Photomedicine

Volume I

Editors

Ehud Ben-Hur, Ph.D.
Department of Radiobiology
Nuclear Research Center-Negev
Beer-Sheva, Israel

Ionel Rosenthal, Ph.D.
Head
Division of Food Technology
Agricultural Research Organization, Volcani Center
Bet Dagan, Israel

CRC Press, Inc.
Boca Raton, Florida

Library of Congress-in-Publication Data

Photomedicine.

Includes bibliographies and index.
1. Phototherapy. 2. Photochemotherapy. 3. Light--
Physiological effect. I. Ben-Hur, Ehud. II. Rosenthal,
Ionel. [DNLM: 1. Phototherapy. WB 480 P5745]
RM837.P48 1987 615.8′3 86-34293
ISBN-0-8493-4673-8 (set)

Direct all inquiries to CRC Press, Inc., 2000 Corporate Blvd., N.W., Boca Raton, Florida, 33431.

International Standard Book Number 0-8493-4674-6 (v. 1)
International Standard Book Number 0-8493-4675-4 (v. 2)
International Standard Book Number 0-8493-4676-2 (v. 3)
International Standard Book Number 0-8493-4673-8 (set)

Library of Congress Card Number 86-34293
Printed in the United States

INTRODUCTION

Go outside and play in the sun. It's good for you.

My mother

The use of sunlight and drugs for the treatment of skin diseases has been documented for over 3400 years; for an even longer time, the reddening, blistering, and tanning effects of sunlight have probably been known. With the discovery of lasers a new dimension was added in the study and application of light in medical therapy. Ophthalmologists adopted the laser in the clinic as a photocoagulator for the treatment of detached retina, and the use of laser as a scalpel for noncontact, noninvasive, and even subcellular surgery is at an earlier state of acceptance. In addition to surgical uses, new, promising ideas are continuing to emerge. Thus, laser can be used to diagnose and treat malignant tumors using photoradiation therapy. This renewed interest, stimulated by the mutual interplay of both scientific and technological innovations, is characterized by a multidisciplinary approach involving physicists, chemists, biochemists, and physicians.

Our objective has been to collect in these three volumes the most up-to-date assessment of our understanding of light in medicine. Since *Photomedicine* was defined as an informative guide to practical applications rather than an esoteric study of medical discipline, the level of medical rigor was reasonably relaxed.

Given limitations on length, the chapters are not intended to be all embracing reviews of the field, but rather to present an overview of key ideas and directions with the objective of delineating the most promising and exciting problems. We hope that the text is sufficiently introductory to stimulate the curiosity and interest of a neophyte, and to simultaneously provide the specialist with a rather short, but current summary of the status of this field. Most important, we hope that the volumes will further highlight this rapidly developing science and spur current and new researchers and ideas.

Ehud Ben-Hur
Ionel Rosenthal

THE EDITORS

Ehud Ben-Hur, Ph.D., was born in Israel in 1940. After graduation from the Hebrew University of Jerusalem in 1965, he went on to study biochemistry at the Technion, Israel Institute of Technology at Haifa, where he obtained his M.Sc. and doctorate degrees. He then joined the Biology Department of Brookhaven National Laboratory as Research Associate where he completed postdoctoral work on the radiobiology of cultured mammalian cells under the auspices of Dr. M. M. Elkind. Upon returning to Israel in 1973, he first joined the Department of Cellular Biochemistry at the Hebrew University and then the Nuclear Research Center-Negev, in 1975, where he is currently engaged in studies of biological effects of ionizing and nonionizing radiations.

The main thrust of his research activity in the past was related to radiation-induced damage in DNA and its repair. During the last few years he has become interested in photodynamic therapy of cancer and is actively involved with Dr. I. Rosenthal in developing new and improved photosensitizers for this purpose.

Dr. Ben-Hur is affiliated with the Department of Radiation Biology, Colorado State University. He is also affiliated with Ben-Gurion University, Beer-Sheva, Israel, where he teaches photobiology. Dr. Ben-Hur has published over 80 papers in scientific journals, is a member of the American Society for Photobiology and the Radiation Research Societies of both the U.S. and Israel, and is on the Editorial Board of the *International Journal of Radiation Biology.*

Dr. Ben-Hur is married with two children and lives most of the time in Beer-Sheva.

Ionel Rosenthal, Ph.D., received his degree in Chemical Engineering from the Polytechnic Institute in Bucharest (Romania) and Ph.D. degree from the Freinberg Graduate School of the Weizmann Institute of Science, Rehovoth, Israel. Dr. Rosenthal has had a very colorful professional career which has included Plant Engineer at "Mahteshim" Chemical Co. and Senior Scientist at the Department of Organic Chemistry at the Weizmann Institute of Science and at the Department of Organic Chemistry, Nuclear Research Center-Negev. Currently he is Principal Scientist at the Department of Food Science, Agricultural Research Organization, Bet-Dagan, and Professor in the Department of Agricultural Biochemistry at the Faculty of Agriculture of the Hebrew University, Jerusalem.

His scientific interests in organic photobiochemistry and food chemistry (and its spin-off: cookery) have resulted in more than 100 research publications in these areas.

ACKNOWLEDGMENTS

The editors wish to acknowledge their indebtedness to the authors for their ready response to our request to contribute, and for their kindness and understanding in accepting their editorial efforts.

CONTRIBUTORS

Esther Azizi, M.D.
Senior Dermatologist
Department of Dermatology
Sheba Medical Center
Tel-Hashomer, Israel

Roger W. Barnes, B.S.
Assistant to the Director
Office of Device Evaluation
U.S. Food and Drug Administration
Silver Springs, Maryland

Ehud Ben-Hur, Ph.D.
Department of Radiobiology
Nuclear Research Center-Negev
Beer-Sheva, Israel

Jeffrey D. Bernhard, M.D.
Director, Division of Dermatology
Director, Phototherapy Center
Assistant Professor
Department of Medicine
University of Massachusetts Medical Center
Worcester, Massachusetts

Homer S. Black, Ph.D.
Director
Photobiology Laboratory
Veterans Administration Medical Center
Baylor College of Medicine
Houston, Texas

Larry E. Bockstahler
Division of Life Sciences
Optical Radiation Branch
U.S. Food and Drug Administration
Rockville, Maryland

S. G. Bown
Department of Surgery
University College London
London, England

John A. S. Carruth
Senior Lecturer in Otolaryngology
Southampton University Hospitals
Southampton, England

James E. Cleaver, Ph.D.
Professor
Laboratory of Radiobiology and Environmental Health
University of California
San Francisco, California

Farrington Daniels, Jr., M.D., M.P.H.
Emeritus Professor of Medicine
The New York Hospital-Cornell Medical Center
New York, New York

John H. Epstein, M.D.
Clinical Professor of Dermatology
University of California Medical School
San Francisco, California

I. Farine, M.D.
Professor and Head
Orthopedic Department
Sheba Medical Center
Tel-Hashomer, Israel

Richard P. Felten
Department of Health and Human Services
Division of Life Sciences
U.S. Food and Drug Administration
Rockville, Maryland

Anna Flint, Ph.D.
Department of Dermatology
Sheba Medical Center
Tel-Hashomer, Israel

Orna Geyer, M.D.
Department of Ophthalmology
Tel Aviv Medical Center
Tel Aviv, Israel

Barbara A. Gilchrest, M.D.
Professor and Chairman
Department of Dermatology
Boston University School of Medicine
Boston, Massachusetts

Philip C. Hanawalt, Ph.D.
Professor and Chairman
Department of Biological Sciences
Stanford University
Stanford, California

M. Heim, M.D.
Department of Rehabilitation
Sheba Medical Center
Tel-Hashomer, Israel

Kiki B. Hellman, Ph.D.
Senior Scientist
Division of Life Sciences
U.S. Food and Drug Administration
Rockville, Maryland

Victoria M. Hitchins, Ph.D.
Division of Life Sciences
Optical Radiation Branch
U.S. Food and Drug Administration
Rockville, Maryland

H. Horoszowski, M.D.
Head
Department of Orthopedy
Sheba Medical Center
Tel-Hashomer, Israel

Elizabeth D. Jacobson, Ph.D.
Acting Director
Office of Science and Technology
Center for Devices and Radiological Health
U.S. Food and Drug Administration
Rockville, Maryland

W. Patrick Jeeves, Ph.D.
Department of Medical Physics
Ontario Cancer Treatment and Research Foundation
Hamilton, Ontario, Canada

Anthony Lamanna
Division of Life Sciences
Optical Radiation Branch
U.S. Food and Drug Administration
Rockville, Maryland

Robert J. Landry, Ph.D.
Electrooptics Branch
U.S. Food and Drug Administration
Rockville, Maryland

Moshe Lazar, M.D.
Department of Ophthalmology
Ichilov Hospital
Tel-Aviv, Israel

Sidney Lerman, Ph.D.
Professor
Department of Opthalmology
Emory University
Atlanta, Georgia

Jerome I. Levine, Ph.D.
Division of Life Sciences
Optical Radiation Branch
U.S. Food and Drug Administration
Rockville, Maryland

Julia G. Levy, Ph.D.
Professor
Department of Microbiology
University of British Columbia
Vancouver, Canada

C. David Lytle, Ph.D.
Division of Life Sciences
Optical Radiation Branch
U.S. Food and Drug Administration
Rockville, Maryland

I. A. Magnus, M.D., F.R.C.P.
Professor Emeritus
Photobiology Department
Institute of Dermatology
London, England

Micheline M. Mathews-Roth, M.D.
Associate Professor of Medicine
Channing Laboratory
Harvard Medical School
Boston, Massachusetts

Daphne Mew, Ph.D.
Faculty of Medicine
Foothill Hospital
University of Calgary
Calgary, Alberta, Canada

K. Mohan, Ph.D.
Science & Technology
U.S. Food and Drug Administration
Rockville, Maryland

Warwick L. Morison, M.D.
Associate Professor of Dermatology
Johns Hopkins Medical Institutions
Baltimore, Maryland

D. Phillips
Royal Institution
London, England

Maureen B. Poh-Fitzpatrick, M.D.
Associate Professor of Dermatology
Columbia University
New York, New York

Michael K. Reusch, M.D.
Department of Dermatology
Stanford University Medical School
Stanford, California

Ionel Rosenthal, Ph.D.
Head
Division of Food Technology
Agricultural Research Organization, Volcani Center
Bet Dagan, Israel

Stephen M. Sykes
Division of Life Sciences
Optical Radiation Branch
U.S. Food and Drug Administration
Rockville, Maryland

Morris Waxler
Office of Science and Technology
Center for Devices and Radiological Health
U.S. Food and Drug Administration
Rockville, Maryland

Abraham Werner, Ph.D.
Chief Physicist
Department of Oncology
Sheba Medical Center
Tel-Hashomer, Israel

Brian C. Wilson, Ph.D.
Department of Medical Physics
Ontario Cancer Treatment and Research Foundation
Hamilton, Ontario, Canada

Bruce U. Wintroub
Department of Dermatology
University of California
San Francisco, California

TABLE OF CONTENTS

Volume I

Volume II

Chapter 1

BASICS OF PHOTOCHEMISTRY

Ionel Rosenthal

TABLE OF CONTENTS

I. PHOTOCHEMICAL TERMS

The science of photochemistry is concerned with the chemical and physical effects associated with electronic excitation of the chemical compounds. The promoter of excitation, for most practical purposes, is electromagnetic radiation in the UV and visible region of the spectrum (Table 1). Electromagnetic radiation can be regarded as having a dual nature.

• The light propagation phenomena such as reflection, refraction, interference, diffraction, and polarization can be explained in terms of a wave propagation. The wave is composed of oscillating electric and magnetic fields operating in planes which are perpendicular to each other and to the direction of propagation. The time variable strength of the resulting vector could be described by a sinusoidal function (Figure 1). The wave of radiation is characterized by wavelength (λ) which is the distance between identical positions on successive waves, or by frequency (υ) which is the number of waves that pass a fixed point per second. These two parameters are related in Equation 1:

$$\nu = \frac{c}{\lambda} \tag{1}$$

where c is the speed of electromagnetic radiation in vacuum (2.9979×10^8 m/sec). The wavelength of radiation determines the color of the light.

• At the same time, the phenomena of absorption or emission of radiation must occur only in discrete units called quanta or photons which have energies E as given by Equation 2:

$$E = h\nu = \frac{hc}{\lambda} \tag{2}$$

where h is a universal constant called Planck's constant (6.6256×10^{-34} J sec). It results that the energy absorbed or emitted by a molecule is inversely proportional to the wavelength of radiation. The energy of one "mole" of light, *viz.*, an Einstein, is given by Equation 3:

$$E = N\frac{hc}{\lambda} \tag{3}$$

where N is Avogadro's constant (6.022×10^{23}).

The first law of photochemistry — the Grotthuss-Draper law — requires that the radiant energy must be absorbed in order to induce a photochemical or photobiological reaction. The energies which can be absorbed by a particular molecule can be determined from the wavelengths in the absorption spectrum. The absorption spectrum of each molecule is, in turn, determined by the kind and arrangement of its component atoms. The first law of photochemistry is true for conventional light sources, but may require modifications for polyphotonic processes initiated by intense light sources such as lasers.

The second fundamental law, which is a result of the quantum theory, was proposed by Stark and Einstein and states that if a molecule absorbs radiation, then only one is excited for each quantum of radiation absorbed. The Stark-Einstein law implies that the efficiency of a photochemical process can be defined by its quantum yield (ϕ) which is the ratio of molecules undergoing a particular change and the number of quanta absorbed. In most photochemical reactions, the quantum yield will range from zero to

Table 1
THE DISTRIBUTION OF THE ELECTROMAGNETIC SPECTRUM

Radiation	Wavelength	Absorption or emission of radiation involves
Gamma rays	0.0005—0.14 nm	Nuclear transition
X-rays	0.01—10 nm	Transitions of inner atomic electrons
Vacuum UV	1—200 nm	Transitions of outer atomic electrons
UV	200—400 nm	Transitions of outer atomic electrons
Visible	400—800 nm	Transitions of outer atomic electrons
Near IR	0.8—1.5 μm	Molecular vibrations
IR	1.5—5.6 μm	Molecular vibrations
Far IR	5.6—1000 μm	Molecular rotations
Micro- and radiowaves	1000 μm—550 m	Oscillations of mobile or free electrons

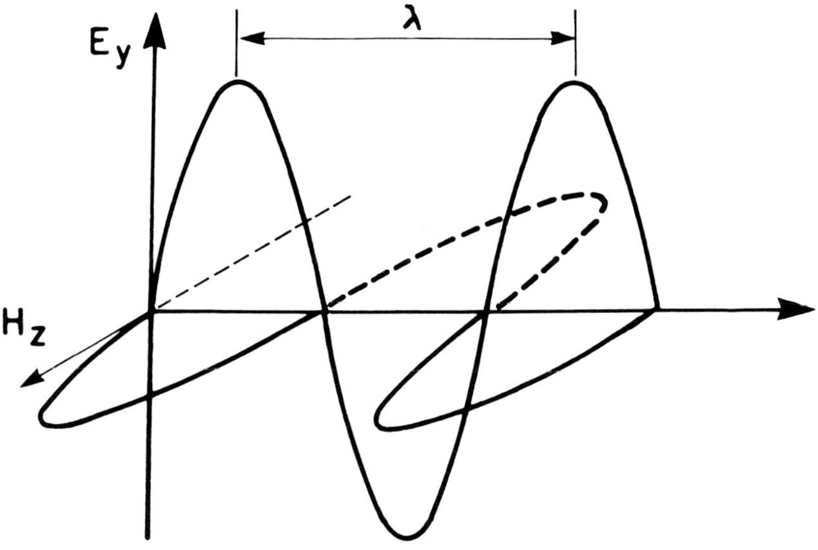

FIGURE 1. The electric (Ey) and magnetic (Hz) vectors of an electromagnetic light wave.

unity. However, in a chain reaction, where the absorption of one photon initiates the reaction, the value of the quantum yield may be several powers of ten. The determination of the quantum yield of a process will depend on the accurate determination of the number of molecules reacting or formed, and the number of photons absorbed. While the first quantity can be obtained by chemical analyses, a measure of the number of photons absorbed is best achieved by comparison with a reaction system, the quantum yield of which has been determined accurately using a photosensitive instrumental technique. Such a reaction system is known as an actinometer.

The intensity of an absorption band follows the empirical Beer-Lambert law, which states that the fraction of light transmitted through an absorbing system is as presented in Equation 4 and, in the log form, in Equation 5:

$$\frac{I_t}{I_o} = 10^{-\xi cd} \quad (4)$$

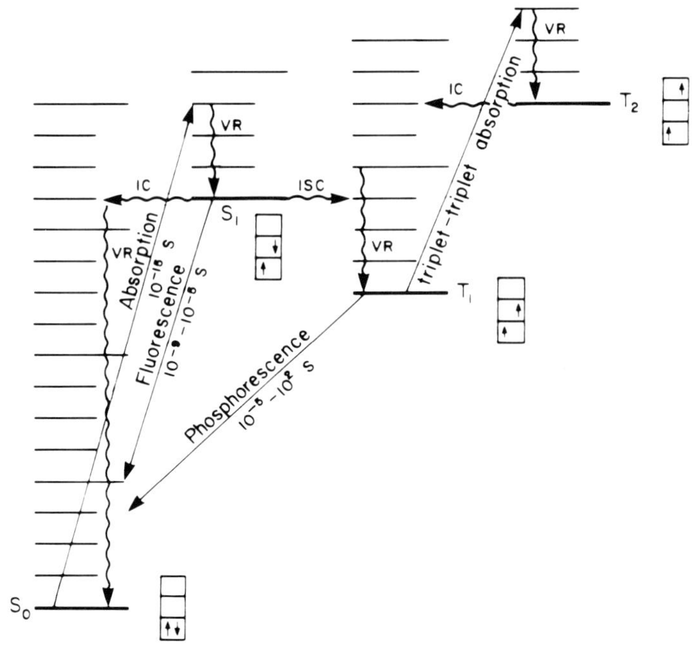

FIGURE 2. Energy level (Jablonski) diagram indicating the principal light-induced molecular photophysical processes. VR = vibrational relaxation, 10^{-10} sec; IC = internal conversion, 10^{-10} sec; ISC = intersystem crossing, 10^{-6} sec.

$$\log \frac{I_o}{I_t} = \xi cd \tag{5}$$

where I_t and I_o are transmitted and incident light intensities, c is the concentration of the absorber, and d is the depth of absorber through which the light has passed. ξ is a constant of proportionality known as the extinction coefficient and is dependent on the wavelength and occasionally on concentration. The product ξcd is also named the optical density (OD) of the system.

When a molecule absorbs energy, the process is referred to as excitation. It is possible to represent the various physical processes involved in the photochemical excitation in the Jablonski diagram (Figure 2). The molecule is raised from its ground state of minimum energy to an excited state of higher energy. In the UV and visible region of the electromagnetic spectrum, this involves excitation to electronic energy levels. In the IR region, the excitation occurs between rotational levels at the lower energies of the far IR, while increase in the vibrational energy occurs in the near and medium IR region. While it is generally true that electronic excited states are the precursors of a photochemical change and, as such, are of primary importance in photobiology, they are inevitably accompanied by some increase in vibrational and rotational energy.

The electronic structure of the absorbing molecule changes as the result of excitation. This change is visualized in simple molecule orbital terms as a change in the occupation pattern of a set of orbitals. The combination of two atomic orbitals of different atoms by overlap along the internuclear axis results in the formation of two molecular orbitals — a bonding orbital (either σ or π), and an antibonding orbital (σ^* or π^*). In addition, the nonbonding electrons, if present as at an oxygen atom, occupy a nonbonding orbital n which by its nature does not possess an antibonding counterpart. Two electrons are assigned to each molecular orbital so that their spins are paired. In general, a molecule in its ground state has all of its electrons spin-paired (a notable exception is molecular oxygen, the ground state of which is a triplet). The absorption of a quanta of radiation in the UV-visible range by a molecule leads to electronic excitation of the absorber. That is the promotion of an electron from one molecular orbital to another, unoccupied, of higher energy, e.g., $\sigma \rightarrow \sigma^*$, $n \rightarrow \sigma$, $\pi \rightarrow \pi^*$, and $n \rightarrow \pi^*$. In general, excitation of an electron to the σ^* orbital requires an amount of energy corresponding to wavelengths below 200 nm, which are not readily accessible to conventional irradiation apparatus and are of little relevance to photobiology. The alternative excitation processes ($\pi - \pi^*$ and $n - \pi^*$) are responsible for the bulk of photochemical reactions, since the wavelengths associated with these two excitations are located in the accessible regions of the UV and visible part of electromagnetic spectrum.

The initial electronic excitation occurs without change in the orbital spin; i.e., the molecule, although in the excited state, still preserves the "singlet" multiplicity. The multiplicity, M, of a system is defined by Equation 6:

$$M = 2S + 1 \tag{6}$$

where S is the total spin of the system. Thus, in a spin-paired system, S = 0 and M = 1 (singlet), while for a spin-parallel system, S = 1 and M = 3 (triplet). The excited singlet state is very short-lived and it may decay to the original ground state by conversion to heat or spontaneous photon emission (fluorescence). Alternatively, the molecule in this state could be destabilized to initiate a chemical reaction or could convert to a triplet state in a process which involves a spin flip of one of the electrons in the highest occupied molecular orbital. In this process, named intersystem crossing, the system becomes spin-parallel. Since transitions between states of different multiplicities are "forbidden" because they involve a spin inversion, the reversal from the excited triplet state to the ground singlet state is slow, and the molecule in the triplet state is long-lived. A triplet state always has a lower energy than the corresponding singlet state because of the repulsive nature of the spin-spin interaction between electrons of the same spin. The spontaneous radiation emission from the triplet state is known as phosphorescence. Because of its long lifetime, the triplet-state molecule is the most important in the initiation of chemical and biochemical reactions in vivo and in vitro. Reactions may also originate from excited singlets, but their much shorter lifetime most often limits their importance. The triplet molecule may react directly, inter- or intramolecularly, or may dispose of its energy by energy transfer to a vicinal suitable system. The concept of energy transfer can be represented in a general form by Equation 7:

$$^3D^* + {}^1A \rightarrow {}^1D + {}^3A^* \tag{7}$$

where D and A are the donor and acceptor, respectively, in the triplet and singlet states. Excitation of a ground-state molecule by energy transfer from another excited species is termed sensitization, and the deactivation of the excited species is termed quenching. The energy-transfer processes play an important role in photobiology. Thus, the photosensitization creates the possibility of generating an excited molecule indirectly, rather than by its irradiation. Furthermore, the excited states of many organic molecules are quenched by molecular oxygen in a diffusion-controlled manner. Consequently, photochemical reactions are often drastically altered by the presence of molecular oxygen.

Chapter 2

MOLECULAR AND CELLULAR PHOTOBIOLOGY

Ehud Ben-Hur

TABLE OF CONTENTS

I. INTRODUCTION

Life evolved on earth under continuous exposure to electromagnetic radiation from the sun, ranging from the far ultraviolet (UVC) to the visible part of the spectrum. UV light can damage many cellular structures, mainly nucleic acids. This results in lowered cell viability and hereditary changes in surviving cells. Thus, the evolving primitive cells were under a selective pressure to protect themselves from the deleterious effects of the UV light of the sun and to develop mechanisms that deal with the damage produced in critical targets, e.g., DNA. Today we have a fairly good, although incomplete, knowledge of the important photodamage in DNA, RNA, proteins, and membranes. We also know the general outlines of most of the repair mechanisms that have been developed to deal with UV-induced damage in DNA. This information is more detailed for prokaryotes than for mammalian cells but new data are being continuously added in this intensively studied field.

Because of the vast scope of the subject matter, it is impossible to give an in-depth state-of-the-art description of molecular and cellular photobiology in one chapter. Rather, what was attempted is to give the reader a general view of this topic with pertinent examples where appropriate. For those wishing to explore this fascinating field further, an extensive list of references is provided.

The first part of this chapter is devoted to the introduction of photodamage and its biological effects. Later parts deal with the biochemical responses to DNA damage, mainly in terms of repair mechanisms, and the way these responses are reflected at the cellular level.

II. MOLECULAR PHOTODAMAGE IN CELLS

A. Photochemistry of Nucleic Acids

Because DNA and RNA have the strongest absorption in the UV of all cellular components, they are also the most sensitive to photochemical changes. The four heterocyclic bases serve as the chromophores, with λ_{max} about 260 nm. The pyrimidines undergo photochemical reactions $\sim$10 times more readily than purines. The following is a list of the photoproducts according to prevalence and importance. In general, those produced in DNA are more important than those produced in RNA because of the unique position of the former in the chain of information transfer within the cell.

1. Cyclobutane-Type Pyrimidine Dimers

The isolation and identification of a cyclobutane-type thymine dimer (Thy <> Thy) from UV-irradiated frozen solution of thymine[1] opened the way for modern molecular photobiology. Thy <> Thy has four isomeric forms, of which two exist in optically active forms[2,3] (Figure 1). In UV-irradiated DNA, dimerization of adjacent pyrimidines is the predominant photochemical reaction. This gives rise to thymine-cytosine and cytosine-cytosine (Thy <> Cyt, Cyt <> Cyt) dimers, in addition to Thy <> Thy, all of which are in *cis-syn* configuration. The photochemical reactivity of DNA depends upon the primary base composition and sequence and DNA secondary structure and its environment during irradiation.[4] For example, the rate of Thy <> Thy formation in denatured DNA is twice that of DNA irradiated in the native state. Also, because of the greater freedom of movement of the bases in denatured DNA, the *trans-syn* isomer is also produced.[5]

Pyr <> Pyr formation in DNA proceeds as follows. Absorption of a photon excites a pyrimidine base from its ground state (S_0) to a singlet excited state (S_1). In the new electronic configuration the van der Waals' radius of the base is reduced, allowing an

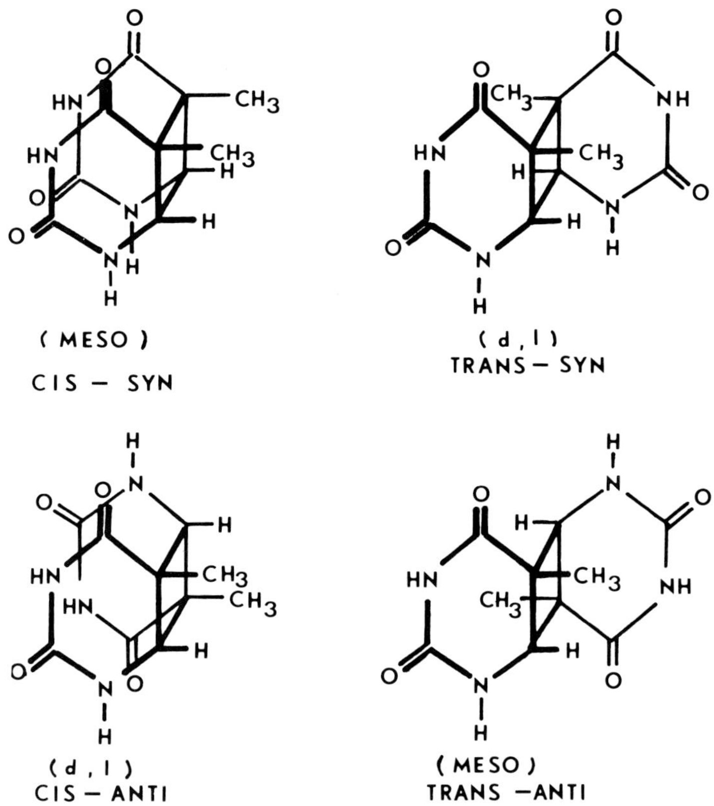

FIGURE 1. Thy <> Thy isomers.

adjacent pyrimidine to move closer and share the excitation energy. This "excimer" can undergo intersystem crossing into the more stable triplet state (T_1) from which a Pyr <> Pyr may arise.

$$Pyr_{S_0} \xrightarrow{h\nu} Pyr_{S_1} + Pyr_{S_0} \rightarrow [Pyr \cdot Pyr]_{S_1} \xrightarrow{isc} [Pyr \cdot Pyr]_{T_1} \rightarrow Pyr <> Pyr$$

DNA undergoes a change from a B conformation to A when its degree of hydration is reduced. Because Pyr <> Pyr yield is drastically reduced and a new photoproduct is formed upon UV irradiation of DNA in the dry state,[6] it was assumed that DNA photochemistry is modified by its conformation. More recent data have shown that the degree of hydration rather than the conformational state is the critical factor in determining which of the photoproducts will form in native DNA.[7]

The absolute yield of Pyr <> Pyr is twofold greater in DNA irradiated (254 nm) in vitro than in vivo.[8] This "shielding" effect is not due to cytoplasmic absorption, although in mammalian cells this may be a contributing factor. Rather, it reflects the dense packing of DNA in cells, as opposed to its more linear, extended form in solution. The ratios of the different Pyr <> Pyr formed in vivo depend on the base content of the DNA. In *E. coli* there are 1 Cyt <> Cyt:5 Cyt <> Thy:10 Thy <> Thy and the yield per genome molecular weight per unit fluence is 2.3 Pyr <> Pyr/10^8 daltons/J/m^2 at 254 nm. At longer wavelengths relatively more Cyt-containing dimers are formed, although the total yield decreases above 265 nm. The presence of Pyr <> Pyr produces some distortion in DNA structure, as evidenced by a lowered melting tem-

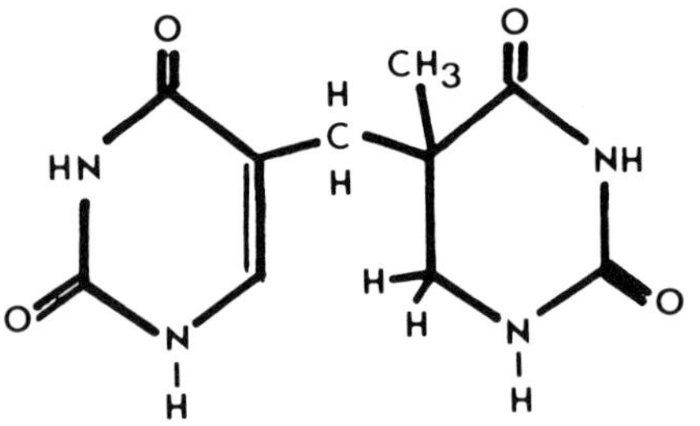

FIGURE 2. Molecular structure of Thy (α-5)hThy.

perature[9] and increased reactivity to formaldehyde.[10] Melting of DNA (4.3 A-T bp are broken per dimer[9]) would reduce the twisting of one strand around its complement and should therefore affect the superhelicity of the entire molecule. By measuring the untwisting in negatively supercoiled DNA it was estimated that the unwinding angle per dimer is −14°.[11] Alterations in superhelicity can affect the binding of proteins to DNA and the processes of transcription, recombination, and replication.[12] In torsionally stressed superhelical DNA, unwinding by dimers will affect the helix structure in more distant regions and thus UV radiation could affect cellular function at a distance from the lesions themselves.[13]

Pyr <> Pyr are nonrandomly distributed in DNA.[10,14] Although this observation was contested,[15,16] it was later shown to be real and that dimerization occurs preferentially in long thymine tracts.[17] However, at the biological dose-range, the probability of forming a dimer "cluster" in DNA of moderate A-T content is rather low, and whether or not this phenomenon has a biological significance remains to be seen.

Detecting the presence and amount of Pyr <> Pyr in DNA can be achieved in two ways: (1) by breaking down the nucleic acid to individual components, separating them by some type of chromatographic procedure, and detecting their presence by absorbance or radioactivity measurements; and (2) by detecting the dimer *in situ* using a specific property (e.g., immunological reaction, the susceptibility to an enzyme). Some of these methods are very sensitive and can detect one dimer per 10^9 daltons DNA. These approaches are also used to detect other photoproducts. For a detailed description see Reference 4.

2. 5-Thyminyl-5,6-Dihydrothymine

When DNA is irradiated in the dry state,[6,18] at subzero temperatures[19] or in bacterial spores,[18] Pyr <> Pyr formation is greatly reduced and a new thymine-derived photoproduct appears. This was identified as 5-thyminyl-5,6-dihydrothymine, Thy(α-5)hThy (Figure 2).[20] It is usually referred to as "spore photoproduct" and was thought to be responsible for spore death following UV irradiation. However, recently another photoproduct was isolated from spores and identified as 6-4′-(pyrimidine-2′-one)-thymine (see next section).[21] Its contribution to the biological effects of UV on bacterial spores has yet to be determined.

The absence of water is necessary for Thy(α-5)hThy formation, which is consistent with the free-radical mechanism proposed.[20] Therefore, this photoproduct is biologically significant only for bacteria exposed to UV under extreme conditions of cold and dryness (e.g., in outer space).

3. Pyrimidine Adducts

A minor photoproduct (about 10% of the total) containing thymine was first isolated from DNA by Varghese and Wang.[22] This was identified as the acid-stable prod-

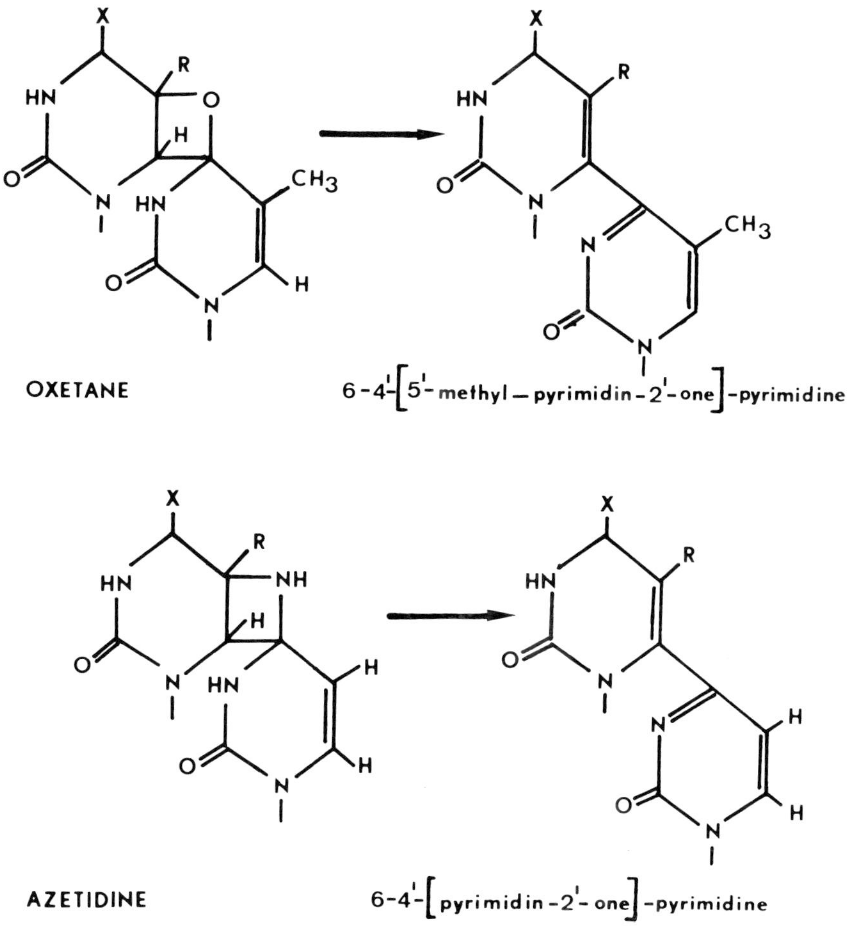

FIGURE 3. Molecular structure of pyrimidine adducts isolated from irradiated DNA. When X is C=O then R is CH_3; X is NH_2 when R is H.

uct 6-4′-(pyrimidine-2′-one)-thymine (Thy(6-4)Pyo) resulting from dehydration of the Thy-Thy adduct containing an oxetane ring[23] (Figure 3). Similar adducts are formed via an azetidine ring between Thy-Cyt, and Cyt-Cyt.

The action spectrum for Pyr-Pyr adduct formation is essentially the same as that for Pyr <> Pyr in *Escherichia coli* DNA,[24] with a maximum at 265 nm. The quantum yield for Cyt-Thy (0.0015) is about half that of Cyt <> Thy (0.0024) and an order of magnitude lower than that for Thy <> Thy formation (0.019). The relative rates of formation of Cyt-Thy and Pyr<> Pyr depend strongly on the percent of Ade + Thy in DNA in an opposite manner. Cyt-Thy adducts formation increases with decreasing Thy contents of the DNA and with increasing dehydration. Also in contrast to Pyr <> Pyr, the formation of Pyr adducts is greater in native than in denatured DNA.[24]

The mechanism of Pyr-Pyr adducts formation does not involve the excited triplet state. It could involve phototautomerization in the G · C bp, serving as traps for singlet excitation energy, with subsequent addition of the imino double bond of the excited Cyt tautomer to the 5,6-double bond of a neighboring Thy.[4]

The most striking feature of Pyr adducts is the absorbance maximum at 315 nm. As a result, they can be eliminated from DNA by irradiation at λ = 310 to 340 nm, which has no effect on Pyr <> Pyr. The quantum yield for Cyt-Thy adduct photolysis is

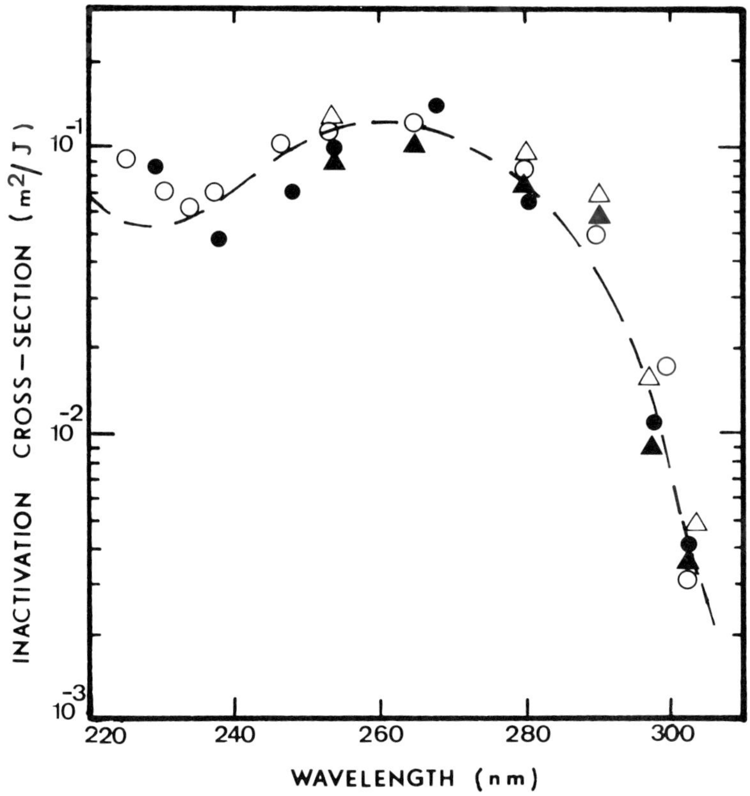

FIGURE 4. Action spectra for killing of *E. coli* (•) and *S. aureus* (○), Chinese hamster cells (△), and induction of Pyr <> Pyr in Chinese hamster cells (▲). The dashed curve shows the relative absorption of DNA, normalized to its value at 260 nm. The data for Chinese hamster cells were normalized relative to 265 nm.[39,53]

This prediction is borne out by action spectra (an action spectrum is the relative response of a system to different wavelengths) obtained before DNA was known to be genetic material. Thus, the action spectra for killing of the bacteria *E. coli* and *Staphylococcus aureus* and of mammalian cells correspond closely to the absorption spectra of the nucleic acids and not to those of the aromatic amino acids of proteins (Figure 4). In addition, a close correspondence also exists between the above and action spectrum of Pyr dimers formation. These spectra were confirmed, for both killing and mutation, by many workers.

Because three of the four bases are the same, absorption spectra of DNA and RNA are very similar; therefore, action spectra do not permit differentiation. Moreover, both DNA and RNA are photoreactivable, although animal cells may not contain an RNA photolyase (see Section V). Nevertheless, because in most biological systems DNA is the primary genetic material, *a priori* DNA must be the main target to far-UV effects on cells. This interpretation is strengthened by experiments in which the effects of UV on transforming principles were studied. These are DNA molecules containing several genes from one organism, which are capable of infecting another organism and combining with its genome, thus transferring to the infected organism heritable genetic changes represented by the genes in the transforming principle. Inactivation of the transforming activity of such DNA in vitro by UV is caused primarily (∿90%) by Pyr <> Pyr.[40] Furthermore, the UV-induced lesions in transforming DNA can kill the host cell into whose genome it is integrated.[41] Another line of evidence implicating DNA as the target is the use of uracil derivatives substituted by a halogen at the 5-position. The 5-chloro, -bromo, and -iodo derivatives are analogs of Thy and are incorporated into DNA. As a result, the organism becomes more sensitive to UVC and UVB.[42] The 5-fluorouracil analog is incorporated into RNA without affecting the response to UV.

Of course, some biological systems do not possess DNA (e.g., mature erythrocytes). In others, RNA serves as the carrier of the genetic information (all plant viruses and many animal viruses). In these systems RNA is expected to be the primary target. In mammalian cells some contribution to the UV response of protein cross-linked to DNA may be envisaged. This has yet to be determined. A notable exception among microorganisms is the highly UV-resistant bacterium *Micrococcus radiodurans.* Because of its highly efficient repair of UV-induced damage in DNA (see Section V), protein damage becomes important for cell killing.

C. Photosensitization

While the phenomena that are called photosensitization are manifold, a general definition is "the action of a component (photosensitizer) of a system that causes another component of the system to react to light". The photosensitizer serves as the chromophore, absorbing the light, and upon excitation transfering the energy to an acceptor molecule. For details see Chapter 1.

Photosensitization of biological systems to visible light involves various dyes and usually requires molecular oxygen. These are the so-called photodynamic reactions. In general, photosensitized oxidations proceed via the triplet state of the sensitizer. In type I photooxidation reactions, the triplet sensitizer molecule can react directly with a substrate molecule by an electron transfer process to give a semireduced form of the sensitizer, $S^{\cdot -}$, and a semioxidized form of the substrate, $A^{\cdot +}$, both of which are free radicals. The semioxidized substrate can react with molecular oxygen to give a fully oxidized product. Other reactions can also be involved.[43] The semireduced form of the sensitizer can also react with oxygen to produce ground-state sensitizer and $O_2^{\cdot -}$. The superoxidc radical can in turn oxidize another molecule in the system.

The type II process involves the interaction of triplet photosensitizer with ground-state oxygen, giving ground-state sensitizer and a highly reactive, singlet-state of oxygen, $^1O_2$:

$$^3S + {}^3O_2 \rightarrow S_o + {}^1O_2$$

$$^1O_2 + A \rightarrow A_{ox}$$

$^1O_2$ can react with a wide variety of substrates to give a fully oxidized form, A_{ox}. This process is common in many biologically important photosensitization reactions involving sensitizers such as acridines, porphyrins, xanthenes, and thiazines.

While the potential substrates in the cell for photooxidation are varied, the actual target molecules depend mainly on the localization of the photosensitizer within the cell. Thus, porphyrins concentrate in the lysosomes. Subsequent illumination causes damage to the lysosomal membranes and rupture of the lysosomes.[44] Other sensitizers, such as rose bengal, accumulate selectively in the plasma membrane of cells while acridines accumulate in chromosomes. Illumination in the latter case produces chromosome breakage. Guanine residues are selectively destroyed in DNA under these conditions, but cross-linking of DNA and proteins, and DNA strand breakage, also occur.[45]

Some photosensitization reactions do not involve molecular oxygen. The most important biologically are the furocoumarins. These naturally occurring photosensitizers intercalate between the DNA bases and form covalent bonds with the pyrimidines upon exposure to near UV.[46] They are discussed extensively in Volume II, Chapter 1.

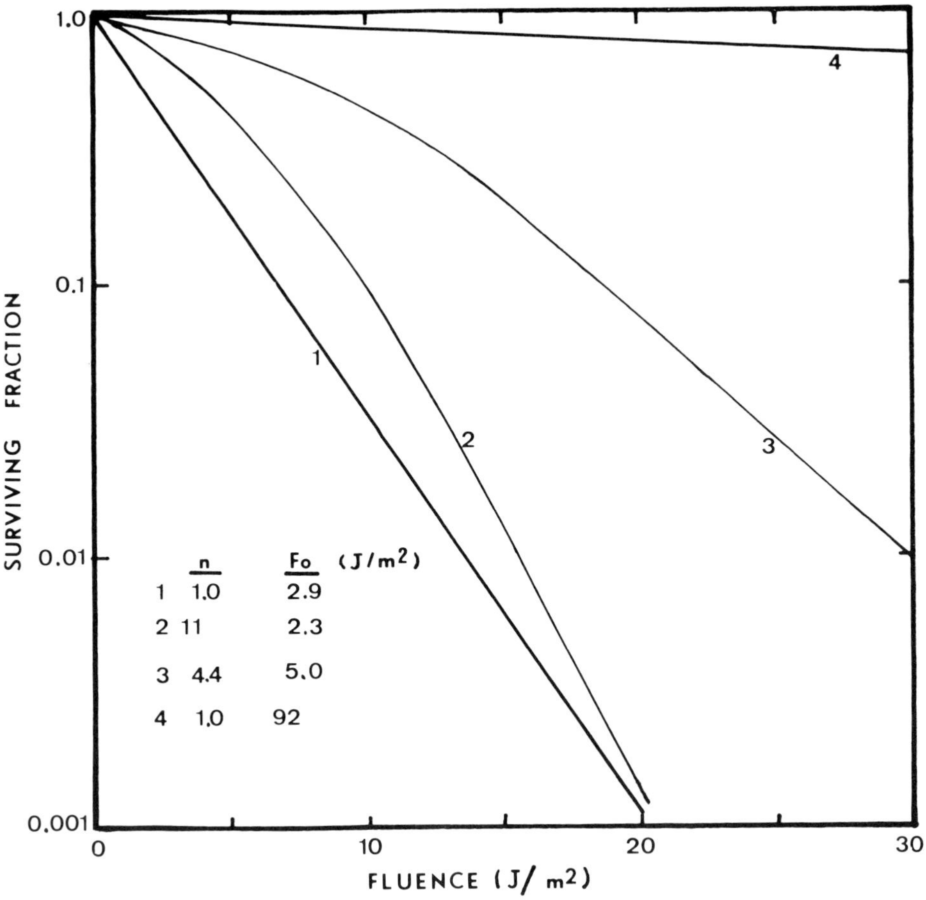

FIGURE 5. Survival curves of UV-irradiated (254-nm) cultured mammalian cells. Fibroblasts obtained from humans (1),[56] bovine (2),[56] and Chinese hamster (3).[57] Curve 4 is for Chinese hamster cells exposed to 265-nm light from a pulsed laser (23 J/m²/15 nsec pulse).[58] 1 and 4 are single-hit curves while 2 and 3 are multitarget survival curves.

survival behavior of simple systems only. In more complex systems, such as the mammalian cell, kinetics of inactivation which fit some target theory kinetics should not be taken to imply a specific biophysical mechanism. This reservation is less serious in repair-deficient mutants.

The lesions leading to the lethal effects of far UV are photoproducts in DNA, mainly Pyr <> Pyr. The evidence for this in bacteria comes mainly from the ability of the photoreactivating enzyme (see Section V) to reverse 50 to 90% of the lethal effect. This enzyme specifically monomerizes Pyr <> Pyr. In mammalian cells, the evidence is less conclusive since they contain very low levels of this enzyme. Nevertheless, studies using cells deficient in DNA repair suggest an important role for dimers in UV-induced lethality of human cells (see Chapter 2, Volume II). In the near-UV region, the role of Pyr <> Pyr in cell killing is reduced progressively with longer wavelengths. Concomitantly the importance of DNA strand breakage increases.[50] This is inferred mainly from the similarity between the oxygen dependence of cell killing and DNA strand breakage. Such a photodynamic effect is mediated via an endogenous photosensitizer, which has not been identified so far. Candidates for the chromophores that could be involved are many. A recently published action spectrum in the visible for the induction of DNA strand breaks in human fibroblasts is similar to the absorption spectrum for riboflavin, a known photosensitizing agent, suggesting this molecule as the absorbing chromophore.[59]

The mechanism by which DNA damage kills a cell is complex and is not completely understood. It will be further discussed in the following sections. The efficiency of a specific lesion in causing lethality varies from one organism to another. Thus, although a typical bacteria, e.g., *E. coli*, is about twice as resistant as a mammalian cell to far UV, because the genome of the latter is $\sim 10^3$ times larger, ~ 300 times more Pyr <> Pyr are produced by a lethal hit (F_{37}) in mammalian cell than in a bacterium. Since DNA repair in mammalian cells is not more efficient than that operating in bacteria it appears that the former can tolerate large numbers of modified bases without losing their proliferative ability. Exactly how this is accomplished is not known, but it should be noted that most of the DNA in a mammalian cell is not required for storing information.

In addition to cell killing in dividing cells discussed earlier, also known as mitotic death, mammalian cells can suffer interphase death after a very large UV fluence. Interphase death is characterized by the development of nuclear pyknosis. Morphologically, this resembles nuclear pyknosis induced in lymphatic cells by low doses of ionizing radiation. The mechanisms, however, are different, in that respiratory metabolism is not required for the development of UV-induced degeneration.[60]

An additional insight into the mechanism of UV-induced cell killing may be obtained by using synchronized cells. Such studies reveal that mammalian cells are most resistant in mitosis (M), resistance decreases as cells move into G_1 and S phase and reaches a minimum in mid S. It then increases as cells progress out of S and into G_2 phase.[61,62] These changes in sensitivity of cell lethality are correlated with enhanced formation of Pyr <> Pyr as cells progress from M into G_1 and S phase.[62] This cycle dependence of photoproduct formation is most probably due to changes in the target geometry and consequent changes in the specific absorption of UV by DNA. However, the cycle dependence of UV killing is only partly explicable by changes in dimer yield. Some of it could be due to cycle-related formation of DNA-protein cross-links[63] and to change in DNA repair capacity through the cell-cycle (see Section V).

B. Mutagenicity

1. General Considerations

The mutagenic potential of UV light in microorganisms has been known for many years. Research on mutation induction accelerated after World War II, but the modern period of study of UV mutagenesis began after the discovery in 1964 of DNA repair (see Section V). Most of our knowledge comes from studies with prokaryotes. However, during the last decade appropriate mammalian cell systems have been developed and information from them is rapidly accumulating.

Since mutations arise as a result of changes in the DNA bases, it is not surprising that the action spectrum for UV-induced mutagenesis mimics the absorption spectrum for DNA. Alterations in DNA that lead to mutations are of two types: (1) Substitution of one base pair by another. This can be due to *transition* in which a purine base is replaced by another purine, or *transversion*, i.e., the replacement of a purine by a pyrimidine or vice versa. (2) *Frame shift* mutation is the deletion or insertion of a base pair, altering the continuity of genetic code. This results in either a *nonsense* codon (a triplet of nucleotides that does not code for an amino acid), which terminates the growing polypeptide chain, or incorrect amino acids would be inserted into the enzyme, making it inactive *(missense)*.

The types of mutations that have been studied have been those that are easily assayed. In bacteria, most studies have followed the acquisition of resistance to killing

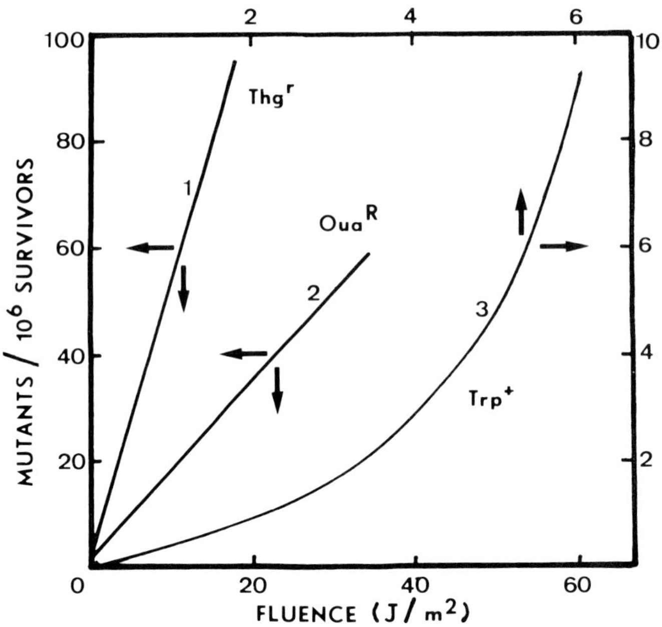

FIGURE 6. UV (254 nm)-induced mutagenesis in *E. coli* B/r Hcr (curve 3) measured as reversion to tryptophan independence ($trp^- \rightarrow trp^+$),[50] and in Chinese hamster cells (curves 1 and 2) assayed as induced resistance to 6-thioguanine (Thg^r) and ouabain (Oua^r).[64] Note the tenfold difference in scale for bacteria and mammalian cells.

by antibiotics and the reversion to prototrophy (an auxotroph requires an amino acid for growth but a prototropic revertant does not). In mammalian cells the most widely used genetic markers have been (1) the acquisition of 6-thioguanine resistance (Thg^r) due to loss of hypoxanthine-guanine phosphoribosyltransferase (HGPRT) activity, and (2) the induced resistance to ouabain (Oua^r) which involves alteration of the plasma membrane Na^+/K^+ ATPase.

An important point to consider when studying mutagenesis is the time needed to express the new phenotype. This time is in a wide range of a few hours in bacteria to a few days in mammalian cells. During that time the cells divide and the premutational lesions in DNA are fixed and then expressed in the daughter cells. Therefore, transfer into the selection medium must be done only at the end of the expression time. When comparing the mutagenic response of two cell lines that differ in their sensitivity to killing by UV, the problem is whether to compare it at equal fluence or at equal survival. Although in most studies the former approach is used, it is not certain that this is most appropriate.

2. Fluence-Response Curves

In bacteria, the mutant frequency response curve after UV irradiation is bending upward, following fluence-square kinetics (Figure 6). This suggests a two-event process, which is supported by other findings. The fluence response in mammalian cells is linear (Figure 6) and therefore the mechanism of mutagenesis is probably not the same as that in bacteria. Also, at low fluences mammalian cells are more sensitive to the mutagenic effect of UV than bacteria. This is actually more pronounced than apparent from Figure 6, since the bacterial strain shown is deficient in excision repair (see Section V) and therefore more susceptible to UV mutagenesis. It is also evident from

Figure 6 that the amount of mutagenic response depends on the genetic locus under study. One reason for this is that target theory dictates that to hit a locus requires fewer photons as its size increases. In addition, inactivation of a function that is not essential for growth, as is Thg^r, would be accomplished by various types of genetic lesions. Oua^r on the other hand involves an essential function and would therefore show mutagenic response only to specific types of changes in DNA (e.g., specific base substitutions) that do not inactivate this function.

3. The Premutational Lesion

Most far UV-induced mutations probably result from the formation of Pyr <> Pyr in bacterial DNA. The strongest evidence for this comes from studies involving photoreactivation. With the *E. coli* mutant strain lacking the enzyme for splitting of Pyr <> Pyr, no photoreactivation of UV-induced mutation lesions was seen, while such photoreactivation did occur with the parental strain possessing this enzyme.[65] However, as will be discussed later, this evidence is not conclusive.

In the case of suppressor mutations for reversion of auxotrophic bacteria, the mutagenic process appears to require two types of DNA-damaging events, an induction/indexing signal and a premutational photoproduct that targets the mutation. While the first of these could be any lesion that disrupts DNA synthesis to stimulate SOS induction or index-critical overlapping daughter-strand gaps[66] (see Section V), the targeting photoproduct could be unique. Using acetophenone as a triplet photosensitizer to change the ratio of Pyr <> Pyr to Thy <> Cyt, it was shown that for class 2 suppressor mutations (those resulting from G·C → A·T transitions) any type of dimer could serve as an induction/indexing signal, but only Thy <> Cyt is a premutational lesion at the target site.[67] Targeted mutagenesis was also shown to occur in mammalian cells, probed with SV40 DNA, through a base pair substitution pathway.[199]

Pyrimidine adducts of the type Pyr(6-4)Pyo can also be mutagenic. This was deduced from a correlation between the UV-induced distribution of nonsense mutations hotspots along the *lac I* gene of *E. coli* and that of UV-induced DNA damage.[68] The correlation is better for Pyr(6-4)Pyo than for Pyr <> Pyr. Using the same system, it was also shown that each of these hotspots occur at a site in the potential hairpin loop of quasi-palindromic sequences.[69] These observations suggest an important role for DNA structure in determining the fate of UV-induced premutational lesions. This is not surprising, since secondary structure has been predicted to play a role in a number of processes, including the initiation of DNA replication and the regulation of transcription.

Currently, there is no information with regard to the mutagenicity of other photolesions in DNA, except strand breaks. These do not appear to be mutagenic.[50]

4. The Mutation Process

Mutagenesis due to photodamage in DNA does not arise simply by the introduction of errors when the lesions are being replicated. This was realized when it was discovered by the pioneering work of Weigle[70] that irradiation of bacteriophages is not sufficient to induce mutation. Irradiation of both the bacterial virus and the host cell is necessary for mutagenesis of the virus. This so-called "Weigle mutagenesis" does not occur when the host cells are *rec A*$^-$ or *lex A*$^-$,[66] in which SOS repair is deficient (see Section V). Such bacteria are not mutagenized by UV, indicating that active participation of the SOS system is required to convert premutational lesions into actual mutations. This system is induced by the inhibitory effect of UV on DNA replication under conditions which allow RNA and protein synthesis. The activity that is induced facilitates DNA synthesis on a damaged template, inserting bases opposite Pyr <> Pyr nonrandomly.[71] A bias toward specific error could result because the chemical nature of Pyr <> Pyr in template DNA is significantly different due to changes in tautomeric configurations. This could result in cytosine bonded in a dimer pairing with adenine. Alternatively, insertion of adenine residues at a noninstructive dimer may be an inherent characteristic of DNA polymerase in transdimer synthesis. The final result is a *de facto* bias for G·C to A·T transitions at cytosine-containing pyrimidine dimers. Modulation of this process can be effected by DNA structural isomers that may transiently form during DNA replication, just when mutations are being fixed by the error-prone repair.[69] Expression of the induced mutations occurs after the DNA containing a fixed mutation is replicated.

The requirement for two events for the mutagenic response in bacteria complicates the interpretation of photoreactivation experiments as an indication of the role of Pyr <> Pyr as premutational lesions. Exposure to UV light followed shortly by photoreactivation removes most of the damage that induces the SOS functions. Under these conditions the SOS functions might never be induced, resulting in drastic reduction in the yield of mutations. Therefore, photoreactivation phenomena may not reflect removal of premutagenic lesions, but rather reflect reversal of the signals necessary to induce the genes required for mutation fixation.

The information regarding the mutation process in mammalian cells is scant. Unlike bacteria, mutagenesis of UV-irradiated viruses can be expressed in unirradiated human cells. The yield of viral mutants can be increased by twofold when the host cells are irradiated 24 hr prior to infection.[72] This, and the linear fluence-response of mutation induction (Figure 6), suggest that mutagenesis in mammalian cells does not require the induction of an error-prone DNA repair. Further evidence for this comes from experiments in which inhibition of protein synthesis in Chinese hamster cells failed to affect UV mutagenesis.[73] Changes in postirradiation DNA synthesis, however, do affect the yield of mutants.[73] This finding can be explained by the fact that imbalanced nucleotide pools produced by UV facilitate errors of replication of damaged DNA,[74] due to decreased fidelity of DNA synthesis by eukaryotic DNA polymerases α and β.[75]

A functional DNA excision repair is important in determining the mutagenic response in both bacteria and mammalian cells. Thus, Chinese hamster[76] and human fibroblasts[77] that are deficient in the initial step of excision are hypermutable as compared with normal fibroblasts. This is most pronounced when comparison is made at the same fluence, although a significant difference remains when cells are compared at equitoxic fluences. Presumably, removal of premutational lesions by excision repair before they can be fixed is the cause of lowered mutability in normal cells.

As for cell killing, the mutagenic response of mammalian cells to UV is cell-cycle dependent. However, the exact age-response structure for mutagenesis has not been defined and is somewhat more difficult to interpret. It will depend, for example, on the time during S that the genetic locus under study is replicated. A detailed action spectrum for mutagenesis in mammalian cells is available for the range 235 to 313 nm.[79] It is similar to that of cell killing and Pyr <> Pyr production, suggesting DNA as the target for UV-induced mutagenesis. Using continuous spectrum near UV with various cutoffs at the shorter wavelengths it was shown that mutagenicity falls off rapidly with increased wavelength at 290 nm.[78] On the basis of induced dimers, at $\lambda >$ 310 nm, 50 to 60 times as many mutants were induced per dimer than by $\lambda >$ 290 nm UV; the latter appeared approximately as mutagenic as 254 nm. These results suggest that UVA induces mutagenic lesions other than Pyr <> Pyr.

C. Oncogenic Transformation

The in vitro transformation of normal mammalian cells into cells that are able to form tumors is a model system for studying carcinogenesis in a quantitative manner. The first, and still most widely used such system is 10T$^{1}/_{2}$ cells derived from a C3H mouse embryo, introduced in 1973 by the late Charles Heidelberger. These are fibroblasts that demonstrate contact inhibited growth in monolayer culture. Morphologically transformed cells are identified as a colony that overgrows the confluent sheet of cells. Two types of such colonies produce fibrosarcomas when inoculated into appropriately prepared C3H mice. The transformation of a cell by an initiator like UV light can be expressed as a frequency per surviving cell. The first fluence-response curve published for mouse cells exposed to 254 nm[80] showed a linear increase in transformation frequency, up to 7.5 J/m^2. No further increase was observed at higher fluences. Surprisingly, the yield of transformants is tenfold higher than that of mutation induction to Thg^r. This must be due to a tenfold larger target for the former process and suggests that a hit in any one of several genes can initiate the transformation process. Such assumption is consistent with the recent observations that there exist in mammalian cells several oncogenic genes that are involved in the transformation of a normal cell into a cancer cell.

Evidence that UV-induced transformation is due to damage in DNA, mainly Pyr <> Pyr, comes from two kinds of experiments. First, the action spectra for neoplastic transformation, production of Pyr <> Pyr, and lethality between 240 and 313 nm in Syrian hamster cells, were essentially the same.[81] These spectra correspond also to the absorption spectrum of DNA. Second, UV-irradiated thyroid cells of the fish *Poecilia formosa*, when injected into isogeneic recipients, give rise to thyroid tumors. If the UV light is followed, but not preceded, by photoreactivation, the yield of thyroid tumors is markedly decreased. Because of the specificity of photoreactivation to Pyr <> Pyr, these data are strong evidence that production of dimers in DNA can give rise to tumors.

Whether or not oncogenic transformation in rodent cells is mediated via an inducible, SOS-like process is not known. Such a process does not appear to operate in the plateau phase (G_o), where enhanced transformation during split fluence UV irradiation is due to plateau-phase holding (see Section VI.B) rather than to the influence of fractionation per se.[82] Since transformation is cell-cycle dependent, it remains to be seen if it can be potentiated by split fluence in other phases of the cell cycle.

In contrast to rodent cells which readily undergo spontaneous as well as UV-induced malignant transformation in culture, cells from normal human donors are difficult to transform by carcinogenic agents. A small fraction of transformed progeny can be obtained in human cells by infection with simian virus 40 (SV40). This fraction can be greatly enhanced by prior exposure to UV.[84] Since this UV-enhanced transformation is not observed in repair-deficient human cells (XP), it is probably mediated by cellular DNA repair functions. XP patients are highly susceptible to cancer caused by UV light (see Chapter 2, Volume II). Viruses similar to SV40, therefore, do not play a major role in carcinogenesis of XP patients. Because of the above, it is not clear what, if any, relationship exists between transformation in vitro and UV-induced carcinogenesis in man. This is said in spite of the apparent relationship between the two processes in mice.[95] For further discussion see Chapter 4.

D. Chromosomal Damage

Morphologically, an early manifestation of DNA damage, such as that caused by exposure to UV light, is the appearance of cells containing chromosome and chromatid breaks and rearrangements. It is therefore not surprising that the action spectrum of the frequency of chromatid aberrations per cell is similar to the absorption spectrum for DNA,[85] and that Pyr <> Pyr are implicated as the main initiating lesions for both chromosomal aberrations and sister-chromatid exchanges.[85-88] For the latter, about 2×10^4 Pyr <> Pyr have to be introduced in Chinese hamster cells for each additional exchange, and an even larger number of dimers are necessary to produce a chromosomal aberration. One of the reasons for the substantial number of dimers is that only those dimers that are not removed by excision repair can participate in producing aberrations. Thus, repair-deficient cells from XP patients are more susceptible than normal cells to the clastogenic effect of UV-mimetic agents.[89]

Further evidence for the role of repair processes comes from studies employing inhibitors of repair. The most widely used, although its mode of action is not clear, is caffeine.[90] When added after UV irradiation, the most striking effect of caffeine was the appearance of a high incidence of cells with multiple aberrations or "shattered chromosomes", also called pulverized chromosomes.[91] A probable explanation is the prevention by caffeine of the G_2 arrest produced by UV. As a result, cells enter mitosis before repair is completed, the damaged chromosomes are shattered, and the cells die. Indeed, there is a pronounced enhancement of UV-induced cell killing by caffeine.

Most recently, new observations indicate that some of the UV-induced chromosome aberrations are produced indirectly.[92] In the case of sister-chromatid exchanges it is 25%. The evidence comes from fusing UV-irradiated mouse cells with unirradiated Chinese hamster cells. The results suggest that unirradiated chromosomes in heterokaryons are damaged by a stable, nondiffusible cytoplasmic component contributed by the irradiated cell.[93] This active factor could be a nucleolytic enzyme released by radiation-induced rupture of lysosomes.[94]

IV. BIOCHEMICAL RESPONSES TO UV LIGHT

A. Inhibition of DNA Synthesis

An immediate and pronounced biochemical response to UV light-irradiation is inhibition of DNA synthesis. This is to be expected if DNA is the main target to UV in the cell. Damage in DNA is likely to interfere with the ability of the DNA to serve as a template for its replication, as well as for RNA synthesis.

In vitro, the residual replications detected when UV-irradiated templates are used proceed semiconservatively and result in the production of pieces of duplex DNA approximately the same size as the average distance between Pyr <> Pyr.[96] This is taken as evidence that dimers are blocks to the progression of the replication fork. In the case of the single-stranded ϕx174 phage DNA as a substrate, Pyr <> Pyr act as an absolute block for chain elongation by DNA polymerase I and III.[97] Nucleotide turnover due to idling by DNA polymerase (i.e., incorporation and subsequent excision of nucleotides opposite the dimer, by the $3' \rightarrow 5'$ "proofreading" exonuclease) prevent replication past the dimer and the potentially mutagenic event that should result. Mammalian DNA polymerase-α that is devoid of exonuclease activity is only partially inhibited by UV irradiation of the template, accompanied by an increased incorporation of noncomplementary nucleotides.

In repair-proficient bacteria, UV fluences below 20 J/m^2 induce short, fluence-dependent delays in DNA replication. Larger fluences do not cause further delay, but the rate of DNA replication is decreased.[98] The immediate stoppage of DNA synthesis after a small UV fluence is controlled by the *rec A* gene product (see Section V.D).[99] Presumably, it enhances the efficiency of repair by allowing the excision enzymes to act on the damaged DNA without interference with DNA replication. Dimers are not permanent blocks to the replication of bacterial DNA. The replication fork can skip past a dimer and continue its normal operation, leaving a single-strand gap behind (see Section V.C).

DNA replication is also depressed in UV-irradiated mammalian cells, but the mechanism is not as clear as it is in bacteria. To begin with, the mammalian chromosome is much larger and its organization more complex than that in bacteria. Its replication proceeds by a hierarchical mechanism in which individual replicating units (replicons) are replicated bidirectionally from central origins and grouped into clusters of ten or more that initiate replication synchronously. Adjacent clusters replicate independently, but are under some higher control that regulates the time in S phase in which replication takes place. Replication forks from adjacent origins approach one another until termination occurs. The size of an individual replicon varies over a wide range (10 to 100 μm). In human cells, the average rate of DNA chain growth is 0.5 to 1.0 μm/min at each fork.

Following exposure to low, noncytotoxic fluences (1 J/m^2, producing less than one Pyr <> Pyr per replicon) initiation of DNA synthesis in new replicons is inhibited. This effect is exerted at the replicon-cluster level. Higher fluences inhibit DNA synthesis in operating replicons presumably because the elongation of nascent strands is blocked when the replication fork arrives at a dimer. The net result is inhibition of chain elongation.[100] This effect is more pronounced in XP cells which are unable to remove Pyr <> Pyr ahead of DNA growing points. Dimers, however, are not permanent blocks to replication and are bypassed sooner or later. After fluences 10 J/m^2, reinitiation beyond a dimer does not occur for at least 1 hr.[101] Indeed, in frog cells, Pyr <> Pyr can act as blocks to nascent DNA synthesis even 24 hr after UV irradiation.[102]

Cells recover their capacity to replicate DNA as time passes after irradiation. The rate at which cells recover their ability to synthesize more and larger-size DNA is a function of both replicon size and of excision-repair capacity.[103] Cells with small replicons recover more rapidly than cells with large replicons, and excision repair-deficient cells recovered less rapidly than excision-competent cells. Recovery of replicon initiation rate after a low fluence occurs within several hours.[100] In part, this recovery is due to completion of excision repair and reinitiation of replication in previously suppressed replicon clusters. However, since recovery is faster than the actual rate of dimer excision, it probably depends on more factors.

B. Inhibition of RNA Synthesis

The major effect of UV photoproducts in bacterial DNA on RNA synthesis is to cause premature termination of RNA chain growth and release of the growing RNA chains and RNA polymerase at the site of DNA photodamage.[104] As a consequence, the rate of transcription is depressed and, on a given transcription unit, the expression of promoter-distal genes appears to be more sensitive to irradiation than that of promoter-proximal genes. Promoter binding and RNA chain initiation steps are relatively insensitive to UV light.

The probability of causing a UV lesion within a transcription unit is directly proportional to its length, making it possible to use UV fluence response to measure the distance between a gene and its promoter. In human cells for example, the synthesis of low molecular weight nuclear RNAs C and D (less than 200 nucleotides long) is highly sensitive to UV because their transcription units are as long as 5 kb.[105] The RNA chains transcribed in UV-irradiated cells are reduced in size. This effect is fluence dependent. The ability of mammalian cells to synthesize full-length RNA transcripts is restored during postirradiation incubation of both primate[106] and murine[107] cells. In the latter, Pyr <> Pyr are not excised and recovery therefore is likely to take place due to the synthesis of DNA daughter strands by a postreplication repair process. In human cells, RNA synthesis recovers quite rapidly after UV irradiation.[108] This recovery is too rapid to be accounted for by excision repair, and, since it occurs in the absence of DNA synthesis, it can neither be explained by postreplication repair. Presumably, for recovery to occur, the transcription machinery is altered in such a way that Pyr <> Pyr no longer cause termination of growing RNA chains and RNA polymerase is able to proceed past the lesions. It should be pointed out that in nondividing human fibroblasts, the lethal action of UV light is due to inhibition of required protein synthesis by preventing RNA transcription.[109] Under these conditions, recovery appeared to depend on the excision-repair capacity of the cells.

C. Effects on Protein and Inducible Enzymes Synthesis

The protein-synthesizing machinery is composed of mRNA, tRNA, ribosomes, and various enzymes and co-factors. RNA in the first three components can be photodamaged by UV light, leading to inhibition of protein synthesis.[110] However, because of target size and multiplicity of the components, inhibition of protein synthesis immediately after UV irradiation requires considerably higher fluences than for inhibition of DNA and RNA syntheses. This is the case for cells growing in culture.[111] The situation in vivo, however, is more complicated. Thus in granular cells of human skin (differentiated epidermal cells) protein synthesis was markedly inhibited within 1 to 3 hr after a mild erythemal dose of UV light.[112]

Some enzyme proteins are not synthesized continuously but only when the need arises, as a result of an inducing signal (inducible enzymes). The induction process usually requires that the appropriate mRNA be first synthesized (transcriptionally controlled induction). In such a case, exposure to UV would be expected to inhibit mRNA synthesis, resulting in reduced amounts of enzyme. Such inhibition of induced enzyme synthesis occurs in human cells following exposure to UV fluences in the biological range (3 to 10 J/m^2). In hepatoma cells, inhibition of tyrosine aminotransferase is inhibited by UV due to inhibition of transcription while pre-existing cytoplasmic mRNA is not inactivated nor its translation inhibited.[113] In human breast carcinoma cells, induction of ornithine decarboxylase is similarly inhibited by low UV fluences. The cells are able to recover from this inhibition by a process that depends on excision repair of DNA damage.[114] These effects of UV on inducible enzyme synthesis should be taken into consideration when searching for a UV-induced DNA repair pathway in mammalian cells (see Section V).

It should be noted that the opposite can also occur. Many genes in mammalian cells are quiescent and can be converted into the fully expressed state by gene activation followed by activation of transcription. Exposure to low UV doses was shown to induce metallothionein-I gene activation in mouse cells.[195] The process is associated with extensive demethylation of the DNA region containing the gene. The extent of methylation of cytosine is thought to regulate gene activity, and exposure to UV can produce demethylation of the repaired sites of DNA.[196] This provides a model for UV carcinogenesis unrelated to point mutagenesis. According to this model, under appropriate conditions UV or chemical carcinogens might activate known cellular oncogenes by causing demethylation. The activated oncogenes then cause transformation of the normal cell into a cancer cell.

V. REPAIR OF DNA PHOTODAMAGE

The fact that organisms have the potential of counteracting the damage caused by solar radiation was recognized only a few decades ago. The idea that cells might be able to prevent radiation-induced damage by repair mechanisms first arose due to observations that cell survival depended on postradiation treatment of the cell. Because of the central role of DNA in both the life of the cell and as a target for UV light, it is no surprise that these mechanisms evolved primarily to repair DNA damage. The elucidation of repair pathways and the enzymes mediating them is critical to our understanding of the effects of sunlight on man.

A. Photoreactivation

The phenomenon of biological recovery when UV-irradiated cells are subsequently exposed to longer wavelength light was first recognized in fungi.[115] Photoreactivation is an enzymatic repair process that occurs in almost all organisms, from bacteria to man.[116] Rupert[117,118] was the first to show that the photoreactivating enzyme (the exact term is DNA photolyase) repairs DNA in a light-dependent reaction. The enzyme is highly specific to Pyr <> Pyr (*cis-syn*).[5,119] It first binds to the dimer-containing region of DNA, forming an enzyme-substrate complex. The complex can absorb light of wavelengths 300 to 600 nm; the dimer is monomerized,[120] yielding repaired DNA and releasing the enzyme for further repair.

$$\text{Pyr} <> \text{Pyr} + \text{E} \rightarrow (\text{Pyr} <> \text{Pyr} \cdot \text{E}) \xrightarrow{h\nu} 2\ \text{Pyr} + \text{E}$$

In eukaryotes, this scheme may involve an additional step, requiring an activating dose of light to produce an activated enzyme-substrate complex.[121]

Photoreactivation is an efficient and economical repair pathway, using the external visible photon as an energy source. It is an error-free, nonmutagenic repair that involves no intermediate structural changes in DNA. Because of its simplicity it is the best characterized repair process. In bacteria, the number of enzyme molecules per cell is small (five to ten in *E. coli*). This number is higher in eukaryotes in order to compensate for a lowered efficiency and a larger amount of DNA per cell.[121] The enzymatic activity is mediated by rather different proteins in the range of 35 to 130 k mol wt. The wavelength of maximum efficiency varies from 365 to 435 nm, and so does the wavelength range over which the various enzymes are active. The short wavelength extent, 300 nm, is probably limited by the ability of these wavelengths to produce dimers directly in DNA. The long wavelength edge varies from about 450 nm for the *E. coli* and yeast enzymes to 600 nm for human[122] and marsupial[123] enzymes. These differences are probably due to the presence of various prosthetic groups serving as chromophores to the various DNA photolyases. Thus, in baker's yeast, the enzyme is a flavoprotein, containing flavin adenine dinucleotide (oxidized) as a prosthetic group.[124] In *Streptomyces griseus* the chromophore appears to be a 5-deazaflavin derivative.[125] However, there is another class of photoreactivating enzymes that appear to lack an intrinsic chromophore (judged from the absence of absorption of the purified enzyme in the visible or near-UV region). During the complex formation of these enzymes with irradiated DNA, a charge-transfer type of absorption band is induced, thus providing the opportunity to absorb photoreactivating light.

The amount of DNA photolyase per cell depends not only on the species of the organism but also on the growth phase and the medium in which the cells are grown. For complex organisms, it may vary with the organ and the animal age or developmental stage.[126] All these results point to cellular regulation of the enzyme levels by unknown effectors, necessitating a careful control of experimental variables when measuring biological or biochemical photoreactivation. In vivo, photoreactivation (monomerization of Pyr <> Pyr) was found to occur in the dermis but not in epidermis of neonatal mouse skin. No photoreactivation could be detected in the skin of adult mice.[127] In human skin, after suberythemal doses, photoreactivation is highly efficient, leading to the disappearance of one half of the dimers in 1.3 min.[128] Thus, during *normal* sunlight exposure, photoreactivation may be the major repair path. The importance of this repair pathway is emphasized by the low levels of DNA photolyase in cells from XP patients.[129] The molecular cause of the lowered photoreactivating enzyme activity in XP patients is not known. It could be a mutation in the structural gene for the enzyme or a mutation in a control region leading to a reduced rate of enzyme production. In any case, the defect appears to be inherited in a simple Mendelian pattern where the heterozygote XP parents are *Pp* and the XP children are *pp*.

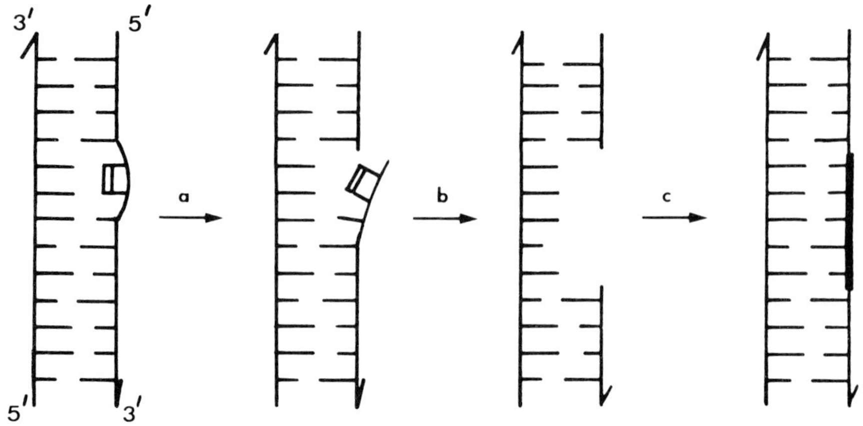

FIGURE 7. Schematic representation of excision repair of Pyr <> Pyr in DNA. (a) UV-irradiated DNA containing a dimer is attacked by an endonuclease specific for such damage, or merely recognizing the distortion in the double helix, to produce a nick adjacent to the dimer; (b) an exonuclease removes the dimer together with a few additional nucleotides on the 3′ side; (c) DNA polymerase inserts the missing nucleotides using the opposing DNA strand as a template (repair synthesis), followed by DNA ligase which joins the newly synthesized DNA piece to the pre-existing strand. The arrows indicate the direction at which the strands are synthesized.

B. Excision Repair

Excision repair is the main pathway by which photodamage in DNA is repaired in the dark. The existence of this repair was discovered in *E. coli* over 20 years ago[6,130,131] and has since been shown to occur in all organisms that were studied. Figure 7 shows the major elements of excision repair, but the details, not all of them known, vary among organisms. The first step is recognition of the damage (Pyr <> Pyr or other photoproducts) or the distortion it causes in the DNA double helix. This is usually performed by a UV-specific endonuclease which then cleaves the phosphodiester backbone 5′ to the dimer. The incision event can be more complicated, involving two steps, as in *Micrococcus luteus* and phage T4.[132,133] In these organisms the UV endonuclease has two activities, an *N*-glycosylase activity that cleaves the *N*-glycosyl bond between the 5′ pyrimidine of a dimer and the corresponding sugar, and an apyrimidinic (AP) endonuclease activity that cleaves a phosphodiester bond 3′ to the newly created AP site.

Incision is followed by excision of the dimer and adjacent nucleotides. In vitro, this step can be performed by various exonucleases (phosphodiesterases that require a terminus to initiate hydrolysis), including the 5′ → 3′ exonucleolytic activity of DNA polymerases I and III of *E. coli*. In *E. coli*, however, most excision is probably mediated by the *uvr ABC* enzyme which cuts the DNA strand on both sides of a Pyr <> Pyr (or other bulky adducts in DNA).[134] In human cells, two exonucleases whose substrate specificities suggest repair roles are DNAse VII and VIII. These enzymes release the dimer and a short oligonucleotide, generating a gap of about 10 nucleotides in length (for a recent review of enzymes involved in repair see Reference 135).

The step of reinserting the excised nucleotides is carried out by various DNA polymerases. In *E. coli*, pol I and III are able to do this, at least in vitro, in coordination with excision because of their associated 5′ → 3′ exonuclease capabilities. However, despite the lack of DNA polymerase-associated exonuclease activity in mammalian cells, it is not certain the excision resynthesis is without coordination in eukaryotic systems. Coordinated repair is indicated by findings that Pyr <> Pyr excision is stimulated by the four deoxynucleoside triphosphates and Mg^{2+} in sonicated human cells[135] and that DNA polymerase-β is stimulated by the incision of AP DNA by a $5' \rightarrow 3'$-exonuclease-AP endonuclease combination from human lymphoblasts.[137] Also, the UV excision-repair system appears to involve a complex of protein subunits in the order of 1 million daltons,[138] which is probably composed of a few enzymatic activities working in concert. Of the best-characterized mammalian polymerases, α, β, and γ, the latter is mitochondrial DNA polymerase, and has no role in nuclear functions such as replication or repair. Pol α has primarily a replicative role but can be involved, together with pol β, in DNA repair synthesis. The latter is primarily a repair enzyme. The fraction of repair mediated by each of the two polymerases is dependent on the DNA-damaging agent and on the amount of damage.[139] Pol α participation increases with increasing DNA damage. The maximal level of its involvement is $\sim$80% following UV radiation (12 J/m^2).

The final integrity of repaired strands is restored by a single enzyme species in bacteria. When nucleotide juxtaposition has been achieved by the repair polymerase, polynucleotide ligase restores strand continuity.[140] In mammalian cells, there are two distinct forms of the enzyme activities.[141] Both activities require Mg^{2+} and ATP as cofactors; however, DNA ligase I is more stable to heat and has a broader pH optimum than DNA ligase II.

The use of UV-sensitive mutants which are deficient in various repair enzymes has allowed a detailed description of the excision-repair pathway in bacteria. Excision-repair deficient mutants are designated *uvr*$^-$. Because incision is the rate-limiting step in excision, the deficiency in *uvr* mutants is usually in the incising endonuclease. Unfortunately, very few UV-sensitive mutants were isolated from mammalian cells. In humans, excision-repair deficiency was observed in XP patients[142] which is analogous to the *uvr* mutation in bacteria (see Chapter 2, Volume II).

Excision repair can be measured by following the various steps of its pathway. Because the process is coordinated, incision breaks are present only transiently and do not accumulate. However, by using inhibitors of the polymerization and/or ligation steps, breaks can be accumulated and the rate of excision measured at any time after UV irradiation.[143] The reason only the rate of excision can be measured and not its total amount is that the inhibition of late steps has a feedback effect on incision. The first inhibitor employed was hydroxyurea,[144] which inhibits the enzyme ribonucleotide reductase, thus preventing the synthesis of precursors for DNA replication. Additional inhibitors now in use are arabinosyl cytosine and aphidicolin, which are specific for pol α, and dideoxythymidine, a pol β specific inhibitor.

The excision step can be measured by determining the reduced dimer content of the DNA with time after UV exposure. Using this approach, it can be shown that human cells remove all the dimers from DNA within 24 hr, both in vitro[145] and in vivo[145] following 10 J/m^2 of 254 nm or 2 minimal erythema doses (MED) of UVB, respectively. Excision becomes progressively saturated after higher fluences. About 30 nucleotides are removed per Pyr <> Pyr dimer excised[147] but there are no clues as to why the cell removes so many nucleotides. One hypothesis that is consistent with the measurements maintains that excision is terminated when a specific base sequence is encountered.[148]

Insertion of new nucleotides into the excision gap is termed DNA repair synthesis. It was first measured by supplying cells undergoing repair with bromodeoxyuridine, an analog which is heavier than thymidine, and measuring the density shift of the DNA.[149] Bromodeoxyuridine incorporated into DNA during repair replication can be subsequently photolysed with 313 m light, resulting in DNA strand breaks which are easily quantitated.[150] This is an extremely sensitive method. For some as yet unexplained reason it yields an estimate of average patch size which is about twice as large as that estimated from the density-shift method. Another often-used method is termed unscheduled DNA synthesis (UDS) in which cells are tested for their ability to incorporate radioactive precursors into DNA in response to UV at times of the cell cycle other than the normal synthetic period.[151] Quantitation is made by means of autoradiography. Its main disadvantage is that it does not measure repair synthesis in S-phase cells.

The completion of excision repair by DNA ligase cannot be measured directly in the cell because intermediate structures for ligation do not accumulate. Interestingly, DNA ligase is the only enzyme participating in repair which was shown to be induced *de novo* in UV-irradiated cells.[152] The activity of this enzyme is further regulated in irradiated mammalian cells by synthesis of the polymer poly (ADP-ribose).[153] Thus, it appears important for the cell that very few breaks are present in DNA at any time during repair.

DNA in mammalian chromatin is organized as nucleosomes consisting of a core region and a linker region. The core region is composed of the inner histones around which is wrapped $\sim$145 bp of DNA. The cores are joined together by shorter stretches of linker DNA. In this arrangement, the nucleosomes appear in electron micrographs like beads on a string. During excision repair, gap filling (and presumably dimer excision) occurs initially faster in the linker regions.[154] In time, however, nucleosomal rearrangement leads to homogeneous distribution of excision in all the chromatin regions. Since induction of Pyr <> Pyr does not occur preferentially in linker regions, it appears that structural constraints placed upon DNA in chromatin also place constraints upon the excision-repair process. Presumably, the DNA in the core regions is less accessible to repair enzymes.

Another important parameter that appears to affect the rate of excision repair is age. Although this is still controversial, a negative correlation was obtained between the age of an individual and the capacity of his peripheral leukocytes to perform UV-induced repair synthesis.[155] A significant decline in DNA repair after 14 months of age was also observed in UV-irradiated rat hepatocytes.[200] Moreover, there is a positive correlation between the ability of fibroblasts to perform UDS and the life span of the species.[156] However, more extensive studies reveal unexplained deviations from this rule.[157] Probably, alternative repair systems, especially postreplication repair, must also be considered.

The repair pathway discussed so far is strictly termed nucleotide excision repair, to distinguish it from base excision repair. The latter pathway operates to remove a damaged base by means of DNA glycosylase.[176] This enzyme, which is specific for various damaged bases, catalyzes hydrolysis of the glycosidic bond linking the base to the sugar in DNA. This is followed by an AP endonuclease that incises the AP site. The rest of the scheme is excision resynthesis as in nucleotide excision repair. Most glycosylases are specific for alkylated bases and only one in *M. luteus*, mentioned earlier, recognizes Pyr <> Pyr and initiates excision after UV irradiation. It is quite likely that there are glycosylases specific for photoproducts other than dimers but none has been described so far.

C. Postreplication Repair

When growing cells are exposed to UV, there is a certain probability that the DNA replication fork will encounter a dimer before it is excised. Bacteria handle such dimers by a bypass-recombinational mechanism,[158] also called postreplication repair since the actual repair occurs after replication of the damaged DNA. Figure 8 shows an outline of the main features of this repair. In the course of DNA replication, the polymerase leaves a gap opposite the dimer. Subsequently, a recombinational mechanism yields a new configuration of the strands so that the dimer-containing region is annealed to a

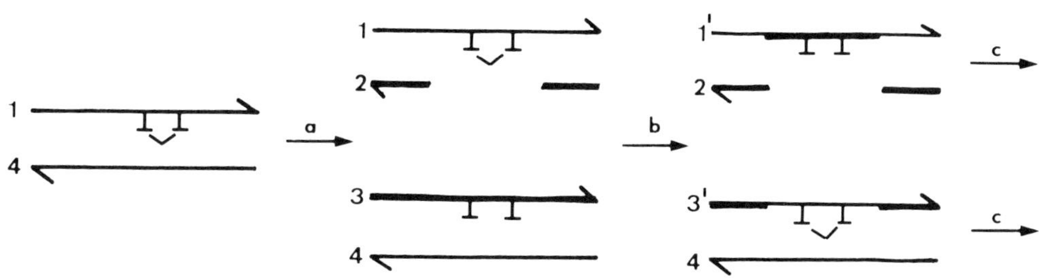

FIGURE 8. A scheme showing postreplication repair in bacteria. (a) DNA containing a Pyr <> Pyr undergoes replication to yield two types of daughter duplexes. Arrows on strands indicate polarity of DNA. (b) By recombination, dimer in strand 1 is transferred to strand 3 and TT sequence from strand 3 is transferred to strand 1. (c) Duplex 1′-2 now requires only gap filling to complete repair, while duplex 3′-4 can be repaired by excision repair (see Figure 7).

region of DNA containing the correct information. The two fundamental reactions involved are homologous pairing and strand exchange, both of which are performed by the *rec A* protein in the presence of ATP.[163,164] The new configuration can now be repaired by excision. The gapped DNA is also opposed by a strand containing complete information and can be filled in by repair replication. In bacteria, this repair pathway is controlled by the *rec A* gene product, mutations in which render the cells sensitive to UV.

Recently, a new mechanism, controlled by the *rec B* gene product has been proposed.[194] This mechanism involves double-strand exchanges between two DNA molecules and was termed sister duplex recombination to distinguish it from gap-filling recombinational repair.

In mammalian cells, the information with regard to postreplication repair is still sketchy. As in bacteria, DNA synthesized in mammalian cells after UV irradiation is shorter in size,[159] but whether it contains gaps and are these gaps opposite dimers has not been settled.[160] Also unsettled is how the gaps, if they exist, are filled. Initially it was thought that this is accomplished by *de novo* DNA synthesis.[159] More recently it was demonstrated that 1 to 3% of the dimers in human lymphocytes can be detected in the newly synthesized daughter strands at 34 to 45 hr after UV irradiation,[161] indicating the occurrence of recombination. Unlike bacteria, very little, if any, DNA can be detected by density-shift experiments to exchange between parent and daughter strands in mammalian cells.[162] Therefore the exchanges must involve relatively short segments of DNA. The lack of mammalian cells mutants analogous to bacterial *rec*$^-$ mutants has also hindered clarification of postreplication repair in mammals. An exception may be a class of XP patients, the XP variants, who show all clinical signs of XP but have normal capacity for excision repair (see Chapter 2, Volume II). There is also a lack of inhibitors specific for enzymes involved in postreplication repair, although caffeine is sometimes considered such an inhibitor.[159] In conclusion, this repair path appears to be of minor importance in ameliorating the cytocidal action of UV light in mammalian cells. However, since *de novo* filling of gaps is expected to be error prone, postreplication repair probably contributes to UV-induced mutagenesis.

D. SOS Repair and Other Inducible Repair Functions

Damage in DNA is normally produced at a very low rate. It would seem advantageous, therefore, for the cell to maintain a low level of repair enzymes constitutively, which can be induced by UV exposure and increased to a higher level. In bacterial cells, the existence of inducible repair paths has been well documented and is covered under the general term SOS repair.[165] This repair is induced by UV light and other DNA-damaging agents, prevented by the administration of protein-synthesis inhibitors after UV, and is error prone (i.e., mutagenic). The system regulating the SOS response in *E. coli* involves the action of two proteins: the *lex A* protein, which represses a set of unlinked genes (among them excision-repair and postreplication-repair genes), and the *rec A* protein. When the DNA is damaged, an inducing signal is produced which reversibly activates a specific protease activity of the *rec A* protein, and this protease cleaves and inactivates the *lex A* repressor. In consequence, the products of the SOS target genes are expressed at much higher levels. When DNA damage is repaired, the level of the signal molecule drops, the protease activity of the *rec A* protein disappears, and *lex A* protein accumulates and repress the SOS genes once again. At low levels of DNA damage, a subinduced state exists in which repressor level is reduced by a low level of cleavage.[166] A mutagenic activity, error-prone repair pathway, controlled by the *umuC* gene product, is also induced as a part of the SOS system.[167] This function is responsible for Weigle reactivation and mutagenesis (see Section III) but its biochemical mechanism has not been established. One possibility is that under SOS-induced conditions, fidelity requirements for DNA polymerase are relaxed and incorrect nucleotides can be incorporated opposite damaged sites in DNA.[168]

Interestingly, with increasing wavelength ($\lambda > 290$ nm), damage induced by solar UV radiations becomes increasingly less susceptible to SOS-induced repair systems.[169] Damage produced by 313 nm no longer induces, and is not being repaired by, the SOS system. Instead a new repair system is induced in *E. coli* by UVA.[170] Its nature is still unknown, but its fluence response is within the range of physiological exposure to UVA, suggesting that this type of repair may be an adaptation to environmental conditions.

Is DNA repair inducible in mammalian cells? A major experimental approach has been the use of viral probes. The results indicate that prior UV exposure of human[72] and monkey cells[171] induces enhanced survival and mutagenesis of UV-irradiated infecting viruses. This strongly suggests the presence of an inducible, error-prone DNA repair system. Although this repair system has not been firmly identified, dose-fractionation studies suggest enhancement of postreplication repair in Chinese hamster cells by a conditioning UV exposure.[172] Painter[173] has questioned the interpretation of such experiments but later studies seem to support it.[174] Of particular interest are the observations that UV light induces the synthesis of plasminogen activator in human cells.[175] Lower fluence (1 J/m^2) is required for induction in XP fibroblasts compared to XP heterozygotes (20 J/m^2). These results suggest that plasminogen activator induction in human cells is caused by unrepaired DNA damage and represents a mammalian SOS-like function. More recently it was shown that exposure of human fibroblasts to low doses of UV induces synthesis of a protein involved in a metabolic pathway that transiently enhances the capacity of cells to repair potentially lethal damage resulting from a subsequent dose of UV.[201]

VI. CELLULAR RECOVERY FROM UV DAMAGE

A. Repair of Sublethal Damage

Sublethal damage (SLD) is a type of damage which must be accumulated for a lethal effect to occur. Its presence in UV-irradiated mammalian cells in culture is evident from the shoulder which is often part of the survival curve (Figure 5). Such damage is subject to repair which is studied by splitting a UV fluence with a time interval between the two exposures, or by exposing the cells at a low fluence rate.[177] The operation of repair under these conditions results in enhanced survival. During a split-dose experiment, the repair of SLD is inferred from the reappearance of the shoulder on the survival curve. Continuous exposure at a low fluence rate results in a survival curve that is less steep than that obtained after acute exposure. In bacteria, enhanced survival due to fluence fractionation is partly due to excision repair, since the effect is reduced in *uvr*$^-$ strains.[178] However, the principal factor apparently is that the number of lesions present at any time remains low, allowing for a more efficient repair. In principle, SOS repair could be induced under these conditions; however, its possible contribution has not been determined.

In mammalian cells, repair of UV-induced SLD was first observed in a Chinese hamster cell-line whose survival curve possessed a pronounced shoulder.[179] Another Chinese hamster cell-line with a small shoulder did not display SLD repair. In order to repair SLD, rodent cells have to pass through S phase after the first exposure to UV.[179-181] This suggests the involvement of postreplication repair in the recovery from SLD, but a definite proof is lacking. Similar results were obtained with human cells.[182] There are no experiments in which mammalian cells were exposed to UV at a low fluence rate during a prolonged period. Also, information with regard to SLD repair following exposure to UVA and UVB is not available. Such information is relevant to the response of human skin to sunlight exposure.

B. Repair of Potentially Lethal Damage

When cells are held under nongrowth conditions following exposure to UV light and before plating to allow colony formation, their survival usually increases. This phenomenon is termed liquid holding recovery (LHR) in bacteria and repair of potentially lethal damage (PLD) in mammalian cells. LHR occurs only in *uvr*$^+$ bacterial strains,[178,183] suggesting that removal of Pyr <> Pyr by excision repair is responsible for LHR. A complicating factor in bacteria is the temporary inhibition of respiration 1 hr after UV irradiation, controlled by UV damage in DNA.[184] Heat treatment of UV-irradiated *E. coli* prevents the turn-off of respiration and is accompanied by an early increase in viability (thermal reactivation).[185] This implies that continuous energy flow is necessary for LHR, a conclusion which is supported by studies in yeast cells.[186]

Repair of PLD in mammalian cells, observed by delayed plating, occurs only in plateau phase. In mouse fibroblasts, it was suggested that PLD repair is associated with excision repair of UV damage in DNA.[187] The elimination of PLD repair in UV-irradiated human[114] and monkey cells[188] by hydroxyurea and aphidicolin, respectively, strongly supports this view. The strongest evidence that PLD repair in human cells depends on excision of Pyr <> Pyr comes from the work of Maher et al.[189] They showed that skin fibroblasts from normal individuals eliminate potentially cytotoxic (and mutagenic) lesions induced by UVC if held ~20 hr in plateau phase. In contrast, fibroblasts from XP patients with no excision-repair activity, showed no increase in survival even when held for 7 days in confluence after UV irradiation. However, PLD repair may depend on factors in addition to excision. Thus, fibroblasts from a patient with Gardner's syndrome exhibit no PLD repair[190] although their excision-repair capacity is indistinguishable from that of normal cells.[191]

Repair of PLD in log-phase cells can be demonstrated by postirradiation treatments that enhance UV-induced cell killing. Such treatments presumably cause expression of PLD by inhibiting its repair. Caffeine is the most potent drug in this respect. However, it affects only rodent cells,[192] while human cells appear to be refractory to its action.[193] This difference may be explained by assuming that caffeine inhibits postreplication repair but not excision. The latter is more important for survival in human cells than in rodent cells. However, caffeine has many pleiotropic effects in mammalian cells and exactly how it potentiates UV-induced cytotoxicity is still not clear.

VII. CONCLUDING REMARKS

From the foregoing it is evident that lesions in DNA are the main cause of the biological damage inflicted on cells exposed to UV from germicidal lamp, sunlamp, or sunlight. The DNA photoproduct of major biological importance is Pyr <> Pyr, although other photoproducts contribute more or less, depending on the wavelength. Photodamage in DNA is repaired enzymatically by various mechanisms which are subject to regulation. The level of repair enzymes varies with the species, the age or developmental stage of the organism, the tissue, and even the growth state of cells in culture. It is thus dangerous to rely on measurements of repair in cultured cells without actually testing repair in the intact organism.

DNA damage that fails to be corrected may not lead to cell death at low levels and may not be detectable on a cellular or organismal level. It may be in a noncoding DNA segment; ambiguity in the genetic code may allow insertion of the correct amino acid into the resulting protein; and even if the coding change results in the insertion of a new amino acid, it may be a conservative substitution allowing adequate enzyme function. In spite of all this, it is also clear that UV does induce cancers in animals and humans. This and other effects of light in man will be discussed in detail in the other chapters of this volume, using the information in the previous and present chapters as a basis.

REFERENCES

1. Beukers, R. and Berends, W., Isolation and identification of the irradiation product of thymine, *Biochim. Biophys. Acta*, 41, 550, 1960.
2. Weinblum, D. and Johns, H. E., Isolation and properties of isomeric thymine dimers, *Biochim. Biophys. Acta*, 114, 450, 1966.
3. Ben-Hur, E., Elad, D., and Ben-Ishai, R., The photosensitized dimerization of thymidine in solution, *Biochim. Biophys. Acta*, 149, 355, 1967.
4. Patrick, M. H. and Rahn, R. O., Photochemistry of DNA and polynucleotides: photoproducts, in *Photochemistry and Photobiology of Nucleic Acids*, Vol. 2, Wang, S. Y., Ed., Academic Press, New York, 1976, chap. 2.
5. Ben-Hur, E. and Ben-Ishai, R., *Trans-syn* thymine dimers in ultraviolet-irradiated denatured DNA: identification and photoreactivability, *Biochim. Biophys. Acta*, 166, 9, 1968.
6. Riklis, E., Studies on the mechanism of repair of ultraviolet-irradiated viral and bacterial DNA *in vivo* and *in vitro*, *Can. J. Biochem.*, 43, 1207, 1965.
7. Patrick, M. H. and Gray, D. M., Independence of photoproduct formation on DNA conformation, *Photochem. Photobiol.*, 24, 507, 1976.
8. Unrau, P., Wheatcroft, R., Cox, B., and Olive, T., The formation of pyrimidine dimers in the DNA of fungi and bacteria, *Biochim. Biophys. Acta*, 312, 626, 1973.
9. Hayes, F. N., Williams, D. L., Ratliff, R. L., Varghese, A. J., and Rupert, C. S., Effect of a single thymine photodimer on the oligodeoxythymidilate-polydeoxyadenilate interaction, *J. Am. Chem. Soc.*, 93, 4940, 1971.
10. Shafranovskaya, N. N., Trifonov, E. N., Lazurkin, Y. S., and Frank-Kamenetskii, M. D., Clustering of thymine dimers in ultraviolet irradiated DNA and the long-range transfer of electronic excitation along the molecule, *Nature (London) New Biol.*, 241, 58, 1973.
11. Ciarrocchi, G. and Pedrini, A. M., Determination of pyrimidine dimer unwinding angle by measurement of DNA electrophoretic mobility, *J. Mol. Biol.*, 155, 177, 1982.
12. Cozzarelli, N. R., DNA gyrase and the supercoiling of DNA, *Science*, 207, 953, 1980.
13. Benham, C. J., Theoretical analysis of competitive conformational transitions in torsionally stressed DNA, *J. Mol. Biol.*, 150, 43, 1981.
14. Brunk, C. F., Distribution of dimers in ultraviolet-irradiated DNA, *Nature (London) New Biol.*, 241, 74, 1973.

15. Rahn, R. O. and Stafford, R. S., Measurements of defects in ultraviolet-irradiated DNA by the kinetic formaldehyde method, *Nature (London)*, 248, 52, 1974.
16. Yonker, L. S. and Blok, J., Evidence against clustering of thymine dimers in ultraviolet-irradiated DNA, *Nature (London)*, 255, 245, 1975.
17. Wilkins, R. J., The non-random distribution of pyrimidine dimers in ultraviolet-irradiated DNA, *Int. J. Radiat. Biol.*, 35, 381, 1979.
18. Donnelan, J. E. and Setlow, R. B., Thymine photoproducts but not thymine dimers found in UV-irradiated bacterial spores, *Science*, 149, 308, 1965.
19. Smith, K. C. and Yoshikawa, H., Variation in the photochemical reactivity of thymine in the DNA of *B. subtilis* spores, vegetative cells and spores germinated in chloramphenicol, *Photochem. Photobiol.*, 5, 777, 1966.
20. Varghese, A. J., 5-Thyminyl-5,6-dihydrothymine from UV-irradiated DNA, *Biochem. Biophys. Res. Commun.*, 38, 484, 1970.
21. Lindsay, J. A. and Murrell, W. G., A comparison of UV induced DNA photoproducts from isolated and non-isolated developing bacterial forespores, *Biochem. Biophys. Res. Commun.*, 113, 618, 1983.
22. Varghese, A. J. and Wang, S. Y., UV-irradiation of DNA produces a third thymine-derived product, *Science*, 156, 955, 1967.
23. Varghese, A. J. and Wang, S. Y., Thymine-thymine adduct as a photoproduct of thymine, *Science*, 160, 186, 1968.
24. Patrick, M. H., Studies on thymine-derived UV photoproducts in DNA-I. Formation and biological role of pyrimidine adducts in DNA, *Photochem. Photobiol.*, 25, 357, 1977.
25. Vanderhoek, J. Y. and Cerutti, P. A., The stability of deoxycytidine photohydrates in the mononucleotide, oligodeoxynucleotides and DNA, *Biochem. Biophys. Res. Commun.*, 52, 1156, 1973.
26. Remsen, J. F., Mattern, M., Miller, N., and Cerutti, P. A., Photohydration of uridine in the ribonucleic acid of coliphage R17. Lethality of uridine photohydrates and nonlethality of cyclobutane-type photodimers, *Biochemistry*, 10, 524, 1971.
27. Yamane, T., Wyluda, B. J., and Shulman, R. G., Dihydrothymine from UV-irradiated DNA, *Proc. Natl. Acad. Sci. U.S.A.*, 58, 439, 1967.
28. Hariharan, P. V. and Cerutti, P. A., Formation of products of the 5,6-dihydroxydihydrothymine type by ultraviolet light in HeLa cells, *Biochemistry*, 16, 2791, 1977.
29. Porschke, D., A specific photoreaction in polydeoxyadenylic acid, *Proc. Natl. Acad. Sci. U.S.A.*, 70, 2683, 1973.
30. Ben-Ishai, R., Green, M., Graff, E., Elad, D., Steinmaus, H., and Salomon, J., Photoalkylation of purines in DNA, *Photochem. Photobiol.*, 17, 155, 1973.
31. Bose, S. N., Davies, R. J. H., Sethi, S. K., and McCloskey, J. A., Formation of an adenine-thymine photoadduct in the deoxynucleoside monophosphate d(TpA) and in DNA, *Science*, 220, 723, 1983.
32. Tyrrell, R. M., Ley, R. D., and Webb, R. B., Induction of single-strand breaks (alkali-labile bonds) in bacterial and phage DNA by near-UV (365 nm) radiation, *Photochem. Photobiol.*, 20, 395, 1974.
33. Bujard, H., Electron microscopy of single-stranded DNA, *J. Mol. Biol.*, 49, 125, 1970.
34. Rahn, R. O. and Patrick, M. H., Photochemistry of DNA, secondary structure, photosensitization, base substitution, and exogenous molecules, in *Photochemistry and Photobiology of Nucleic Acids*, Vol. 2, Wang, S. Y., Ed., Academic Press, New York, 1976, chap. 3.
35. Alexander, P. and Moroson, H., Cross-linking of deoxyribonucleic acid to protein following ultraviolet irradiation of different cells, *Nature (London)*, 194, 882, 1962.
36. Varghese, A. J., Properties of photoaddition products of thymine and cysteine, *Biochemistry*, 12, 2725, 1973.
37. Todd, P. and Han, A., UV-induced DNA to protein cross-linking in mammalian cells, in *Aging, Carcinogenesis, and Radiation Biology*, Smith, K. C., Ed., Plenum Press, New York, 1976, 83.
38. Smith, K. C., The radiation-induced addition of proteins and other molecules to nucleic acids, in *Photochemistry and Photobiology of Nucleic Acids*, Vol. 2, Wang, S. Y., Ed., Academic Press, New York, 1976, chap. 5.
39. Jagger, J., Ultraviolet inactivation of biological systems, in *Photochemistry and Photobiology of Nucleic Acids*, Vol. 2, Wang, S. Y., Ed., Academic Press, New York, 1976, chap. 4.
40. Setlow, J. K., Effects of UV on DNA: correlations among biological changes, physical changes and repair mechanisms, *Photochem. Photobiol.*, 3, 405, 1964.
41. Beattie, K. L. and Setlow, J. K., Killing of *Haemophilus influenzae* by integrated ultraviolet-induced lesions from transforming deoxyribonucleic acid, *J. Bacteriol.*, 100, 1284, 1969.
42. Hutchinson, F. and Kohnlein, W., The photochemistry of 5-bromouracil and 5-iodouracil in DNA, in *Progress in Molecular and Subcellular Biology*, Vol. 7, Hahn, F. E., Kersten, H., Kersten, W., and Szybalski, W., Eds., Springer-Verlag, Berlin, 1980, 1.
43. Foote, C. S., Photosensitized oxidation and singlet oxygen: consequences in biological systems, in *Free Radicals in Biology*, Vol. 2, Pryor, W. A., Ed., Academic Press, New York, 1976, 85.

44. Allison, A. E., Magnus, I. A., and Young, M. R., Role of lysosomes and of cell membranes in photosensitization, *Nature (London),* 209, 874, 1966.
45. Spikes, J. D. and Livingston, R., The molecular biology of photodynamic action: sensitized photoautoxidations in biological systems, *Adv. Radiat. Biol.,* 3, 29, 1969.
46. Ben-Hur, E. and Song, P. S., The photochemistry and photobiology of furocoumarins (psoralens), *Adv. Radiat. Biol.,* 11, 131, 1984.
47. Ben-Ishai, R., Ben-Hur, E., and Horenfeld, Y., Photosensitized dimerization of thymine and cytosine in DNA, *Isr. J. Chem.,* 6, 769, 1968.
48. Tyrrell, R. M., Induction of pyrimidine dimers in bacterial DNA by 365 nm radiation, *Photochem. Photobiol.,* 17, 69, 1973.
49. Niggli, H. J. and Cerutti, P. A., Temperature dependence of induction of cyclobutane-type pyrimidine photodimers in human fibroblasts by 313 nm light, *Photochem. Photobiol.,* 37, 467, 1983.
50. Webb, R. B., Lethal and mutagenic effects of near-ultraviolet radiation, in *Photochemical and Photobiological Reviews,* Vol. 2, Smith, K. C., Ed., Plenum Press, New York, 1977, 169.
51. Sutherland, J. and Griffin, K., Monomerization of pyrimidine dimers in DNA by tryptophan-containing peptides: wavelength dependence, *Radiat. Res.,* 83, 529, 1980.
52. Gantt, R., Jones, G. M., Stephens, E. N., Baeck, A. E., and Sanford, K. K., Visible light-induced DNA crosslinks in cultured mouse and human cells, *Biochim. Biophys. Acta,* 565, 231, 1979.
53. Rothman, R. H. and Setlow, R. B., An action spectrum for cell killing and pyrimidine dimer formation in Chinese hamster V-79 cells, *Photochem. Photobiol.,* 29, 57, 1979.
54. Moss, S. H. and Smith, K. C., Membrane damage can be a significant factor in the inactivation of *Escherichia coli* by near-ultraviolet radiation, *Photochem. Photobiol.,* 33, 203, 1981.
55. Ito, A. and Ito, T., Possible involvement of membrane damage in the inactivation by broad-band near-UV radiation in *Saccharomyces cerevisiae* cells, *Photochem. Photobiol.,* 37, 395, 1983.
56. Cleaver, J. E., Excision repair: our current knowledge based on human (xeroderma pigmentosum) and cattle cells, in *Molecular and Cellular Repair Processes,* Beers, R. F., Herriott, R. M., and Tilghman, R. C., Eds., Johns Hopkins University Press, Baltimore, 1972, 195.
57. Ben-Hur, E. and Elkind, M. M., DNA damage and its repair in hyperthermic mammalian cells: relation to enhanced cell killing, in *Radiation Research — Biomedical, Chemical and Physical Perspectives,* Nygaard, O. F., Adler, H. I., and Sinclair, W. K., Eds., Academic Press, New York, 1975, 703.
58. Stroud, A. N., Rounds, D. E., Creter, S., and El Kadi, S. S., Reproductive integrity of cells after ultraviolet laser irradiation, *Int. J. Radiat. Biol.,* 19, 157, 1971.
59. Rosenstein, B. S. and Ducore, J. M., Induction of DNA strand breaks in normal human fibroblasts exposed to monochromatic ultraviolet and visible wavelengths in the 240—546 nm range, *Photochem. Photobiol.,* 38, 51, 1983.
60. Scaife, J. F. and Brohee, H., Pyknotic interphase death of cultured human kidney cells after high doses of ultraviolet radiation, *Int. J. Radiat. Biol.,* 13, 531, 1968.
61. Han, A. and Sinclair, W. K., Sensitivity of synchronized Chinese hamster cells to ultraviolet light, *Biophys. J.,* 9, 1171, 1969.
62. Downes, C. S., Collins, A. R. S., and Johnson, R. T., DNA damage in synchronized HeLa cells irradiated with ultraviolet, *Biophys. J.,* 25, 129, 1979.
63. Han, A. M., Korbelick, M., and Ban, J., DNA-to-protein crosslinking in synchronous HeLa cells exposed to ultraviolet light, *Int. J. Radiat. Biol.,* 27, 63, 1975.
64. Singh, B. and Gupta, R. S., Mutagenic response to ultraviolet light and x-rays at five independent genetic loci in Chinese hamster ovary cells, *Environ. Mutagenesis,* 4, 543, 1982.
65. Kondo, S. and Jagger, J., Action spectra for photoreactivation of mutation to prototrophy in *Escherichia coli* phr^+ and phr^-, *Photochem. Photobiol.,* 5, 189, 1966.
66. Witkin, E. M., Ultraviolet light mutagenesis and inducible DNA repair in *Escherichia coli, Bacteriol. Rev.,* 40, 869, 1976.
67. Fix, D. and Bockrath, R., Targeted mutation at cytosine-containing pyrimidine dimers: studies of *Escherichia coli* B/r with acetophenone and 313-nm light, *Proc. Natl. Acad. Sci. U.S.A.,* 80, 4446, 1983.
68. Brash, D. E. and Haseltine, W. A., UV-induced mutation hotspots occur at DNA damage hotspots, *Nature (London),* 198, 189, 1982.
69. Todd, P. A. and Glickman, B. W., Mutational specificity of UV light in *Escherichia coli*: indication for a role of DNA secondary structure, *Proc. Natl. Acad. Sci. U.S.A.,* 79, 4123, 1982.
70. Weigle, J., Induction of mutations in a bacterial virus, *Proc. Natl. Acad. Sci. U.S.A.,* 39, 628, 1953.
71. Lawrence, C. W., Are pyrimidine dimers non-instructive lesions?, *Mol. Gen. Genet.,* 192, 511, 1981.
72. Lytle, C. D. and Knott, D. C., Enhanced mutagenesis parallels enhanced reactivation of herpes virus in a human cell line, *EMBO J.,* 1, 701, 1982.

73. Stone-Wolff, D. S. and Rossman, T. G., Effects of inhibitors of de novo protein synthesis on UV-mutagenesis in Chinese hamster cells. Evidence against mutagenesis via inducible, error-prone DNA repair, *Mutat. Res.*, 82, 147, 1981.
74. Peterson, A. R., Landolph, J. R., Peterson, H., and Heidelberger, C., Mutagenesis of Chinese hamster cells is facilitated by thymidine and deoxycytidine, *Nature (London)*, 276, 508, 1978.
75. Kunkel, T. A., Silber, J. R., and Loeb, L. A., The mutagenic effect of deoxynucleotide substrate imbalances during DNA synthesis with mammalian DNA polymerases, *Mutat. Res.*, 94, 413, 1982.
76. Thompson, L. H., Brookman, K. W., Dillehay, L. E., Mooney, C. L., and Carrano, A. V., Hypersensitivity to mutation and sister-chromatid-exchange induction in CHO cell mutants defective in incising DNA containing UV lesions, *Somat. Cell Genet.*, 8, 759, 1982.
77. Grosovsky, A. J. and Little, J. B., Mutagenesis and lethality following S phase irradiation of xeroderma pigmentosum and normal human diploid fibroblasts with ultraviolet light, *Carcinogenesis*, 4, 1389, 1983.
78. Zelle, B., Reynolds, R. J., Kottenhagen, M. J., Schuite, A., and Lohman, P. H. M., The influence of the wavelength of ultraviolet radiation on survival, mutation induction and DNA repair in irradiated Chinese hamster cells, *Mutat. Res.*, 72, 491, 1980.
79. Jacobson, E. D., Krell, K., and Dempsey, M. H., The wavelength dependence of ultraviolet light-induced cell killing and mutagenesis in L5178Y mouse lymphoma cells, *Photochem. Photobiol.*, 33, 257, 1981.
80. Chan, G. L. and Little, J. B., Induction of oncogenic transformation *in vitro* by ultraviolet light, *Nature (London)*, 264, 442, 1976.
81. Doniger, J., Jacobson, E. D., Krell, K., and DiPaolo, J. A., Ultraviolet light action spectra for neoplastic transformation and lethality of Syrian hamster embryo cells correlate with spectrum for pyrimidine dimer formation in cellular DNA, *Proc. Natl. Acad. Sci. U.S.A.*, 78, 2378, 1981.
82. Chan, G. L. and Little, J. B., Effects of fluence fractionation on UV-induced malignant transformation and cell killing in non-cycling plateau phase mouse cells, *Photochem. Photobiol.*, 36, 409, 1982.
84. Hall, J. D., Transformation of ultraviolet-irradiated human fibroblasts by simian virus 40 is enhanced by cellular DNA repair functions, *Biochim. Biophys. Acta*, 652, 314, 1981.
85. Chu, E. H. Y., Effect of ultraviolet radiation on mammalian cells. I. Induction of chromosome aberrations, *Mutat. Res.*, 2, 75, 1965.
86. Bender, M. A., Griggs, H. G., and Walker, P. L., Mechanisms of chromosomal aberration production. I. Aberration induction by ultraviolet light, *Mutat. Res.*, 20, 387, 1973.
87. Reynolds, R. J., Natarajan, A. T., and Lohman, P. H. M., *Micrococcus lutcus* UV-endonuclease-sensitive sites and sister-chromatid exchanges in Chinese hamster ovary cells, *Mutat. Res.*, 64, 353, 1979.
88. Natarajan, A. T., Van Zeeland, A. A., Verdegaal-Immerzeel, E. A. M., and Filon, A. R., Studies on the influence of photoreactivation on the frequencies of UV-induced chromosomal aberrations, sister-chromatid exchanges and pyrimidine dimers in chicken embryonic fibroblasts, *Mutat. Res.*, 69, 307, 1979.
89. Sasaki, M. S., DNA repair capacity and susceptibility to chromosome breakage in xeroderma pigmentosum cells, *Mutat. Res.*, 20, 291, 1973.
90. Timson, J., Caffeine, *Mutat. Res.*, 47, 1, 1977.
91. Cremer, C., Gremer, T., and Simickova, M., Induction of chromosome shattering and micronuclei by ultraviolet light and caffeine. I. Temporal relationship and antagonistic effects of the four deoxyribonucleotides, *Environ. Mutagenesis*, 2, 338, 1980.
92. Kellow, G. N. and Graves, J. A. M., Evidence for an indirect effect of radiation on mammalian chromosomes. II. UV-induced isochromosome formation in cell hybrids, *Cancer Genet. Cytogenet.*, 8, 297, 1983.
93. Graves, J. A. M. and Kellow, G. N., Evidence for an indirect effect of radiation on mammalian chromosomes. III. UV- and x-ray-induced sister-chromatid exchanges in heterokaryons, *Cancer Genet. Cytogenet.*, 8, 307, 1983.
94. Allison, A. C. and Paton, G. R., Chromosome damage in human diploid cells following activation of lysosomal enzymes, *Nature (London)*, 207, 1170, 1965.
95. Fisher, M. S., Kripke, M. S., and Chan, G. L., Antigenic similarity between cells transformed by ultraviolet radiation *in vitro* and *in vivo*, *Science*, 223, 593, 1984.
96. Masker, W. E. and Kuemmerle, N. B., In vitro replication of bacteriophage T7 DNA damaged by ultraviolet radiation, *Biochim. Biophys. Acta*, 609, 61, 1980.
97. Villani, G., Boiteaux, S., and Radman, M., Mechanism of ultraviolet-induced mutagenesis: extent and fidelity of *in vitro* DNA synthesis on irradiated templates, *Proc. Natl. Acad. Sci. U.S.A.*, 75, 3037, 1978.
98. Trgovevic, Z., Petranovic, D., Petranovic, M., and Salajsmic, E., *rec A* gene product is responsible for inhibition of DNA synthesis after ultraviolet irradiation, *J. Bacteriol.*, 143, 1506, 1980.

100. Kaufmann, W. K. and Cleaver, J. E., Mechanisms of inhibition of DNA replication by ultraviolet light in normal human and xeroderma pigmentosum fibroblasts, *J. Mol. Biol.*, 149, 171, 1981.
101. Edenberg, J. H., Inhibition of DNA replication by ultraviolet light, *Biophys. J.*, 16, 849, 1976.
102. Rosenstein, B. S. and Setlow, R. B., DNA repair after ultraviolet irradiation of ICR 2A frog cells. Pyrimidine dimers are long acting blocks to nascent DNA synthesis, *Biophys. J.*, 31, 195, 1980.
103. Cleaver, J. E., Kaufmann, W. K., Kapp, L. N., and Park, S. D., Replicon size and excision repair as factors in the inhibition and recovery of DNA synthesis from ultraviolet damage, *Biochim. Biophys. Acta*, 739, 207, 1983.
104. Michalke, H. and Bremer, H., RNA synthesis in *Escherichia coli* after irradiation with ultraviolet light, *J. Mol. Biol.*, 41, 1, 1969.
105. Eliceriri, G. L., Sensitivity of low molecular weight RNA synthesis to UV radiation, *Nature (London)*, 179, 80, 1979.
106. Nocentini, S., Inhibition and recovery of ribosomal RNA synthesis in ultraviolet-irradiated mammalian cells, *Biochim. Biophys. Acta*, 454, 114, 1976.
107. Ali, R. and Sauerbier, W., Effects of ultraviolet irradiation and postirradiation incubation on heterogeneous nuclear RNA size in murine cells, *Biophys. J.*, 22, 393, 1978.
108. Mayne, L. V. and Lehmann, A. R., Failure of RNA synthesis to recover after UV irradiation: an early defect in cells from individuals with Cockayne's syndrome and xeroderma pigmentosum, *Cancer Res.*, 42, 1472, 1982.
109. Kantor, G. J. and Hull, D. R., An effect of UV light on RNA and protein synthesis in nondividing human fibroblasts, *Biophys. J.*, 27, 359, 1979.
110. Brunschede, H. and Bremer, H., Protein synthesis in *Escherichia coli* after irradiation with ultraviolet light, *J. Mol. Biol.*, 41, 25, 1969.
111. Takeda, S., Naruse, S., and Yatani, R., Effects of UV microbeam irradiation of various sites in HeLa cells on the synthesis of RNA, DNA and protein, *Nature (London)*, 213, 696, 1967.
112. Fukuyuma, K., Epstein, W. J., and Epstein, J. H., Effect of ultraviolet light on RNA and protein synthesis in differentiated epidermal cells, *Nature (London)*, 216, 1031, 1967.
113. Bushnell, D. E., Yager, J. D., Becker, J. E., and Potter, V. R., Inhibition of messenger RNA accumulation but not translation in ultraviolet irradiated hepatoma cells, *Biochem. Biophys. Res. Commun.*, 57, 949, 1974.
114. Ben-Hur, E., Prager, A., and Buonaguro, F., Recovery from inhibition by UV-irradiation of ornithine decarboxylase induction in human cells: implication of excision repair, *Photochem. Photobiol.*, 35, 671, 1982.
115. Kelner, A., Effect of visible light on the recovery of *Streptomyces griseus* conidia from ultraviolet irradiation injury, *Proc. Natl. Acad. Sci. U.S.A.*, 35, 73, 1949.
116. Sutherland, B. M., Photoreactivation, *Bioscience*, 31, 439, 1981.
117. Rupert, C. S., Photoenzymatic repair of ultraviolet damage in DNA. I. Kinetics of the reaction, *J. Gen. Physiol.*, 45, 703, 1962.
118. Rupert, C. S., Photoenzymatic repair of ultraviolet damage in DNA. II. Formation of an enzyme-substrate complex, *J. Gen. Physiol.*, 45, 725, 1962.
119. Setlow, J. K., Boling, M. E., and Bollum, F. J., The chemical nature of photoreactivable lesions in DNA, *Proc. Natl. Acad. Sci. U.S.A.*, 53, 1430, 1965.
120. Cook, J. S., Direct demonstration of the monomerization of thymine-containing dimers in UV-irradiated DNA by yeast photoreactivating enzyme and light, *Photochem. Photobiol.*, 6, 97, 1967.
121. Bronk, B. V. and Van de Merwe, W. P., The shape of the blacklight dose photoreactivation curve for chick embryo fibroblasts, *Mutat. Res.*, 112, 109, 1983.
122. Sutherland, J. C. and Sutherland, B. M., Human photoreactivation enzyme: action spectrum and safelight conditions, *Biophys. J.*, 15, 435, 1975.
123. Chiang, T. and Rupert, C. S., Action spectrum for photoreactivation of ultraviolet irradiated marsupial cells in tissue culture, *Photochem. Photobiol.*, 30, 525, 1979.
124. Iwatsuki, N., Joe, C. O., and Werbin, H., Evidence that deoxyribonucleic acid photolyase from baker's yeast is a flavoprotein, *Biochemistry*, 19, 1172, 1980.
125. Eker, A. P. M., Dekker, R. H., and Berends, W., Photoreactivating enzyme from *Streptomyces griseus*. I. On the nature of the chromophoric cofactor in *Streptomyces griseus* photoreactivating enzyme, *Photochem. Photobiol.*, 33, 65, 1981.
126. Cook, J. S. and McGrath, J. R., Photoreactivating enzyme activity in metazoa, *Proc. Natl. Acad. Sci. U.S.A.*, 58, 1359, 1967.
127. Ananthaswamy, H. N. and Fisher, M. S., Photoreactivation of ultraviolet radiation-induced pyrimidine dimers in neonatal BALBc mouse skin, *Cancer Res.*, 41, 1829, 1981.
128. D'Ambrosio, S. M., Whetsone, J. W., Slazinski, L., and Lowney, E., Photorepair of pyrimidine dimers in human skin *in vivo*, *Photochem. Photobiol.*, 34, 461, 1981.
129. Sutherland, B. M. and Oliver, R., Inheritance of photoreactivating enzyme deficiencies in human cells, *Photochem. Photobiol.*, 24, 449, 1976.

130. Setlow, R. B. and Carrier, W. L., The disappearance of thymine dimers from DNA: an error correcting mechanism, *Proc. Natl. Acad. Sci. U.S.A.*, 51, 226, 1964.
131. Boyce, R. P. and Howard-Flanders, P., Release of ultraviolet light induced thymine dimers from DNA of *Escherichia coli* K-12, *Proc. Natl. Acad. Sci. U.S.A.*, 51, 293, 1964.
132. Haseltine, W. A., Gordon, L. K., Lindan, C. P., Grafstrom, R. H., Shaper, J. L., and Grossman, L., Cleavage of pyrimidine dimers in specific DNA sequences by a pyrimidine dimer DNA-glycosylase of *M. luteus*, *Nature (London)*, 285, 240, 1980.
133. Gordon, L. K. and Haseltine, W. A., Comparison of the cleavage of pyrimidine dimers by the bacteriophage T4 and *Micrococcus luteus* UV-specific endonucleases, *J. Biol. Chem.*, 255, 12047, 1980.
134. Sancar, A. and Rupp, W. D., A novel repair enzyme: UVRABC excision nuclease of *E. coli* cuts a DNA strand on both sides of the damaged region, *Cell*, 33, 249, 1983.
135. Grossman, L., Enzymes involved in the repair of damaged DNA, *Arch. Biochem. Biophys.*, 211, 511, 1981.
136. Friedberg, E. C., Rude, J. M., Cook, K. H., Ehmann, U. K., Mortelman, K., Cleaver, J. E., and Slor, H., in *DNA Repair Processes*, Nichols, W. W. and Murphey, D. G., Eds., Miami Symposia Specialists, Miami, 1977, 21.
137. Strauss, B., Rose, K., Alfimirano, M., Sklar, P., and Tatsumi, K., in *DNA Repair Mechanisms*, Hanawalt, P. C., Friedberg, E. C., and Fox, C. F., Eds., Academic Press, New York, 1978, 621.
138. Park, S. D., Choi, K. H., Hong, S. W., and Cleaver, J. E., Inhibition of excision-repair of ultraviolet damage in human cells by exposure to methyl methanesulfonate, *Mutat. Res.*, 82, 365, 1981.
139. Dresler, S. L. and Lieberman, M. W., Identification of DNA polymerases involved in excision repair in diploid human fibroblasts, *J. Biol. Chem.*, 258, 9990, 1983.
140. Youngs, D. A. and Smith, K. C., The involvement of polynucleotide ligase in the repair of UV-induced DNA damage in *Escherichia coli* K-12 cells, *Mol. Gen. Genet.*, 152, 37, 1978.
141. Soderhall, S. and Lindahl, T., Mammalian DNA ligases, *J. Biol. Chem.*, 250, 8438, 1975.
142. Cleaver, J. E., Defective repair replication of DNA in xeroderma pigmentosum, *Nature (London)*, 218, 652, 1968.
143. Collins, A. R. S. and Johnson, R. T., The inhibition of DNA repair, *Adv. Radiat. Biol.*, 11, 71, 1984.
144. Ben-Hur, E. and Ben-Ishai, R., DNA repair in ultraviolet light irradiated HeLa cells and its reversible inhibition by hydroxyurea, *Photochem. Photobiol.*, 13, 337, 1971.
145. Ahmed, F. E. and Setlow, R. B., Saturation of DNA repair in mammalian cells, *Photochem. Photobiol.*, 29, 983, 1979.
146. Eggset, G., Volden, G., and Krokan, H., UV-induced DNA damage and its repair in human skin *in vivo* studied by sensitive immunohistochemical methods, *Carcinogenesis*, 4, 745, 1983.
147. Edenberg, H. and Hanawalt, P., Size of repair patches in the DNA of ultraviolet-irradiated HeLa cells, *Biochim. Biophys. Acta*, 272, 361, 1971.
148. Collier, I. E., Olson, W. H., and Regan, J. D., A mechanism for the control of patch size in mammalian cell DNA excision repair, *J. Theor. Biol.*, 102, 487, 1983.
149. Smith, C. A., Cooper, P. K., and Hanawalt, P. C., Measurement of repair replication by equilibrium sedimentation, in *DNA Repair: A Laboratory Manual of Research Procedures*, Vol. 1 (Part B), Friedberg, E. C. and Hanawalt, P. C., Eds., Marcel Dekker, New York, 1981, chap. 24.
150. Setlow, R. B. and Regan, J. D., Measurement of repair synthesis by photolysis of bromouracil, in *DNA Repair: A Laboratory Manual of Research Procedures*, Vol. 1 (Part B), Friedberg, E. C. and Hanawalt, P. C., Eds., Marcel Dekker, New York, 1981, chap. 25.
152. Mezzina, M. and Nocentini, S., DNA ligase activity in UV-irradiated monkey kidney cells, *Nucleic Acid Res.*, 5, 4317, 1978.
153. Ben-Hur, E., The involvement of poly (ADP-ribose) in the radiation response of mammalian cells, *Int. J. Radiat. Biol.*, 46, 659, 1984.
154. Smerdon, M. J. and Lieberman, M. W., Nucleosome rearrangement in human chromatin during UV-induced DNA repair synthesis, *Proc. Natl. Acad. Sci. U.S.A.*, 75, 4238, 1978.
155. Lambert, B., Ringborg, U., and Skoog, L., Age-related decrease of ultraviolet light-induced DNA repair synthesis in human peripheral leukocytes, *Cancer Res.*, 39, 2792, 1979.
156. Hart, R. W. and Setlow, R. B., Correlation between DNA excision-repair and life-span in a number of mammalian species, *Proc. Natl. Acad. Sci. U.S.A.*, 71, 2169, 1974.
157. Francis, A. A., Lee, W. H., and Regan, J. D., The relationship of DNA excision repair of ultraviolet-induced lesions to the maximum life span of mammals, *Mech. Ageing Dev.*, 16, 181, 1981.
158. Rupp, W. D. and Howard-Flanders, P., Discontinuities in the DNA synthesized in an excision-defective strain of *Escherichia coli* following ultraviolet irradiation, *J. Mol. Biol.*, 31, 291, 1968.
159. Lehman, A. R., Postreplication repair of DNA in mammalian cells, *Life Sci.*, 15, 2005, 1975.
160. Park, S. D. and Cleaver, J. E., Postreplication repair: questions of its definition and possible alteration in xeroderma pigmentosum cell strains, *Proc. Natl. Acad. Sci. U.S.A.*, 76, 3927, 1979.

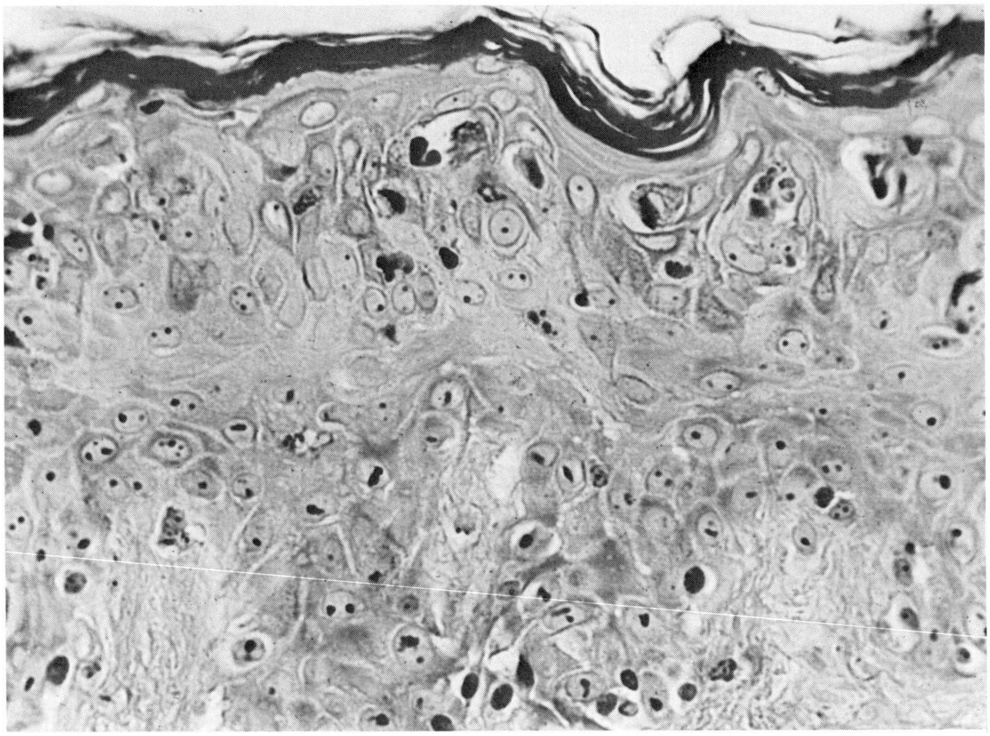

FIGURE 2. Giemsa stain. Severe UVB burn at 48 hr. Rounded, presumably spherical collections of dead or dying keratinocytes, suggesting a distribution around Langerhans cells. Mitotic figures and small and multiple nucleoli at basal level. Melanocytes have deep-stained nuclei and pale cytoplasm.

of severe sunburn. Instead of isolated sunburn cells, there are now nests of necrotic cells giving the impression of a globular configuration. The epidermis is markedly thickened. Figure 3 shows a biopsy taken at the same time and in the same individual as in Figure 2. However, this is 8-methoxypsoralen photosensitization. The destruction is so different that the author has never accepted the idea of referring to PUVA as an exaggerated sunburn. Finally, Figure 4 shows the histology 72 hr after UVB exposure. The old upper third of the epidermis is being cast off and a new granular layer is forming under it. By 5 days, the main finding is epidermal thickening as the remainder of acanthosis. An increase in the number of melanocytes at the basal level is evident.

III. TARGETS IN CUTANEOUS PHOTOBIOLOGY

A. Nucleic Acids

DNA is probably the most studied of possible UVR targets in skin under the general heading of photobiology. Its damage and repair are discussed in Chapter 2.

In serial biopsy studies by Daniels et al.,[4,6,27] the important features related to DNA were blanching or swelling of upper-level prickle cell nuclei at 4 and 7 hr. The methods used included the Feulgen reaction. Beginning between 24 and 48 hr, there was an increase in basal-level mitoses, reaching a maximum at 72 hr. That DNA is damaged by UVR is further indicated by the fact that it becomes antigenic.[29,30] Volden et al.[31] showed that DNA was damaged by UVB at half of MED, and that the erythema was better correlated with DNA damage than with the dose of UVB. The information

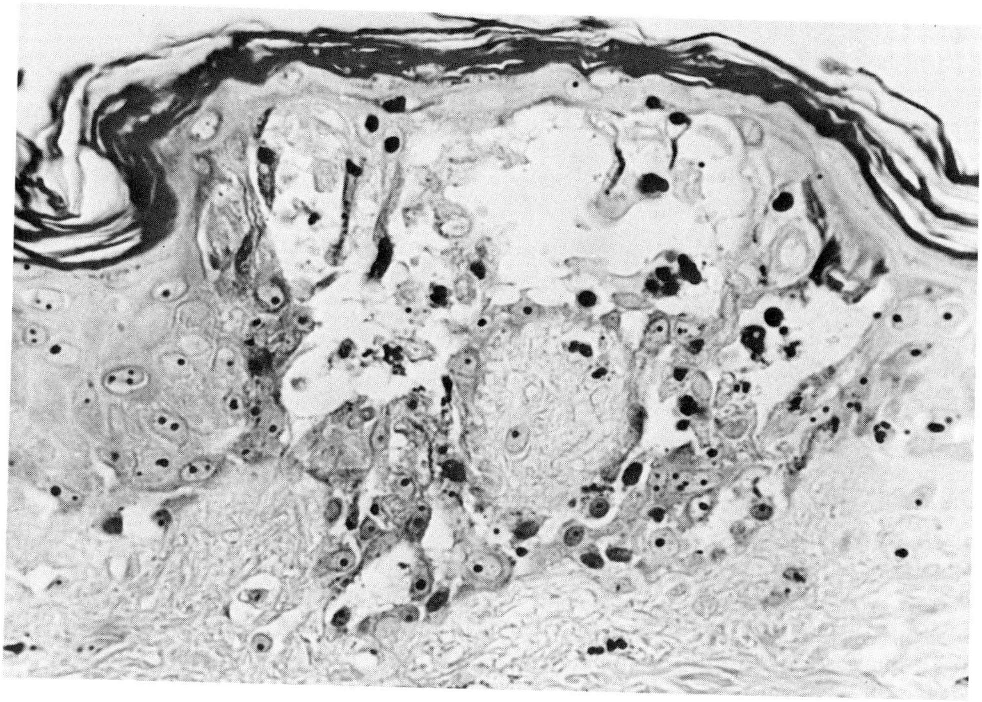

FIGURE 3. Same subject as Figure 2, 48 hr after UVA exposure with 8-methoxypsoralen as a photosensitizer.

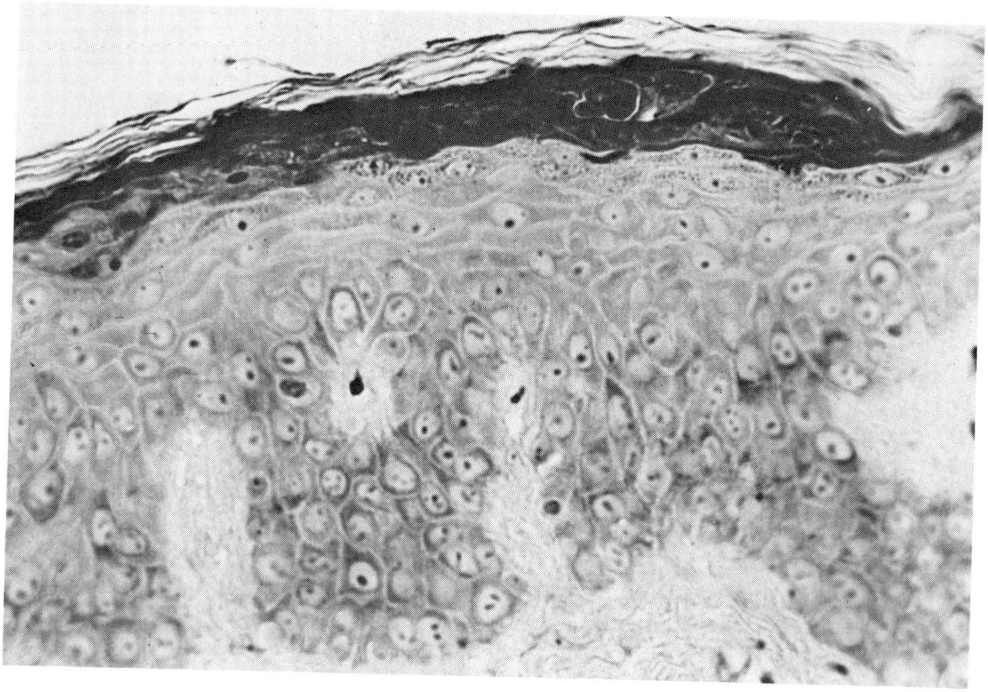

FIGURE 4. Histological appearance at 72 hr after UVB exposure.

vitiligo areas at 24 hr after 1, 4, and 8 MED UVB radiation are strikingly different, with fewer sunburn cells, with much less reaction around them, and little increase in their number with increasing dose. In the vitiliginous areas, the granular layer is intact and thicker with one to three cells. Is it possible that vitiligo is a disease of the granular layer rather than of melanocytes?

C. Membrane Constituents

Routine light microscopy with its treatment of tissue with organic solvents and paraffin for sectioning, has hidden the fact that cells and organelles are organized by lipid membranes. Electron microscopy not only increased magnification but the fixation techniques using osmium emphasized the membranes. Tickner[52] found that after UV radiation of skin there was no change in nucleic acid content, but a decrease in phospholipids with a concomitant increase in acid-soluble phosphate. Johnson[53] found a decrease in phospholipids in mouse skin after UVR. Norins[54] demonstrated free-radical formation in UV-irradiated skin. A paper presented at the 1st International Congress on Photobiology[55] described lipid peroxidation following UVR. Changes in membrane metabolism in cultured keratinocytes exposed to UVR are another indication of the importance of membranes.[56] There are many reviews on lipid membranes, including the one edited by Kummerow.[57] Even though the evidence was very strong that lipid membrane peroxidation was an important feature of UV damage,[58-62] there remained the question of the chromophore. Though the fatty acids contain double bonds they do not absorb in the UVB range.

Lysosomes were originally described as single-membrane bound organelles containing a number of acid hydrolases, and were originally considered to be intracellular ''suicide packets''.[63-65] Part of the rationale in studying lysosomes in sunburn cells was that it would appear they would have the ideal arrangement for the digestion of DNA, RNA, and proteins, in the general process of keratinocyte cornification.[66] Lysosome function has gone beyond the suicide packet, and the lysosomes are now believed to have roles in the uptake, storage, and discharge of materials from outside the cell; in effect serving as the gastrointestinal and urinary tracts of the cell.

It was established earlier[64] that lysosomes could be disrupted by UVR, and that the membranes were stabilized by adrenocortical steroids. It was also found that the lysosome membrane was stabilized by 7-dehydrocholesterol, and lysed by vitamins A and D. Lysosomal damage with the azo-coupling method for acid phosphatase was found first in mice, then in men.[67-69] By differential incubation, an initial increase in staining intensity was interpreted as sufficient membrane damage to increase substrate penetration. Later, diffusion, disorganization, and then fading were found.[8]

An interesting feature of the sunburn cell is that it frequently has more ingested melanin than the adjacent keratinocytes, tending to confirm that phagolysosomes are more vulnerable to UV damage than primary lysosomes, or that melanin-containing lysosomes have more acid phosphatase.[24]

Conversion of 7-dehydrocholesterol in the epidermis to previtamin D_3 is due to UVR exposure. The lysosomal membrane is stabilized by 7-dehydrocholesterol and not by vitamin D.[68,69] It is noteworthy[70] that the vitamin D metabolite produced by single hydroxylation in the liver is toxic and that it may influence the structure of lipid bilayer membranes. The circulating level of 25-hydroxy vitamin D is influenced by UV exposure of the skin. The final and most active metabolite (hormone), 1,25-dihydroxy vitamin D_3 is end-product regulated. Targets for this metabolite have been identified in epidermal keratinocytes,[71] and in line with the other information on the granular layer it is of interest that in vitamin D-deficient rats the basal cells are elongated and there is a marked reduction in the granular layer.[72]

Histamine[73] and kinins[74] had been ruled out as mediators of erythema but as each new group of mediators was identified, it was considered a factor in UV erythema. Thus, prostaglandins (PGs) were also implicated.[56,58-62] The recent medical literature is replete with reviews of the cyclooxygenase and lipoxygenase pathway of products derived from arachidonic acid (AA). Many of these have offered hope of explaining the UV erythema, but the time course and the persistence of the erythema long after the bolus of mediator would have passed have made them identified as involved; however, this does not explain the entire picture.

The role of leukotrienes as mediators, particularly of allergic reactions, has received its interest in turn. Soter et al.[75] have probably spared us a long period of speculation by injecting a series of leukotrienes into human skin and clearly producing effects that differ markedly from a UVB burn. Of interest were the different sensory effects in different subjects.

Tryptophan and its photoproducts have been recognized as key items in the effects of UVA on the lens,[76-78] and as the dominant chromophore present in cell membranes.[79] The finding that tryptophan is the chromophore in lipid membranes brings amino acids back into the picture after almost being eclipsed by DNA. Conversion of tryptophan to tryptamine by UVR was described.[80]

IV. DYNAMICS OF THE UV ERYTHEMAS AND TANNING

The MED or minimal perceptible erythema (MPE) is the minimum dose to produce an erythema. The minimum dose to produce melanization, or the minimum dose to produce vesiculation can also be used. While the use of a threshold of visual detection makes investigators feel uncomfortable because of its subjective nature, the MED has been and continues to be a useful clinical standard. Daniels et al. studied visual grading in evaluating sunburn and tanning in comparison to reflectance readings and found that in large samples the grading of +, ++, +++, and ++++ fell on a straight line on a log scale,[81] and when a field experiment added a +++++ category,[82] it fell on the same line. With television techniques, nothing could be seen below an MED.[83]

Hausser and Vahle[12] in a widely reproduced graph based on color standard grading of erythema, showed that the erythema increased almost vertically at 297 nm at a ++ erythema. With UVC the slope is much flatter and the maximum reached is ++, while the UVB goes to +++++ on their scale. The UVC erythema stays pink, whereas at longer wavelengths erythema go to deep red. Erythema is only one manifestation of injury and attention must be paid to edema, vesiculation, desquamation, and pigmentation. Buettner[84] suggested that the subvesicular dose was less variable than the minimal erythema. Adams et al.[85] compared the data of Hausser and Vahle[12] to the process of photography. They presented an interesting plot of the different degrees of erythema with energy of different wavelengths. By combining the erythema curve and the spectrum of sunlight reaching the earth they found a "peak effectiveness" at 306 nm.[86] If the square root of dye concentration in the standard is plotted against the log of the exposure, straight lines are obtained, and at 334 and 365 nm, the slope of erythema as a function of dose is so large that "taking it as infinite would not greatly change the calculated exposure."[85] Recording reflectance spectrophotometers have been used to measure skin reflectance at IR, visible, and UV wavelengths.[87-89] Melanin has a broad absorption band but no specific peaks. Aside from slight absorption from carotene at 482 nm, of interest are the absorption peaks of oxyhemoglobin at 542 and 576 nm. The reduced hemoglobin maximum at 556 nm is not usually separable from the two oxyhemoglobin bands. Usually a green filter (or an interference filter with peak transmission at 542 or 576 nm) is used to measure erythema, and the red filter (or a 610-nm peak transmission filter), to sample melanin. Reflectance spectra for Caucasian, black, sunburned, and suntanned skin are given elsewhere.[4]

Reflectance meters have been used extensively by anthropologists to measure skin color in different human populations,[90,91] and by skin photobiologists in following the effects of different wavelengths of UVR. Interference filters of 544 and 661 nm were used in a reflectance meter to follow the erythemas produced by cold quartz (254-nm) lamps, fluorescent sunlamps, three sources of predominantly UVB, and sunlight.[92] The 254-nm erythema appeared earliest and faded in 5 days for 1 MED and in 10 days with 4 MED. The other three sources had a much slower decrease from the peak erythema. Wan et al.[93] calculated the blood content of the skin based on reflectance spectroscopy measurements. In line with other observations that UVA erythema requires 1000 to 3000 times the dose of UVB, their data are presented as J/cm^2 for UVA, and mJ/cm^2 of UVB. It is, however, reiterated that there may be as much as 100 times the intensity of UVA in sunlight as compared to UVB.[94]

Reciprocity is the characteristic of photochemical reactions in which the multiplication of intensity and time is constant. Claesson[95] demonstrated that UVB effects on skin exhibit reciprocity over a 10 million-fold range. Two[96,97] independent studies have put the upper limit of reciprocity for human UVB erythema at about 8 hr.

The usual method for identifying a target in a photobiological system is by comparing an action spectrum with the absorption spectrum of potential molecular targets. An action spectrum is defined by determining the reciprocal of the dose of each wavelength required to produce a certain reaction, which in the case of sunburn has usually been erythema. The question of erythema action spectrum constitutes a major portion of reviews on the cutaneous effects of UVR.[4,51,98-100] However, with the finding that dermal and epidermal effects of UVR are largely separated, there is less reason to try to explain erythema by epidermal chromophores.

V. EPIDERMAL LANGERHANS CELLS (ELC) IN THE SKIN REACTION TO UVR

Figure 1 shows sunburn cells in a pattern resembling the distribution of ELC. In 1962, Daniels[6] suggested that ELC might be UV targets. In 1964, Daniels et al. found damage to the dendrites and body of the ELC in a 24-hr UVB burn using gold chloride and osmium zinc iodide techniques. Gilchrest[102,103] found that sunburn cells were keratinocytes in contact with UV-damaged ELC. The ELC disappeared at 24 hr, at which time only a minority of keratinocytes displayed marked vacuolization or dyskeratosis. The spreading of damage in a globular cluster of keratinocytes[104] around the initial sunburn cell is best explained by the linking of the keratinocytes in the "Langerhans cell-epidermal unit".[105,106] In the dose range 1 to 30 mJ/cm^2 for UVB and 1 × 40 to 2 × 100 J/cm^2 of UVA, ELC were the only epidermal cells to show damage.[107] In mice, 40 mJ/cm^2 of UVB or 300 J/cm^2 of UVA had no discernible effect on keratinocytes, but 40 to 60 mJ/cm^2 of UVB produced an almost complete elimination of ATPase-positive cells.[108] ATPase and Ia-reactive cells were markedly reduced by 100 J/cm^2 of UVA. It is not yet known why ELC damage in UVA does not lead to sunburn cells. Potter[109] has argued that the inverse relationship between epidermal hyperplasia and the presence of ELC suggest that the ELC might be the source of the G_1 chalone. In describing the histologic and histochemical changes following UVB Lobitz et al.[110] suggested that the keratinocyte chalone production was interrupted at the time the sunburn cells appeared, because the number of basal level mitoses was maximal at 72 hr, as compared to the 48 hr following tape stripping. This suggestion was supported by Iverson.[111]

ELC disappear for about 2 weeks after UVB irradiation, and return as the hyperplasia subsides. As indicated,[108] the upper dendrites of Langerhans cells are located in the granular layer, with an appearance suggesting termination at the nucleus of the granular-layer cells. In their own 15-μm sections with osmium zinc iodide Daniels et al. have observed by focusing up and down with phase contrast that the appearance of terminating at a granular-layer cell nucleus is often seen. After UVB irradiation, a new granular layer is formed, and with the return of ELC, the conspicuous link between basal cells and the granular layer is reestablished. In spite of the identification of the ELC as the front line of the immune system, this does not exclude other functions. With ATPase and osmium zinc iodide there is the appearance of wires or pipes in ELC, connecting the granular layer with the basal layer. Because it is known that a message is sent to regulate basal-cell mitoses, it is hard to ignore the appearance of these connections provided by the ELC.

VI. MELANOCYTES AND MELANIN SYSTEM IN THE ACUTE REACTIONS OF THE SKIN TO UVR

This subject could easily require an entire chapter, so only a few highlights will be given. For reviews see References 112 to 117.

Paling of melanocytes has been noted on histologic sections[102] but has not been studied as systematically as other cells. The 72-hr findings are visible in Figure 4.

Some subjects, when exposed to UVA or even visible light, develop darkening as a result of oxidation of preformed melanin.[118-121] More recent work emphasized the role of cytoskeletal elements. Since the immediate tanning peaks in 1 to 3 hr and fades variably, the cosmetic appeal of "tan without burn" has to be watched, particularly during the current craze for UVA boxes.

Rosdahl[113] has shown that local UVR causes a general increase in the number of skin melanocytes. Is this due to the "circulating solar factors" that appear in the epidemiology of malignant melanomas?

VII. EFFECTS OF UVA, UVB, AND UVC ON THE DERMIS

Erythema is produced by engorgement of small blood vessels, mostly in the subpapillary venous plexus. The term "small vessels" is used to avoid "capillaries", since capillary loops would not have a sufficient coverage to produce uniform erythema. Van der Luen[51,122] suggested that the greater warmth of a UVB erythema compared to a UVC erythema is probably due to arteriolar involvement. He has also presented evidence that UVC and UVA erythemas involve direct action on blood vessels, while that of UVB involves a diffusable mediator. Sams and Winkelman[123] have shown direct effects of UV on isolated blood vessels, and Epstein et al.[124] have shown that there must be direct vessel damage, because of unscheduled DNA synthesis following UVR.

Rosario et al.[104] studied the different erythemas with equally erythemogenic doses. In general, UVB and UVC produced more damage in the epidermis, whereas UVA produced more dermal effects. With UVA, papillary edema was evident at 72 hr and at 7 days. In UVB erythema, endothelial cell enlargement followed the UVB irradiation in a matter of minutes, which suggests direct action and not a diffusing mediator. There is also a discrepancy between the clinical and the histopathologic pictures. The clinical picture is of blood-filled vessels; the pathologic picture is of occlusion of blood-vessel lumen by endothelial cells protruding into it.[103] Merrill Chase[125] has told the author that in his studies of allergic contact dermatitis in guinea pigs he always clamped off perforating arteries and veins before fixing the tissue, so that he could see the erythema in the pathology. Daniels[126] saw vessels occluded by red cells in frostbite or the equivalent cryosurgery, where extravasation of the fluid following mast-cell rupture is believed to leave a thrombus of red cells. In the normal skin or in erythema one might expect the vessels to contain not only the red cells but plasma as well, and hence one would expect more evident blood vessels. A punch biopsy of a sunburned area is a bloody biopsy and Daniels suggests that the erythema has drained out of the biopsy specimen. If true, this has considerable general importance because the clinician sees erythema and the pathologist does not.

A suggestion[127] that histamine might be involved in the UVR reaction was settled by the finding that a histamine liberator did not prevent subsequent UV erythema.[128] Mast cells were found to be important in inflammation but not in sunburn cell production.[129]

Hawk and Parrish[130] found some polymorphonuclear (PMN) leukocytes in UVA erythema, which they considered might be due to a thermal effect. They also noted that neutrophils are not found in UVB burn in man but they are found in rodents. Potter[109] suggests that the shedding epidermis is not an invasive foreign body or an enclosed area of necrosis, and that epidermal cells release their own interleukins and other mediators without the intervention of the neutrophils. The author does not believe sunburn is a typical inflammation any more than he believes that it is an example of a first-degree thermal burn.

VIII. CONCLUDING REMARKS

There are some incompletely answered questions: (1) Why is there a wavelength-dependent time lag in the onset of erythema? (2) What is/are the mediator(s) of the erythema? A diffusable mediator is an easy explanation for the latent period between UV irradiation and the appearance of erythema. However, what is the reason for the persistence of the erythema, and particularly its prolongation in the skin cancer-prone subject? Whatever causes the erythema has to include the finding that the erythema does not blanch with injected epinephrine. Why does the slope of response to dose vary so consistently with wavelength? (3) What is the adaptive value of the basic skin melanization? There is a strong negative correlation between human skin color and latitude. Is this due to an almost-closed melanin shutter near the equator, and an almost-open melanin shutter toward the North Pole for the regulation of vitamin D synthesis? If not vitamin D, what?

A. Areas of Current Rapid Progress

The Langerhans cell has turned out to be even more interesting than when the reviewer suggested it as a UVR target in 1962. It organizes a unit of keratinocytes, and in serial biopsies it appears that this unit is damaged by UVB burn, as well as the ELC itself. ELC acts as the front line of immunological defense, and the fact remains that morphologically it should regulate epidermal thickness by communicating between the granular layer and the basal cells. That a cell can have more than one function is evident from the survival of cells in culture.

REFERENCES

1. Rothman, S., Untersuchungen über die Physiologie der Lichtwurkungen, *Z. Gesamte Exp. Med.*, 36, 398, 1923.
2. Daniels, F., Jr., Man and radiant energy: solar radiation, in *Handbook of Physiology, Section 4*, Dill, D. B., Ed., Williams & Wilkins, Baltimore, 1964, 969.
3. Daniels, F., Jr., Ultraviolet radiation and dermatology, in *Therpeutic Electricity and Ultraviolet Radiation*, Vol. 4, 2nd ed., Licht, S., Ed., Licht Publishing, New Haven, Conn., 1967, 370.
4. Johnson, B. E., Daniels, F., Jr., and Magnus, I. A., Responses of the human skin to ultraviolet light, in *Photophysiology*, Vol. 4, Giese, A. C., Ed., Academic Press, New York, 1968, 139.

5. Daniels, F., Jr., The physiological effects of sunlight, *J. Invest. Dermatol.*, 32, 147, 1959.
6. Daniels, F., Jr., Ultraviolet carcinogenesis in man, in The 1st Conference on Biology of Cutaneous Cancer, National Cancer Institute Monogr. No. 10, Urbach, F., Ed., NCI, Bethesda, Md., 1962, 407.
7. Daniels, F., Jr., Solar radiation, in *Environmental Physiology*, Slonim, B., Ed., C. V. Mosby, St. Louis, 1974, 276.
8. Daniels, F., Jr. and Johnson, B. E., Normal physiologic and pathologic effects of solar radiation on the skin, in *Sunlight and Man*, Fitzpatrick, T. B., Pathak, M. A., Harber, L. C., Seiji, M., and Kukita, A., Eds., University of Tokyo Press, Tokyo, 1974, 117.
9. Mayer, E., *Clinical Application of Sunlight and Artificial Radiation*, Williams & Wilkins, Baltimore, 1926.
10. Laurens, H., *The Physiological Effects of Radiant Energy*, Chemical Catalog, New York, 1933.
11. Blum, H. F., *Photodynamic Action and Disease Caused by Light*, (reprinted ed.), Hafner, New York, 1964.
12. Hausser, K. W. and Vahle, W., Sunburn and suntanning, *Wiss. Veroeff. Siemens Konzern*, 6, 101, 1927; translated by Urbach, F., in *The Biologic Effects of Ultraviolet Radiation*, Urbach, F., Ed., Pergamon Press, Oxford, 1969, 3.
13. Van der Leun, J. C., Personal communication.
14. Birge, S. and Haddad, J. G., 25-hydroxycholecalciferol stimulation on muscle metabolism, *J. Clin. Invest.*, 56, 1100, 1975.
15. Stumpf, W. E., Madhabananda, S., Reid, F. A., Tanaka, U., and DeLuca, H. F., Target cells for 1,25-dihydroxy vitamin D_3 in intestinal tract, stomach, kidney, skin, pituitary, and parathyroid, *Science*, 206, 1188, 1979.
16. Stumpf, W. E., Sar, M., Clarck, S. A., and DeLuca, H. F., Brain target sites for 1,25-dihydroxy vitamin D_3, *Science*, 215, 1403, 1982.
17. Snyder, D. S., Cutaneous effects of topical indomethacin, an inhibitor of prostaglandin synthesis on UV-damaged skin, *J. Invest. Dermatol.*, 64, 322, 1975.
18. Kaidbey, K. H. and Kurban, A. K., The influence of corticosteroids and topical indomethacin on sunburn erythema, *J. Invest. Dermatol.*, 66, 153, 1976.
19. Pearse, A. D., Wolska, H., and Marks, R., Do sunscreens prevent premalignant change in the epidermis?, *Br. J. Dermatol.*, 101(Suppl. 17), 1979.
20. Sambuco, C. P., Forbes, P. D., Davies, R. D., and Urbach, F., An animal model to determine sunscreen protectiveness against both vascular injury and epidermal cell damage, *J. Am. Acad. Dermatol.*, 10, 737, 1984.
21. Green, H., The keratinocyte as differentiated cell type, *Harvey Lect.*, 74, 101, 1979.
22. Kligman, A. M., Comments on the stratum corneum, in *The Biologic Effects of Ultraviolet Radiation*, Urbach, F., Ed., Pergamon Press, Oxford, 1969, 165.
23. Maizelis, M. Y., Effect of solar radiation upon skin permeability, *Vestn. Dermatol. Venerol.*, 3, 10, 1957.
24. Johnson, B. E., Mandell, G., and Daniels, F., Jr., Melanin and cellular reactions to ultraviolet radiation, *Nature (London) New Biol.*, 235, 147, 1972.
25. Dane, E. T. and Herman, D. L., Haematoxylin-phloxine-alcian blue-orange G differential staining of prekeratin, keratin and mucin, *Stain Technol.*, 38, 97, 1963.
26. Daniels, F., Jr., Van der Leun, J. C., and Johnson, B. E., Sunburn, *Sci. Am.*, 219, 38, 1968.
27. Daniels, F., Jr., Brophy, D., and Lobitz, W. C., Jr., Histochemical responses of human skin following ultraviolet irradiation, *J. Invest. Dermatol.*, 37, 351, 1961.
28. Lobitz, W. C., Jr., Brophy, D., Larner, A. E., and Daniels, F., Jr., Glycogen response in epidermal basal cells, *Arch. Dermatol.*, 86, 207, 1962.
29. Tan, E. M. and Stoughton, R. B., Ultraviolet light induced damage to deoxyribonucleic acid in human skin, *J. Invest. Dermatol.*, 52, 537, 1969.
30. LeFeber, W. P., Norris, D. A., Huff, J. C., Kub, M., Ryan, S. R., and Weston, W. L., Induction of nuclear antigen expression in human keratinocytes by ultraviolet light, *J. Invest. Dermatol.*, 80, 328, 1983.
31. Volden, G., Eggset, G., Krokan, H., and Kavli, G., UV-induced DNA damage and its repair in human skin *in vivo* correlated to erythema, tanning and light penetration, *J. Invest. Dermatol.*, 82, 550, 1984.
32. Nix, T. E., Jr., Nordquist, R. E., and Scott, J. R., An ultrastructural study of nucleolar enlargement following ultraviolet irradiation of human epidermis, *J. Invest. Dermatol.*, 45, 114, 1965.
33. Baden, H. F. and Pearlman, C., The effect of ultraviolet light on protein and nucleic acid synthesis in the epidermis, *J. Invest. Dermatol.*, 43, 7, 1964.
34. Fukuyama, K., Epstein, W. L., and Epstein, J. H., Effect of ultraviolet light on RNA and protein synthesis in differentiated epidermal cells, *Nature (London)*, 216, 1031, 1967.

35. Epstein, J. H., Fukuyama, K., and Fye, K., Effects of ultraviolet radiation on the mitotic cycle and DNA, RNA, and protein synthesis in mammalian epidermis *in vivo, Photochem. Photobiol.,* 12, 57, 1970.
36. Hanke, W., Die Bedeutung von Latenzzeiten für U.V.-Strahlenbedingte Veranderungen am lebenden Gewebe, in *Progress in Photobiology, The Finsen Memorial Congress,* Christensen, B. and Buchmann, B., Eds., Elsevier, Amsterdam, 1961, 336.
37. Daniels, F., Jr., Johnson, B. E., Beaghen, S., and Binford, R. T., Histology and histochemistry of acute sunburn, Exhibit at American Academy of Dermatology, Bal Harbour, Fla., December 3 to 8, 1966.
38. Argyris, T. S., Ribosomal RNA synthesis through epidermal hyperplasia induced by abrasion or treatment with tetradecanoyl phorbol-13-acetate, *J. Invest. Dermatol.,* 76, 388, 1981.
39. Fukuyama, K., Nakamura, T., and Bernstein, I. A., Differentially localized incorporation of amino acids in relation to epidermal keratinization in the newborn rat, *Anat. Rec.,* 152, 525, 1965.
40. Voorhees, J. D., Chakrabarti, S. G., and Bernstein, I. A., The metabolism of "histidine-rich" protein in normal and psoriatic keratinization, *J. Invest. Dermatol.,* 51, 344, 1968.
41. Zenisek, A. and Kral, J. A., Sunscreening effect of urocanic acid, *Biochem. Biophys. Acta,* 18, 589, 1959.
42. Hais, I. M. and Strych, A., Increase in urocanic acid concentration in human epidermis following insolation, *Coll. Czech. Commun.,* 34, 649, 1969.
43. Brusilow, S. L. and Ikai, K., Urocanic acid in sweat: an artifact of elution from the epidermis, *Science,* 160, 1257, 1968.
44. Baden, H. P. and Pathak, M. A., The metabolism and function of urocanic acid in skin, *J. Invest. Dermatol.,* 48, 11, 1967.
45. Anglin, J. G. and Batten, W. H., Studies on *cis* urocanic acid, *J. Invest. Dermatol.,* 50, 463, 1968.
46. DeFabo, E., Noonan, F., Fisher, M., Burns, J., and Kacser, H., Further evidence that the photoreceptor mediated UV-induced systemic immune suppresion is urocanic acid, *J. Invest. Dermatol.,* 80, 319, 1983.
47. DeFabo, E., Noonan, F., Fisher, M., Burns, J., and Kacser, H., Mice deficient in skin urocanic acid are unable to be systemically immunosuppressed by UV radiation, *Photochem. Photobiol.,* 37, S36, 1983.
48. Shelley, W. B. and Juhlin, L., The Langerhans cell: its origin, nature, and function, *Acta Derm.,* Suppl. 79, 7, 1978.
49. Daniels, F., Jr., Optics of the skin as related to ultraviolet radiation, in *The Biologic Effects of Ultraviolet Radiation,* Urbach, F., Ed., Pergamon Press, Elmsford, N.Y., 1969, 151.
50. Nix, T. E., Jr., Nordquist, R. E., and Everett, M. A., Ultrastructural changes induced by ultraviolet light in human epidermis: granular and transitional cell layers, *J. Ultrastruct. Res.,* 12, 547, 1965.
51. Van der Luen, J. C., Ultraviolet Erythema: a Study of Diffusion Processes in Human Skin, Ph.D. thesis, University of Utrecht, Utrecht, the Netherlands, 1966.
52. Tickner, A., Changes in phospholipids of mouse skin after ultraviolet irradiation *in vitro, Biochem. J.,* 88, 80, 1963.
53. Johnson, B. E., A Study of the Physiological Effects of Ultraviolet Radiation on Mammalian Skin, Ph.D. thesis, University of London, 1965.
54. Norins, A. L., Free radical formation in the skin following exposure to ultraviolet light, *J. Invest. Dermatol.,* 39, 445, 1962.
55. Dubouloz, P. and Dumas, J., Sur la formation de peroxide lipideques dan la peau sous l'action des radiation, in *Proc. 1st Int. Photobiological Congr.,* Veenman and Zonen, Wageningen, the Netherlands, 1954, 247.
56. De Leo, V. A., Horlick, H., Hanson, D., Eisinger, M., and Harber, L. C., Ultraviolet radiation induces changes in membrane metabolism of human keratinocytes in culture, *J. Invest. Dermatol.,* 83, 323, 1984.
57. Kummerow, F. A., Benga, G., and Holmes, R. P., Biomembranes and cell function, *Ann. N.Y. Acad. Sci.,* 414, 1983.
58. Eaglstein, W. H. and Weinstein, G. D., Prostaglandin and DNA synthesis in human skin: possible relationship to ultraviolet light effects, *J. Invest. Dermatol.,* 64, 386, 1975.
59. Snyder, D. S., Cutaneous effects of topical indomethacin, an inhibitor of prostaglandin synthesis on UV damaged skin, *J. Invest. Dermatol.,* 64, 322, 1975.
60. Camp, R. D., Greaves, M. W., Hensby, C. N., Plummer, M. A., and Warin, A. P., Irradiation of human skin by short wavelength ultraviolet radiation (200—290 nm) (UVC): increased concentrations of arachidonic acid and prostaglandins E_2 and $F_{2\alpha}$, *Br. J. Clin. Pharmacol.,* 6, 145, 1978.
61. Zibon, V. A., Casebolt, T. L., Marcelo, C. L., and Voorhees, J. J., Lipoxygenation of arachidonic acid by subcellular preparation from murine keratinocytes, *J. Invest. Dermatol.,* 83, 248, 1984.

62. Hawk, J. L. M., Black, A. K., Jaenicke, K. F., Barr, R. M., Soter, N., Mallett, A. I., Gilchrest, B. A., Hensby, C. N., Parrish, J. A., and Greaves, M., Increased concentrations of arachidonic acid, prostaglandin E_2, D_2, and 6-oxo-F_1 and histamine in human skin following UVA irradiation, *J. Invest. Dermatol.*, 80, 496, 1983.
63. DeDuve, C. and Wattiaux, R., Functions of lysosomes, *Annu. Rev. Physiol.*, 28, 435, 1966.
64. Weissmann, G. and Dingle, J. T., Release of lysosomal protease by UV irradiation and inhibition by hydrocortisone, *Exp. Cell Res.*, 25, 207, 1961.
65. Weissmann, G. and Fell, H. B., The effect of hydrocortisone on the response of fetal rat skin in culture to ultraviolet irradiation, *J. Exp. Med.*, 365, 1968.
66. Frienkel, R. K. and Traczyk, T. N., Acid hydrolases of the epidermis: subcellular localization and relationship to cornification, *J. Invest. Dermatol.*, 80, 441, 1983.
67. Johnson, E. B., Ultraviolet radiation and lysosomes in skin, *Nature (London)*, 219, 1258, 1968.
68. Johnson, E. B. and Daniels, F., Jr., Lysosomes and reactions of skin to ultraviolet radiation, *J. Invest. Dermatol.*, 53, 85, 1969.
69. Johnson, E. B., Lysosomal damage by UV-irradiation of the skin, in 6th Int. Congr. Photobiology, Schenk, G. O., Ed., Bochum, West Germany, 1972.
70. Kummerow, F. A., Modification of cell membrane composition by dietary lipids and its implications for atherosclerosis, *Ann. N.Y. Acad. Sci.*, 414, 29, 1983.
71. Feldman, D., Chen, T., and Hirst, M., Demonstration of 1,25 dihydroxy vitamin D receptors in human skin biopsies, *J. Clin. Endocrinol.*, 51, 1463, 1980.
72. Pavlovitch, J. H. and Delescluse, C., Vitamin D and the metabolism of epidermis, *Int. J. Dermatol.*, 22, 98, 1983.
73. Weber, G., Excited states in proteins, in *Light and Life*, McElroy, W. D. and Glass, B., Eds., Johns Hopkins University Press, Baltimore, 1961, 82.
74. Yeargers, E., Ultraviolet light effects on proteins, in *The Biologic Effects of Ultraviolet Radiation*, Urbach, F., Ed., Pergamon Press, Oxford, 1969, 37.
75. Soter, N. A., Lewis, R. A., Corey, E. J., and Austen, K. F., Local effects of synthetic leukotrienes (LTC_4, LTD_4, LTE_4, LTB_4) in human skin, *J. Invest. Dermatol.*, 80, 115, 1983.
76. Walrant, P., Santis, R., and Grossweiner, L. I., Photosensitizing properties of N_1-formylkynurenine, *Photochem. Photobiol.*, 22, 63, 1975.
77. Zigman, S., Schultz, J., Yolo, T., and Griess, G., Possible roles of near UV light in the cataractous process, *Exp. Eye Res.*, 15, 201, 1973.
78. Zigman, S., The role of sunlight in human cataract formation, *Surv. Ophthalmol.*, 27, 317, 1983.
79. Kochevar, I. E. and Yoon, M., Photosensitization of methyl linoleate oxidation by tryptophan in peptides, *Photochem. Photobiol.*, 37, 279, 1983.
80. Wiskemann, A., Die heutigen Kenntnisse über die Mechanismen der UV-Wirkung auf die menschliche haut, *Strahlentherapie*, 122, 463, 1963.
81. Daniels, F., Jr. and Van der Luen, J. C., Problems in quantifying skin reactions, *Arch. Dermatol.*, 97, 553, 1968.
82. Daniels, F., Jr. and Imbrie, J. D., Comparison between visual grading and reflectance measurements of erythema produced by sunlight, *J. Invest. Dermatol.*, 30, 295, 1958.
83. Van der Luen, J. C., Presentation at the 6th Int. Congr. Photobiology, Bochum, West Germany, 1972.
84. Buettner, K. J. K., The effects of natural sunlight on human skin, in *The Biologic Effects of Ultraviolet Radiation*, Urbach, F., Ed., Pergamon Press, Oxford, 1969, 237.
85. Adams, E. Q., Barnes, B. T., and Forsythe, W. E., Erythema due to ultraviolet radiation, *J. Opt. Soc. Am.*, 21, 207, 1931.
86. Schulze, R. and Grafe, K., Consideration of sky ultraviolet radiation in the measurement of solar ultraviolet radiation, in *The Biologic Effects of Ultraviolet Radiation*, Urbach, F., Ed., Pergamon Press, Oxford, 1969.
87. Edwards, E. A. and Duntley, S. Q., An analysis of skin pigment changes after exposure to sunlight, *Science*, 90, 235, 1939.
88. Jansen, M. T., A reflection spectrophotometric study of ultraviolet erythema and pigmentation, *J. Clin. Invest.*, 32, 1053, 1953.
89. Jacquez, J. A., Kuppenheim, H. F., Dimitroff, J. M., McKeehan, W., and Huss, J., Spectral reflectance of human skin in the region 235—700 nm, *J. Appl. Physiol.*, 8, 212, 1955.
90. Lasker, G. W., Photoelectric measurement of skin color in a Mexican Mestizo population, *Am. J. Phys. Anthropol.*, 12, 115, 1954.
91. Weiner, J. S., Harisson, G. A., Singer, R., Harris, R., and Jopp, W., Skin color in southern Africa, *Hum. Biol.*, 36, 294, 1964.
92. Breit, R. and Kligman, A. M., Measurement of erythemal and pigmentary responses to ultraviolet radiation of different spectral properties, in *The Biologic Effects of Ultraviolet Radiation*, Urbach, F., Ed., Pergamon Press, Oxford, 1969, 267.

93. Wan, S., Jaenicke, K. F., and Parrish, J. A., Comparison of the erythemogenic effectiveness of ultraviolet-B (290—320 nm) and ultraviolet-A (320—400 nm) radiation by skin reflectance, *Photochem. Photobiol.,* 37, 547, 1983.
94. Parrish, J. A., Anderson, R. R., Urbach, F., and Pitts, D., *UVA: Biological Effects of Ultraviolet Radiation with Emphasis on Human Responses to Longwave Ultraviolet,* Plenum Press, New York, 1978, 7—14 and 65—73.
95. Claesson, S., Juhlin, L., and Wettermark, G., The reciprocity law of UV-irradiation effects, *Acta Derm.,* 38, 123, 1958.
96. Berger, D., Personal communication, 1979.
97. Colsky, L. D., Halprin, K. M., Taylor, J. R., Comerford, M., and Kramer, J. R., UVB is additive when repeated within an 8-hour interval, *J. Am. Acad. Dermatol.,* 8, 760, 1983.
98. Everett, W. A., Sayer, R. M., and Olson, R. L., Physiologic response of human skin to UV light, in *The Biologic Effects of Ultraviolet Radiation,* Urbach, F., Ed., Pergamon Press, Oxford, 1969, 181.
99. Van der Leun, A. C., On the action spectrum of ultraviolet erythema, in *Research Progress in Organic, Biological and Medicinal Chemistry,* North-Holland, Amsterdam, 1972, 711.
100. Parrish, J. A., Ying, C. Y., Pathak, M. A., and Fitzpatrick, T. B., Erythemogenic properties of long-wave ultraviolet light, in *Sunlight and Man,* Pathak, M. A., Harber, L. C., Seiji, M., Kukita, A., and Fitzpatrick, T. B., Eds., University of Tokyo Press, Tokyo, 1974, 131.
101. Johnson, B. E., Reactions of normal skin to solar radiation, in *The Physiology and Pathophysiology of the Skin,* Vol. 8, Jarrett, A., Ed., 1984, 2413.
102. Gilchrest, B. A., Soter, N. A., Stoff, J. S., and Mihm, M. C., Human sunburn reaction: histologic and biochemical studies, *J. Am. Acad. Dermatol.,* 5, 411, 1981.
103. Gilchrest, B. A., Murphy, G. F., and Soter, N. A., Effect of chronological aging and ultraviolet irradiation on Langerhans cells in human epidermis, *J. Invest. Dermatol.,* 79, 85, 1982.
104. Rosario, R., Mark, G. J., Parrish, J. A., and Mihm, M. C., Histologic changes produced in skin by equally erythemogenic doses of UVA, UVB, UVC, and UVA with psoralens, *Br. J. Dermatol.,* 101, 299, 1979.
105. Wolff, K. and Winklemann, R. F., Ultrastructural localization of nucleoside triphosphatase in Langerhans cells, *J. Invest. Dermatol.,* 48, 50, 1967.
106. Potten, C. S. and Allen, T. D., A model implicating the Langerhans cell in keratinocyte proliferation control, *Differentiation,* 5, 43, 1976.
107. Aberer, W., Schuler, G., Stingl, G., and Wolff, K., Ultraviolet light depletes surface markers of Langerhans cells, *J. Invest. Dermatol.,* 76, 202, 1981.
108. Stingl, G., New aspects of Langerhans cell function, *Int. J. Dermatol.,* 19, 289, 1980.
109. Potter, G., Langerhans Cell: Studies in Epidermal Cell Culture and Clinical Applications, Ph.D. thesis, Cornell University, Ithaca, N.Y., 1981.
110. Lobitz, W. C., Jr., Brophy, D., Larner, A. E., and Daniels, F., Jr., Glycogen response in human epidermal basal cell, *Arch. Dermatol.,* 86, 207, 1962.
111. Iverson, O., Discussion, in 1st Int. Conf. Biology of Cutaneous Cancer, National Cancer Institute Monograph No. 10, Urbach, F., Ed., NCI, Bethesda, Md., 1963, 418.
112. Riley, V., Ed., *Pigmentation: Its Genesis and Biologic Control,* Appleton-Century-Crofts, New York, 1972.
113. Kawamura, T., Ed., *Biology of Normal and Abnormal Melanocytes,* University of Tokyo Press, Tokyo, 1974.
114. Jimbow, K., Pathak, M. A., Szabo, G., and Fitzpatrick, T. B., Ultrastructural changes in human melanocytes after ultraviolet radiation, in *Sunlight and Man,* Pathak, M. A., Harber, L. C., Seiji, M., Kukita, A., and Fitzpatrick, T. B., Eds., University of Tokyo Press, Tokyo, 1974.
115. Riley, V., Ed., *Pigment Cell,* Vol. 3, S. Karger, Basel, 1975.
116. Carter, D. M., The Yale conference on pigment cell biology, *Yale J. Biol. Med.,* 46, 331, 1973.
117. Fitzpatrick, T. B., Szabo, G., Seiji, M., and Quevedo, W. C., Jr., Biology of the melanin pigmentary system, in *Dermatology in General Medicine,* 2nd ed., Fitzpatrick, T. B., Eisen, A. Z., Wolff, K., Freedberg, I. M., and Austen, K. F., Eds., McGraw-Hill, New York, 1979.
118. Miescher, G. and Minder, H., Untersuchungen über die durch langwelliges Ultraviolett hervorgerufene Pigmentdunkelun, *Strahlentherapie,* 66, 6, 1939.
119. Wiskemann, A. and Winser, H., Die Beziehung der direkten Pigmentiertung zur Konstitution, *Strahlentherapie,* 99, 594, 1956.
120. Lorincz, A. L., Pigmentation, in *Physiology and Biochemistry of the Skin,* Rothman, S., Ed., University of Chicago Press, Chicago, 1954, 515.
121. Rosdahl, I. K., Local and systemic effects on the epidermal melanocyte population in UV-irradiated mouse skin, *J. Invest. Dermatol.,* 73, 306, 1979.
122. Van der Luen, A. C., On the action spectrum of ultraviolet erythema, in *Research Progress in Organic, Biological, and Medicinal Chemistry,* North-Holland, Amsterdam, 1972, 711.

123. Sams, W. J. and Winkelmann, R. K., The effect of ultraviolet light on isolated cutaneous blood vessels, *J. Invest. Dermatol.*, 53, 79, 1969.
124. Epstein, W. L., Fukuyama, K., and Epstein, J. H., Early effects of ultraviolet light on DNA synthesis in human skin *in vivo, Arch. Dermatol.*, 100, 84, 1969.
125. Chase, M., Personal communication.
126. Daniels, F., Jr., Some of the cryobiology behind cryosurgery, *Cutis,* 16, 421, 1975.
127. Lewis, T., *The Blood Vessels of the Human Skin and their Responses,* Shaw & Sons, London, 1927.
128. Partington, M. W., The vascular response of the skin to ultraviolet light, *Clin. Sci.*, 13, 425, 1954.
129. Ikai, T., Danno, K., Horop, T., and Narumya, S., Effect of ultraviolet radiation on mast cell deficient mice, *J. Invest. Dermatol.*, 82, 391, 1984.
130. Hawk, J. L. M. and Parrish, J. A., Responses of normal skin to ultraviolet radiation, in *The Science of Photomedicine,* Regan, J. D. and Parrish, J. A., Eds., Plenum Press, New York, 1982, 219.

Table 1
EVIDENCE FOR INVOLVEMENT OF FREE-RADICAL MEDIATED REACTIONS IN UV CARCINOGENESIS

1. Free radicals are formed in skin upon response to UV radiation.[56]
2. UV-induced oxidation of skin sterols results in formation of cholesterol epoxide[77] — a putative carcinogen, the formation of which is a sensitive indicator of lipid peroxidation.[78]
3. Dietarily administered antioxidants inhibit the formation of cholesterol epoxide — its formation being inversely related to the skin antioxidant content.[57]
4. UV radiation promotes the formation of stable lipid-peroxidation products in skin.[74]
5. Antioxidants inhibit epidermal malonaldehyde formation — an empirical test for lipid peroxidation.[73]
6. Biological effects of potential free-radical-mediated reactions, i.e., membrane labilization, have been shown to occur in acute and chronically irradiated skin.[75,76]
7. Intake of free-radical stimulators, i.e., polyunsaturated fats, exacerbates UV carcinogenesis.[71,72]
8. Antioxidants inhibit UV carcinogenesis at a level roughly equivalent to the degree of exacerbation induced by unsaturated lipid.[71]

those in which immunosuppressive agents have been shown to enhance UV carcinogenesis,[90,91] provide convincing evidence of an immunologic component to photocarcinogenesis and a link between UV radiation and the immune system.

Just as T-cell function has been implicated in the process of UV carcinogenesis, it is T-cell populations which primarily reflect age-associated immunologic change. Based upon the process of involution and atrophy through which the thymus begins to undergo shortly after puberty, proponents of the "biological clock" theory of aging suggest that the T-dependent lymphoid system is genetically programmed to decline in effectiveness.[79] In theory, the thymus would be under control of a genetically programmed clock mechanism set to self-destruct after having undergone a fixed number of cell divisions.

Regardless of the intrinsic nature of potential control mechanisms of immunocompetence, it has been shown that extrinsic factors, among them diet, can counteract age-related immunologic insufficiency. Just as reduced caloric intake has been demonstrated to inhibit carcinogenesis, Good et al.[92] observed that not only was life span increased in certain strains of animals receiving restricted dietary calories but that immunologic decline was forestalled. Dietary fat also played a significant role in ameliorating age-related autoimmune processes. It has been suggested that dietary restriction causes a delay in maturation of the immune system, with a subsequent increase in time span of immunocompetence.[93]

There is some evidence that endogenous free-radical reactions have an adverse effect on the immune system. Age-associated increases in random free-radical reactions are reflected in elevated serum oxidase activities and declining concentrations of mercaptans,[94,95] both conditions occurring concomitantly with increasing age and declining immunocompetence. Although there is no evidence that antioxidants slow the decline of age-related immune responses, these agents are capable of enhancing both humoral and cell-mediated immunologic parameters. Harman et al.[96] found that several free-radical inhibitors were as effective in eliciting a humoral response from spleen cells as levamisole, a known immune-system stimulant.

As yet, there have been no investigations into the possible role antioxidants might play in potentiating UV carcinogenesis by modulation of immune responses. However, considering the influence of UV upon immune response, its relation to UV carcinogenesis, and the photoprotective effects of antioxidants, it would be surprising if these agents did not diminish at least some of the known deleterious effects UV has upon skin-associated lymphoid tissues.[97,98]

VI. CONCLUSIONS

The involvement of UV radiation in either carcinogenic or aging processes of skin has been documented in numerous reviews.[99-103] Those cited are intended to be representative rather than exhaustive. Here, an attempt has been made to explore the mechanistic relationship between UV-influenced carcinogenesis and aging through general theories upon which both aging and carcinogenesis may converge. The areas of potential congruence through which homologies of these two processes have been sought include: (1) perturbations in the maintenance of genetic integrity, (2) formation and/or accumulation of harmful materials, and (3) an intrinsically timed, but extrinsically influenced, biological clock that shuts down control of essential life processes.

Before recapping the evidence regarding each of these theories, a general comment regarding the nature of UV aging/carcinogenesis relationships should be noted. Epidemiologic evidence indicates that given equal actinic exposure, older individuals are more likely to develop skin cancer than are younger ones.[6] Indeed, data of this nature form the basis for seeking potential relationships between aging and cancer. Experimentally, however, it has been demonstrated that older animals are less susceptible to UV-induced tumor formation.[7,9,10] This paradox is further confounded by the observation that skin from old animals, when grafted to young recipients, is more susceptible to UV carcinogenesis but that more tumors develop in skin grafted to young recipients than in old, regardless of the age of the skin donor.[10] These observations led to the supposition that predisposition to UV carcinogenesis was not attributable to inherent properties of the skin per se, but was dependent upon physiological factors in the host that change as a function of age. If this were so, conventional wisdom might predict that older animals would be more susceptible to carcinogenesis when immunocompetence or other physiological controls were in decline — a circuitous argument that appears in conflict with the experimental data.

In addition, it must be stressed that most recognized aging effects attributable to UV are manifested in the dermis.[104-106] It seems reasonable that relationships between aging and carcinogenesis should be sought in the tissue where carcinogenesis occurs. If direct relationships exist at the mechanistic level, carcinogenesis might be considered as an acute manifestation of the underlying bases of aging. Certainly much more is known of the photocarcinogenic process in the epidermis than is known of aging. Aside from the general observation of age-associated epidermal atrophy, no biochemical indices are known which represent epidermal aging. Whereas the epidermis would be the logical tissue in which to pursue aging/carcinogenic relationships, this is not to say that the dermis does not influence those processes as they may be expressed in the epidermis. Indeed, some view the actinic changes that occur in the dermis to be of paramount importance to epidermal carcinogenesis.[107] Nor would one imagine that changes that occur in the epidermis be without effect upon the dermis.

With respect to mechanisms, considerable evidence points to a decline in DNA repair capacity as a factor in both aging and carcinogenesis. Impaired ability to repair DNA has been demonstrated in cells from both progeria and XP patients, disease states representing accelerated aging and extreme susceptibility to UV-induced skin cancer, respectively. Corroboration of these results has not always been forthcoming.[30,40] Evidence that UV plays a role in aging comes from cell-culture studies in which the number of population doublings decline in cells derived from chronically exposed sites. However, cells derived from these sites demonstrate an increased plating efficiency.[23] Further, while decreased DNA excision repair has been related to increased age, fibroblasts derived from chronically sun-exposed areas demonstrate a higher capacity for macromolecular synthesis and DNA repair.[108,109] Thus, the role of DNA repair in UV-mediated aging remains clouded and information concerning the fidelity of actinically impaired repair mechanisms is lacking.

Evidence for the formation and accumulation of toxic by-products and their involvement in aging and carcinogenesis is based largely upon the free-radical theory. By-products of free-radical reactions, such as malonaldehyde, are themselves mutagenic and affect DNA repair.[110] Whereas a considerable body of evidence, albeit indirect, points to involvement of free radicals in UV carcinogenesis, evidence for their involvement in aging is less convincing. Some have suggested a "wear and tear" mechanism for aging which would be reflected in a diminished capacity to safely cope with ever-present endogenously formed free radicals.[111] The result would be gradual accumulation of damage — an effect not incompatible with the error-accumulation or biological-clock theories. However, in at least one study, free-radical inhibitors were less effective in preventing UV carcinogenesis in older animals than in young ones — a time when wear and tear would be greatest, when the host would be least capable of coping with continuous insult, and when one would predict antiradicals would be most effective.[11] On the other hand, the wear and tear version of the free-radical theory would predict that older animals would be most susceptible to carcinogenesis, again, a prediction not borne out by the experimental evidence.

The immunologic influence upon carcinogenesis and aging perhaps provides the best associations in support of the biological-clock theory. The decline in immunocompetence with age, and consequent increase in cancer incidence, coincides with lymphoid cell decline. The latter is apparently intrinsically controlled, although it is obvious that extrinsic factors, while not altering the final outcome, can modify the response within time limits. Photoimmunology represents one of the most promising areas for future investigation into relationships of actinically mediated aging and carcinogenesis. However, the conflict between epidemiological and experimental evidence, with regard to age-associated carcinogenic susceptibility, should again be emphasized. This is a question that must be resolved! In addition, one should be mindful that much of our knowledge concerning general cellular and molecular parameters attributable to either aging or carcinogenic processes has been gleaned from studies of cell-culture systems, where systemic immunologic effect is absent.

The difficulty in seeking an unified theory with respect to the mechanistic relation of actinic aging and carcinogenesis will have become obvious. In areas where insight has been gained into specific influences of UV on carcinogenesis, knowledge is lacking for aging processes, and vice versa.

Further, Kligman[112] has cautioned about using cutaneous changes to support theoretical generalizations about the aging process, as it is unlikely that a single mechanism could account for the numerous and varied alterations that have been observed. If this be the case, the odds that a single theory could explain actinic-mediated mechanisms for both processes are even more remote. The French essayist Michel Montaigne's argument that the study of life is no more than one of decay may be more prophetic than realized. These two processes for which a relationship has been sought seem to be linked only by the general expression of dysfunction, dysfunction manifested by loss of physiologic control, i.e., decay (a change from a state of perfection). However, even in this general relationship it should be obvious that there are a number of mechanistic parallels between both processes that provide a starting point for more intense investigation.

ACKNOWLEDGMENT

The author is grateful to Mrs. Wanda Lenger for assistance in preparing the manuscript.

REFERENCES

1. Scott, J., Fears, T., Lisiecki, E., and Radosevich, S., Incidence of nonmelanoma skin cancer in the United States, 1977—78, Preliminary Report, Publ. No. (NIH) 80-2154, National Cancer Institute, U.S. Department of Health, Education and Welfare, Bethesda, Md., 1980.
2. Bergstresser, P. R. and Halprin, K. M., Multiple sequential skin cancers. The risk of skin cancer in patients with previous skin cancer, *Arch. Dermatol.*, 111, 995, 1975.
3. Johnson, M. L. T. and Robert, J., Relevance of Dermatological Disease Among Persons 1—74 years of age. Advance data from vital and health statistics of the National Center for Health Statistics, No. 4, U.S. Department of Health, Education and Welfare, Bethesda, Md., 1977.
4. Pitot, H. C., Carcinogenesis and aging — two related phenomena?, *Am. J. Pathol.*, 87, 440, 1977.
5. Peto, R., Roe, F. J. C., Lee, P. N., Levy, L., and Clack, J., Cancer and aging in mice and men, *Br. J. Cancer*, 32, 411, 1975.
6. Vitaliano, P. P. and Urbach, F., The relative importance of risk factors in nonmelanoma skin carcinoma, *Arch. Dermatol.*, 116, 454, 1980.
7. Forbes, P. D., Davies, R. E., and Urbach, F., Aging, environmental influences, and photocarcinogenesis, *J. Invest. Dermatol.*, 73, 131, 1979.
8. Doll, R., Incidence of cancer in humans, in *Origins of Human Cancer*, Hiatt, H. H., Watson, J. D., and Winsten, J. A., Eds., Cold Spring Harbor Laboratory, Cold Spring Harbor, N.Y., 1977, 1.
9. Blum, H. F., Grady, H. G., and Kirby-Smith, J. S., Relationships between dosage and rate of tumor induction by ultraviolet radiation, *J. Natl. Cancer Inst.*, 3, 91, 1942.
10. Ebbesen, P. and Kripke, M. L., Influences of age and anatomical site on ultraviolet carcinogenesis in BALB/c mice, *J. Natl. Cancer Inst.*, 68, 691, 1982.
11. Black, H. S., McCann, V., and Thornby, J. I., Influence of animal age upon antioxidant-modified UV carcinogenesis, *Photobiochem. Photobiophys.*, 4, 107, 1982.
12. Strehler, B. L., Aging theories and immunological senescence: perspects and prospects, in *Handbook of Immunology in Aging*, Kay, M. M. B. and Makinodan, T., Eds., CRC Press, Boca Raton, Fla., 1981, 3.
13. Becker, F. F., Evolution, chemical carcinogenesis, and mortality: the cycle of life, in *Carcinogens: Identification and Mechanisms of Action*, Griffin, A. C. and Shaw, C. R., Eds., Raven Press, New York, 1979, 5.
14. Orgel, L. E., The maintenance of the accuracy of protein synthesis and its relevance to ageing, *Proc. Natl. Acad. Sci. U.S.A.*, 49, 517, 1963.
15. Hart, R. W., D'Ambrosio, S. M., and Ng, K. J., Longevity, stability and DNA repair, *Mech. Ageing Dev.*, 9, 203, 1979.
16. Hart, R. W. and Setlow, R. B., Correlation between deoxyribonucleic acid excision-repair and life-span in a number of mammalian species, *Proc. Natl. Acad. Sci. U.S.A.*, 71, 2169, 1974.
17. Hayflick, L. and Moorhead, P. S., The serial cultivation of human diploid cell strains, *Exp. Cell Res.*, 25, 585, 1961.
18. Hayflick, L., The limited *in vitro* lifetime of human diploid cell strains, *Exp. Cell Res.*, 37, 614, 1965.
19. Hayflick, L., The cell biology of aging, *J. Invest. Dermatol.*, 73, 8, 1979.
20. Martin, G. M., Sprague, C. A., and Epstein, C. J., Replicative life-span of cultivated human cells. Effects of donor's age and genotype, *Lab. Invest.*, 23, 86, 1970.
21. Schneider, E. L. and Mitsui, Y., The relationship between in vitro cellular aging and in vivo human age, *Proc. Natl. Acad. Sci. U.S.A.*, 73, 3584, 1976.
22. Gilchrest, B. A., Relationship between actinic damage and chronologic aging in keratinocyte cultures of human skin, *J. Invest. Dermatol.*, 72, 219, 1979.
23. Gilchrest, B. A., Prior chronic sun exposure decreases the lifespan of human skin fibroblasts *in vitro*, *J. Gerontol.*, 35, 537, 1980.
24. Cristofalo, V. T. and Sharf, B. B., Cellular senescence and DNA synthesis: thymidine incorporation as a measure of population age in human diploid cells, *Exp. Cell Res.* 76, 419, 1973.
25. Little, J. B., Relationships between DNA repair capacity and cellular aging, *Gerontology*, 22, 28, 1976.
26. Chan, A. C. and Walker, I. C., Loss of DNA repair capacity during successive subcultures of primary rat fibroblasts, *J. Cell Biol.*, 74, 365, 1977.
27. Epstein, J., Williams, J. R., and Little, J. B., Rate of DNA repair in progeric and normal human fibroblasts, *Biochem. Biophys. Res. Commun.*, 59, 850, 1974.
28. Goldstein, S., The role of DNA repair in aging of cultured fibroblasts from Xeroderma Pigmentosum and normals, *Proc. Soc. Exp. Biol. Med.*, 137, 730, 1971.
29. Hart, R. W. and Setlow, R. B., DNA repair in late-passage human cells, *Mech. Ageing Dev.*, 5, 67, 1976.
30. Regan, J. D. and Setlow, R. B., DNA repair in human progeroid cells, *Biochem. Biophys. Res. Commun.*, 59, 858, 1974.

31. Robbins, J. H., Kraemer, K. H., Lutzner, M. A., Festoff, B. W., and Coon, H. G., An inherited disease with sun sensitivity, multiple cutaneous neoplasms, and abnormal DNA repair, *Ann. Intern. Med.*, 80, 221, 1974.
32. Grossman, L., Braun, A., Feldberg, R., and Mahler, I., Enzymatic repair of DNA, *Annu. Rev. Biochem.*, 44, 19, 1975.
33. Epstein, J. H., Fukuyama, K., and Epstein, W. L., UVL induced stimulation of DNA synthesis in hairless mouse epidermis, *J. Invest. Dermatol.*, 51, 445, 1968.
34. Setlow, R. B., Regan, J. D., German, J., and Carrier, W. L., Evidence that Xeroderma Pigmentosum cells do not perform the first step in the repair of ultraviolet damage to their DNA, *Proc. Natl. Acad. Sci. U.S.A.*, 64, 1035, 1969.
35. Cleaver, J. E., Xeroderma Pigmentosum: a human disease in which an initial stage of DNA repair is defective, *Proc. Natl. Acad. Sci. U.S.A.*, 63, 428, 1969.
36. Cleaver, J. E., DNA repair and radiation sensitivity in human (Xeroderma Pigmentosum) cells, *Int. J. Radiat. Biol.*, 18, 557, 1970.
37. Kleijer, W. J., Lohman, P. H. M., Mulder, M. P., and Bootsma, D., Repair of X-ray damage in DNA of cultivated cells from patients having Xeroderma Pigmentosum, *Mutat. Res.*, 9, 517, 1970.
38. Epstein, W. L., Fukuyama, K., and Epstein, J. H., Ultraviolet light, DNA repair and skin carcinogenesis in man, *Fed. Proc.*, 30, 1766, 1971.
39. Cleaver, J. E., DNA damage and repair in light-sensitive human skin disease, *J. Invest. Dermatol.*, 54, 181, 1970.
40. Cleaver, J. E., Xeroderma Pigmentosum: variants with normal DNA repair and normal sensitivity to ultraviolet light, *J. Invest. Dermatol.*, 58, 124, 1972.
41. Kohn, R. R., Evidence against cellular aging theories, in *Testing the Theories of Aging*, Adelman, R. C. and Roth, G. S., Eds., CRC Press, Boca Raton, Fla., 1982, 221.
42. Cutler, R. G., Evolutionary biology of aging and longevity in mammalian species, in *Aging and Cell Function*, Johnson, J. E., Ed., Plenum Press, New York, 1984, 1.
43. Harman, D., Role of free radicals in mutation, cancer, aging and the maintenance of life, *Radiat. Res.*, 16, 753, 1962.
44. Harman, D., The free radical theory of aging, in *Free Radicals in Biology V*, Pryor, W. A., Ed., Academic Press, New York, 1982, 255.
45. Black, H. S., Chemoprevention of cutaneous carcinogenesis, *Cancer Bull.*, 35, 252, 1983.
46. Horgan, V. J., Philpot, J. St. L., Porter, B. W., and Roodyn, D. B., Toxicity of autoxidized squalene and linoleic acid and of simpler peroxides, in relation to toxicity of radiation, *Biochem. J.*, 67, 551, 1957.
47. Shamberger, R. J., Andereone, T. L., and Willis, C. E., Antioxidants and cancer. IV. Initiating activity of malonaldehyde as a carcinogen, *J. Natl. Cancer Inst.*, 53, 1771, 1974.
48. Shamberger, R. J., Corlett, C. L., Beaman, K. D., and Kasten, B. L., Antioxidants reduce the mutagenic effect of malonaldehyde and β-propiolactone. IX. Antioxidants and cancer, *Mutat. Res.*, 66, 349, 1979.
49. Chio, K. S. and Tappel, A. L., Synthesis and characterization of the fluorescent products derived from malonaldehyde and amino acids, *Biochemistry*, 7, 2821, 1969.
50. Donato, H., Jr., Lipid peroxidation, cross-linking reactions, and aging, in *Age Pigments*, Sohal, R. S., Ed., Elsevier/North-Holland, Amsterdam, 1981, 63.
51. Berg, B. N. and Simms, H. S., Nutrition and longevity in the rat. II. Longevity and onset of disease with different levels of food intake, *J. Nutr.*, 71, 255, 1960.
52. Harman, D., The free radical theory of aging: effect of the amount and degree of unsaturation of dietary fat on mortality rate, *J. Gerontol.*, 26, 451, 1971.
53. Harman, D., Free radical theory of aging: effect of free radical reaction inhibitors on the mortality rate of male LAF_1 mice, *J. Gerontol.*, 23, 476, 1968.
54. Daniels, F., Jr., The physiological effects of sunlight, *J. Invest. Dermatol.*, 32, 147, 1959.
55. Daniels, F., Jr. and Johnson, B. E., Normal, physiologic and pathologic effects of solar radiation on the skin, in *Sunlight and Man*, Fitzpatrick, T. B., Ed., University of Tokyo Press, Tokyo, 1974, 117.
56. Pathak, M. A. and Stratton, K., Free radicals in human skin before and after exposure to light, *Arch. Biochem. Biophys.*, 123, 468, 1968.
57. Lo, W. B. and Black, H. S., Inhibition of carcinogen formation in skin irradiated with UV, *Nature (London)*, 246, 489, 1973.
58. Black, H. S., Effects of dietary antioxidants on actinic tumor induction, *Res. Commun. Chem. Pathol. Pharmacol.*, 7, 783, 1974.
59. Black, H. S. and Chan, J. T., Suppression of ultraviolet light-induced tumor formation by dietary antioxidants, *J. Invest. Dermatol.*, 65, 412, 1975.
60. DeRios, G., Chan, J. T., Black, H. S. et al., Systemic protection by antioxidants against UVL-induced erythema, *J. Invest. Dermatol.*, 70, 123, 1978.

61. Black, H. S., Chan, J. T., and Brown, G. E., Effects of dietary constituents on ultraviolet light-mediated carcinogenesis, *Cancer Res.*, 38, 1384, 1978.
62. Mathews-Roth, M. M., Pathak, M. A., Fitzpatrick, T. B. et al., Beta-carotene therapy for Erythopoietic Protoporphyria and other photosensitivity diseases, *Arch. Dermatol.*, 113, 1229, 1977.
63. Epstein, J. H., Effects of Beta-carotene on UV induced cancer formation in the hairless mouse skin, *Photochem. Photobiol.*, 25, 211, 1977.
64. Mathews-Roth, M. M., Antitumor activity of β-carotene, canthaxanthin and phytoene, *Oncology*, 39, 33, 1982.
65. Krinsky, N. I. and Deneke, S. M., Interaction of oxygen and oxyradicals with carotenoids, *J. Natl. Cancer Inst.*, 69, 205, 1982.
66. Pauling, L., Willoughby, R., Reynolds, R., et al., Incidence of squamous cell carcinoma in hairless mice irradiated with ultraviolet light in relation to intake of ascorbic acid (vitamin C) and dl-α-tocopheryl acetate (vitamin E), in *Vitamin C:* new clinical applications in Immunology, lipid metabolism and cancer, Hanck, A., Ed., *Int. J. Vitam. Nutr. Res.*, 23(Suppl.), 53, 1982.
67. Forbes, P. D., Photocarcinogenesis: an overview, *J. Invest. Dermatol.*, 77, 139, 1981.
68. Black, H. S., Henderson, S. V., Kleinhans, C. M. et al., The effect of dietary cholesterol on ultraviolet-light carcinogenesis, *Cancer Res.*, 39, 5022, 1979.
69. Baumann, C. A. and Rusch, H. P., Effect of diet on tumors induced by ultraviolet light, *Am. J. Cancer*, 35, 213, 1939.
70. Black, H. S., Lenger, W., Phelps, A. W. et al., Influence of dietary lipid upon ultraviolet-light carcinogenesis, *Nutr. Cancer*, 5, 59, 1983.
71. Black, H. S., Lenger, W., MacCallum, M. et al., The influence of dietary lipid level on photocarcinogenesis, *Photochem. Photobiol.*, Suppl. 37, 539, 1983.
72. Mathews-Roth, M. M. and Krinsky, N. I., Effect of dietary fat level on UV-B induced skin tumors, and anti-tumor action of β-carotene, *Photochem. Photobiol.*, 40, 671, 1984.
73. Black, H. S. and Lenger, W., Inhibition of epidermal lipid peroxidation by dietarily administered antioxidants, *Proc. Am. Assoc. Cancer Res.*, 25, 132, 1984.
74. Meffert, H., Diezel, W., and Sonnichsen, N., Stable lipid peroxidation products in human skin: detection, ultraviolet light-induced increase, pathogenic importance, *Experientia*, 32, 1397, 1976.
75. Johnson, B. E. and Daniels, F., Jr., Lysosomes and the reactions of skin to ultraviolet radiation, *J. Invest. Dermatol.*, 53, 85, 1969.
76. Grossie, V. B., Jr. and Black, H. S., The effect of ultraviolet light (UVL) on the lysosomes of hairless mouse epidermis, *Experientia*, 33, 425, 1977.
77. Black, H. S. and Lo, W. B., Formation of a carcinogen in UV-irradiated human skin, *Nature (London)*, 234, 306, 1971.
78. Sevanian, A., Mead, J. F., and Stein, R. A., Epoxides as products of lipid autoxidation in rat lungs, *Lipids*, 14, 623, 1979.
79. Kay, M. M. B., The thymus: clock for immunologic aging?, *J. Invest. Dermatol.*, 73, 29, 1979.
80. Gross, L., Immunologic defect in aged populations and its relation to cancer, *Cancer*, 18, 201, 1965.
81. Gardner, I. D. and Remington, J. S., Age-related decline in the resistance of mice to infection with intracellular pathogens, *Infect. Immunol.*, 16, 593, 1977.
82. Kripke, M. L., Antigenicity of murine skin tumors induced by ultraviolet light, *J. Natl. Cancer Inst.*, 53, 1333, 1974.
83. Kripke, M. L. and Fisher, M. S., Immunologic parameters of ultraviolet carcinogenesis, *J. Natl. Cancer Inst.*, 57, 211, 1976.
84. Kripke, M. L. and Fidler, I. J., Enhanced experimental metastasis of ultraviolet light-induced fibrosarcomas in ultraviolet light-irradiated syngeneic mice, *Cancer Res.*, 40, 625, 1980.
85. Fisher, M. S. and Kripke, M. L., Further studies on the tumor-specific suppressor cells induced by ultraviolet radiation, *J. Immunol.*, 121, 1139, 1978.
86. Spellman, C. W. and Daynes, R. A., Properties of ultraviolet light-induced suppressor cells by ultraviolet radiation, *Cell Immunol.*, 31, 182, 1977.
87. Fisher, M. S. and Kripke, M. L., Systemic alteration induced in mice by ultraviolet light irradiation and its relationship to ultraviolet carcinogenesis, *Proc. Natl. Acad. Sci. U.S.A.*, 74, 1688, 1977.
88. de Gruijl, F. R. and van der Leun, J. C., Systemic influence of pre-irradiation of a limited skin area on UV-carcinogenesis, *Photochem. Photobiol.*, 35, 379, 1982.
89. Fisher, M. S. and Kripke, M. L., Suppressor T lymphocytes control the development of primary skin cancers in UV-irradiated mice, *Science*, 216, 1133, 1982.
90. Nathanson, R. B., Forbes, P. D., and Urbach, F., Modification of photocarcinogenesis by two immunosuppressive agents, *Cancer Lett.*, 1, 243, 1976.
91. Daynes, R. A., Harris, C. C., Connor, R. J., et al., Skin cancer development in mice exposed chronically to immunosuppressive agents, *J. Natl. Cancer Inst.*, 62, 1075, 1979.

92. Good, R. A., Fernandes, G., and West, A., Nutrition, immunologic aging, and disease, in *Aging and Immunity,* Singhal, S. K., Sinclair, N. R., and Stiller, C. R., Eds., Elsevier/North-Holland, New York, 1979, 141.
93. Watson, R. R. and Safranski, D. V., Dietary restrictions and immune responses in the aged, in *Handbook of Immunology in Aging,* Kay, M. M. B. and Makinodan, T., Eds., CRC Press, Boca Raton, Fla., 1981, 125.
94. Harman, D., The free radical theory of aging: effect of age on serum copper levels, *J. Gerontol.,* 20, 151, 1965.
95. Leto, S., Yiengst, M. J., and Barrows, C. H., Jr., The effect of age and protein deprivation on the sulfhydryl content of serum albumin, *J. Gerontol.,* 25, 4, 1970.
96. Harman, D., Heidrick, M. L., and Eddy, D. E., Free radical theory of aging: effect of free-radical reaction inhibitors on the immune response, *J. Am. Geriatr. Soc.,* 25, 400, 1977.
97. Streilein, J. W. and Tigelarr, R. E., SALT: skin associated lymphoid tissues, in *Photoimmunology,* Parrish, J. A., Kripke, M. L., and Morison, W. L., Eds., Plenum Press, New York, 1983, 95.
98. Bergstresser, P. R., Elmets, C. A., and Streilein, J. W., Local effects of ultraviolet radiation on immune function in mice, in *The Effect of Ultraviolet Radiation on the Immune System,* Parrish, J. A., Ed., Johnson & Johnson, Raritan, N.J., 1983, 73.
99. Emmett, E. A., Ultraviolet radiation as a cause of skin tumors, *CRC Crit. Rev. Toxicol.,* 2, 211, 1973.
100. Black, H. S. and Chan, J. T., Experimental ultraviolet light-carcinogenesis, *Photochem. Photobiol.,* 26, 183, 1977.
101. Montagna, W., Kligman, A. M., Wuepper, K. D., and Bentley, J. P., Eds., Special issue on aging, Proceedings of the 28th symposium on the biology of skin, *J. Invest. Dermatol.,* 73, 1, 1979.
102. Epstein, J. H., Photocarcinogenesis, skin cancer, and aging, *J. Am. Acad. Dermatol.,* 9, 487, 1983.
103. Gilchrest, B. A., *Skin and Aging Processes,* CRC Press, Boca Raton, Fla., 1984, 136.
104. Montagna, W., Morphology of aging skin: the cutaneous appendage, in *Advances in Biology of Skin,* Montagna, W., Ed., Pergamon Press, Oxford, 1965, 1.
105. Montagna, W. and Carlisle, K., Structural changes in aging human skin, *J. Invest. Dermatol.,* 73, 47, 1979.
106. Lavker, R. M., Structural alterations in exposed and unexposed aged skin, *J. Invest. Dermatol.,* 73, 59, 1979.
107. Mackie, B. S. and McGovern, V. J., The mechanism of solar carcinogenesis, *Arch. Dermatol.,* 78, 218, 1958.
108. Schneider, E. L., Aging and cultured human skin fibroblasts, *J. Invest. Dermatol.,* 73, 15, 1979.
109. Sbano, E., DNA repair after UV irradiation in skin fibroblasts from patients with actinic keratoses, *Arch. Dermatol. Res.,* 262, 55, 1978.
110. Munkres, K. D., Biochemical genetics of aging of *Neurospora crassa* and *Podospora anserina,* in *Age Pigments,* Sohal, R. S., Ed., Elsevier/North-Holland, New York, 1981, 83.
111. Pryor, W. A., The formation of free radicals and the consequences of their reactions *in vivo, Photochem. Photobiol.,* 28, 787, 1978.
112. Kligman, A. M., Perspectives and problems in cutaneous gerontology, *J. Invest. Dermatol.,* 73, 39, 1979.

Chapter 5

EFFECTS OF SUNLIGHT ON THE EYE

Sidney Lerman

TABLE OF CONTENTS

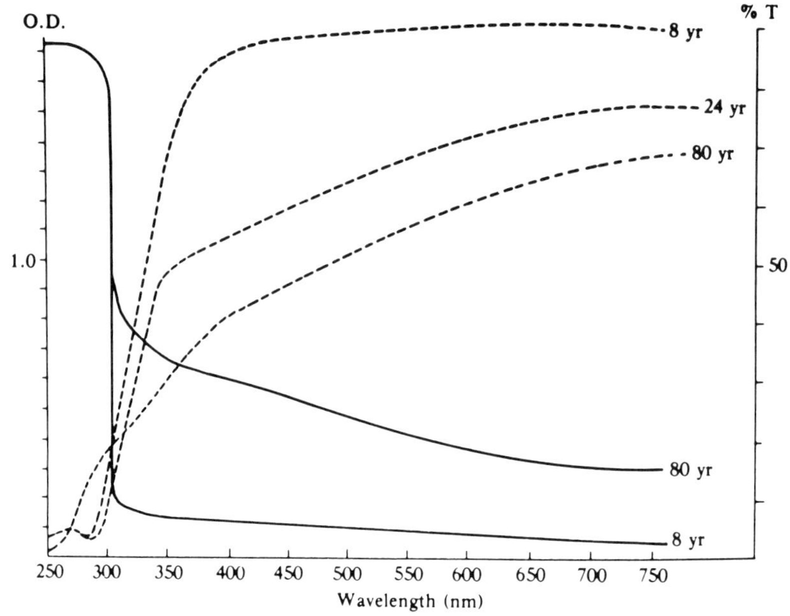

FIGURE 2. Absorption (solid line) and transmission (dotted line) of the human cornea. (From Lerman, S., *Radiant Energy and the Eye,* Macmillan, New York, 1980. With permission.)

tors.[17-19] These investigators noted that the maximum efficiency for experimental photokeratitis and photokeratoconjunctivitis peaks sharply at 288 to 290 nm, with a smaller peak at 254 nm. This effect can be ascribed to absorption by the nucleotides (at 254 nm) and by tyrosine and tryptophan in the corneal epithelium. Studies on free-radical generation in human corneas exposed to 290 nm radiation show an interesting correlation with the UV action spectrum for photokeratitis reported by Sherashov[19] as shown in Figure 3. Clinically, corneal photodamage from UV radiation has long been appreciated, a typical example being snow blindness experienced by polar explorers. This type of photokeratitis is due to the relatively high levels of UV radiation which can be reflected by snow compared with less than 5% from earth or grass. Aside from the well-known polar and industrial photokeratitis, UV radiation has also been implicated in a variety of conjunctiveal and corneal lesions. These include pingueculae and pterygiums, exposure keratosis (which involves epithelial changes related to actinic radiation and is analogous to actinic keratosis of the skin), nodular band-shaped keratopathy, experimentally induced tumors in animals, and the relatively rare dysplasia and intra-epithelial carcinoma. Certain corneal diseases can also be triggered by exposure to UV radiation; for example, herpes simplex keratitis and recurrent erosions of the cornea.

Since the cornea filters out UV radiation shorter than 295 nm, we are left with the question regarding the effect of longer wavelength UV radiation (300 to 400 nm) on the eye. The cornea transmits almost all of this radiation into the interior of the globe and the same can be said for the aqueous humor. The ocular lens absorbs a significantly increasing amount of this radiation, as well as visible light, as it ages.

B. Lens

A knowledge of the morphology and composition of the lens will not only explain its transparency to visible light but will also indicate why this organ is susceptible to UV-induced photochemical changes as it ages.

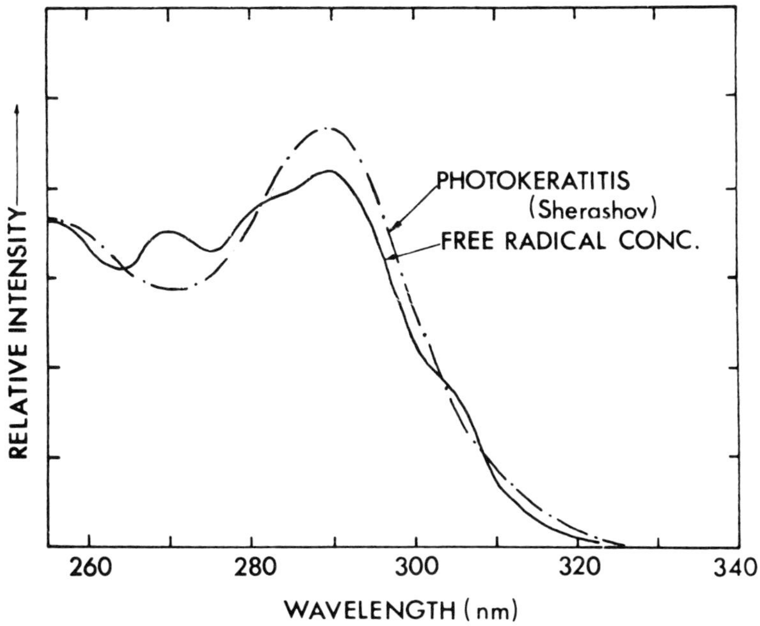

FIGURE 3. Action spectrum for photokeratitis[19] and free radical generation in normal cornea exposed to monochromatic UV radiation.

The human crystalline lens, derived from the surface ectoderm at about the 4.5-mm stage of embryonic life, is a unique, biconvex, transparent, and avascular organ that represents a pure cell line and preserves the complete population of cells from fetal life until death. The mature strap-like lens-fiber cells lose their nuclei but remain viable with only scanty organelles. The lens fibers grow from all around the equatorial region and meet at radiating lines or sutures. During the first 8 weeks of life, the lens is a sphere. Since further growth of the lens depends on the continuous division of cells located in the germinative zone, there will be a gradual increase in the equatorial diameter of the lens in comparison with its axial diameter, and the organ will take on the more characteristic oval shape seen in the adult.

The adult human lens is a precisely formed structure containing about 65% water and 35% organic matter, the latter being chiefly structural proteins. Since the structural proteins (alpha, beta, and gamma crystallins and the insoluble protein fraction) constitute most of the dry weight of the lens, they must play a major role in the transmission as well as in the absorption and refraction of light through this organ. The degree of hydration and perhaps the action of the mucopolysaccharides as cement substances between the lens fibers may also play a role in lenticular transparency. In the infant, this organ does not contain any chromophores capable of absorbing a significant amount of visible radiation (400 to 750 nm), and the potential for absorbing UV radiation longer than 295 nm resides mainly in the tryptophan residues present in the lens proteins (and the relatively small amount of free tryptophan within the lens amino-acid pool). Aside from the minute amount of cytochromes in the lens epithelial cells, there are no other chromophores in the young lens capable of absorbing radiation wavelengths longer than 320 nm. The metabolic activity of the lens is mainly directed toward maintaining normal hydration (60 to 65%) and protein synthesis. The constituent lens proteins are densely packed and arranged in a high degree of spatial order within the intact fiber. There is little change in the density of scattering particles as one moves from point to point within the lens. According to Benedek[13,14] the changes in the index of refraction within specific anatomic zones of the lens still permit transparency. That is, "the scattered waves add up to give zero intensity in all directions except for those directions predicted for the bending of the primary beam by the laws of refraction in geometrical optics".[14] Refraction itself, therefore, is a consequence of scattering, and the refracted beam is the result of constructive interference of the scattered light. As Benedek[13] has pointed out "when lens proteins are extracted and mixed with water the constituent proteins are now dispersed throughout the solution. The proteins are no longer tightly packed and the density of proteins fluctuates markedly about its average value, hence, the solution can appear milky or turbid." In a simplified analogy, the accumulation of water within the lens results in a decreased transparency. Localized alterations in the density of packing of lens proteins, because of aggregation and/or configurational changes developing in various portions of the lens, will lead to changes in transparency; i.e., the development of opacities which could be due to localized changes in the refractive index.

Aging of the lens is reflected by pronounced changes at the molecular level, particularly with respect to the structural lens proteins, which constitute most of its dry weight. It seems that postsynthetic modifications mainly are involved, although errors at the translational level cannot be excluded. A variety of chemically or physically defined processes have been observed in lens proteins with aging. At birth, >95% of the human structural lens proteins are water soluble. One well-defined aging parameter is the progressive age-related increase in the amount of insoluble protein (to 40% or more of the total crystallin content), with a concomitant decrease in the total soluble protein (TSP) fraction.[1] The generation of a high molecular weight (HMW) fraction has been implicated in the increase in light scattering which develops as the lens ages.[20] There is also a marked change in the relative concentration of at least one of the soluble crystallin fractions. At birth, the gamma crystallins constitute over 50% of the TSP and then decrease progressively with age to about 10% in the old human lens (60+ age group) and in the 2-year-old rat lens.[1] The aggregation/insolubilization processes occurring with age appear to be associated with the formation of disulfide bridges and other covalent cross links. The latter may (at least in part) be due to chronic cumulative exposure to ambient UV radiation (over a 30 to 50 year span) resulting in the photochemical generation of a series of progressively longer wavelength UV and visible radiation-absorbing chromophores in the lens.[1,15,16,21] Thus, as the lens ages it displays an increasing absorbance in the 300- to 400-nm UV region, and in the visible region, which is associated with increased pigmentation of the lens core (nucleus). This is associated with increased lenticular fluorescence.

1. Intrinsic Fluorescence of the Lens

Before discussing the visible fluorescence of the lens, it is important to note that most proteins are endowed with an intrinsic UV-excited fluorescence because they contain aromatic amino acids (phenylalanine, tyrosine, and tryptophan). Of these three aromatic amino acids, phenylalanine has the lowest fluorescence quantum yield (approximately 1/10 to 1/20 of the latter two). Protein fluorescence spectra are thus generally considered only with respect to tyrosine and tryptophan. The presence of even one tryptophan residue in a protein (for example, horse serum albumin which has 17 tyrosines and 1 tryptophan) will result in an intrinsic fluorescence essentially that of tryptophan alone. Thus in most proteins containing at least one tryptophan residue, tyrosine appears to have a very low quantum efficiency so that the emission from this residue is markedly overshadowed by tryptophan fluorescence. However, tryptophan fluorescence in proteins appears to have quantum yields which in some cases are equivalent to the free amino acids.[22] It is also known that the tryptophan fluorescence maximum in proteins varies from 330 to 352 nm depending on the location of this amino acid within the protein, whereas free tryptophan has a characteristic fluorescence emission maximum at approximately 350 nm. It should be noted that tryptophan residues in native proteins are not necessarily present in identical locations nor are they equally influenced by their environment.[23,24]

The microenvironment of every residue is characterized by a particular set of physiochemical conditions such as polarizability, microviscosity, availability of charged groups, and possible specific interactions, that can influence chromophore fluorescence. Thus, protein (tryptophan) fluorescence is conditioned by the sum of fluorescent contributions of individual tryptophan residues which can vary over a relatively wide range. Studies using spectral band width, spectral maximum position, fluorescence quenching by external ionic quenchers, fluorescence lifetime, quantum yield, and changes upon denaturation of a variety of proteins indicate that fluorescence properties of various tryptophan residues in proteins can be delineated into three discrete spectral classes.[23,24] The Type I tryptophan residues are buried in nonpolar regions of the protein, and the Type II and III tryptophan residues are located on the protein surface, Type III being fully exposed to water and Type II being in limited contact with water. The following discussion will describe how these ideas can be applied to tryptophan fluorescence in lens proteins.

Lens proteins, in general, have a considerably higher content of tyrosine than tryptophan residues, but the presence of the latter results in an intrinsic protein fluorescence in the lens due to tryptophan alone.[21] The data show that the tryptophan fluorescence maximum in whole lenses is at 332 ± 2 nm with a band width of 47 nm. These results indicate that the majority of tryptophan residues in mammalian lenses belong to Type I of the Burstein classification,[23] i.e., they are buried in nonpolar regions of the proteins.

Fluorescence-lifetime measurements on intact lenses support the conclusions that lens protein tryptophan residues belong to Type I. Analysis of the decay data produces a fluorescence lifetime of 2.3 nsec, compared to a value of 2.8 nsec which is obtained for free tryptophan in aqueous solution. The value for tryptophan in the lens is consistent with these residues being located primarily in hydrophobic environments. Recent studies on bovine and human lens protein provide further substantiation that the tryptophan residues in the gamma crystallins are all buried or in crevice positions.[25-27] In the alpha and beta crystallins, most of the tryptophan residues are likewise buried, although one of the beta fractions may contain several surface tryptophans.

2. *Extrinsic Fluorescence of the Lens*

The visible fluorescence of the ocular lens was first noted well over 100 years ago,[28] but it was not until the last decade that this phenomenon was associated with increased yellow coloration of the lens nucleus by the increased visible and near-UV absorption. François et al.[29] noted the presence of a fluorescent peptide in primate lenses which could be activated by UV light. Specific fluorescent proteins in the human and other mammalian lenses have been demonstrated by a variety of investigators.[15,16,30-42] In addition to the fluorescence due to tryptophan, these proteins contain one or more fluorescent chromophores (fluorogens) with activation wavelengths of approximately 340 to 360 nm and 420 to 435 nm and emission maximums at 420 to 440 nm and 500 to 520 nm.[21] This type of fluorescence has been shown to be present in the soluble, HMW and insoluble fractions of the human lens (mainly in the latter) and to increase with age, particularly in the insoluble-protein fraction. These fluorogens are tightly bound to peptides derived from the insoluble lens protein.

In vitro UV-induced changes in human lenses include increased visible fluorescence intensities at 440 and 520 nm and increased pigmentation.[1,15,16] Similar fluorescence enhancement can be observed in UV-irradiated animal lenses both in vitro and in

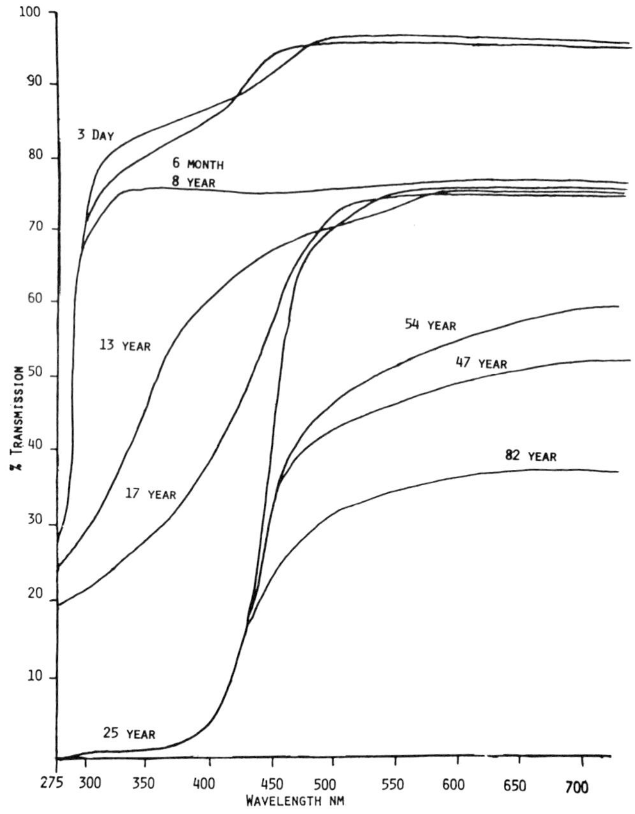

FIGURE 4. Transmission changes in the aging ocular lens.

vivo.[1,15,43] These studies indicate that the photochemical changes are similar to the increasing fluorescence seen in normal human lenses as they age.[1,15,16,21] The cumulative effects of chronic exposure to ambient UV radiation include the generation of a series of fluorescent chromophores (absorbing at increasingly longer wavelengths), a deepening yellow color of the lens nucleus, and a progressive increase in the level of insoluble lens proteins. The latter phenomenon may be related to the cross linking of previously soluble polypeptides by some of the fluorescent pigments as well as by disulfide bond formation.

Since the human ocular lens is constantly exposed to ambient UV radiation (300 to 400 nm) throughout life, the consequence of cumulative photochemical damage is an increasing absorption of UV radiation and some visible light due to the presence of photochemically generated fluorescent chromophores which increase in concentration and in number as the lens ages. UV and visible-light transmission studies on normal human lenses ranging in age from 3 days to 82 years have shown that over 75% of the UV radiation (300 to 400 nm) is transmitted by lenses under 10 years of age, while the corresponding UV transmission in lenses above 25 years of age drops markedly (Figure 4). These data are consistent with the relative lack of UV-absorbing chromophores in the young lens and an increase in their concentration as the lens ages.[1] Accompanying the measured increase in these chromophores and decrease in light transmission, there is a concomitant increase in the yellow color of the lens nucleus as it ages. These pigments are mainly confined to the nucleus and may play a role in the development of nuclear sclerosis. However, nuclear sclerosis can also develop without significant discoloration. The discoloration is mainly confined to the lens nucleus since the cortex has much higher levels of glutathione and other compounds capable of aborting most of the photochemical reactions.[1,15,16] In about 10% of our population, this process progresses at a more rapid pace, resulting in the development of the brown (nuclear) cataract which is an extreme example of these age-related photochemical changes. This type of discoloration, in moderation, is actually beneficial since it enables the lens to become a very effective filter for UV and short-wavelength visible radiation (by the second to third decade) thus protecting the retina from cumulative photochemical damage which could occur during our lifetime. Recent studies have shown that such radiation can cause irreversible retinal photodamage in the aphakic Rhesus monkey and even in man.[2-6,44-47] It is interesting that nature has provided us with the ability to develop a lenticular UV filter to protect the retina from continuous radiation exposure which could be harmful, particularly in the older individual where the retinal metabolism and repair processes are no longer as effective as in the young. We are now beginning to see confirmation of the hypothesis that long-wavelength UV radiation may play a role in certain retinal diseases (e.g., cystoid macular edema) which tend to occur in older patients following removal of their cataractous lenses and even in degenerative processes (macular degeneration and retinitis pigmentosa).

Although only a small amount of UV radiation from the sun enters the eye under normal circumstances, the cumulative effect of many years exposure is significant, particularly when one considers man's ever-increasing life span. Epidemiologic surveys provide some support for the thesis that sunlight plays a role in lenticular aging and in the generation of senile cataracts.[48-51] For example, cataracts and the rate of cataract extraction are much higher in India, Pakistan, and certain areas of Africa than in the temperate zones. An epidemiologic investigation into the relationship between sunlight and cataract in the U.S. reported that " . . . cataract to control ratios for persons aged 65 years or older were significantly larger in locations with large amounts of sunlight . . . "[48]

Two recent epidemiologic studies involving Australian aborigines exposed to relatively intense solar radiation are of interest. Detailed personal histories were obtained on 350 individuals, and the possible role of a variety of environmental factors in cataractogenesis were investigated. Of the 350 individuals studied by Taylor,[49] all of whom were over 30 years of age, 116 had lens opacities as determined by slit-lamp examination. These findings indicate a relationship between cataracts, increased hours of sunlight, and higher annual UV radiation levels. Aside from age, no other factors studied appeared to be associated with cataractogenesis. Hollows and Moran's study,[50] involving a much greater geographical area and over 50% of the total aborigine population, found a statistically significant correlation between environmental UV irradiation and the prevalence of cataract. Cataracts occurred much more frequently in the younger age group (40 to 59 years) in zones with high UV radiation than in zones with low UV radiation. Brilliant et al.[51] studied the relationship between the prevalence of cataracts, altitude, and hours of sunlight exposure in Nepal. Exposure to an average of 12 hr of sunlight resulted in 3.8 times the number of cataracts compared with individuals exposed to 7 hr or less of sunlight. These epidemiologic studies support the hypothesis that senile cataracts are associated with higher exposure to sunlight. Obviously other factors play a role in cataractogenesis, including heredity, nutrition, metabolism, etc.

Aside from the effects of chronic exposure to ambient UV radiation, exposure to higher radiation levels can produce cortical opacities in human, rat, and rabbit lenses in vitro and in vivo.[52-54] Thus, more intense UV radiation (300 nm and longer) can induce lens changes involving the cortex, while chronic exposure affects mainly the lens nucleus. These effects appear to be dose and time related.[55] There is also the possibility that more intense UV radiation may affect the protein/water order within the lens, struction of tryptophan, producing small amounts of anthranilic acid which may be involved in the production of one or more of the lens fluorogens.[31,42,66,67] H_2O_2 can itself be produced by near-UV radiation action on tryptophan.[68] Thus, a build up of H_2O_2 may also be significant in potential lenticular damage and the presence of adequate levels of GSH would also serve to protect the ocular lens in such a situation. Superoxide dismutase (SOD) and catalase are present in the normal lens in order to protect it from such oxidative damage.[69] H_2O_2 inhibits SOD activity in the lens extracts, and this inhibition is potentiated by 3-AT, while 3-AT alone has no effect on lens SOD; thus, it is postulated that endogenous catalase protects lens SOD from inactivation by H_2O_2.[69]

b. Singlet Oxygen

The photosensitizing action of the psoralen compounds has been correlated with the generation of cyclobutane photoaddition reactions with thymine in the DNA molecule.[70-72] A similar mechanism has been proposed for 8-methoxypsoralen (8-MOP) lens damage.[73] It has been demonstrated that 8-MOP can enter the lens, and following photic stimulation, it can bind to lens proteins thereby enhancing whatever UV damage is already occurring in that individual, i.e., an increase in lenticular fluorescence and phosphorescence.[74] 8-MOP can also exert a photodynamic action in lens proteins by forming new photoproducts with certain aromatic amino acids in which singlet oxygen is involved.[75-77] That is, the triplet state of 8-MOP can react with ground-state oxygen to generate $^1O_2$.[78] The recent demonstration of an oxygen requirement for the photoreaction of 8-MOP with proteins[75-80] indicates that this compound can undergo the Type-II reaction as well as the anaerobic Type-I reaction with DNA, resulting in its binding to lens proteins and the permanent retention of the photoproduct(s) which can also function as new longer wavelength-absorbing photosensitizers in the lens. Many reagents (antioxidants) are capable of quenching $^1O_2$ and doubtless the lens contains its share of them.[81]

One of the immediate effects of radiation on biologic tissue is the production of strongly oxidizing species such as the hydroxyl radical. In order to obtain significant biological effects from the OH radicals (or similar strongly oxidizing species such as a singlet oxygen, the peroxide, or the hydrogen peroxide radicals), they must be produced at or near the target, thereby directly inducing molecular damage; or there may be an energy transfer within the molecule (e.g., the OH radical attacks a protein and the energy is transferred along the peptide chain frequently terminating at an SH group in the chain); or the energy could migrate via lipid peroxidation. In all instances, the SH groups in biologic tissues play an important reparative as well as protective role. These groups may be small SH compounds (GSH) or part of a protein. Thus biologic reactions with strongly oxidizing agents appear to be involved with as well as determined by sulfhydryl groups. The GSH level within the lens is of extreme importance in ionizing radiation damage; a significant fall occurs almost immediately after an experimental animal is exposed,[82] while catalase and SOD may also play an important role in protecting the lens against H_2O_2 and the superoxide radicals.[83]

C. Vitreous

As previously noted, the vitreous is normally protected from UV radiation by the filtering action of the cornea (up to 295 nm) and by the lens as it ages (>295 nm). However, patients who have had their cataracts removed (aphakes and pseudophakes) have lost a significant and protective intraocular filter. The normal vitreous is a gel-like material composed mainly of water, collagenous protein, and long-chain carbohydrates. These compounds do not contain any significant chromophores absorbing above 250 to 260 nm, although a small number of aromatic amino acids are present.

In this respect, the cornea plays a much more significant role as a UV filter since it prevents UV radiation below 295 nm from entering the eye. However, the vitreous does contain some tryptophan residues and several cell types — the hyalocytes and fibrocytes. Thus, some chromophores are present that are capable of absorbing UV radiation longer than 295 nm, provided that the filtering action of the ocular lens is removed. There is some experimental evidence that exposure of the vitreous to UV radiation up to 320 nm results in shrinkage of the vitreous gel and denaturation of the collagen network.[84] There is also a decrease in the viscosity of hyaluronic acid preparations derived from UV-exposed vitreous which can be attributed mainly to a decrease in molecular weight. UV irradiation causes an increase in the reducing power of the polysaccharide solution due to a breakdown of the glucosidic linkages, an increase of the reducing end groups, and the formation of smaller fragments.[84] The absorption spectra of hyaluronic acid subjected to UV irradiation shows significant alterations with the development of new chromophores absorbing at approximately 267 nm. Other studies in which human vitreous gels were subjected to monochromatic UV radiation (280 to 300 nm) resulted in the generation of one or more fluorescent compounds. A 320- to 340-nm excitation and a 420-fluorescence emission peak could be demonstrated following 4 to 6 hr of irradiation, suggesting the possible photodegradation of certain amino acids.

It would thus appear that the cornea, which filters all UV radiation shorter than 295 nm, plays a major role in protecting the vitreous from UV damage. There is still insufficient experimental evidence to assess the effect of UV radiation longer than 295 nm, but there are indications that the vitreous is also sensitive to longer-wavelength UV radiation.[84] Thus the 295- to 400-nm filtering action of the ocular lens may also be of significance in protecting the vitreous.

D. Retina

The fact that visible light is required in the cyclic process of shedding and renewal of the outer membrane disks, which contain the visual pigments, might explain the finding that even moderate but prolonged exposure to visible light, at thresholds of illumination well below those capable of causing thermal damage to the retina, can result in retinal pathology in a variety of experimental animals.[85-93] Aging is known to be characterized by a loss of rod and cone cells.[94] The recent observation that photic trauma can damage the receptors suggests a potential cumulative action of light resulting in an enhanced loss of visual cells over a period of years. That is, phototoxic effects may be cumulative in the normal aging process of the retina. Photon energies in the electromagnetic spectrum increase as the wavelength decreases, from 1.6 eV at 750 nm to 3.3 eV at 400 nm and higher energies in the UV wavelengths capable of penetrating to the retina in the aphakic or pseudophakic eyes. One would anticipate that photic damage would be greatest for UV radiation (320 to 400 nm) and shortwave visible light in the blue region (400 to 475 nm), and decrease with increasing wavelengths of light, with the least photic damage occurring with red light. A variety of reports strongly implicate longwave UV and shortwave visible radiation (320 to 450 nm) as significant factors in retinal photodamage in primates as well as other experimental animals, and even in man.[1-6,95-100] These data are of particular concern in young patients as well as aphakes and pseudophakes who are on photosensitizing drugs and to all patients exposed to prolonged or above-ambient levels of UVA radiation (e.g., occupational-exposure industries where UV polymerization is employed, and certain activities such as sailing, snow skiing, etc.).

The spectral sensitivity of the human retina plays a role with respect to the efficiency of a specific wavelength in producing retinal damage. The ocular lens protects the retina from visible as well as UV radiation since it filters more of the shorter wave-

4. Ophthalmoscopy, the Operating Microscope, and Retinal Damage

One should also consider the potential problems with respect to indirect ophthalmoscopy and retinal damage. Several workers have demonstrated that indirect ophthalmoscopy is capable of producing retinal damage in primate eyes.[98,99] It should be noted that the standard method employed in indirect ophthalmoscopy involves the use of a 20-diopter convex lens placed between the light source and the patient's eye, which in effect will serve to act as an additional focusing element, thereby concentrating the light and energy per unit area onto the patient's retina (in addition to concentrating properties already inherent in the patient's own ocular media). It has been estimated that the amount of focal energy applied to the retina and choroid by the indirect ophthalmoscope is approximately 0.1 to 0.2 mW/cm^2, which approximates the amount of irradiance received at the retinal surface from the sun. However, it should also be noted that the indirect ophthalmoscope has 90% of its power in the IR region (longer than 750 nm). The retinal irradiance levels with an indirect vs. a direct ophthalmoscope have been measured and the data indicate a tenfold increase in retinal irradiance with the indirect ophthalmoscope.[108] It is estimated that at the patient's retina the total irradiance is composed of one third visible light to two thirds IR radiation (750 to 1400 nm). Some workers have proposed that IR filters should be incorporated into all ophthalmoscopes to prevent potential thermal damage from extensive indirect ophthalmoscopy. They also proposed that reasonably short exposures should be used when examining the posterior pole of the eye. Intense illumination and frequent reexamination at short intervals should be avoided. It should be pointed out that the energy delivered during indirect ophthalmoscopy is only 200 times less than the energy that is capable of producing a retinal burn.

Lesions resembling central serious retinopathy have been demonstrated in primate eyes exposed to long-wave UV radiation,[90,99] and recent reports have described clinically demonstrable retinal lesions in patients undergoing intraocular surgery with the operating microscope.[4,6] One must therefore consider the potential for retinal damage with these clinical techniques, particularly in certain conditions, such as retinitis pigmentosa, which might be accelerated by light exposure.[93] Photic damage has also been shown to result in pathologic changes in the pigment epithelium as well as the outer segments; this suggests the possibility that such damage would weaken the adhesions between the retina and pigment epithelium and potentiate retinal detachment.

5. Summary and Conclusions

Recent reviews have considered the possibility of ocular hazards resulting from ambient light exposure.[93,109,110] They indicate that ambient light levels are currently approaching the threshold level for permanent retinal damage (following chronic exposure). The levels of artificial illumination that we are exposed to at present vary between 20 and 50 fc (3.2 to 8.1×10^{-5} W/cm^2). Although these levels of irradiance are well below the levels that are capable of producing retinal photopathology in primates and man, they have been shown to be of sufficient intensity to produce permanent as well as reversible retinal damage in certain experimental animals, particularly albino rats. The trend by lighting engineers toward higher levels of illumination (e.g., 100 to 1000 fc) should be approached with extreme caution in view of the potential photic damage that could be incurred at these levels of illumination. It should also be noted that the lowest reported retinal damage threshold in primates occurs at wavelengths in the blue region of the spectrum, and the potential adverse effects of UV radiation on the retina must also be considered, particularly in the aphakic eye. Another point to bear in mind is the fact that the pigment epithelium is one of the major sites of primary photic damage; thus the melanosomes in the pigment epithelium may play a significant role as transducers of photic energy in the eye. These organelles seem to play a central role in both the thermal and chemical mechanisms of primary photic damage to the retina and may play a role in the pathogenesis of senile macular degeneration. One must also consider the role of photosensitizing agents in retinal photobiology. The fact that the psoralens absorb mostly in the longer-wavelength UV spectrum (between 320 and 360 nm) has allayed concern about the risk of these drugs to the retina. However, the ocular hazard from such photosensitizing drugs should be of concern in aphakic eyes that have lost their natural UV filter (the ocular lens), in pseudophakes, and in young children (whose own lens has yet to develop as an effective UV filter).

III. PHOTOSENSITIZED RADIATION EFFECTS ON OCULAR TISSUES

A photosensitizing agent can be defined as a compound whose chemical structure endows it with the ability to absorb optical radiation (UV and visible) and undergo a primary photochemical reaction resulting in the generation of highly reactive and relatively long-lived intermediates (triplets, radicals, and ions) that can cause chemical modifications in other (nearby) molecules of the biologic system. Regardless of subsequent events, the primary event of photosensitization in ocular tissues appears to be the absorption of light by the photosensitizer molecule, eventually resulting in a complex of photosensitizers with nucleic acid or protein. Only the pyrimidine bases in the nucleic acids are susceptible to photosensitization, and in the proteins, histidine is the most readily altered of all the amino acid residues (either alone or as part of the protein) followed by tryptophan, tyrosine, cystine, and methionine.[1] Sensitized photooxidation reactions in proteins generally do not involve disulfide and peptide bonds, with chemical alterations occurring most frequently in the aforementioned five amino acid residues. Hundreds of chemical compounds are capable of acting as photosensitizing agents.[1,111-116] Such compounds include a wide variety of drugs used in routine medical practice as well as in industry, agriculture, and the home. The most common photosensitizing reactions involve the skin, although internal organs are also susceptible to phototoxic reactions.

There are several parameters that could influence the possibility and extent of photosensitized damage in the eye. These include (1) the chemical structure of the drug in question and its absorption spectra, (2) the role of the blood-aqueous and blood-retinal barriers, (3) the age and/or the health of the patient, and (4) the oxygen concentration in the affected tissue. Certain drugs having a tricyclic (three fused aromatic rings), heterocyclic, or porphyrin ring system are known to be efficient photosensitizers. The reasons for this are (1) they have long-lived triplet states, (2) they have a low oxidation potential (Type I reaction), and (3) their triplet-state energy is such that transfer of energy to ground-state oxygen is possible (Type II reaction).

In addition to the demonstrated direct photochemical action of UV radiation on ocular tissues, particularly the cornea, lens, and retina, there is the possibility of photobiologic damage by means of photosensitized reactions due to the accumulation and retention of certain drugs within these tissues. For example, after the 13-mm stage of development, the ocular lens is completely encapsulated and never sheds its cells throughout life. Thus, photobinding a drug to the lens proteins and nucleic acids ensures its lifelong retention within the lens with the potential for enhanced photodamage if these compounds are capable of acting as photosensitizing agents. A similar situation exists in the neural retina which does not regenerate.

Only two groups of compounds (the phenothiazines and the psoralens) have been clearly identified as intraocular photosensitizing agents that are capable of causing photochemical damage to the choroid, the retina, and the lens in man as well as in experimental animals. This may be due to the screening effect of the blood-aqueous and the blood-retinal barriers. It is well known, for example, that it is difficult if not and minocycline can result in pseudotumor cerebri.[116] Since extraocular paresis and/or paralysis and papilledema can be caused by pseudotumor cerebri, these signs and symptoms are probably secondary to the pseudotumor and not directly related to any photosensitizing or toxic effects of the drug.

Griseofulvin is an antifungal agent which can cause severe allergic reactions with involvement of the eyelids and erythema multiforme-like rashes.[111] Though rare, these side effects can occur when the medication is given in heavy doses over extended periods of time, and are probably due to a photoallergic mechanism.[111] Transient macular edema has been reported with this drug.

Adriamycin is an anthracycline antibiotic that is also used as a chemotherapeutic agent in treating cancer. Adriamycin can affect cells by intercalating with DNA, membrane binding, and lipid peroxidation. The most thoroughly documented action of adriamycin is DNA intercalation.[136] The consequences of this intercalation include blockage of DNA, RNA, and protein synthesis, inhibition of DNA repair, and fragmentation of DNA. Adriamycin also binds to cell surfaces and alters membrane function at concentrations below those required for DNA intercalation. Adriamycin is reduced to a semiquinone free radical by cytochrome P_{450}. The semiquinone transfers an electron to molecular oxygen, producing the superoxide radical ion.[137] This leads to a free-radical chain reaction that results in peroxidation of unsaturated fatty acids. Superoxide generation may also lead to single-stranded scission of DNA.

Nalidixic acid therapy can also be associated with mild visual disturbances and changes in color perception, with reversible papilledema and sixth-nerve palsies.[116]

Other drugs with photosensitizing photoallergic and/or phototoxic properties include the sulfonamides, the oral hypoglycemic agents, antimalarial agents (chloroquine), and some of the oral contraceptives.[116]

The increasing interest in using HPD for phototherapy (and photodiagnostic procedures) merits careful evaluation with respect to their phototoxic (or sensitizing) potential. Because the HPD absorb over a wide range in the UV and visible region of the electromagnetic spectrum and the fact that they appear to be slowly metabolized (HPD can be retained for 3 months after injection)[138] necessitates that adequate precautions should be taken when such therapy is employed. These compounds can exert a photosensitizing action via the type-I and type-II reactions. The photodynamic action is mediated via singlet oxygen and has been shown to polymerize lens proteins in vitro.[139] The porphyrin-induced photochemical reactions also involve OH radicals and have been implicated in the generation of H_2O_2;[138] thus the potential for ocular damage certainly exists. Patients undergoing HPD therapy must be monitored for ocular side effects, and they should also be informed about the potential for severe sunburns, particularly within the first 3 months following therapy.

There has been a recent resurgence of interest in the therapeutic uses of topical vitamin A acid (retinoic acid) and closely related compounds[140-142] particularly for skin conditions such as acne and related disorders and psoriasis. In addition, such compounds have been tested as a method for treating corneal xerophthalmia, an ocular condition caused by a severe vitamin A deficiency.[143,144] Considerable success has been claimed with such a therapeutic regime, however, the oral administration of isotretinoin has resulted in some side effects; corneal opacities have been reported in a patient receiving this drug.[145] Topical retinoic acid in treating experimental corneal xerophthalmia has met with mixed results. While little is known about the specific mode of action of such drugs, the possibility of phototoxicity (and/or photosensitization) should be considered, particularly with respect to the eye. Patients undergoing clinical trials with these drugs should have careful ocular examinations prior to instituting therapy and be re-evaluated at specific intervals to assess their ocular status.

IV. CLINICAL LENS-FLUORESCENCE STUDIES

Since laboratory studies have demonstrated enhanced fluorescence in the ocular lens associated with aging and drug therapy, and human photosensitized cataracts have recently been reported, a method to monitor lens fluorescence in vivo has been developed.[146,147] A new slit lamp densitographic apparatus (based on the Scheimpflug principle) of accurately and reproducibly recording visible changes in lens density as the lens ages was recently introduced.[148,149] This apparatus has been modified to utilize UV radiation (300 to 400 nm) to measure and quantitate the age-related fluorescence levels in the normal lens in vivo and correlate them with in vitro data.[1,15,16,21,146,147] Visible and UV slit-lamp photographs taken with the Scheimpflug (Topcon SL45) camera on normal eyes and corresponding densitograms show increased lens fluorescence with age.[146,147] A series of UV and visible slit-lamp photographs of normal patients ranging in age from 5 to 82 years demonstrate a lack of fluorescence in the young lens and a progressive increase in fluorescence with age (Figures 6 to 8). These data can be expressed in graphic form (Figure 9) showing the normal age-related increase in lens fluorescence (in vivo) which corresponds well with the in vitro data (Figures 10 and 11) previously reported.[15,16,21] The in vitro studies were performed on lenses from normal eye bank eyes and represent two (nontryptophan) fluorescent peaks obtained by fluorescence spectroscopy.

Aside from demonstrating the normal age-related increase in lens fluorescence, abnormal enhanced fluorescence caused by occupational (or accidental) exposure to higher levels of UV radiation can also be detected. This is shown in Figure 12, obtained on a 40-year-old patient exposed to excessive UV radiation in his workplace. The increased fluorescence can easily be appreciated by comparing this lens with a normal 40-year-old eye (Figure 13). Enhanced fluorescence and/or abnormal fluorescence emission can also occur in patients on PUVA therapy and failure to properly protect such patients from all UV radiation exposure for at least 24 hr following ingestion of the drug can result in cataract formation as shown in Figure 14. This 52-year-old psoriatic patient received 4 years of intermittent PUVA treatment (without proper eye protection). Although many dermatology clinics now provide all PUVA patients with proper UV absorbing or reflecting spectacles, data obtained on a series of patients who were treated prior to 1977 (when the potential for photosensitized lens damage from psoralen therapy was first demonstrated)[74] show a significant elevation of one of the lens fluorescence peaks (Figure 15). In contrast, patients who have been on D-penicillamine therapy (for a variety of diseases) tend to have lower lens-fluorescence intensities (Figure 15). This is due to the fact that D-penicillamine (which is an excellent free-radical scavenger as well as a chelating agent) is capable of entering the lens, both in vivo as well in vitro.[43,58] As a free-radical scavenger, D-penicillamine aborts the UV photodamage.

These studies demonstrate the feasibility of obtaining in vivo lens-fluorescence data which are objective, reproducible, and which can be quantified. Thus UV slit-lamp densitography can be used to objectively monitor one parameter of lens aging (fluorescence), as well as photosensitized lens damage, at a molecular level months to years before visible opacities become manifest by conventional slit-lamp examination, and measures can be instituted to at least retard if not prevent such lens opacities.

In addition to performing in vivo lens-fluorescence measurements, the same photographs can also be utilized (with proper software) to obtain biometric measurements on these eyes including: (1) radius of curvature of the anterior and posterior cornea and corneal thickness, (2) depth of the anterior chamber, and (3) radius of curvature of the anterior and posterior lens surfaces and lens thickness. Thus the clinician can

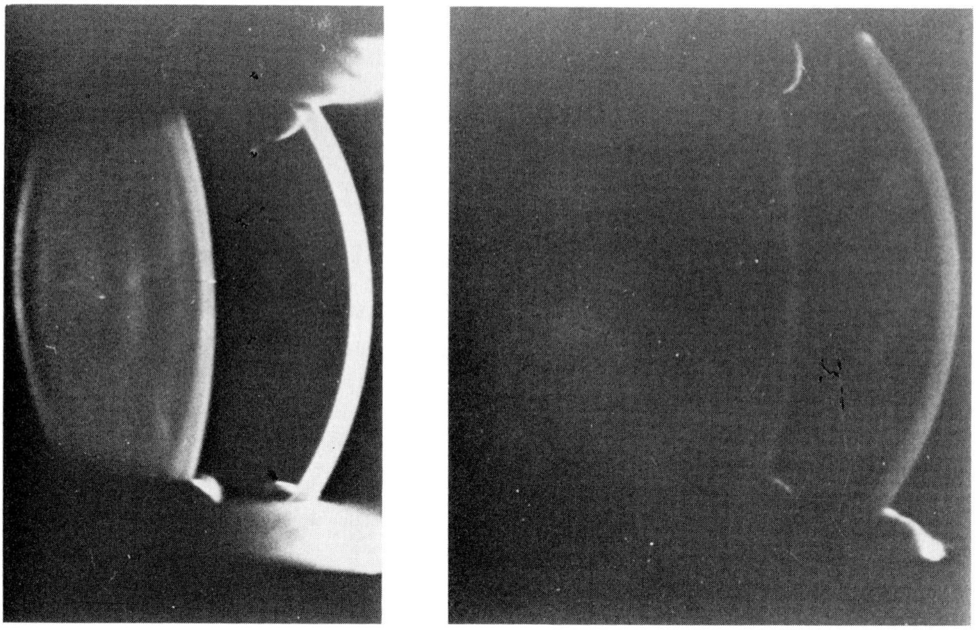

A

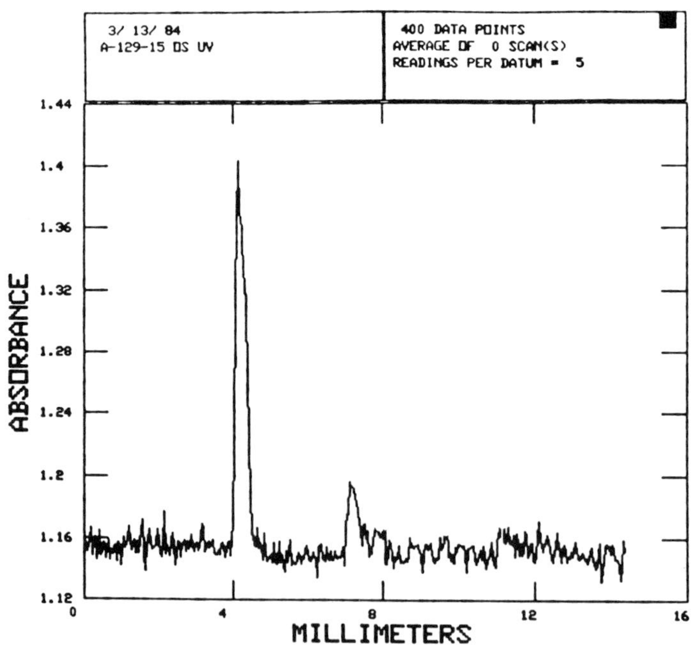

B

FIGURE 6. (A) Visible (L) and UV (R) photos of a 5-year-old normal eye; (B) UV (fluorescence) densitogram.

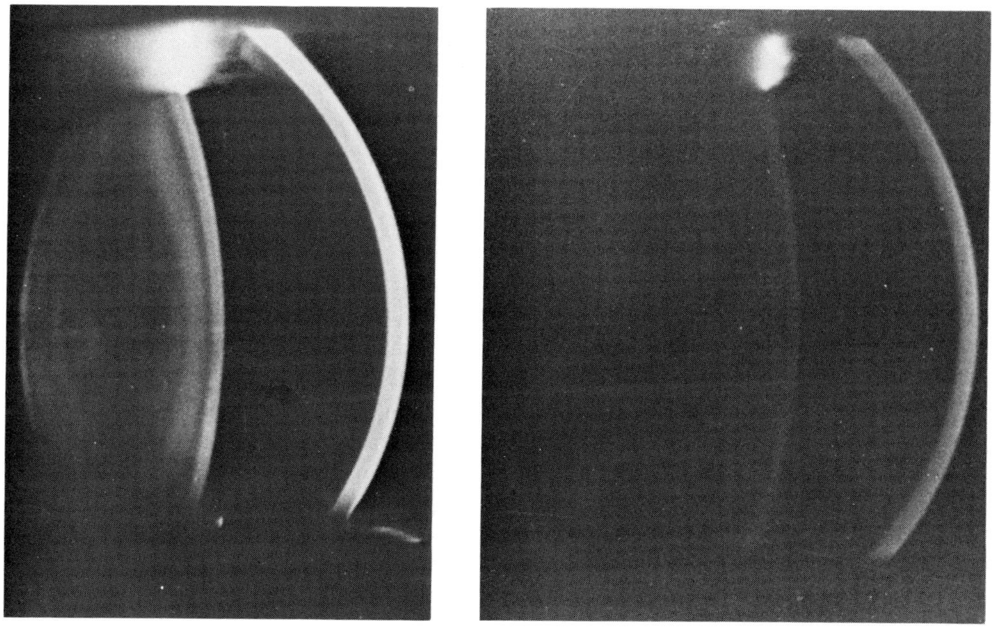

A

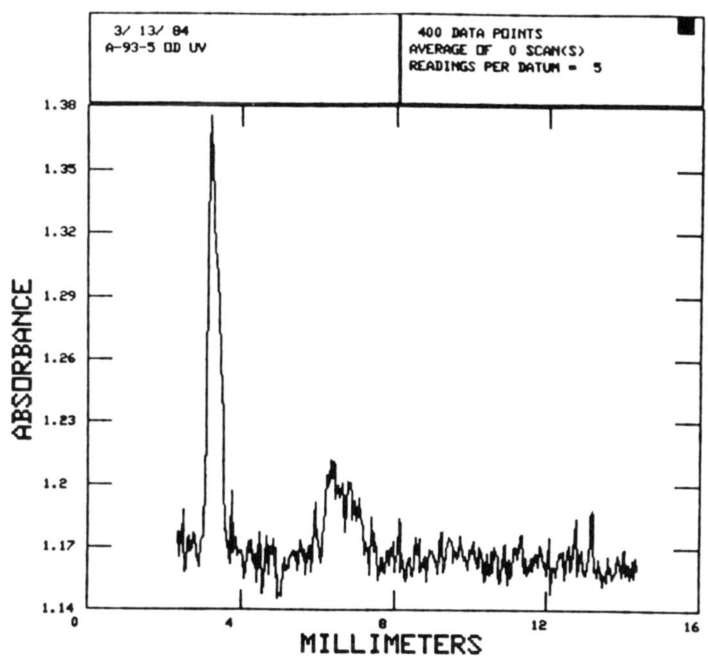

B

FIGURE 7. (A) Visible (L) and UV (R) photos of a 28-year-old normal eye; (B) UV (fluorescence) densitogram.

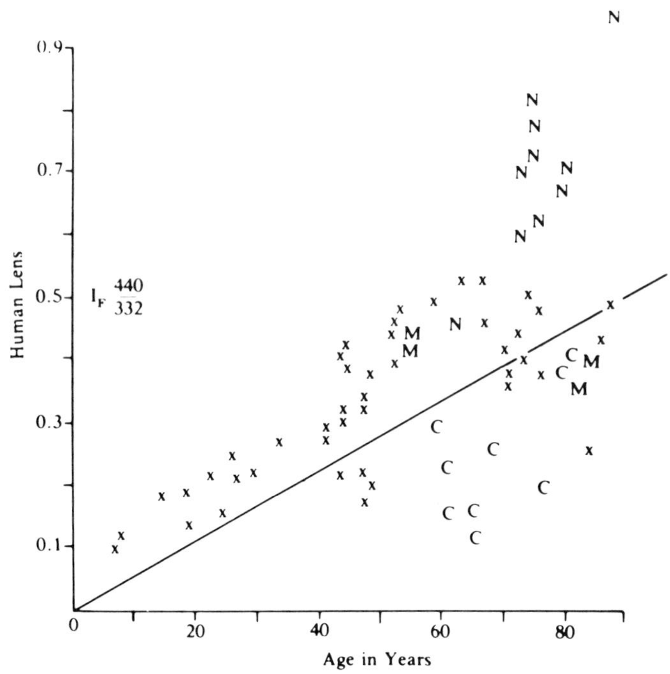

FIGURE 10. I_F 440/332 ratios representing whole-lens fluorescence intensity at 440 nm (360-nm excitation) divided by tryptophan intensity (in whole lens) at 332 nm (290-nm excitation). The I_F ratio shows age-related increase in the normal lens (X and solid line), a marked increase in brown nuclear cataracts (N), relatively normal or below-normal levels in cortical cataracts (C), and high normal values in mixed cortical and nuclear cataracts (M). Each point represents a single lens.

nate'' — thus the term nuclear magnetic resonance. Current studies utilizing ^{31}P NMR spectroscopy enable the investigator to monitor organophosphate metabolism in the lens in vivo (surface scanning) as well as in vitro.[152-156] Such data provide dynamic rather than static information about the general metabolic state of this organ and quickly reflect changes in high-energy phosphates (ATP, ADP, etc.), sugar phosphates, inorganic phosphates, and so forth. For example, in experimental sugar cataracts (glucose, galactose, xylose) one can measure the accumulation of sugar alcohols and the concomitant alterations in organophosphate levels as soon as these metabolic changes develop and well before any opacities become manifest. Other forms of NMR spectroscopy permit us to monitor changes in sodium and amino acids and detect very early changes in molecular aggregate formation as reflected by alterations in the lens water compartments.[157-159] Although such studies are still undergoing laboratory evaluation, sufficient data have accumulated to demonstrate the feasibility of applying one or more of these techniques to human studies in vivo. In essence, one will obtain a more sensitive type of CT scan capable of detecting molecular dysfunction within the eye without exposing it to ionizing radiation.

V. THE ROLE OF COMMERCIAL SUNGLASSES

Because of the confusion and controversial claims regarding the efficacy of commercially available sunglasses in protecting the eye from UV photodamage, a large series of such lenses have been analyzed to determine their transmission characteristics.

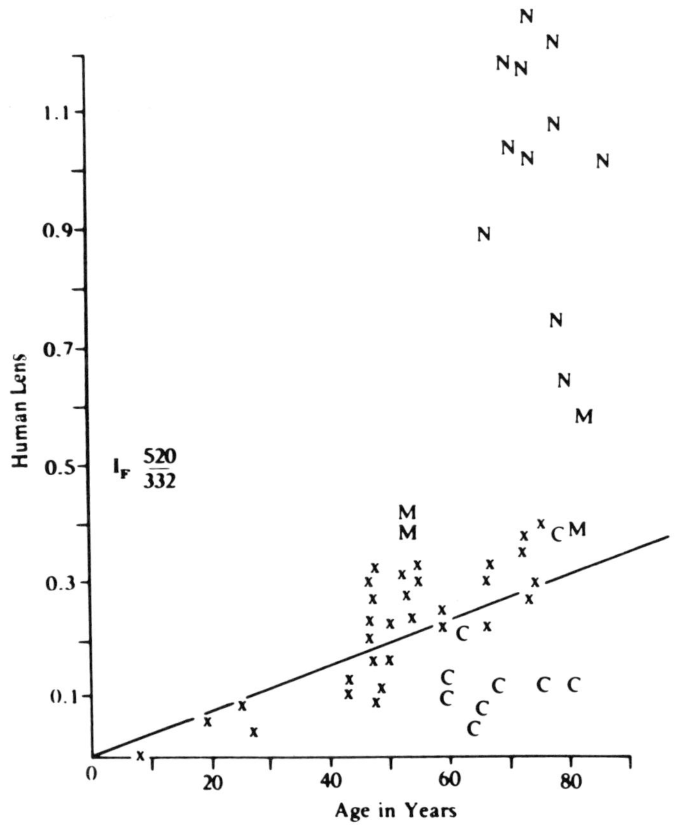

FIGURE 11. I_F 520/332 ratios representing fluorescence intensity of second fluorescent region in the lens at 520 nm (420- to 435-nm excitation) divided by tryptophan fluorescence intensity in the lens at 332 nm (295-nm excitation). Interference filters (295 and 435 nm) were employed in order to decrease the light scattering when cortical and mixed cataracts were examined. Each point represents a single lens.

These studies[160] are in general agreement with an earlier report[161] and demonstrate a wide variation in the UV-transmission characteristics of the sunglasses evaluated, ranging from 1.5 to 40%, with similar transmission values noted when tested for more discrete wavelengths (340 to 380 nm). Only the NOIR, Spectra-Shield, Silor, Univis, and UV 400 lenses were >99% effective in filtering all the UV radiation. It should also be noted that visible radiation is significantly decreased in darkly tinted sunglasses, while still permitting some long wavelength UV transmission. In patients with blue, gray, and hazel eyes, a 50% or more decrease in visible radiation can result in pupillary enlargement of 0.25 mm, thereby increasing the effective dose of radiation incident on the intraocular tissues.

VI. VDTs AND THE EYE

The recent controversy regarding ocular problems which have been reported in workers using VDTs may turn out to be beneficial and opportune. In contrast with the indifference and ignorance with which the development of the X-ray was embraced (in the not-too-distant past), the community at large, as well as the scientific specialists, are now well aware of possible dangers in misusing electromagnetic radiation. We now

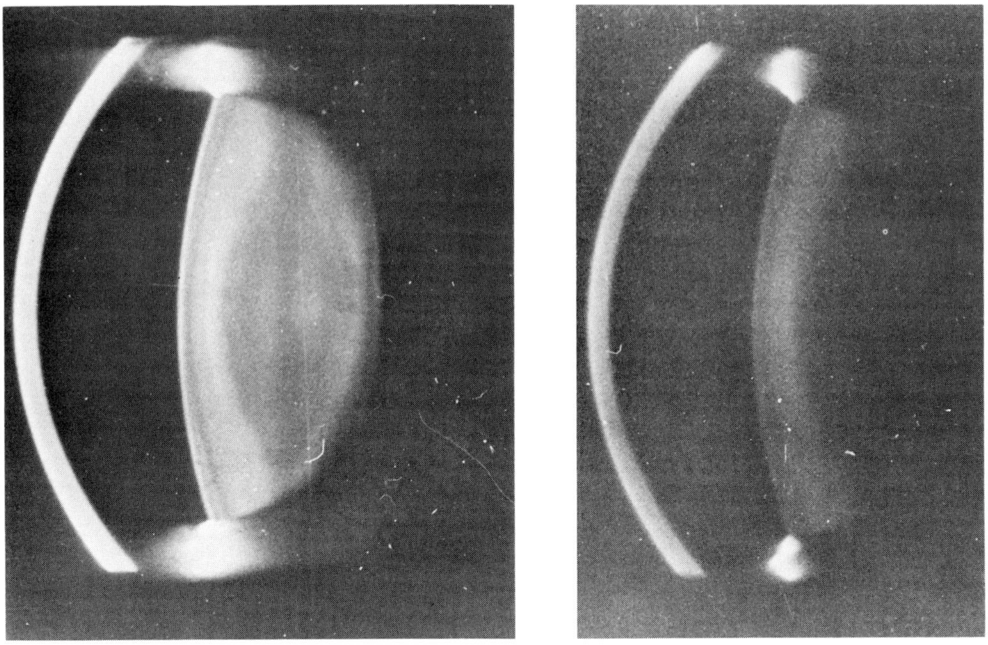

A

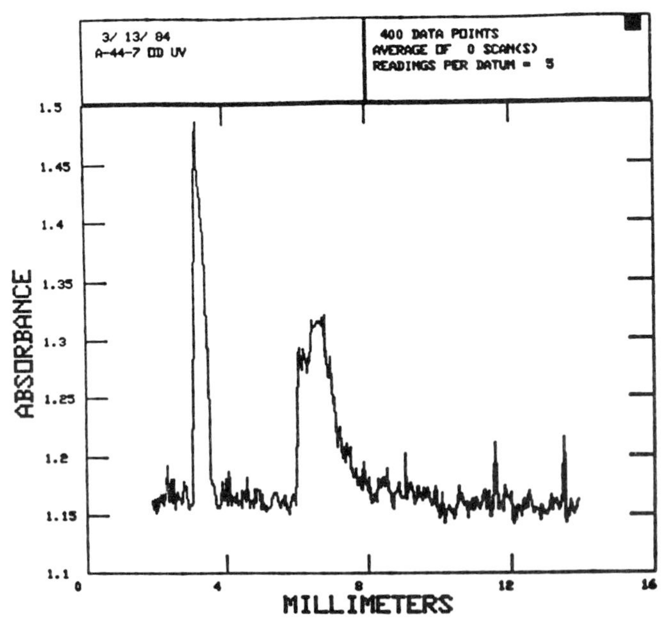

B

FIGURE 12. (A) Visible (L) and UV (R) photos of a 40-year-old eye showing abnormal fluorescence due to occupational excessive UV exposure; (B) UV (fluorescence) densitogram. Note enhanced fluorescence compared with normal values (see Figure 13B).

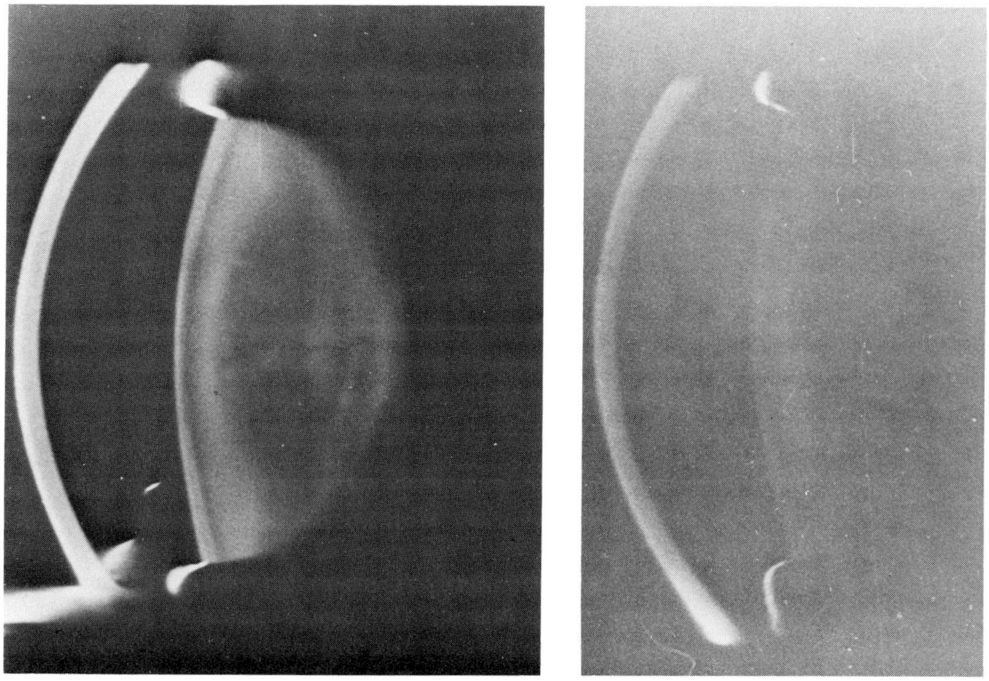

A

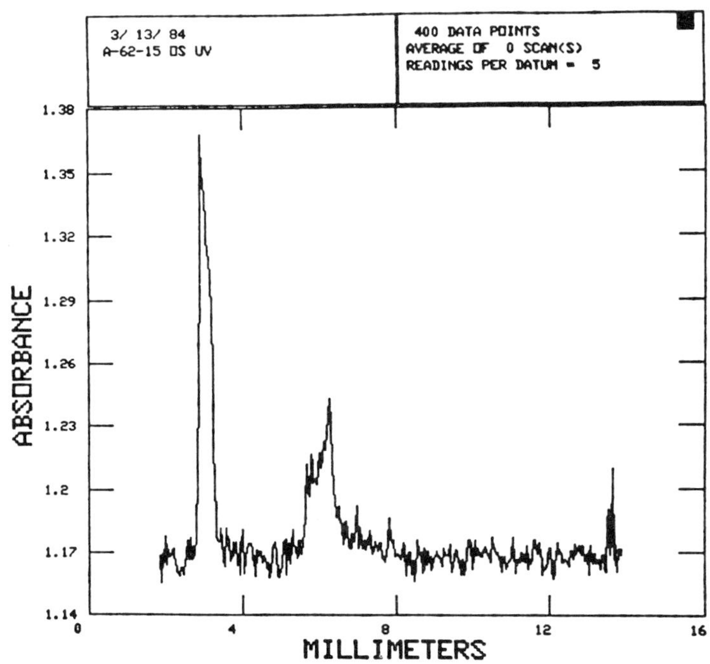

B

FIGURE 13. (A) Visible (L) and UV (R) photos of a normal 40-year-old eye; (B) UV (fluorescence) densitogram.

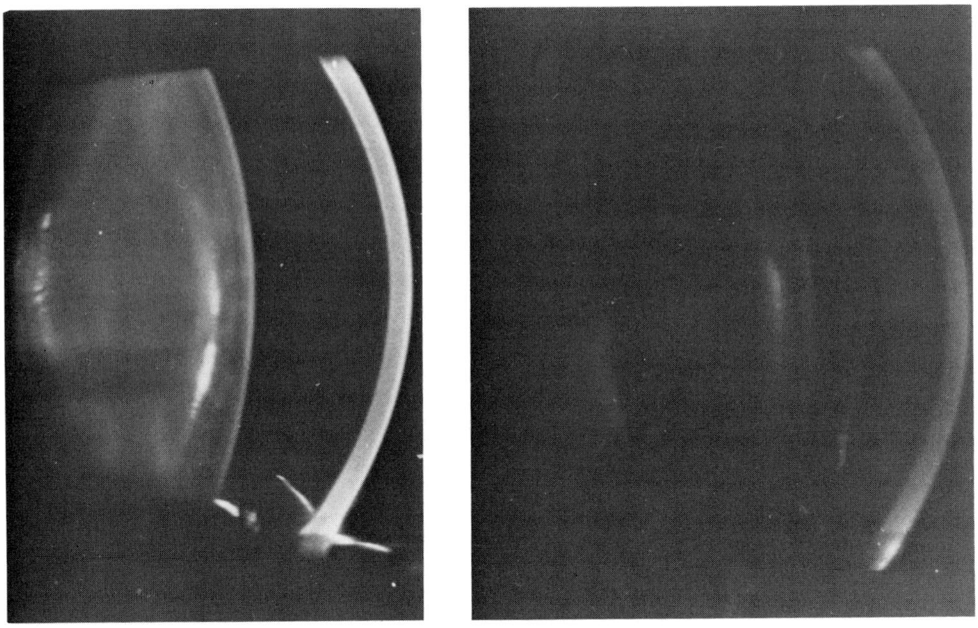

A

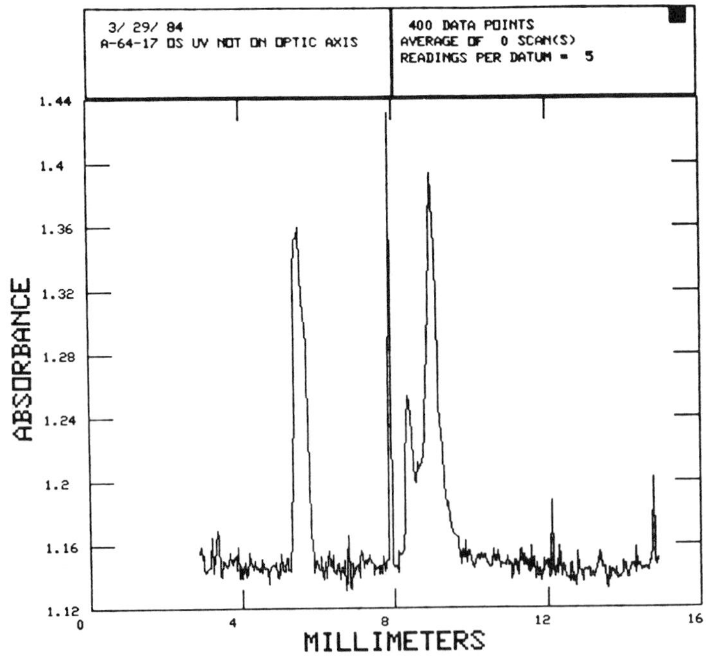

B

FIGURE 14. (A) Visible (L) and UV (R) photos of a PUVA-induced cataract with abnormal fluorescence in a 52-year-old patient on PUVA therapy for 4 years; (B) UV (fluorescence) densitogram.

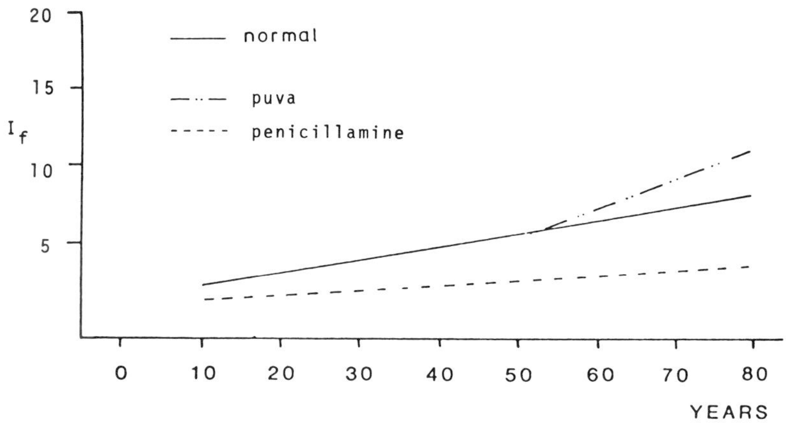

FIGURE 15. Densitographic analyses from in vivo photos showing enhanced fluorescence in PUVA patients compared with normal individuals. Note decreased fluorescence level in patients on chronic D-penicillamine (over 1 year).

have the ability to monitor radiation emission levels from our technological marvels and obtain reasonably accurate measurements on most wavelength emissions of interest. We also have a considerable body of ever-expanding scientific knowledge which enables us to correlate specific radiation emission levels with threshold doses for possible photobiologic damage. Such measurements have recently been performed by scientists at the Bureau of Radiological Health (BRH), Bell Laboratories, and NIOSH.[162-164] Scientists at the BRH measured ionizing radiation (e.g., X-ray) emission levels on 125 VDTs; of these, 34 VDTs were also measured for nonionizing radiation emission (UV, visible, IR, microwave, and other radiofrequency [RF] emissions). A smaller number of VDTs were similarly evaluated by Bell Laboratory scientists. These data indicate that the VDTs "emit little or no harmful radiation under normal operating conditions".[162] The specific emissions that were detectable were all well below the current national and international safety standards.

Aside from "ocular fatigue", questions have been raised regarding a possible relationship between VDT radiation emission and "organic" ocular photodamage, particularly to the ocular lens. Of the various wavelengths in the electromagnetic spectrum, UV radiation and ocular photodamage have received considerable attention during the last two decades. The best current experimental and epidemiologic evidence does not indicate that the level of UV radiation emitted by the VDTs is capable of exerting any deleterious effects on the ocular lenses of personnel using these terminals. We should, however, pay more attention to the workplace with respect to the types of lighting (their UV emission levels) used in these environments and the reflectance level[165] of work surfaces (including reflectance of painted walls, desks, etc.) in order to maintain total ambient UV radiation exposure at minimum levels. Our own ocular lens serves us well as a natural filter for removing the UV radiation that penetrates into the eye, thereby protecting the underlying retina which has been shown to be quite sensitive to rather low levels of UV radiation. Thus, particular care must be taken to protect the aphakic and pseudophakic individuals from potential low-level UV radiation damage. A similar situation pertains to some patients who are receiving photosensitizing drugs. Fortunately, the levels of UV radiation emitted by the VDTs are approximately 1000-fold lower than the amount shown to cause retinal photodamage in experimental primate studies.

Further research should be encouraged, particularly in the areas relating to nonionizing photobiologic radiation damage in order to clarify and better define the molecular mechanisms resulting from exposure to UV, IR, microwave, and RF radiation. It is also important to determine whether damage can result from cumulative low-level exposure over a lifetime, and whether a synergistic effect could accrue from low-level exposure to broadband radiation (e.g., UV + microwave, ionizing + nonionizing).

Finally, it should be noted that we now have the capability of protecting the eye from potential UV photodamage by special eyeglasses and similar UV-absorbing materials could be employed at the manufacturing end to prevent UV radiation emission from our ever-increasing array of technological products.

VII. THERAPY

A. Radiation Filters

The section on the effect of UV and visible radiation on the aphakic or pseudophakic eye has considered the mechanisms involved in ocular photodamage caused by optical radiation, i.e., nonionizing radiant energy (280 to 1400 nm) which is present in our environment. As noted, the simplest and most effective preventive therapy for UV-induced cataracts involves the use of spectacles which are capable of filtering out 99+% of the UV radiation (300 to 400 nm) which penetrates through the cornea. The development of UV-visible slit-lamp densitography now enables us to detect lenticular UV photodamage at a molecular level, months to years before the opacities become manifest with conventional slit-lamp examination. Such a screening procedure will detect enhanced or abnormal lens fluorescence for any particular age group (by decade), thereby alerting the clinician that the patient has had excessive UV exposure. Studies indicate that any patient whose lens-fluorescence levels are >30% above the amount present for his age group (by decade), should be considered a potential candidate for the development of UV cataracts, particularly the brown nuclear cataract.[1,16,128] Such enhanced lens fluorescence can be caused by occupational or recreational exposure to prolonged solar UV radiation in those people living in the "sun belt" areas of the U.S., or by exposure in industries where UV radiation is employed in research and development or manufacturing procedures. Further progression of this type of lens fluorescence can easily be prevented by prescribing proper UV-filtering glasses to such patients.

Since a certain amount of ambient UV exposure exerts a positive photochemical influence on the developing lens during childhood, inducing the generation of pigments so that the lens becomes an effective filter for UV and short wavelength visible radiation (thereby protecting the aging and metabolically less efficient retina from potential photodamage), UV-protective spectacles are not recommended for the general population, nor are they required. Such spectacles are only required when enhanced or abnormal fluorescence is present. If the UV slit-lamp densitography screening procedure demonstrates abnormal lens fluorescence, it may be indicative of current or past ingestion of photosensitizing drugs or chemicals. Such patients should also be given UV-absorbing spectacles with sidepieces (to prevent reflected radiation from entering the eye) and, depending on the mode and frequency with which these patients are ingesting the photosensitizer, they may only need to wear them for 24 hr after ingesting the drug.

Since the aging retina is particularly sensitive to UV and shortwave visible radiation, such spectacles are also recommended for all aphakes and pseudophakes irrespective of their site of domicile (i.e., not just those living in sunbelt areas) or occupation. It is important to note that none of the contact lenses are capable of filtering UV radiation; in fact, all the plastic contact lenses transmit much more UV radiation shorter than 320 nm than does ordinary glass. The same applies to most of the intraocular lenses currently employed, although many manufacturers are now eagerly attempting to develop and market UV-filtering intraocular lenses. Once again, a word of caution is advisable. To be truly effective and protective, such intraocular lenses must filter 99+% of all UV radiation between 300 to 400 nm and at least 50 to 75% of the shortwave visible radiation between 400 to 475 nm.

B. Medication

Aside from the use of UV and shortwave visible radiation filters, the potential exists for a medical approach to protect the lens from direct and photosensitized optical radiation damage. A variety of investigators have noted that D-penicillamine exerts a significant protective action both in vivo and in vitro and prevents experimental UV radiation photodamage and/or cataracts from developing in spite of direct UV exposure.[15] This drug is an effective free-radical scavenger as well as chelating agent. Its latter action is the basis for its clinical use in treating Wilson's disease. It efficacy as a free-radical scavenger within the lens has been demonstrated by in vitro and in vivo experiments in animals[16,58] where it aborts radiation-induced free-radical damage and prevents the subsequent photochemical reactions from being promulgated. UV slit-lamp densitography studies on a series of patients who had been on 1 or more years of D-penicillamine therapy (for a variety of nonocular conditions) demonstrated that all had significantly decreased lens-fluorescence values compared with their control peer groups. This suggests that medical therapy to inhibit or prevent progression of radiation-induced lenticular photodamage is a distinct possibility in the future. Such therapy could be applied locally in the form of drops, ointment, or a prolonged-release form of medication, in much lower concentrations than are currently being employed for treating Wilson's disease or certain forms of arthritis (i.e., in microgram rather than milligram quantities) thereby diminishing if not obviating some of the side effects currently seen with D-penicillamine therapy. Aside from this specific drug, there are a significant number of other free-radical quenchers or scavengers which, at least in theory, should be effective as possible forms of medical therapy for radiation cataractogenesis. However, prior to such therapy becoming available, it must be proved effective as well as safe.

REFERENCES

1. Lerman, S., *Radiant Energy and the Eye,* Macmillan, New York, 1980.
2. Mainster, M. A., Solar retinitis, photic maculopathy and the pseudophakic eye, *Am. Intra-Ocular Implant Soc. J.,* 4, 84, 1978.
3. Mainster, M. A., Spectral transmittance of intraocular lenses and retinal damage from intense light sources, *Am. J. Ophthalmol.,* 85, 167, 1978.
4. Berler, D. K. and Peyser, R., Light intensity and visual acuity following cataract surgery, *Ophthalmology,* 90, 933, 1983.
5. Mainster, M. A., Ham, W. T., Jr., and Delori, F. C., Potential retinal hazards: instrument and environmental light sources, *Ophthalmology,* 90, 927, 1983.
6. Boldrey, E. E., Ho, B. T., and Griffith, R. D., Retinal burns occurring at cataract extraction, *Opthalmology,* 91, 1297, 1984.
7. Lerman, S., NMR and fluorescence spectroscopy on the normal, aging and cataractous lens, *Lens Res.,* 1, 175, 1983.
8. Lerman, S., Observations on the use of higher power lasers in ophthalmology, *IEEE J. Quantum Electron.,* 20, 1465, 1984.
9. Lerman, S., Thrasher, B., and Moran, M., Vitreous changes after posterior lens capsule or mid-vitreous Nd:YAG laser irradiation, *A. J. Ophthalmol.,* 97, 470, 1984.
10. Dougherty, T. J., Thoma, R. E., Boyle, D. G., and Weishaupt, K. R., Interstitial photoradiation therapy for primary solid tumors in pet cats and dogs, *Cancer Res.,* 41, 401, 1981.

11. Menon, I. A., Persad, S., Haberman, H. F., Kurian, C. J., and Basu, P. K., A qualitative study of the melanins from blue and brown human eyes, *Exp. Eye Res.*, 34, 531, 1982.
12. Maurice, D., The structure and transparency of the cornea, *J. Physiol. (London)*, 136, 263, 1957.
13. Benedek, G. B., Theory of transparency of the eye, *Appl. Opt.*, 10, 459, 1971.
14. Miller, D. and Benedek, G., *Intraocular Light Scattering*, Charles C Thomas, Springfield, Ill., 1973.
15. Lerman, S. and Borkman, R. F., Photochemistry and lens aging, in *Interdisciplinary Topics in Gerontology*, Vol. 13, S. Karger, Basel, 1978, 154.
16. Lerman, S., Lens transparency and aging, in *Aging of the Lens*, Regnault, F., Hockwin, O., and Courtois, Y., Eds., Elsevier, Amsterdam, 1980, 263.
17. Cogan, D. G. and Kinsey, V. E., Action spectrum of keratitis produced by ultraviolet radiation, *Arch. Ophthalmol.*, 35, 670, 1946.
18. Bachem, A., Ophthalmic ultraviolet action spectra, *Am. J. Ophthalmol.*, 41, 969, 1956.
19. Sherashov, S. G., Spectral sensitivity of the cornea to ultraviolet radiation, *Biofizika*, 15, 543, 1970.
20. Jedziniak, J. A., Nicoli, D. F., Baram, H., and Benedek, G. B., Quantitative verification of the existence of high-molecular-weight protein aggregates in the intact normal human lens by light scattering spectroscopy, *Invest. Ophthalmol. Vis. Sci.*, 17, 51, 1978.
21. Lerman, S. and Borkman, R. F., Spectroscopic evaluation and classification of the normal, aging and cataractous lens, *Ophthalmic Res.*, 8, 335, 1976.
22. Teale, F. W. J., The ultraviolet fluorescence of proteins in neutral solution, *Biochem. J.*, 76, 381, 1960.
23. Burstein, E. A., Vedenkina, N. S., and Ivkova, M. N., Fluorescence and the location of tryptophan residues in protein molecules, *Photochem. Photobiol.*, 18, 263, 1973.
24. Grinwald, A. and Steinberg, I. A., The fluorescence decay of tryptophan residues in native and denatured proteins, *Biochim. Biophys. Acta*, 427, 663, 1976.
25. Lerman, S., Megaw, J. M., and Moran, M. N., NMR analyses of the cold cataract. III. ^{13}C acrylamide studies, *Invest. Ophthal. Vis. Sci.*, submitted, 1984.
26. Blundell, T. L., Lindley, P. F., Miller, L. R., Moss, D. S., Slingsby, C., Turnell, W. G., and Wistow, G. T., Interactions of gamma crystallin in relation to eye-lens transparency, *Lens Res.*, 1, 109, 1983.
27. Summers, L., Slingsby, C., White, H., Narebor, M., Moss, D., Miller, L., Mahadevan, D., Lindley, P., Driessen, H., Blundell, T., Den Dunnen, J., Moorman, R., Van Leen, R., and Schoenmaker, J., The molecular structures and interactions of bovine and human gamma crystallins, in *Ciba Foundation Symp. 106 Human Cataract Formation*, Nugent, J. and Whelen, J., Eds., Pitman Press, London, 1984, 219.
28. Regnauld, J., Sur la fluorescence des milieux de l'oeil chez l'homme et quelques mammiferes, *Institut*, 26, 410, 1958.
29. François, J., Rabaey, M., and Recoules, N., A fluorescent substance of low molecular weight in the lens of primates, *Arch. Ophthalmol.*, 65, 118, 1961.
30. Lerman, S., Tan, A. T., Louis, D., and Hollander, M., Anomalous absorptivity of lens proteins due to a fluorogen, *Ophthal. Res.*, 1, 338, 1970.
31. Lerman, S., Lens proteins and fluorescence, *Isr. J. Med. Sci.*, 8, 1583, 1972.
32. Spector, A., Roy, D., and Stauffer, J., Isolation and characterization of age-dependent polypeptide from human lens with non-tryptophan fluorescence, *Exp. Eye Res.*, 21, 9, 1975.
33. Pirie, A., Color and solubility of the proteins of human cataract, *Invest. Ophthalmol.*, 7, 634, 1968.
34. Pirie, A., Formation of *N'*-formylkynurenine in proteins from lens and other sources by exposure to sunlight, *Biochem. J.*, 125, 203, 1971.
35. van Heynigen, R., Fluorescent derivatives of 3-hydroxy-l-kynurenine in the lens of man, the baboon and the grey squirrel, *Biochem. J.*, 123, 30, 1971.
36. Pirie, A., Photo-oxidation of proteins and comparison of photo-oxidized proteins with those of the cataractous human lens, *Isr. J. Med. Sci.*, 8, 1583, 1972.
37. van Heyningen, R., The glucoside of 3-hydroxy-kynurenine and other fluorescent compounds in the human lens, *Exp. Eye Res.*, 15, 121, 1973.
38. Zigman, S., Griess, G., Yulo, T., and Schultz, J., Ocular protein alteration by near UV light, *Exp. Eye Res.*, 15, 255, 1973.
39. Satoh, K., Bando, M., and Nakajima, A., Fluorescence in human lens, *Exp. Eye Res.*, 16, 167, 1973.
40. Dilley, K. J. and Pirie, A., Changes to the proteins of the human lens nucleus in cataract, *Exp. Eye Res.*, 19, 59, 1974.
41. Augusteyn, R. C., Distribution of fluorescence in the human cataractous lens, *Ophthal. Res.*, 7, 217, 1975.
42. Dillon, J., Spector, A., and Nakanishi, A., Identification of σ-carbolines isolated from fluorescent human lens proteins, *Nature (London)*, 259, 422, 1976.
43. Lerman, S., Kuck, J. F., Borkman, R., and Saker, E., Induction, acceleration and prevention (in vitro) of an aging parameter in the ocular lens, *Ophthal. Res.*, 8, 213, 1976.

44. Ts'o, M. O. M., Fine, B. S., and Zimmerman, L. E., Photic maculopathy produced by indirect ophthalmoscope. I. Clinical and histopathic study, *Am. J. Ophthalmol.*, 73, 686, 1972.
45. Ham, W. T., Muller, H. A., Ruffolo, J. J., Guerry, D., and Guerry, R. K., Action spectrum for retinal injury from near ultraviolet radiation in the aphakic monkey, *Am. J. Ophthalmol.*, 93, 299, 1982.
46. Hochheimer, B., A possible cause of chronic cystic maculopathy: the operating microscopy, *Ann. Ophthalmol.*, 13, 153, 1981.
47. Berler, D. and Peyser, R., Light intensity and visual acuity following cataract surgery, *Ophthalmology*, Suppl. 89, 117, 1982.
48. Hiller, R., Giacometti, L., and Yuen, K., Sunlight and cataract: an epidemiologic investigation, *Am. J. Epidemiol.*, 105, 450, 1977.
49. Taylor, H. R., The environment and the lens, *Br. J. Ophthalmol.*, 64, 303, 1980.
50. Hollows, F. and Moran, D., Cataract — the ultraviolet risk factor, *Lancet*, 1, 1249, 1981.
51. Brilliant, L. B., Grasset, N. C., Pokhrel, R. P., Kolstad, A., Lepkowski, J. M., Brilliant, G. E., and Hawks, W. N., Associations among cataract prevalence, sunlight hours and altitude in the Himalayas, 1983 NEI Epidemiol. Symp., *Am. J. Ophthalmol.*, 113, 250, 1983.
52. Zigman, S. and Vaughn, W., Near UV light effects on the lenses and retinas of mice, *Invest. Ophthalmol.*, 13, 462, 1974.
53. Pitts, D., Hacker, P., and Parr, W. H., Ocular Ultraviolet Effects from 295 nm to 400 nm in the Rabbit Eye, Publication No. 77, National Institute for Occupational Safety and Health, U.S. Department of Health, Education and Welfare, Washington, D.C., 1977, 175.
54. Lerman, S., Human UV radiation cataracts, *Ophthal. Res.*, 12, 303, 1980.
55. Lerman, S., Gardner, K., Megaw, J., and Borkman, R., Prevention of direct and photosensitized UV radiation damage to the ocular lens, *Ophthal. Res.*, 13, 284, 1981.
56. Thomas, D. M. and Schepler, K. L., Raman spectra of normal and ultraviolet-induced cataractous rabbit lens, *Invest. Ophthalmol. Vis. Sci.*, 19, 904, 1980.
57. Kurzel, R. B., Wolbarsht, M. L., and Yamanashi, B. S., UV radiation effects on the human eye, *Photochem. Photobiol. Rev.*, 2, 133, 1977.
58. Borkman, R. F. and Lerman, S., Evidence for a free radical mechanism in aging and UV irradiated ocular lenses, *Exp. Eye Res.*, 25, 303, 1977.
59. Weiter, J. J. and Finch, E. D., Paramagnetic species in cataractous human lenses, *Nature (London)*, 254, 536, 1975.
60. Zigman, S., Groff, J., and Yulo, T., Enhancement of the non-tryptophan fluorescence of human lens proteins after near-UV light exposure, *Photochem. Photobiol.*, 26, 437, 1977.
61. Lerman, S., Ocular phototoxicity and PUVA therapy: an experimental and clinical evaluation, FDA Photochemical Toxicity Symp., *J. Natl. Cancer Inst.*, 69, 287, 1982.
62. Goosey, J. D., Zigler, S., Matheson, I. B. C., and Kinoshita, J. H., Effects of singlet oxygen on human lens crystallins, *Invest. Ophthalmol. Vis. Sci.*, 20, 679, 1981.
63. Varma, S. D., Superoxide and lens of the eye: a new theory of cataractogenesis, *Int. J. Quantum Chem.*, 20, 479, 1981.
64. Jernigan, H. M., Jr., Fukui, H. N., Goosey, J. D., and Kinoshita, J. H., Photodynamic effects of rose bengal or riboflavin on carrier-mediated transport systems in rat lens, *Exp. Eye Res.*, 32, 461, 1981.
65. Varma, S. D., Srivastava, V. K., and Richards, R. D., Photoperoxidation in lens and cataract formation: preventive role of superoxide dismutase, catalase and vitamin C, *Ophthal. Res.*, 14, 167, 1982.
66. van Haard, P. M. M., Hoenders, H. J., and Ketelaars, H. C. J., Human lens nuclear cataract: hydrogen peroxide action and anthranilic acid association with lens proteins, *Ophthal. Res.*, 12, 252, 1980.
67. Truscott, R. J. W., Faull, K., and Augusteyn, R. C., The identification of anthranilic acid in proteolytic digests of cataractous lens proteins, *Ophthal. Res.*, 9, 263, 1977.
68. McCormick, J. P., Fischer, J. R., Pachlatko, J. P., and Eisenstark, A., Characterization of a cell-lethal product from the photo oxidation of tryptophan: hydrogen peroxide, *Science*, 191, 468, 1976.
69. Bhuyan, K. C. and Bhuyan, D. K., Superoxide dismutase of the eye: relative functions of superoxide dismutase and catalase in protecting the ocular lens from oxidative damage, *Biochim. Biophys. Acta*, 542, 28, 1978.
70. Musajo, L. and Rodighiero, G., Mode of photosensitizing action of furocoumarins, in *Photophysiology*, Vol. 7, Giese, A. C., Ed., Academic Press, New York, 1972, 115.
71. Mantulin, W. W. and Song, P. S., Excited states of skin-sensitizing coumarins and psoralens. Spectroscopic studies, *J. Am. Chem. Soc.*, 95, 5112, 1973.
72. Parrish, J. A., Fitzpatrick, T. B., Tanenbaum, L., and Pathak, M. A., Photochemotherapy of psoriasis with oral methoxsalen and longwave ultraviolet light, *N. Engl. J. Med.*, 291, 1207, 1974.

73. Jose, J. J. and Yielding, K. L., Photosensitive cataractogens, chlorpromazine and methoxypsoralen, cause DNA repair synthesis in lens epithelial cells, *Invest. Ophthalmol. Vis. Sci.*, 17, 687, 1978.
74. Lerman, S., Jocoy, M., and Borkman, R. F., Photosensitization of the lens by 8-methoxypsoralen, *Invest. Ophthalmol. Vis. Sci.*, 16, 1065, 1977.
75. Lerman, S., Megaw, J., and Willis, I., The photoreaction of 8-MOP with tryptophan and lens proteins, *Photochem. Photobiol.*, 31, 519, 1980.
76. Lerman, S., Megaw, J., and Willis, I., Potential ocular complications of PUVA therapy and their prevention, *J. Invest. Dermatol.*, 74, 197, 1980.
77. Megaw, J., Lee, C., and Lerman, S., NMR analyses of tryptophan-8-methoxypsoralen photoreaction products, *Photochem. Photobiol.*, 32, 265, 1980.
78. Poppe, W. and Grossweiner, L. I., Photodynamic sensitization by 8 methoxypsoralen via singlet oxygen mechanism, *Photochem. Photobiol.*, 22, 217, 1975.
79. Yoshikawa, K., Mori, N., Sakakibara, S., Mizuno, N., and Song, P. S., Photo-conjugation of 8-methoxypsoralen with proteins, *Photochem. Photobiol.*, 29, 1127, 1979.
80. Dall'Acqua, S., Marciani, M., Zambon, F., and Rodighiero, G., Kinetic analysis of the photoreaction (365 nm) between psoralen and DNA, *Photochem. Photobiol.*, 29, 489, 1979.
81. Bessems, G. J. H., de Man, B. M., Hoenders, H. J., and Wollensak, J., Alpha-tocopherol in the normal and nuclear-cataractous human lens, *Lens Res.*, in press.
82. Hockwin, O., Early changes of lens metabolism after X-irradiation, *Exp. Eye Res.*, 1, 422, 1962.
83. Armstrong, D. A. and Buchanan, J. D., Reactions of O_2, H_2O_2 and other oxidants with sulfhydryl enzymes, *Photochem. Photobiol.*, 28, 741, 1978.
84. Balazs, E. A., Laurent, T. C., Howe, A. F., and Varga, L., Irradiation of mucopolysaccharides with ultraviolet light and electrons, *Radiat. Res.*, 11, 149, 1959.
85. Lauber, J. K., Schutze, J. B., and McGinnis, J., Effects of exposure to continuous light on the eye of the growing chicken, *Proc. Soc. Exp. Biol. Med.*, 106, 871, 1961.
86. Dowling, J. E. and Sidman, R. L., Inherited retinal dystrophy in the rat, *J. Cell Biol.*, 14, 73, 1962.
87. Noell, W. K., Aspects of experimental and hereditary retinal degeneration, in *Biochemistry of the Eye*, Graymore, C. N., Ed., Academic Press, New York, 1963, 51.
88. Noell, W. K., Delmelle, M. C., and Albrecht, R., Vitamin A deficiency effect on retina: dependence on light, *Science*, 172, 72, 1971.
89. Noell, W. K. and Albrecht, R., Irreversible effects of visible light on the retina: role of vitamin A, *Science*, 171, 76, 1971.
90. Ts'o, M. O. M., Photic maculopathy in Rhesus monkey: a light and electron microscopic study, *Invest. Ophthalmol. Vis. Sci.*, 12, 17, 1973.
91. Noell, W. K., Hereditary retinal degeneration and damage by light, in Estratto dagli Atti del Somposio di Oftalmologia Pediatrica, Parma, October, 1974, 332.
92. Organisciak, D. T. and Noell, W. K., The rod outer segment phospholipid/opsin ratio of rats maintained in darkness or cyclic light, *Invest. Ophthalmol. Vis. Sci.*, 16, 188, 1977.
93. Lanum, J., The damaging effects of light on the retina. Empirical findings, theoretical and practical applications, *Surv. Ophthalmol.*, 22, 221, 1978.
94. Marshall, J., Grindle, C. F. J., Ansell, P. L., and Borwein, B., Convolution in human rods: an aging process, *Br. J. Ophthalmol.*, 63, 181, 1979.
95. Clark, B., Johnson, M. L., and Dreher, R., The effect of sunlight on dark adaptation, *Am. J. Ophthalmol.*, 29, 828, 1946.
96. Hecht, S., Hendley, C. D., Ross, H., and Richmond, P. N., The effect of exposure to sunlight on night vision, *Am. J. Ophthalmol.*, 31, 1573, 1948.
97. Penner, R. and McNair, J. N., Eclipse blindness, *Am. J. Ophthalmol.*, 61, 1452, 1966.
98. Dawson, W. W. and Herron, W. L., Retinal illumination during indirect ophthalmoscopy: subsequent dark adaptation, *Invest. Ophthalmol.*, 9, 89, 1970.
99. Ts'o, M. O. M., Fine, B. S., and Zimmerman, L. E., Photic maculopathy produced by the indirect ophthalmoscope. I. Clinical and histopathic study, *Am. J. Ophthalmol.*, 73, 686, 1972.
100. Ham, W. T., Jr., Muller, H. A., Williams, R. C., and Geeraets, W. J., Ocular hazards from viewing the sun unprotected and through various windows and filters, *Appl. Opt.*, 12, 21, 1973.
101. Persad, S., Menon, A., and Haberman, H. F., Comparison of the effects of UV visible irradiation of melanins and melanin-hematoporphyrin complexes from human black and red hair, *Photochem. Photobiol.*, 37, 63, 1983.
102. Ham, W. T. and Mueller, H. A., Retinal sensitivity to damage from short wavelength light, *Nature (London)*, 260, 153, 1976.
103. Ham, W. T., Ruffolo, J. R., Jr., Mueller, H. A., Clark, A. M., and Moon, M. E., Histologic analysis of photochemical lesions produced in rhesus retina by short-wavelength light, *Invest. Ophthalmol. Vis. Sci.*, 17, 1029, 1978.
104. Ham, W. T., Mueller, H. A., Ruffolo, J. R., Jr., Clarke, A. M., Sensitivity of the retina to radiation damage as a function of wavelength, *Photochem. Photobiol.*, 29, 735, 1979.

105. Kuwabara, T., Retinal recovery from exposure to light, *Am. J. Ophthalmol.*, 70, 187, 1970.
106. Lawwill, T., Three major pathologic processes caused by light in the primate retina: a search for mechanisms, *Trans. Am. Ophthalmol. Soc.*, 80, 517, 1982.
107. Noell, W. K., Walker, V. S., Kang, B. S., and Berman, S., Retinal damage by light in rats, *Invest. Ophthalmol.*, 5, 450, 1966.
108. Pomerantzeff, P., Govignon, J., and Schapens, C. L., Indirect ophthalmoscopy: is the illumination level dangerous?, *Trans. Am. Acad. Ophthalmol. Otolaryngol.*, 73, 246, 1961.
109. Sliney, D. H., The ambient light environment and ocular hazards, *Adv. Exp. Med. Biol.*, 77, 211, 1977.
110. Sliney, D. H., Biohazards of ultraviolet, visible and infrared radiation, *J. Occup. Med.*, 25, 203, 1983.
111. Baer, R. L. and Harber, L. C., Photosensitivity induced by drugs, *JAMA*, 192, 989, 1965.
112. Gallo, U. and Santa Maria, L., Eds., *Research Progress in Organic, Biological and Medicinal Chemistry*, Vol. 3, North-Holland, Amsterdam, 1972.
113. Meier-Ruge, W., Drug induced retinopathy, *CRC Crit. Rev. Toxicol.*, 352, 1972.
114. Grant, W. M., *Toxicology of the Eye*, Charles C Thomas, Springfield, Ill., 1974.
115. Hanna, C. and Fraunfelder, F. T., Drugs and toxins, in *Pathobiology of Ocular Disease: A Dynamic Approach*, Garner, A. and Klintworth, G. T., Eds., Marcel Dekker, New York, 1982, 1125.
116. Fraunfelder, F. T., *Drug-Induced Ocular Side Effects and Drug Interactions*, Lea & Febiger, Philadelphia, 1982.
117. Parrish, J. A., Fitzpatrick, T. B., Shea, C., and Pathak, M. A., Photochemotherapy of vitiligo. Use of orally administered psoralens and a high intensity longwave ultraviolet light (UV-A) system, *Arch. Dermatol.*, 112, 1531, 1976.
118. Cloud, T. M., Hakim, R., and Griffin, A. C., Photosensitization of the eye with methoxalen. I. Acute effect, *Arch. Ophthalmol.*, 64, 346, 1960.
119. Cloud, T. M., Hakim, R., and Griffin, A. C., Photosensitization of the eye with methoxsalen. II. Chronic effects, *Arch. Ophthalmol.*, 66, 689, 1961.
120. Freeman, R. G. and Troll, D., Photosensitization of the eye by 8-methoxypsoralen, *J. Invest. Dermatol.*, 53, 449, 1969.
121. Crylin, M. N., Pedvis-Leftick, A., and Sugar, J., Cataract formation in association with ultraviolet photosensitivity, *Ann. Ophthalmol.*, 12, 786, 1980.
122. Lerman, S., Megaw, J., Gardner, K., Takei, Y., and Willis, I., Localization of 8-methoxypsoralen in ocular tissues, *Ophthal. Res.*, 13, 106, 1981.
123. Wulf, H. C. and Andreasen, M. P., Distribution of ^{3}H-8-MOP and its metabolites in rat organs after a single oral administration, *J. Invest. Dermatol.*, 76, 252, 1981.
124. Wulf, H. C. and Andreasen, M. P., Concentration of ^{3}H-8-methoxypsoralen and its metabolites in the rat lens and eye after a single oral administration, *Invest. Ophthalmol. Vis. Sci.*, 22, 32, 1982.
125. Lerman, S., Psoralens and ocular effects in animals and man: in vivo monitoring of human ocular and cutaneous manifestations, in Photochemotherapeutic Aspects of Psoralens, Monogr. no. 66, National Cancer Institute, Bethesda, Md., 1984, 227.
126. Lerman, S., Megaw, J., and Gardner, K., P-UVA therapy and human cataractogenesis, *Invest. Ophthalmol. Vis. Sci.*, 23, 801, 1982.
127. Lerman, S., Takei, Y., Franks, Y., Megaw, J., Gardner, K., and Gammon, A., Photobinding of ^{3}H-8-Methoxypsoralen to monkey intraocular tissues, *Invest. Ophthalmol. Vis. Sci.*, 25, 1267, 1984.
128. Lerman, S., Psoralens and ocular effects in animals and man: in vivo monitoring of human ocular and cutaneous manifestations, in Photochemotherapeutic Aspects of Psoralens, Monogr. no. 66, National Cancer Institute, Bethesda, Md., 1984, 227.
129. Dayhow-Barker, P. and Barker, F. M., II, Retinal effects of short term exposure to 8-MOP and UV-A, *Photochem. Photobiol.*, Suppl. 37, 283, 1983.
130. Dayhaw-Barker, P., Barker, F. M., II, and Diebert, K., Effects of three drugs on the retinal threshold to damage, *Am. J. Opt. Phys. Opt.*, 59, 10, 1982.
131. Potts, A. and Gonasum, L. M., Toxicology of the eye, in *Toxicology*, Casarett, J. J. and Doull, J., Eds., Macmillan, New York, 1975, 275.
132. Fraunfelder, F. T., Hanna, C., Dreis, M. W., and Cogrove, K. W., Possible lens changes associated with allopurinol therapy, *Am. J. Ophthalmol.*, 94, 137, 1982.
133. Lerman, S., Megaw, J., and Gardner, K., Allopurinol therapy and human cataractogenesis, *Am. J. Ophthalmol.*, 94, 141, 1982.
134. Lerman, S., Megaw, J., and Fraunfelder, F., Further studies on allopurinol and human cataractogenesis, *Am. J. Ophthalmol.*, 97, 205, 1984.
135. Krejci, L., Brettschneider, I., and Triska, J., Tetracycline hydrochloride and lens changes, *Ophthal. Res.*, 10, 30, 1978.
136. Zunino, F., Gambetts, R., and DiMarco, A., Interaction of daunomycin and its derivatives with DNA, *Biochim. Biophys. Acta*, 277, 489, 1972.

137. Berlin, V. and Haseltine, W. A., Reduction of adriamycin to a semiquione free radical by cytochrome P-450 reductase produces DNA cleavage in reaction modulated by molecular oxygen, *J. Biol. Chem.*, 256, 4747, 1981.
138. Dayhaw-Barker, P., Forbes, D., Fox, D., Lerman, S., Metgaw, S., McGinniss, J., Waxler, M., and Felten, R., Drug phototoxicity and visual health, in Long Term Visual Health Risks of Optical Radiation, U.S. Food and Drug Administration Symp., Bethesda, Md., September 24 to 27, 1983.
139. Roberts, J. E., The photodynamic effect of chlorpromazine, promazine and hematoporphrin on lens protein, *Invest. Ophthalmol. Vis. Sci.*, 25, 748, 1984.
140. Thomas, J. R., III, and Doyle, J. A., The therapeutic uses of topical vitamin A acid, *J. Am. Acad. Dermatol.*, 4, 505, 1981.
141. Ward, A., Brogden, R. N., Heel, R. C., Speight, T. M., and Avery, G. S., Etretinate, a review of its pharmacological properties and therapeutic efficacy in psoriasis and other skin disorders, *Drugs*, 26, 9, 1983.
142. Shalita, A. R., Cunningham, W. J., Leyden, J. J., Pochi, P. E., and Strauss, J. S., Isotretinoin treatment of acne and related disorders: an update, *J. Am. Acad. Dermatol.*, 9, 629, 1983.
143. Sommer, A., Treatment of corneal xerophthalmia with topical retinoic acid, *Am. J. Ophthalmol.*, 95, 349, 1983.
144. Hatchell, D. L., Faculjak, M., and Kubicek, D., Treatment of xerophthalmia with retinol, tretinoin, and etretinate, *Arch. Ophthalmol.*, 102, 926, 1984.
145. Weiss, J., Degnan, M., Leupold, R., and Lumpkin, L. R., Bilateral corneal opacities: occurrence in a patient treated with oral isotretinoin, *Arch. Dermatol.*, 117, 182, 1981.
146. Lerman, S. and Hockwin, O., UV-visible slit lamp densitography of the human eye, *Exp. Eye Res.*, 33, 587, 1981.
147. Lerman, S., Dragomirescu, V., and Hockwin, O., In vivo monitoring of direct and photosensitized UV radiation damage to the lens, *Acta 24th Int. Congr. Ophthalmol.*, 1, 354, 1983.
148. Dragomirescu, V., Hockwin, O., and Koch, H. R., Development of a new equipment for rotating slit image photography according to Scheimpflug's principle, in *Interdisciplinary Topics in Gerontology*, Vol. 13, S. Karger, Basel, 1978, 118.
149. Dragomirescu, V., Hockwin, O., and Koch, H. R., Photo-cell device for slit-beam adjustment to the optical axis of the eye in Scheimpflug photography, *Ophthal. Res.*, 12, 78, 1980.
150. Weiss, J. N., Rand, L. I., Gleason, R. E., and Soeldner, J. S., Laser light scattering spectroscopy of in vivo human lenses, *Invest. Ophthalmol. Vis. Sci.*, 25, 594, 1984.
151. Nishio, I., Weiss, J. N., Tanaka, T., et al., Early detection of in vivo cataractogenesis using laser light scattering spectroscopy, *Curr. Eye Res.*, in press.
152. Schleich, T., Matson, G. B., Acosta, G. et al., Surface coil ^{31}P NMR studies of the intact eye, *Exp. Eye Res.*, in press.
153. Schleich, T., Willis, G. A., and Matson, G. V., Longitudinal (T_1) relaxation times of the phosphorous metabolites in the bovine and rabbit lens, *Exp. Eye Res.*, 39, 455, 1984.
154. Lerman, S., Moran, M., and Megaw, J., The UV and acrylamide effect on the human lens, *Lens Res.*, in press.
155. Lerman, S., Megaw, J. M., and Moran, M. N., Further studies on the effects of UV radiation on the human lens, *Ophthal. Res.*, in press.
156. Schleich, T. and Willis, J. A., The effect of ultraviolet-A and visible light irradiation and the phosphorus metabolite levels of the rabbit lens in organ culture, *Lens Res.*, in press.
157. Gonzalez, R. G., Willis, J., Aguayo, J., Campbell, P., Chylack, L. T., Jr., and Schleich, T., ^{13}C-nuclear magnetic resonance studies of sugar cataractogenesis in the single intact rabbit lens, *Invest. Ophthalmol. Vis. Sci.*, 22, 808, 1982.
158. Lerman, S., Megaw, J., Gardner, K., Long, R., Ashley, D., and Goldstein, J. H., NMR analysis of the cold cataract: whole lens studies, *Invest. Ophthalmol. Vis. Sci.*, 23, 218, 1982.
159. Lerman, S., Megaw, J., Gardner, K., Ashley, D., Long, R., and Goldstein, J. H., NMR analyses of the cold cataract. II. Studies on protein solutions, *Invest. Ophthalmol. Vis. Sci.*, 24, 99, 1983.
160. Lerman, S. and Megaw, J., Transmission characteristics of commercially available sunglasses, *J. Ocular Cutaneous Toxicol.*, 2, 47, 1983.
161. Anderson, W. J. and Grabel, R. K. H., Ultraviolet windows in commercial sunglasses, *Appl. Opt.*, 16, 515, 1977.
162. Anon., An evaluation of radiation emission from video display terminals, in *Radiological Health*, Food and Drug Administration, U.S. Department of Health and Human Services, Washington, D.C., 1981.
163. Weiss, M. M. and Peterson, R. C., Electromagnetic radiation emitted from video computer terminals, Bell Telephone Laboratories, Inc., *Am. Ind. Hyg. Assoc. J.*, 40, 300, 1979.
164. Anon., Potential Health Hazards of Video Display Terminals, NIOSH Res. Rep. Publ. No. 81-129, National Institute for Occupational Safety and Health, U.S. Department of Health and Human Services, Washington, D.C., 1981.

165. Ulrich, O. A. and Evans, R. M., Ultraviolet reflectance of paints, American Welding Society Report, August 6, 1976.

Chapter 6

PHOTOIMMUNOLOGY

Warwick L. Morison

TABLE OF CONTENTS

I. INTRODUCTION

Photoimmunology is an old field of investigation with a new name. The first suggestions that nonionizing radiation might affect immune function were made early in the 20th century. Dermatologists were recognizing various types of photosensitivity and hypothesizing on the pathogenesis of these disorders. Solar urticaria, photoallergy, and lupus erythematosus (LE) were conditions produced or exacerbated by exposure to sunlight and in which a disturbance of immune function was considered central to the pathogenesis. Therefore, some interaction between sunlight and the immune system was thought to be involved in the development of these diseases. For two reasons it is not surprising that sunlight-induced immunologic effects were considered many years ago. First, it is obvious that sunlight penetrates deeply into the body. Any person who has held a flashlight in the palm of the hand and observed red light coming through the back of the hand is aware that nonionizing radiation penetrates to a considerable depth. More energetic UV radiation is less penetrating but still reaches the dermis and subcutaneous tissue. Second, it has long been known that the immune system is represented in skin in blood vessels and tissue spaces. Thus, the target was present within range of a known toxin, sunlight.

The name photoimmunology crept, or some people might say leapt, into usage about 10 years ago and is now generally used to encompass all phenomena involving interactions between nonionizing radiation and the immune system. Introduction of the name coincided with greatly increased interest in the field, which was the result of three stimuli. First, it was discovered that UV radiation, besides causing neoplastic transformation of cells in mice, also produced very specific alterations of immune function which played an integral role in photocarcinogenesis. This finding has revolutionized our approach to the pathogenesis of nonmelanoma skin cancer and has provided a focus for much of the recent investigation in photoimmunology. It also has resulted in much new information about immunosurveillance of cancer and has led to the use of UV radiation as a tool for unraveling immunoregulatory pathways in the mouse. The second stimulus for photoimmunology came from clinical observations. About 10 years ago a new treatment for certain skin diseases, in particular psoriasis, was introduced which involved oral ingestion of a photosensitizer called methoxsalen and subsequent exposure to UVA (320 to 400 nm) radiation, now usually referred to by the acronym PUVA therapy. It was soon found that this treatment was effective in a wide variety of skin diseases which appeared to be related only in possibly having an immunologic pathogenesis, and this suggested that PUVA therapy might be effective via a photoimmunologic mechanism. Much study of the effects of PUVA treatment on the immune system in humans and experimental animals ensued and this has in many ways meshed with studies of photocarcinogenesis, particularly since PUVA treatment appears to be carcinogenic in humans. Finally, the third stimulus to research in photoimmunology came from the greatly increased understanding of the skin as an immune organ. The demonstration that Langerhans cells have antigen-presenting (AP) cell function, that keratinocytes produce substances capable of modifying the function of lymphocytes, and that a population of lymphocytes is closely associated with the skin has revolutionized our approach to the integument. Obviously, since the skin is constantly bathed with nonionizing radiation, this has provided a great stimulus to photoimmunology.

Several motivations have prompted studies in photoimmunology. Some people have approached directly questions of how and why nonionizing radiation affects the immune system. An example of this approach is the study of interactions between UV radiation and immunity in photocarcinogenesis. Other people have used UV radiation as a tool for investigating immune function. An example of this approach is the early studies of the effects of UV radiation on Langerhans cells. Most recently, UV radiation has been used as a tool for deliberately manipulating immune function for possible therapeutic benefit. An example of this is the irradiation of islets of Langerhans prior to transplantation across a histocompatibility barrier in the treatment of experimental diabetes. One or two successes in this direction will probably provide a further stimulus for studies in photoimmunology.

II. EXPERIMENTAL STUDIES IN ANIMALS

Investigation of the effects of nonionizing radiation on immune function in animals has been confined largely to a narrow focus. Most work has concentrated on a few immune responses, in particular contact hypersensitivity (CHS) and rejection of UV-induced tumors. In addition, the fluorescent sunlamp bulb has been the source of radiation in most studies. By convention, radiation from this source is referred to as UVB (280 to 320 nm) radiation but its emission spectrum is continuous from the UVC (mainly 254-nm) region through the UVB region and includes much UVA radiation. It is important to remember that sunlight at the surface of the earth does not contain radiation shorter than 295 nm and is rich in UVA, visible, and IR radiation. Obviously the term "sunlamp bulb" is a misnomer.

A. Contact Hypersensitivity (CHS)

CHS is a term used to refer to an immune response in animals that is analogous to contact allergy or allergic contact dermatitis in humans; poison ivy dermatitis is probably the most common example. The response follows topical application to the skin of a chemical, called a hapten or contact sensitizer, which then binds to a protein or cell in the skin, called a carrier, to form the complete antigen. AP cells in the skin process the antigen and aid in stimulating the proliferation of T lymphocytes specifically sensitized to the antigen. Langerhans cells are a population of dendritic cells in the epidermis which act as AP cells, but there appear to be other populations of cells that also perform this function. These steps are the afferent limbs of the immune response and result in generation of a population of circulating, sensitized, long-lived T lymphocytes. This process takes 5 to 10 days. The state of sensitivity to the specific antigen can be adoptively transferred to another animal by injection of T lymphocytes from the sensitized animal. The efferent limb of the response is its expression in the skin upon further contact with the antigen and involves an influx of macrophages and T lymphocytes. In humans, the response is manifest as eczema while in animals the main features are swelling or edema of the skin with or without erythema, crusting, and scaling. The response reaches a peak 24 to 48 hr or later after exposure to the antigen.

Exposure of animals to UVB radiation suppresses the development of CHS, with two types of suppression being observed. So-called local suppression occurs when the sensitizing dose of hapten is applied to skin at the site of exposure to radiation. The suppression is called systemic when the sensitizing dose of hapten is applied to nonirradiated skin. In fact, both types of suppression are associated with the generation of circulating suppressor T lymphocytes that can be adoptively transferred, and therefore both can be systemic phenomena. However, the distinction between the two effects is important because the mechanisms leading to the generation of suppressor cells are probably quite different.

1. Local Suppression of CHS

Suppression of CHS in irradiated skin was first reported by Haniszko and Suskind[1]

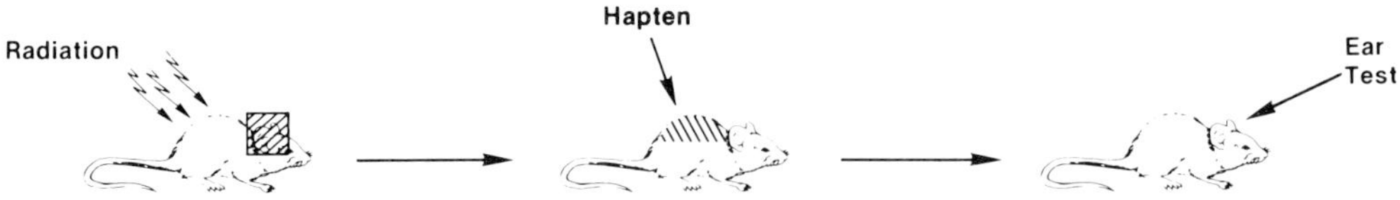

FIGURE 1. System for demonstrating local suppression of CHS by UV radiation in the mouse.

when they found that 14 daily exposures to UVB radiation suppressed the response to dinitrochlorobenzene (DNCB) in guinea pigs. The suppression was similar when the irradiation was given 2 weeks pre- or post-sensitization, which indicates that the effect was mediated on both the induction and elicitation phases of CHS. In a few subsequent studies, all in outbred guinea pigs, PUVA,[2] UVB,[3,4] and UVC[5] radiation have been found to suppress the elicitation of CHS in previously sensitized animals, but the mechanism of this effect is unknown. All other studies have focused on suppression of the induction of CHS.

Interest in local suppression of CHS by UV radiation was reawakened by studies of the function of epidermal Langerhans cells. Toews et al.[6] found that elimination of Langerhans cells in mice by exposure to UVB radiation was associated with suppression of CHS, which supported their hypothesis of Langerhans cells being an essential element in the development of this immune response. This local effect, in which the site of sensitization and irradiation are the same but the site of elicitation is not irradiated (Figure 1), requires only a low dose of UVB radiation. The site of elicitation is the ear and the response is evaluated by measuring specific swelling of the ear 24 hr after application of the hapten. The suppression is hapten-specific and is associated with the generation of hapten-specific Lyt-1$^+$ T cells that suppress the induction phase of the immune response.[7] Local suppression of CHS cannot be induced in all strains of mice that have been tested, but recently Glass et al.[8] found that suppressor cells are present in all strains, including those that do not exhibit suppression of CHS. Presumably, a positive immunoregulatory pathway is activated in nonsuppressible strains that overrides the tolerogenic signal elicited by UVB radiation. Local suppression of CHS by UV radiation is not restricted to mice; the phenomenon also occurs in hamsters following exposure to UVB radiation[9] and in guinea pigs following treatment with PUVA.[10] In both instances, there is some evidence that, as in the mouse, the suppression is active and associated with suppressor cells.

Several UV-induced alterations of cutaneous immune function have been identified and linked to the mechanism of local suppression of CHS. Initially, it was thought that UV-induced changes in the function of Langerhans cells might be the key to suppression of CHS. Langerhans cells are very sensitive to damage by UV radiation, and the AP cell function of the epidermis is impaired after exposure to this wave band of radiation.[11-13] However, although alterations in Langerhans cells are probably involved in local suppression of CHS, changes in other cells also play a role. Recent studies have shown that several other components of the epidermis are involved in immune function and are altered by exposure to radiation, and these alterations are possibly important in the mechanism of local suppression of CHS.

The findings of Granstein et al.[14] suggest that the UV-induced alteration of AP cell function may not involve a simple loss of function, but rather that it may involve an active process. Functional studies revealed that a UV-resistant population of I-J$^+$ AP cells in the epidermis is active in the generation of suppressor-cell pathways that lead to suppression of CHS rather than help after exposure to UV radiation. The I-J$^+$ cell has not been identified morphologically, and it may be a Langerhans cell or another

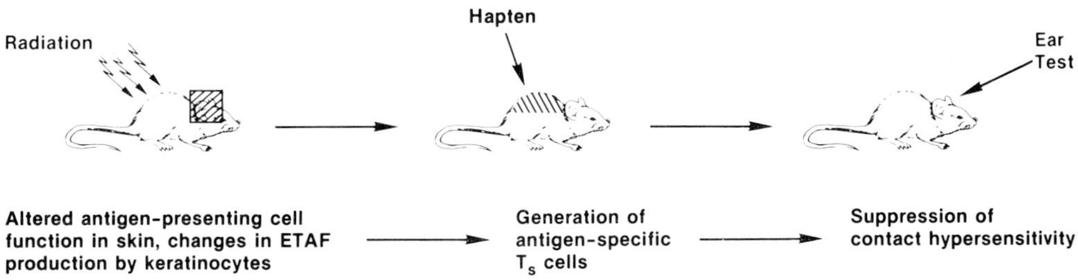

FIGURE 2. Proposed mechanism of UV-induced local suppression of CHS in the mouse.

type of cell. Thus, it can be hypothesized that normal immunologic balance is disturbed after exposure to UV radiation because one population of AP cells, responsible for generation of helper cells, is UV-sensitive while a second population, responsible for generation of suppressor cells, is UV-resistant.

UV-induced changes in the function of keratinocytes may also be involved in local suppression of CHS. Keratinocytes produce epidermal cell-derived thymocyte-activating factor (ETAF),[15] which is functionally and biochemically very similar to interleukin-1 (IL-1), a nonspecific helper factor necessary for activation of T cells by antigen. Sauder et al.[16] found that epidermal cells from mice that had been exposed to monochromatic 270-nm radiation were defective in stimulating allogeneic T cells in vitro and that this defect could be partially corrected by the addition of ETAF. This finding was supported by in vitro studies in which a UV-induced impairment of AP cell function was partially restored by addition of IL-1[17,18] or partially purified ETAF.[19] Furthermore, Stingl et al.[19] found that after exposure to UV radiation, the time course for development of the AP cell defect in vitro matched the time course of reduction in ETAF production by epidermal cells. However, in two studies,[20,21] exposure of epidermal cells to broadband UV radiation from sunlamp bulbs either in vitro or in vivo resulted in increased production of ETAF. Differences in the radiation sources, the doses administered, and the model systems probably explain the discrepancies between the results of these studies. In either case, an alteration in ETAF production might be a key factor in UV-induced immunosuppression.

Finally, a population of dendritic cells in murine epidermis has been recently identified. These cells are Thy-1$^+$, are derived from bone marrow, and are distinct from Langerhans cells.[22,23] A preliminary study[24] found that these dendritic cells were morphologically altered by PUVA treatment, but their sensitivity to UVB radiation is unknown. The role of these cells in immune function and its suppression by UV radiation has not been determined.

In summary (Figure 2), the mechanism of local suppression of CHS in the mouse involves an alteration in AP cell function in the skin at the site of exposure to radiation. This altered function might be due to UV-induced damage to one cell population, possibly Langerhans cells, but may also involve defective function of another AP cell population and of keratinocytes. In either case, the result is preferential activation of a suppressor-cell pathway so that there is diminution of the normal immune response to the particular antigen.

2. Systemic Suppression of CHS

UVB radiation also produces suppression of the development of CHS to chemicals subsequently applied to nonirradiated areas of skin. In this model, illustrated in the mouse in Figure 3, the sites of sensitization and elicitation are both on skin that has not been exposed to radiation. Systemic suppression requires a much higher exposure

FIGURE 3. System for demonstrating systemic suppression of CHS by UV radiation in the mouse.

dose of radiation for its induction as compared with local suppression. The phenomenon was first described in the mouse,[25,26] but it has also been observed recently in inbred guinea pigs.[27] In both species, the mechanism involves the generation of hapten-specific T lymphocytes which suppress the induction, but not the elicitation, of CHS. In the mouse, the phenotype of the T lymphocytes is Lyt-1^+2^-.[28]

Apart from UVB radiation, other wave bands can also affect CHS. PUVA treatment and sunlight produce suppression of CHS in mice and guinea pigs which is associated with generation of suppressor cells.[27,29,30] Large doses of visible radiation produce a small but consistent degree of suppression of CHS, while large doses of UVA radiation produce some enhancement of this immune response;[30] the mechanisms of these effects are unknown.

Attempts have been made to identify the chromophore molecule that absorbs radiation and leads to suppression of CHS. Urocanic acid is one candidate. In support of this possibility, DeFabo and Noonan[31] have presented evidence that the chromophore is superficially located in the epidermis, as is urocanic acid, and can be removed by stripping the epidermis with cellophane tape. They have also determined that wavelengths in the 260- to 270-nm region, which are within the absorption spectrum of urocanic acid, are most effective in producing systemic suppression of CHS. These findings would be strengthened by direct measurements of levels of urocanic acid to show that it has been removed from tape-stripped skin and followed by the demonstration that urocanic acid applied to the skin restores susceptibility to UV-induced suppression. In a recent study Morison and co-workers found in a different strain of mice that tape stripping did remove most of the urocanic acid from skin, but it did not abolish UV-induced suppression of CHS. Tape stripping itself produces much nonspecific immunosuppression which tends to make interpretation of data very difficult. Another candidate as the chromophore molecule is DNA. Its absorption spectrum also includes the relevant wavelengths. Furthermore, physicochemical agents that interact with DNA produce systemic suppression of CHS that resembles UVB-induced suppression in that it can be adoptively transferred to normal animals by injection of lymphoid cells obtained from suppressed animals.[32] Agents tested and found to have this ability include methoxsalen/UVA, 5-methyl isopsoralen/UVA, and superficial X-ray. In contrast, phototoxic treatments such as eosin/visible light and rose bengal/visible light that do not produce lesions in DNA also do not produce suppression of CHS that can be adoptively transferred. Of course, it is possible that DNA is involved in the pathway to suppression but is not the chromophore molecule.

There has been much interest in how UVB radiation causes systemic suppression of CHS. A first step in understanding the mechanism was the finding that exposure of mice to UVB radiation produces a defect in AP activity of splenic adherent cells and fewer Ia positive cells can be recovered from the spleens of these mice.[33,34] Furthermore, the impaired reactivity in these mice can be overcome by immunization of the irradiated animals with hapten coupled to normal AP cells.[26,33,35] These studies were interpreted to indicate that the UV-induced impairment of AP cell activity in the spleen was a central step in the suppression of CHS, and it was suggested that the splenic defect led to activation of a T suppressor cell pathway.[35] This may be true, but it is important to note that the immune response restored by immunization with hapten-coupled normal AP cells was delayed hypersensitivity and not CHS. The mechanisms of suppression of these two responses by UV radiation might be quite different, as will be discussed later. Suffice it to say that a defect of AP cell function is present in the spleens of irradiated mice and this defect might be causally related to the suppression of CHS.

The mechanism whereby absorption of radiation by skin leads to a systemic AP cell defect is unknown. One possible mechanism is that migration of damaged Langerhans cells from the irradiated skin to internal organs might give rise to the AP cell defect in the spleen and elsewhere. This possibility has been explored by asking whether alterations in Langerhans cells are essential for systemic suppression of CHS by UV radiation. Several pieces of evidence indicate that the answer is no. First, Langerhans cells are normal at the unirradiated sites of sensitization and challenge with hapten.[36-38] Second, elimination of Langerhans cells with high doses of UVA radiation is not associated with suppression of CHS and does not prevent the development of UVB-induced suppression.[38] Third, UV-induced suppression can be produced by doses of radiation of certain wavelengths that do not alter Langerhans cells.[36] Therefore, although migration of damaged Langerhans cells might occur after exposure to UV radiation, this is probably not an essential step in systemic suppression of CHS.

Circulating monocytes might be directly damaged by UVB radiation as they pass through cutaneous capillaries and then lodge in the spleen to give rise to the AP cell defect in that organ. This hypothesis has been tested by supplying alternative sources of precursors of splenic AP cells. UVB-irradiated mice were given exogenous bone marrow cells, spleen fragments, and splenic adherent cells in attempts to reconstitute the AP cell defect but none of these procedures restored normal CHS.[39] Furthermore, immunization of UVB-irradiated mice with hapten-coupled splenic adherent cells also failed to restore CHS.

A third possible mechanism could involve an alteration of cell trafficking that results in the AP cell defect and, through that, the systemic suppression of CHS. There is some evidence that lymphocyte trafficking is perturbed in UV-irradiated animals and that lymphocytes migrate preferentially to regional lymph nodes.[35,40] In contrast, treatment of lymphoid cells with UV radiation in vitro reduces their tendency to localize in lymph nodes after injection into normal animals.[40,41] An extension of this hypothesis is that the splenic AP cell defect is due to passive accumulation of inflammatory cells in the skin. The main points against this concept are that the splenic defect cannot be reconstituted by alternative sources of cells, and nonspecific trauma to the skin, which causes an influx of inflammatory cells, does not produce a specific suppression of CHS.[39]

Finally, factors may be released from irradiated skin and circulate to produce the AP cell defect and systemic suppression of CHS. A recent study suggests that this may occur. Serum obtained from mice 3 hr after exposure to UVB radiation produced suppression of the development of CHS in recipient mice that was associated with the generation of suppressor cells.[42] The factor in serum that produces this effect has not yet been identified, but certain of its properties indicate that it is not one of the known mediators of inflammation. This is supported by recent studies in which we have found that maximally tolerated doses of inhibitors of various mediators of inflammation are without effect on the suppression of CHS by UVB radiation. Inhibitors of histamine, serotonin, prostaglandins, and the leukotrienes had no effect, and complement-deficient animals exhibited suppression. In addition, corticosteroids did not inhibit the suppression. Gahring et al.[21] used a different approach and found that elevation of the level of ETAF/IL-1 activity in serum of UV-irradiated mice correlated with elevation

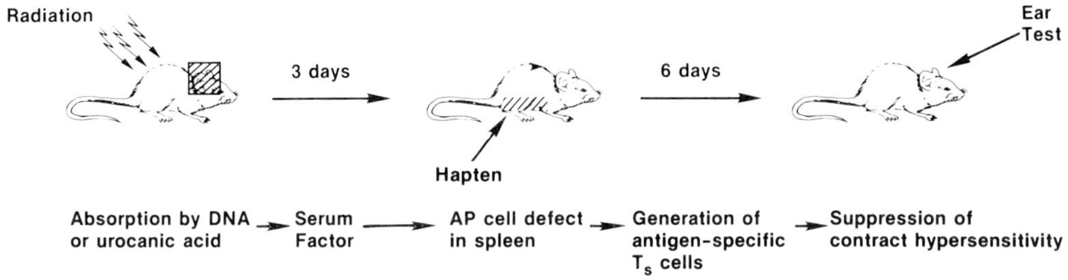

FIGURE 4. Proposed mechanism of UV-induced systemic suppression of CHS in the mouse.

of the levels of the acute-phase reactants, amyloid, complement, and fibrinogen after a single exposure to radiation. However, after multiple exposures, the serum concentrations of acute-phase reactants were not elevated, despite sustained elevation of the serum level of ETAF/IL-1 activity. The authors suggest that these changes might at least partially explain UV-induced immunosuppression: the suppression after a single exposure could be mediated by acute-phase reactants and the suppression after multiple exposures could be due to lack of a necessary signal for effective immune function because of desensitization of target cells to the elevated levels of ETAF/IL-1. Somewhat at variance with these observations are those of Tominaga et al.[18] who found that UVB-induced systemic suppression of delayed hypersensitivity could be prevented by administration of IL-1.

In summary (see Figure 4), UVB radiation can produce systemic suppression of CHS in mice and guinea pigs and the suppression is due to generation of hapten-specific suppressor T lymphocytes. A defect of splenic AP cell function is associated with the suppression, but the mechanism by which this is produced is unknown. Furthermore, it is not known whether the two defects are interrelated. One possible scenario is that radiation is absorbed by a chromophore such as urocanic acid or DNA in skin, a mediator or photoproduct is released into the circulation and causes the AP cell defect in the spleen, and this results in preferential activation of a suppressor-cell pathway and consequent suppression of CHS.

B. Photocarcinogenesis

Much of the impetus for recent work in photoimmunology has come from observations of the role of the immune system in photocarcinogenesis. This interest began with the observation that tumors induced in mice by exposure to UV radiation have the unusual feature of being highly antigenic. Thus, when these tumors are transplanted in syngeneic, i.e., genetically identical, normal mice they fail to grow, but they do grow progressively in immunosuppressed mice (Figure 5).[43] Rejection of the tumors in normal mice is immunologic and mediated by T lymphocytes. This observation raised the obvious question of why such antigenic tumors were able to escape the normal immunologic surveillance mechanism and grow in the primary host. Investigation of the immune status of the primary host revealed susceptibility to growth of transplants of other UV-induced tumors but this defect was selective since most other immune responses in these animals were normal.[44-47] Furthermore, a subcarcinogenic dose of UV radiation was sufficient to produce susceptibility to the growth of transplanted UV-induced tumors, and this susceptibility only applied to UV-induced tumors since tumors produced by chemicals did not show accelerated growth in irradiated mice.[48]

Therefore, UV radiation in mice produces two alterations in the process of tumor development (Figure 6): (1) neoplastic transformation of cells in the skin, and (2) a specific, systemic, immunologic alteration which favors the growth of tumors. Subse-

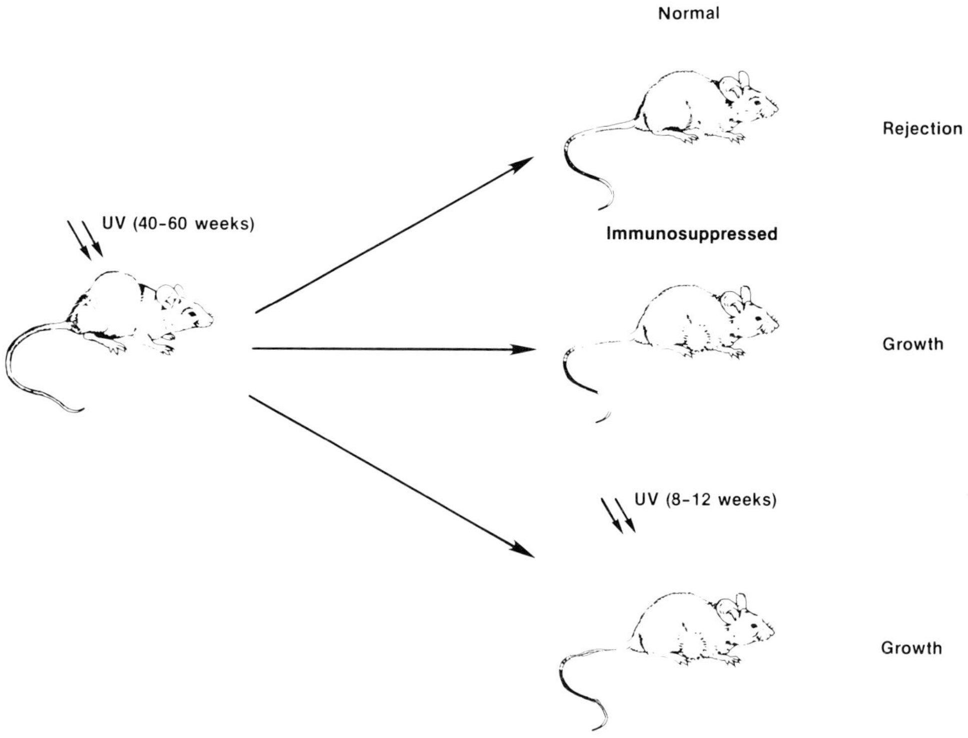

FIGURE 5. Transplantation and patterns of growth of UV-induced tumors in syngeneic mice.

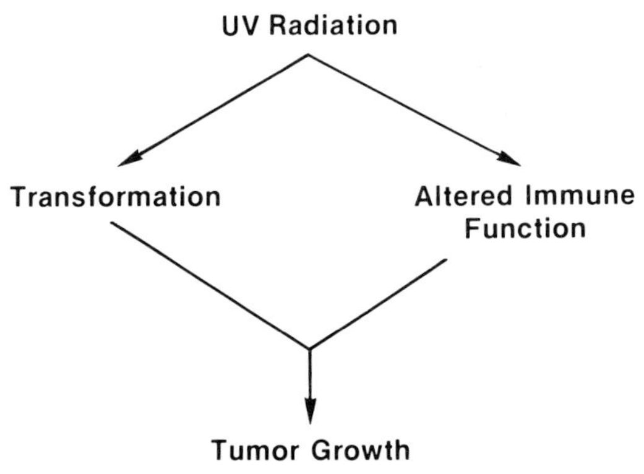

FIGURE 6. Mechanism of photocarcinogenesis in the mouse.

quent studies revealed that the immunologic alteration involved the generation of suppressor T lymphocytes which inhibit the rejection of UV-induced tumors.[49,50] The suppressor cells have the same phenotype as the CHS suppressor cells in that they are Lyt-1^+2^-, Ia^-.[28] The dose of radiation required to induce formation of the suppressor cells is much less than a carcinogenic dose, and they appear early in the development of neoplasia. The early and pivotal role of the suppressor cells was demonstrated by the finding that if suppressor cells are present at the time of commencement of irradiation, tumors appear earlier and in greater numbers.[51] This extended an observation that irradiation of one site markedly accelerated the induction of tumors by subsequent irradiation of another site.[52]

The stimulus for the formation of suppressor cells is unknown. However, presumably a putative tumor antigen appears on transformed cells and initiates the immune response. There is some evidence to support this concept. First, cells of the $10T_{1/2}$ mouse fibroblast line transformed in vitro by UV radiation are antigenically similar to those from skin cancers produced in vivo by UV radiation, and the transformed cells are recognized by UV-induced suppressor cells.[53] Second, UV-irradiated skin contains the elements necessary to induce a state of tumor susceptibility. Grafts of UV-irradiated skin measuring 5 × 2.5 cm transplanted onto normal mice induced tolerance to the growth of UV-induced tumors.[54] This observation suggests that the UV-associated antigen is present in skin at an early stage of neoplasia, but the possibility that the effect is due to the presence of passenger leukocytes in the grafted skin has not been excluded.

A further question is why the putative tumor antigen stimulates suppression and not help. Studies in the CHS system provide some clues to the answer. Exposure of mice to UV radiation alters antigen presentation in the skin so that an $I\text{-}J^+$ AP cell, which is resistant to damage, has a dominant influence and this $I\text{-}J^+$ cell preferentially activates a suppressor-cell pathway.[14] More direct evidence to support this concept comes from the finding that the function and/or development of tumor suppressor cells can be inhibited by administration of anti-$I\text{-}J^k$ antibody.[55] Of course this finding is not only important for understanding the mechanism of neoplasia but also raises the possibility of specific therapy to prevent the development of skin cancer.

The FS40 sunlamp bulb has been the main source of radiation used in these studies of the immunobiology of skin cancer. Wavelengths between 275 and 315 nm from this source were most effective in producing suppression of tumor rejection.[56] Recently, sunlight has also been found effective in producing tumor susceptibility in mice.[57] The 295- to 320-nm wave band appears to be responsible for this effect of sunlight because it was prevented by filtering the radiation through Mylar® and by application of a sunscreen containing para-aminobenzoic acid.

Interestingly, PUVA-induced tumors in mice are not highly antigenic and do not show preferential growth in UVB-irradiated animals.[58] Furthermore, treatment of animals with PUVA does not prevent their normal immunologic rejection of UVB-induced tumors.[59]

C. Other Immune Responses

The finding that exposure of mice to UVB radiation suppressed the rejection of UV-induced tumors led to an investigation of other possible immunologic alterations in irradiated animals.

Delayed hypersensitivity to injected antigens is impaired in some circumstances. Both PUVA treatment and UVB radiation produce local and systemic impairment of delayed hypersensitivity to an injected antigen in the guinea pig.[60,61] UVB radiation also impairs delayed hypersensitivity in mice and this effect appears to involve the generation of antigen-specific T suppressor cells. However, the mechanism of this effect appears to be different from UV-induced suppression of CHS since it can be reconstituted with antigen-coupled, normal splenic adherent cells,[35] whereas these cells do not reconstitute suppression of CHS.[39]

The local graft vs. host reaction is also reduced in irradiated mice. There is less enlargement of the popliteal lymph node after injection of parental lymphocytes into the footpads of irradiated F_1 hybrid mice than occurs in unirradiated recipients.[45,62] This effect is due to a reduction in the host component of the reaction, possibly through diversion of cells from the site of the response, and there is no evidence that it involves activation of a suppressor-cell pathway.[62]

Allogeneic reactions are usually unaffected by UV radiation. Rejection of skin allografts in mice and rabbits is unaffected by UVB radiation,[45,63] although PUVA treatment of the donor and recipient sites produced some prolongation of allograft survival in rabbits.[63] Tumor allografts are rejected normally by irradiated mice,[44] and lymphocytes from such animals have normal cytotoxic responses against allogeneic cells in vitro.[46,47]

Obviously, further studies are required to fully define the effect of nonionizing radiation on various immune responses. The effect upon antibody responses and cell-mediated responses to microbial antigens are two areas awaiting study. However, with our present knowledge it is possible to say that the action of radiation is selective; only certain immune responses are affected while others are quite normal in the irradiated animal. Furthermore, it is becoming clear that the mechanisms by which UV radiation affects immune function are diverse, with formation of suppressor cells in some cases, alteration in the trafficking of cells in others, and selective cell killing in yet other cases. The selectivity and diversity of the effects of UV radiation may well have value for selective, targeted manipulation of immune function.

III. PHOTOIMMUNOLOGY IN HUMANS

Investigation of the effects of nonionizing radiation on immune function in humans has lagged far behind experimental studies in animals, but already enough is known to suggest that photoimmunology will be important in human health and disease. Only a few investigations have been designed to answer a specific question about the effects of radiation, and most studies have been directed toward determining the pathogenesis of various diseases. These two approaches will be discussed separately since they have quite different goals.

A. Experimental Studies

1. In Vivo Immune Responses

There are several reports of inhibition of contact allergy in humans by exposure to UV radiation. The response to DNCB was diminished in normal volunteers as compared with control subjects after daily exposures for 2 weeks in a solarium emitting mainly UVA but also a small amount of UVB radiation.[64] UVB radiation also delayed the development of contact allergy to mechlorethamine in patients being treated topically with this agent.[65,66] In these studies it is not possible to distinguish between an effect on the induction vs. the elicitation phase of the response. However, in several studies sunlight, UVB, or PUVA therapy inhibited the elicitation of contact allergy in previously sensitized subjects,[67-70] and in one study PUVA therapy possibly affected both induction and elicitation of contact allergy.[71]

The mechanism of suppression of contact allergy by UV radiation in humans is unknown, but since in all instances the effect has been detected locally at the site of exposure to radiation, it is reasonable to suspect it will be similar to the mechanism of local suppression in the mouse. Although the latter mechanism has not been completely defined, th early suggestion of a central involvement of Langerhans cells focused attention on the effect of UV radiation on these cells in humans. Langerhans cells are reduced in number in sun-damaged skin[72] following exposure to UVB and UVA radiation,[73,74] and after a course of PUVA therapy.[75-77] However, these cells are not completely eliminated by these treatments and a normal population is restored after a brief period of recovery. The relationship between the morphologic alteration in Langerhans cells and suppression of contact allergy has not been studied in humans.

2. Effect of UV Radiation on Lymphocytes

Alterations in the distribution and function of circulating lymphocytes can be detected after exposure to UV radiation. Two weeks of daily exposure in a solarium or to sunlight produced a decrease in OKT8 suppressor/cytotoxic cells and of suppressor T-cell activity, and depression of natural killer-cell activity.[64,78] These changes were still evident in some subjects 2 weeks after completion of the exposures. A single whole-body exposure to UVB radiation or PUVA treatment also altered circulating lymphocytes with a decrease in identifiable T and B cells and an increase in null cells; these changes lasted less than 72 hr.[79,80] The mechanism of these changes is unknown, but possibly lymphocytes are damaged as they percolate through cutaneous capillaries and are then sequestered out of the circulation.

Apart from studies in normal subjects, there has also been much interest in the effect of radiation during treatment of various diseases of the skin. Lymphocytes from patients with psoriasis treated with PUVA therapy were found to have suppression of normal DNA synthesis[81] and some impairment of mitogen responses[82,83] early in the course of treatment, but after months of treatment, lymphocyte function was found to be normal.[84] Longer-term follow-up of patients who had been treated for years and who had evidence of aging of the skin as a result of treatment revealed a reduction of circulating helper/inducer T cells and impairment of the response of lymphocytes to mitogens.[85,86] Studies are required to determine if the patients with these changes are those most likely to develop skin cancer which can result from this treatment.

3. In Vitro Exposure of Lymphocytes to UV Radiation

In contrast to the paucity of literature on in vivo effects of radiation on immune function in humans, there are many studies of the effects on in vitro exposure of lymphocytes to various wave bands of radiation. Some of these effects are of more interest from a photobiologic rather than an immunologic viewpoint since they provide information about the relative toxicity of the various wave bands without indicating how this toxicity might be important in vivo. For example, when viability of lymphocytes or their response to a mitogen are tested after exposure, UVC is more toxic than UVB, which in turn is more toxic than UVA radiation, but the toxicity of UVA radiation is greatly increased by the addition of psoralen.[87-89]

In vitro treatment of lymphocytes with PUVA also inhibits the mixed-lymphocyte-culture reaction when either the stimulator or the responder cells are exposed,[90,91] and increases chromosomal aberrations and the frequency of sister-chromatid exchanges.[92,93] However, the frequency of sister-chromatid exchanges is not increased in lymphocytes obtained from patients during treatment.[94] The disparity between these findings after in vivo and in vitro exposure is probably due to the much higher exposure doses used in the in vitro model; this of course is one problem in the interpretation of most of the in vitro studies.

Although in vitro studies involving irradiation of human lymphocytes have not been very contributory to our understanding of photoimmunology, they could provide a good avenue for further studies. For example, development of in vitro correlates of in vivo immune responses is one way in which some of the limitations upon studies in humans can be overcome.

B. Nonmelanoma Skin Cancer

Demonstration of a central role of the immune system in the pathogenesis of nonmelanoma skin cancer in mice has of course focused interest on whether sunlight-associated skin cancer in humans might involve similar immunologic alterations. At present the evidence is at best suggestive.

The strongest evidence for immune surveillance being important in human skin cancer is that the incidence of tumors on sun-exposed skin is greatly increased in renal transplant recipients treated with immunosuppressive drugs.[95-97] The incidence rises shortly after transplantation and reached 44% 9 years post-transplant in one study from a sunny climate.[98] Furthermore, the incidence appears to be dose-related and highest in patients exposed to most sunlight.[99] However, it must be noted that the drugs used in these patients have other effects apart from immunosuppression, and therefore this clinical observation is persuasive but by no means conclusive evidence of involvement of the immune system in the pathogenesis of skin cancer. Detection of human papillomavirus in skin cancers of an immunosuppressed renal transplant recipient raises the possibility that sunlight might be acting as a cofactor for the expression of viral oncogenesis in these patients.[100]

The number of circulating T lymphocytes and the degree of lymphocytic infiltration of nonmelanoma skin cancers were both found to be inversely proportional to the size of the tumor in one study.[101] These observations were interpreted as indicating that the extent of the immune response was important in determining the size of a tumor, but it is equally plausible that a reduced response is a consequence of the presence of a large tumor.

In another study, humoral and cell-mediated immunologic reactivity toward autologous tumor tissue was detected in most patients with squamous cell carcinoma that were tested.[102] The specificity and quantitation of the reactivity requires further study, but the results do suggest that human tumors are antigenic and can provoke an immunologic response.

A somewhat different approach was taken in a recent study which examined the ability of patients with xeroderma pigmentosum to develop contact allergy on sun-exposed skin.[103] The reasoning behind this study was that if immunologic abnormalities are important in human skin cancer then they should be most obvious in patients with the greatest susceptibility to the development of these tumors. It was found that in comparison to normal control subjects, most patients with xeroderma pigmentosum were grossly defective in their capacity to be sensitized to a common allergen, and the degree of this impairment was closely related to the severity of the cutaneous manifestations of the disease. These findings are consistent with the possibility that immunologic factors, presumably induced by exposure to sunlight, could be important in the development of skin cancer in these people. More direct evidence awaits the demonstration that human skin tumors are antigenic, that the patient does mount an immune response to the tumor, and that suppressor cells are important in blocking the expression of that response. A suitable in vitro model will be required for these studies.

C. Photosensitivity Disorders

Photoimmunology began with the investigation of several diseases which predominantly affect the skin and are characterized by photosensitivity. The term "photosensitivity" is used here in a clinical sense and may be defined as an altered reaction of the skin or other organs following exposure to nonionizing radiation. Immunologic mechanisms have been postulated to explain the pathogenesis of a number of these disorders, but strong evidence for this is only available in a few.

1. Solar Urticaria

Solar urticaria, which is simply the development of hives following exposure to sunlight, was the first photosensitivity disorder to be categorized as immunologic or "allergic" in nature. The suggestion arose because all types of hives were considered to have an immunologic pathogenesis. We now know that this is not true, and similarly, we now know that only some cases of solar urticaria have an immune basis. In other cases, although the pathogenesis is unknown, there is no evidence for an immune response. Solar urticaria usually develops in young adults without apparent cause and probably persists for life. Most clinicians consider it to be an uncommon disease, but it seems probable that it is fairly common in a mild form, so that most patients, instead of seeking medical advice, just avoid exposing themselves to the amount of sunlight which they know will trigger the problem.

The mechanism of development of solar urticaria in cases which have an immunologic basis appears to involve an antigen-antibody reaction in the skin at the site of exposure to sunlight. The putative antigen is a photoproduct and is either a molecule produced in normal skin or a molecule that is formed only in patients with solar urticaria. The nature of the antigen is unknown and the main evidence for an immune mechanism comes from serum-transfer tests. Two types of tests have been used in solar urticaria to study the nature of the reaction. The passive serum-transfer test is analogous to the Prausnitz-Kustner reaction in that the patient's serum is injected intradermally into a normal subject and the site is exposed to radiation within 24 hr. A positive response is indicated by an immediate wheal. In the reversed passive-transfer reaction, the serum is injected after exposure to radiation and a positive reaction eliminates the possibility of the response being due to a phototoxic agent in serum. When both tests are positive, the antigen presumably is a photoproduct formed in normal skin. In some patients, the passive-transfer test is positive but the reversed passive-transfer test is negative. A recent study of these patients found that a positive result in the reversed test was obtained when the patient's serum was irradiated, which suggests formation of a photoproduct that is unique to the patient.[104]

Certain properties of the serum factor, which is presumably an antibody, have been characterized: (1) it is thermolabile, (2) it progressively loses activity in a passively sensitized site after exposure to radiation, and (3) it retains reactivity in the sensitized site several days after injection.[105] These properties are consistent with the serum factor being an IgE reagenic antibody.

The mast cell is thought to be the target for the urticarial reaction and the source of mediators. Degranulation of mast cells has been observed after exposure to radiation.[106,107] The levels of histamine, eosinophil chemotactic factor, and neutrophil chemotactic factor are elevated in venous blood draining the site of the reaction.[106-108] These findings, combined with the results of serum-transfer tests, are all consistent with a pathogenesis for solar urticaria in which IgE antibody is bound to mast cells, exposure to radiation alters a component of skin to form an antigen, and mast cell degranulation and release of mediators is triggered by an antigen-antibody reaction on the cell surface.

2. Photoallergy

Photoallergy is a state of specifically altered reactivity toward an exogenous chemical in which photons play a role. Two types of photoallergy are recognized. Contact photoallergy follows epicutaneous application of a chemical and exposure to radiation, and the immunologic basis of this reaction is established. Systemic photoallergy involves oral or parenteral administration of the chemical, and there is no definite evidence that the pathogenesis of this disorder is immunologic; instead it is assumed to be immunologic by analogy to contact photoallergy.

The role of nonionizing radiation in contact photoallergy is in the formation of the antigen. The antigen is composed of a low molecular weight compound called a hapten which is conjugated to a carrier protein to form the complete antigen. There are three mechanisms by which radiation could be involved in production of the antigen: (1) formation of a photoproduct which then serves as the hapten, (2) photochemical conjugation of the photosensitizing compound to protein, and (3) photochemical alteration of a protein. Experimental evidence indicates that the first two mechanisms apply in different cases, but there is no evidence for the third mechanism being important.

Once the antigen has been formed, radiation plays no further part in the reaction.

The antigen triggers a cell-mediated immune response manifested as eczema and thus, apart from the early role of radiation, this immune response is very similar to ordinary contact allergy. The immunologic nature of the reaction has been established in guinea pigs[109] and mice[110-112] by adoptively transferring immune reactivity by injection of lymphoid cells from sensitized animals; the reaction is mediated by T lymphocytes. Contact photoallergy is much less common than ordinary contact allergy, and a recent study possibly has provided an explanation for the different frequencies of the reactions. To obtain a vigorous photoallergic response in mice, it is necessary to pretreat the animals with cyclophosphamide, and only a minimal response develops in the absence of this treatment. Granstein and co-workers[112] found considerable net splenic suppressor-cell activity in mice that were not pretreated with cyclophosphamide. Furthermore, these suppressor cells, which are antigen specific, inhibited the induction of the immune response. Thus, we may be protected from developing photoallergy to environmental agents by this immunoregulatory imbalance.

3. Lupus Erythematosus (LE)

Various diseases have been classified as autoimmune because responses to self-antigens appear to be involved in their pathogenesis. One of the most common of these diseases is LE which is a chronic inflammatory disease affecting multiple organs including skin. Photosensitivity is a feature of LE, and nonionizing radiation often precipitates or aggravates the disease. However, it is not clear whether this is a nonspecific effect of radiation or a specific effect due to a selective alteration of immune function.

LE is characterized by the appearance of a whole array of autoantibodies against individual constituents of cells, whole cells, and other proteins. The most important of these antibodies are directed against DNA and other nuclear constituents, and these appear to play a central role in the disease process. DNA is only weakly antigenic, but exposure to UV radiation results in the formation of a photoproduct, UV-DNA, which is antigenic and in animals stimulates the formation of antibodies that react with UV-DNA and native DNA.[113] UV-DNA is formed in vivo after exposure to UV radiation and it can be detected in skin[114] and serum.[115] When mice with high titers of anti-UV-DNA were exposed to UVC radiation, immunoglobulin and complement were deposited in the skin in a pattern similar to that seen in patients with LE,[116] suggesting that circulating antibody to UV-DNA may play a role in the pathogenesis of the skin lesions in LE. According to this theory, UV-DNA formed as a result of exposure to radiation is released from skin and immunizes the host, who then forms antibodies to UV-DNA. On subsequent exposure to radiation, the anti-UV-DNA reacts with the photoproduct in the skin, fixes complement, and gives rise to an inflammatory response. However, somewhat at variance with this theory is the finding that NZB mice, which spontaneously develop a disease resembling LE, do not show acceleration of the process when exposed to UV radiation.[117,118]

Antibodies to UV-DNA have been found in the serums of patients with LE.[119] These antibodies show immunologic specificity for UV-DNA and are distinct from antibodies that react with native DNA.[120] However, in the latter study, there was a poor correlation between the presence of antibodies to UV-DNA and a history of photosensitivity.

IV. THERAPEUTIC PHOTOIMMUNOLOGY

Photoimmunology is still very much in its infancy as a field of research, so at first glance it might appear premature to consider any practical applications, such as in therapy of disease. However, two considerations suggest this possibility might not be too far in the future. First, much interest in photoimmunology has resulted from the use of PUVA therapy in diseases thought to have an immunologic pathogenesis, and

25. Jessup, J. M., Hanna, N., Palaszysnki, E., and Kripke, M. L., Mechanisms of depressed reactivity to dinitrochlorobenzene and ultraviolet-induced tumors during ultraviolet carcinogenesis in BALB/c mice, *Cell. Immunol.,* 38, 105, 1978.
26. Noonan, F. P., Kripke, M. L., Pedesen, G. M., and Greene, M. I., Supression of contact hypersensitivity in mice by ultraviolet irradiation is associated with defective antigen presentation, *Immunology,* 43, 527, 1981.
27. Morison, W. L. and Kripke, M. L., Systemic suppression of contact hypersensitivity by ultraviolet b radiation or methoxsalen/ultraviolet a radiation in the guinea pig, *Cell. Immunol.,* 85, 270, 1984.
28. Ullrich, S. E. and Kripke, M. L., Mechanisms in the suppression of tumor rejection produced in mice by repeated UV irradiation, *J. Immunol.,* 133, 2786, 1984.
29. Kripke, M. L., Morison, W. L., and Parrish, J. A., Systemic suppression of contact hypersensitivity in mice by psoralen plus UVA radiation (PUVA), *J. Invest. Dermatol.,* 81, 87, 1983.
30. Morison, W. L., Pike, R. A., and Kripke, M. L., Effect of sunlight and its component wavebands on contact hypersensitivity in mice and guinea pigs, *Photodermatology,* 2, 195, 1985.
31. DeFabo, E. C. and Noonan, F. P., Mechanism of immune suppression by ultraviolet irradiation *in vivo.* I. Evidence for the existence of a unique photoreceptor in skin and its role in photoimmunology, *J. Exp. Med.,* 157, 84, 1983.
32. Morison, W. L., Systemic suppression of contact hypersensitivity associated with suppressor lymphocytes: is a lesion in DNA an essential step in the pathway?, *J. Invest. Dermatol.,* 82, 421, 1984.
33. Greene, M. I., Sy, M. S., Kripke, M., and Benacerraf, B., Impairment of antigen-presenting cell function by ultraviolet radiation, *Proc. Natl. Acad. Sci. U.S.A.,* 76, 6591, 1979.
34. Letvin, N. L., Nepom, J. T., Greene, M. I., Benacerraf, B., and Germain, R. N., Loss of Ia-bearing splenic adherent cells after whole body ultraviolet irradiation, *J. Immunol.,* 125, 2550, 1980.
35. Fox, I. J., Sy, M., Benacerraf, B., and Greene, M. I., Impairment of antigen-presenting cell function by ultraviolet radiation, *Transplantation,* 31, 262, 1981.
36. Noonan, F. P., Bucana, C., Sauder, D. N., and DeFabo, E. C., Mechanism of systemic immune suppression by UV irradiation *in vivo.* II. The UV effects on number and morphology of epidermal Langerhans cells and the UV-induced suppression of contact hypersensitivity have different wavelength dependencies, *J. Immunol.,* 132, 2408, 1984.
37. Lynch, D. H., Gurish, M. F., and Daynes, R. A., The effects of high-dose UV exposure on murine Langerhans cell function at exposed and unexposed sites as assessed using *in vivo* and *in vitro* assays, *J. Invest. Dermatol.,* 81, 336, 1983.
38. Morison, W. L., Bucana, C., and Kripke, M. L., Systemic suppression of contact hypersensitivity by UVB-radiation is unrelated to the UVB-induced alterations in the morphology and number of Langerhans cells, *Immunology,* 52, 299, 1984.
39. Kripke, M. L. and Morison, W. L., Modulation of immune function by UV-radiation, *J. Invest. Dermatol.,* 85, 625, 1985.
40. Spangrude, G. J., Bernhard, E. J., Ajioka, R. S., and Daynes, R. A., Alterations in lymphocyte homing patterns within mice exposed to ultraviolet radiation, *J. Immunol.,* 130, 2974, 1983.
41. Milton, J. D., The effect of ultraviolet irradiation on the localization of syngeneic and allogeneic lymphoid cells and on alloimmunogenicity, *Immunology,* 44, 281, 1981.
42. Swartz, R. P., Role of UVB-induced serum factor(s) in suppression of contact hypersensitivity in mice, *J. Invest. Dermatol.,* 83, 305, 1984.
43. Kripke, M. L., Antigenicity of murine skin tumors induced by ultraviolet light, *J. Natl. Cancer Inst.,* 53, 1333, 1974.
44. Kripke, M. L. and Fisher, M. S., Immunologic parameters of ultraviolet carcinogenesis, *J. Natl. Cancer Inst.,* 57, 211, 1976.
45. Kripke, M. L., Lofgreen, J. S., Beard, J., Jessup, J. M., and Fisher, M. S., *In vivo* immune responses of mice during carcinogenesis by ultraviolet irradiation, *J. Natl. Cancer Inst.,* 59, 1227, 1977.
46. Spellman, C. W., Woodward, J. G., and Daynes, R. A., Modification of immunological potential by ultraviolet radiation, *Transplantation,* 24, 112, 1977.
47. Norbury, K. C., Kripke, M. L., and Budmen, M. B., *In vitro* reactivity of macrophages and lymphocytes from ultraviolet-irradiated mice, *J. Natl. Cancer Inst.,* 59, 1231, 1977.
48. Kripke, M. L., Thorn, R. M., Lill, P. H., Civin, C. I., Pazmino, N. H., and Fisher, M. S., Further characterization of immunological unresponsiveness induced in mice by ultraviolet radiation, *Transplantation,* 28, 212, 1979.
49. Fisher, M. S. and Kripke, M. L., Systemic alteration induced in mice by ultraviolet light irradiation and its relationship to ultraviolet carcinogenesis, *Proc. Natl. Acad. Sci. U.S.A.,* 74, 1688, 1977.
50. Spellman, C. W. and Daynes, R. A., Modification of immunological potential by ultraviolet radiation. II. Generation of suppressor cells in short-term UV-irradiated mice, *Transplantation,* 24, 120, 1977.
51. Fisher, M. S. and Kripke, M. L., Suppressor T lymphocytes control the development of primary skin cancers in ultraviolet-irradiated mice, *Science,* 216, 1133, 1982.

52. De Gruijl, F. R. and van der Leun, J. C., Systemic influence of pre-irradiation of a limited skin area on UV-tumorigenesis, *Photochem. Photobiol.*, 35, 379, 1982.
53. Fisher, M. S., Kripke, M. L., and Chan, G. L., Antigenic similarity between cells transformed by ultraviolet radiation *in vitro* and *in vivo, Science*, 223, 593, 1984.
54. Palaszynski, E. W. and Kripke, M. L., Transfer of immunological tolerance to ultraviolet-radiation-induced skin tumors with grafts of ultraviolet-irradiated skin, *Transplantation*, 36, 465, 1983.
55. Granstein, R. D., Parrish, J. A., McAuliffe, D. J., Waltenbaugh, C., and Greene, M. I., Immunologic inhibition of ultraviolet radiation-induced tumor suppressor cell activity, *Science*, 224, 615, 1984.
56. Fabo, E. C. and Kripke, M. L., Wavelength dependence and dose-rate independence of UV radiation-induced immunologic unresponsiveness of mice to a UV-induced fibrosarcoma, *Photochem. Photobiol.*, 32, 183, 1980.
57. Morison, W. L. and Kelley, S. P., Sunlight suppresses the rejection of UVB-induced skin tumors in mice, *J. Natl. Cancer Inst.*, 74, 525, 1985.
58. Kripke, M. L., Morison, W. L., and Parrish, J. A., Induction and transplantation of murine skin cancers induced by methoxsalen plus ultraviolet (320—400 nm) radiation, *J. Natl. Cancer Inst.*, 68, 685, 1982.
59. Kripke, M. L., Morison, W. L., and Parrish, J. A., Differences in the immunologic reactivity of mice treated with UVB or methoxsalen plus UVA radiation, *J. Invest. Dermatol.*, 76, 4445, 1981.
60. Briffa, D. V., Parker, D., Tosca, N., Turk, J. L., and Greaves, M. W., The effect of photochemotherapy (PUVA) on cell mediated immunity in guinea pig, *J. Invest. Dermatol.*, 77, 377, 1981.
61. Morison, W. L., Parrish, J. A., Woehler, M. E., Krugler, J. I., and Bloch, K. J., Influence of PUVA and UVB radiation on delayed hypersensitivity in the guinea pig, *J. Invest. Dermatol.*, 76, 484, 1981.
62. Morison, W. L. and Pike, R. A., Suppression of graft-versus-host reactivity in the mouse popliteal nody by UVB radiation, *J. Invest. Dermatol.*, 84, 483, 1985.
63. Morison, W. L., Parrish, J. A., Woehler, M. E., and Bloch, K. J., The influence of PUVA and UVB radiation on skin-graft survival in rabbits, *J. Invest. Dermatol.*, 75, 331, 1980.
64. Hersey, P., Hasic, E., Edwards, A., Bradley, M., Haran, G., and McCarthy, W. H., Immunological effects of solarium exposure, *Lancet*, 1, 548, 1983.
65. Halprin, K. M., Comerford, M., Presser, S. E., and Taylor, J. R., Ultraviolet light treatment delays contact sensitization to nitrogen mustard, *Br. J. Dermatol.*, 105, 71, 1981.
66. Nusbaum, B. P., Edwards, E. K., Horwitz, S. N., and Frost, P., The effect of UV radiation on sensitization to mechlorethamine, *Arch. Dermatol.*, 119, 117, 1983.
67. Volden, G., Molin, L., and Thomsen, K., PUVA-induced suppression of contact sensitivity to mustine hydrochloride in mycosis fungoides, *Br. Med. J.*, 2, 865, 1978.
68. O'Dell, B. L., Jessen, R. T., Becker, L. E., Jackson, R. T., and Smith, E. B., Diminished immune response in sun-damaged skin, *Arch. Dermatol.*, 116, 559, 1980.
69. Kalimo, K., Koulu, L., and Jansen, C. T., Effect of a single UVB or PUVA exposure on immediate and delayed skin hypersensitivity reactions in humans, *Arch. Dermatol.*, 275, 374, 1983.
70. Strains, G. H., Greaves, M., Price, M., Bridges, B. A., Hall-Smith, P., and Vella-Briffa, D., Inhibition of delayed hypersensitivity reaction in skin (DNCB test) by 8-methoxypsoralen photochemotherapy, *Lancet*, 2, 556, 1980.
71. Moss, C., Friedmann, P. S., and Shuster, S., Impaired contact hypersensitivity in untreated psoriasis and the effects of photochemotherapy and dithranol/UV-B, *Br. J. Dermatol.*, 105, 503, 1981.
72. Thiers, B. H., Maize, J. C., Spicer, S. S., and Cantor, A. B., The effect of aging and chronic sun exposure on human Langerhans cell populations, *J. Invest. Dermatol.*, 82, 223, 1984.
73. Gilchrest, B. A., Murphy, G. F., and Soter, N. A., Effect of chronologic aging and ultraviolet irradiation on Langerhans cells in human epidermis, *J. Invest. Dermatol.*, 79, 85, 1982.
74. Aberer, W., Schuler, G., Stingl, G., Honigsmann, H., and Wolff, K., Ultraviolet light depletes surface markers of Langerhans cells, *J. Invest. Dermatol.*, 76, 202, 1981.
75. Friedmann, P. S., Disappearance of epidermal Langerhans cells during PUVA therapy, *Br. J. Dermatol.*, 105, 219, 1981.
76. Friedman, P. S., Ford, G., Ross, J., and Diffey, B. L., Reappearance of epidermal Langerhans cells after PUVA therapy, *Br. J. Dermatol.*, 109, 301, 1983.
77. Ree, K., Reduction of Langerhans cells in human epidermis during PUVA therapy: a morphometric study, *J. Invest. Dermatol.*, 78, 488, 1982.
78. Hersey, P., Haran, G., Hasic, E., and Edwards, A., Alteration of T cell subsets and induction of suppressor T cell activity in normal subjects after exposure to sunlight, *J. Immunol.*, 31, 1, 1983.
79. Morison, W. L., Parrish, J. A., Bloch, K. J., and Krugler, J. I., *In vivo* effect of UV-B on lymphocyte function, *Br. J. Dermatol.*, 101, 513, 1979.
80. Morison, W. L., Parrish, J. A., Bloch, K. J., and Krugler, J. I., *In vivo* effect of PUVA on lymphocyte function, *Br. J. Dermatol.*, 104, 405, 1981.

81. Kraemer, K. H. and Weinstein, G. D., Decreased thymidine incorporation in circulating leukocytes after treatment of psoriasis with psoralen and long-wave ultraviolet light, *J. Invest. Dermatol.*, 69, 211, 1977.
82. Cormane, R. H., Hamerlinck, F., and Siddiqui, A. H., Immunologic implications of PUVA therapy in psoriasis vulgaris, *Arch. Dermatol. Res.*, 265, 245, 1979.
83. Morison, W. L., Parrish, J. A., Bloch, K. J., and Krugler, J. I., Transient impairment of peripheral blood lymphocyte function during PUVA therapy, *Br. J. Dermatol.*, 101, 391, 1979.
84. Harper, R. A., Tam, D. W., Vonderheid, E. C., and Urbach, F., Normal T-lymphocyte function in psoriatic patients undergoing methoxsalen photochemotherapy, *J. Invest. Dermatol.*, 72, 323, 1979.
85. Moscicki, R. A., Morison, W. L., Parrish, J. A., Bloch, K. J., and Colvin, R. B., Reduction of the fraction of circulating helper-inducer T cells identified by monoclonal antibodies in psoriatic patients treated with long-term psoralen/ultraviolet-A radiation (PUVA), *J. Invest. Dermatol.*, 79, 205, 1982.
86. Morison, W. L., Wimberly, J., Parrish, J. A., and Bloch, K. J., Abnormal lymphocyte function following long-term PUVA therapy for psoriasis, *Br. J. Dermatol.*, 108, 445, 1983.
87. Morison, W. L., Parrish, J. A., Anderson, R. R., and Bloch, K. J., Sensitivity of mononuclear cells to UV radiation, *Photochem. Photobiol.*, 29, 1047, 1979.
88. Morison, W. L., Parrish, J. A., McAuliffe, D. J., and Bloch, K. J., Sensitivity of mononuclear cells to UV radiation: effect on subsequent stimulation with phytohemagglutinin, *Photochem. Photobiol.*, 32, 99, 1980.
89. Morison, W. L., Parrish, J. A., McAuliffe, D. J., and Bloch, K. J., Sensitivity of mononuclear cells to PUVA: effect on subsequent stimulation with mitogens and on exclusion of trypan blue dye, *Clin. Exp. Dermatol.*, 6, 273, 1981.
90. Morhenn, V. B., Benike, C. J., and Engleman, E. G., Inhibition of cell mediated immune responses by 8-methoxypsoralen and long-wave ultraviolet light: a possible explanation for the clinical effects of photoactivated psoralen, *J. Invest. Dermatol.*, 75, 249, 1980.
91. Kraemer, K. H., Levis, W. R., Cason, J. C., and Tarone, R. E., Inhibition of mixed leukocyte culture reaction by 8-methoxypsoralen and long-wavelength ultraviolet radiation, *J. Invest. Dermatol.*, 77, 235, 1981.
92. Carter, D. M., Wolff, K., and Schnedl, W., 8-methoxypsoralen and UVA promote sister-chromatid exchanges, *J. Invest. Dermatol.*, 67, 548, 1976.
93. Swanbeck, G., Thyresson-Hök, M., Bredberg, A., and Lambert, B., Treatment of psoriasis with oral psoralens and longwave ultraviolet light, *Acta Derm. Venerol.*, 55, 367, 1975.
94. Wolff-Schreiner, E. C., Carter, D. M., Schwarzacher, H. G., and Wolff, K., Sister chromatid exchanges in photochemotherapy, *J. Invest. Dermatol.*, 69, 387, 1977.
95. Marshall, V. C., Skin tumours in immunosuppressed patients, *Aust. N.Z. J. Surg.*, 43, 214, 1973.
96. Koranda, F. C., Dehmel, E. M., Kahn, G., and Penn, I., Cutaneous complications in immunosuppressed renal homograft recipients, *JAMA*, 229, 419, 1974.
97. Hoxtell, E. O., Mandel, J. S., Murray, S. S., Schuman, L. M., and Goltz, R. W., Incidence of skin carcinoma after renal transplantation, *Arch. Dermatol.*, 113, 436, 1977.
98. Hardie, I. R., Strong, R. W., Hartley, L. C. J., Woodruff, P. W., and Clunie, G. J. A., Skin cancer in caucasian renal allograft recipients living in a subtropical climate, *Surgery*, 177, 1980.
99. Boyle, J., Briggs, J. D., MacKie, R. M., and Junor, B. J. R., Cancer, warts, and sunshine in renal transplant patients, *Lancet*, 1, 702, 1984.
100. Letzner, M. A., Orth, G., Detronquay, V., Ducasse, M., Kreis, H., and Crosnier, J., Detection of human papillomavirus type 5 DNA in skin cancers of an immunosuppressed renal allograft recipient, *Lancet*, 2, 422, 1983.
101. Dellon, A. L., Potvin, C., Chretien, P. B., and Rogenntine, C. N., The immunobiology of skin cancer, *Plast. Reconstr. Surg.*, 55, 343, 1975.
102. Nairn, R. C., Nind, A. P. P., Guli, E. P. G., Muller, H. K., Rolland, J. M., and Minty, C. C. J., Specific immune response in human skin carcinoma, *Br. Med. J.*, 2, 702, 1971.
103. Morison, W. L, Bucana, C., Hashem, N., Kripke, M. L., Cleaver, J. E., and German, J. L., III, Impaired immune function in patients with xeroderma pigmentosum, *Cancer Res.*, 45, 3929, 1985.
104. Horio, T. and Minami, K., Solar urticaria, photoallergen in a patient's serum, *Arch. Dermatol.*, 113, 157, 1977.
105. Sams, W. M., Jr., Solar urticaria: studies of the active serum factor, *J. Allergy*, 45, 295, 1970.
106. Hawk, J. L. M., Eady, R. A. J., Challoner, A. V. J., Kobza-Black, A., Keahey, T. M., and Greaves, M. W., Elevated blood histamine levels and mast cell degranulation in solar urticaria, *Br. J. Clin. Pharmacol.*, 9, 183, 1980.
107. Keahey, T. M., Lavker, R. M., Kaidbey, K. H., Atkins, P. C., and Zweiman, B., Studies on the mechanism of clinical tolerance in solar urticaria, *Br. J. Dermatol.*, 110, 327, 1984.
108. Soter, N. A., Wasserman, S. I., Pathak, M. S., Parrish, J. A., and Austen, K. F., Solar urticaria: release of mast cell mediators into the circulation after experimental challenge, *J. Invest. Dermatol.*, 72, 282, 1974.

109. Harber, L. C. and Baer, R. L., Mechanisms of drug photosensitivity reactions, *Toxicol. Appl. Pharmacol. Suppl.*, 3, 58, 1969.
110. Maguire, H. C. and Kaidbey, K., Experimental photoallergic contact dermatitis: a mouse model, *J. Invest. Dermatol.*, 79, 147, 1982.
111. Takigawa, M. and Miyachi, Y., Mechanisms of contact photosensitivity in mice. I. T cell regulation of contact photosensitivity of tetrachlorosalicylanilide under the genetic restrictions of the major histocompatibility complex, *J. Invest. Dermatol.*, 78, 108, 1982.
112. Granstein, R. D., Morison, W. L., and Kripke, M. L., The role of suppressor cells in the induction of murine photoallergic contact dermatitis and in its suppression by prior UVB irradiation, *J. Immunol.*, 130, 2099, 1983.
113. Levine, L., Seaman, E., Hammerschlag, E., and van Vunakis, H., Antibodies to photoproducts of deoxyribonucleic acids irradiated with ultraviolet light, *Science*, 153, 1666, 1966.
114. Tan, E. M. and Stoughton, R. B., Ultraviolet light alteration of cellular deoxyribonucleic acid *in vivo*, *Proc. Natl. Acad. Sci. U.S.A.*, 62, 708, 1969.
115. Tan, E. M., Production of potentially antigenic DNA in cells, in *Immunopathology 6th Int. Symp.*, Miescher, P. A., Ed., Grune & Stratton, New York, 1971, 346.
116. Natali, P. G. and Tan, E. M., Experimental skin lesions in mice resembling systemic lupus erythematos, *Arthritis Rheum.*, 16, 579, 1973.
117. Davis, P. and Percy, J. S., Effect of ultraviolet light on disease characteristics of NZB/W mice, *J. Rheumatol.*, 5, 126, 1978.
118. Strickland, P. T., Photocarcinogenesis and influence of UV radiation on autoimmune disease in NZB/N mice, *J. Natl. Cancer Inst.*, 73, 537, 1984.
119. Tan, E. M., Antibodies to deoxyribonucleic acid irradiated with ultraviolet light: detection by precipitins and immunofluorescence, *Science*, 161, 1353, 1968.
120. Davis, P., Russell, A. S., and Percy, J. S., Antibodies to UV light denatured DNA in systemic lupus erythematosus: detection by filter radioimmunoassay and clinical correlations, *J. Rheumatol.*, 3, 375, 1976.
121. Morison, W. L., Parrish, J. A., and Fitzpatrick, T. B., Oral psoralen photochemotherapy of atopic eczema, *Br. J. Dermatol.*, 98, 25, 1978.
122. Lau, H., Reemtsma, K., and Hardy, M. A., Prolongation of rat islet allograft survival by direct ultraviolet irradiation of the graft, *Science*, 223, 607, 1984.
123. Glazier, A., Morison, W. L., Bucana, C., Hess, A. D., and Tutschka, P. J., Suppression of epidermal graft-versus-host disease with ultraviolet radiation, *Transplantation*, 37, 211, 1984.
124. Hymes, S. R., Morison, W. L., Farmer, E. V., Walters, L. L., Tutschka, P. J., and Santos, G. W., Methoxysalen and ultraviolet a radiation in the treatment of chronic cutaneous graft-versus-host reaction, *J. Am. Acad. Dermatol.*, in press, 1984.
125. Hauser, S. L., Weiner, H. L., Che, M., Shapiro, M. E., Gilles, F., and Letvin, N. L., Prevention of experimental allergic encephalomyelitis (EAE) in the SJL/J mouse by whole body ultraviolet irradiation, *J. Immunol.*, 132, 1276, 1984.

Chapter 7

PHOTOSENSITIVITY TO DRUGS

John H. Epstein and Bruce U. Wintroub

TABLE OF CONTENTS

I. INTRODUCTION

Photosensitivity reactions have become increasingly common in the past several decades due not only to the general public's desire to develop a "good sun tan" but also due to the increasing amounts of photosensitizing chemicals in our environment. A significant number of these chemicals are therapeutic agents, and these will be the primary concern of this discussion.

Before we examine the problems of photochemical toxicity in human skin, however, perhaps a brief discussion of certain basic concepts might be in order. Let us start with the basic law of photobiology. This law states that nonionizing radiation must be absorbed to produce a photochemical, and thus the resultant photobiological, reaction. This law then stipulates that the photosensitizer, chromophore, radiation absorber, or whatever term one wishes to use must be present at the time of radiation for a reaction to occur.

The action spectrum of a photochemical reaction and the resultant photobiological reaction consists of the wavelengths absorbed by a molecule which leads to a specific response. The action spectrum usually, but not always, simulates the absorption spectrum of a molecule. However, it must always be included in the absorption spectrum, since this radiation must be absorbed to produce a reaction.

"Photosensitivity" is the broad term used to connote adverse reactions to this radiation.[1] Two types of reactions may occur — phototoxic or photoallergic. The phototoxic response can be divided into 2 subgroups — photodynamic and nonphotodynamic. Simplistically, the differentiation relates to the need of oxygen in the photodynamic process, though the process is much more complex.[2]

Phototoxicity is common. It will occur in everybody if enough of the chromophore or photosensitizers are present and the skin is exposed to a sufficient amount of the appropriate wavelengths. Clinically, the responses usually are comprised of a delayed erythema with or without edema, followed by hyperpigmentation and desquamation. Thus they resemble the ordinary UVB radiation (between 280 or 290 nm and 315 or 320 nm) sunburn response. In fact, the ordinary UVB sunburn is by far the most common phototoxic reaction that is seen.

In contrast, photoallergy is uncommon.[3-5] It is an acquired, altered reactivity, presumably dependent on an antigen-antibody or a cell-mediated hypersensitivity. Clinically, it is characterized by unusual reactions. These reactions include immediate urticarial responses and delayed papular to eczematous lesions.

Almost all of the well-documented photobiological events that occur in the skin are induced by UVB radiation. Such radiation can inhibit DNA, RNA, and protein synthesis and mitosis formation; labilize lysosomal membranes; stimulate formation or release of vasoactive substances such as prostaglandins and kinins; induce a delayed erythema and new pigment formation; cause cancer; and produce vitamin D precursors in the epidermis. For practical purposes, this radiation does not pass through window glass. Thus, a good sturdy piece of window glass will eliminate all of these cutaneous responses.

UVA radiation (315 or 320 to 400 nm) also produces a few minor photobiological events, such as immediate pigment darkening, due to photo-oxidation of melanin, new pigment formation, and a transient immediate and a delayed erythema response. In addition, UVA radiation is important for two reasons. UVA radiation markedly accentuates the injury produced by the UVB radiation and is responsible for the vast majority of exogenously photosensitized reactions that occur in the skin.

It is with this last issue (i.e., chemical photosensitivity) that this chapter is concerned.

II. EXOGENOUS PHOTOSENSITIZERS

Exogenous chemical photosensitizers may arrive on the skin topically or systemically through the vascular system.

The topical exogenous photosensitizers may be placed in the following categories:

1. Cosmetics: perfumes, creams, soaps, etc.
2. Medications
   A. Advertent: psoralens, tars
   B. Inadvertent: phenothiazines, sulfonamides, halogenated salicylanilides (and related compounds), sunscreens, etc.
3. Plants: gas plant, lime rinds, celery, carrots, etc.

Topical exogenous photosensitizers may be applied to the skin in the form of cosmetics, perfumes, moisturizing creams, soaps, and the like. Certain medications, some of which are intended to produce photoreactions, e.g., the psoralen compounds and occasionally tar, are applied to the skin.

Other medications arrive on the skin without the intent to photosensitize. These include phenothiazines, sulfonamides, antibacterial agents such as the halogenated salicylanilides and related compounds, and sunscreens.

A number of plants contain photosensitizing chemicals, primarily psoralen compounds, which can and do induce photosensitive reactions. These plants include carrots, celery (especially celery infected with pink rot disease), gas plant, lime rinds, and the like. The primary photosensitizers in these plants are psoralen compounds, though other photosensitizers are found in other plants.[6]

Industrial emissions constitute another source of topical photoactive chemicals. These agents are primarily polycyclic hydrocarbons that can act as photosensitizers as well as additive carcinogens.[2,7] In addition, potential promoters of photocarcinogenesis may be found in industrial emissions.

The commonly used systemic photosensitizers may be placed in the following categories:

1. Antibacterial sulfonamides
2. Thiazide diuretic medications
3. Sulfonylurea antidiabetic medications
4. Phenothiazines
5. Antibiotics — especially demeclocycline

Photosensitizing chemicals that reach the skin through the vasculature are generally medications. The number of systemic photosensitizers is immense. They include potent photosensitizers such as the psoralen compounds that are usually employed to produce photoreactions, to weak photosensitizers such as the antihistamine Benadryl®. Included in this list of commonly used systemic photosensitizers are antibacterial sulfonamides and their relatives, the thiazide diuretics, and the sulfonylurea antidiabetic drugs. The phenothiazines, especially chlorpromazine and certain antibiotics, are also included in this group. More complete lists of potential topical and systemic photosensitizers are available.[8,9]

III. PHOTOTOXICITY

As noted, the most common photosensitivity reactions to exogenous chemicals are phototoxic in nature. A significant number of these chemicals are therapeutic agents, i.e., drugs. Drug-induced phototoxic reactions can occur in 100% of people following the first exposure to the chemical and a dose response can be demonstrated with the incidence of phototoxicity being related to the concentration of the sensitizers and the amount of the appropriate radiation. This section will focus on the clinical expression and mechanism of phototoxicity due to topically or systemically administered drugs and will consider only briefly phototoxicity associated with certain metabolic disorders, such as the porphyrias, in which a product with photosensitizing properties accumulates in the skin as a result of an inborn error of metabolism.

A. Clinical Reactions

Acute phototoxic responses are by far the most common reactions to these agents. These responses are usually characterized by a delayed erythema and edema followed by hyperpigmentation and desquamation or peeling. If the chemical is applied to the skin, the epidermis may be severely damaged, such as is seen in topical psoralen photosensitization. In contrast, if the agent reaches the skin through the vasculature, the primary site of injury may be in the dermis with little or no visible epidermal change. In both instances, the inflammatory cellular response is very slight in the dermis of human skin.[8]

This is similar to the lack of cellular infiltration noted in human cutaneous acute UVB sunburn reactions. When the tissue is examined at the submicroscopic level, vascular injury can be detected in responses to systemic photosensitizers such as protoporphyrins.[10,11]

Host effects may influence the natural history of drug-induced phototoxic reactions. Several factors, including hair, pigment, and thickness of the stratum corneum, influence radiation penetration. Topical photosensitizers are dependent on percutaneous absorption and metabolism in the skin.[12] Systemic chromophores are influenced by gastrointestinal absorption, distribution, and metabolism of the chemicals. In addition, increased humidity, temperature, and wind may also enhance phototoxic responses.[13,14] However, recent studies suggest that heat may inhibit phototoxic reactions.[15,16] The reasons for these discrepancies are unknown.

Certain exogenous chemicals such as tetracycline hydrochloride, nalidixic acid, and the sulfones can produce low-grade phototoxic injuries that are expressed as blistering, fragility, denudation, scarring, and milium formation.[17,18] These changes are identical (clinically and microscopically) with those found in a number of the porphyrias.[19]

Chronic actinic damage and skin cancer formation are induced primarily by the UVB rays from the sun. In general, this process in humans has not been associated with drugs. However, such a relationship with certain forms of photochemotherapy has recently been established. In a large series of psoriatic patients treated with psoralens and UVA radiation, Stern et al.[20] noted a definite increase in cutaneous cancer formation. This increase was even more clinically significant because the cancers were found primarily on parts of the body usually not chronically exposed to the sun. In addition, there is histologic evidence of focal dystrophy of epidermal cells, atypical changes in melanocytes, and deposition of colloid bodies and amyloid at the dermoepidermal junction.[21]

Stern et al.[22] also reported an increased skin cancer incidence in psoriatic patients who had received very extensive treatments with coal tar and UVB radiation (the Goeckerman regimen). This may represent an additive carcinogenic effect between the tar and UVB radiation energy, since the action spectrum for coal tar photosensitization is in the UVA radiation range. Thus, patients with psoriasis may be at increased risk of developing skin cancer because of certain treatment modalities in wide use at the present time.

B. Mechanisms

The interactions of light, phototoxic chemicals, and their biologic targets may result from several types of photochemical reactions. In each case, damage to biological substrates is initiated by the absorption of light energy by a phototoxic compound. Historically, phototoxic reactions have been divided into two categories: oxygen-dependent (photodynamic) and oxygen-independent (nonphotodynamic).[2,8] In the case of drug phototoxicity, recent studies have resulted in a more detailed understanding of the various mechanisms of phototoxicity.

The action spectrum for these reactions usually includes at least the UVA range, though certain molecules such as porphyrins and a number of dyes photosensitize to visible light.[2] In addition, a limited number of chemicals are activated by UVB rays.[23,24]

Photodynamic processes appear to be responsible for reactions to certain dyes, coal tar, polycyclic hydrocarbons, and porphyrin molecules.[2] Photodynamic reactions may induce an excited triplet state which reacts with oxygen, forming a singlet oxygen or superoxide anion.[25-27] This activated oxygen can then damage cell components. The available evidence indicates that this process occurs in the endothelial cells of the superficial dermal blood vessels of the skin in the porphyrias.[10,11,19,28] In addition, there is evidence that porphyrin photosensitization can lead to activation of the classical complement pathway and generate chemotactic activity derived from the fifth component of complement.[29,30] It should be noted that demethylchlortetracycline photosensitization can also cause complement activation and generate chemotactic factors.[31]

Nonphotodynamic reactions have been described for psoralen photosensitization.[32-34] In this instance, certain psoralen compounds intercalate into the DNA helix. On photoactivation with UVA radiation, mono- and bifunctional adducts may be formed in the DNA, depending on the structure of the psoralen molecule and the wavelengths utilized for irradiation. Photoreactions with proteins may also occur. It should be noted that psoralens can produce photodynamic responses as well.[34] Psoralen photoinduced singlet oxygen formation has been correlated with erythema production by several psoralens, suggesting that these photodynamic responses may be responsible for some of the effects of psoralen photosensitization in human skin.

Photoinduced DNA cross-links have been reported with coal-tar extracts and anthracene.[32] Also, lysosomal labilization has been noted with anthracene as well as porphyrins.[28] How this injury translates into erythema, edema, and eventual hyperpigmentation remains to be determined. Another established mechanism concerns the production of toxic photoproducts which can cause membrane damage and result in erythematous cutaneous responses. Recent studies indicate that photoreactions to certain chemicals, such as chlorpromazine and protryptyline, may be due at least in part to this type of mechanism.[35]

The erythrogenic mediators of these reactions are not well understood. Unlike UVB phototoxicity, prostaglandins do not appear to be involved in 8-methoxypsoralen (8-MOP) and UVA photoreactions. Anthracene phototoxic-induced hyperemia in mouse skin appears to be mediated by histamine.[36] Recently, phototoxic-induced immediate erythema and edema reactions to hematoporphyrin[37] and benoxaprofen[38] have been shown to be related to mast-cell degranulation. In the case of benoxaprofen, the degranulation appears to be due to photosensitization of the membrane by a lipophilic photoproduct of the parent chemical.[39] How these findings relate to the erythematous responses to photoinjury induced by other molecules remains to be determined. In general, much more information is needed to clarify the mechanisms which lead to the changes we see during phototoxic cutaneous responses.

C. Phototoxic Agents of Special Interest

Certain frequently used orally and topically administered agents are known to produce phototoxicity. In certain cases phototoxicity is employed for therapeutic benefit, as in the case of psoralen and tar, while in other cases, phototoxicity is a distinctly adverse event. In this section, those drugs which are commonly used will be reviewed with respect to clinical problems and mechanisms of phototoxicity.

1. Psoralen Phototoxicity

Furocoumarins are a class of compounds to which psoralens belong, and these compounds consist of a double-ringed coumarin moiety to which a furan ring is attached. At least 28 different furocoumarins have been isolated from natural sources, but only 4 are clinically useful: psoralen, 8-MOP, 5-methoxypsoralen, and 4,5,8-trimethylpsoralen. The absorption maximums of psoralens range between 210 and 330 nm, but the erythema action spectrum is in the UVA range from 320 to 370 nm.

Psoralens combined with sun exposure have been used to treat vitiligo for centuries. The availability of high intensity, artificial, UVA light sources in the past decade has permitted the use of psoralens in combination with controlled doses of UVA light.[40] The acronym PUVA (psoralen and UVA) photochemotherapy is used to indicate this form of dermatologic therapy. Three major dermatologic diseases are frequently treated with PUVA — vitiligo, psoriasis, and mycosis fungoides.

The mechanism of psoralen photosensitization has been intensively studied. Nonphotodynamic reactions have been described for psoralen photosensitization.[32] In this instance, certain psoralen compounds intercalate into the DNA helix.[33,34]

On photoactivation with UVA radiation, mono- and bifunctional adducts may be formed in the DNA, depending on the structure of the psoralen molecule and the wavelengths utilized for irradiation.[32-34] Photoreactions with proteins may also occur. It should be noted that psoralens can produce photodynamic responses as well.[34] Psoralen photoinduced singlet oxygen formation has been correlated with erythema production by several psoralens, suggesting that these photodynamic responses may be responsible for some of the effects of psoralen photosensitization in human skin.[34]

2. Chlorpromazine Phototoxicity

Chlorpromazine is a major tranquilizer frequently utilized in psychiatric patients. Clinical photosensitivity to chlorpromazine was reported in the 1950s[41] and the drug is now known to be phototoxic in a variety of in vitro systems including RBC, bacteria, mammalian cells, bacteriophage, and viruses.[42] Chlorpromazine causes several photodermatoses when taken systemically. An exaggerated sunburn reaction (phototoxicity) and a hyperpigmentation of purple or slate-gray tone in light-exposed skin have been described. The wavelength range required to elicit a clinical reaction has been a topic of controversy. Chlorpromazine has an absorption maximum at 305 nm in aqueous solution, but the action spectrum for phototoxicity has been reported to be below[41] and above 320 nm in man.[43] For phototoxicity in mice, the action spectrum has been reported to be 320 to 340 nm with a maximum of 330 nm.[44]

The mechanisms of the chlorpromazine phototoxicity have been studied extensively and at least two possible mechanisms have been demonstrated in vitro. Photoaddition of chlorpromazine to protein, DNA, and cell membranes has been reported.[45] In addition, it appears that stable photoproducts of chlorpromazine elicit cutaneous toxicity in animals and hemolyse RBC.[35]

In both systems, toxicity was detected after induction by preirradiated chlorpromazine. In the case of red cell hemolysis, the stable lysis-producing product did not require oxygen for activity nor cause lipid oxidation in the presence of oxygen. Thus, chlorpromazine has been shown to be phototoxic in many systems, but the route of translation of these in vitro effects to human phototoxicity is unclear.

3. Tetracycline Phototoxicity

Of the tetracyclines, demethylchlortetracyline (Declomycin®), is the most potent phototoxic agent. It causes immediate and delayed phototoxic reactions in most individuals, according to dose.[46,47] Human studies indicate that the action spectrum for demethylchlortetracycline phototoxicity is above 320 nm, and studies in mice demonstrated an effective radiation range of 350 to 450 nm with greatest response at 400 nm. Recent experiments clearly have demonstrated a requirement for complement and polymorphonuclear leukocytes as amplifiers or effectors of demethylchlortetracycline phototoxicity.[31]

4. Nalidixic Acid Phototoxicity

Nalidixic acid is a bacteriostatic antibiotic frequently used to treat urinary tract infections. In association with extensive sun exposure, nalidixic acid caused a characteristic bullous eruption, fragility, scarring, and milium formation in light-exposed areas, particularly the lower legs and feet.[17] Similar reactions have been noted for tetracycline hydrochloride and sulfones.[18] While the mechanisms of phototoxicity are not known, the clinical and histopathologic changes are identical with those found in a number of the porphyrias.[19] Recently Keane et al.[48] reported inducing such lesions in CF-1 female mice with repeated UVA exposures following photosensitization by nalidixic acid.

5. Phototoxicity due to Nonsteroidal Anti-Inflammatory Agents

In the past 10 years, many new nonsteroidal anti-inflammatory drugs have been introduced and are widely used to treat a variety of inflammatory and symptomatic musculoskeletal disorders. Though photosensitivity reactions have been reported to a number of these agents,[49,50] two of these agents, benoxaprofen (Oraflex®) and piroxicam (Feldene®) commonly produce phototoxic reactions.

Benoxaprofen is a proprionic-acid derivative which was introduced into European and U.S. markets for treatment of rheumatoid arthritis, osteoarthritis, ankylosing spondylitis, and psoriasis. The drug was suspended because it caused fatal cholestatic hepatitis, especially in the elderly. Unlike other nonsteroidal anti-inflammatory drugs which act principally by inhibiting prostaglandin synthetase, benoxaprofen is a 5-lipoxygenase inhibitor. Benoxaprofen caused side effects in 65% of patients, and cutaneous reactions were the most common, accounting for 70% of all side effects.[51] The most common cutaneous side effect was photosensitivity, which occurred in 29% of patients on drug. The action spectrum of the phototoxic reaction is 310 to 340 nm. The reaction is manifested as immediate itching, burning, and erythema.[52] The mechanism of this phototoxicity is unknown; however, mast-cell degranulation is a prominent feature of this process.[38,39]

Piroxicam (Feldene®) is an oxicam which inhibits prostaglandin synthetase and is used for treatment of rheumatoid arthritis, osteoarthritis, gout, and ankylosing spondylitis. While the incidence of cutaneous side effects is approximately 2.4%, a unique photosensitivity due to piroxicam has been reported in 23 patients.[50] The photosensitivity reaction is usually vesiculobullous and is often pruritic. The mechanism of piroxicam photosensitivity is unknown.

D. Assays for Phototoxicity to Drugs

Agents capable of producing phototoxic effects have been identified by use of a variety of in vitro and in vivo systems. However, the ultimate issue concerns the effect these molecules have on human tissue in vivo.[53,54] In general, the following four techniques have been utilized: systemic administration, topical application with occlusion, topical application after the stratum corneum is stripped off, and intradermal injection of the molecules. Each of the techniques has disadvantages, among which are lack of distribution to the skin of the active photosensitizing molecule, metabolism of the chemical to a nonphotosensitizing structure, lack of metabolism to an active photosensitizing structure, lack of penetration to the appropriate target tissue, lack of irradiation with the appropriate wavelengths, and lack of irradiation at the appropriate time.

IV. PHOTOALLERGY

A. Clinical Expression

Photoallergic reactions are uncommon and are acquired altered reactivities dependent on antibody-dependent or cell-mediated hypersensitivity. Clinically, photoallergic reactions present as an immediate wheal and flare or a delayed papular to eczematous dermatitis.[3,4,8,55,56] Delayed reactions are microscopically characterized by a dense dermal perivascular round-cell infiltrate similar to that seen in allergic contact dermatitis. Though vasodilitation and edema are present in the immediate reaction, these are difficult to discern histologically.

B. Photoallergy to Exogenous Chemicals

The skin is the site of photoallergic reactions to exogenous chemicals which may be applied topically or administered systemically. As with phototoxicity, in the photoallergic reactions to these chemicals the action spectrum usually falls in the UVA range. The immunologic basis for these reactions is supported primarily by studies on contact photoallergy. Photoallergic reactions to drugs and other exogenous chemicals are almost always of a cell-mediated type.

1. Photoallergic Contact Dermatitis

a. Clinical Description

Photoallergic contact reactions are clinically identical to any other type of allergic contact dermatitis. Though the most common picture is eczematous in nature, the spectrum of the possible responses may range from a simple erythema to a severe vesiculobullous eruption. When the process is chronic, lichenification results from repeated mechanical trauma (rubbing and scratching). The sun-exposed areas of the skin are involved primarily, as would be expected. However, the eruption may extend to unexposed parts of the body and even become generalized. Even when this occurs, the dermatitis is most marked in light-exposed sites.

Adults are affected much more commonly than children, and reactions occur in populations that are exposed to sunlight and photocontactants. For example, reactions to the optical brighteners more likely would be seen in persons who use them, and those to chlorpromazine more likely would occur in people who work in mental institutions. The most extensive statistical data have been compiled on the photocontact reactions induced by the halogenated salicylanilides and related antibacterial compounds because of the large number of people photosensitized by these agents between 1960 and 1970.[57] For unknown reasons, reactions occurred predominantly in men past the age of 40 years. Skin color and race apparently have little influence on the problem, which occurs readily in blacks and Asians as well as in Caucasians.

b. Persistent Light Reactions

In general, removal of the offending photosensitizer and related compounds will eliminate the problem. A small percentage of patients continue to develop the dermatitis on sun-exposed areas without apparent further contact with the offending agent or related structures. These patients are persistent light reactors.[58] Up to 25% of patients with photocontact reactions to the halogenated salicylanilides have become persistent light reactors. In studies with antibacterial halogenated salicylanilides, two types of persistent light reactors were found: a mild variety that loses its reactivity within 1½ years and a severe type that persists indefinitely.[59]

Chemicals other than the antibacterial halogenated salicylanilides induce persistent light reactions. These include the antifungal compounds buclosamide (Jadit),[60,61] chlorpromazine,[62-64] promethazine,[65] musk ambrette,[66] and epoxy resins.[67]

The mechanism or mechanisms of persistent light reactivity are not clear. Cross reaction to unknown photocontactants has been considered. Willis and Kligman[68,69] presented evidence suggesting that the reactions could result from the retention of small amounts of the photosensitizer in the dermis. Photoinduced covalent binding of the halogenated salicylanilide to albumin has been demonstrated.[5] Other possibilities include the persistence of a sensitized mononuclear infiltrate in the dermis, which will react on minimal antigenic exposure; hypersensitivity to the protein component of the complete antigen, which then becomes an independent photoallergen; or the development of clones of cells that are persistently sensitive to a number of photoallergens.[57] A possible relationship of persistent light reaction to the chronic eczematous type of polymorphous light eruption (PMLE) has been reported.[59] The histology of persistent light reactions is nonspecific and resembles allergic contact dermatitis.

c. *Differential Diagnosis*

The differential diagnosis includes any eruption that may involve the sun-exposed skin. These include allergic contact dermatitis, the eczematous type of polymorphous light reactions, and phototoxic reactions. Airborne allergens contact exposed skin, but unlike photocontactant reactions, the airborne contact eruption will be accentuated in skin-fold areas such as the upper eyelids, antecubital fossae, and flexural areas of the wrist. Differentiation from eczematous polymorphous light reactions and allergic contact reactions to sunscreens, "sun tanning" lotions and creams, and medications used to relieve the discomfort of a sunburn is more difficult since the reactions are confined to the sun-exposed areas and are indistinguishable clinically and histologically. Therefore, the most useful diagnostic tools are the patch, photopatch, and phototesting procedures.

The diagnosis of photocontact dermatitis is suspected by the clinical picture, and confirmation and identification of the offending chemical depend on photopatch testing. This is accomplished by the application in duplicate of nonirritating concentrations of the potential photosensitizers in appropriate vehicles (i.e., a 1% concentration of the halogenated salicylanilides in petrolatum). The use of an extra layer of black paper over the patches will help prevent a "masked" positive reaction to the photopatch test.[70] This is unnecessary with the use of aluminum patches. Twenty-four hours after application one set of patches is irradiated with the UVA light. Any light source that emits sufficient amounts of UVA light can be utilized.[71] Window-glass filtration is necessary if there is a significant amount of UVB emitted by the lamp, as in the case of the hot-quartz source. Twenty-four hours after irradiation, the unexposed and exposed sites are compared. A positive reaction to the photopatch tests reproduces the clinical lesions morphologically and histologically.

Difficulties with patch testing may occur. If the patient is contact allergic to a chemical, it may be difficult to determine if he or she is photocontact allergic as well. However, in general, the photopatch test site will be much more reactive than the patch-test area in a patient with dual sensitivity. Another problem is identification of the potential photoallergen. Although the history may be helpful in this determination, the patients often are not aware of the contact (except for suntan oils or the like). This is often true of preservatives or ingredients in soaps. Therefore, clinicians use the following series of compounds which commonly cause photocontact allergy, and add whatever can be determined from the history: 5% amyl dimethyl para-aminobenzoic acid

(PABA), glyceryl PABA, 5% musk ambrette, octyl dimethyl PABA, oxybenzone, dioxybenzone, 0.25% tetrachlorosalicylanilide (TCSA) and 1% TBS. They also routinely patch test with the screening materials suggested by the North American Contact Dermatitis Groups (NACDG)[72] to evaluate other potential contactants and phototest to examine for PMLE.[73]

d. Photocontactants

Photoallergic reactions to many topically contacted chemicals have been reported. These include sulfonamides,[55,56,74] phenothiazines,[43,56,62,63,65,75-79] sulfonylcarbamides,[80] men's colognes and aftershave lotions,[81] blankophores (optical brighteners),[82] Persian lime rind,[83] ragweed,[84] sunscreens,[85-93] diphenhydramine,[94] psoralens,[95,96] epoxy resins,[67] and quindoxin.[97,98] The halogenated salicylanilides and related antibacterial and antifungal compounds represented the most important group of allergic contact photosensitizers. In the 1960s, these chemicals were responsible for numerous photoallergic reactions. Perhaps TCSA was the most potent photosensitizer of this group. Between 1960 and 1961 it was responsible for an estimated 10,000 cases in England,[99] and it was therefore removed from general use. Subsequently, a number of related phenolic compounds were incorporated into soaps and other vehicles to combat infection, reduce body odor, act as preservatives, and destroy fungi. Photocontact reactions were induced by many of these agents, including bithionol, the brominated salicylanilides, hexachlorophene, dichlorophene, fentichlor (bis [2-hydroxy-5-chlorophenyl] sulfide), Multifungin (bromochlorosalicylanilide), Jadit (a mixture of buclosamide and salicylic acid),[3,57] and chloro-2-phenylphenol.[100]

There has been a rapid decline in the induction of photocontact dermatitis by the halogenated salicylanilides and related compounds since 1968.[101] This is due to the removal of the more potent of these photosensitizers from general use. However, within the past 5 years, there have been two new contactants which appear to produce photoallergic responses. These are the two widely used fragrance compounds 6-methylcoumarin[102,103] and musk ambrette.[66,104,105] In addition, musk ambrette has been reported to be responsible for the induction of a persistent light-reactor state similar to that noted with the halogenated salicylanilides.[66] The action spectrum for the photoallergic reactions to these fragrances falls in the UVA range.[102,103,106]

e. Mechanism

The clinical appearance, histology, and photopatch test responses suggest that the photoreactions are dependent on cell-mediated immunity (CMI). Experimental studies utilizing animal and human models support this concept.

Photoallergic reactions in human skin characteristic of CMI responses have been induced using multiple agents including sulfonamides,[107] phenothiazines,[63] and the halogenated salicylanilides.[69] Perhaps the most extensive studies have been accomplished with the halogenated salicylanilides. Willis and Kligman[69] utilized UVB as well as UVA exposures plus the chemicals to induce photosensitivity, but only UVA and the halogenated salicylanilides elicited the photocontact allergy. No evidence of antibody-mediated hypersensitivity has been detected in patients with photocontact reactions.

Animal and human studies have confirmed the CMI mechanism for photocontact allergy reactions; however, the nature of the antigen has not been settled. Two mechanisms are possible, and both may be correct. The studies of Burckhardt and Schwarz-Speck,[80] Jung and Schwarz,[108] and Schwarz and Speck[74] with sulfonamide and related compounds; Epstein and Enta,[109] Jung and Schultz,[110] and Willis and Kligman[69,111] with the halogenated salicylanilides; and Fulton and Willis[95] with methoxypsoralen suggested that the haptens were stable photoproducts of these chemicals. An alternative concept is that the photoproducts might well be shortlived free radicals, which attach the protein carrier within microseconds to form the complete antigen.[112] Support for this theory has developed from clinical observations.[3,57,113] In addition, Jung's studies of postirradiation free-radical formation and subsequent binding to albumin and β-globulin of chlorpromazine,[114] and triplet-state induction by radiation of triacetyldiphenolisatin (TDI) with deactivation by binding to albumin, β-globulin, and skin protein,[115] present further evidence in favor of the latter concept. The studies of Jung et al.[116,117] with Jadit were even more supportive. These workers demonstrated protein binding in vitro after irradiation. The protein complex then acted as a full antigen on plain patch testing of Jadit-photosensitive patients.

f. Experimental Predictive Testing Studies

Determination of the potential for a chemical to induce allergic photocontact sensitization has depended on demonstration of such reactions in patients who clinically have developed photocontact allergies to the chemical. Though animal models are readily available for phototoxicity studies, their use in evaluation of possible photoallergic reactions has presented a complicated problem. In fact, no animal models have proved to be valuable for predictive screening of photoallergic potential of chemicals.

Animal models are of great value in the study of photocontact reactions. Schwarz and Speck[74] first reported the induction of photoallergic contact dermatitis to sulfanilamide in guinea pigs. Subsequently, photoallergic contact reactions in guinea pigs have been induced in TCSA and related compounds, chlorpromazine, musk ambrette, and 6-methylcoumarin.[118,119] In vitro and in animals, immunologic studies have supported the cell-mediated nature of the hypersensitivity.[57,116,117,120] More recently, mouse models have been shown to be most efficient for experimentally examining this problem.[121,122] Using such animals, demonstration of the necessity of Langerhans cells, apparently to process the photoallergen appropriately; the need for genetically dependent T cells for the development of the reactivity; and the inhibiting influence of UVB-induced suppressor T cells have been established under experimental conditions.[123-125]

g. Human Experimental Model

Predictive testing in human skin is not definitive. In the 1960s, identification of TCSA and related phenolic compounds was accomplished by photopatch testing clinically involved patients. Subsequently Willis and Kligman[69] induced contact photoallergy to certain of these agents in normal human subjects using a modification of the maximization test which was developed for evaluating the potential for chemicals to produce contact dermatitis.[126] Recently, Kaidbey and Kligman[127,128] further modified the photomaximization procedure, and were able to photoallergic contact sensitize normal human volunteers to certain methylated coumarin derivatives, TCSA, 3.5-DBS, chlorpromazine, and sodium omadine. A lesser number of positive-induction responses were noted with TBS contaminated with 47% DBS, 4,5-DBS, Jadit, and bithionol. However, using the same test, negative results were noted with PABA and musk ambrette which have produced photoallergic contact reactions clinically. The authors considered them weak photosensitizers. Thus, to date, there is no proven effective predictive testing model for photoallergic contact dermatitis.

2. Systemic Drug Photoallergy

Photoallergic reactions to systemic photosensitizers are less common than those induced by contactants, and they are less well understood. However, a few apparent photoallergic reactions have been reported to a number of agents including the antibacterial sulfonamides,[74,107,129-131] quinethazone (Hydromox),[132] sulfonylurea antidiabetic agents,[133] the chlorothiazide diuretics,[134] phenothiazines,[135,136] oral contraceptive hormones,[137] griseofulvin,[138] chlordiazepoxide hydrochloride (Librium®),[139] several antihistamines,[140] triacetyldiphenolisatin,[141] the artificial sweetener calcium cyclamate,[142-144] chloroquine,[145] quinidine,[146,147] carprofen[148] and 5-fluorocytosine.[149,150]

The reaction time of these responses is delayed in nature, and the clinical eruptions range from lichenoid papules to eczematous changes. Though the problems are generally short-term, occasionally persistent light reactivity may be induced by systemic photosensitizers.[55]

V. THERAPY OF DRUG-INDUCED PHOTOSENSITIVITY

Therapy of acute photoallergic or phototoxic responses is identical to that used for any such inflammatory reactions. This includes the use of topical cool wet dressings, soothing shake lotions, topical corticosteroids, and systemic antipruritus agents. If the process is severe enough, systemic corticosteroids may be indicated. As with any topical or systemic drug-induced eruption, removal of the offending agent is essential to the cure of the allergic process. In case of photoallergy or phototoxicity, either the sun or the chemical could be removed. Because exposure to radiant energy is difficult to control, removal of the chemical is the most appropriate approach. Rarely, the medication is indispensible and in such cases avoidance of the sun is essential.

Some patients develop persistent light eruptions which persist after discontinuing use of the offending agent. When this occurs, the patient usually becomes markedly photosensitive. Long-term anti-inflammatory therapy and avoidance of any sun exposure may be necessary to control such a process.

REFERENCES

1. Epstein, J. H., Phototoxicity and photoallergy, in *Sunlight and Man,* Pathak, M. A., Harber, L. C., Seiji, M., and Kukita, A., Eds., University of Tokyo Press, Tokyo, 1974, 459.
2. Blum, H. F., *Photodynamic Action and Diseases Caused by Light,* Reinhold, New York, 1941.
3. Epstein, J. H., Photoallergy. A review, *Arch. Dermatol.,* 106, 741, 1972.
4. Morison, W. L., Parrish, J. A., and Epstein, J. H., Photoimmunology, *Arch. Dermatol.,* 115, 350, 1979.
5. Kochevar, I. E., Photoallergic responses to chemicals, *Photochem. Photobiol.,* 30, 437, 1979.
6. Pathak, M. A., Phytophotodermatitis, in *Sunlight and Man,* Pathak, M. A., Harber, L. C., Seiji, M., and Kukita, A., Eds., University of Tokyo Press, Tokyo, 1974, 495.
7. Epstein, J. H., Photocarcinogenesis, skin cancer and aging, *J. Am. Acad. Dermatol.,* 9, 487, 1983.
8. Epstein, J. H., Advances in cutaneous reactions to the sun, in *Yearbook of Dermatology,* Malkinson, F. D. and Pearson, R. W., Eds., Year Book Medical Publishing, Chicago, 1971, 5.
9. Parrish, J. A. and Pathak, M. A., Photomedicine, in *Dermatology in General Medicine,* 2nd ed., Fitzpatrick, T. B., Eisen, A. Z., Wolff, K., Freedberg, I. M., and Austen, K. F., Eds., McGraw-Hill, New York, 1979, 942.
10. Honigsman, H., Scnait, G. S., Konrad, R. A., Stingl, G., and Wolff, K., Mouse model for protoporphyria, *J. Invest. Dermatol.,* 66, 188, 1976.
11. Schnait, F. G., Wolff, K., and Konrad, K., Erythropoietic protoporphyria — submicroscopic events during the acute photosensitivity flare, *Br. J. Dermatol.,* 92, 545, 1975.
12. Anderson, R. R. and Parrish, J. A., Optics of human skin, *J. Invest. Dermatol.,* 77, 13, 1981.
13. Levine, G. M. and Harber, L. C., The effect of humidity on the phototoxic response to 8-methoxypsoralen in guinea pigs, *Acta Derm. Venereol.,* 49, 82, 1969.
14. Owens, D. W. and Knox, J. M., Influence of heat, wind, and humidity on ultraviolet radiation injury, in *Ultraviolet Carcinogenesis,* Natl. Cancer Institute Monogr. no. 50, Kripke, M. L. and Sass, E. R., Eds., NCI, Bethesda, Md., 1978, 161.
15. Kaidbey, K. H., Witkowski, T. A., and Kligman, M. A., The influence of infrared radiation on short-term ultraviolet-radiation-induced injuries, *Arch. Dermatol.,* 118, 315, 1982.
16. Walter, J. F., Gange, R. W., and Mendelson, I. R., Psoralen-containing sunscreen induces phototoxicity and epidermal ornithine decarboxylase activity, *J. Am. Acad. Dermatol.,* 6, 1022, 1982.

17. Ramsay, C. A. and Obreashkova, E., Photosensitivity from nalidixic acid, *Br. J. Dermatol.*, 91, 523, 1974.
18. Epstein, J. H., Tuffanelli, D. I., Seibert, J. S., and Epstein, W. L., Porphyria-like cutaneous changes induced by tetracycline hydrochloride photosensitization, *Arch. Dermatol.*, 112, 661, 1976.
19. Epstein, J. H., Tuffanelli, D. L., and Epstein, W. L., Cutaneous changes in the porphyrias: a microscopic study, *Arch. Dermatol.*, 107, 689, 1973.
20. Stern, R. S., Laird, N., Melskin, J., Parrish, J. A., Fitzpatrick, T. B., and Bleich, H. L., Cutaneous squamous cell carcinoma in patients treated with PUVA, *N. Engl. J. Med.*, 310, 1156, 1984.
21. Abel, E. A. and Farber, E. M., Photochemotherapy, in *Recent Advances in Dermatology,* Rook, A. and Savin, J., Eds., Churchill Livingstone, Edinburgh, 1980, 259.
22. Stern, R. S., Zibler, S., and Parrish, J. A., Skin carcinoma in patients treated with topical tar and artificial ultraviolet radiation, *Lancet,* 1, 732, 1980.
23. Breza, T. S., Halprin, K. M., and Taylor, J. R., Photosensitivity to vinblastine, *Arch. Dermatol.*, 111, 1168, 1975.
24, Kunse, J., Roeber, H., and Kollakowski, M., Phototoxic dermatitis with DTIC treatment, *Z. Hautkr.*, 55, 100, 1979.
25. Stratigos, J. D. and Magnus, I. A., Photosensitivity by demethylchlortetracycline and sulphanilamide, *Br. J. Dermatol.*, 80, 391, 1968.
26. Foote, C. S., Photosensitized oxidation and singlet oxygen: consequences in biological systems, in *Free Radicals in Biology,* Pryor, W. A., Ed., Academic Press, New York, 1976, 85.
27. Ito, T. and Kobayashi, K., A survey of in vivo photodynamic activity of xanthenes, thiazines and acridines in yeasts cells, *Photochem. Photobiol.*, 26, 591, 1978.
28. Harber, L. C. and Bickers, D. R., The porphyrias: basic science aspects, clinical diagnosis and management, in *Year Book of Dermatology,* Malkinson, F. D. and Pearson, R. W., Eds., Year Book Medical Publishing, Chicago, 1975, 9.
29. Lim, H. W., Perez, H. D., Poh-Fitzpatrick, M., and Gigli, I., Generation of chemotactic activity in serum from patients with erythropoietic protoporphyria and porphyria cutanea tarda, *N. Engl. J. Med.*, 304, 212, 1981.
30. Lim, H. W., Perez, H. D., Goldstein, I. M., and Gigli, I., Complement-derived chemotactic activity is generated in human serum containing uroporphyrin after irradiation with 405 nm light, *J. Clin. Invest.*, 67, 1072, 1981.
31. Lim, H. W., Novotny, H., and Gigli, I., Role of complement and polymorphonuclear cells in demethylchlortetracycline-induced phototoxicity in guinea pigs, *J. Clin. Invest.*, 71, 1326, 1983.
32. Pathak, M. A., Kramer, D. M., and Fitzpatrick, T. B., Photochemistry and photobiology of furocoumarins (psoralens), in *Sunlight and Man,* Pathak, M. A., Harber, L. C., Seiji, M., and Kukita, A., Eds., University of Tokyo Press, Tokyo, 1974, 335.
33. Song, P. S. and Tapley, K. J., Jr., Photochemistry and photobiology of psoralens, *Photochem. Photobiol.*, 29, 1177, 1979.
34. Ben-Hur, E. and Song, P. S., The photochemistry and photobiology of furocoumarins (psoralens), *Adv. Radiat. Biol.*, 11, 131, 1984.
35. Kochevar, I. and Lamola, A. A., Chlorpromazine and protriptyline phototoxicity: photosensitized, oxygen-independent red cell hemolysis, *Photochem. Photobiol.*, 29, 1177, 1979.
36. Argenbright, L. W., Forbes, P. D., and Steward, G. J., Quantitation of phototoxic hyperemia and permeability. II. Inhibition of histamine (H_1 and H_2) receptor antagonists in mouse skin, *J. Invest. Dermatol.*, 75, 417, 1980.
37. Kamide, R., Gigli, I., and Lim, H. W., Participation of mast cells and complement in the immediate phase of hematoporphyrin-induced photosensitivity, *J. Invest. Dermatol.*, 82, 485, 1984.
38. Sik, R. H., Paschall, C. S., and Chignell, C. F., The phototoxic effect of benoxaprofen and its analogs on human erythrocytes and rat peritoneal mast cells, *Photochem. Photobiol.*, 38, 411, 1983.
39. Kochevar, I. E., Hoover, K. W., and Grulenouski, M., Benoxaprofen photosensitization of cell membrane disruption, *J. Invest. Dermatol.*, 82, 214, 1984.
40. Parrish, J. A., Fitzpatrick, T. B., Pathak, M. A., and Tanenbaum, L., Photochemotherapy of psoriasis with oral methoxsalen and longwave ultraviolet light, *N. Engl. J. Med.*, 291, 1207, 1974.
41. Epstein, J. H., Brunsting, L. A., Petersen, M. C., and Schwarz, B. E., A study of photosensitivity occurring with chlorpromazine therapy, *J. Invest. Dermatol.*, 28, 329, 1957.
42. Kochevar, I. E., Phototoxicity mechanisms: chlorpromazine photosensitized damage to DNA and cell membranes, *J. Invest. Dermatol.*, 77, 59, 1981.
43. Epstein, S., Chlorpromazine photosensitivity, phototoxic and photoallergic reactions, *Arch. Dermatol.*, 98, 354, 1968.
44. Hunter, J. A., Bhutani, L. K., and Magnus, I. A., Chlorpromazine photosensitivity in mice: its action spectrum and effect of anti-inflammatory agents, *Br. J. Dermatol.*, 82, 157, 1970.
45. Rosenthal, I., Ben-Hur, E., Prager, A., and Riklis, E., Photochemical reactions of chlorpromazine: chemical and biochemical implications, *Photochem. Photobiol.*, 28, 591, 1978.

46. Kligman, A. M., Declomycin Compendium, Lederle Laboratories, Division of American Cyanamide Company, New York, 1962.
47. Maibach, H. I., Sams, N. M., Jr., and Epstein, J. H., Screening for drug toxicity by wavelengths greater than 310 nm, *Arch. Dermatol.*, 95, 12, 1967.
48. Keane, J. T., Pearson, R. W., and Malkinson, F. D., Nalidixic acid-induced photosensitivity in mice: a model for pseudoporphyria, *J. Invest. Dermatol.*, 82, 210, 1984.
49. Diffey, B. L., Daymond, T. J., and Fairgreaves, H., Phototoxic reactions to piroxicam, maproxen and tiaprofenic acid, *Br. J. Rheumatol.*, 22, 239, 1983.
50. Stern, R. S. and Bigby, M., An expanded profile of cutaneous reactions to nonsteroidal anti-inflammatory drugs, *JAMA*, 252, 1433, 1984.
51. Halsey, J., Benoxaprofen: side-effects profile in 300 patients, *Br. J. Med.*, 284, 1365, 1982.
52. Ferguson, J., A study of benoxaprofen-induced photosensitivity, *Br. J. Dermatol.*, 107, 429, 1982.
53. Emmett, E. A., Phototoxicity from exogenous agents, *Photochem. Photobiol.*, 38, 429, 1979.
54. Kligman, A. M. and Kaidbey, K. H., Human models for identification of photosensitizing chemicals, *J. Natl. Cancer Inst.*, 69, 264, 1982.
55. Epstein, S., Photoallergy versus phototoxicity, in *Dermatoses Due to Environmental and Physical Factors*, Rees, R. B., Ed., Charles C Thomas, Springfield, Ill., 1962, 119.
56. Storck, H., Photoallergy and photosensitivity: due to systemically administered drugs, *Arch. Dermatol.*, 91, 469, 1965.
57. Herman, P. and Sams, W. M., Jr., *Soap Photodermatitis*, Charles C Thomas, Springfield, Ill., 1972.
58. Jillson, V. F. and Baughman, R. D., Contact dermatitis from bithionol, *Arch. Dermatol.*, 88, 409, 1963.
59. Epstein, J. H., Wuepper, K. D., and Maibach, H. I., Photocontact dermatitis to halogenated compounds and related compounds, *Arch. Dermatol.*, 97, 236, 1968.
60. Burry, J. N., Persistent light reactions from buclosamide, *Arch. Dermatol.*, 101, 95, 1970.
61. Burry, J. N. and Hunter, G. A., Photocontact dermatitis from Jadit, *Br. J. Dermatol.*, 82, 244, 1970.
62. Amblard, P., Beani, J. C., and Raymond, J. L., Persistent light reaction due to phenothiazines in atopic disease, *Ann. Dermatol. Venereol.*, 109, 225, 1982.
63. Burdick, K. H., Prolonged sensitivity to intradermal chlorpromazine, *Cutis*, 5, 1113, 1969.
64. Wiskemann, A. and Wulf, K., Untersuchungen über den auslosenden Spektralbereich und die direkte Lichtpigmentierung bei chronischen und akuten Lichtauschlagen, *Arch. Klin. Exp. Dermatol.*, 209, 443, 1959.
65. Sidi, E., Hincky, M., and Gervais, A., Allergic sensitization and photosensitization to Phenergan cream, *J. Invest. Dermatol.*, 24, 345, 1955.
66. Giovinazzo, V. J., Harber, L. C., Bickers, D. R., Armstrong, R. B., and Silvers, D. N., Photoallergic contact dermatitis to musk ambrette, *J. Am. Acad. Dermatol.*, 3, 384, 1980.
67. Allen, H. and Kaidbey, K. H., Persistent photosensitivity following occupational exposure to epoxy resin, *Arch. Dermatol.*, 115, 1307, 1979.
68. Willis, I. and Kligman, A. M., The mechanism of the persistent light reactor, *J. Invest. Dermatol.*, 51, 385, 1968.
69. Willis, I. and Kligman, A. M., The mechanism of photoallergic contact dermatitis, *J. Invest. Dermatol.*, 51, 378, 1968.
70. Epstein, S., "Masked" photopatch tests, *J. Invest. Dermatol.*, 41, 369, 1963.
71. Harber, L. C., Bickers, D. R., Epstein, J. H., Pathak, M. A., and Urbach, F., Report on ultraviolet light sources, *Arch. Dermatol.*, 109, 833, 1974.
72. North American Contact Dermatitis Group of the National Program for Dermatology, The Role of Patch Testing in Allergic Contact Dermatitis, Johnson & Johnson, New Brunswick, N.Y., 1974.
73. Epstein, J. H., Polymorphus light eruption, *Ann. Allergy*, 24, 397, 1966.
74. Schwarz, K. and Speck, M., Experimentelle Untersuchungen zun Frage der Photoallergie der Sulfonamide, *Dermatologica*, 114, 232, 1957.
75. Epstein, S., Allergic photocontact dermatitis from promethazine (Phenergan), *Arch. Dermatol.*, 81, 175, 1960.
76. Epstein, S. and Rowe, R. J., Photoallergy and photocross-sensitivity to Phenergan, *J. Invest. Dermatol.*, 29, 319, 1957.
77. Ertle, T., Work-related contact and photocontact allergy in a farmer caused by chlorpromazine, *Derm. Beruf. Umwelt.*, 30, 120, 1982.
78. Polano, J. K., Photosensitivity due to drugs, *Excerpta Med. Int. Congr. Ser.*, 85, 102, 1964.
79. Torinuki, W., Kumai, N., and Miura, T., Chronic photosensitive dermatitis due to phenothiazines, *Tohoku J. Exp. Med.*, 138, 223, 1982.
80. Burckhardt, W. and Schwarz-Speck, M., Photoallergische Ekzeme durch Nadisan, *Schweiz. Med. Wochenschr.*, 87, 954, 1954.
81. Epstein, E., Persome dermatitis in men, *JAMA*, 209, 911, 1969.

82. Burckhardt, W., Photoallergy eczema due to blankophores (optic brightening agents), *Hautarzt,* 8, 486, 1957.
83. Epstein, S., Discussion of Reference 92, 1956.
84. Epstein, S., Role of dermal sensitivity in ragweed contact dermatitis, *Arch. Dermatol.,* 82, 48, 1960.
85. Davies, M. G., Hawk, J. L., and Rycroft, R., Acute photosensitivity from the sunscreen 2-ethoxy-ethyl-*p*-methoxycinnamate, *Contact Dermatitis,* 8, 190, 1982.
86. Fagerlund, V. L., Kalimo, K., and Hansen, C., Photocontact allergy from sunscreens, *Duodecin,* 99, 146, 1983.
87. Fitzpatrick, T. B., Pathak, M. A., Magnus, I. A. et al., Abnormal reactions of man to light, *Annu. Rev. Med.,* 14, 195, 1963.
88. Goldman, G. C. and Epstein, E., Jr., Contact photosensitivity dermatitis from sun-protective agent, *Arch. Dermatol.,* 100, 447, 1969.
89. Holzle, E. and Plewig, G., Photoallergic contact dermatitis by benzophenone containing sunscreening preparations, *Hautarzt,* 33, 391, 1982.
90. Kaidbey, K. H. and Allen, H., Photocontact allergy to benzocaine, *Arch. Dermatol.,* 117, 77, 1981.
91. Mathias, C. G. T., Maibach, H. I., and Epstein, J. H., Allergic contact photodermatitis to para-aminobenzoic acid, *Arch. Dermatol.,* 114, 1665, 1978.
92. Sams, W. M., Contact photodermatitis, *Arch. Dermatol.,* 73, 142, 1956.
93. Satulsky, E. M., Photosensitization induced by monoglyceron para-aminobenzoate, *Arch. Dermatol.,* 62, 711, 1950.
94. Emmett, E. A., Diphenydramine photoallergy, *Arch. Dermatol.,* 110, 249, 1974.
95. Fulton, J. E. and Willis, I., Photoallergy to methoxsalen, *Arch. Dermatol.,* 98, 445, 1968.
96. Sidi, E. and Bourgeois-Cavardin, J. Mise au point due traitement du vitiligo par l'Ammi Majus, *Presse Med.,* 61L, 436, 1953.
97. Scott, K. W. and Dawson, T. A. J., Photocontact dermatitis arising from the presence of quindoxin in animals feeding stuffs, *Br. J. Dermatol.,* 90, 543, 1974.
98. Zanoun, S., Johnson, B. E., and Frain-Bell, W., The investigation of quindoxin photosensitivity, *Contact Dermatitis,* 2, 342, 1976.
99. Wilkinson, D. S., Photodermatitis due to tetrachlorosalicylanilide, *Br. J. Dermatol.,* 73, 213, 1961.
100. Adams, R. M., Photoallergic contact dermatitis to chloro-2-phenyl-phenol, *Arch. Dermatol.,* 106, 711, 1972.
101. Smith, S. Z. and Epstein, J. H., Photocontact dermatitis to halogenated salicylanilides and related compounds, *Arch. Dermatol.,* 113, 1372, 1977.
102. Jackson, R. T., Nesbit, L. T., and DeLeo, V. A., 6-Methyl coumarin photocontact dermatitis, *J. Am. Acad.. Dermatol.,* 2, 124, 1980.
103. Kaidbey, K. H. and Kligman, A. M., Contact photoallergy to 6-methyl coumarin in proprietary sunscreens, *Arch. Dermatol.,* 114, 1709, 1978.
104. Galosi, A. and Plewig, G., Photoallergic eczema caused by musk ambrette, *Hautarzt,* 33, 589, 1982.
105. Raugi, G. J., Storrs, F. J., and Larsen, W. G., Photoallergic contact dermatitis to men's perfume, *Contact Dermatitis,* 5, 251, 1979.
106. Giovinazzo, V. J., Harber, L. C., Bickers, D. R., and Armstrong, R. B., Photoallergic contact dermatitis to musk ambrette: action spectrum in guinea pigs and man, *Photochem. Photobiol.,* 33, 773, 1981.
107. Epstein, S., Photoallergy and primary phototoxicity to sulfanilamide, *J. Invest. Dermatol.,* 2, 43, 1939.
108. Jung, E. G. and Schwarz, K., Photoallergy to "Jadit" with photocrossreactions to derivatives of sulfanilamide, *Int. Arch. Allergy Appl. Immunol.,* 27, 313, 1965.
109. Epstein, S. and Enta, T., Photoallergic content dermatitis, *JAMA,* 194, 1016, 1965.
110. Jung, E. G. and Schultz, R., Kontakt und Photoallergien durch Desinfizien, *Dermatologica,* 137, 216, 1968.
111. Willis, I. and Kligman, A. M., Photocontent allergic reactions: elicitation by low dose ultraviolet rays, *Arch. Dermatol.,* 110, 535, 1969.
112. Jenkins, F. P., Welti, D., and Baines, D., Photochemical reactions of tetrachlorosalicylanilide, *Nature (London),* 210, 827, 1964.
113. Osmundsen, P. E., Contact photoallergy to tribromsalicylanide, *Br. J. Dermatol.,* 81, 429, 1969.
114. Jung, E. G., In vitro-Untersuchungen zur Chlorpromazine (CPZ) Photoallergie, *Arch. Klin. Exp. Dermatol.,* 237, 501, 1970.
115. Jung, E. G., Photoallergic durch Triacetyldiphenolisatin (TDI). II. Photochemische Untersuchungen zur Pathogenese, *Arch. Klin. Exp. Dermatol.,* 231, 39, 1967.
116. Jung, E. G., Dummler, U., and Immich, H., Photoallergie durch 4-chlor-2-hydroxy-benzoesaure-*n*-butylmid. I. Lichtbiologische Untersuchungen zur Antigenbildung, *Arch. Klin. Exp. Dermatol.,* 232, 403, 1968.

117. Jung, E. G., Hornke, J., and Hajede, P., Photoallergic durch 4-chlor-2-hydrobenzoesaure-*n*-butyl-mid. II. Photochemische Untersuchungen, *Arch. Klin. Exp. Dermatol.*, 233, 287, 1968.
118. Harber, L. C., Current status of mammalian and human models for predicting drug photosensitivity, *J. Invest. Dermatol.*, 77, 65, 1981.
119. Jordan, W. P., Jr., The guinea pig model for predicting photoallergic contact dermatitis, *Contact Dermatitis*, 8, 109, 1982.
120. Harber, L. C., Targovnik, S. E., and Baer, R. L., Contact photosensitivity patterns to halogenated salicylanilides in man and guinea pigs, *Arch. Dermatol.*, 96, 646, 1967.
121. Maguire, H. C., Jr. and Kaidbey, K. H., Experimental photoallergic contact dermatitis: a mouse model, *J. Invest. Dermatol.*, 79, 147, 1982.
122. Miyachi, Y. and Takigawa, M., Mechanisms of contact photosensitivity in mice. III. Predictive testing of chemicals with photoallergic potential in mice, *Arch. Dermatol.*, 119, 736, 1983.
123. Granstein, R. D., Morison, W. L., and Kripke, M. L., The role of suppressor cells in the induction of murine photoallergic contact dermatitis and in its suppression by prior UVB irradiation, *J. Immunol.*, 130, 2099, 1983.
124. Miyachi, Y. and Takigawa, M., Mechanisms of contact photosensitivity in mice. II. Langerhans cells are required for successful induction of contact photosensitivity to TCSA, *J. Invest. Dermatol.*, 78, 363, 1982.
125. Takigawa, M. and Miyachi, Y., Mechanisms of contact photosensitivity in mice. I. T cell regulation of contact photosensitivity to tetrachlorsalicylanilide under genetic restrictions of the major histocompatibility complex, *J. Invest. Dermatol.*, 79, 108, 1982.
126. Kligman, A. M., Identification of contact allergens by human assay. III. The maximization test: a procedure for screening and rating contact sensitizers, *J. Invest. Dermatol.*, 47, 393, 1966.
127. Kaidbey, K. H. and Kligman, A. M., Photo-maximization test for identifying photoallergic contact sensitizers, *Contact Dermatitis*, 6, 161, 1980.
128. Kaidbey, K. H., The evaluation of photoallergic contact sensitizers in humans, in *Dermatotoxicology*, 2nd ed., Marzulli, F. N. and Maibach, H. I., Eds., Hemisphere Publishers, Washington, D.C., 1983, 405.
129. Blum, H. F., Studies of photosensitivity due to sulfanilamide, *J. Invest. Dermatol.*, 4, 159, 1941.
130. Burkhardt, W., Untersuchungen über die Photoaktivitat einiger Sulfanilamide, *Dermatologica*, 83, 63, 1941.
131. Stevanovic, D. V., Photosensitivity due to certain drugs, *Br. J. Dermatol.*, 73, 233, 1961.
132. Miller, R. C. and Beltrani, V. S., Quinethazone photosensitivity dermatitis, *Arch. Dermatol.*, 93, 346, 1966.
133. Hitselberger, J. F. and Foxnaugh, R. P., Photosensitivity due to chlorpropamide, *JAMA*, 180, 142, 1962.
134. Harber, L. C., Lashinsky, A. M., and Baer, R. L., Photosensitivity due to chlorothiazide and hydrochlorothiazide, *N. Engl. J. Med.*, 261, 1378, 1959.
135. Calnan, C. D., Frain-Bell, W., and Cuthbert, W. I., Occupational dermatitis from chlorpromazine, *Trans. St. John's Hosp. Dermatol. Soc.*, 48, 49, 1962.
136. Schultz, K. H., Wickemann, A., and Wulf, K., Kliniche und experimentelle Untersuchungen über die photodynamiche Wirksamkeit von Phinothiazin Derivaten, insbesondere von Magaphen, *Arch. Klin. Exp. Dermatol.*, 202, 285, 1956.
137. Erickson, L. R. and Peterka, E. S., Sunlight sensitivity from oral contraceptives, *JAMA*, 203, 980, 1968.
138. Chang, T. W., Cold urticaria and photosensitivity due to griseofulvin, *JAMA*, 193, 848, 1965.
139. Luton, E. F. and Finchum, R. N., Photosensitivity reaction to chlordiazepoxide, *Arch. Dermatol.*, 91, 362, 1965.
140. Schreiber, M. M. and Naylor, L. Z., Antihistamine photosensitivity, *Arch. Dermatol.*, 86, 58, 1962.
141. Kasuistik, I. and Jung, E. G., Photoallergic durch Triacetydiphenolisatin (TDI), *Arch. Klin. Exp. Dermatol.*, 229, 170, 1967.
142. Kobori, T. and Araki, H., Photoallergy in dermatology, *J. Asthma Res.*, 3, 213, 1966.
143. Lamberg, S. I., A new photosensitizer: the artificial cyclamate, *JAMA*, 201, 121, 1967.
144. Tatsuji, K. and Toshie, A., Photoallergic dermatitis probably due to artificial sweetening agents, *Med. Cult.*, 5, 795, 1963.
145. Van Weelden, H., Bolling, H. H., Baart de la Faille, H., and van der Leun, J. C., Photosensitivity caused by chloroquine, *Arch. Dermatol.*, 118, 290, 1982.
146. Lang, P. G., Jr., Quinidine-induced photodermatitis confirmed by photopatch testing, *J. Am. Acad. Dermatol.*, 91, 124, 1983.
147. Marx, J. L., Eisenstat, B. A., and Gladstein, A. H., Quinidine photosensitivity, *Arch. Dermatol.*, 119, 39, 1983.
148. Merot, Y., Harms, M., and Saurat, J. H., Photosensitivity associated with carprofen (Imadyl) a new non-steroidal anti-inflammatory drug, *Dermatologica*, 116, 310, 1983.

149. Beardmore, G. L., Recalcitrant sporotrichosis: a report of a patient treated with various therapies, including oral miconozole and 5-fluorocytosine, *Aust. J. Dermatol.*, 20, 10, 1979.
150. Shelley, W. B. and Sica, P. A., Disseminate sporotrichosis of skin and bone cured with 5-fluorocytosine: photosensitivity as a complication, *J. Am. Acad. Dermatol.*, 8, 229, 1983.

Chapter 8

PORPHYRIAS

Maureen B. Poh-Fitzpatrick

TABLE OF CONTENTS

I. INTRODUCTION

The porphyrias are a group of related clinical disorders that arise from several different enzymatic errors of heme biosynthesis. These errors are partial enzyme blocks that result in accumulation of abnormal amounts of intermediaries and by-products of the heme biosynthetic pathway in patterns that differ according to the locus of the defective enzyme in the sequence. Each unique porphyria has been associated with defective activity of a particular enzyme. They are listed in Table 1.

The porphyrias are usually classified into "erythropoietic" or "hepatic" categories based on whether the bulk of the total excess porphyrin production occurs in the heme-synthesizing cells of the erythron or in hepatocytes. More recent information about these disorders suggests that in some disorders (particularly protoporphyria) abnormal amounts of heme precursors may be produced in both organs of heme synthesis and to a lesser extent in other tissues as well. The clinical characteristics of each distinct porphyria are determined by the properties of the particular array of porphyrins or porphyrin precursors that accumulate in the tissues of origin, are liberated into the blood and disseminate into other tissues, and are excreted by renal or hepatobiliary mechanisms.

II. ENZYME DEFECTS IN PORPHYRIAS

An outline of heme biosynthesis is shown in Figure 1. In normal individuals, heme synthesis is closely regulated so that very few of the intermediaries escape conversion into heme. The rate-limiting enzyme for the pathway is δ-aminolevulinic acid synthase (ALAS). The control of overall pathway activity is exerted by negative feedback repression of either ALAS synthesis or of its activity by the end-product heme. When heme supplies are adequate, ALAS activity is geared to maintain the level constant. Heme depletion or deficiency results in induction of new enzyme synthesis or in direct derepression of ALAS activity, thus driving the production of increased amounts of precursors to supply new heme.

In hepatic heme synthesis, ALAS regulation is affected by numerous agents including many drugs, selected steroid hormones, and other compounds that rapidly stimulate new ALAS synthesis by several different mechanisms having the common result of reducing available free heme.[1-3] Repletion of heme or heme analogues from exogenous sources can abort or prevent this effect.[4] High concentrations of dietary glucose also appear to retard induction of ALAS in clinical settings.[5,6]

Erythropoietic heme synthesis appears to have a different set of factors that stimulate ALAS function, and the mechanism of heme feedback repression appears to be directly on enzyme activity rather than on rate of new enzyme synthesis.[7] Factors that promote or inhibit ALAS activity in in vitro experimental systems may have clinical relevance in preventive and interventive management of patients.

A specific enzyme defect for each of the porphyrias has been determined. Accumulation of the substrates for the partially blocked enzymes results in the biochemical abnormalities that characterize each distinct porphyric entity. Since the array of clinical manifestations demonstrated by each of these disorders is related to the types of porphyrin precursors accumulated, knowledge of the site of the defect associated with each porphyria aids greatly in the understanding of both the biochemical and clinical features of each syndrome.

The partial enzyme defect associated with a unique porphyria that occurs earliest in the pathway is in uroporphyrinogen synthase (US), otherwise termed porphobilinogen deaminase. Deficient activity of this enzyme is linked with acute intermittent porphyria

Table 1
CLASSIFICATION OF PORPHYRIAS AND THEIR ENZYMATIC DEFECTS

Disorder	Defective enzyme
Erythropoietic	
Congenital erythropoietic porphyria	Uroporphyrinogen cosynthase
Protoporphyria	Heme synthase
Hepatic	
Porphyria cutanea tarda	Uroporphyrinogen decarboxylase
Acute intermittent porphyria[a]	Uroporphyrinogen synthase
Variegate porphyria[a]	Protoporphyrinogen oxidase
Hereditary coproporphyria[a]	Coproporphyrinogen oxidase

[a] Acute attack forms of porphyria.

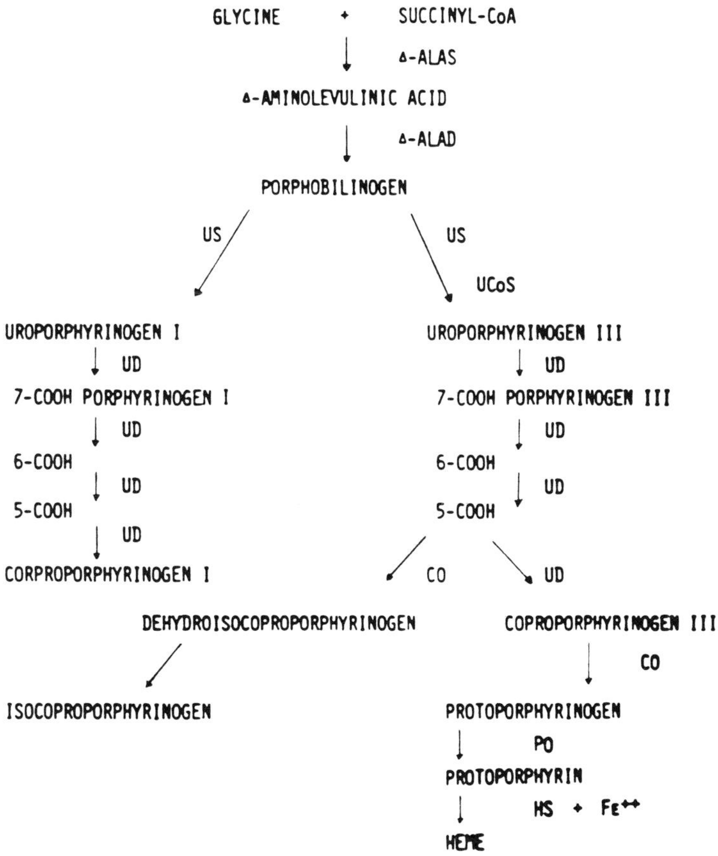

FIGURE 1. Heme biosynthetic pathway. Δ-ALAS = delta-aminolevulinic acid synthase; Δ-ALAD = delta-aminolevulinic acid dehydrase; US = uroporphyrinogen synthase (porphobilinogen deaminase); UCoS = uroporphyrinogen cosynthase; UD = uroporphyrinogen decarboxylase; CO = coproporphyrinogen oxidase; PO = protoporphyrinogen oxidase; HS = heme synthase (ferrochelatase).

(AIP).[8] US utilizes porphobilinogen as substrate; thus this precursor (and its predecessor aminolevulinic acid) accumulate and are subsequently excreted in abnormal amounts. None of the subsequent intermediaries or photosensitizing by-products of heme synthesis are accumulated in AIP.

Partial deficiency of the next enzyme in the sequence, uroporphyrinogen cosynthase (UCoS) results in the inefficent conversion of the product of the uroporphyrinogen synthase reaction to urcporphyrinogen of the isomer III configuration. Only series III isomers can undergo complete conversion into heme. The accumulated substrate of this UCoS deficiency undergoes spontaneous (nonenzymic) conversion to the series I isomer of uroporphyrin. Series I precursors can be further utilized as substrates only by one additional pathway enzyme, thereby being metabolized through the 7-, 6-, 5-, and 4-carboxylic series I porphyrinogens, resulting in accumulation of their corresponding spontaneously oxidized, photoactive series I porphyrin by-products of the pathway. Accumulation and excretion of large amounts of the isomeric series I porphyrins are the major biochemical indicators of congenital erythropoietic porphyria, the disorder linked to a defect in UCoS activity.[9]

Uroporphyrinogen decarboxylase (UD) is the next enzyme in the sequence, and is responsible for the stepwise removal of 4-carboxyl groups from the uroporphyrinogen I or III substrate. Decreased UD activity has been established as the enzymic abnormality in all forms of porphyria cutanea tarda.[10] Partial deficiency of UD results in accumulation of all of the 8-, 7-, 6-, and 5-carboxylic porphyrinogens of both III and I isomeric series and their corresponding spontaneously oxidized photosensitizing porphyrins. The excess accumulation of the 5-carboxylic porphyrinogen, however, can serve as substrate for another enzyme, coproporphyrinogen oxidase (CO). CO converts this 5-carboxyl compound to dehydroisocoproporphyrinogen, a by-product of the pathway occurring in large amounts only in the presence of deficient UD activity. Spontaneous oxidation of this tetracarboxylic porphyrinogen yields isocoproporphyrin, a marker molecule for UD deficiency disorders that is readily detectable in both urine and feces when present in large amounts.[11]

Deficient activity of CO levels leads to accumulation of its substrate coproporphyrinogen III. Defective CO and overproduction and excretion of coproporphyrin III are associated with hereditary coproporphyria.[12,13]

The next enzyme in the sequence is protoporphyrinogen oxidase (PO). Deficient conversion of protoporphyrinogen to protoporphyrin is the result of this partial block. This defect is associated with variegate porphyria (VP).[14] The array of heme precursors that accumulate and are excreted in abnormal amounts in VP includes not only oxidized protoporphyrin, considered to be the nonenzymatic oxidation product of accumulated protoporphyrinogen rather than the protoporphyrin normally formed by enzymatic conversion,[14] but also lesser amounts of coproporphyrin and hydrophilic porphyrinpeptide complexes.[15] The accumulation of coproporphyrin could be explained by the observation that protoporphyrinogen is an effective competitive inhibitor of CO,[16] leading to a secondary backup of the CO substrate, coproporphyrinogen III. Some investigators have also given evidence that at least in some cases, a defect in heme synthase activity may also be associated with this disorder.[17,18]

The last enzyme, heme synthase (or ferrochelatase), is responsible for incorporation of a ferrous iron molecule at the central position of the tetrapyrrole ring structure of protoporphyrin. This results in heme, a nonphotoactive compound. Partial deficiency of heme synthase results in accumulation of the dicarboxylic, hydrophobic protoporphyrin, the only photosensitizing molecule serving as a substrate in the pathway. Excess protoporphyrin and defective heme synthase are linked with the disorder photoporphyria.[19,20]

The solubility properties of the intermediaries and porphyrin by-products of heme synthesis determine the route of their excretion. The more water-soluble group — aminolevulinic acid (ALA), porphobilinogen (PBG), uroporphyrin, and heptacarboxylic porphyrins — are primarily excreted into the urine, while the dicarboxylic hydrophobic protoporphyrin molecule is excreted by biliary secretion into the feces. Coproporphyrin and isocoproporphyrin with 4-carboxyl side groups and the penta- and hexacarboxylic porphyrins have intermediate solubility properties, and can be found in both urine and feces.

Whether patients with different forms of porphyria exhibit cutaneous photosensitivity and the different types of photocutaneous lesions that occur appear to be determined by the physiochemical properties of the particular metabolites accumulated and transported to the skin in each disorder. All forms of porphyria, except AIP, may have light-induced skin lesions as part of the clinical symptom complex. The explanation for the exclusion of AIP appears to be that the porphyrin precursors that accumulate in AIP (e.g., ALA and PBG) are not photosensitizing molecules. Patients with all of the other porphyrias do accumulate molecules that can absorb light radiation in the visible spectrum, and re-emit this energy by several mechanisms including fluorescence and performance of photochemistry.

The reduced tetrapyrrolic porphyrinogens that are the true intermediaries of the pathway are not photoactive molecules. However, as these molecules accumulate, nonenzymatic oxidation converts them irreversibly to the corresponding photosensitizing porphyrins. Oxidized porphyrins are purple pigments, and discolor the tissues or fluids in which they are heavily deposited.

III. CLINICAL PATTERNS IN PORPHYRIAS

Patients with any of the porphyrias that are biochemically characterized by accumulation and excretion of the more water-soluble porphyrins exhibit similar cutaneous lesions. The skin of these patients is so fragile in sun-exposed areas that the epidermis shears away from the underlying dermis with minimal mechanical trauma. Blisters or bullae form spontaneously, with the fluid-filled cavity between the epidermis and dermis (Figure 2). These lesions are slow to heal, and form heavy crusts and residual scars (Figure 3). Hypertrichosis, pigmentary disturbances (Figure 4), sclerodermoid changes with loss of dermal appendages and calcification, and scarring of potentially mutilating severity are the major chronic manifestations of this type of porphyric phototoxicity. Although the chronic changes may be severely deforming, very often there is little or no acute discomfort in the skin at the time of light exposure, and some patients remain unaware of the association of skin lesions and light.

In contrast, patients with protoporphyria who have accumulated hydrophobic protoporphyrin exhibit a different acute and chronic phototoxicity response pattern. Within minutes of sunlight exposure, patients with protoporphyria experience sensations of stinging, tingling, or burning deep within their skin. With sufficient exposure, the skin becomes massively edematous within several hours and then develops petechial erythema (Figure 5). The skin remains very painful during the 2 to 3 days of evolution of these lesions. However, after the petechiae resolve, the residual scarring is usually limited to shallow, pitted depressions of the skin of the face, and a roughened or leathery texture of the dorsum of the hands, and occasionally of the forehead, nose, and malar skin (Figure 6).

Three of the hepatic porphyrias are clinically similar in the occurrence of acute attack episodes which may be of life-threatening severity. These three are AIP, VP, and hereditary coproporphyria (HCP). Each of these hepatic porphyrias is characterized

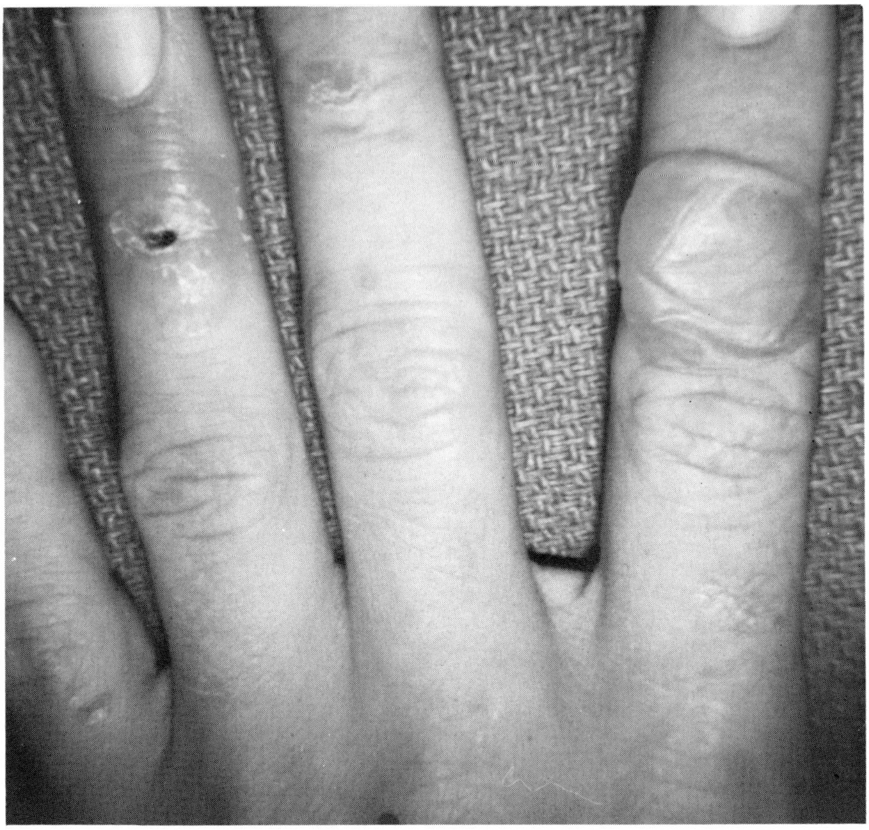

FIGURE 2. Large bulla on the index finger of a patient with porphyria cutanea tarda. Note also a crusted erosion over the middle knuckle and the healing site of a previous bulla on the fourth finger.

by the excretion of large amounts of ALA and PBG, either constantly (AIP), or primarily during the episode (VP and HCP). The same factors are responsible for triggering onset of the attacks in all of these disorders, and the treatment of such attacks is also the same for all.

A. Congenital Erythropoietic Porphyria (CEP)

CEP is the least-common porphyria, and the only one inherited as an autosomal recessive defect. The onset of signs and symptoms may occur at birth or shortly thereafter. Copious excretion of porphyrins of the more water-soluble classes into the urine causes pink discoloration of urine and diapers.

As described earlier, reduced activity of UCoS results in accumulation of series I isomers of the porphyrinogens and the corresponding spontaneously oxidized porphyrin molecules with 8- to 4-carboxylic side groups. Only series III porphyrinogen isomers can be fully processed to complete the heme sequence beyond coproporphyrin formation. Reduction in heme synthesized is often profound and reflected clinically as an anemia detectable at birth. Heme deficiency also results in derepression of the rate-controlling enzyme of the pathway, ALAS, so that the sequence is driven to produce more intermediaries. In addition to the characteristically heavy accumulation of the series I porphyrin by-products, the corresponding series III isomers may be found in increased amounts as well.

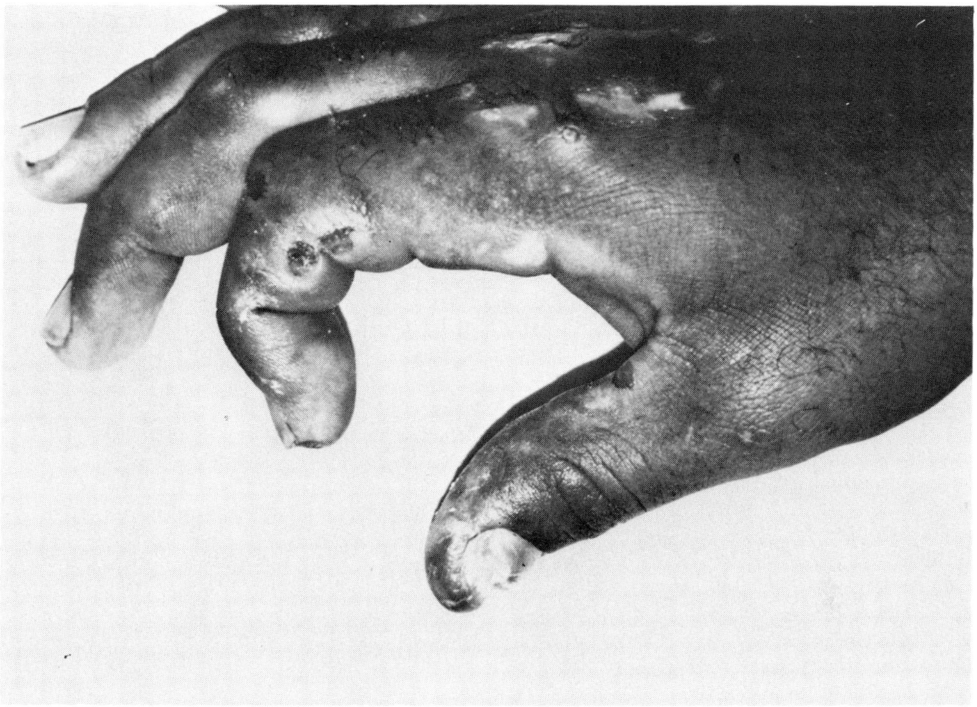

FIGURE 3. Crusted erosions and depigmented scars forming on the hand of a patient with porphyria cutanea tarda.

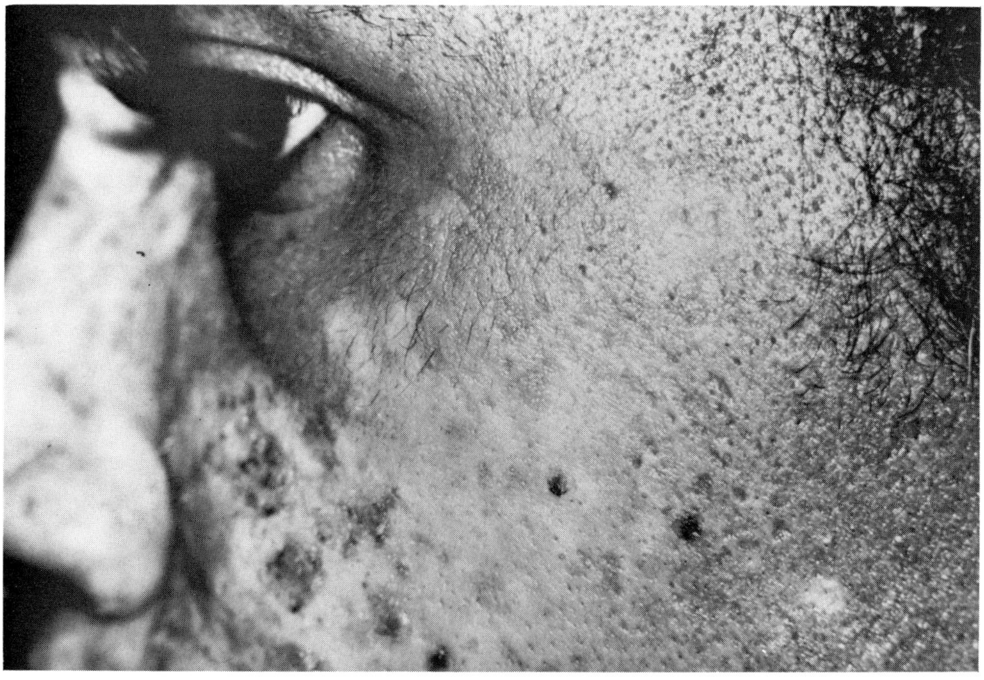

FIGURE 4. Crusted facial erosions and marked depigmentation in a patient with porphyria cutanea tarda. Several coarse hairs are present over the zygomatic arch.

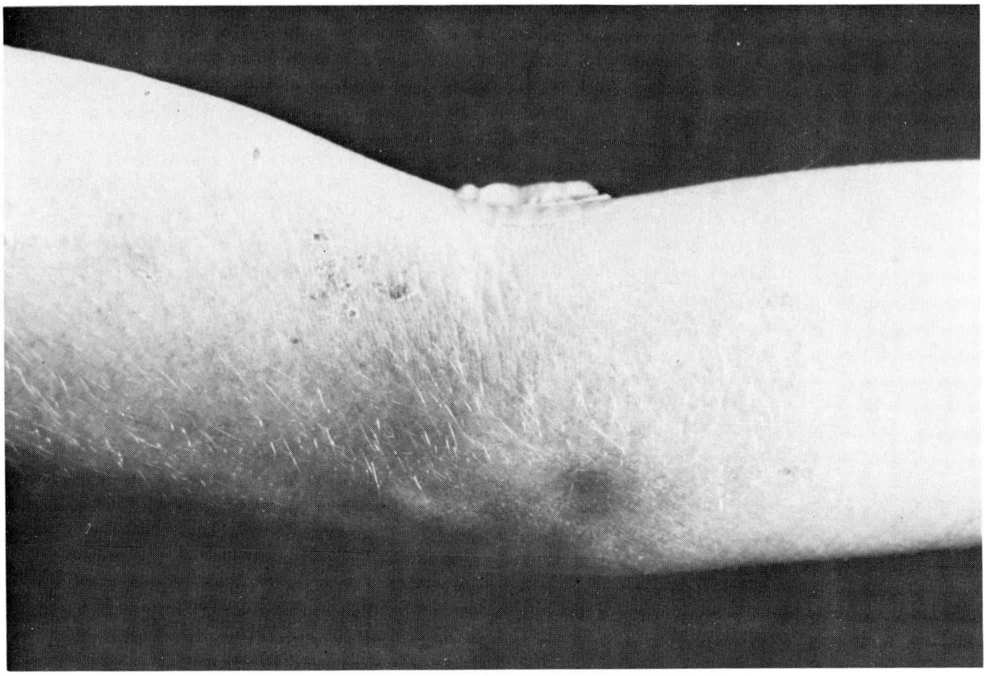

FIGURE 5. Petechial erythema in a patient with protoporphyria.

Photosensitivity is severe, and may be present at birth. Infants exposed to sunlight will form blisters. If not protected rigorously from light exposure, the chronic sequelae of CEP can be terribly deforming. Deep scars, destruction of cartilage of ala nasi and auricular pinnae, scarring alopecia, pigmentary changes, ectropion, eclabium, resorption of distal phalanges, and teeth stained brown-red by deposited porphyrin pigment all contribute to a clinical morbidity of potentially horrific proportions. The associated anemia may be severe, with secondary splenomegaly and stunting of growth. The hyperactive bone marrow, attempting to compensate the anemia and shortened erythrocyte life span, may encroach on the bony cortices causing retarded growth and thinned bones that may fracture easily.

Treatment for these patients includes measures to physically protect the skin from offensive wavelengths of light, oral β-carotene photoprotection (see Section III.B), splenectomy in selected cases to reduce premature destruction of circulating erythrocytes, and transfusion to provide adequate heme in circulation to reduce erythropoiesis. Recent clinical trials of hypertransfusion have had good success, combined with vigorous cutaneous photoprotection, in limiting the usual clinical morbidity associated with CEP.[22,23]

B. Protoporphyria (EPP)

Inherited in an autosomal dominant mode, human protoporphyria is one of the more common disorders of porphyrin metabolism. A partial block of the enzyme heme synthase (ferrochelatase) reduces conversion of protoporphyrin to heme, resulting in accumulation of the photoactive, hydrophobic protoporphyrin molecule. In the majority of cases, this defect appears to be expressed chiefly in the erythropoietic bone marrow tissue.[24,25] Thus, this disorder is often referred to as "erythropoietic protoporphyria". In selected cases, an important contribution to the total excess of protopor-

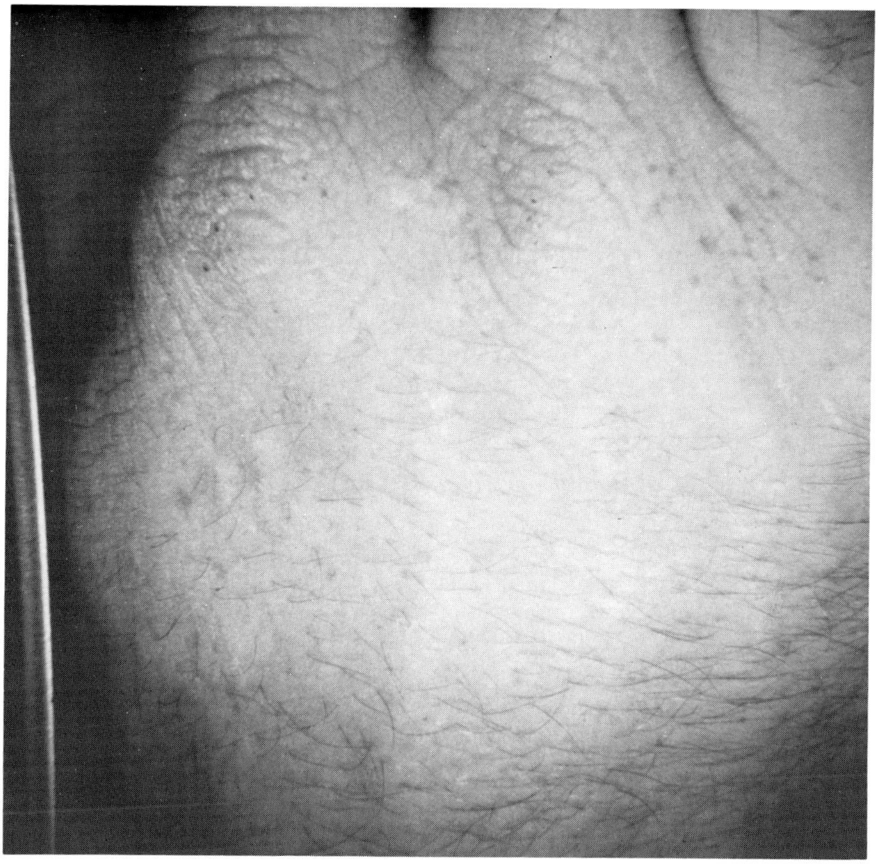

FIGURE 6. Roughened leathery skin over the knuckles of a patient with protoprophyria.

phyrin may also derive from hepatocytic protoporphyrin production.[25,26] Therefore, many investigators now refer to this syndrome simply as "protoporphyria".

Despite the defective enzyme activity, sufficient heme is usually formed, so that clinically detectable anemia is typically mild or may be absent. The most frequent clinical manifestations are the cutaneous symptoms that typically appear sometime during childhood. Apart from painful acute phototoxicity, patients with EPP have an increased incidence of early-onset cholelithiasis.[27] This appears related to the very large excesses of protoporphyrin cleared from the blood by the liver and secreted into the bile. Fecal protoporphyrin excretion rates are affected by the enterohepatic reabsorption of some fraction of the secreted protoporphyrin.[28]

The livers of patients with EPP are characterized by the presence of pigmented protoporphyrin deposits in hepatocytes, macrophages, and biliary canaliculi. Crystals of protoporphyrin and abnormal mitochondria can be seen in hepatocytes with electron microscopy.[29,31] and with polarizing lenses, the porphyrin deposits are seen as doubly refractile "Maltese cross" bodies.[32] Although liver biopsy specimens from virtually all patients show these abnormalities to a variable extent, in a minority of unknown percentage, protoporphyrin deposition is heavy and associated with inflammatory changes, fibrosis, and even cirrhosis. Some patients with EPP will develop signs and symptoms of hepatic dysfunction, and some 20 cases of progressive liver failure and death from dysfunction or hemorrhage from esophageal varices have been recorded in the medical literature.

Treatment for the cutaneous phototoxicity includes avoidance of sunlight exposure and ingestion of β-carotene. The naturally occurring yellow pigment, β-carotene, was noted to have photoprotective effects in bacterial and animal model systems.[33,34] Subsequently, this provitamin was administered to human volunteers with EPP and was found to increase cutaneous tolerance to light exposure.[35,36] Ingestion of synthetic β-carotene in amounts sufficient to raise serum carotene levels to the 400 to 800 μg/dℓ range produces clinical carotenoderma. As the skin yellows, patients note decreasing symptom levels compared with pretreatment exposures. Synthetic β-carotene has now been available for use for several years without recognition of serious side effects. Safety for use in pregnancy has not been determined.

The mechanism of action by which the β-carotene molecule provides photoprotection is considered to reside chiefly in its avid singlet-oxygen quenching and free-radical scavenging properties rather than from a physical filter effect of the yellow pigment.[37] However, all of these mechanisms may contribute to the total effect. The efficacy of carotene in EPP has suggested its use in the porphyrias in which the more water-soluble porphyrin molecules are the major photosensitizers. In these other disorders, benefit is less readily evident than in EPP.

C. Porphyria Cutanea Tarda (PCT)

Probably the most frequently occurring disorder of porphyrin metabolism is PCT. The disorder is an hepatic porphyria which may occur in familial or sporadic forms.

An autosomal dominant transmission of the genetic defect for deficiency in UD results in the familial form of PCT. Heterozygous inheritance results in the classical adult-onset form of PCT (only rarely seen in children), while homozygous inheritance of the same defect apparently causes a much more severe, childhood-onset disorder whose clinical picture resembles that of CEP. In addition to familial PCT, cases have been observed to occur in sporadic instances or epidemic outbreaks apparently related to exposure to hepatotoxic drugs or environmental toxins. PCT as a symptom of hepatic tumors has been well described,[38,39] and PCT-manifesting symptoms after institution of maintenance dialysis for chronic renal failure has also recently been recognized.[40,41]

UD mediates the sequential decarboxylation of 4-carboxyl groups from uroporphyrinogen I or III, yielding the hepta-, hexa-, and pentacarboxylic porphyrinogens and coproporphyrinogens. Partial deficiency of the enzyme (approximately 50% of normal activity in heterozygous inherited PCT) results in accumulation of 8-, 7-, 6-, and 5-carboxylic spontaneously oxidized porphyrin by-products. Although the normal conversion of pentacarboxylic porphyrinogen to coproporphyrinogen is compromised, the next enzyme in the pathway, CO, causes conversion of this substrate to dehydroisocoproporphyrinogen which oxidizes to isocoproporphyrin, a tetracarboxylic molecule that accumulates in large amounts characteristically in PCT.[42] Isocoproporphyrin appears, along with coproporphyrin, in feces as well as in urine. A fecal isocoproporphyrin to coprocoproporphyrin ratio >0.1 is associated with PCT.[43]

In familial PCT, the enzyme defect can be measured in erythrocytes as well as in hepatocytes, while in sporadic cases the defect appears to be present only in hepatocytes.[44,45] In cases of homozygous inheritance of the genetic defect for PCT, the enzyme activity is further reduced to approximately 10% of normal.[46]

Inheritance of the defect in the heterozygous mode is often insufficient to produce clinically, or even biochemically manifest disease. PCT may be silent, or clinically latent, until the predisposed individual encounters an additional factor that induces the unmasking of the disorder. Factors associated with clinical appearance of PCT include iron overload, alcohol, estrogenic hormones, or exposure to environmental hepatotoxins. In the last instance sufficient exposure appears able to produce PCT even in persons not genetically predisposed. The class of compounds that have produced epidemic PCT are the polychlorinated aromatic hydrocarbons. The classical example of hepatotoxin-induced nongenetic epidemic PCT occurred in Turkey.[47] Famished peoples there consumed seed wheat that had been treated with hexachlorobenzene as a fungicide. Large numbers of them subsequently developed cutaneous fragility, bullae, hyperpigmentation, and hypertrichosis, and urinary porphyrins were excreted in amounts and types similar to genetic PCT. Children breast fed by poisoned mothers developed skin lesions of a peculiar type. Short stature, shortened digits, increased pigmentation, and hypertrichosis have persisted for two decades after the acute events.[48] Occurrences of PCT have also been linked to exposure to polychlorinated biphenyls[49] and to dioxin.[50]

PCT has developed in association with benign and malignant liver tumors and with metastatic liver disease.[39] In addition to numerous individuals who have developed PCT-like blisters while receiving chronic hemodialysis treatment for renal failure, a situation associated with increased incidence of mildly elevated plasma porphyrin levels,[51,52] there are also numerous reports of similar patients with extraordinarily high levels of plasma porphyrins of the PCT type who had never had any cutaneous symptoms until after several months' dialysis.[53]

Patients with all of these types of PCT demonstrate similar skin lesions of the vesiculobullous, scarring pattern. Those with the largest accumulations (homozygous children or dialysis-related forms) have the most profound photosensitivity. Children with homozygous PCT may be as severely affected as children with CEP, and may exhibit erythrodontia. Homozygous PCT is also associated with anemia and with hepatosplenomegaly and abnormal liver function in some cases.[54]

Treatment of familial heterozygous or sporadic cases includes avoidance of sunlight exposure, removal of factors that are known inducers, and use of phlebotomy and/or chloroquine phosphate regimens to try to induce remission. In uncomplicated familial heterozygous PCT or in sporadic cases, removal of the inducer may have a beneficial effect even without additional treatment, but in most cases biochemical and clinical improvement is hastened by use of either or a combination of the two therapeutic approaches.

Patients with PCT often have associated increased body iron stores and a tendency toward polycythemia. Increased hepatic iron may play a role in facilitating the enzyme defect.[55,56] Removal of whole blood (several units over a period of several weeks to several months) often produces both biochemical and clinical improvement.[57-61] Although the rationale for this is usually considered to be the beneficial result of removing excess tissue iron,[57-60,62] it has also been suggested that removal of a plasma factor may play a role.[63] Supporting the role of iron mobilization is the observation that replacement of the iron is accompanied by worsening of the disease.[64,65] Supporting the role of the postulated plasma factor is the response of some patients to plasmapheresis.[66,67] Plasmapheresis has also been effective therapeutically for hemodialysis patients with PCT,[68] in whom anemia of chronic disease is a relative contraindication to phlebotomy as a method of treatment.

Chloroquine phosphate has been shown to complex with hepatocyte porphyrin stores in PCT, forming a water-soluble complex that is apparently discharged from the cells and excreted in the urine.[69] This appears to be a hepatotoxic event, for if large doses of the drug are given, patients with PCT may become quite ill, with markedly elevated serum liver enzymes, indicating liver-cell necrosis. Therefore, the dosages used for this indication are much smaller than the amounts used for lupus erythematosus or for malaria. It cannot be used in anephric patients, as dialysis does not remove the complex from the plasma, and patients are worsened.

Patients with PCT as a sequela of hepatic tumor may improve following resection of the lesion. There is no current experience with successful treatment of homozygous PCT.

β-Carotene oral photoprotection is not as obviously beneficial in PCT as in EPP, but may be used in cases in which other treatment is not possible or is ineffective.

Patients with PCT have an increased incidence of glucose intolerance and may have overt diabetes.[70] PCT has also been associated with systemic lupus erythematosus.[70]

D. Acute Intermittent Porphyria (AIP)

AIP is the only porphyria with solely systemic symptomatology. An autosomal-dominant disorder, AIP is characterized biochemically by the constant urinary excretion of increased levels of ALA and PBG.

ALA and PBG are colorless, nonphotosensitizing compounds that accumulate due to a partial block in US as discussed earlier. This defect is clinically significant in hepatocytes, where heme synthesis is active to provide the prosthetic heme for cytochromes, catalases, and peroxidases. Inheritance of the gene defect remains silent during childhood, becoming biochemically manifest after puberty. Clinical expression of the disease may remain latent in adulthood unless the affected individual encounters an inducing factor that triggers systemic symptoms. These symptoms are neurovisceral in nature, and are identical in AIP, VP, and HCP. These symptoms may include one or more chronic complaints (neurosis, psychosis, vague abdominal distress, constipation, neurological dysfunctions) or one or more of the more severe "acute attack" manifestations (coma, paralysis, seizures, psychosis, severe abdominal pain, hypertension, tachycardia, fluid and electrolyte disturbances) that may be life threatening in severity.

Acute attacks may be induced by several factors including dietary carbohydrate restriction, infections, hormonal fluctuations, stress, and an extensive list of drugs,[71,72] some commonly used (i.e., barbiturates, sulfonamides, griseofulvin). Patients subject to these attacks must be informed of their diagnosis and carefully counseled in avoidance of these factors.

Patients recognized to be suffering from an attack are often encountered in hospital intensive care settings. Early recognition of an acute attack is usually treated by infusion of glucose as well as general supportive measures dictated by the condition of the individual patient. High levels of glucose appear to inhibit the rate-controlling enzyme ALAS, thus slowing overproduction of ALA and PBG. The precise molecular mechanisms by which these precursors elicit the neurovisceral effects seen in AIP, VP, and HCP remain uncertain. High-dose propranolol has also been administered with beneficial effect during acute attack episodes,[73] but has significant side-effect potential and must be used with caution.[74] If the attack continues, intravenous infusion of hematin may be tried and may reverse the deregulation of heme synthesis.[75-78] Hematin appears to affect hepatic heme synthesis by providing an increased supply of a heme analogue that exerts negative feedback repression on the rate-limiting enzyme similar to that of heme. This repression slows production of ALA and PBG which then accumulate to a lesser extent at the partially blocked level of US.

Patients with AIP have no apparent defect in erythropoietic heme production. Nevertheless, deficient activity of the enzyme can be detected in erythrocytes as well as in other nonliver cell lines.

E. Variegate Porphyria (VP)

VP is another autosomal-dominant genetic defect resulting in an hepatic porphyria characterized by acute attack episodes.

As in AIP, symptoms are rarely apparent until after puberty, and may also remain latent in adulthood unless triggered by the same factors that are associated with AIP. The acute attacks and chronic systemic disorders are identical to those of AIP and are treated in the same manner.

In contrast to AIP, patients with VP may develop cutaneous photosensitivity, either alone, or in association with any of the systemic complaints. Some VP patients, however, have only systemic symptoms, and VP must be distinguished from AIP by laboratory analyses of porphyrins and precursors in urine, plasma, and stool. The enzyme defective appears to be PO, although in some cases heme synthase may be deficient as well.[17,18] The defect in PO, although demonstrable in several cell lines, is clinically manifested only in hepatocytes. Erythropoietic heme synthesis appears to be adequate, and there is no anemia or accumulation of erythrocyte porphyrins. Treatment for the acute attack and preventive measures are as outlined for AIP.

The cutaneous lesions, if present, are identical to those of PCT. Therapy of these lesions is difficult if the hepatic heme synthetic pathway is induced. Sun avoidance is the mainstay of treatment. β-carotene may be tried, but its efficacy in VP is unclear. Phlebotomy or chloroquine treatments are not helpful in VP.

F. Hereditary Coproporphyria (HCP)

HCP is encountered rarely, but is an autosomal-dominant, acute attack type hepatic porphyria with an occurrence of PCT-like skin lesions in occasional cases.[79]

As in AIP and VP, symptoms are absent in childhood, and are triggered in adult life by the set of agents noted for AIP. The attack symptoms, and treatment thereof, are identical to those of AIP and VP.

Cutaneous lesions are infrequent, but when observed, are similar to those of PCT. Phlebotomy and chloroquine are not effective. There is no associated anemia or accumulation of erythrocyte porphyrins.

REFERENCES

1. Granick, S., The induction in vitro of the synthesis of δ-aminolaevulinic acid synthetase in chemical porphyria: a response to certain drugs, sex hormones, and foreign chemicals, *J. Biol. Chem.*, 241, 1359, 1966.
2. Tschudy, D. P., Porphyrin metabolism and porphyrias, in *Metabolic Control and Disease*, Bondy, P. K. and Rosenbery, L. E., Eds., W. B. Saunders, Philadelphia, 1980, 939.
3. Giger, V. and Meyer, V. A., Role of haem in the induction of cytochrome P-450 by phenobarbitone, *Biochem. J.*, 198, 321, 1981.
4. Granick, S., Sinclair, P., Sassa, S., and Grieninger, G., Effects by heme, insulin, and serum albumin on heme and protein synthesis in chick embryo liver cells cultured in a chemically defined medium, and a spectrofluorometric assay for porphyrin composition, *J. Biol. Chem.*, 250, 9215, 1975.
5. Tschudy, D. P., The influence of hormonal and nutritional factors on the regulation of liver haem biosynthesis, in *Heme and Hemoproteins*, Springer-Verlag, Berlin, 1973, 255.
6. Redeker, A. G. and Sterling, R. E., The "glucose effect" in erythropoietic protoporphyria, *Arch. Intern. Med.*, 121, 466, 1968.
7. Elder, G. H., Haem synthesis and breakdown, in *Iron in Biochemistry and Medicine*, Vol. 2, Jacobs, A. J. and Worwood, M., Eds., Academic Press, New York, 1980.
8. Strand, L. J., Felsher, B. F., Redeker, A. G., and Marver, H. S., Heme biosynthesis in intermittent acute prophyria: decreased hepatic conversion of porphobilinogen to porphyrins and increased δ-aminolaevulinic acid synthetase activity, *Proc. Natl. Acad. Sci. U.S.A.*, 67, 1315, 1970.
9. Romeo, G. and Levin, E. Y., Uroporphyrinogen III cosynthetase in human congenital erythropoietic porphyria, *Proc. Natl. Acad. Sci. U.S.A.*, 63, 856, 1969.
10. Pimstone, N. R., Porphyria cutanea tarda, *Semin. Liver Dis.*, 2, 132, 1982.
11. Elder, G. H., Porphyrin metabolism in porphyria cutanea tarda, *Semin. Hematol.*, 14, 227, 1977.

12. Elder, G. H., Evans, J. O., Thomas, N., Cox, R., Brodie, M. J., Moore, M. R., Goldberg, A., and Nicholson, D. C., The primary enzyme defect in hereditary corproporphyria, *Lancet*, 2, 1217, 1976.
13. Grandchamp, B. and Nordmann, Y., Decreased lymphocyte coproporphyrinogen III oxidase in hereditary coproporphyria, *Biochem. Biophys. Res. Commun.*, 74, 1089, 1977.
14. Brenner, D. A. and Bloomer, J. R., The enzymatic defect in variegate porphyria, *N. Engl. J. Med.*, 302, 765, 1980.
15. Rimington, C., Lockwood, W. H., and Belcher, R. V., The excretion of porphyrin-peptide conjugates in porphyria variegata, *Clin. Sci.*, 35, 211, 1968.
16. Elder, G. H. and Evans, J. O., A radiochemical method for the measurement of coproporphyrinogen oxidase and the utilization of substrates other than coproporphyrinogen III by the enzyme from rat liver, *Biochem. J.*, 169, 205, 1978.
17. Becker, D. M., Viljoen, D. J., Katz, J., and Kramer, S., Reduced ferrochelatase activity: a defect common to porphyria variegata and protoporphyria, *Br. J. Haematol.*, 36, 171, 1977.
18. Viljoen, D. J., Cyanis, E., and Becker, D. M., Reduced ferrochelatase activity in fibroblasts from patients with porphyria variegata, *Am. J. Hematol.*, 6, 185, 1979.
19. Bottomley, S. S., Tanaka, M., and Everett, M. A., Diminished erythroid ferrochelatase activity in protoporphyria, *J. Lab. Clin. Med.*, 86, 126, 1975.
20. deGoeij, A. F. P. M., Christianse, K., and van Steveninck, J., Decreased haem synthetase activity in blood cells of patients with erythropoietic protoporphyria, *Eur. J. Clin. Invest.*, 5, 397, 1975.
21. Bloomer, J. R., Bonkowski, H. L., Ebert, P. S., and Mahoney, M. J., Inheritance in protoporphyria — comparison of heme synthetase activity in skin fibroblasts with clinical features, *Lancet*, 2, 226, 1976.
22. Haining, R. G., Cowger, M. L., Labbe, R. F., and Finch, C. A., Congenital erythropoietic porphyria. II. The effects of induced polycythemia, *Blood*, 36, 297, 1970.
23. Piomelli, S., Poh-Fitzpatrick, M., Skolnick, L., Graziano, J., Berdon, W., and Seaman, C., Supression of porphyrin production in congenital erythropoietic porphyria by hypertransfusion, *Blood*, 63(Abstr.), 49, 1983.
24. Schwartz, S., Johnson, J. A., Stephenson, B. D., Anderson, A. S., Edmondson, P. R., and Fusaro, R. M., Erythropoietic defects in protoporphyria: a study of factors involved in labelling of porphyrins and bile pigments from ALA-^{3}H and glycine-^{14}C, *J. Lab. Clin. Med.*, 78, 411, 1971.
25. Lamon, J. M., Poh-Fitzpatrick, M. B., and Lamola, A. A., Hepatic protoporphyrin production in human protoporphyria: alteration of protoporphyrin levels in blood and feces with intravenous hematin and analysis of red cell protoporphyrin distribution, *Gastroenterology*, 79, 115, 1980.
26. Scholnick, P., Marver, H. S., and Schmid, R., Erythropoietic protoporphyria: evidence for multiple sites of excess protoporphyrin formation, *J. Clin. Invest.*, 50, 203, 1971.
27. Cripps, D. J. and Scheuer, P. J., Hepatobiliary changes in erythropoietic protoporphyria, *Arch. Pathol.*, 80, 500, 1965.
28. Ibrahim, G. W. and Watson, C. J., Enterohepatic circulation and conversion of protoporphyrin to bile pigment in man, *Proc. Soc. Exp. Biol. Med.*, 127, 890, 1968.
29. Wolff, K., Wolff-Schreiner, E., and Gschnait, F., Liver inclusions in erythropoietic protoporphyria, *Eur. J. Clin. Invest.*, 5, 21, 1975.
30. Bruguera, M., Esquerda, J. E., Mascaro, J. M., and Pinol, J., Erythropoietic protoporphyria, *Arch. Pathol. Lab. Med.*, 100, 587, 1976.
31. MacDonald, D. M., Germain, D., and Perrot, H., The histopathology and ultrastructure of liver disease in erythropoietic protoporphyria, *Br. J. Dermatol.*, 104, 7, 1981.
32. Klaskin, G. and Bloomer, J., Birefringence of hepatic pigment deposits in erythropoietic protoporphyria. Specificity and sensitivity of polarization microscopy in the identification of hepatic protoporphyrin deposits, *Gastroenterology*, 67, 294, 1974.
33. Mathews, M. M. and Sistrom, W. R., Function of carotenoid pigments in non-photosynthetic bacteria, *Nature (London)*, 184, 1892, 1959.
34. Mathews, M. M., Effect of beta-carotene against lethal photosensitization by hematoporphyrin, *Nature (London)*, 203, 1092, 1964.
35. Mathews-Roth, M. M., Pathak, M. A., Fitzpatrick, T. B., Harber, L. C., and Kass, E. H., Beta-carotene as an oral photoprotective agent in erythropoietic protoporphyria, *JAMA*, 228, 104, 1974.
36. DeLeo, V. A., Poh-Fitzpatrick, M. B., Mathews-Roth, M. M., and Harber, L. C., Erythropoietic protoporphyria: ten years experience, *Am. J. Med.*, 60, 8, 1976.
37. Lamola, A. A. and Blumberg, W. E., The effectiveness of beta-carotene and phytoene as systemic sunscreens, in Program 4th Annu. Meet. American Society for Photobiology, 1976, 109.
38. Tio, T. H., Leijnse, B., Jarrett, A., and Rimington, C., Acquired porphyria from a liver tumor, *Clin. Sci.*, 16, 517, 1957.
39. Grossman, M. E. and Bickers, D. R., Porphyria cutanea tarda, a rare cutaneous manifestation of hepatic tumors, *Cutis*, 21, 782, 1978.

40. Poh-Fitzpatrick, M. B., Bellet, N., DeLeo, V. A., Grossman, M. E., and Bickers, D. R., Porphyria cutanea tarda in two patients treated with hemodialysis for chronic renal failure, *N. Engl. J. Med.*, 299, 292, 1978.
41. Poh-Fitzpatrick, M. B., Masullo, A. S., and Grossman, M. E., Porphyria cutanea tarda associated with chronic renal disease and hemodialysis, *Arch. Dermatol.*, 116, 191, 1980.
42. Elder, G. H., The metabolism of porphyrins of the isocaproporphyrin series, *Enzyme*, 17, 61, 1974.
43. Elder, G. H., Differentiation of porphyria cutanea tarda symptomatica from other types of porphyria by measurement of isocoproporphyrin in feces, *J. Clin. Pathol.*, 28, 601, 1975.
44. Kushner, J. P., Barbuto, A. J., and Lee, G. R., An inherited enzymatic defect in porphyria cutanea tarda: decreased uroporphyrinogen decarboxylase activity, *J. Clin. Invest.*, 58, 1089, 1976.
45. Elder, G. H., Lee, G. B., and Tovey, J. A., Decreased activity of hepatic uroporphyrinogen decarboxylase in sporadic porphyria cutanea tarda, *N. Engl. J. Med.*, 299, 274, 1978.
46. Elder, G. H., Smith, S. G., Herrero, C., Mascaro, J. M., Lecha, M., Muniesa, A. M., Czarnecki, D. B., Brenan, J., Poulos, V., and DeSalamanca, R. E., Hepatoerythropoietic porphyria: a new uroporphyrinogen decarboxylase defect or homozygous porphyria cutanea tarda?, *Lancet*, 1, 916, 1981.
47. Cam, C. and Nigogosyan, G., Acquired toxic porphyria due to hexachlorobenzene, *JAMA*, 183, 88, 1963.
48. Cripps, D. J., Gocmen, A., and Peters, H. A., Porphyria turcica. Twenty years after hexachlorobenzene intoxication, *Arch. Dermatol.*, 116, 46, 1980.
49. Bleiberg, J., Wallen, M., Brodkin, R., and Applebaum, I. L., Industrially acquired porphyria, *Arch. Dermatol.*, 89, 793, 1964.
50. Poland, A. P., Smith, D., Metter, G., and Possick, P., A health survey of workers in a 2,4-D and 2,4,5-T plant, *Arch. Environ. Health*, 22, 316, 1971.
51. Day, R. S. and Eales, L., Porphyrins in chronic renal failure, *Nephron*, 26, 90, 1980.
52. Poh-Fitzpatrick, M. B., Sossin, A. E., and Bemis, J., Porphyrin levels in plasma and erythrocytes of chronic hemodialysis patients, *J. Am. Acad. Dermatol.*, 7, 100, 1982.
53. Hannon, R. and Callen, J. P., Porphyria cutanea tarda as a cause of bullous dermatosis of hemodialysis. A case report and review of the literature, *Cutis*, 28, 261, 1981.
54. Czarnecki, D. B., Hepatoerythropoietic porphyria, *Arch. Dermatol.*, 116, 307, 1980.
55. Kushner, J. P., Lee, G. R., and Nacht, S., The role of iron in the pathogenesis of porphyria cutanea tarda. An in vitro model, *J. Clin. Invest.*, 51, 3044, 1972.
56. Kushner, J. P., Steinmuller, D. P., and Lee, G. H., The role of iron in the pathogenesis of porphyria cutanea tarda. II. Inhibition of uroporphyrinogen decarboxylase, *J. Clin. Invest.*, 56, 661, 1975.
57. Ippen, H., Allgemeinsymptome der spaten Hautporphyrie (porphyria cutanea tarda) als Hinweise für deren Behandlung, *Dtsch. Med. Wochenschr.*, 86, 127, 1961.
58. Epstein, J. H. and Redeker, A. G., Porphyria cutanea tarda, a study of the effect of phlebotomy, *N. Engl. J. Med.*, 279, 1301, 1968.
59. Kalivas, J. T., Pathak, M. A., and Fitzpatrick, T. B., Phlebotomy and iron-overload in porphyria cutanea tarda, *Lancet*, 1, 1184, 1969.
60. Walsh, J. R., Lobitz, W. C., Mahler, D. J., and Kingery, F. A. J., Phlebotomy therapy in cutaneous porphyria, *Arch. Dermatol.*, 101, 167, 1970.
61. Ippen, H., Treatment of porphyria cutanea tarda by phlebotomy, *Semin. Hematol.*, 14, 253, 1977.
62. Felsher, B. F. and Kushner, J. P., Hepatic siderosis and porphyria cutanea tarda: relation of iron excess to the metabolic defect, *Semin. Hematol.*, 14, 243, 1977.
63. Rifkind, A. B., Sassa, S., Merkatz, I. R., Winchester, R., Harber, L. C., and Kappas, A., Stimulators and inhibitors of hepatic porphyrin formation in human sera, *J. Clin. Invest.*, 53, 1167, 1974.
64. Lundvall, O., The effect of replenishment of iron stores after phlebotomy therapy in porphyria cutanea tarda, *Acta Med. Scand.*, 189, 51, 1971.
65. Felsher, B. J., Jones, M. L., and Redeker, A. G., Iron and hepatic uroporphyrin synthesis. Relation in porphyria cutanea tarda, *JAMA*, 226, 663, 1973.
66. Allen, B. R., Parker, S., Thompson, G. G., Moore, M. R., Darby, F. J., and Hunter, J. A. A., The effect of treatment on plasma uroporphyrin levels in cutaneous hepatic porphyria, *Br. J. Dermatol.*, 93, 37, 1975.
67. Miyachi, S., Shiraichi, S., and Miki, Y., Small volume plasmaphoresis in the management of porphyria cutanea tarda, *Arch. Dermatol.*, 119, 752, 1983.
68. Disler, P., Day, R., Burman, N., Blekkenhorst, G., and Eales, L., Treatment of hemodialysis-related porphyria cutanea tarda with plasma exchange, *Am. J. Med.*, 72, 989, 1982.
69. Scholnick, P. L., Epstein, J. H., and Marver, H. S., The molecular basis of chloroquine in porphyria cutanea tarda, *J. Invest. Dermatol.*, 61, 226, 1973.
70. Grossman, M. E., Bickers, D. R., Poh-Fitzpatrick, M. B., DeLeo, V. A., and Harber, L. C., Porphyria cutanea tarda: clinical features and laboratory findings in forty patients, *Am. J. Med.*, 67, 277, 1979.

71. Moore, M. R., International review of drugs in acute porphyria — 1980, *Int. J. Biochem.*, 12, 1089, 1980.
72. Disler, P. B., Blekkenhorst, G. H., Eales, L., Moore, M. R., and Straughan, J., Guidelines for drug prescription in patients with the acute porphyrias, *S. Afr. Med. J.*, 61, 656, 1982.
73. Atsmon, A., Blum, I., and Fischl, J., Treatment of an acute attack of porphyria variegata with propranolol, *S. Afr. Med. J.*, 46, 331, 1972.
74. Bonkowsky, H. H. and Tschudy, D. P., Hazards of propranolol in treatment of acute porphyria, *Br. Med. J.*, 4, 47, 1974.
75. Watson, C. J., Pierach, C. A., Bossenmaier, I., and Cardinal, R., Use of hematin in the acute attack of the "inducible" hepatic porphyrias, *Adv. Intern. Med.*, 23, 265, 1978.
76. Lamon, J. M., Frykholm, B. C., Bennett, M., and Tschudy, D. P., Prevention of acute porphyric attacks by hematin, *Lancet*, 2, 492, 1978.
77. Lamon, J. M., Frykholm, B. C., Hess, R. A., and Tschudy, D. P., Hematin therapy for acute porphyria, *Medicine (Baltimore)*, 58, 252, 1979.
78. Pierach, C. A., Hematin therapy for the porphyric attack, *Semin. Liver Dis.*, 2, 125, 1982.
79. Goldberg, A., Rimington, C., and Lochhead, A. C., Hereditary coproporphyria, *Lancet*, 1, 632, 1967.

INDEX

A

B

C

D

F

G

H

I

N

O

P

Q

R

S

T

U

V

W

X

Photomedicine

Volume II

Editors

Ehud Ben-Hur, Ph.D.
Department of Radiobiology
Nuclear Research Center-Negev
Beer-Sheva, Israel

Ionel Rosenthal, Ph.D.
Head
Division of Food Technology
Agricultural Research Organization, Volcani Center
Bet Dagan, Israel

CRC Press, Inc.
Boca Raton, Florida

Library of Congress-in-Publication Data

Photomedicine.

Includes bibliographies and index.
1. Phototherapy. 2. Photochemotherapy. 3. Light--
Physiological effect. I. Ben-Hur, Ehud. II. Rosenthal,
Ionel. [DNLM: 1. Phototherapy. WB 480 P5745]
RM837.P48 1987 615.8′3 86-34293
ISBN-0-8493-4673-8 (set)

Direct all inquiries to CRC Press, Inc., 2000 Corporate Blvd., N.W., Boca Raton, Florida, 33431.

International Standard Book Number 0-8493-4674-6 (v. 1)
International Standard Book Number 0-8493-4675-4 (v. 2)
International Standard Book Number 0-8493-4676-2 (v. 3)
International Standard Book Number 0-8493-4673-8 (set)

Library of Congress Card Number 86-34293
Printed in the United States

INTRODUCTION

Go outside and play in the sun. It's good for you.

My mother

The use of sunlight and drugs for the treatment of skin diseases has been documented for over 3400 years; for an even longer time, the reddening, blistering, and tanning effects of sunlight have probably been known. With the discovery of lasers a new dimension was added in the study and application of light in medical therapy. Ophthalmologists adopted the laser in the clinic as a photocoagulator for the treatment of detached retina, and the use of laser as a scalpel for noncontact, noninvasive, and even subcellular surgery is at an earlier state of acceptance. In addition to surgical uses, new, promising ideas are continuing to emerge. Thus, laser can be used to diagnose and treat malignant tumors using photoradiation therapy. This renewed interest, stimulated by the mutual interplay of both scientific and technological innovations, is characterized by a multidisciplinary approach involving physicists, chemists, biochemists, and physicians.

Our objective has been to collect in these three volumes the most up-to-date assessment of our understanding of light in medicine. Since *Photomedicine* was defined as an informative guide to practical applications rather than an esoteric study of medical discipline, the level of medical rigor was reasonably relaxed.

Given limitations on length, the chapters are not intended to be all embracing reviews of the field, but rather to present an overview of key ideas and directions with the objective of delineating the most promising and exciting problems. We hope that the text is sufficiently introductory to stimulate the curiosity and interest of a neophyte, and to simultaneously provide the specialist with a rather short, but current summary of the status of this field. Most important, we hope that the volumes will further highlight this rapidly developing science and spur current and new researchers and ideas.

Ehud Ben-Hur
Ionel Rosenthal

THE EDITORS

Ehud Ben-Hur, Ph.D., was born in Israel in 1940. After graduation from the Hebrew University of Jerusalem in 1965, he went on to study biochemistry at the Technion, Israel Institute of Technology at Haifa, where he obtained his M.Sc. and doctorate degrees. He then joined the Biology Department of Brookhaven National Laboratory as Research Associate where he completed postdoctoral work on the radiobiology of cultured mammalian cells under the auspices of Dr. M. M. Elkind. Upon returning to Israel in 1973, he first joined the Department of Cellular Biochemistry at the Hebrew University and then the Nuclear Research Center-Negev, in 1975, where he is currently engaged in studies of biological effects of ionizing and nonionizing radiations.

The main thrust of his research activity in the past was related to radiation-induced damage in DNA and its repair. During the last few years he has become interested in photodynamic therapy of cancer and is actively involved with Dr. I. Rosenthal in developing new and improved photosensitizers for this purpose.

Dr. Ben-Hur is affiliated with the Department of Radiation Biology, Colorado State University. He is also affiliated with Ben-Gurion University, Beer-Sheva, Israel, where he teaches photobiology. Dr. Ben-Hur has published over 80 papers in scientific journals, is a member of the American Society for Photobiology and the Radiation Research Societies of both the U.S. and Israel, and is on the Editorial Board of the *International Journal of Radiation Biology.*

Dr. Ben-Hur is married with two children and lives most of the time in Beer-Sheva.

Ionel Rosenthal, Ph.D., received his degree in Chemical Engineering from the Polytechnic Institute in Bucharest (Romania) and Ph.D. degree from the Freinberg Graduate School of the Weizmann Institute of Science, Rehovoth, Israel. Dr. Rosenthal has had a very colorful professional career which has included Plant Engineer at "Mahteshim" Chemical Co. and Senior Scientist at the Department of Organic Chemistry at the Weizmann Institute of Science and at the Department of Organic Chemistry, Nuclear Research Center-Negev. Currently he is Principal Scientist at the Department of Food Science, Agricultural Research Organization, Bet-Dagan, and Professor in the Department of Agricultural Biochemistry at the Faculty of Agriculture of the Hebrew University, Jerusalem.

His scientific interests in organic photobiochemistry and food chemistry (and its spin-off: cookery) have resulted in more than 100 research publications in these areas.

ACKNOWLEDGMENTS

The editors wish to acknowledge their indebtedness to the authors for their ready response to our request to contribute, and for their kindness and understanding in accepting their editorial efforts.

CONTRIBUTORS

Esther Azizi, M.D.
Senior Dermatologist
Department of Dermatology
Sheba Medical Center
Tel-Hashomer, Israel

Roger W. Barnes, B.S.
Assistant to the Director
Office of Device Evaluation
U.S. Food and Drug Administration
Silver Springs, Maryland

Ehud Ben-Hur, Ph.D.
Department of Radiobiology
Nuclear Research Center-Negev
Beer-Sheva, Israel

Jeffrey D. Bernhard, M.D.
Director, Division of Dermatology
Director, Phototherapy Center
Assistant Professor
Department of Medicine
University of Massachusetts Medical Center
Worcester, Massachusetts

Homer S. Black, Ph.D.
Director
Photobiology Laboratory
Veterans Administration Medical Center
Baylor College of Medicine
Houston, Texas

Larry E. Bockstahler
Division of Life Sciences
Optical Radiation Branch
U.S. Food and Drug Administration
Rockville, Maryland

S. G. Bown
Department of Surgery
University College London
London, England

John A. S. Carruth
Senior Lecturer in Otolaryngology
Southampton University Hospitals
Southampton, England

James E. Cleaver, Ph.D.
Professor
Laboratory of Radiobiology and Environmental Health
University of California
San Francisco, California

Farrington Daniels, Jr., M.D., M.P.H.
Emeritus Professor of Medicine
The New York Hospital-Cornell Medical Center
New York, New York

John H. Epstein, M.D.
Clinical Professor of Dermatology
University of California Medical School
San Francisco, California

I. Farine, M.D.
Professor and Head
Orthopedic Department
Sheba Medical Center
Tel-Hashomer, Israel

Richard P. Felten
Department of Health and Human Services
Division of Life Sciences
U.S. Food and Drug Administration
Rockville, Maryland

Anna Flint, Ph.D.
Department of Dermatology
Sheba Medical Center
Tel-Hashomer, Israel

Orna Geyer, M.D.
Department of Ophthalmology
Tel Aviv Medical Center
Tel Aviv, Israel

Barbara A. Gilchrest, M.D.
Professor and Chairman
Department of Dermatology
Boston University School of Medicine
Boston, Massachusetts

Philip C. Hanawalt, Ph.D.
Professor and Chairman
Department of Biological Sciences
Stanford University
Stanford, California

M. Heim, M.D.
Department of Rehabilitation
Sheba Medical Center
Tel-Hashomer, Israel

Kiki B. Hellman, Ph.D.
Senior Scientist
Division of Life Sciences
U.S. Food and Drug Administration
Rockville, Maryland

Victoria M. Hitchins, Ph.D.
Division of Life Sciences
Optical Radiation Branch
U.S. Food and Drug Administration
Rockville, Maryland

H. Horoszowski, M.D.
Head
Department of Orthopedy
Sheba Medical Center
Tel-Hashomer, Israel

Elizabeth D. Jacobson, Ph.D.
Acting Director
Office of Science and Technology
Center for Devices and Radiological Health
U.S. Food and Drug Administration
Rockville, Maryland

W. Patrick Jeeves, Ph.D.
Department of Medical Physics
Ontario Cancer Treatment and Research Foundation
Hamilton, Ontario, Canada

Anthony Lamanna
Division of Life Sciences
Optical Radiation Branch
U.S. Food and Drug Administration
Rockville, Maryland

Robert J. Landry, Ph.D.
Electrooptics Branch
U.S. Food and Drug Administration
Rockville, Maryland

Moshe Lazar, M.D.
Department of Ophthalmology
Ichilov Hospital
Tel-Aviv, Israel

Sidney Lerman, Ph.D.
Professor
Department of Opthalmology
Emory University
Atlanta, Georgia

Jerome I. Levine, Ph.D.
Division of Life Sciences
Optical Radiation Branch
U.S. Food and Drug Administration
Rockville, Maryland

Julia G. Levy, Ph.D.
Professor
Department of Microbiology
University of British Columbia
Vancouver, Canada

C. David Lytle, Ph.D.
Division of Life Sciences
Optical Radiation Branch
U.S. Food and Drug Administration
Rockville, Maryland

I. A. Magnus, M.D., F.R.C.P.
Professor Emeritus
Photobiology Department
Institute of Dermatology
London, England

Micheline M. Mathews-Roth, M.D.
Associate Professor of Medicine
Channing Laboratory
Harvard Medical School
Boston, Massachusetts

Daphne Mew, Ph.D.
Faculty of Medicine
Foothill Hospital
University of Calgary
Calgary, Alberta, Canada

K. Mohan, Ph.D.
Science & Technology
U.S. Food and Drug Administration
Rockville, Maryland

Warwick L. Morison, M.D.
Associate Professor of Dermatology
Johns Hopkins Medical Institutions
Baltimore, Maryland

D. Phillips
Royal Institution
London, England

Maureen B. Poh-Fitzpatrick, M.D.
Associate Professor of Dermatology
Columbia University
New York, New York

Michael K. Reusch, M.D.
Department of Dermatology
Stanford University Medical School
Stanford, California

Ionel Rosenthal, Ph.D.
Head
Division of Food Technology
Agricultural Research Organization, Volcani Center
Bet Dagan, Israel

Stephen M. Sykes
Division of Life Sciences
Optical Radiation Branch
U.S. Food and Drug Administration
Rockville, Maryland

Morris Waxler
Office of Science and Technology
Center for Devices and Radiological Health
U.S. Food and Drug Administration
Rockville, Maryland

Abraham Werner, Ph.D.
Chief Physicist
Department of Oncology
Sheba Medical Center
Tel-Hashomer, Israel

Brian C. Wilson, Ph.D.
Department of Medical Physics
Ontario Cancer Treatment and Research Foundation
Hamilton, Ontario, Canada

Bruce U. Wintroub
Department of Dermatology
University of California
San Francisco, California

TABLE OF CONTENTS

Volume I

Volume II

Volume III

Chapter 1

THE IDIOPATHIC PHOTODERMATOSES

I. A. Magnus

TABLE OF CONTENTS

I. PHOTODERMATOSES OF UNKNOWN ETIOLOGY

Earlier recent reviews on the photodermatoses include the relevant chapters and sections in the books of Harber and Bickers[1] and of Johnson.[2] This chapter will be dealing with only four types of photodermatoses (Table 1). These may be called the "primary" or idiopathic photodermatoses. See Table 2 for a summary of their main features; this table is necessarily oversimplified and the full text must be consulted.

The secondary photodermatoses, which are dealt with elsewhere, are briefly listed in Table 3. In addition (not shown in these tables), chronic, life-long exposure to sunlight is associated with (1) aging changes and (2) skin cancer. These may be considered "normal" responses in that probably all normal people with a pale skin would be subject to these changes given sufficient radiant exposure. The relative absence of melanin pigment in white skin must lead to impaired photoprotection against aging and cancer as compared with Asiatic or Negroid skin, and this is surely important. However, other genetic factors, as yet undetermined, also play a part in "normal white populations". These seem to include, as well as having a pale skin, being of "Celtic" heritage. As yet, what the relevant qualities of being Celtic are have not been isolated, let alone quantified or scientifically characterized.

Table 1
THE PRIMARY PHOTODERMATOSES

1. Polymorphic light eruption (Rasch)
2. Actinic or summer prurigo (Hutchinson)
3. Chronic actinic dermatitis (actinic reticuloid and photosensitive eczema)
4. Hydroa vacciniforme (Bazin)

II. POLYMORPHIC LIGHT ERUPTION

A. Introduction

More or less minor rashes of the skin, thought to be associated with solar exposure, had been described from the start of what may be called modern dermatological literature, viz., Robert Willan[4] wrote in London, just before the turn of the 18th century, on ''eczema solaris''. Willan is credited as the pioneer of a proper classification of skin eruptions, and it is to him that credit is probably due for first associating solar exposure with a skin disorder. Other early writers were Rayer[5] in Paris and Veiel[6] in Vienna. Presumably, all these observers were writing about what is here termed polymorphic light eruption, but this is not always certain. Uncertainty has long bedeviled early medical classifications. The position is not much better today, where the main difficulty remains putting an agreed terminology to an agreed morphological and clinical state, where mechanism is unknown.

1. Terminology and Relation to Other Photodermatoses (Especially Actinic Prurigo)

The story of this has been reviewed previously by several writers, viz., Rasch[7] and also Magnus.[8] The term polymorphic light eruption (PLE) was introduced by Carl Rasch in 1900 in a brief note describing two patients, mother and daughter, who had what he called an ''eczema-like polymorphic light eruption''. In a later review, Rasch[7] tells how earlier, his pupil Haxthausen[9] in 1918 had proposed the forming into one disease group the eczematous condition with another pruriginous condition described by Hutchinson.[10] This has turned out to be an unfortunate disservice to clear nomenclature.

Thus, seemingly was the beginnings of a widespread and continuing confusion of two separate dermatoses. In retrospect, it is tempting to ask how such distinct skin disease pictures could possibly be confused for so long.[2]

In spite of the confusion, there had always been some who had emphasized the distinctions between the various solar-induced rashes on the basis of morphology, e.g., Epstein,[11] who wrote about ''solar erythema and eczema'' and ''summer prurigo'' as separate clinical entities.

With the more recent observation of sun-induced rashes in indigenous races of the New World, the muddle of nomenclature has returned. The solar-induced prurigo in patients of North, Central, and South America has been called PLE. This has continued the confusion.[12–16]

Magnus strongly supports the view that summer (or actinic) prurigo and PLE must be separated on the grounds of their (1) having different morphology, (2) taking a different clinical course, (3) responding to a different therapy, and (4) reacting with possible differences in T cells.[17–19]

B. General Features

1. Racial and Geographical Distribution

The racial and geographical distribution of PLE has received little or no notice. This is, presumably, partly because of differences in terminology. Such an endeavor, to

Table 2
SUMMARY OF CHIEF CHARACTERISTICS OF THE FOUR PRIMARY PHOTODERMATOSES

	Prevalence	Characteristic primary lesion	Diagrammatic view from above and profile	Progression of lesion	Other skin lesions	Body surface distribution	Demonstrable photosensitivity	Characteristic lesion histology
Polymorphic light eruption	Common	Papule, 1—2 mm	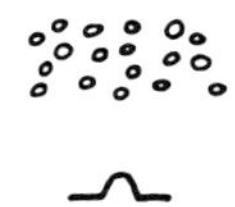	Resolves 1—2 weeks, without scars	Rare, urticarial, target-like	Diffusely; exposed limbs, V of neck, trunk; not face	+ or ±	±
Chronic actinic dermatitis	Less common	Varied and not specific		Persistent	Reticulosis-like papules (2—5 mm) or plaque (1—3 cm)	Diffusely; not strictly on exposed areas	+++ or ±	+++ or ±
Actinic prurigo	Uncommon	Prurigo papule, 1—3 cm, with surmounting scratched vesicle 1—2 mm		Slowly resolves, ± superficial scars	Not usual, except in exacerbation, edema, and papules	Diffusely; mostly exposed limbs and face; also covered areas	+ or ±	±
Hydroa vacciniforme	Very rare	Closely grouped tense vesicles, 1—2 mm each; umbilicated larger edematous papule, 1 cm		Lesions coalesce, form scabs; resolves 2—4 weeks; depressed scars, 3—6 mm	Lesions often in different stages, some umbilicated, some scabbed	Lesions in restricted number of groups; exposed limbs and face	±	+

Table 3
THE SECONDARY PHOTODERMATOSES

I. Photodermatoses associated with known external or exogenous chemicals or chemical photosensitizers
   A. Therapeutic drugs, which include the following
      1. Certain antibiotics, especially tetracyclines and in particular demethyloxytetracycline
      2. Tranquilizers, especially chlorpromazine and protriptyline
      3. Antibacterials, namely nalidixic acid and sulfonamides, especially with trimethoprim
      4. Oral diuretics such as hydrochlorothiazide and furosemide
      5. Oral antidiabetics such as chlorpropamide
      6. Other miscellaneous drugs[3]
   B. Industrial chemicals or products, which include the following
      1. The products of the destructive distillation of coal, e.g., pitch, anthracene, and related substances
      2. Dyes, e.g., of the hydroquinone and triphenylmethane families
   C. Botanical products: some plants of the parsley, celery, figs, citrous families. The known active photochemical agents are nearly always psoralens as in bergamot oil. Of unknown importance is hypericin, an established photosensitizer in cattle but very doubtful in humans
   D. Cosmetics: these include ingredients incorporated into soaps, perfumes, etc. of which the incriminating agents may include halogenated salicylanilides, certain nitromusks, *p*-aminobenzoic acid, cinnamates, and psoralens (bergamot oil)

II. Photodermatoses associated with endogenous photosensitizers or metabolic abnormalities
   A. The porphyrias
   B. Certain genodermatoses, associated mostly with dwarfism, including xeroderma pigmentosum, Cockayne, Bloom, and Rothmund-Thomson syndromes, ataxia telangiectasia, Hartnup disease, and dyskeratosis congenita. In the first five of these, there is evidence of either abnormal mechanisms associated with DNA repair or synthesis, or RNA synthesis, or of features such as highly abnormal spontaneous sister chromatid exchange
   C. Pellagra

III. Dermatoses, usually not associated with UVR, but liable to exacerbation or precipitation by solar exposure[2,3]

determine world prevalence in a relatively uncommon dermatosis, would necessitate a multicenter cooperative project. Some suggestive points, however, can be made.

PLE occurs essentially in places of mild temperate climate, e.g., in Europe, mostly in regions higher than about latitude 50° N, as in Holland, England, and Scotland, southern Scandinavia, and northern Germany; and in North America, near latitude 42° N upwards on the West and East Coasts. However, PLE does not seem to occur nearly as often in regions of similar latitudes in the Southern Hemisphere, as amongst the Europeans of South Africa. PLE is rare also in Australia and New Zealand.

PLE seems to be more common in the relatively pale-skinned people, as is evidenced by case collections written up by Haxthausen[9] of Copenhagen and by Wiskemann and Wulf[20] of Hamburg. It would be surprising if genetic pigmentation did not play a decisive protective role, and indeed, PLE is almost unknown in the natives of India.[21] Wiskemann and Wulf[20] had noted that their white-skinned patients failed to show the normal immediate pigmentation reaction in response to irradiation with UVA and wondered whether this had a causative role. However, PLE can occur in the very dark skinned, e.g., it occasionally has been described in Black Americans,[22] and in Central Africa, Verhagen et al.[23] noted a number of cases. In St. John's Hospital, London, nine cases of PLE among Asian immigrants have been seen in a period of about 15 years. The clinical picture in the cases the author saw is identical to that of the pale-skinned native.

An important part is played by seasonal changes in sunshine. Magnus[8] previously discussed this point when the apparent absence of PLE had been noted in parts of the

Table 4
AGE OF ONSET OF PLE IN TWO LARGE STUDIES

	0—10	11—20	21—30	31—40	41 and over
Dundee, Scotland	50%	23%	27% (21 years and over)		
London and southern England	9%	27%	33%	16%	15%

world that did not have marked seasonal changes in sunlight. Generally, the sun is more or less uniformly strong. This condition occurs in the southern U.S., e.g., Texas, parts of southern Africa, and Australasia. A proper survey of the prevalence of PLE in such geographical sites, where there is an Anglo-Saxon or Nordic population, would be valuable. If, as one suspects, PLE is rare or unknown in these regions, the relationship of this disease to the phenomenon of seasonal ''hardening'' or ''habituation'' in patients with PLE is of interest. Some earlier writers had noted the tendency of PLE to be less severe in many patients toward the middle and end of the sunny season; this suggests some kind of hardening process.

2. Prevalence

In the majority of cases, PLE does not seem to cause significant disability and probably many cases are too mild to attract dermatological consultation. It has been claimed, though seemingly on no ascertainable published authority, that including the very mildest cases, PLE may affect about one in ten of the population as a whole. This very high figure applies supposedly only to a racial stock predominantly Caucasian in mild, northern temperate climates. A proper survey on this would be valuable.

3. The Seasonal and Long-Term Course of the Disorder

PLE, as defined here, is commonly a relatively mild recurrent condition, occurring in short attacks each lasting several days. Sometimes there are individual attacks which reduce in number or cease before the middle or end of the sunny season, Thus, there may be season amelioration earlier than would be expected from the amount of sunlight available.

The long-term tendency for the disorder is for it to persist indefinitely; this has been shown by a follow-up study[24] on some 300 cases of PLE observed for varying times up to 30 years at St. John's Hospital, London. There was almost no predilection to the disease undergoing natural termination. Smaller surveys by others have given essentially the same result.

4. Age of Onset

This is very variable and is probably mostly due to variation in diagnostic criteria for PLE (see Table 4). Onset appears to be in an earlier group in Scotland, compared with the more general spread noted in England.

5. Sex Distribution

Wiskemann and Wulf,[20] Frain-Bell et al.,[25] Magnus,[8] and Jansén[26] all found female patients to be much commoner, approximately two to one. However, this is to be contrasted with the findings of Epstein,[22] who alone found the sex incidence about equal in his study of about 80 patients. There seems to be very little other data available in the literature on large studies of cases.

6. Family Prevalence

This seems to occur not uncommonly, but with what frequency it is difficult to say because of the different criteria for diagnosis of PLE and whether there is a distinction made with actinic prurigo. Johnson[27] points out that out of four different groups of papers on so-called PLE patients, the frequency is from 14 to 75%. One of these papers is by Birt and Davies, who, it is clear, call PLE what most other people name actinic prurigo (see Section IV.A).

7. Underlying Mechanism

This is generally assumed to be due to an allergic process of some kind. Evidence of this has not been easy to find, and what evidence exists is circumstantial and inconsistent. Thus morphologically, the rash is similar to a contact dermatitis. Data supporting a cell-mediated immunological reaction were found by Horkay and Meszáros,[28] but this was not confirmed by Raffle et al.[29] or by Jansén and Helander.[30] The excision-repair systems for correcting UV-induced damage in DNA seem to be normal in PLE.[31]

C. Clinical Features

1. Precipitation of an Attack

First, patients usually connect an outbreak of erythemato-papular lesions with their being exposed to sunlight. Usually an individual exposure is linked to a particular attack of skin lesions. This is the more common pattern of events. In other patients, there is a gradual build up of the number of lesions after a series of daily solar exposures.

Second, patients usually note that lesions appear after a definite latent period, which separates the solar exposure from the onset of the rash. This usually is about 6 to 18 hr, but less commonly may be prolonged to 2 to 3 days. A common event is for solar exposure just before midday to yield a rash near bedtime. A very short latent period of seconds or minutes is, properly, a feature of solar urticaria; such a urticarial or urticated type of response may be seen with the more usual erythemato-papular eruption developing later, viz., after 12 hr or so. Thus, in the majority of instances solar urticaria seems to be part of the PLE syndrome.[8]

Third, PLE lesions of the erythemato-papulo-vesicular type generally begin to fade in 24 to 36 hr, begin to clear in 2 to 3 days, and usually clear completely within 1 week, provided aggravation from inappropriate treatment and further exposure of the skin to sunlight has not occurred. The short course of individual attacks may frequently result in the dermatologist not seeing the patient while lesions are still present, for the typical PLE attack subsides without sequelae. The author found it worthwhile to encourage patients to have their lesions photographed by spouse, relations, friends, etc. Amateur clinical photography frequently helps to confirm a history suggesting PLE (Figures 1 and 2).

Fourth, precipitation of lesions in temperate climates is usually by sunlight on a clear day, 2 or 3 hr before or after midday. Cloudy conditions may tend to reduce the power of the sun, and the patient may escape trouble.

2. Quality of Sunlight

This is determined by geographical, meteorological, and environmental factors. Many discerning patients learn that summer sunlight during midday is more likely to precipitate attacks than sunlight in the early morning or late evening. Some may note worsening of attacks when in high altitudes. Some patients may also note that sunlight through window glass does not provoke lesions. (However, aggravation of existing lesions by the heat of sunlight coming through glass windows should not be confused with the precipitation of new attacks.) Many patients claim that sunlight through car windows or windshield will provoke lesions, and such lesions may be unilateral, corresponding with the position of the sun during a more or less long drive. (This is in

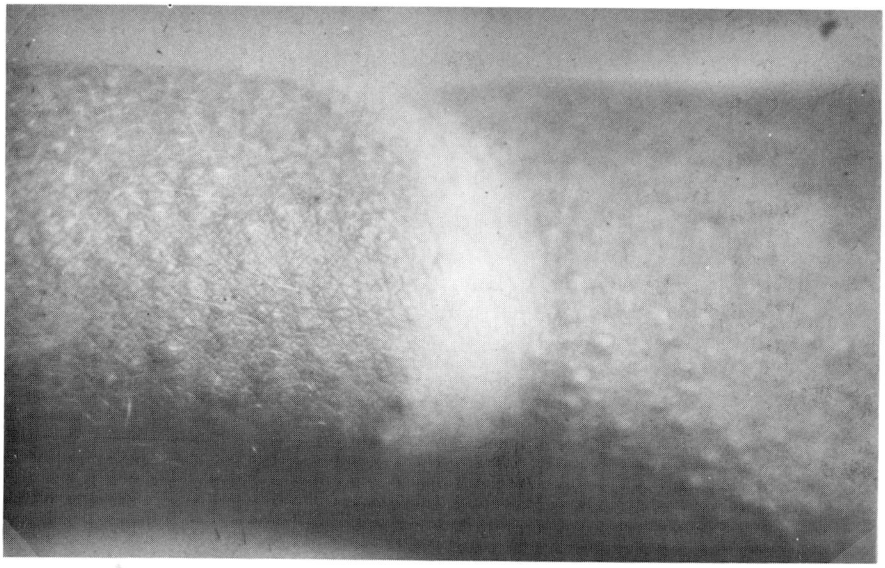

FIGURE 1. PLE, papular lesions, 1- to 2-mm diameter. Close-up of lesions on exposed part of back of hand.

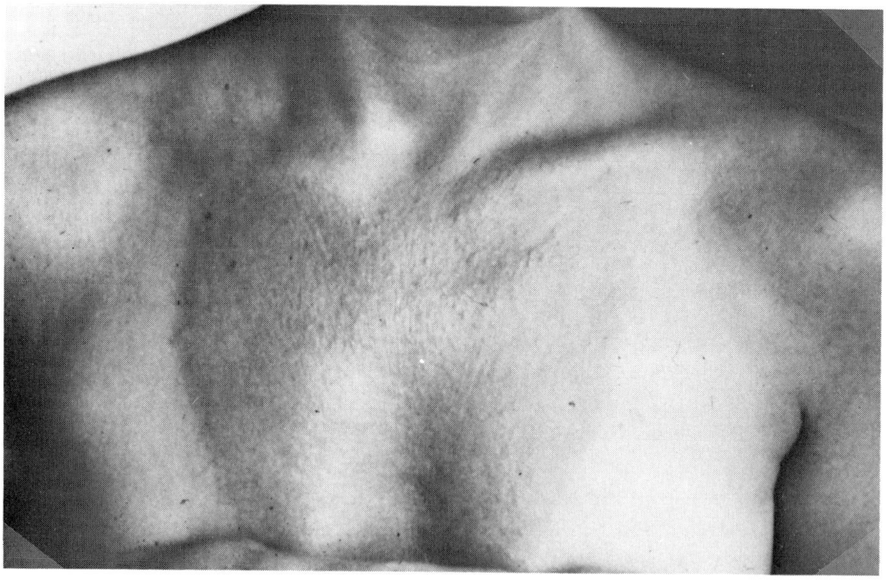

FIGURE 2. PLE. General view of diffuse distribution of lesions on exposed shoulders and adjoining chest. Note areas spared by protective effect of clothing. (Photograph taken during vacation by a member of patient's family.)

spite of the glass concerned supposedly stopping the transmission of UVB wavelengths.) The aggravation by heat from the sun or from a radiant electric file is, one supposes, a nonspecific thermal effect leading to warming of the skin, vasodilatation, and itching. Subsequently, there may well be aggravation of the rash by scratching.

Lesions provoked by sunlight through clothing have been noted by some observers and have been taken to indicate an allergic mechanism (humoral or cell mediated) is in operation "at a distance", but experience with patients in London points to such a happening as rare. Clothes seem to give excellent protection against UVR-provoked lesions; full coverage would afford very effective preventive therapy, were it not that present fashion dictates otherwise (see Figure 2). One may speculate as to whether such a rash on a clothes-covered area is more part of a generalized irritability of the skin, viz., so-called "angry back" syndrome, sometimes seen in patients undergoing multiple patch-test investigations.

3. Symptoms

Itching is a principal feature and may precede the rash. Thus, scratching is often an important aggravating factor, especially when lesions are excoriated by the sleeping patient. In some children, scratching leads to impetiginous complications and the prolonging of an attack.

Patients frequently describe the rash as "nettle rash", "hives", or "sweat rash". This is confusing, but a careful clinical history should clarify matters.

4. The Morphological Features

The most common lesions are erythema and papules. The papules are small, 2 to 5 mm in diameter, usually discrete, and are dispersed more or less regularly on exposed sites on an erythematous background. Sometimes there is a tendency to form confluent sheets of more densely packed lesions (Figure 1). Erythematomaculo-papules are as far as the lesions in an attack will progress in most instances and constitute the complete morphological picture. The development of vesicles from the papules is less common, though in some instances the lesions have vesicular or urticated features from the start; these are often particularly itchy and are often scratched.

It must be admitted that it is not usual for the dermatologist to have the opportunity to see the progress of an individual attack of lesions from beginning to end. Either he sees the final stages of resolving papules or more often nothing, as the attack has cleared.

a. Varieties of Lesions

Certain observers[32] have described PLE papules as being of large as well as small type. These large lesions seem to be rare, but the category "large papule" seems sometimes to be used as a handy term for a nodule or plaque.[33]

True plaque-like PLE lesions, originally noted by Lamb et al.,[34,35] are another morphological example of PLE; again, these have been most unusual in the London patients. This clinical picture may lead to diagnostic difficulties, for the appearance may resemble lupus erythematosus.

Small, round "target"-type lesions, more or less reminiscent of erythema multiforme, may occur in some patients, but are rare. Proper erythema multiforme is not to be confused with the previously described lesions and should not be included under the "umbrella" of PLE. Sunlight-induced erythema multiforme is best treated as a separate entity from PLE.

Solar urticaria is one of the rarest types of lesion seen in PLE (for further information on solar urticaria, see Reference 27). It may be possible to elicit from the history that a patient has *two distinct types* of skin reaction to sunlight, viz., a more or less immediate response of solar urticaria, followed later by a delayed response, the more usual papular type of reaction of PLE.

5. Areas of the Body Affected

In PLE it is characteristically only sunlight-exposed skin areas that are affected; clothed areas, areas protected by long hair, e.g., the ears, and relatively shaded areas,

Table 5
SPECTRAL DISTRIBUTION IN 180 PATIENTS WITH PLE, FOR PRODUCTION OF ABNORMAL RESPONSES TO MONOCHROMATOR LIGHT TESTS

Wavelength (nm)	307.5	320	340	360	380	400
Males	13	15	10	2	0	0
Females	49	59	15	12	4	1

Note: Figures indicate number of patients with upper limit at wavelength indicated. All patients reacted at 300 nm, usually the shortest wavelength region tested. Tests were conducted between January 1979 and December 1983.

as under the nose and the chin, escape. The upper eyelid, retracted during waking hours, is also relatively unaffected. Other areas that excape are the flexures of the elbows and knees. Curiously, also, the face seldom suffers in PLE. In fact, where facial lesions have occurred, in the experience of the author, the diagnosis has proved to be not PLE but actinic prurigo (AP) (see Section IV.A). Diffuse rash on the end of the nose is particularly characteristic of AP and rules strongly against a diagnosis of PLE.

Sites of predilection are the extensor surface of the forearms, the backs of the hands and knuckles, the vee of the neck (especially in females), the fronts of the thighs and tibial regions, and the dorsum of the feet should the lower limbs be unclothed, as may occur in some leisure activities.

D. Histological Appearance

Skin biopsies of lesions examined with routine histological techniques, viz., stained with hematoxylin and eosin, show only what are largely nonspecific changes, and are helpful only in excluding other conditions such as chronic actinic dermatitis and erythropoietic protoporphyria.[36]

E. Investigations

In the clear-cut case, no special tests are required. However, where facilities allow, it may be desirable to perform irradiation tests, preferably monochromatic. Most observers seem to think that these tests should be done on skin not normally exposed to sun and clinically showing no lesions. In many instances the most convenient area is on the back. These show, in varying numbers of patients, abnormal reactions predominantly in the UVB spectrum. Occasionally adjoining UVA wavelengths, viz., up to 360 nm, also cause abnormal experimental responses (see Table 5). However, entirely normal results may also be found in patients who are undoubtedly photosensitive on clinical grounds. It is not clear why results are so variable.[25,27]

There is another school of thought which calls for testing patients on abnormal or exposed skin, i.e., that showing natural lesions. There are others who give repeated irradiation on the same skin site. These procedures were started by Epstein,[37] and were later used occasionally by Rottier,[38] and most recently by Hölzle et al.[39] The results of Hölzle and co-workers, who claim *only* UVA to be causative, not UVB, pose a paradox. Possibly, their results may be due to their testing procedure in which large doses of UVA are given repeatedly to exposed-skin sites, seemingly affected with lesions. This is quite different from the tests done by investigators on what is clinically normal skin.

It would appear that the validity of results using abnormal skin needs to be further substantiated in controlled investigations on further test subjects, i.e., on the abnormal skin of patients with other nonphotosensitive dermatoses, discoid eczema and seborrheic dermatitis, not merely on the skin of nonphotosensitive subjects. Much further work is certainly merited here, and for the present, the significance of phototesting, whatever the methods used, must remain unclear. Until proper control studies are done using the procedures where repeated irradiation is made of natural lesions, one must remain uneasy as to the meaning to attach to these procedures.

F. Treatment

This depends on the severity of the disorder. The very mildest attacks will clear rapidly on their own; further attacks can be avoided by simple preventive measures such as wearing protective clothing and scrupulously avoiding exposure to sunlight during the 4 hr or so around midday when the UVB in sunlight is sufficient to be troublesome.

When exposure is unavoidable, sun-barrier preparations have to be used. Again, these have to be used meticulously. The higher the "sun protective factor", or SPF, the more successful the prevention of lesions. SPFs as high as 15 to 20 are claimed, and if confirmed could be significant; but SPFs are probably only approximate guides to efficacy as far as their use in photodermatoses is concerned. Details of sun-barrier preparations are dealt with elsewhere (see also Murphy and Hawk[40]). It must suffice to mention here that even an efficient sun barrier has to be applied much more frequently in a patient with a photodermatosis than is usually advised by the manufacturer for normal persons — hourly instead of two to three times daily. Furthermore, the efficacy of sun barriers does not seem to survive the washing-off effects of bathing and heavy perspiration as well as is claimed in normal persons.

Thus it is necessary to emphasize that sun-barrier preparations do not seem to work as well on patients as they do on normal subjects. Endeavoring to avoid a rash precipitated in ordinary day-to-day circumstances is quite different from the vacationer seeking to control the extent of his or her deliberate sun exposure.

Apart from their relative inadequacies in successful preventive therapy, other than in the milder cases, sun-barrier preparations have a number of disadvantages. For practical purposes, they are only more or less effective in the UVB spectral range; effective agents in the UVA are constantly being sought, but so far without proven efficacy in human photodermatoses. Fortunately, most patients with PLE are sensitive mostly in the UVB range, the UVA wavelengths being predominantly unimportant. Very fortunately, PLE patients are not sensitive to visible radiation, and indeed if they were, the sun-barrier preparation would necessarily have to contain colored active ingredients or white reflectants.

Treatment with various oral drugs has been advocated for some 40 years, starting with the synthetic antimalarials, originally mepacrine. Next chloroquine and/or hydroxychloroquine were recommended for a number of years. All these drugs seemed to be of more or less therapeutic benefit, though how they worked was quite uncertain. When (more recently) these drugs were submitted to controlled therapeutic trials, most were shown only to have relatively minor, somewhat disappointing efficacy. Hydroxychloroquine appears to be the best, with a "drug protection factor" (analogous with the SPF of a sun barrier) as high as eight, according to the results of one particular trial. Unfortunately, the results of all these controlled trials and values of the protective factors must be treated with caution. They are only approximate guides to efficacy, because it is impossible to arrange a trial on sufficient numbers of patients to yield a statistical result of satisfactory power. If the results are somewhat indecisive, this probably indicates that the drugs tested are not outstandingly efficacious — not much more than a placebo.

Nonsteroidal anti-inflammatory drugs have been tried in PLE, and one of them, ketoprofen, in a controlled trial, was found to have no significant benefit over the placebo.[41] Interestingly, the data for efficacy of the placebo was better than that for ketoprofen, which seems to suggest that ketoprofen may make the PLE worse, i.e., it may be a photosensitizer. Although this result with ketoprofen was not statistically significant, soon afterwards yet another nonsteroidal anti-inflammatory drug, benoxaprofen, was introduced with undeniable photosensitizing action.

β-Carotene, alone or combined with canthoxanthin has some vogue in the treatment of PLE, though it is much more popular in the prevention of photosensitive symptoms in porphyria, especially in erythropoietic protoporphyria. Nearly all reports on the use of carotenoids in therapy, whatever the disease, have been uncontrolled and thus are not decisive. The practice usually followed is to give the oral preparation of β-carotene and/or canthoxanthin in sufficient dose to produce a definite yellowing of the skin (serum or plasma concentration of carotenoids should reach about 10 to 12 μmol carotene per liter[42]). Again, controlled therapeutic trials in PLE patients failed to show any benefit in one instance, and a benefit of little more than marginal degree in another instance. Once more, the number of patients tested in both trials was somewhat small.[41] Probably there is no harm in prescribing carotenoids and they may be regarded as placebos; most patients do not seem to mind yellowing of the skin.

The most effective treatment may be so-called PUVA.[43] The results of a controlled therapeutic trial were clear cut statistically, though there were, as before, too small a number of patients examined.[44] In particular, there was not much difference between the PUVA-treatment group of patients and a similar group undergoing a multiexposure regimen receiving simple UVB irradiation. Possibly, the latter is to be preferred to PUVA; certainly more work is required in this field.

III. CHRONIC ACTINIC DERMATITIS

A. Introduction

1. Definition

This is a chronic photodermatosis of varying degrees of severity, sometimes serious and debilitating. The clinical, photobiological, and histological manifestations may be present in markedly pathological form. The condition was orignally recognized in its severe state; later a milder disorder was reported. Now the two are thought to represent one disease process.[45,46] The condition has been subject to controversy and confusion, partly due to lack of agreement as to how to define the clinical picture, but particularly because of poor understanding of and mistaken attitudes toward the underlying mechanisms. Nevertheless, the fact that convincing reports of cases have recently appeared from various quarters,[47,48] suggests that some consensus of opinion is developing.

2. Synonyms

The rather curious but quite apt title actinic reticuloid, introduced by Ive et al.,[49] brings out two peculiar features. On the one hand, there is an extensive and severe dermatosis, sometimes amounting to an erythroderma, associated with a severe skin photosensitivity, often of spectacular severity and spectral extent. On the other hand, there is a histological and clinical picture resembling cutaneous lymphoma, in particular mycosis conditions. A milder disease, which had at one time been assumed to be a separate entity, unrelated to actinic reticuloid, was termed "photosensitive eczema" by Ramsay and Kobza Black.[50] However, these two conditions are now thought to be aspects of the same disorder, and there is evidence that one condition may pass into the other, i.e., actinic reticuloid may turn into photosensitive eczema, and vice versa.[46] The whole picture had already been termed "photosensitive dermatitis-actinic reticuloid syndrome'', shortened to PD/AR syndrome. This was used by Frain-Bell et al.,[45] who particularly wanted to indicate that they were one disease. The shorter, more pithy ''chronic actinic dermatitis'' is what is preferred by Hawk and Magnus[46] and by Ramsay,[51] and is used here in this chapter. See Figure 3 for the various clinical, photobiological, and histological combinations that are possible.

3. Etiology

As mentioned earlier, this remains largely unclear. There are some seven causal mechanisms that have been proposed. Enumerated in approximate order of published appearance, they are

1. A manifestation of photocontact dermatitis, as is suggested by the occasional positive photopatch tests to materials to which certain patients may have been exposed — mostly ingredients such as germicides or cosmetics (salicylanilides and nitromusks[51])
2. Due to a humoral immunological mechanism, presumably type 1 immunological reaction of Gell and Coombs because of positive serum transfer tests, reported by Degos et al.[52]
3. A neoplastic condition related to the lymphomas[53]
4. Related to contact sensitivity, possibly to chromium and to the sesquiterpene lactones of oleoresins[54,55]
5. A manifestation of granulomatous reactions associated with or related to the Sézary syndrome[56,57]
6. Related to tryptophan and kynurenic acid metabolism[58-60]
7. Due to a failure of normal free-radical neutralization processes[61]

B. General Features

1. Prevalence

When first described, the condition was assumed to be rare, but this is clearly not so now. The diagnosis is seemingly being made more often as dermatologists become more aware of the disease and the facilities for investigating it increase in availability. In the 2 years and 3 months of October 1981 to December 1983, 107 new cases have been seen at St. John's Hospital for Diseases of the Skin, London; the average rate of reaching a firm diagnosis is currently about 1/week. Case reporting of CAD, mostly in more or less isolated instances, seems to continue and is probably on the increase.

2. Sex Distribution

Males predominate in patients based in the London area, approximately 10:1; similar findings occur elsewhere.[45]

3. Onset

The disease starts during later adult life (see Table 6). It progresses insidiously or it may undergo a series of more or less severe episodes over months or years. The earlier stages usually are not associated with overt skin photosensitivity. Frequently, the history of skin trouble commences with some kind of chronic or recurrent dermatosis, not specifically connected with photosensitivity, such as discoid eczema, seborrheic dermatitis, or psoriasis. It is usually difficult to know exactly when abnormal photosensitivity starts or how it progresses because both patient and medical attendants are very often unaware of its presence. Likewise, in the mature stages of the clinical picture, the role of solar exposure as a precipitating factor may be quite difficult or impossible to establish on evaluating the clinical history. Such is often appreciated in retrospect, i.e., worsening in the sunny season.

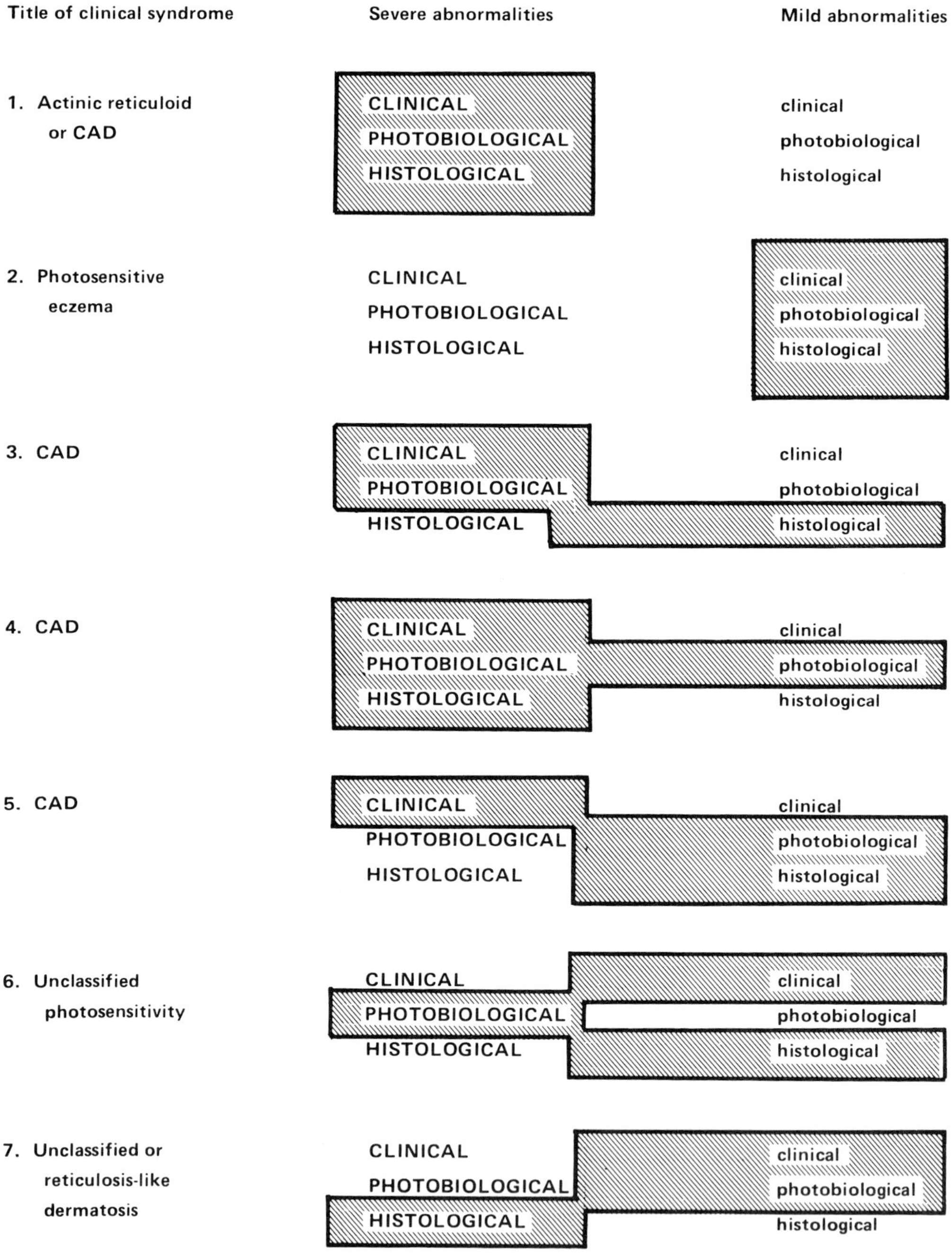

FIGURE 3. Summarization of the various clinical, photobiological, and histological forms that CAD may take and illustration how one form may pass into another. The first column shows the various titles that may be used: actinic reticuloid (syndrome 1 in column 1). Columns 2 and 3 show the three features — clinical, photobiological, and histological; where these are severe, they are shown in column 2 in capital letters; where mild in column 3 in lower-case lettering. Compare syndromes 1 and 2, actinic reticuloid and photosensitive eczema, respectively. Syndromes 3, 4, and 5 are other combinations of the three features, severe or mild. Syndrome 6 could be regarded as an unclassified type of abnormal photosensitivity; syndrome 7 a reticulosis-like dermatosis with the main abnormality in its histological features.

Table 6
AGE AT DIAGNOSIS OF 107 PATIENTS WITH CAD SEEN AT ST. JOHN'S HOSPITAL FOR DISEASES OF THE SKIN, OCTOBER 1981— DECEMBER 1983

Sex, no. of patients	Av. age, range
Male, 97	64, 40—81
Female, 10	65, 54—80

4. Course

This is usually prolonged and varying in degree of severity, though exacerbation may be blamed on the sunny season of the year (the spring often ushers in an acute exacerbation), or turns for the worse may be blamed on inappropriate therapy. These exacerbations may include episodes of erythroderma. It may not be until the severity has retreated that abnormal photosensitivity is suspected on the grounds of distribution of skin sites affected.

Spontaneous recovery seems to be most unusual, but a few patients may slowly improve after long periods of careful treatment.[62,63] It is estimated that more patients will eventually improve as experience with treatment increases.

C. Clinical Features

1. Morphology and Distribution of Lesions

The type of lesion in the earliest stages of the disease is not characteristic. Associated conditions such as psoriasis or seborrheic dermatitis may confuse the picture. Developed CAD has no particular morphological hallmark and is often a somewhat nonspecific chronic erythematous, erythemato-papulo-vesicular or erythemato-papulo-vesiculo-squamous dermatitis. It is not necessarily confined to uncovered skin areas. Later, the skin is diffusely lichenified and may show marked lichenoid features, especially around the hands and wrists.

As the condition becomes chronic over the months and years, and especially if there have been several attacks of erythroderma, there is diffuse skin thickening, exaggeration of normal skin creases (see Figure 4). This is especially evident in the temples, cheeks, and forehead. Infiltration of the forehead, neck, and scalp also occurs in the form of large, flat papules and plaques, and later as a diffuse thickening. The clinical appearance of mycosis fungoides may be then closely mimicked.

Although the distribution of the lesions on the body surface may at first not point to sun exposure playing a part, eventually the worst lesions are often on the face, neck, upper part of the back, backs of hands, and on the dorsal surface of the forearms. A helpful diagnostic sign may be the relative sparing of the digital and the deep body clefts, although involvement of the axillae and groins and natal cleft seem to occur in patients in which seborrheic dermatitis had previously predominated.

2. Clinical Investigations

The essentials are (1) test exposures of the skin with UV and visible radiation, preferably with an irradiation monochromator; (2) patch (epicutaneous) tests to the skin for contact and photocontact sensitivity against certain agents; and (3) histological examinations of biopsies of skin lesions. The first two tests may be difficult to perform on a patient with widespread skin lesions and may have to be postponed until an erythematous skin is in a more or less quiescent stage.

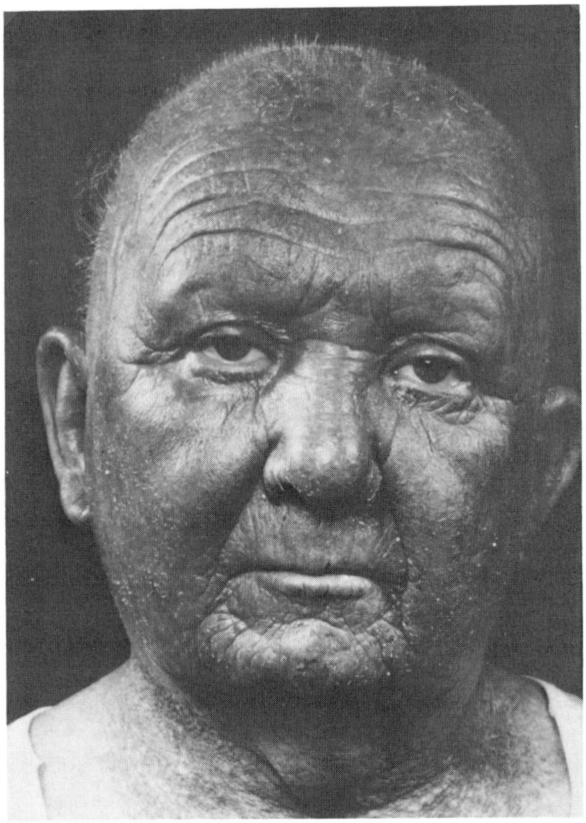

FIGURE 4. CAD. Note thickening of skin, especially on forehead.

3. The Results of Tests for Contact Sensitivity and of Immunological Mechanisms

a. Patch Testing

What agents should be patch tested is not a matter of general agreement. Also, the results have varied from one series of patients to another. In predominantly rural populations, as in parts of India, Canada, and Scotland, the suspicion has arisen that contact sensitivity (rather than photocontact sensitivity) to certain plants may be a factor. It has been hinted that the oleoresins in certain lichens and grasses and other plants lead to photosensitivity.[64-66] Patients living in urban regions do not seem to show any particular pattern of contact sensitivity and those which show any contact seem to be more sensitive to materials such as chromate, nickel, and chemicals associated with rubber or synthetic rubber. A few are contact sensitive to colophony, a material which may contain ingredients related to oleoresins. However, in one consecutive series of 12 London patients with CAD reported by Menter et al.,[67] all showed negative results to all patch (not only photopatch) tests.

b. Other Possible Investigations

Apart from patch and photopatch testing, little in the way of immunological investigations can be done that is ethical.

c. Photobiology Tests

These have been confined mainly to tests with measured doses of more or less narrow-band (''monochromatic'') irradiation exposures of the skin itself. Ideally, a dose-

Table 7
SUMMARY OF ABNORMAL SPECTRAL PHOTOSENSITIVITY IN NEW PATIENTS WITH CAD SEEN OCTOBER 1981 TO DECEMBER 1983 AT ST. JOHN'S HOSPITAL FOR DISEASES OF THE SKIN, LONDON

	No. of patients, upper spectral limit of abnormal photosensitivity (nm)					
No. of patients, sex	320	340	360	380	400	500
97, male	10	18	27	23	14	3
10, female	3	—	3	4	—	—

response test is done with such a monochromatic band. In practice, it is usual to determine a threshold dose throughout the spectrum. Such irradiation monochromator tests very often show spectacular abnormal skin reactivity light, all the more remarkable as patients may be quite unaware of the abnormal photosensitivity their skin has. Patients will often reject the notion of their being solar photosensitive until this is brought to their notice by these tests. The procedure of light testing thus has a very useful therapeutic purpose in educating the patient to avoiding sunlight.

The abnormal features may be as follows: (1) abnormal skin photosensitivity may be to a relatively wide spectrum, from below 300 nm to at least 360 nm in the UVA range (see Table 7); (2) marked lowering of the dose required for minimum erythema in the 300-nm region, e.g., less than 1 mJ/cm^2, i.e., the threshold is reduced by 20-fold or more; this seems to occur in the severest cases, who are also photosensitive well into the visible light spectrum (out to 500 nm; blue-green light); (3) the response to test exposures with UVB and UVA may take the form of strikingly abnormal — fierce, erythematous swellings, which are tender to the touch; these develop 1 to 2 hr after exposure and last several days. Responses to the UVB and UVA may be of this type, but responses to the visible spectrum are milder and may occur as erythema alone. All skin reactions to irradiation monochromator light tests are of relatively delayed appearance; true urticarial responses have not been recorded.

D. Histological Appearances of the Skin

The picture in a biopsy from skin lesions in a patient who has a mild degree of CAD, as exemplified by the title "photosensitive eczema", is not helpful. It is merely that of a mild nonspecific inflammatory change in the epidermis and most superficial dermis. These lesions include a sparse, irregularly disposed infiltration of lymphocytes, sometimes mostly perivascular.[49] By contrast, in patients with the established severe disease, typified by the title "actinic reticuloid", there may be striking skin histological abnormalities, but an infinite variety of intermediate histological states with the mild picture noted earlier may be seen. The fully developed severe disease shows three main features: (1) a dense infiltrate of lymphohistiocytes in the dermis: this may be widespread and extend from the epidermis to into the mid-dermis or even deeper to reach the subcutaneous fatty layer; (2) invasion of the epidermis may be by well-localized focuses of lymphocytic cells reminiscent of the "Pautrier micro-abscess" of mycoses fungoides; (3) the lymphohistiocytes in the dermal infiltrate often show abnormal staining, i.e., hypohromatic nuclei, and abnormal morphological features such as cells which are larger than normal and show a crenated border. Menter et al.[67] have described a band of autofluorescence in CAD skin lesion biopsies that needs to be further investigated.

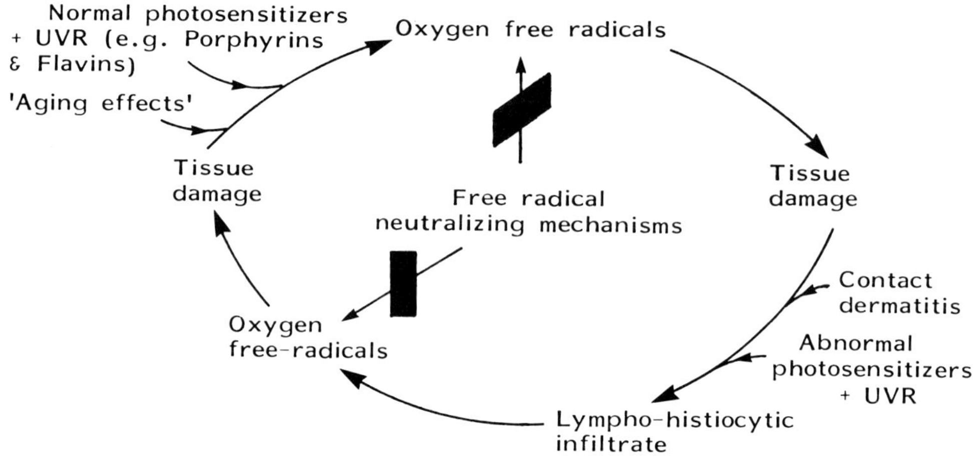

FIGURE 5. To illustrate the hypothesis of the vicious circle in CAD, it is proposed that there is a failure of mechanisms which normally neutralize free radicals liberated by tissue damage, aging effects, photosensitization, and the inflammation resulting from contact and photocontact dermatitis.

vacciniforme, the diagnosis of AP depends on diagnostic acumen and clinical experience. Pathological and experimental studies have contributed almost nothing, and the underlying mechanisms are unknown. The disease was described quite clearly in a well-written paper by Jonathan Hutchinson,[76] whose description is still worth consulting.

A sizeable number of dermatologists, it seems, confuse AP with other conditions such as atopic dermatitis and PLE. Others do not recognize it because, evidently, they never encounter patients with it. Still other dermatologists recognize its existence and see it occasionally; but they treat it as a subgroup within PLE.

In the last 10 years or so, three points with regard to AP have emerged. First, a significant advance has been made in its treatment; this is with thalidomide. Second, a number of review articles have appeared indicating that it is being increasingly realized that AP must be regarded as a true entity and that it deserves more attention.[77-81] The preliminary reports of Muhlbauer et al.[82,83] and Moncada[84] on T cells suggests possible differences between AP and PLE; this merits confirmation and extension. Third, from the point of view of etiology, what is of interest is the outburst of recognition of AP in the American continental Indians.

A. Etiology and Other General Features

1. Racial and Geographical Distribution

AP occurs in patients from European and North American urban areas, but also more significantly in patients from South and Central America and the North American indigenous Indian in urban, rural, and mountain regions. This was perhaps first noted by Schenk in 1960,[85] but seen again by Scheen et al.[80] and Moncada.[84] The large literature, mostly in the Spanish language, is typified by Londoño and Mundi[86] and Domínguez Soto.[87] Additionally, there are the papers in English by Everett et al.,[88] Birt,[89] and Birt and Davies[90] on AP in Canadian Indians and, more recently, that of Orr and Birt[91] on AP in Canadian Eskimos.

2. Course and Natural History

As stated earlier, recently several relatively large studies of patients have been collected and published. These data have contributed considerably to re-establishing this disease as a true entity and clarifying its course and natural history.

AP is a disorder that generally starts early in life, usually in the first decade.[79,81] In one recent study,[92] AP also started in the second and third decades. Once started, the disease seems to continue for an indefinite period, however often with periods of remission, though only temporary. In some of the patients whose disease started early in life, there is considerable improvement by puberty. In general, however, there are no firm data on the long-term prospects of complete remission.

3. Precipitating Factors, Sex Distribution, Family History, and Socioeconomic Status

Precipitating factors — In some patients, it has not been easy to ascertain the identity of the precipitating factors. Others claim their skin is bad all year. Symptoms may or may not be obviously connected to solar exposure, though they are commonly worse in summer.[81] South American AP seems to predominate at higher altitudes and some authors (for example, González Ochoa[93]) in the Spanish literature on AP believe geographical factors are important, hence the special name given: "prurigo solar de altiplanicie".[94]

Sex distribution — Female patients predominate; see reviews by Calnan and Meara[79] and Addo and Frain-Bell.[81]

Prevalence of atopy in the family — AP is not infrequently confused with atopic dermatitis, and its association with atopy is sometimes difficult to ascertain from the literature. However, there seems no reason for this; a clear distinction can often be made easily on clinical grounds between atopy and AP. Also, a history of the eczema-asthma-hay fever syndrome is by no means the rule in the latter. The prevalence of atopy in a recent large study in either the patient himself or his family was about one in five.[81]

Family history of AP — Most writers find a family history of AP is common. Addo and Frain-Bell[81] found a positive family history of AP in about one third of their patients. Similar high incidences are also found in some studies reported from South America.

Socioeconomic status — This had not been remarked on before the percipient review of Calnan and Meara,[79] who noted that AP patients seem to be well represented by the lower socioeconomic groups. The finding is perhaps not surprising in the indigenous American continental Indian patient. Certainly in London, a number of the AP patients had been under care of the social services, on probation, or in prison. The type of medical service the patient is seen in, whether as a private patient or in a clinic organized by state or municipality must have some bearing. An appropriate survey might confirm the role of social and economic status and could go some way toward explaining the curious patchy distribution of AP in the more affluent Western population.

B. Clinical Picture

1. Skin Lesions

The rash of AP is nearly always very distinctive when in its resting or chronic phase, and this is how it is commonly seen. Once this particular morphological picture is borne in mind by the clinician, it is one of the dermatoses that is nearly always instantly recognized, even from afar — diagnosed, as it were, at the door as the patient walks in (see Figure 6).

The pathognomic feature is the prurigo lesion, a class of skin abnormality that is not easy to describe readily in words. In fresh lesions, in the acute stage, AP is perhaps not so clearly recognizable, for the rash may temporarily include erythema, edema, and other papular elements. The typical settled prurigo lesion, however, is very characteristic. This is a slightly raised papule of 1- to 3-cm diameter, or perhaps more, with an excoriated vesicle on its top. The appearance is to be compared with the 1- to 2-mm

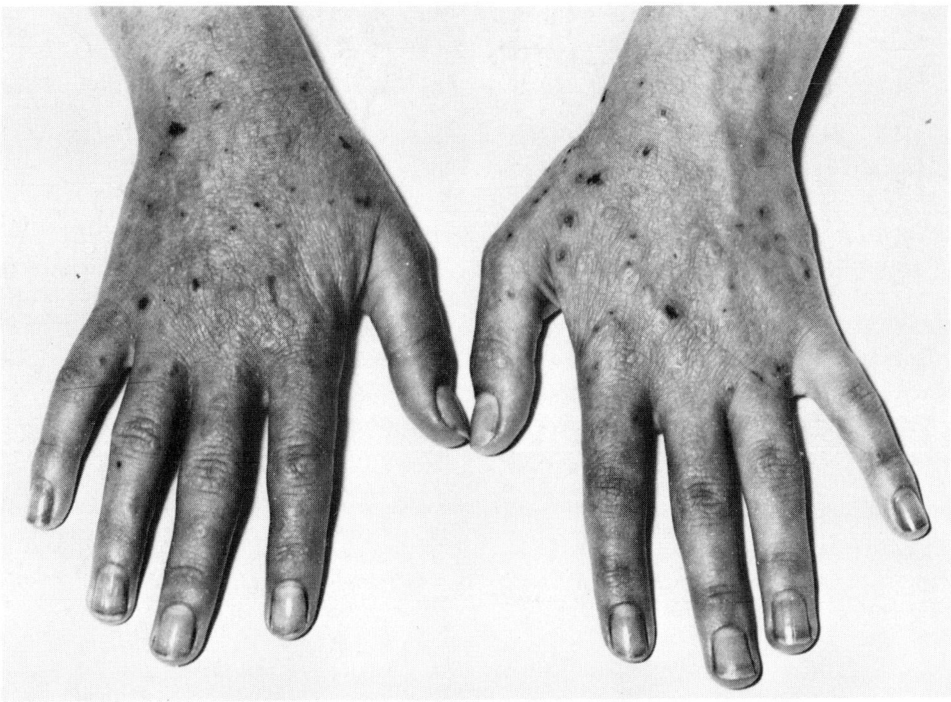

FIGURE 6. AP, prurigo lesions on back of hands. Note large papules approximately 1 to 2 cm in diameter, each surmounted by small (2- to 3-mm) scratched lesion. (Acknowledgments due to Dr. Frain-Bell and editors of *Photodermatology.)*

steeply rising papule of PLE. The AP papule may become confluent with other large papules and become a large plaque, in contrast to the discrete, small PLE papule (see Figure 6).

2. Distribution of Rash

A characteristic feature of AP is its distribution on the body surface. This is mostly on sunlight-exposed skin areas, limbs and face, but also (though to a lesser degree) on covered areas such as the posterior part of the chest wall and buttocks. In fact, Addo and Frain-Bell[81] found about one third of their patients to have lesions on covered skin areas, and Calnan and Meara[73] found about one half of their patients to be similarly affected. An important diagnostic feature is that the face is commonly affected, in contradistinction to PLE where the face escapes.[79] A confluent patch of rash with papular, eczematous, and exudative features, 3 or 4 cm in diameter, on each cheek, is very characteristic of AP. Addo and Frain-Bell[81] also emphasize cheilitis and involvement of the end of the nose as being characteristic (see Figure 7).

C. Investigations

Routine laboratory studies are of little or no help in diagnosing or elucidating mechanisms of AP. In testing for skin photosensitivity, preferably with an irradiation monochromator, the results are neither pathognomonic nor do they have a pattern suggestive of the diagnosis. Meara et al.[77] report that about 50% of patients showed no objective evidence of photosensitive abnormality at all and are entirely normal to irradiation tests for skin photosensitivity. Another 25% (approximate) had abnormal test results to UVB radiation only. The remaining 25% or so patients showed abnormalities in both UVB and UVA, some of them with severe grades of skin reactions

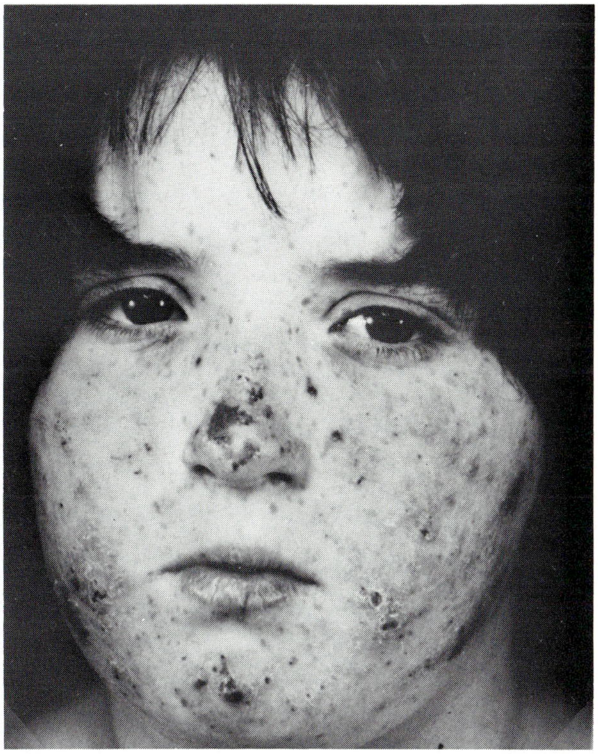

FIGURE 7. AP. View of more or less diffuse confluent prurigo lesions. Note severe appearance of nose.

indicating marked spectral photosensitivity. Addo and Frain-Bell[81] found similar variable spectral photosensitivity and even a handful of patients were sensitive to visible light from an irradiation monochromator.

Overall, these results seem puzzling; they suggest that either AP is a collection of different diseases or that the skin response may vary markedly from one time to another. The latter is perhaps not likely, as in the few patients that have been retested, including those sensitive to UVA, there has been no change in their abnormal spectral photosensitivity.

The histological appearance of biopsies from prurigo lesions are of little help except for excluding other entities. The picture is that of prurigo only.

D. Treatment

Treatment has until recently been along conventional lines and includes exhorting the patient to avoid exposure to sunlight, to wear protective clothing such as long sleeves and wide-brimmed hats, and to use efficient sub-barrier preparations. In most of the author's patients, these measures have only proved of marginal help. In addition, some patients have needed pacifying reassurance and have been given tranquilizers.

The drug that has so improved the life of the AP patient is thalidomide. This was introduced by Londoño et al.[95] in 1968 and reported in full in 1973.[96,97] Londoño originally used relatively large doses, 300 mg orally daily for 3 months or more, similar to one regime for treating lepromatous leprosy. Magnus and his colleagues at St. John's Hospital use smaller doses and shorter courses: 50 to 100 mg at night for 2 to 4 weeks, reducing to a maintenance dose of 25 to 50 mg nightly or 50 mg on alternate nights. Some of their patients need larger doses: 200 mg, exceptionally 300 mg, nightly for control. Because of the very well known possibility of adverse side effects, they prefer to stop treatment entirely as soon as possible, after about 4 weeks. They recommence only when recurrence occurs, but usually only after a drug-free period of 4 to 6 weeks. Remissions may sometimes last several months, even as long as 1 year.

The notorious adverse side effects of thalidomide make it essential that patients are selected carefully and treatment is monitored meticulously. The patient must be made aware of the possibility of side effects and, in the case of fertile females, satisfactory contraception is essential to avoid the teratogenic effects and serious birth defects. Some authors have confined thalidomide to prepubertal and postmenopausal females.[93] Otherwise, peripheral neuropathy is potentially the most serious problem and the patient should be instructed to stop thalidomide at once should persistent symptoms such as paresthesia be noted. The sedative effects of thalidomide are usually circumvented by the patient taking the drug at night time. See Clemmensen et al.[98] for a recent review of thalidomide neurotoxicity.

V. HYDROA VACCINIFORME OF BAZIN

Described by Bazin,[99] hydroa vacciniforme (HV) is the rarest of the well-established idiopathic photodermatoses, but, in a typical case, it has perhaps the most easily diagnosed and distinctive appearance.

The main preoccupation of earlier authors has been to differentiate it from other photodermatoses and especially from congenital porphyria. Present-day clinicians have the advantage of good diagnostic chemical tests for the porphyrias and this is no longer a diagnostic difficulty; however, for the recognition of HV, one depends entirely on diagnostic acumen and clinical experience, as is also the case in AP and PLE.

The literature is somewhat sparse and often unsatisfactory in that some descriptions are insufficient in detail; among the better earlier accounts is that of Rasch,[100] an authority on the photodermatoses. From the 1930s, a number of papers describing single cases or small collections have appeared; these were reviewed by Bielický and Ondráčková,[101] who were able to separate clearly HV from congenital porphyria and from epidermolysis bullosa and other blistering and scarring conditions; satisfactory chemical tests for porphyria were then only in their infancy and unreliable except in the hands of a few experts. McGrae and Perry[102] have presented the most extensive literature review. Since then, only a few case reports have appeared intermittently.[103,104]

A. Mechanism

Very few patients have been investigated from the photobiological point of view. Among these investigators were Schiff and Jillson,[106] Ramsay,[107] and Bickers et al.[103] and together their work, on the whole, indicated abnormal sensitivity in the UVB range. However, most of the results in the literature are indecisive. Most recently, Sonnex and Hawk[105] found the action spectrum for abnormal photosensitivity in HV might also extend into the UVA. Four of their cases were tested with a monochromator and were found to show abnormal photosensitivity as erythema and/or edematous responses in the near UVR (340 to 360 nm). Like most writers, they did not claim to reproduce the natural clinical lesions of HV. Claims that artificial UVB irradiation in large doses can produce the genuine HV vesicles[106] need to be confirmed, and whether the experimental lesions were mere blisters from a normal sunburn response needs to be clarified.

B. Clinical Features

In the ten cases of Sonnex and Hawk,[105] age of onset was from 1 to 11 years. More patients were boys than girls and in the review by McGrae and Perry[102] the ratio was about 2:1.

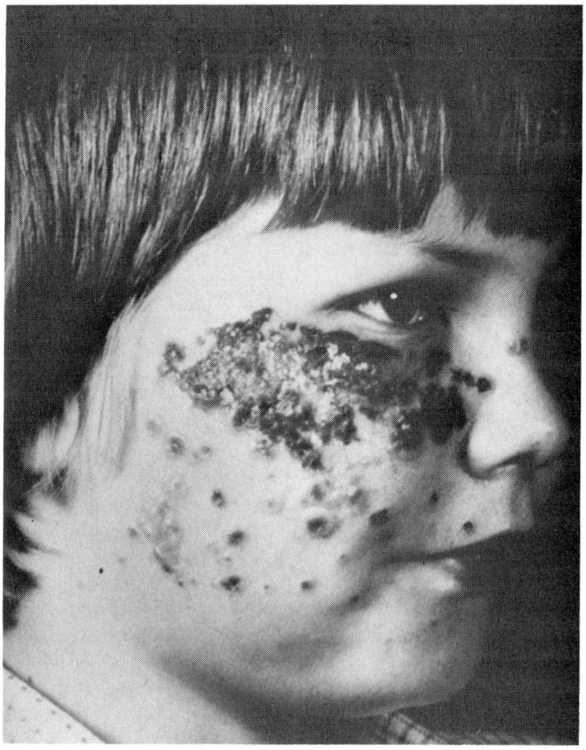

FIGURE 8. HV. Note lesions at different stages of development: umbilicated and scabbed. (Courtesy of Dr. Patrick Hall-Smith and Dr. Margaret Price.)

The condition has only a superficial resemblance to PLE in that recurrent episodes of rash occur during the summer on exposed parts. Occasionally, the history indicates that crops of lesions appear within a few hours of solar exposure, but more often this connection is less clear cut. Otherwise the two diseases differ markedly.

HV differs from PLE in the quite striking appearance of the fresh lesions; in HV they are tense erythemato-papular vesicles 2 to 4 mm in diameter, closely grouped together within a larger edematous or bullous area. The numbers of groups of lesions are usually relatively low. In PLE, on the other hand, the lesions are much more often papules than vesicles, are dispersed more or less diffusely in sheets or over more or less large areas, and heal without complication. In HV, a lesion group becomes hemorrhagic, may take on an umbilicated appearance, coalesce, and form a scab. New lesions are apt to appear at intervals of a few days; thus individual groups of lesions will be at different stages of development, some as fresh vesicles, and other scabbed or more or less healed (see Figure 8).

An individual crop of lesions may remain scabbed for several weeks; when subsided, the crust separates off, and scars remain the size and shape of those of smallpox.

The scleras and conjunctivas may sometimes be mildly inflamed. A minor degree of constitutional disturbance with slight pyrexia may also occur.

C. Histological Examination

Examination of biopsies of lesions shows localized epidermal vesciculation with adjacent dermal focal necrosis, hemorrhage, and thrombosis. This microscopic picture seems to be sufficiently distinctive, in some instances, to be helpful in confirming the clinical diagnosis, though ultimately it is the clinical course and gross morphology of HV that are decisive.

D. Progress, Course, and Treatment

The prognosis is difficult to make, but fortunately some patients may only suffer infrequent attacks, isolated by several years. Fortunately also, the disease appears to be self-limiting, although this may not be for several years into the second decade according to Sonnex and Hawk.[105]

With regard to therapy, it is reported that oral carotenoids, thalidomide, and PUVA are ineffective,[105] though Piñol Aguadé et al.[108] reported a case in which thalidomide seemed to help, and this was also found in subsequent cases in Barcelona. Sonnex and Hawk[105] suggest that efficient sun screens are helpful, and possibly antimalarials; the latter were also advocated by McGrae and Perry.[102]

REFERENCES

1. Harber, L. C. and Bickers, D. R., *Photosensitivity Diseases,* W. B. Saunders, Philadelphia, 1981.
2. Johnson, B. E., *The Physiology and Pathophysiology of the Skin,* Vol. 8, Jarrett, A., Ed., Academic Press, London, 1984, 2514.
3. Magnus, I. A., *Dermatological Photobiology,* Blackwell Scientific, Oxford, 1976, 211.
4. Willan, R., Eczema solare: plate 56, in *Delineations of Cutaneous Diseases,* Batemen, R., Ed., Longman, London, 1817.
5. Rayer, P., *Traité Theorique et Pratique des Maladies de la peau,* 1, 274, J. B. Bailliere, Paris, 1826.
6. Veiel, Th., Ueber einen Fall von Eczema solare, *Vierteljahressch. Dermato. Syphol.,* 14, 1113, 1887.
7. Rasch, C., Some historical and clinical remarks on the effect of light on the skin and skin diseases, *Proc. R. Soc. Med.,* 20(i & ii), 11, 1926.
8. Magnus, I. A., The classification and diagnosis of the photodermatoses, in *Dermatological Photobiology, Clinical and Experimental Aspects,* Magnus, I. A., Ed., Blackwell Scientific, Oxford, 1976, chap. 12.
9. Haxthausen, H., Studier over lysdermatiterne. II. Undersøgelser over pathologiske seras virkning paa fotodemiske processer, *Hospitalstidende,* 19—20, 577, 1918.
10. Hutchinson, J., Summer prurigo, in *Lecture IX on Clinical Surgery,* Vol. 1, Churchill, London, 1879.
11. Epstein, S., Studies in abnormal human sensitivity to light. I. Prurigo aestivalis, eczema solare and urticaria photogenica. Report of 17 cases and review of the literature, *J. Invest. Dermatol.,* 5, 187, 1942.
12. Everett, M. A., Crockett, W., Lamb, J. H., and Minor, D., Light-sensitive eruptions in American Indians, *Arch. Dermatol.,* 83, 243, 1961.
13. Londoño, F., Muvdi, F., Giraldo, F., Rueda, L. A., and Caputo, A., Familial actinic prurigo, *Dermatol. Ibero Lat. Am.,* 3, 61, 1968.
14. Birt, A. R., Photodermatitis in Indians of Manitoba, *Can. Med. Assoc. J.,* 98, 392, 1968.
15. Birt, A. R. and Davis, R. A., Photodermatitis in North American Indians: familial actinic prurigo, *Int. J. Dermatol.,* 10, 107, 1971.
16. Hojyo-Tomoka, M. T. and Domínguez Soto, L., Epidemiology, clinical characteristics and photosensitivity tests in polymorphous light eruption in Mexico, in 3rd Congr. Soc. Int. Dermatologiae Tropicae, São Paulo, Brazil, September 1 to 5, 1974.
17. Calnan, C. D. and Meara, R. H., Actinic prurigo (Hutchinson's summer prurigo), *Clin. Exp. Dermatol.,* 2, 356, 1977.
18. Scheen, S. R., Connolly, S. M., and Dicken, C. H., Actinic prurigo, *J. Am. Acad. Dermatol.,* 5, 189, 1981.
19. Addo, H. A. and Frain-Bell, W., Actinic prurigo — a specific photodermatosis?, *Photodermatology,* 1, 119, 1984.
20. Wiskemann, A. and Wulf, K., Untersuchungen über den auslösenden Spektrlbereich und die direkte Lichtpigmentierung bei chronischen und akuten Lichtausschlägen, *Arch. Klin. Exp. Dermatol.,* 209, 443, 1959.
21. Bhutani, L., Personal communication, 1985.
22. Epstein, J. H., Polymorphous light eruption, *Ann. Allergy,* 24, 397, 1966.
23. Verhagen, A. R. H. B., Koten, J. W., Chaddah, V. K., and Patel, R. I., Skin diseases in Kenya. A clinical and histopathological study of 3,168 patients, *Arch. Dermatol.,* 98, 577, 1968.
24. Dawson, T. A. J., Personal communication, 1968.

25. Frain-Bell, W., Dickson, A., Herd, J., and Sturrock, I., The action spectrum in polymorphic light eruption, *Br. J. Dermatol.*, 89, 243, 1973.
26. Jansén, C. T., The natural history of polymorphic light eruption, *Arch. Dermatol.*, 115, 165, 1979.
27. Johnson, B. E., Abnormal reactions associated with skin disorders, in *The Physiology and Pathophysiology of the Skin*, Vol. 8, Jarrett, A., Ed., Academic Press, London, 1984, 2493.
28. Horkay, I. and Mészáros, Cs., A study on lymphocyte transformation in light dermatoses. Preliminary report, *Acta Derm. Venereol.*, 51, 268, 1971.
29. Raffle, E. J., McLeod, T. M., and Hutchinson, F., *In vitro* lymphocyte studies in chronic polymorphic light eruption, *Br. J. Dermatol.*, 89, 143, 1973.
30. Jansén, C. T. and Helander, I., Cell mediated immunity in chronic PLEs, *Acta Derm. Venereol.*, 56, 121, 1976.
31. Jung, E. G. and Bohnert, E., Chronisch polymorphe Lichtdermatose; Untersuchungen an Lymphozyten *in vitro, Dermatologica*, 148, 209, 1974.
32. Epstein, J. H., Continuing medical education — polymorphic light eruption, *J. Am. Acad. Dermatol.*, 3, 329, 1980.
33. Jansén, C. T., *Monimuotienen Valiohuttuna*, Kirjapaino Polytpos, Turku, 1975.
34. Lamb, J. H., Shelmire, B., Cooper, Z., Morgan, R. J., and Keaty, C., Solar dermatitis, *Arch. Dermatol.*, 62, 1, 1950.
35. Lamb, J. H., Jones, P. E., and Maxwell, T. B., Solar dermatitis, *Arch. Dermatol.*, 75, 171, 1957.
36. Tanenbaum, L., Mark, G. J., Mihm, M. C., and Parrish, J. A., The role of histopathology in polymorphous light eruption light testing, *Clin. Exp. Dermatol.*, 6, 123, 1981.
37. Epstein, J. H., Polymorphous light eruptions. Phototest technique studies, *Arch. Dermatol.*, 85, 502, 1962.
38. Rottier, P. B., Light dermatoses without abnormal reactions to single test irradiations, *Dermatologica*, 129, 286, 1964.
39. Hölzle, E., Plewig, G., Hofmann, C., and Roser-Maas, E., Polymorphic light eruption, *J. Am. Acad. Dermatol.*, 7, 111, 1982.
40. Murphy, G. M. and Hawk, J. L. M., Choosing a suitable sunscreen, *Pharm. J.*, July 14, p. 43, 1984.
41. Corbett, M. F., Hawk, J. L. M., Herxheimer, A., and Magnus, I. A., Controlled therapeutic trials in polymorphic light eruption, *Br. J. Dermatol.*, 107, 571, 1982.
42. Corbett, M. F., Herxheimer, A., Magnus, I. A., Ramsay, C. A., and Kobza Black, A., The long term treatment with β-carotene in erythropoietic protoporphyria: a controlled trial, *Br. J. Dermatol.*, 97, 655, 1977.
43. Gschnait, F., Schwarz, T., and Ladich, I., Treatment of polymorphous light eruption, *Arch. Dermatol. Res.*, 275, 379, 1983.
44. Murphy, G. M., Logan, R., Lovell, C. R., Morris, R., Hawk, J. L. M., and Magnus, I. A., The efficacy of phototherapy in polymorphic light eruption, *Br. J. Dermatol.*, 111(Suppl. 26) (Abstr.), 24, 1984.
45. Frain-Bell, W., Lakshmipathi, T., Rogers, J., and Willock, J., The syndrome of chronic photosensitivity dermatitis and actinic reticuloid, *Br. J. Dermatol.*, 91, 617, 1974.
46. Hawk, J. L. M. and Magnus, I. A., Chronic actinic dermatitis — an idiopathic photosensitivity syndrome including actinic reticuloid and photosensitive eczema, *Br. J. Dermatol.*, 101(Suppl. 17), 24, 1979.
47. Guardiola, A. and Sánchez, J. L., Actinic reticuloid, *Int. J. Dermatol.*, 19, 154, 1980.
48. Haynes, H. A., Bernhard, J. D., and Gange, R. W., Actinic reticuloid, *J. Am. Acad. Dermatol.*, 10, 947, 1984.
49. Ive, F. A., Magnus, I. A., Warin, R. P., and Wilson-Jones, E., "Actinic reticuloid"; chronic dermatosis associated with severe photosensitivity and the histological resemblance to lymphoma, *Br. J. Dermatol.*, 81, 469, 1969.
50. Ramsay, C. A. and Kobza Black, A., Photosensitive eczema, *Trans. St. John's Hosp. Dermatol. Soc.*, 59, 152, 1973.
51. Ramsay, C. A., Personal communication, 1980.
52. Degos, R., Belaich, S., and Civatte, J., Actino-reticulosis, in *Essays on Tropical Dermatology*, Vol. 2, Marshall, J., Ed., Excerpta Medica, Amsterdam, 1972, 18.
53. Jensen, N. E. and Sneddon, I. B., Actinic reticuloid with lymphoma, *Br. J. Dermatol.*, 82, 287, 1970.
54. Frain-Bell, W., Hetherington, A., and Johnson, B. E., Contact allergic sensitivity to chrysanthemum and the photosensitivity dermatitis and actinic reticuloid syndrome, *Br. J. Dermatol.*, 101, 491, 1979.
55. Addo, H. A., Ferguson, J., Johnson, B. E., and Frain-Bell, W., The relationship between exposure to fragrance materials and persistent light reaction in the photosensitivity dermatitis with actinic reticuloid syndrome, *Br. J. Dermatol.*, 107, 261, 1982.
56. Ecker, R. I. and Winkelmann, R. K., Lymphomatoid contact dermatitis, *Contact Dermatitis*, 7, 84, 1981.

57. Winkelmann, R. K., Buechner, S. A., and Díaz-Pérez, J. L., Pre-Sézary syndrome, *J. Am. Acad. Dermatol.*, 10, 992, 1984.
58. Swanbeck, G. and Wennersten, G., Evidence of kynurenic acid as a possible photosensitiser in actinic reticuloid, *Acta Derm. Venereol.*, 53, 109, 1973.
59. Swanbeck, G. and Wennersten, G., Photohaemolytic activity of tryptophan at phenylalamine metabolites, *Acta Derm. Venereol.*, 54, 99, 1974.
60. Nilsson, R., Swanbeck, G., and Wennersten, G., Primary mechanisms of erythrocyte photolysis induced by biological sensitizers and phototoxic drugs, *Photochem. Photobiol.*, 22, 183, 1975.
61. Botcherby, P. K., Magnus, I. A., Marimo, B., and Giannelli, F., Actinic reticuloid — an idiopathic photodermatosis with cellular sensitivity to near-ultraviolet radiation, *Photochem. Photobiol.*, 39, 641, 1984.
62. Frain-Bell, W., Personal communication, 1984.
63. Hawk, J. L. M., Personal communication, 1985.
64. Mitchell, J. C., Allergic contact dermatitis from compositae, *Trans. St. John's Hosp. Dermatol. Soc.*, 55, 174, 1969.
65. Frain-Bell, W., Photosensitivity dermatitis and actinic reticuloid, *Semin. Dermatol.*, 1, 161, 1982.
66. Lonkar, A., Mitchell, J. C., and Calnan, C. D., Contact dermatitis from parthenium hysterophorus, *Trans. St. John's Hosp. Dermatol. Soc.*, 60, 43, 1974.
67. Menter, M. A., McKerron, R. A., and Amos, H. E., Actinic reticuloid: an immunological investigation providing evidence of basement membrane damage, *Br. J. Dermatol.*, 90, 507, 1974.
68. Herd, J., Sturrock, F., and Frain-Bell, W., The use of plastic material for the protection of patients with severe photodermatoses, *Br. J. Dermatol.*, 88, 283, 1973.
69. Leigh, I. M. and Hawk, J. L. M., Treatment of chronic actinic dermatitis with azathioprine, *Br. J. Dermatol.*, 110, 691, 1984.
70. Harman, D., Free radical theory of ageing, *Age*, 3, 64, 1980.
71. Amos, H. E., Photoallergy: a critical survey, *Trans. St. John's Hosp. Dermatol. Soc.*, 59, 147, 1973.
72. Miyachi, Y. and Takigawa, M., Mechanisms of contact photosensitivity in mice. II. Langerhans cells are required for successful induction of contact photosensitivity to TCSA, *J. Invest. Dermatol.*, 78, 363, 1982.
73. Maguire, H. C. and Kaidbey, K., Experimental photoallergic contact dermatitis; a mouse model, *J. Invest. Dermatol.*, 79, 147, 1982.
74. Granstein, R. D., Morison, W. L., and Kripke, M. L., The role of UVB radiation in the induction and elicitation of photocontact hypersensitivity to TCSA in the mouse, *J. Invest. Dermatol.*, 80, 158, 1983.
75. Giannelli, F., Botcherby, P. K., Marimo, B., and Magnus, I. A., Cellular hypersensitivity to UV-A: a clue to the aetiology of actinic reticuloid?, *Lancet*, 2, 88, 1983.
76. Hutchinson, J., Summer prurigo, in *Lecture IX on Clinical Surgery*, Vol. 1, Churchill, London, 1879.
77. Meara, R. H., Magnus, I. A., Grice, K., Wilson-Jones, E., and Johansson, S. G. O., Hutchinson's summer prurigo, *Trans. St. John's Hosp. Dermatol. Soc.*, 57, 87, 1971.
78. Magnus, I. A., *Dermatological Photobiology*, Blackwell Scientific, Oxford, 1976, 183.
79. Calnan, D. and Meara, R. H., Actinic prurigo (Hutchinson's summer prurigo), *Clin. Exp. Dermatol.*, 5, 47, 1977.
80. Scheen, S. R., Connolly, S. M., and Dicken, C. H., Actinic prurigo, *J. Am. Acad. Dermatol.*, 5, 183, 1981.
81. Addo, H. A. and Frain-Bell, W., Actinic prurigo — a specific photodermatosis?, *Photodermatology*, 1, 119, 1984.
82. Muhlbauer, J. E., Bhan, A. K., Harris, T. J., Bernhard, J. D., and Mihm, M. C., Immunopathology of polymorphous light eruption, *J. Invest. Dermatol.*, 78, 347a, 1982.
83. Muhlbauer, J. E., Bhan, A. K., Harris, T. J., et al., Papular polymorphic light eruption: an immunoperoxidase study using mononuclear antibodies, *Br. J. Dermatol.*, 108, 153, 1983.
84. Moncada, B., Actinic prurigo, *Int. J. Dermatol.*, 22, 415, 1983.
85. Schenk, R. R., Controlled trials of methoxalen in solar dermatitis of Chippewa Indians, *JAMA*, 172, 1134, 1960.
86. Londoño, F. and Mundi, F., Prurigo actínico familiar, *Arch. Argent. Dermatol.*, 16, 290, 1966.
87. Domínguez Soto, L., Las dermatoses solares, *Boliviana Dermatol.*, 3, 6, 1971.
88. Everett, M. A., Crockett, W., Lamb, J. H., and Minor, D., Light sensitive eruptions in American Indians, *Arch. Dermatol.*, 83, 243, 1961.
89. Birt, A. R., Photodermatitis in Indians of Manitoba, *Can. Med. Assoc. J.*, 98, 392, 1968.
90. Birt, A. R. and Davies, R. A., Photodermatitis in North American Indians: familial actinic prurigo, *Int. J. Dermatol.*, 10, 107, 1971.
91. Orr, P. H. and Birt, A. R., Hereditary polymorphic light eruption in Canadian Inuit, *Int. J. Dermatol.*, 23, 472, 1984.

92. Lovell, C. R., Hawk, J. L. M., Calnan, C. D., and Magnus, I. A., Thalidomide in actinic prurigo, *Br. J. Dermatol.*, 108, 467, 1983.
93. Gonzâlez Ochoa, A., Solar eczema, *Dermatol. Trop. Ecol. Geogr.*, 1, 137, 1962.
94. Flores, O., Prurigo solar de altiplanicie, *Dermatología (Mexico City)*, 19, 26, 1975.
95. Londoño, F., Muvdi, F., Giraldo, F., Rueda, L. A., and Caputo, A., Familial actinic prurigo, *Dermatol. Ibero Lat. Am.*, 111, 61, 1968.
96. Londoño, F., Thalidomide in the treatment of actinic prurigo, *Int. J. Dermatol.*, 12, 326, 1973.
97. Saul, A., Flores, O., and Novales, J., Polymorphic light eruption: treatment with thalidomide, *Aust. J. Dermatol.*, 17, 17, 1976.
98. Clemmensen, O. J., Olsen, P. Z., and Andersen, K. E., Thalidomide neurotoxicity, *Arch. Dermatol.*, 120, 338, 1984.
99. Bazin, *Leçons Theoriques et Cliniques sur les Affections Cutanées de Nature Arthritique et Dartreuse*, 1860, 197.
100. Rasch, C., Tilfaelde of Hydroa vacciniformis (Bazin), *Hospitalstidende*, 7, 506, 1899.
101. Bielický, T. and Ondráčková, J., Hidroa vacciniformia a porfyrinurie, *Ceska Dermatol.*, 36, 261, 1961.
102. McGrae, J. D. and Perry, H. O., Hydroa vacciniforme, *Arch. Dermatol.*, 87, 618, 1963.
103. Bickers, D. R., Demar, L. K., DeLeo, V., Poh-Fitzpatrick, M. B., Aronberg, M., and Harber, L. C., Hydroa vacciniforme, *Arch. Dermatol.*, 114, 1193, 1978.
104. Goldgeier, M. H., Nordlund, J. J., Lucky, A. W., Sibrack, L. A., McCarthy, M. J., and McGuire, J., Hydroa vacciniforme, *Arch. Dermatol.*, 118, 588, 1982.
105. Sonnex, T. S. and Hawk, J. L. M., Hydroa vacciniforme, *Br. J. Dermatol.*, in press, 1985.
106. Schiff, M. and Jillson, O. F., Photoskin tests in hydroa vacciniforme, *Arch. Dermatol.*, 82, 204, 1960.
107. Ramsay, C., Hydroa vacciniforme, *Br. J. Dermatol.*, 87, 618, 1972.
108. Piñol Aguadé, J., Lecha, M., Castells Rodellas, A., et al., Osservazioni di qualche fotodermatosi con reazione persistente alla luce, *G. Ital. Dermatol. Minerva Dermatol.*, 109, 129, 1975.

Chapter 2

XERODERMA PIGMENTOSUM

James E. Cleaver

TABLE OF CONTENTS

I. INTRODUCTION

A large variety of diseases are known in which cells or tissues exhibit abnormal responses to one or more kinds of DNA-damaging agents, i.e., they are hypersensitive.[1] Only in a few of these diseases, especially the various forms of xeroderma pigmentosum (XP), is the reason for the hypersensitivity known to be a defect in the repair of damaged DNA.[2,3] In other hypersensitive diseases there may be more complex abnormalities: in ataxia-telangiectasia after X irradiation there are alterations in both semiconservative replication[4,5] and repair,[6-8] and in Cockayne syndrome after irradiation by UV light there are alterations in the recovery of DNA replication.[9-12]

XP, therefore, occupies a unique place in radiation biology and medicine. It is a disease in which the capacity of the cell to mend its own genes is severely impaired. In consequence, XP cells and XP individuals are greatly at risk from the deleterious consequences of environmental exposure to mutagens and carcinogens, especially sunlight.

II. THE GENETIC DEFECT IN XP

A. The Disease

XP is an autosomal recessive inherited skin disease in which homozygotes show a marked tendency to develop skin cancers from exposure to sunlight.[13-15] Heterozygotes are generally asymptomatic, although a few cancers and a slight excess of neurological abnormalities have been reported.[16-18] There is no consistently observed linkage to other inherited diseases, HLA antigens,[19] or chromosomal abnormalities.[20] An early reported association with the ABO blood groups has not been supported by later investigations.[21,22]

Many patients are children of consanguine marriages, particularly in areas of the world in which intermarriage between relatives is common: parts of North Africa (especially Egypt),[21,23,24] Israel,[25] and Japan.[26] The disease is found among all racial groups at a frequency of about 1 in 250,000 in the general population, but about 10 times as frequently in Egypt, for unexplained reasons.

The nature and severity of cutaneous, ocular, and neurological symptoms vary widely with genetic makeup and amount of sun exposure.[15] The typical progression of the disease consists of an early stage in the first years of life in which there may be an abnormal erythemal response. The abnormality may involve a prolonged latent period, atypical morphology, edema, and blistering on minimal sun exposure. The minimal erythemal dose is lower than normal, and the erythemal response can be used in early diagnosis before other skin changes develop.[27] After several years of life, other skin changes, including excessive freckling, keratoses, and cancers, develop to varying degrees. The pigmentary changes, which involve disturbances in melanocyte structure and function,[28] and the frequent recurrence of cancers in sun-exposed regions of the skin are the characteristic features on which diagnosis is based. The cancers involve all cell types that receive sun exposure and include squamous and basal cell carcinomas, malignant melanomas, keratocanthomas, angiomas, and sarcomas.[15,29-31] There can also be considerable ocular involvement, including at least mild conjunctivitis and photophobia.[15,32,33]

Patients with severe neurological involvement have been classified as a distinct form of XP, the de Sanctis-Cacchione syndrome.[34] Most patients with neurological involvement fall into complementation groups A, B, C, D, and G, which will be described subsequently. The neurological symptoms include microcephaly, progressive mental deficiency, retarded skeletal development, sensorineural deafness, choriocathetosis, ataxia, and eventual quadriparesis. Not all patients show the full range of the most severe symptoms, but common findings are microcephaly, retardation, and areflexia.

A few patients having a wide spectrum of cutaneous and ocular symptoms and unambiguously diagnosed as having XP may have a normal erythemal response.[27] In many cases this clinical sign has been associated with the form of XP known as the XP variant or pigmented xerodermoid.[27,35,36] This form can be diagnosed fully only by biochemical means.[37,38] Compared to other XP patients, XP variants may show a higher proportion of basal to squamous cell carcinomas.[39]

There appears to be an overlap between XP and another sun-sensitive hereditary disorder, Cockayne syndrome. This disease consists of dwarfism, premature senility, retinal pigment degeneration, optic atrophy, deafness, mental retardation, absence of subcutaneous fat, and sensitivity to sunlight.[40] Cockayne syndrome patients do not show the characteristic malignant changes of the skin seen in XP, but at least two patients have now been reported who appear to suffer from a conjunction of XP and Cockayne syndrome symptoms.[15,41] Both of these patients constitute sole members of distinct complementation groups of both XP and Cockayne syndrome, which raises a difficult diagnostic question. The biochemical abnormality of Cockayne syndrome does not add a distinctive, and experimentally detectable, abnormality beyond that already generated by the XP defects.[9,12,42] These patients could therefore equally well be diagnosed as belonging to unique XP complementation groups that exhibit symptoms that mimic Cockayne syndrome. Their assignment to Cockayne syndrome itself is arguable, but can be resolved when the enzymes or genes are characterized in detail.

Another difficult category is that in which clinical symptoms are diagnosed as XP by clear dermatological criteria, but the biochemistry of DNA repair and replication shows nothing abnormal.[43] Several cases have even been identified among Egyptian XP patients whose clinical symptoms are severe and unambiguous.[331] Whether these patients should remain diagnosed as XP, in anticipation of later characterization (perhaps named "pseudo XP" or "XP, form fruste") or simply regarded as belonging to a completely different set of disorders is a difficult decision that has not yet been made.

XP and three other radiosensitive diseases, ataxia-telangiectasia, Cockayne syndrome, and Fanconi's anemia,[1] have a variety of neurological symptoms. A logical extension of this has been to search among other specifically neurological disorders for evidence of repair deficiencies and sensitivity to radiation or chemical carcinogens. Despite extensive effort, this generally has been unfruitful. Slight increases in sensitivity have been reported in a few disorders,[44,45] but these could be explained as secondary consequences of altered cell cycle parameters, pool sizes, or other nonspecific cellular parameters. No precise identification of repair deficiencies in neurological disorders has been made at present.

The frequency of cancers in XP patients in sites other than the skin is uncertain because of the difficulty in obtaining enough statistically significant data.[31,46,47] A retrospective survey of clinical reports of tumors in skin and other organs indicated a 2000-fold increase in skin (basal and squamous cell carcinoma, and melanoma), ocular, and tongue tumors in comparison to the general population.[31] In comparison, there was a 12-fold increase in cancers in other sites, of which the majority involved the brain and the oral cavity.

Immunological functions can be impaired by exposure to UV light.[48] Several reports have demonstrated more severe immunological dysfunction in XP patients than in normal subjects.[49,50] In a recent study of Egyptian XP patients, the development of contact allergy to dinitrochlorobenzene was very poor compared to comparable normal individuals although the frequency of Langerhans cells in the skin was normal.[50] Increased immunosuppression therefore seems to be associated with the increased sunlight damage found in XP patients.

In general, the response of XP patients to sunlight reflects a cellular abnormality in the capacity to tolerate damage to DNA. Because of this, the characteristic symptoms are concentrated in the skin, but the internal symptoms may reflect exposure to carcinogens at a lower dose rate than that from sunlight.

B. The Biochemical Pathway

1. Mechanisms of DNA Damage

Sunlight contains wavelengths that extend into the near-UV region, the shortest detectable being about 300 nm. This lower limit slightly overlaps the upper region of the absorption spectra of nucleic acids and proteins. Energy in this region of overlap is absorbed in macromolecules in the skin and can cause harmful effects, such as erythema, burns, and actinic carcinogenesis.[51,52] Comparisons between direct sunlight and shortwave UV light (254 nm) indicate that sunlight in the midwestern U.S. is equivalent in germicidal activity to about 0.1 to 0.2 $J/m^2/min$ of 254-nm UV light.[53,54] Since normal human cells in culture have a D_{37}* of only about 3 to 5 J/m^2 of radiation at 254 nm, the direct exposure of human proliferating cells to sunlight can result in significant amounts of cell killing.

Light at the UV end of the sun's spectrum produces its biological effects through absorption of quanta in molecules that have unsaturated chemical bonds, such as aromatic acids in proteins and purine and pyrimidine components of DNA and RNA (ribonucleic acid). The action spectra for production of DNA damage (pyrimidine dimers),[55,56] cell killing,[57-59] chromosome aberration production,[60] and induction of unscheduled DNA synthesis[61] are all similar, exhibiting maximum efficiency from 260- to 280-nm wavelengths. Solar wavelengths (around 313 nm) tend to give a greater ratio of point mutations to deletions than 254 nm.[62] Although there is little energy in this region of the sun spectrum, there is sufficient overlap of the shortest end of the sun spectrum with the longer-wavelength side of the absorption spectrum for DNA for significant photochemical reaction to occur.

One of the best-characterized photoproducts formed in DNA by absorption of UV light is the cyclobutane pyrimidine dimer. This is formed between adjacent pyrimidines in the same strand of DNA by the formation of new bonds between 5-5 and 6-6 positions on the pyrimidine rings to produce the form I (meso) dimer. Another photoproduct that is important for mutagenic effects is the azetidine dimer formed between the 6 and 4 positions of adjacent pyrimidines, usually in the form pyr-cytosine.[63]

2. Excision Repair Defects in XP

Two major types of excision repair can be distinguished on the basis of the initial mechanisms of excision: base excision and nucleotide excision.[1] Pyrimidine dimers and other carcinogen adducts in DNA constitute the major substrates for the nucleotide excision pathway that is important for XP (Figure 1). Base excision acts on singly modified bases to remove them by cleavage of the glycosyl bond to release free damaged bases, leaving apurinic or apyrimidinic sites.[64]

Major chemical changes in DNA such as UV-induced pyrimidine dimers, DNA-protein adducts, and chemical carcinogen adducts require extensive reconstruction of the damaged regions (Figure 2). The initial incision in mammalian cell DNA appears to be endonucleolytic to one or both sides of the damaged sites, and it is this step that is defective in most XP cells.[65]

The distribution of excision repair sites in DNA is likely to be a complex function of the initial number and distribution of damaged sites, the number of repair enzymes, and the mechanism by which the enzymes search for damaged sites in DNA. Damage

* The D_{37} is the dose required to reduce survival to 0.37 from the initial value of 1.0, and corresponds to the dose required to produce an average of one lethal hit on the sensitive target of an irradiated organism when the survival curve is exponential.

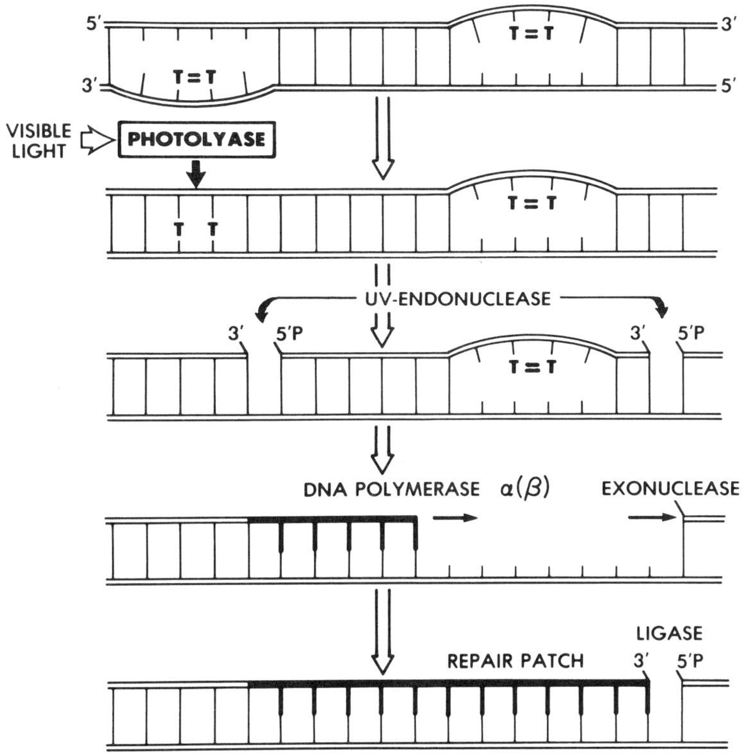

FIGURE 1. Mechanisms of repair of pyrimidine dimers by photoreactivation and by nucleotide excision repair. Photoreactivation involves binding of a DNA photolyase to the dimer, followed by absorption of visible light to cleave the 5-5,6,6-cyclobutane bonds of the dimer, causing release of the enzyme. Nucleotide excision repair of dimers (and other large DNA adducts) proceeds by asymmetric cleavage of the DNA on either side of the damaged sites, followed by release of the damaged oligonucleotide. This leaves a large single-strand gap that may then be enlarged by 5′ → 3′ exonuclease action. Resynthesis proceeds by DNA polymerase α with some contribution from polymerase β, followed by ligation. This mechanism of excision is based on one demonstrated for *E. coli*[129] and may also be inferred from work in eukaryotic cells.[239,330]

from UV light is approximately random, but repair of moderate to high levels of damage (>6 J/m^2) passes through a stage at which the repair sites are clustered.[66] The sites are random when repair is complete at late times (72 hr) after irradiation. This observation fits a model in which the repair enzymes are partially processive or have limited diffusion along the DNA, so that adjacent damaged sites will tend to be repaired sequentially. Repair will therefore proceed rapidly at first, producing clustered repair sites, but will be slower at later times because unrepaired sites will be widely spaced.

In addition to this overall pattern of excision repair, further variations in rates of repair of UV damage are likely according to gene structure and function. General patterns in relation to gene function have yet to emerge, however. The actively transcribed gene family of ribosomal sequences is repaired much more slowly than the average,[67] whereas untranscribed alpha sequences are repaired at rates similar to the average.[68] Additional complications are to be anticipated when different kinds of DNA damage and cell types with different repair capacities are investigated.

In XP cells, the rate of removal of damaged sites, such as pyrimidine dimers[69] or psoralen adducts[70] or other chemical adducts,[71,72] is markedly reduced. As a consequence, later stages of repair, including strand breakage[73-76] and the incorporation of

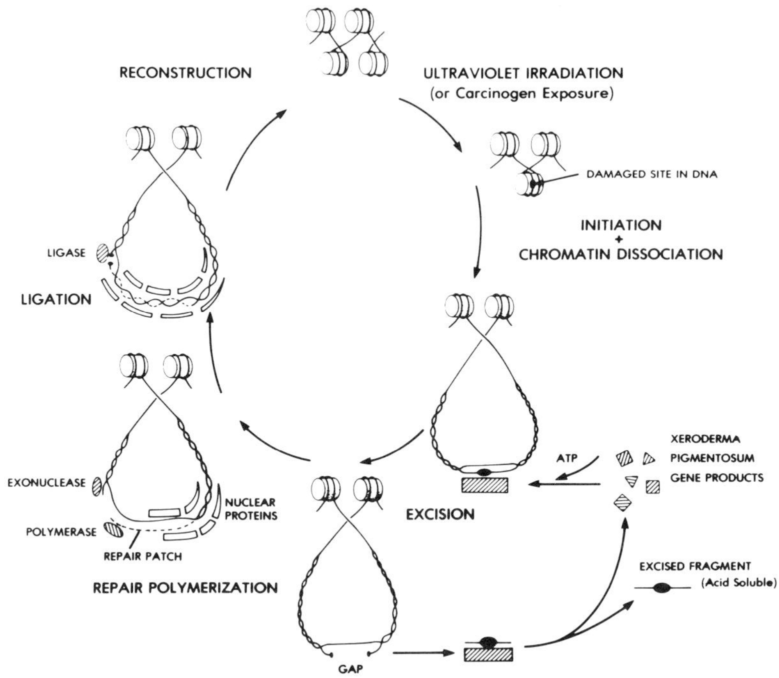

FIGURE 2. Excision repair cycle for human chromatin illustrating changes in nucleosomal conformation. Excision involves temporary loosening of chromatin structure in the vicinity of the damaged site, making the region accessible to repair enzymes. The multiple gene products specified by XP complementation groups form an excision complex that removes an oligonucleotide that initially may be acid precipitable before release and further breakdown. As repair replication proceeds, chromatin proteins rapidly become associated with the new patch. Once ligation is complete, the region rapidly acquires native nucleosomal structure with a half-time of 10 to 20 min.

new bases by repair replication,[2,77,78] are reduced to low or negligible levels. Cells therefore exhibit a large increase (up to approximately tenfold) in their sensitivity to killing by UV light[79] (Figure 3). Any of these end points are sufficient to characterize an XP cell line as belonging to the excision-defective category. This category, however, is genetically complex; its subdivision into multiple genetic subgroups will be described subsequently.

XP cells also show an increased sensitivity to a large variety of chemical carcinogens (Table 1). This indicates that there is cross reaction of the nucleotide excision-repair pathway to most bulky adducts in DNA. Certain chemicals raise problems, however. XP cells are sensitive to asbestos,[80] antipain,[81] and interferon,[82] none of which are known to damage DNA.[83] UV-sensitive mutants of *Escherichia coli* did not, however, show increased sensitivity to antipain,[81] suggesting that XP cells exhibit additional abnormalities unrelated to their repair deficiencies.

Repair deficiencies in XP heterozygotes generally have not been identified. Some heterozygote cells have shown slight reductions in repair, often at high-UV doses.[78,84,85] Hybridization of heterozygous and homozygous cell lines has also made it possible to distinguish heterozygotes from normal cells.[86] Sensitive methods based on adenoviral antigen production in irradiated cells also detected deficits in repair in heterozygous XP cells,[87] although host-cell reactivation of adenovirus[88] and herpesvirus[89] is normal.

3. DNA Replication Abnormalities in XP Variant

One particular category of XP patients appears to have no abnormality in excision

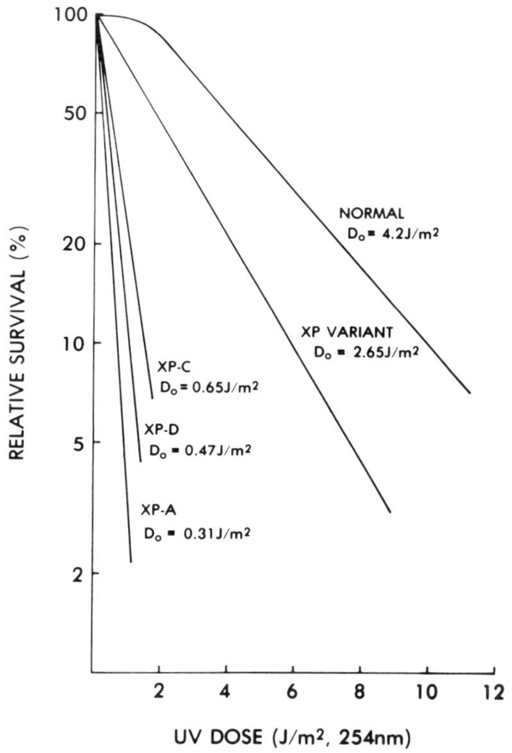

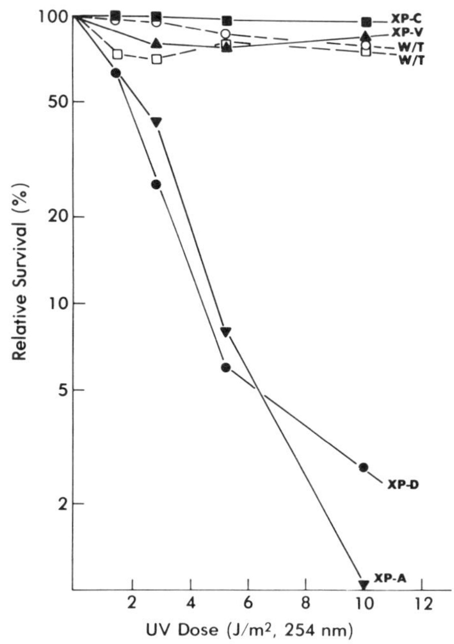

FIGURE 3. Relative survival of normal and XP cells irradiated with UV light and assayed either for colony formation (left) or for survival as confluent cultures (right). Colony formation curves represent the mean curves for each group from the table presented by Andrews et al.[79] Confluent survival was measured either by prelabeling cells with [^{14}C]thymidine (0.01 μCi/mℓ; 54 mCi/mmol) and assaying for ^{14}C activity, or by measuring the capacity of cells to incorporate [^{3}H]hypoxanthine (5 μCi/mℓ; 10 Ci/mmol) into acid-insoluble RNA and DNA.

repair of UV damage, and was originally classified as XP variant.[35,36] Subsequently, an abnormality in the semiconservative replication of damaged DNA was found in these cells. After irradiation, replication forks appear to stop or be interrupted at most sites of damage.[37,38,42,90-92] After irradiation, there is an accumulation of nascent DNA of shorter size than in normal cells (Figure 4). Replication forks in normal cells are rarely stopped at sites of damage and tend to replicate in lengths longer than interdimer distances. Excision-defective XP cells also exhibit replication anomalies, but these are secondary consequences of the increased level of damage resulting from failure in excision.[91-95]

The recovery of DNA synthesis after irradiation is delayed more in the variant than in normal cells, and this phenomenon affords a point for comparison with excision-defective XP, Cockayne syndrome, and ataxia-telangiectasia cells (Figure 5). All of these genetic disorders show abnormal responses to irradiation for different reasons. Excision-defective XP, XP variant, and Cockayne syndrome cells show greater reduction and delayed recovery of DNA synthesis after UV. In excision-defective cells, this is because greater levels of unrepaired damage interrupt DNA synthesis. The variant has normal excision but has a primary abnormality in the replication of damaged DNA. In addition, the variant cells show an extended delay from G_1 to S after irradiation.[96] Cockayne syndrome cells also show a primary abnormality in replication that appears to involve an inability to reinitiate replication after UV irradiation. Ataxia-telangiectasia cells, in contrast, have an opposite kind of abnormality after exposure

Table 1
AGENTS TO WHICH XP CELLS SHOW INCREASED SENSITIVITY OR REDUCED DNA REPAIR

Agent	Survival	Chromosomal aberrations or SCEs	Repair	Ref.
DNA damaging				
Acetoxy-acetyl-aminofluorene	X	X	X	301,302
Aromatic amides	X	—	—	303,304
Hydrogen peroxide	X	—	—	305
4-Nitroquinoline-1-oxide	X	X	X	306—309
Anthramycin	X	—	X	176,310
Polycyclic hydrocarbons	X	—	X	241,311—314
Nitrofuryl acrylamide	X	—	—	315
Tryptophan pyrolysate	X	—	—	316
Aflatoxin	X	X	X	317,318
Nitropyridine	X	—	X	319
Isoniazid	—	—	X	320
Diarylcarbamate	—	—	X	302
Platinum diammine	—	—	X	321
ICR170	—	—	X	302
Psoralens	—	X	X	261,322,323
Diepoxybutane	—	X	—	258
Agents not known to damage DNA				
Asbestos	X	—	—	80
Antipain	X	X	—	81
Interferon	X	—	—	82

Note: The symbol "X" denotes the nature of evidence presented in the cited references. The symbol "—" denotes no evidence in the references.

to ionizing radiation. These cells fail to switch off their DNA synthesis after irradiation.[97]

The recovery rates of XP variant cells fall approximately in the range of XP group C cells (Figure 5). This may be relevant to the phenotypic consequences of the respective defects. Group C and variant show similar clinical symptoms, and both have extensive capacity to recover from similar UV fluences when allowed to remain confluent for extended periods[98,99] (also see Figure 3 and subsequent discussion).

Although the argument has been made that XP variants might have a highly localized excision defect at replication forks,[42,100] this now appears to be unlikely. The absence of any increase in sister chromatid exchanges in XP variant cells above that in normal cells after UV irradiation argues against there being any excess damage at the growing points.[101-103] The most likely interpretation of the XP variant is that these cells lack gene products whose primary role is in the replication of damaged DNA. In normal cells, these gene products would facilitate the rapid passage of replication forks past damaged sites with minimal interruption of daughter-strand continuity.

4. Photoreactivation in XP

Photoreactivation is a distinct repair system that is highly specialized for the repair of the cyclobutane pyrimidine dimers produced by UV light. It is a one-enzyme system that uses visible light as an energy source to cleave dimers without excision of damaged sites. It is widespread among living organisms, although its activity is weak in mammalian cells. Until about 1974, it was generally assumed to be absent from placental mammals.[104,105] A low level of enzymatic activity was then reported in human leuko-

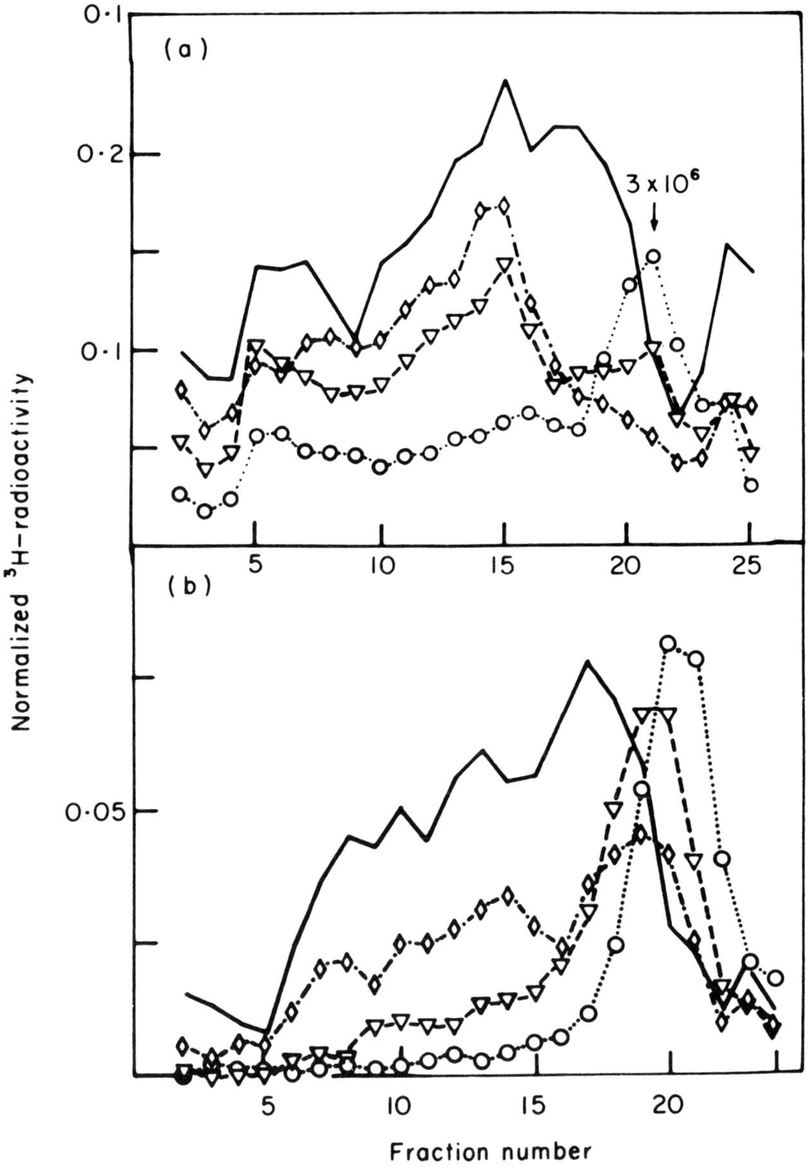

FIGURE 4. Dose-response of UV-induced alterations in steady-state sedimentation profiles of [^{3}H]DNA from normal and XP variant fibroblasts. (a) Cultures of normal fibroblasts (HF1) were incubated in medium containing [^{3}H]thymidine for 10 min beginning 30 min after sham irradiation (—) or irradiation with 1.3-J/m^2 (diamond), 2.6-J/m^2 (inverted triangle) or 5.2-J/m^2 (circle) UV; (b) XP variant cultures (XP30RO) also labeled for 10 min with [^{3}H]thymidine after 1.3-J/m^2 (diamond), 2.6-J/m^2 (inverted triangle), or 5.2-J/m^2 (circle) UV. Cells were harvested and [^{3}H]DNA was sedimented in alkaline sucrose gradients at 25 krev/min for 4 hr. Normalized ^{3}H radioactivity was calculated by dividing the net ^{3}H cpm in each gradient fraction by the total ^{14}C cpm recovered in the gradient. Total ^{14}C cpm were about 5000. The solid line is a trace of profile from the sham-irradiated (control) cultures.[91]

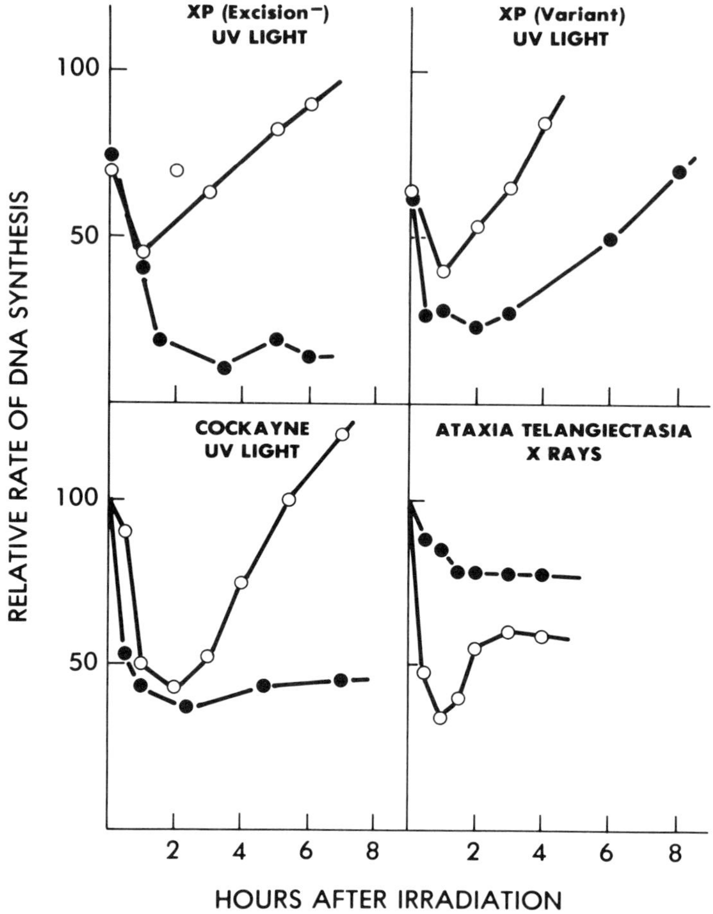

FIGURE 5. Relative rate of DNA synthesis, measured by 10- to 30-min labeling with [^{3}H]thymidine as a function of time after irradiation of various normal (circle) or affected (black circle) cell types. Excision-defective XP, XP variant, and Cockayne syndrome cells were exposed to 5-J/m^2 UV light. Ataxia-telangiectasia cells were exposed to 1-krad X-rays.[239]

cytes, although the enzyme activity in vitro reduced pyrimidine dimer content in DNA substrates only from 5.2 to 4.5%, a barely detectable reduction.[106] Subsequently, there were reports of deficient enzyme activity in XP cells.[107] Continued investigations have not supported claims for a reproducible photoreactivation activity in various mammalian (including human) cell lines.[108] Observed monomerization of pyrimidine dimers generally has been small, inconsistent, and extremely dependent on experimental conditions. In mouse and human skin in vivo, there have been occasional reports of the photoreactivation of pyrimidine dimers, but again the activity has not been observed in all cases, is rarely a large effect, and its absence has not been associated with any pathologic condition.[109-112] Claims for photoreactivation enzyme deficiencies in XP cells therefore have uncertain significance.

C. The Complementation Groups

Genetic heterogenicity in excision repair-defective cells was suggested by the different residual activities of repair replication observed in cells from different patients[2,77,113] and by the varied clinical patterns.[15] These genetic differences have been analyzed by somatic cell hybridization. Cells from different XP patients can be hybrid-

Table 2
XP COMPLEMENTATION GROUPS

Group	Case frequency[a]	CNS symptoms	Excision[b] (%)	UDS[c] (%)	D_0 ratio (XP/ normal)	Ref.
A	26% (all A)	+	0	0—5	0.07	116
A	3 cases	−	0	0—35	0.17—0.41	79
B (Cockayne)	1 case	+	0	3—7	—	15,116
C	38%	−	~15	10—15	0.15—0.25	79,116
D	10%	+	30	25	0.09—0.11	79,116
E	3% (2 cases)	−	55	40—60	0.7	116,324,325
F	3% (2 cases)	−	60	10[d]	0.45	325—327
G	3% (2 cases)	+	0	10—15	0.11	328,329
H (Cockayne)	1 case	+	—	30	—	41
I	1—2 cases			14		332
Variant	15%	−	100	70—100	0.6—1.0	37,38,69

[a] Based on the catalog of the National Institute for Genetic and Medical Research, Camden, N.J., 1983, which contains cases contributed from the U.S., England, Egypt, and Japan.
[b] Percentage of the excision seen in normal cells at 24 hr after irradiation (70%).[69]
[c] Percentage of the repair synthesis seen in normal cells labeled for the first few hours after irradiation.
[d] This apparent discrepancy between excision and repair synthesis can be resolved if excision and labeling are performed over similar time periods.[325]

ized in culture under the influence of inactivated Sendai virus or polyethylene glycol to produce multinucleated cells (heterokaryons) with nuclei from each XP patient. Heterokaryons from some combinations of XP patients exhibit complementation that can be detected as increased repair synthesis and increased UV resistance, whereas other combinations remain repair deficient.[3,15,114-117] Complementation is an indication that the original cell types contained different defects in the initial step of pyrimidine-dimer excision. Numerous studies in which cells from many patients were hybridized in pairs have demonstrated at least nine excision repair-deficient complementation groups, designated A through I.[14] The same groups are obtained when chemicals such as 4-nitroquinoline-1-oxide are used as the damaging agent.[118]

The complementation groups have distinctive clinical and biochemical features (Table 2), although none would permit objective assignment of a cell line to a group without cell-hybridization experiments. Neurological abnormalities are concentrated in patients in groups A, B, D, and G, although there are a few exceptions. Complementation group A contains the most severely affected patients, with neurological abnormalities representative of the de Sanctis-Cacchione syndrome. There are, however, several well-documented cases in group A in which patients exhibit no neurological abnormalities.[18,119,120] Skin and neurological abnormalities are therefore not always coupled in group A.

Cells in complementation group C appear to have a milder defect that affects only rates of repair. If these cells are maintained in confluence, they can recover slowly to near-normal extents (Figure 3). Cells from groups A and D, however, show much less recovery under these conditions.[98,99] Group-C cells could, in consequence, involve a gene that determines the energy requirement in excision repair, possibly the ATP binding site, whereas the other groups could involve more important structural genes that determine the recognition mechanisms for DNA damage.

Some evidence has been presented claiming that excision-repair sites are more clustered in XP group-C cells than in normal cells.[121,122] Similar investigations of group-D cells, however, which have comparable levels of excision repair, have not yet been reported. Although it is attractive to consider this clustering of repair sites to represent repair of functionally defined regions, it might represent no more than a consequence of the partially progressive character of the excision-repair system itself.[66] Repair sites are clustered when repair is still incomplete, and the low frequency of repair enzymes in group C could simply exaggerate this clustering phenomenon, irrespective of the functional states of the mutant repair system or the DNA being mended.

The patterns of complementation between cell lines of each group are not identical. Fusions between groups A and D result in rapid complementation, whereas those involving group C are much slower.[123-125] Similar rapid and slow complementation occurs when enucleated cells (cybrids) are used, implying that the gene products can be found in the cytoplasm.[126] These studies suggest that the group-A gene product is present in excess and rapidly reassociates with other products during hybridization, whereas the group-C gene product may be synthesized slowly. The product of group A, however, may turn over more rapidly than those of groups C and D.[123] The repair complex itself appears to be synthesized slowly and is unaffected by inhibition of protein synthesis with cycloheximide.[127] Complementation is not prevented by cycloheximide, suggesting that the gene products are not induced.[123,128]

The initial step of pyrimidine dimer excision that appears to be merely the action of a single UV endonuclease is therefore much more complex, and is regulated by at least nine different genetic factors. The XP variant constitutes yet another group involving a different pathway. Whether these factors are all different polypeptides, analogous to the three polypeptides involved in the same repair step in *E. coli*,[129] is unresolved, but the system is clearly much more complex than that in microorganisms.[86,123,124] The excision-defective complementation groups in XP could represent intergenic or intragenic complementation. The possibility that two complementation groups correspond to intron and extron regions of a single gene or different functional regions of one polypeptide cannot be discounted, and there may be fewer polypeptides than complementation groups. The affected genes may play a role as both structural and regulatory genes in the formation of one or more enzymes mediating the early steps of excision repair. One hypothesis consistent with the observed kinetics of excision and strand breakage is that the excision, replication, and ligation enzymes act in a coordinated way by means of an enzyme complex.

Each complementation group is not found equally around the world. There are two major groups, A and C, and the remainder are less frequent. Group C, however, is also rare in Japan. The next most frequent groups are D and variant. The remaining groups B, E, F, and G are thus far represented by only one or two families. The low frequency of group C cases among the Japanese[26,130] was prefigured in a study of gene frequencies that suggested a lower frequency for XP in Japanese than in Caucasian populations.[131]

Identification of the chromosomes carrying the XP genes has not been easy, mainly because of the lack of precise analogs to XP among rodent cell lines. Ideally, the human chromosomes bearing XP genes could be identified by hybridization of normal human cells to repair-deficient rodent cells, followed by selective elimination of the human chromosomes. Unfortunately, although many mouse and hamster cells are repair-deficient, including some specifically selected for their UV sensitivity,[132-134] none of these are known to have precisely the same genetic abnormality as XP, despite superficial similarities. Hybridization between normal human cells and repair-deficient rodent cells has, however, identified several chromosomes carrying complementing genes. In fusion with mouse cells, chromosomes 3[135] and 13[136] were identified as important in UV sensitivity. In a similar study using hamster cells, chromosome 1q was also identified as a probable location for XP group A.[137,138] Hybridization between XP group-A cells and normal mouse cells resulted in assignment of the mouse gene homologous to XP group A to mouse chromosome 4.[139] If mouse and human linkage groups are assumed to be similar,[140] this work would imply that the gene for XP group A is on chromosome 1.

Experiments designed to clone the genes involved in XP have encountered serious and unexpected technical problems. The direct approach involves transfection of genomic DNA from normal human cells in an appropriate vector with coselectable markers into recipient XP cells. Successful transfection of genes for UV repair should permit selection of cells by their resistance to UV light. This approach appeared to show some initial promise,[141,142] but subsequent experiments have failed.[143,144] Human DNA sequences, approximately 7 kb in size, that complement UV-sensitive hamster cells have, however, been identified and cloned.[145,146] The relationship of these genes to XP has not yet been clarified; neither has the reason for consistent failure in similar experiments using XP cells. One unexpected observation in attempts to use XP cells has been the production of a low frequency of UV-resistant XP revertants, independent of the DNA used in attempted gene transfer.[144] Perhaps the genes that are important in XP are larger and more complex than those in Chinese hamster cells, or maybe they are *cis*-acting regulatory genes that must be integrated in the correct position to be effective.

D. The Gene Products

Enzymes for the first step of repair should have unique properties related to their action on UV photoproducts. The enzymes for the later stages of excision repair (exonucleases, polymerases, and ligase) are less likely to have unique repair properties because they operate on molecules in which specific damage has been excised. The later stages of repair are performed by polymerases[147,148] and ligases[149] already identified and characterized in animal tissues. Polymerase α is the main polymerase involved in repair of UV and many other forms of damage in proliferating cells,[150-155] although some participation of polymerase β is also likely.[153,154,156] The situation in human cells may resemble that in *E. coli,* in which several different polymerases can play a role in excision repair depending on the genetic and physiological state of the cell, although the preferred mode involves polymerase I.[157] Polymerases and ligases appear normal in XP cells.[158,159]

Identification of a UV-specific endonuclease involved in an early step of repair in human cells has been much more difficult than anticipated. This difficulty is due, in part, to uncertainty in defining the expected activities in assays using crude or partially purified extracts with nonspecific nuclease activities. Another difficulty is the instability of many enzyme extracts. Indirect evidence suggested that the enzymatic function may lie in an extremely large aggregate of polypeptides.[123,160,161]

Bacterial repair enzymes have been isolated through their ability to make nicks in UV-irradiated supercoiled DNA.[157,162] Experiments based on this kind of substrate have identified a similar activity in human cells,[163-165] rat liver,[166] XP fibroblast cell lines,[162,164] and calf thymus.[167,168] This nicking activity did not appear to be involved in the repair of pyrimidine dimers because it made nicks at a frequency of only about 1 to 2% of the number of the dimers present in the UV-irradiated substrates.[162,163,165,167] Specific monomerization of dimers by photoreactivation in vitro also did not alter the nicking activity of the extracts on the substrate.[162] Recently, however, similar experiments identified a nicking activity that was present in normal human cells but not XP cells, but that could be restored by mixing extracts from XP cells of different complementation groups.[165] Although this represents an important advance, the activity was weak and the extract made single-strand breaks at only a small percentage of the total pyrimidine dimer sites in the substrate. The enzyme activities identified in most of these kinds of experiments therefore do not appear to nick close to a large proportion of pyrimidine dimers. They may not, therefore, represent the enzymes associated with the repair deficiency in XP.

A more specific assay has been devised that is more closely related to the function in vivo. This was based on the ability of a cell extract to excise pyrimidine dimers in vitro.[169,170] Whether this assay will detect only an endonuclease or both an endonuclease and an exonuclease depends on one's view of the mechanism of excision. Extracts from normal human cells excised dimers from DNA in vitro, but the activity was easily destroyed by freezing and thawing or excessive sonication during preparation of the extracts.[171]

Unfrozen extracts from both normal and XP cells can excise dimers from UV-irradiated DNA.[169,170] On this basis, even XP cells that cannot perform excision repair in vivo still perform normal repair in vitro.[169,170] This suggests either that the defect is more complex than first envisaged and is not simply a missing UV-specific endonuclease, or that the excision observed in vitro is due to other enzymes that are prevented from excising in the cell.

When cell extracts were compared for their ability to excise dimers from both exogenous, purified DNA and their own DNA still associated with nucleoprotein, extracts from XP cells (complementation groups A and C) excised dimers only from exogenous DNA and not from DNA in chromatin. Combined extracts from XP cells of different complementation groups could, however, excise dimers from chromatin, indicating that complementation between existing polypeptides in the extracts could occur.[170] The defect in XP cells revealed by these experiments therefore may be some factor that governs the accessibility of damaged sites in chromatin to repair enzymes.

None of these studies bear on the nature of the gene product in the XP variant. This appears to be a homogeneous complementation group[172] in which cells lack a gene product required for replication of UV-damaged DNA, but further characterization has not been possible.

A number of other enzyme deficiencies have been reported in XP, some of which appear to be more closely related to a base excision-repair pathway than the nucleotide-excision pathway. Deficiencies in apurinic endonuclease have been reported for XP groups A and D,[173,174] and in a single-strand binding protein for group A.[175] Other deficiencies have been found in activities acting on anthramycin-treated DNA[176] and alkylated or apurinic DNA.[177-179] A coherent picture of the enzyme deficiencies in XP has yet to be achieved, but the disease clearly displays a bewildering variety of deficiencies at the cellular level.

E. Damage-Inducible Cellular Processes

Damage to DNA causes an overall depression in RNA and protein synthesis. Transcription of RNA appears to be terminated at damaged sites, and this phenomenon has been used to estimate transcript sizes and gene sequences in multicistronic messages.[180] Recovery of RNA synthesis is a genetically determined process dependent at least on prior excision of damage, because recovery is delayed in XP cells.[11] Recovery is also delayed in the XP variant and in the three known complementation groups of Cockayne syndrome, which implies that several additional gene products regulate recovery of RNA synthesis after irradiation.

Against this background of general depression in macromolecular synthesis, there are reports of a disparate array of novel proteins that appear to be induced by UV irradiation.[181] In addition, semiconservative DNA synthesis can be stimulated by UV damage in a small proportion of a confluent cell population.[182] Well-characterized proteins such as ornithine decarboxylase can be induced by irradiation of epidermal cells.[183] To what extent these changed or increased activities represent degradative or pathological sequelae of cell death caused by irradiation, and to what extent they may be associated with cellular recovery, is difficult to resolve. Studies in which synthesis of plasminogen activator in XP cells[184] or synthesis of a variety of unspecified proteins[181] was induced by UV are relevant to this question. These studies all observed much greater induction by UV in excision-defective XP, intermediate levels of induction in Cockayne and XP variant cells, and least induction in normal fibroblasts. Similarly, reactivation of UV-damaged herpes virus is increased by irradiation of the host cells; the dose required for maximal induction is least in XP groups A and D, slightly greater in XP group C, intermediate in Cockayne syndrome, and greatest in XP variant, XP heterozygotes, and normal fibroblasts.[185] These relative efficiencies of induction correspond precisely to the relative sensitivity to cell killing by UV light, despite the widely varied biochemical abnormalities in these cells. It is likely, therefore, that these responses are correlated with pathological rather than recovery processes.

A similar mechanism involving cell death may underlie the observation of an induction of semiconservative DNA synthesis[182] and of DNA ligase[186] 24 hr after irradiation. These processes could resemble a "wound response". Death of a small proportion of the population could relieve contact inhibition and permit cells to commence progress through the cell cycle.

The common factor in these experiments therefore seems to be the degree of cell killing, which would argue against the results being caused by an inducible process analogous to the bacterial SOS system.[187-189] In *E. coli,* damage generates a signal, probably associated with blocked DNA replication forks and DNA breakdown, that initiates a *recA/lexA*-dependent system involving protease cleavage of *lexA* gene product to abrogate its repressor function.[187] A large variety of genes are then coordinately induced, new protein synthesis ensues, and many gene products specifically associated with mutagenesis and cell survival are produced *de novo.* Evidence for a similar system in mammalian cells[190] and yeast[191] is weak and controversial. Some attempts to force a close parallel have employed protease inhibitors (e.g., antipain) in mammalian cells to generate phenomena that would be used as evidence for a protease-dependent step in transformation.[192,193] However, experiments involving transformation are sufficiently complex that they are susceptible to several interpretations other than inducibility, especially because antipain does not influence mutagenesis in mammalian cells.[192,194]

Two distinct problems are involved when inducible processes are discussed. One is the nature of error-prone systems that generate permanent genetic changes, and the other is the role of enzymes whose synthesis is induced by DNA damage. In *E. coli* these two problems coalesce; mutations are generated by the inducible *umuC* system.[187] In eukaryotic cells, error-prone systems clearly exist but they are not necessarily inducible.[95,190,191]

One line of evidence for an inducible system in mammalian cells comes from experiments in which increased survival and mutation frequencies are observed in viruses as a consequence of irradiation of the host cell. This process, known as Weigle reactivation, implies that survival is increased and mutagenic events are created in the viral DNA by an inherently inaccurate cellular system induced *de novo* by irradiation of the host. Substantial differences exist, however, between the phenomenon of Weigle reactivation in *E. coli* and in mammalian cells. Direct irradiation of phage has only a weak mutagenic effect, whereas irradiation of host bacteria produces a high rate of mutagenesis in the phage.[195] In contrast, in mammalian cells and virus systems, irradiation of the virus alone is sufficient to produce high levels of viral mutagenesis.[196-199] The observations in mammalian cells are therefore more representative of modulation of a constitutive system by irradiation than induction of a completely new system.

To fully interpret these increases in viral survival that occur after irradiation of host cells, complete viral inactivation curves are important. Misleading impressions can be created if the increased survival is merely expressed at a few discrete levels of damage to the host cell. Since viral inactivation curves are exponential, large increases in viral survival can occur from very slight changes in the slope of the inactivation curves. For example, at 1% survival on an exponential curve, a fivefold increase in survival corresponds to a reduction in the effective dose of only about one third. Therefore, quite small changes in cellular physiology could cause slight alterations in the slope of viral survival curves, which could be misread as large inducible effects if the increased survival is considered only at discrete doses.

Irradiation of the host cell increases mutagenesis of the virus less markedly than in *E. coli*. The largest effects have been observed with SV40-infected monkey cells.[198,199] Increases are weaker or nonexistent with herpes virus.[200,201] Mutagenesis of adenovirus is also not detectable as a result of host-cell irradiation.[202] In experiments using a shuttle vector in which transfected DNA shows high mutation frequencies,[203] prior exposure of cells to ethyl methanesulfonate increased mutagenesis by a factor of ten.[204] The interpretation of these experiments is much more complicated than in *E. coli,* because in mammalian cells only a very small proportion of infecting viral particles (10^{-2} to 10^{-3}) produce plaques.

Implicit in these viral studies is the assumption that viruses and viral-like shuttle vectors are probes for cellular biochemistry. The reduced survival of irradiated viruses grown in repair-deficient XP cells supports this view.[205-209] However, phenomena such as multiplicity reactivation[208] and the high mutation rates of infecting viruses[196,198] and shuttle vectors[203,204] and high rates of recombination[210] indicate that extrachromosomal DNA is exposed to an environment that creates mutations at a higher rate and qualitatively different from those in chromosomal DNA. Also, the large structure of chromosomal DNA, which is replicated from multiple origins of replications, implies that chromosomal DNA has mechanisms for modulating replication in response to damage that are unavailable to extrachromosomal DNA.[92,93]

It is possible that at least some of the responses that have been interpreted as induced error-prone repair systems could be the consequence of alterations in pool sizes as a result of exposure to radiations and carcinogens.[211-215] Pool imbalance has been shown to have a profound effect on cellular toxicity and mutagenesis,[216-218] and the time course of pool size changes in cells exposed to UV or chemical carcinogens is similar to that expected for induced processes. In part, pool changes could occur because of continued nucleotide synthesis in cells in which DNA synthesis is inhibited. Since the patterns of DNA synthesis inhibition are different in XP, XP variant, Cockayne syndrome, and ataxia-telangiectasia, these cell types could all show different changes in pool size that could masquerade as differences in hypothetical inducible processes.

A uniquely eukaryotic process that enters into a variety of cellular responses to DNA damage is the polymerizing system by which nicotinamide adenine dinucleotide (NAD) and adenosine triphosphate pools are depleted by the formation of poly(ADP-ribose).[219-221] This polymer covalently modifies many nuclear proteins and inhibits their function. This system is activated by breaks in DNA, and prevention of its activity is associated with increased cell killing. In XP cells, there are very few single-strand breaks after irradiation because of the repair deficiency, and much less poly(ADP-ribose) synthesis occurs.[222] Abnormalities in the poly(ADP-ribose) system have been reported in Cockayne syndrome,[223] ataxia-telangiectasia,[224] and Fanconi's anemia,[225] but their significance is unclear.[226,227]

Another unique mammalian process that can be observed after irradiation or carcinogen exposure is an alteration in DNA methylation.[228] Repair patches are initially undermethylated compared to the bulk DNA of the cell.[228] Some genes can become fully demethylated and activated, although this is still a rare event occurring at a frequency of about 10^{-4} after 10 J/m^2.[229]

Another induced process that occurs preferentially in XP and ataxia-telangiectasia cells is that of collagenase secretion.[230] Using an inhibitor of calmodulin, XP groups A and D and, to a much lesser extent, group C, could be induced to secrete collagenase, suggesting that the ability of these cells to interact with an extracellular matrix may be altered.

Damage-inducible systems clearly exist in mammalian cells, but whether they play a major role in mutagenesis is uncertain. The dose responses for mutagenesis in normal and XP cells are close to linear, which argues for mutations to be governed by a single dose-dependent event, i.e., the amount of damage. Additional dose-dependent induction of a mutagenic SOS process would be expected to create nonlinear dose-response curves. Experiments using fractionated doses to investigate possible induction of a process associated with DNA replication[231,232] produced distinctly different responses in normal XP and XP variant cells, but the interpretation of these results does not require an inducible process. Different inhibitory effects of second doses on DNA replication profiles that had been markedly disturbed by the first dose, together with inappropriate normalization of sucrose gradient profiles, satisfactorily explains the original observations.[233] Similarly, fractionated UV doses give no indication of enhanced mutagenesis.[234]

In conclusion, although the concept of damage-inducible events is versatile, evidence is inconclusive. Mammalian cells contain such a rich diversity of biochemical processes that any attempt to force their responses to fit a bacterial model is likely to be distracting rather than helpful.

III. THE ORIGINS OF CANCER IN XP

One theory of carcinogenesis[235,236] invokes an accumulation of genetic damage to some critical region(s) of the genome by exposure to environmental mutagens.[237] Damaged regions could become sites for somatic mutation, chromosomal rearrangement, gene amplification, or other alterations in the genome.[236] Evidence bearing on some of these possible mechanisms in carcinogens has been obtained from studies of mutagenesis, chromosomal abnormalities, and transformation in XP cells. The life spans of XP patients[31] and of XP cells in culture[238,239] appear normal (excepting early deaths due to the malignancies), which implies that aging, unlike cancer, is unrelated to the kinds of repair abnormalities seen in XP.

A. Mutagenesis

Few gene loci are available for studying quantitative mutagenesis in vitro. Of these, resistance to 6-thioguanine or 8-azaguanine denotes loss of function of the hypoxanthine-guanine phosphoribosyl transferase enzymic pathway, which could result from point mutations, frame shifts, and deletions. Resistance to ouabain and to diphtheria toxin are, however, much more restrictive and can be produced in viable cells only if there is a point mutation that leaves a functional protein with altered binding sites.[240]

Experiments involving these loci have all shown that the mutation rates in excision-defective XP cells are elevated approximately in proportion to their increase in sensitivity to killing by DNA-damaging agents and their decrease in repair.[241-246] Although this would suggest that mutagenesis and cell killing are altered in the same quantitative sense by excision-repair deficits, other studies point to important differences in mechanisms of these two processes.

In normal fibroblasts, for example, irradiation in early G_1 leaves cells with a long period for repair to occur before DNA synthesis. The consequence of this time delay before replication is a reduction in the mutation rates, as compared to cells irradiated late in G_1, but no change in cell killing.[247] Recovery from the lethal effects of UV occurs later in the cell cycle than recovery from the mutagenic effects, and is generally expressed by passage through S phase.[248-250]

The XP variant cells are very sensitive to the induction of mutations by UV light, even though their sensitivity to killing is only slightly increased.[251-253] The defect in these cells, therefore, is one that preferentially affects mutagenesis.

B. Chromosomal Aberrations and Sister Chromatid Exchanges

The repair deficiencies in XP have a major effect on the chromosome stability in irradiated cells. Unexposed cells do not, however, exhibit spontaneous chromosome abnormalities or sister chromatid exchanges (SCE), unlike some other disorders such as Bloom's syndrome, Fanconi's anemia, or ataxia-telangiectasia.[254]

Aberrations of several types have been reported in XP cells after UV irradiation. These often appear to be the kinds that would be associated with lethal events and may be related to the increased cell killing in irradiated XP cells.[255-261]

After exposure to DNA-damaging agents, increased frequencies of SCE have been observed in XP fibroblasts, transformed cells, and lymphoid cells.[81,101-103,262-267] Chemicals that form O^6-alkylguanine in DNA produce cellular effects that can often be related to an abnormality in O^6-alkyltransferase that develops in culture (the mer^- or mex^- phenotype).[268-270] Early work reporting increased SCE in an SV40-transformed XP cell line (XP12RO) that failed to repair O^6-ethylguanine is attributable to the transformed phenotype, not the UV repair defect of XP.[271-273] Untransformed XP fibroblasts have normal capacities for O^6-methylguanine repair.[274]

Exposure to UV light causes increased SCE in both XP fibroblasts[114] and lymphoblastoid cells.[103] The excision-defective XP cells all showed increases in SCE frequencies, but the XP variant cells showed normal frequencies. This difference between the two forms of XP constitutes an important clue to the mechanism of production and biological significance of SCE. The presence of increased damage during DNA synthesis, as a consequence of repair deficiency in the excision-defective XP groups, is a likely mechanism for stimulation of SCE, but the interruptions in DNA chain growth in the XP variant clearly do not generate SCE.[275] In contrast, increased SCE frequencies are observed in Cockayne syndrome cells, which have a DNA replication abnormality different from that of the XP variant.[103,276] Since the excision-defective and variant XP groups all show increased mutagenesis and carcinogenesis, the normal frequency of SCE in XP variant cells implies that SCE have little relevance to mutation or carcinogenesis in this system. Similarly, elevated SCE frequencies in Cockayne syndrome, which does not exhibit increased carcinogenesis, support this view. Since there is a strong correlation between mutagenic and carcinogenic mechanisms, and SCE are used extensively for carcinogen screening, the lack of correlation between alterations in SCE frequencies and malignancy in XP variant and Cockayne syndrome constitutes important evidence that deserves greater attention.

C. Transformation

Radiations and chemical carcinogens have been used extensively in studies of morphological transformation of rodent cell types in vitro.[277-279] Transformation studies using human cell types have been more difficult, however, because the experimental systems are more difficult to employ routinely for quantitative purposes.[280-284] Also, although transformed cells produced from mouse and hamster cells in vitro can form invasive tumors in vivo in immunosuppressed animals, human transformed cells tend to form benign sarcomas that regress. Transformation in vitro, therefore, mimics some but not all aspects of carcinogenesis in vivo.

Several studies have shown enhanced transformation of XP cells in vitro from either short wavelength (254 nm)[281] or medium wavelength ($\sim$300 nm)[285] UV light. These were produced in normal cells when a period of DNA synthesis followed promptly after irradiation, but not when cells had an extended period of G_1 during which excision repair could occur. Maher et al.[281] observed that the yield of transformants was higher (2.5 times) than the frequency of single gene mutations (6-thioguanine resistance). Higher yields of transformants than single gene mutants have often been reported. This implies that there is either a large genomic target for transformation, or very efficient activation of a process in many cells that favors spontaneous mutations during subsequent growth.[286,287]

D. Carcinogenesis

XP constitutes a strong example that implicates mutagenesis and chromosomal abnormalities as mechanisms for an environmentally induced cancer. This view has been strengthened by the association of many of these mechanisms with activation of oncogenes.[288] Point mutations have been identified in the *ras* series of oncogenes.[289-294] Translocation has been identified for the *myc* oncogene in lymphoma.[295,296] Amplification of *myc*-related sequences has also been identified in a series of human tumors.[297,298] Although there are no published studies on gene amplification in XP, preliminary evidence suggests that repair deficiency also leads to an elevated frequency of methotrexate resistance, which is often associated with gene amplification after irradiation.[299]

Incomplete, inefficient, or inaccurate repair in some hereditary diseases therefore should be correlated with increased carcinogenesis. Carcinogenesis induced by radiation or chemicals in normal individuals would be then due to the normal amount of inaccuracy present in repair, or to replication of a damaged region before repair was complete. XP would be an example of increased damage remaining in DNA during replication, with a resulting accumulation of mutations and other radiation-induced genetic changes. The XP variant provides additional insight into this problem. Although patients in this group appear very similar to those in excision-defective groups, there is an apparent excess of basal cell cancers.[39] This correlates well with the replication anomaly in the XP variant because most DNA replication in the skin is concentrated in the basal layer. It is therefore probable that replication of excess unrepaired DNA damage from solar radiation, or aberrant replication of normal levels of DNA damage before time for repair, are mechanisms that converge to produce similar cellular and clinical consequences. The extreme organ specificity of tumors in XP patients[31,47] points to a direct mutagenic effect of sunlight and other carcinogens on the organs at risk. The rapid onset of tumors[31] without the many decades of lag commonly observed for skin cancer in normal individuals suggests that sunlight might also have a strong promoting effect on skin carcinogenesis in XP.

Substantial evidence from microorganisms and mammalian cells supports a theory in which initiating events in carcinogenesis are akin to mutagenic events. We would therefore expect XP tumors to contain mutations or other changes in some of the oncogenes recently identified, especially the *ras* family of oncogenes.[293,294] The 12th and 61st codons of the *ras* gene appear to be the only codons that can lead to transforming activity in these genes.[294,300] These codons are GGT and CAA, respectively.[294] Analysis of these codons in XP tumors would be of particular interest because codon 12 does not contain adjacent pyrimidines subject to UV mutagenesis, but codon 61 has adjacent thymines in the complementary strand. Therefore, if mutagenesis arises in *ras* gene codons from direct damage to the bases within these sequences, we would expect a bias to mutations in the 61st codon, or participation of flanking bases in adjacent codons.

These considerations suggest that the various cellular parameters studied experimentally, such as DNA damage, repair, mutation, and transformation, are but a few of the complex mechanisms involved in carcinogenesis. Further elucidation of details of the biochemical, cellular, and clinical characteristics of XP should provide a better understanding of one kind of carcinogenesis.

ACKNOWLEDGMENT

This work was supported by the Office of Health and Environmental Research, U.S. Department of Energy (DE-ACO3-76-SF01012).

REFERENCES

1. Cleaver, J. E., DNA damage, repair systems and human hypersensitive diseases, *J. Environ. Pathol. Toxicol.*, 3, 53, 1980.
2. Cleaver, J. E., Defective repair replication of DNA in xeroderma pigmentosum, *Nature (London)*, 218, 652, 1968.
3. Cleaver, J. E. and Bootsma, D., Xeroderma pigmentosum: biochemical and genetic characteristics, *Annu. Rev. Genet.*, 9, 19, 1975.
4. Painter, R. B. and Young, B. R., Radiosensitivity in ataxia-telangiectasia: a new explanation, *Proc. Natl. Acad. Sci. U.S.A.*, 77, 7315, 1980.
5. Painter, R. B., Radioresistant DNA synthesis: an intrinsic feature of ataxia-telangiectasia, *Mutat. Res.*, 84, 183, 1981.
6. Paterson, M. C., Smith, B. P., Lohman, P. H. M., Anderson, A. K., and Fishman, L., Defective excision repair of γ-ray-damaged DNA in human (ataxia-telangiectasia) fibroblasts, *Nature (London)*, 260, 444, 1976.
7. Paterson, M. C., Anderson, A. K., Smith, B. P., and Smith, P. J., Enhanced radiosensitivity of cultured fibroblasts from ataxia-telangiectasia heterozygotes manifested by defective colony-forming ability and reduced DNA repair replication after hypoxic γ-irradiation, *Cancer Res.*, 39, 3725, 1979.
8. Taylor, A. M. R., Metcalfe, J. A., Oxford, J. M., and Harnden, D. G., Is chromatid-type damage in ataxia-telangiectasia after irradiation at G_0 a consequence of defective repair?, *Nature (London)*, 260, 441, 1976.
9. Lehmann, A. R., Kirk-Bell, S., and Mayne, L., Abnormal kinetics of DNA synthesis in ultraviolet light-irradiated cells from patients with Cockayne's syndrome, *Cancer Res.*, 39, 4237, 1979.
10. Lehmann, A. R., Three complementation groups in Cockayne syndrome, *Mutat. Res.*, 106, 347, 1982.
11. Mayne, L. V. and Lehmann, A. R., Failure of RNA synthesis to recover after UV irradiation: an early defect in cells from individuals with Cockayne's syndrome and xeroderma pigmentosum, *Cancer Res.*, 42, 1473, 1982.
12. Cleaver, J. E., Normal reconstruction of DNA supercoiling and chromatin structure in Cockayne syndrome cells during repair of damage from ultraviolet light, *Am. J. Hum. Genet.*, 34, 566, 1982.
13. McKusick, V. A., Ed., *Mendelian Inheritance in Man, Catalogs of Autosomal Dominant, Autosomal Recessive, and X-linked Phenotypes*, 3rd ed., Johns Hopkins University Press, Baltimore, 1971, 294.
14. Cleaver, J. E., Xeroderma pigmentosum, in *Metabolic Basis of Inherited Disease*, 5th ed., Stanbury, J. B., Wyngaarden, J. B., Fredrickson, D. S., Goldstein, J. L., and Brown, M. S., Eds., McGraw-Hill, New York, 1983, 1227.
15. Robbins, J. H., Kraemer, K. H., Lutzner, M. A., Festoff, B. W., and Coon, H. G., Xeroderma pigmentosum: an inherited disease with sun sensitivity, multiple cutaneous neoplasms, and abnormal DNA repair, *Ann. Intern. Med.*, 80, 221, 1974.
16. Swift, M. and Chase, C., Cancer in families with xeroderma pigmentosum, *J. Natl. Cancer Inst.*, 62, 1415, 1979.
17. Welshimer, K. and Swift, M., Congenital malformations and developmental disabilities in ataxia-telangiectasia, Fanconi's anemia, and xeroderma pigmentosum families, *Am. J. Hum. Genet.*, 34, 781, 1982.
18. Cleaver, J. E., Zelle, B., Hashem, N., El-Hefnawi, M. H., and German, J., Xeroderma pigmentosum patients from Egypt. II. Preliminary correlations of epidemiology, clinical symptoms and molecular biology, *J. Invest. Dermatol.*, 77, 96, 1981.
19. Hodge, S. E., Berkel, A. I., Gatti, R. A., Boder, E., and Spence, M. A., Ataxia-telangiectasia and xeroderma pigmentosum: no evidence of linkage to HLA, *Tiss. Antigens*, 15, 313, 1980.
20. Harnden, D. G. and Taylor, A. M. R., Chromosomes and neoplasia, *Adv. Hum Genet.*, 9, 1, 1979.

21. El-Hefnawi, H., Maynard Smith, S., and Penrose, L. S., Xeroderma pigmentosum — its inheritance and relationship to the ABO blood-group system, *Ann. Hum. Genet.*, 28, 273, 1965.
22. German, J., Hashem, N., El-Hefnawi, M., and Cleaver, J. E., Xeroderma pigmentosum in Egypt. III. ABO blood grouping in 22 affected families, *Ann. Hum. Genet.*, 48, 61, 1984.
23. Hashem, N., Bootsma, D., Keijzer, W., Greene, A., Coriell, L., Thomas, G., and Cleaver, J. E., Clinical characteristics, DNA repair, and complementation groups in xeroderma pigmentosum patients from Egypt, *Cancer Res.*, 40, 13, 1980.
24. Marshall, J., *Skin Diseases in Africa, an Essay in Epidemiology,* Maskew Miller, Cape Town, 1964, 91.
25. Berlin, C. and Tager, A., Dermatitis solaris in children. A herald manifestation of xeroderma pigmentosum, *Arch. Dermatol.*, 85, 41, 1962.
26. Takebe, H., Miki, Y., Kozuka, T., Furuyama, J. I., Tanaka, K., Sasaki, M. S., Fujiwara, Y., and Akiba, H., DNA repair characteristics and skin cancers of xeroderma pigmentosum patients in Japan, *Cancer Res.*, 37, 490, 1977.
27. Ramsay, C. A. and Giannelli, F., The erythemal action spectrum and deoxyribonucleic acid repair synthesis in xeroderma pigmentosum, *Br. J. Dermatol.*, 92, 49, 1975.
28. Guerrier, C. J., Lutzner, M. A., Devico, V., and Prunieras, M., An electron microscopical study of the skin in 18 cases of xeroderma pigmentosum, *Dermatologica,* 146, 211, 1973.
29. Lynch, H. T., Anderson, D. E., Smith, J. L., Jr., Howell, J. B., and Krush, A. J., Xeroderma pigmentosum, malignant melanoma, and congenital ichthyosis. A family study, *Arch. Dermatol.*, 96, 625, 1967.
30. Lynch, H. T., Krush, A. J., and Larsen, A. L., Heredity and multiple primary malignant neoplasms: six cancer families, *Am. J. Med. Sci.*, 254, 322, 1967.
31. Kraemer, K. H., Lee, M. M., and Scotto, J., DNA repair protests against cutaneous and internal neoplasia: evidence from xeroderma pigmentosum, *Carcinogenesis,* 5, 511, 1984.
32. Stenson, S., Ocular findings in xeroderma pigmentosum: report of two cases, *Ann. Ophthalmol.*, 14, 580, 1982.
33. de Wolff-Rouendaal, D., Xeroderma pigmentosum with ophthalmological symptoms, *Ophthalmologica,* 173, 290, 1976.
34. de Sanctis, C. and Cacchione, A., L'idiozia xerodermica, *Riv. Sper. Freniatr. Med. Leg. Alienazioni Ment.*, 56, 269, 1932.
35. Burk, P. G., Yuspa, S. H., Lutzner, M. A., and Robbins, J. H., Xeroderma pigmentosum and D.N.A. repair, *Lancet,* 2, 601, 1971.
36. Cleaver, J. E., Xeroderma pigmentosum: variants with normal DNA repair and normal sensitivity to ultraviolet light, *J. Invest. Dermatol.*, 58, 124, 1972.
37. Cleaver, J. E., Arutyunyan, R. M., Sarkisian, T., Kaufmann, W. K., Greene, A. E., and Coriell, L., Similar defects in DNA repair and replication in the pigmented xerodermoid and the xeroderma pigmentosum variants, *Carcinogenesis,* 1, 647, 1980.
38. Lehmann, A. R., Kirk-Bell, S., Arlett, C. F., Paterson, M. C., Lohman, P. H. M., deWeerd-Kastelein, E. A., and Bootsma, D., Xeroderma pigmentosum cells with normal levels of excision repair have a defect in DNA synthesis after UV-irradiation, *Proc. Natl. Acad. Sci. U.S.A.*, 72, 219, 1975.
39. Kobayashi, M., Satoh, Y., Irimajiri, T., Mitoh, Y., Kozuka, T., Sato, K., Ichihashi, M., and Nakanishi, T., Skin tumors of xeroderma pigmentosum, *J. Dermatol. (Tokyo),* 9, 319, 1982.
40. Schmickel, R. D., Chu, E. H. Y., Trosko, J. E., and Chang, C. C., Cockayne syndrome: a cellular sensitivity to ultraviolet light, *Pediatrics,* 60, 135, 1977.
41. Dupuy, J.-M., Moshell, A. N., Lutzner, M. A., and Robbins, J. H., A new patient with both xeroderma pigmentosum and Cockayne syndrome is not in complementation group B, *J. Invest. Dermatol.*, 78(Abstr.), 356, 1982.
42. Cleaver, J. E., Thomas, G. H., and Park, S. D., Xeroderma pigmentosum variants have a slow recovery of DNA synthesis after irradiation with ultraviolet light, *Biochim. Biophys. Acta,* 564, 122, 1979.
43. Cleaver, J. E., Greene, A. E., Coriell, L., and Riccardi, V. M., An autosomal dominant inheritance for multiple sunlight-induced malignancy in a patient without abnormalities in DNA repair or replication, *Cytogenet. Cell Genet.*, 29, 122, 1981.
44. Scudiero, D. A., Meyer, S. A., Clatterbuck, B. E., Tarone, R. E., and Robbins, J. H., Hypersensitivity to *N*-methyl-*N'*-nitro-*N*-nitrosoguanidine in fibroblasts from patients with Huntington disease, familial dysautonomia, and other primary neuronal degenerations, *Proc. Natl. Acad. Sci. U.S.A.*, 78, 6451, 1981.
45. Scudiero, D. A., Moshell, A. N., Scarpinato, R. G., Meyer, S. A., Clatterbuck, B. E., Tarone, R. E., and Robbins, J. H., Lymphoblastoid lines and skin fibroblasts from patients with tuberous sclerosis are abnormally sensitive to ionizing radiation and to a radiomimetic chemical, *J. Invest. Dermatol.*, 78, 234, 1982.

46. Cairns, J., The origin of human cancers, *Nature (London)*, 289, 353, 1981.
47. Feinberg, A. P. and Coffey, D. S., Organ site specificity for cancer in chromosomal instability disorders, *Cancer Res.*, 42, 3252, 1982.
48. Kripke, M. L., Immunology of UV-induced skin cancer, *Photochem. Photobiol.*, 32, 837, 1980.
49. Dupuy, J. M. and Lafforet, D., A defect of cellular immunity in xeroderma pigmentosum, *Clin. Immunol. Immunopathol.*, 3, 52, 1974.
50. Morison, W. L., Bucana, C., Hashem, N., Kripke, M. L., Cleaver, J. E., and German, J. L., Impaired immune function in patients with xeroderma pigmentosum, *Cancer Res.*, 45, 3929, 1985.
51. Setlow, R. B., The wavelengths in sunlight effective in producing skin cancer: a theoretical analysis, *Proc. Natl. Acad. Sci. U.S.A.*, 71, 3363, 1974.
52. Epstein, J. H., Ultraviolet carcinogenesis, *Photophysiology*, 5, 235, 1970.
53. Harm, W., Use of an *E. coli uvr⁻ rec⁻* mutant for monitoring the germicidal activity of sunlight, *Radiat. Res.*, 39(Abstr.), 517, 1969.
54. Trosko, J. E., Krause, D., and Isoun, M., Sunlight-induced pyrimidine dimers in human cells *in vitro*, *Nature (London)*, 228, 358, 1970.
55. Rothman, R. H. and Setlow, R. B., An action spectrum for cell killing and pyrimidine dimer formation in Chinese hamster V-79 cells, *Photochem. Photobiol.*, 29, 57, 1979.
56. Pathak, M. A., Kramer, D. M., and Gungerich, U., Formation of thymine dimers in mammalian skin by ultraviolet radiation *in vivo*, *Photochem. Photobiol.*, 15, 177, 1972.
57. Todd, P., Coohill, T. P., and Mahoney, J. A., Responses of cultured Chinese hamster cells to ultraviolet light of different wavelengths, *Radiat. Res.*, 35, 390, 1968.
58. Keyse, S. M., Moss, S. H., and Davies, D. J. G., Action spectra for inactivation of normal and xeroderma pigmentosum human skin fibroblasts by ultraviolet radiations, *Photochem. Photobiol.*, 37, 307, 1983.
59. Kantor, G. J., Sutherland, J. C., and Setlow, R. B., Action spectra for killing non-dividing normal human and *xeroderma pigmentosum* cells, *Photochem. Photobiol.*, 31, 459, 1980.
60. Chu, E. H. Y., Effects of ultraviolet radiation on mammalian cells. I. Induction of chromosome aberrations, *Mutat. Res.*, 2, 75, 1965.
61. Ichihasi, M. and Ramsay, C. A., Excision repair of DNA in some photodermatoses, *Br. J. Dermatol.*, 95, 13, 1976.
62. Tyrrell, R. M., Mutagenic action of monochromatic UV radiation in the solar range on human cells, *Mutat. Res.*, 129, 103, 1984.
63. Lippke, J. A., Gordon, L. K., Brasch, D. E., and Haseltine, W. A., Distribution of UV light-induced damage in a defined sequence of human DNA: detection of alkaline-sensitive lesions at pyrimidine nucleoside-cytidine sequences, *Proc. Natl. Acad. Sci. U.S.A.*, 78, 3388, 1981.
64. Lindahl, T., DNA repair enzymes, *Annu. Rev. Biochem.*, 51, 61, 1982.
65. Cleaver, J. E., Xeroderma pigmentosum: a human disease in which an initial stage of DNA repair is defective, *Proc. Natl. Acad. Sci. U.S.A.*, 63, 428, 1969.
66. Cohn, S. M. and Lieberman, M. W., The distribution of DNA excision-repair sites in human diploid fibroblasts following ultraviolet irradiation, *J. Biol. Chem.*, 259, 12463, 1984.
67. Cohn, S. M. and Lieberman, M. W., The use of antibodies to 5-bromo-2′-deoxyuridine for the isolation of DNA sequences containing excision-repair sites, *J. Biol. Chem.*, 259, 12456, 1984.
68. Zolan, M. E., Cortopassi, G. A., Smith, C. A., and Hanawalt, P. C., Deficient repair of chemical adducts in α DNA of monkey cells, *Cell*, 28, 613, 1982.
69. Zelle, B. and Lohman, P. H. M., Repair of UV-endonuclease susceptible sites in the 7 complementation groups of xeroderma pigmentosum A through G, *Mutat. Res.*, 62, 363, 1979.
70. Cleaver, J. E. and Gruenert, D. C., Repair of psoralen adducts in human DNA: differences among xeroderma pigmentosum complementation groups, *J. Invest. Dermatol.*, 82, 311, 1984.
71. Cleaver, J. E., DNA repair with purines and pyrimidines in radiation- and carcinogen-damaged normal and xeroderma pigmentosum human cells, *Cancer Res.*, 33, 362, 1973.
72. Regan, J. D. and Setlow, R. B., Two forms of repair in the DNA of human cells damaged by chemical carcinogens and mutagens, *Cancer Res.*, 34, 3318, 1974.
73. Fornace, A. J., Jr., Kohn, K. W., and Kann, H. E., Jr., DNA single-strand breaks during repair of UV damage in human fibroblasts and abnormalities of repair in xeroderma pigmentosum, *Proc. Natl. Acad. Sci. U.S.A.*, 73, 39, 1976.
74. Fornace, A. J., Jr. and Seres, D. S., Detection of DNA single-strand breaks during the repair of UV damage in xeroderma pigmentosum cells, *Radiat. Res.*, 93, 107, 1983.
75. Dunn, W. C. and Regan, J. D., Inhibition of DNA excision repair in human cells by arabinofuranosyl cytosine: effect on normal and xeroderma pigmentosum cells, *Mol. Pharmacol.*, 15, 367, 1979.
76. Cleaver, J. E., Sedimentation of DNA from human fibroblasts irradiated with ultraviolet light: possible detection of excision breaks in normal and repair-deficient xeroderma pigmentosum cells, *Radiat. Res.*, 57, 207, 1974.

77. Bootsma, D., Mulder, M. P., Pot, F., and Cohen, J. A., Different inherited levels of DNA repair replication in xeroderma pigmentosum cell strains after expsoure to ultraviolet irradiation, *Mutat. Res.,* 9, 507, 1970.
78. Kleijer, W. J., de Weerd-Kastelein, E. A., Sluyter, M. L., Keijzer, W., de Wit, J., and Bootsma, D., UV-induced DNA repair synthesis in cells of patients with different forms of xeroderma pigmentosum and of heterozygotes, *Mutat. Res.,* 20, 417, 1973.
79. Andrews, A. D., Barrett, S. F., and Robbins, J. H., Xeroderma pigmentosum neurological abnormalities correlate with colony-forming ability after ultraviolet radiation, *Proc. Natl. Acad. Sci. U.S.A.,* 75, 1984, 1978.
80. Yang, L. L., Kouri, R. E., and Curren, R. D., Xeroderma pigmentosum fibroblasts are more sensitive to asbestos fibers than are normal human fibroblasts, *Carcinogenesis,* 5, 291, 1984.
81. Ishizaki, K., Yagi, T., and Takebe, H., Cytotoxic effects of protease inhibitors on human cells. I. High sensitivity of xeroderma pigmentosum cells to antipain, *Cancer Lett.,* 10, 199, 1980.
82. Suzuki, N., Kojima, T., Kuwata, T., Nishimaki, J., Takakubo, Y., and Miki, T., Cross sensitivity between interferon and uv in human cell strains: IF^r, HEC-1, and CRL1200, *Virology,* 135, 20, 1984.
83. Borek, C. and Cleaver, J. E., Protease inhibitors neither damage DNA nor interfere with DNA repair or replication in human cells, *Mutat. Res.,* 82, 373, 1981.
84. Cleaver, J. E. and Carter, D. M., Xeroderma pigmentosum variants: influence of temperature on DNA repair, *J. Invest. Dermatol.,* 60, 29, 1973.
85. Ritter, M. A., Unpublished observations.
86. Giannelli, F. and Pawsey, S. A., DNA repair synthesis in human heterokaryons. II. A test for heterozygosity in xeroderma pigmentosum and some insight into the structure of the defective enzyme, *J. Cell Sci.,* 15, 163, 1974.
87. Rainbow, A. J., Reduced capacity to repair irradiated adenovirus in fibroblasts from xeroderma pigmentosum heterozygotes, *Cancer Res.,* 40, 3945, 1980.
88. Day, R. S., III, Studies on repair of adenovirus 2 by human fibroblasts using normal, xeroderma pigmentosum, and xeroderma pigmentosum heterozygous strains, *Cancer Res.,* 34, 1965, 1974.
89. Selsky, C. A. and Greer, S., Host-cell reactivation of UV-irradiated and chemically-treated herpes simplex virus-1 by xeroderma pigmentosum, XP heterozygotes and normal skin fibroblasts, *Mutat. Res.,* 50, 395, 1978.
90. Lehmann, A. R., The relationship between pyrimidine dimers and replicating DNA in UV-irradiated human fibroblasts, *Nucl. Acids Res.,* 7, 1901, 1979.
91. Kaufmann, W. K. and Cleaver, J. E., Mechanisms of inhibition of DNA replication by ultraviolet light in normal human and xeroderma pigmentosum fibroblasts, *J. Mol. Biol.,* 149, 171, 1981.
92. Cleaver, J. E., Kaufmann, W. K., Kapp, L. N., and Park, S. D., Replicon size and excision repair as factors in the inhibition and recovery of DNA synthesis from ultraviolet damage, *Biochim. Biophys. Acta,* 739, 207, 1983.
93. Park, S. D. and Cleaver, J. E., Postreplication repair: questions of its definition and possible alteration in xeroderma pigmentosum cell strains, *Proc. Natl. Acad. Sci. U.S.A.,* 76, 3927, 1979.
94. Park, S. D. and Cleaver, J. E., Recovery of DNA synthesis after ultraviolet irradiation of xeroderma pigmentosum cells depends on excision repair and is blocked by caffeine, *Nucl. Acids Res.,* 6, 1151, 1979.
95. Cleaver, J. E., DNA repair and its coupling to DNA replication in eukaryotic cells, *Biochim. Biophys. Acta,* 516, 489, 1978.
96. Imray, P., Mangan, T., Saul, A., and Kidson, C., Effects of ultraviolet irradiation on the cell cycle in normal and UV-sensitive cell lines with reference to the nature of the defect in xeroderma pigmentosum variant, *Mutat. Res.,* 112, 301, 1983.
97. Painter, R. B. and Young, B. R., Effect of hypertonicity and X radiation on DNA synthesis in normal and ataxia-telangiectasia cells, *Radiat. Res.,* 92, 552, 1982.
98. Maher, V. W., Dorney, D. J., Mendrala, A. L., Konze-Thomas, B., and McCormick, J. J., DNA excision-repair processes in human cells can eliminate the cytotoxic and mutagenic consequences of ultraviolet irradiation, *Mutat. Res.,* 62, 311, 1979.
99. Chan, G. L. and Little, J. B., Resistance of plateau-phase human normal and xeroderma pigmentosum fibroblasts to the cytotoxic effect of ultraviolet light, *Mutat. Res.,* 63, 401, 1979.
100. Dingman, C. W. and Kakunaga, T., DNA strand breaking and rejoining in response to ultraviolet light in normal and xeroderma pigmentosum cells, *Intl. J. Radiat. Biol.,* 30, 55, 1976.
101. Tohda, H. and Oikawa, A., Differential features of sister-chromatid exchange responses to ultraviolet radiation and caffeine in xeroderma pigmentosum lymphoblastoid cell lines, *Mutat. Res.,* 107, 387, 1983.
102. de Weerd-Kastelein, E. A., Keijzer, W., Rainaldi, G., and Bootsma, D., Induction of sister chromatid exchanges in xeroderma pigmentosum cells after exposure to ultraviolet light, *Mutat. Res.,* 45, 253, 1977.

103. Cheng, W.-S., Tarone, R. E., Andrews, A. D., Whang-Peng, J. S., and Robbins, J. H., Ultraviolet light-induced sister chromatid exchanges in xeroderma pigmentosum and in Cockayne's syndrome lymphocyte cell lines, *Cancer Res.*, 38, 1601, 1978.
104. Cook, J. S. and McGrath, J. R., Photoreactivating-enzyme activity in metazoa, *Proc. Natl. Acad. Sci. U.S.A.*, 58, 1359, 1967.
105. Cleaver, J. E., Photoreactivation: a radiation repair mechanism absent from mammalian cells, *Biochem. Biophys. Res. Commun.*, 24, 569, 1966.
106. Sutherland, B. M., Photoreactivating enzyme from human leukocytes, *Nature (London)*, 248, 109, 1974.
107. Sutherland, B. M., Rice, M., and Wagner, E. K., Xeroderma pigmentosum cells contain low levels of photoreactivating enzyme, *Proc. Natl. Acad. Sci. U.S.A.*, 72, 103, 1975.
108. Mortelmans, K., Cleaver, J. E., Friedberg, E. C., Paterson, M. C., Smith, B. P., and Thomas, G. H., Photoreactivation of thymine dimers in UV-irradiated human cells: unique dependence on culture conditions, *Mutat. Res.*, 44, 433, 1977.
109. Sutherland, B. M., Harber, L. C., and Kochevar, I. E., Pyrimidine dimer formation and repair in human skin, *Cancer Res.*, 40, 3181, 1980.
110. d'Ambrosio, S. M., Whetstone, J. W., Slazinski, L., and Lowney, E., Photorepair of pyrimidine dimers in human skin *in vivo*, *Photochem. Photobiol.*, 34, 461, 1981.
111. Ananthaswamy, H. N. and Fisher, M. S., Photoreactivation of ultraviolet radiation induced pyrimidine dimers in neonatal BALB/c mouse skin, *Cancer Res.*, 41, 1829, 1981.
112. Eggset, G., Volden, G., and Krokan, H., U.V.-induced DNA damage and its repair in human skin *in vivo* studied by sensitive immunohistochemical methods, *Carcinogenesis*, 4, 745, 1983.
113. Cleaver, J. E., DNA damage and repair in light-sensitive human skin disease, *J. Invest. Dermatol.*, 54, 181, 1970.
114. de Weerd-Kastelein, E. A., Keijzer, W., and Bootsma, D., Genetic heterogeneity of xeroderma pigmentosum demonstrated by somatic cell hybridization, *Nature (London) New Biol.*, 238, 80, 1972.
115. de Weerd-Kastelein, E. A., Kleijer, W. J., Sluyter, M. L., and Keijzer, W., Repair replication in heterokaryons derived from different repair-deficient xeroderma pigmentosum strains, *Mutat. Res.*, 19, 237, 1973.
116. Kraemer, K. H., de Weerd-Kastelein, E. A., Robbins, J. H., Keijzer, W., Barrett, S. F., Petinga, R. A., and Bootsma, D., Five complementation groups in xeroderma pigmentosum, *Mutat. Res.*, 33, 327, 1975.
117. Cleaver, J. E., Rapid complementation method for classifying excision repair-defective xeroderma pigmentosum cell strains, *Somat. Cell Genet.*, 8, 801, 1982.
118. Zelle, B. and Bootsma, D., Repair of DNA damage after exposure to 4-nitroquinoline-1-oxide in heterokaryons derived from xeroderma pigmentosum cells, *Mutat. Res.*, 70, 373, 1980.
119. de Weerd-Kastelein, E. A., Keijzer, W., Sabour, M., Parrington, J. M., and Bootsma, D., A xeroderma pigmentosum patient having a high residual activity of unscheduled DNA synthesis after UV is assigned to complementation group A, *Mutat. Res.*, 37, 307, 1976.
120. Stefanini, M., Keijzer, W., Dalpra, L., Elli, R., Porro, M. N., Nicoletti, B., and Nuzzo, F., Differences in the levels of UV repair and in clinical symptoms in two sibs affected by xeroderma pigmentosum, *Hum. Genet.*, 54, 177, 1980.
121. Mansbridge, J. N. and Hanawalt, P. C., Domain-limited repair of DNA in ultraviolet irradiated fibroblasts from xeroderma pigmentosum complementation group C, in *Cellular Responses to DNA Damage*, Friedberg, E. C. and Bridges, B. A., Eds., Alan R. Liss, New York, 1983, 195.
122. Mullenders, L. H. F., van Kesteren, A. C., Bussmann, C. J. M., can Zeeland, A. A., and Natarajan, A. T., Preferential repair of nuclear matrix associated DNA in xeroderma pigmentosum complementation group C, *Mutat. Res.*, 141, 75, 1984.
123. Giannelli, F., Pawsey, S. A., and Avery, J. A., Differences in patterns of complementation of the more common groups of xeroderma pigmentosum: possible implications, *Cell*, 29, 451, 1982.
124. Giannelli, F. and Pawsey, S. A., DNA repair synthesis in human heterokaryons. III. The rapid and slow complementing varieties of xeroderma pigmentosum, *J. Cell Sci.*, 20, 207, 1976.
125. Matsukuma, S., Zelle, B., Keijzer, W., Berends, F., and Bootsma, D., Different rates of restoration of the repair capacity in complementing xeroderma pigmentosum cells after fusion, *Exp. Cell Res.*, 134, 103, 1981.
126. Keijzer, W., Verkerk, A., and Bootsma, D., Phenotypic correction of the defect in xeroderma pigmentosum cells after fusion with isolated cytoplasts, *Exp. Cell Res.*, 140, 119, 1982.
127. Gautschi, J. R., Young, B. R., and Cleaver, J. E., Repair of damaged DNA in the absence of protein synthesis in mammalian cells, *Exp. Cell Res.*, 76, 87, 1973.
128. Giannelli, F., Croll, P. M., and Lewin, S. A., DNA repair synthesis in human heterokaryons formed by normal and UV-sensitive fibroblasts, *Exp. Cell Res.*, 78, 175, 1973.
129. Sancar, A. and Rupp, W. D., A novel repair enzyme: UVRABC excision nuclease of *Escherichia coli* cuts a DNA strand on both sides of the damaged region, *Cell*, 33, 249, 1983.

130. Takebe, H., Difference in distribution of genetic complementation groups of xeroderma pigmentosum between Europe, USA, and Japan, *J. Radiat. Res.*, 17(Abstr.), 29, 1976.
131. Neel, J. V., Kodani, M., Brewer, R., and Anderson, R. C., The incidence of consanguineous matings in Japan, with remarks on the estimation of comparative gene frequencies and the expected rate of appearance of induced recessive mutations, *Am. J. Hum. Genet.*, 1, 156, 1949.
132. Thompson, L. H., Rubin, J. S., Cleaver, J. E., Whitmore, G. F., and Brookman, K., A screening method for isolating DNA repair-deficient mutants of CHO cells, *Somat. Cell Genet.*, 6, 391, 1980.
133. Thompson, L. H., Busch, D. B., Brookman, K., Mooney, C. L., and Glaser, D. A., Genetic diversity of UV-sensitive DNA repair mutants of Chinese hamster ovary cells, *Proc. Natl. Acad. Sci. U.S.A.*, 78, 3734, 1981.
134. Busch, D. B., Cleaver, J. E., and Glaser, D. A., Large-scale isolation of UV-sensitive clones of CHO cells, *Somat. Cell Genet.*, 6, 407, 1980.
135. Lalley, P. A., Diaz, J. A., Francis, A. A., Dunn, W. C., and Regan, J. D., The expression and chromosomal assignments of genes required for repair of UV-induced DNA damage, *Cytogenet. Cell Genet.*, 37(Abstr.), 516, 1984.
136. Hori, T.-A., Shiomi, T., and Sato, K., Human chromosome 13 compensates a DNA repair defect in UV-sensitive mouse cells by mouse-human cell hybridization, *Proc. Natl. Acad. Sci. U.S.A.*, 80, 5655, 1983.
137. Stefanini, M., Keijzer, W., Westerveld, A., Geurts van Kessel, A., Jongkind, J. F., and Bootsma, D., Complementation and mapping of genes involved in deficient DNA repair in xeroderma pigmentosum cells, *Cytogenet. Cell Genet.*, 32(Abstr.), 321, 1982.
138. Keijzer, W., Stefanini, M., Westerveld, A., and Bootsma, D., Mapping of the XPAC gene involved in complementation of the defect in xeroderma pigmentosum group A cells, *Cytogenet. Cell Genet.*, 37(Abstr.), 508, 1984.
139. Lin, P.-F. and Ruddle, F. H., Murine DNA repair gene located on chromosome 4, *Nature (London)*, 289, 191, 1981.
140. Lalley, P. A., Minna, J. D., and Francke, U., Conservation of autosomal gene synteny groups in mouse and man, *Nature (London)*, 274, 160, 1978.
141. Takano, T., Noda, M., and Tamura, T.-A., Transfection of cells from a xeroderma pigmentosum patient with normal human DNA confers UV resistance, *Nature (London)*, 296, 269, 1982.
142. Ishizaki, K., Shima, A., Sekiguchi, T., and Takebe, H., Microcell-mediated transfer of human DNA repair gene(s) into xeroderma pigmentosum cells, *Jpn. J. Genet.*, 56, 105, 1981.
143. Cleaver, J. E., Workshop summary: DNA repair in normal and repair-defective human cells, in *Cellular Responses to DNA Damage*, Friedberg, E. C. and Bridges, B. A., Eds., Alan R. Liss, New York, 1983, 327.
144. Royer-Pokora, B. and Haseltine, W. A., Isolation of UV-resistant revertants from a xeroderma pigmentosum complementation group A cell line, *Nature (London)*, 311, 390, 1984.
145. Rubin, J. S., Joyner, A. L., Bernstein, A., and Whitmore, G. F., Molecular identification of a human DNA repair gene following DNA-mediated gene transfer, *Nature (London)*, 306, 206, 1983.
146. Westerveld, A., Hoeijmakers, J. H. J., van Duin, M., de Wit, J., Odijk, H., Pastink, A., Wood, R. D., and Bootsma, D., Molecular cloning of a human DNA repair gene, *Nature (London)*, 310, 425, 1984.
147. Hubscher, U., Kuenzle, C. C., and Spadari, S., Functional roles of DNA polymerases β and γ, *Proc. Natl. Acad. Sci. U.S.A.*, 76, 2316, 1979.
148. Weissbach, A., Vertebrate DNA polymerases, *Cell*, 5, 101, 1975.
149. Lindahl, T. and Edelman, G. M., Polynucleotide ligase from myeloid and lymphoid tissues, *Proc. Natl. Acad. Sci. U.S.A.*, 61, 680, 1968.
150. Ciarrocchi, G., Jose, J. G., and Linn, S., Further characterization of a cell-free system for measuring replicative and repair DNA synthesis with cultured human fibroblasts and evidence for the involvement of DNA polymerase α in DNA repair, *Nucl. Acids Res.*, 7, 1205, 1979.
151. Seki, S. and Oda, T., Effects of 2′,3′-dideoxythymidine triphosphate on replicative DNA synthesis and unscheduled DNA synthesis in permeable mouse sarcoma cells, *Biochim. Biophys. Acta*, 606, 246, 1980.
152. Cleaver, J. E., Sensitivity of excision repair in normal human, xeroderma pigmentosum variant and Cockayne's syndrome fibroblasts to inhibition by cytosine arabinoside, *J. Cell. Physiol.*, 108, 163, 1981.
153. Cleaver, J. E., Structure of repaired sites in human DNA synthesized in the presence of inhibitors of DNA polymerases alpha and beta in human fibroblasts, *Biochim. Biophys. Acta*, 739, 301, 1983.
154. Dresler, S. L. and Lieberman, M. W., Identification of DNA polymerases involved in DNA excision repair in diploid human fibroblasts, *J. Biol. Chem.*, 258, 9990, 1983.
155. Snyder, R. D. and Regan, J. D., Aphidicolin inhibits repair of DNA in UV-irradiated human fibroblasts, *Biochem. Biophys. Res. Commun.*, 99, 1088, 1981.

156. Smith, C. A. and Okumoto, D. S., Nature of DNA repair synthesis resistant to inhibitors of polymerase α in human cells, *Biochemistry,* 23, 1383, 1984.
157. Grossman, L., Enzymes involved in the repair of DNA, *Adv. Radiat. Biol.,* 4, 77, 1974.
158. Bertazzoni, U., Scovassi, A. I., Stefanini, M., Guilotto, E., Spadari, S., and Pedrini, A. M., DNA polymerases alpha, beta and gamma in inherited diseases affecting DNA repair, *Nucl. Acids Res.,* 5, 2189, 1978.
159. Parker, V. P. and Lieberman, M. W., Levels of DNA polymerases alpha, beta, and gamma in control and repair-deficient human diploid fibroblasts, *Nucl. Acids Res.,* 4, 2029, 1977.
160. Cleaver, J. E., Inactivation of ultraviolet repair in normal and xeroderma pigmentosum cells by methyl methanesulfonate, *Cancer Res.,* 42, 860, 1982.
161. Park, S. D., Choi, K. H., Hong, S. W., and Cleaver, J. E., Inhibition of excision-repair of ultraviolet damage in human cells by exposure to methyl methanesulfonate, *Mutat. Res.,* 82, 365, 1981.
162. Bacchetti, S., van der Plas, A., and Veldhuisen, G., A UV-specific endonucleolytic activity present in human cell extracts, *Biochem. Biophys. Res. Commun.,* 48, 662, 1972.
163. Brent, T. P., Repair enzyme suggested by mammalian endonuclease activity specific for ultraviolet-irradiated DNA, *Nature (London) New Biol.,* 239, 172, 1972.
164. Duker, N. J. and Teebor, G. W., Different ultraviolet DNA endonuclease activity in human cells, *Nature (London),* 255, 82, 1975.
165. Helland, D., Kleppe, R., Lillehaug, J. R., and Kleppe, K., Xeroderma pigmentosum: *in vitro* complementation of DNA repair endonuclease, *Carcinogenesis,* 5, 833, 1984.
166. Van Lancker, J. L. and Tomura, T., Purification and some properties of a mammalian repair endonuclease, *Biochim. Biophys. Acta,* 353, 99, 1974.
167. Bacchetti, S. and Benne, R., Purification and characterization of an endonuclease from calf thymus acting on irradiated DNA, *Biochim. Biophys. Acta,* 390, 285, 1975.
168. Waldstein, E. A., Peller, S., and Setlow, R. B., UV-endonuclease from calf thymus with specificity toward pyrimidine dimers in DNA, *Proc. Natl. Acad. Sci. U.S.A.,* 76, 3746, 1979.
169. Mortelmans, K., Friedberg, E. C., Slor, H., Thomas, G., and Cleaver, J. E., Defective thymine dimer excision by cell-free extracts of xeroderma pigmentosum cells, *Proc. Natl. Acad. Sci. U.S.A.,* 73, 2757, 1976.
170. Kano, Y. and Fujiwara, Y., Defective thymine dimer excision from xeroderma pigmentosum chromatin and its characteristic catalysis by cell-free extracts, *Carcinogenesis,* 4, 1419, 1983.
171. Slor, H., Lev-Sobe, T., and Friedberg, E. C., Evidence for inactivation of DNA repair in frozen and thawed mammalian cells, *Mutat. Res.,* 45, 137, 1977.
172. Jaspers, N. G. J., Jansen-Van de Kuilen, G., and Bootsma, D., Complementation analysis of xeroderma pigmentosum variants, *Exp. Cell Res.,* 136, 81, 1981.
173. Kuhnlein, U., Penhoet, E. E., and Linn, S., An altered apurinic DNA endonuclease activity in group A and group D xeroderma pigmentosum fibroblasts, *Proc. Natl. Acad. Sci. U.S.A.,* 73, 1169, 1976.
174. Kuhnlein, U., Lee, B., Penhoet, E. E., and Linn, S., Xeroderma pigmentosum fibroblasts of the D group lack an apurinic DNA endonuclease species with a low apparent K_m, *Nucl. Acids Res.,* 5, 951, 1978.
175. Kuhnlein, U., Tsang, S.-S., Lokken, O., Tong, S., and Twa, D., Cell lines from xeroderma pigmentosum complementation group A lack a single stranded DNA binding activity, *Biosci. Rep.,* 3, 667, 1983.
176. Lee, D. E., Okorodudu, A. O., Lambert, W. C., and Lambert, M. W., Defective DNA endonuclease activity on anthramycin treated DNA in xeroderma pigmentosum and mouse melanoma cells, *Biochem. Biophys. Res. Commun.,* 107, 395, 1982.
177. Lambert, M. W., Lambert, W. C., and Okorodudu, A. O., Nuclear DNA endonuclease activities on partially apurinic/apyrimidinic DNA in normal human and xeroderma pigmentosum lymphoblastoid and mouse melanoma cells, *Chem. Biol. Interact.,* 46, 109, 1983.
178. Okorodudu, A. O., Lambert, W. C., and Lambert, M. W., Nuclear deoxyribonuclease activities in normal and xeroderma pigmentosum lymphoblastoid cells, *Biochem. Biophys. Res. Commun.,* 108, 576, 1982.
179. Thielmann, H. W. and Witte, I., Correlation of the colony-forming abilities of xeroderma pigmentosum fibroblasts with repair-specific DNA incision reactions catalyzed by cell-free extracts, *Arch. Toxicol.,* 44, 197, 1980.
180. Sauerbier, W., UV damage at the transcriptional level, *Adv. Radiat. Biol.,* 6, 49, 1976.
181. Schorpp, M., Mallick, U., Rahmsdorf, H. J., and Herrlich, P., UV-induced extracellular factor from human fibroblasts communicates the UV response to nonirradiated cells, *Cell,* 37, 861, 1984.
182. Cohn, S. M., Krawisz, B. R., Dresler, S. L., and Lieberman, M. W., Induction of replicative DNA synthesis in quiescent human fibroblasts by DNA damaging agents, *Proc. Natl. Acad. Sci. U.S.A.,* 81, 4828, 1984.

183. Lichti, U., Bowden, G. T., Patterson, E., Ben, T., and Yuspa, S. H., Germicidal ultraviolet light induces ornithine decarboxylase in mouse epidermal cells and modifies the induction caused by phorbol ester tumor promoters, *Photochem. Photobiol.*, 32, 177, 1980.
184. Miskin, R. and Ben-Ishai, R., Induction of plasminogen activator by UV light in normal and xeroderma pigmentosum fibroblasts, *Proc. Natl. Acad. Sci. U.S.A.*, 78, 6236, 1981.
185. Coppey, J. and Menezes, S., Enhanced reactivation of ultraviolet-damaged herpes virus in ultraviolet pretreated skin fibroblasts of cancer prone donors, *Carcinogenesis*, 2, 787, 1981.
186. Mezzina, M. and Nocentini, S., DNA ligase activity in UV-irradiated monkey kidney cells, *Nucl. Acids Res.*, 5, 4317, 1978.
187. Walker, G. C., Mutagenesis and inducible responses to deoxyribonucleic acid damage in *Escherichia coli*, *Microbiol. Rev.*, 48, 60, 1984.
188. Radman, M., Is there SOS induction in mammalian cells?, *Photochem. Photobiol.*, 32, 823, 1980.
189. Strauss, B. S., Cellular aspects of DNA repair, *Adv. Cancer Res.*, 45, 45, 1985.
190. Herrlich, P., Mallick, U., Ponta, H., and Rahmsdorf, H. J., Genetic changes in mammalian cells reminiscent of an SOS response, *Hum. Genet.*, 67, 360, 1984.
191. Seide, W. and Eckardt, F., Inducibility of error-prone DNA repair in yeast?, *Mutat. Res.*, 129, 3, 1984.
192. Kuroki, T. and Drevon, C., Inhibition of chemical transformation in C3H/10T1/2 cells by protease inhibitors, *Cancer Res.*, 39, 2755, 1979.
193. Kennedy, A. R., Antipain, but not cycloheximide, suppresses radiation transformation when present for only one day at five days post-irradiation, *Carcinogenesis*, 3, 1093, 1982.
194. Kinsella, A. R. and Radman, M., Inhibition of carcinogen-induced chromosomal aberrations by an anticarcinogenic protease inhibitor, *Proc. Natl. Acad. Sci. U.S.A.*, 77, 3544, 1980.
195. Bleichrodt, J. F. and Verheij, W. S. D., Mutagenesis by ultraviolet radiation in bacteriophage ϕX174: on the mutation stimulating processes induced by ultraviolet radiation in the host bacterium, *Mol. Gen. Genet.*, 135, 19, 1974.
196. Cleaver, J. E. and Weil, S., UV-induced reversion of a temperature sensitive late mutant of simian virus 40 to a wild-type phenotype, *J. Virol.*, 16, 214, 1975.
197. Cornelis, J. J., Lupker, J. H., and van der Eb, A. J., Error-prone repair processes in mammalian cells, *Mutat. Res.*, 74(Abstr.), 178, 1980.
198. Sarasin, A. and Benoit, A., Induction of an error-prone mode of DNA repair in UV-irradiated monkey kidney cells, *Mutat. Res.*, 70, 71, 1980.
199. Bourre, F. and Sarasin, A., Targeted mutagenesis of SV40 DNA induced by UV light, *Nature (London)*, 305, 68, 1983.
200. Das Gupta, U. B. and Summers, W. C., Ultraviolet reactivation of herpes simplex virus is mutagenic and inducible in mammalian cells, *Proc. Natl. Acad. Sci. U.S.A.*, 75, 2378, 1978.
201. Lytle, C. D., Goddard, J. G., and Lin, C.-H., Repair and mutagenesis of herpes simplex virus in UV-irradiated monkey cells, *Mutat. Res.*, 70, 139, 1980.
202. Day, R. S., III and Ziolkowski, C., Studies on UV-induced viral reversion, Cockayne's syndrome, and MNNG damage using adenovirus 5, in *DNA Repair Mechanisms*, Hanawalt, P. C., Friedberg, E. C., and Fox, C. F., Eds., Academic Press, New York, 1978, 535.
203. Lebkowski, J. S., DuBridge, R. B., Antell, E. A., Greisen, K. S., and Calos, M. P., Transfected DNA is mutated in monkey, mouse, and human cells, *Mol. Cell. Biol.*, 4, 1951, 1984.
204. Sarkar, S., Das Gupta, U. B., and Summers, W. C., Error-prone mutagenesis detected in mammalian cells by a shuttle vector containing the *supF* gene of *Escherichia coli*, *Mol. Cell. Biol.*, 4, 2227, 1984.
205. Day, R. S., III, Cellular reactivation of ultraviolet-irradiated human adenovirus 2 in normal and xeroderma pigmentosum fibroblasts, *Photochem. Photobiol.*, 19, 9, 1974.
206. Day, R. S., III, Giuffrida, A. S., and Dingman, C. W., Repair by human cells of adenovirus-2 damaged by psoralen plus near ultraviolet light treatment, *Mutat. Res.*, 33, 311, 1975.
207. Lytle, C. D., Aaronson, S. A., and Harvey, E., Host-cell reactivation in mammalian cells. II. Survival of herpes simplex virus and vaccinia virus in normal human and xeroderma pigmentosum cells, *Int. J. Radiat. Biol.*, 22, 159, 1972.
208. Lytle, C. D., Day, R. S., III, Hellman, K. B., and Bockstahler, L. E., Infection of UV-irradiated xeroderma pigmentosum fibroblasts by herpes simplex virus: study of capacity and Weigle reactivation, *Mutat. Res.*, 36, 257, 1976.
209. Aaronson, S. A. and Lytle, C. D., Decreased host cell reactivation of irradiated SV40 virus in xeroderma pigmentosum, *Nature (London)*, 228, 359, 1970.
210. Das Gupta, U. B. and Summers, W. C., Genetic recombination of *herpes simplex* virus, the role of the host cell and UV-irradiation of the virus, *Mol. Gen. Genet.*, 178, 617, 1980.
211. Eadie, J. S., Conrad, M., Toorchen, D., and Topal, M. D., Mechanism of mutagenesis by O^6-methylguanine, *Nature (London)*, 308, 201, 1984.

212. Dinsart, C., Cornelis, J. J., Klein, B., van der Eb, A. J., and Rommelaere, J., Transfection with extracellularly UV-damaged DNA induces human and rat cells to express a mutator phenotype towards parvovirus H-1, *Mol. Cell. Biol.,* 4, 324, 1984.
213. Das, S. K. and Loeb, L. A., UV irradiation alters deoxynucleoside triphosphate pools in *Escherichia coli, Mutat. Res.,* 131, 97, 1984.
214. Das, S. K., Benditt, E. P., and Loeb, L. A., Rapid changes in deoxynucleoside triphosphate pools in mammalian cells treated with mutagens, *Biochem. Biophys. Res. Commun.,* 114, 458, 1983.
215. Weinberg, G., Ullman, B., and Martin, D. W., Jr., Mutator phenotypes in mammalian cell mutants with distinct biochemical defects and abnormal deoxyribonucleoside triphosphate pools, *Proc. Natl. Acad. Sci. U.S.A.,* 78, 2447, 1981.
216. Meuth, M., Role of deoxynucleoside triphosphate pools in the cytotoxic and mutagenic effects of DNA alkylating agents, *Somat. Cell Genet.,* 7, 89, 1981.
217. Meuth, M., Sensitivity of a mutator gene in Chinese hamster ovary cells to deoxynucleoside triphosphate pool alterations, *Mol. Cell. Biol.,* 1, 652, 1981.
218. Meuth, M., Deoxycytidine kinase-deficient mutants of Chinese hamster ovary cells are hypersensitive to DNA alkylating agents, *Mutat. Res.,* 110, 383, 1983.
219. Sugimura, T., Miwa, M., Saito, H., Kanai, Y., Ikejima, M., Terada, M., Yamada, M., and Utakoji, T., Studies of nuclear ADP-ribosylation, *Adv. Enzyme Regul.,* 18, 195, 1980.
220. Hayaishi, O. and Ueda, K., Poly(ADP-ribose) and ADP-ribosylation of proteins, *Annu. Rev. Biochem.,* 46, 95, 1977.
221. Jacobson, E. L., Antol, K. M., Juarez-Salinas, H., and Jacobson, M. K., Poly(ADP-ribose) metabolism in ultraviolet irradiated human fibroblasts, *J. Biol. Chem.,* 258, 103, 1983.
222. McCurry, L. S. and Jacobson, M. K., Poly(ADP-ribose) synthesis following DNA damage in cells heterozygous or homozygous for the xeroderma pigmentosum genotype, *J. Biol. Chem.,* 256, 551, 1981.
223. Fujiwara, Y., Goto, K., and Kano, Y., Ultraviolet hypersensitivity of Cockayne's syndrome fibroblasts. Effects of nicotinamide adenine dinucleotide and poly(ADP-ribose) synthesis, *Exp. Cell Res.,* 139, 207, 1982.
224. Edwards, M. J. and Taylor, A. M. R., Unusual levels of $(ADP\text{-ribose})_n$ and DNA synthesis in ataxia-telangiectasia cells following γ-ray irradiation, *Nature (London),* 287, 745, 1980.
225. Klocker, H., Auer, B., Hirsch-Kauffmann, M., Altmann, H., Burtscher, H. J., and Schweiger, M., DNA repair dependent NAD^+ metabolism is impaired in cells from patients with Fanconi's anemia, *EMBO J.,* 2, 303, 1983.
226. Zwelling, L. A., Kerrigan, D., and Mattern, M. R., Ataxia telangiectasia cells are not uniformly deficient in poly(ADP-ribose) synthesis following X-irradiation, *Mutat. Res.,* 120, 69, 1983.
227. Mayne, L. V., Broughton, B. C., and Lehmann, A. R., The ultraviolet sensitivity of Cockayne syndrome cells is not a consequence of reduced cellular NAD content, *Am. J. Hum. Genet.,* 36, 311, 1984.
228. Kastan, M. B., Gowans, B. J., and Lieberman, M. W., Methylation of deoxycytidine incorporated by excision-repair synthesis of DNA, *Cell,* 30, 509, 1982.
229. Lieberman, M. W., Beach, L. R., and Palmiter, R. D., Ultraviolet radiation-induced metallothionein-I gene activation is associated with extensive DNA demethylation, *Cell,* 35, 207, 1983.
230. Aggeler, J. and Murnane, J. P., Unpublished observations.
231. d'Ambrosio, S. M. and Setlow, R. B., Enhancement of postreplication repair in Chinese hamster cells, *Proc. Natl. Acad. Sci. U.S.A.,* 73, 2396, 1976.
232. d'Ambrosio, S. M. and Setlow, R. B., Defective and enhanced postreplication repair in classical and variant xeroderma pigmentosum cells treated with *N*-acetoxy-2-acetylaminofluorene, *Cancer Res.,* 38, 1147, 1978.
233. Painter, R. B., Does ultraviolet light enhance postreplication repair in mammalian cells?, *Nature (London),* 275, 243, 1978.
234. Chang, C. C., d'Ambrosio, S. M., Schultz, R., Trosko, J. E., and Setlow, R. B., Modification of UV-induced mutation frequencies in Chinese hamster cells by dose fractionation, cycloheximide and caffeine treatments, *Mutat. Res.,* 52, 231, 1978.
235. Burnet, M., Cancer — a biological approach. I. The processes of control, *Br. Med. J.,* April 6, 779, 1957.
236. Straus, D. S., Somatic mutation, cellular differentiation, and cancer causation (editorial), *J. Natl. Cancer Inst.,* 67, 233, 1981.
237. Ames, B. N., Dietary carcinogens and anticarcinogens, *Science,* 221, 1256, 1983.
238. Goldstein, S., The role of DNA repair in aging of cultured fibroblasts from xeroderma pigmentosum and normals, *Proc. Soc. Exp. Biol. Med.,* 137, 730, 1971.
239. Cleaver, J. E., DNA repair deficiencies and cellular senescence are unrelated in xeroderma pigmentosum cell lines, *Mech. Ageing Dev.,* 27, 189, 1984.

240. Cleaver, J. E., Induction of thioguanine- and ouabain-resistant mutants and single-strand breaks in the DNA of Chinese hamster ovary cells by ^{3}H-thymidine, *Genetics*, 87, 129, 1977.
241. Maher, V. M., McCormick, J. J., Grover, P. L., and Sims, P., Effect of DNA repair on the cytotoxicity and mutagenicity of polycyclic hydrocarbon derivatives in normal and xeroderma pigmentosum human fibroblasts, *Mutat. Res.*, 43, 117, 1977.
242. Gupta, R. S. and Goldstein, S., Diphtheria toxin resistance in human fibroblast cell strains from normal and cancer-prone individuals, *Mutat. Res.*, 73, 331, 1980.
243. Deluca, J. G., Kaden, D. A., Komives, E. A., and Thilly, W. G., Mutation of xeroderma pigmentosum lymphoblasts by far-ultraviolet light, *Mutat. Res.*, 128, 47, 1984.
244. Grosovsky, A. J. and Little, J. B., Mutagenesis and lethality following S phase irradiation of xeroderma pigmentosum and normal human diploid fibroblasts with ultraviolet light, *Carcinogenesis*, 4, 1389, 1983.
245. Maher, V. M., Dorney, D. J., Mendrala, A. L., Konze-Thomas, B., and McCormick, J. J., DNA excision-repair processes in human cells can eliminate the cytotoxic and mutagenic consequences of ultraviolet irradiation, *Mutat. Res.*, 62, 311, 1979.
246. Glover, T. W., Chang, C.-C., Trosko, J. E., and Li, S. S.-L., Ultraviolet light induction of diphtheria toxin-resistant mutants in normal and xeroderma pigmentosum human fibroblasts, *Proc. Natl. Acad. Sci. U.S.A.*, 76, 3982, 1979.
247. Konze-Thomas, B., Hazard, R. M., Maher, V. M., and McCormick, J. J., Extent of excision repair before DNA synthesis determines the mutagenic but not the lethal effect of UV radiation, *Mutat. Res.*, 94, 421, 1982.
248. Todd, P., Fractionated ultraviolet light irradiation of cultured Chinese hamster cells, *Radiat. Res.*, 55, 93, 1973.
249. Todd, P., Dalen, H., and Schroy, C. B., Survival of synchronized cultured human liver cells following single and fractionated exposures to ultraviolet light, *Radiat. Res.*, 69, 573, 1977.
250. Domon, M. and Rauth, A. M., Cell cycle specific recovery from fractionated exposures of ultraviolet light, *Radiat. Res.*, 55, 81, 1973.
251. Maher, V. M., Ouellette, L. M., Curren, R. D., and McCormick, J. J., Frequency of ultraviolet light-induced mutations is higher in xeroderma pigmentosum variant cells than in normal human cells, *Nature (London)*, 261, 593, 1976.
252. Maher, V. M., Ouellette, L. M., Curren, R. D., and McCormick, J. J., Caffeine enhancement of the cytotoxic and mutagenic effect of ultraviolet irradiation in a xeroderma pigmentosum variant strain of human cells, *Biochem. Biophys. Res. Commun.*, 71, 228, 1976.
253. Myhr, B. C., Turnbull, D., and DiPaolo, J. A., Ultraviolet mutagenesis of normal and xeroderma pigmentosum variant human fibroblasts, *Mutat. Res.*, 62, 341, 1979.
254. German, J., Genes which increase chromosomal instability in somatic cells and predispose to cancer, *Prog. Med. Genet.*, 8, 61, 1972.
255. Parrington, J. M., Delhanty, J. D. A., and Baden, H. P., Unscheduled DNA synthesis, u.v.-induced chromosome aberrations and SV40 transformation in cultured cells from xeroderma pigmentosum, *Ann. Hum. Genet.*, 35, 149, 1971.
256. Marshall, R. R. and Scott, D., The relationship between chromosome damage and cell killing in UV-irradiated normal and xeroderma pigmentosum cells, *Mutat. Res.*, 36, 397, 1976.
257. Stich, H. F., Stich, W., and San, R. H. C., Chromosome aberrations in xeroderma pigmentosum cells exposed to the carcinogens, 4-nitroquinoline-1-oxide and *N*-methyl-*N*′-nitro-nitrosoguanidine, *Proc. Soc. Exp. Biol. Med.*, 142, 1141, 1973.
258. Auerbach, A. D. and Wolman, S. R., Susceptibility of Fanconi's anaemia fibroblasts to chromosome damage by carcinogens, *Nature (London)*, 261, 494, 1976.
259. Hartley-Asp, B., The influence of caffeine on the mitomycin C-induced chromosome aberration frequency in normal human and xeroderma pigmentosum cells, *Mutat. Res.*, 49, 117, 1978.
260. Lo, L. W. and Stich, H. F., DNA damage, DNA repair and chromosome aberrations of xeroderma pigmentosum cells and controls following exposure to nitrosation products of methylguanidine, *Mutat. Res.*, 30, 397, 1975.
261. Natarajan, A. T., Verdegaal-Immerzeel, E. A. M., Ashwood-Smith, M. J., and Poulton, G. A., Chromosomal damage induced by furocoumarins and UVA in hamster and human cells including cells from patients with ataxia telangiectasia and xeroderma pigmentosum, *Mutat. Res.*, 84, 113, 1981.
262. Kato, H. and Stich, H. F., Sister chromatid exchanges in ageing and repair-deficient human fibroblasts, *Nature (London)*, 260, 447, 1976.
263. Wang, X., Li, X., Jiang, Z., and Hu, Y., Study of DNA repair enzyme system 2 observation on the sister chromatid exchanges induced by UV light and mitomycin C in xeroderma pigmentosum peripheral lymphocytes, *Acta Genet. Sin.*, 9, 14, 1983.
264. Pant, G. S. and Kamada, N., Sister chromatid exchanges and their relevance to defective DNA repair in xeroderma pigmentosum, *Indian J. Exp. Biol.*, 16, 1194, 1978.

265. Bartram, C. R., Koske-Westphal, T., and Passarge, E., Chromatid exchanges in ataxia-telangiectasia, Bloom syndrome, Werner syndrome, and xeroderma pigmentosum, *Ann. Hum. Genet.*, 40, 79, 1976.
266. Schonwald, A. D. and Passarge, E., UV-light induced sister chromatid exchanges in xeroderma pigmentosum lymphocytes, *Hum. Genet.*, 36, 213, 1977.
267. Fujiwara, Y., Kano, Y., Paul, P., Goto, K., Yamamoto, K., and Miyazaki, N., Excision and cross-link repair of DNA and sister chromatid exchanges in cultured human fibroblasts with different repair capacities, in *Mutation, Promotion and Transformation in Vitro,* Inui, N. et al., Eds., Japan Scientific Societies Press, Tokyo, 1981, 33.
268. Pegg, A. E., Alkylation and subsequent repair of DNA after exposure to dimethylnitrosamine and related carcinogens, in *Review of Biochemical Toxicology,* Vol. 5, Hodgson, E., Bend, J. R., and Philpot, R. M., Eds., Elsevier, Amsterdam, 1983, 83.
269. Day, R. S., III, Ziolkowski, C. H. J., Scudiero, D. A., Meyer, S. A., and Mattern, M. R., Human tumor cell strains defective in the repair of alkylation damage, *Carcinogenesis,* 1, 21, 1980.
270. Sklar, R. and Strauss, B., Removal of O^6-methylguanine from DNA of normal and xeroderma pigmentosum-derived lymphoblastoid lines, *Nature (London),* 289, 417, 1981.
271. Wolff, S., Bodycote, J., Thomas, G. H., and Cleaver, J. E., Sister chromatid exchange in xeroderma pigmentosum cells that are defective in DNA excision repair or post-replication repair, *Genetics,* 81, 349, 1975.
272. Wolff, S., Rodin, B., and Cleaver, J. E., Sister chromatid exchanges induced by mutagenic carcinogens in normal and xeroderma pigmentosum cells, *Nature (London),* 265, 347, 1977.
273. Goth-Goldstein, R., Repair of DNA damaged by alkylating carcinogens is defective in xeroderma pigmentosum-derived fibroblasts, *Nature (London),* 267, 81, 1977.
274. Medcalf, A. S. C. and Lawley, P. D., Time course of O^6-methylguanine removal from DNA of *N*-methyl-*N*-nitrosourea-treated human fibroblasts, *Nature (London),* 289, 796, 1981.
275. Cleaver, J. E., Correlations between sister chromatid exchange frequencies and replicon sizes, *Exp. Cell Res.*, 136, 27, 1981.
276. Marshall, R. R., Arlett, C. F., Harcourt, S. A., and Broughton, B. A., Increased sensitivity of cell strains from Cockayne's syndrome to sister-chromatid-exchange induction and cell killing by UV light, *Mutat. Res.*, 69, 107, 1980.
277. Borek, C., Radiation oncogenesis in cell culture, *Adv. Cancer Res.*, 37, 159, 1982.
278. Barrett, J. C. and Ts'o, P. O. P., Relationship between somatic mutation and neoplastic transformation, *Proc. Natl. Acad. Sci. U.S.A.*, 75, 3297, 1978.
279. Reznikoff, C. A., Bertram, J. S., Brankow, D. W., and Heidelberger, C., Quantitative and qualitative studies of chemical transformation of cloned C3H mouse embryo cells sensitive to postconfluence inhibition of cell division, *Cancer Res.*, 33, 3239, 1973.
280. Borek, C., X-ray-induced *in vitro* neoplastic transformation of human diploid cells, *Nature (London),* 283, 776, 1980.
281. Maher, V. M., Rowan, L. A., Silinskas, K. C., Kateley, S. A., and McCormick, J. J., Frequency of UV-induced neoplastic transformation of diploid human fibroblasts is higher in xeroderma pigmentosum cells than in normal cells, *Proc. Natl. Acad. Sci. U.S.A.*, 79, 2613, 1982.
282. Sutherland, B. M., Delihas, N. C., Oliver, R. P., and Sutherland, J. C., Action spectra for ultraviolet light-induced transformation of human cells to anchorage-independent growth, *Cancer Res.*, 41, 2211, 1981.
283. Milo, G. E., Weisbrode, S. A., Zimmerman, R., and McCloskey, J. A., Ultraviolet radiation-induced neoplastic transformation of normal human cells, in vitro, *Chem. Biol. Interact.*, 36, 45, 1981.
284. Kakunaga, T., Neoplastic transformation of human diploid fibroblast cells by chemical carcinogens, *Proc. Natl. Acad. Sci. U.S.A.*, 75, 1334, 1978.
285. Andrews, A. D. and Borek, C., UVB light-induced *in vitro* transformation of skin fibroblasts from xeroderma pigmentosum and Bloom's syndrome patients, *J. Invest. Dermatol.*, 78(Abstr.), 355, 1982.
286. Kennedy, A. R., Cairns, J., and Little, J. B., Timing of the steps in transformation of C3H 10T1/2 cells by X-irradiation, *Nature (London),* 307, 85, 1984.
287. Crawford, B. D., Barrett, J. C., and Ts'o, P. O., Neoplastic conversion of preneoplastic Syrian hamster cells: rate estimation by fluctuation analysis, *Mol. Cell. Biol.*, 3, 931, 1983.
288. Klein, G. and Klein, E., Oncogene activation and tumor progression, *Carcinogenesis,* 5, 429, 1984.
289. Tabin, C. J., Bradley, S. M., Bargmann, C. I., Weinberg, R. A., Papageorge, A. G., Scolnick, E. M., Dhar, R., Lowy, D. R., and Chang, E. H., Mechanism of activation of a human oncogene, *Nature (London),* 300, 143, 1982.
290. Reddy, E. P., Reynolds, R. K., Santos, E., and Barbacid, M., A point mutation is responsible for the acquisition of transforming properties by the T24 human bladder carcinoma oncogene, *Nature (London),* 300, 149, 1982.

291. Yuasa, Y., Srivastava, S. K., Dunn, C. Y., Rhim, J. S., Reddy, E. P., and Aaronson, S. A., Acquisition of transforming properties by alternative point mutations within c-*bas/has* human proto-oncogene, *Nature (London)*, 303, 775, 1983.
292. Guerrero, I., Villasante, A., Corces, V., and Pellicer, A., Activation of a c-K-*ras* oncogene by somatic mutation in mouse lymphomas induced by gamma radiation, *Science*, 225, 1159, 1984.
293. Shimizu, K., Goldfarb, M., Suard, Y., Perucho, M., Li, Y., Kamata, T., Feramisco, J., Stavnezer, E., Fogh, J., and Wigler, M. H., Three human transforming genes are related to the viral *ras* oncogenes, *Proc. Natl. Acad. Sci. U.S.A.*, 80, 2112, 1983.
294. Taparowsky, E., Shimizu, K., Goldfarb, M., and Wigler, M., Structure and activation of the human N-*ras* gene, *Cell*, 34, 581, 1983.
295. Shen-Ong, G. L. C., Keath, E. J., Piccoli, S. P., and Cole, M. D., Novel *myc* oncogene RNA from abortive immunoglobulin-gene recombination in mouse plasmacytomas, *Cell*, 31, 443, 1982.
296. Adams, J. M., Gerondakis, S., Webb, E., Corcoran, L. M., and Cory, S., Cellular *myc* oncogene is altered by chromosome translocation to an immunoglobulin locus in murine plasmacytomas and is rearranged similarly in human Burkitt lymphomas, *Proc. Natl. Acad. Sci. U.S.A.*, 80, 1982, 1983.
297. Collins, S. and Groudine, M., Amplification of endogenous *myc*-related DNA sequences in a human myeloid leukaemia cell line, *Nature (London)*, 298, 679, 1982.
298. Schwab, M., Ellison, J., Busch, M., Rosenau, W., and Varmus, H. E., Enhanced expression of the human gene N-*myc* consequent to amplification of DNA may contribute to malignant progression of neuroblastoma, *Proc. Natl. Acad. Sci. U.S.A.*, 81, 4940, 1984.
299. Cleaver, J. E., Defective DNA repair and cancer-prone disorders of man, in *Genes and Cancer*, Bishop, J. M., Rowley, J. D., and Greaves, M., Eds., Alan R. Liss, New York, 1984, 117.
300. Seeburg, P. H., Colby, W. W., Capon, D. J., Goeddel, D. V., and Levinson, A. D., Biological properties of human c-Ha-*ras* 1 genes mutated at codon 12, *Nature (London)*, 312, 71, 1984.
301. Stich, H. F., San, R. H. C., Miller, J. A., and Miller, E. C., Various levels of DNA repair synthesis in xeroderma pigmentosum cells exposed to the carcinogens *N*-hydroxy and *N*-acetoxy-2-acetylaminofluorene, *Nature (London) New Biol.*, 238, 9, 1972.
302. Francis, A. A., Snyder, R. D., Dunn, W. C., and Regan, J. D., Classification of chemical agents as to their ability to induce long- or short-patch DNA repair in human cells, *Mutat. Res.*, 83, 159, 1981.
303. Heflich, R. H., Hazard, R. M., Lommel, L., Scribner, J. D., Maher, V. M., and McCormick, J. J., A comparison of the DNA binding, cytotoxicity, and repair synthesis induced in human fibroblasts by reactive derivatives of aromatic amide carcinogens, *Chem. Biol. Interact.*, 29, 43, 1980.
304. Maher, V. M., Birch, N., Otto, J. R., and McCormick, J. J., Cytotoxicity of carcinogenic aromatic amides in normal and xeroderma pigmentosum fibroblasts with different DNA repair capabilities, *J. Natl. Cancer Inst.*, 54, 1287, 1975.
305. Hoffmann, M. E. and Meneghini, R., Action of hydrogen peroxide on human fibroblast in culture, *Photochem. Photobiol.*, 30, 151, 1979.
306. Takebe, H., Furuyama, J.-I., Miki, Y., and Kondo, S., High sensitivity of xeroderma pigmentosum cells to the carcinogen 4-nitroquinoline-1-oxide, *Mutat. Res.*, 15, 98, 1972.
307. Stich, H. F. and San, R. H. C., Reduced DNA repair synthesis in xeroderma pigmentosum cells exposed to the oncogenic 4-nitroquinoline 1-oxide and 4-hydroxyaminoquinoline 1-oxide, *Mutat. Res.*, 13, 279, 1971.
308. Ikenaga, M., Takebe, H., and Ishii, Y., Excision repair of DNA base damage in human cells treated with the chemical carcinogen 4-nitroquinoline 1-oxide, *Mutat. Res.*, 43, 415, 1977.
309. Mirzayans, R. and Waters, R., DNA damage and its repair in human normal or xeroderma pigmentosum fibroblasts treated with 4-nitroquinoline 1-oxide or its 3-methyl derivative, *Carcinogenesis*, 2, 1359, 1981.
310. Petrusek, R. L., Uhlenhopp, E. L., Duteau, N., and Hurley, L. H., Reaction of anthramycin with DNA: biological consequences of DNA damage in normal and xeroderma pigmentosum cell lines, *J. Biol. Chem.*, 257, 6207, 1982.
311. Yang, L. L., Maher, V. M., and McCormick, J. J., Relationship between excision repair and the cytotoxic and mutagenic effect of the "anti" 7,8-diol-9,10-epoxide of benzo(a)pyrene in human cells, *Mutat. Res.*, 94, 435, 1982.
312. McCaw, B. A., Dipple, A., Young, S., and Roberts, J. J., Excision of hydrocarbon DNA adducts and consequent cell survival in normal and repair defective human cells, *Chem. Biol. Interact.*, 22, 139, 1978.
313. Slor, H., Induction of unscheduled DNA synthesis by the carcinogen 7-bromomethylbenz(a)anthracene and its removal from the DNA of normal and xeroderma pigmentosum lymphocytes, *Mutat. Res.*, 19, 231, 1973.
314. Regan, J. D., Francis, A. A., Dunn, W. C., Hernandez, D., Yagi, H., and Jerina, D. M., Repair of DNA damaged by mutagenic metabolites of benzo(a)pyrene in human cells, *Chem. Biol. Interact.*, 20, 279, 1978.

I. INTRODUCTION

Exposure to UV or visible light can have profound effects on the skin of a susceptible individual. The harmful effects of the sunburn spectrum (290 to 320 nm) are well known. In addition, there are several diseases associated with sensitivity to visible light. An example of one of these diseases is erythropoietic protoporphyria (EPP), in which sensitivity to visible light (380 to 560 nm) is a cardinal symptom. Until recently, no satisfactory way existed for preventing the sensitivity to visible light characteristic of EPP other than to avoid light exposure. Sunscreens effective in preventing abnormal reactions to light of the sunburn spectrum are ineffective for the prevention of sensitivity to visible light. Observations that the carotenoid pigments of green plants and photosynthetic bacteria played an important role as protective agents against photosensitization by the organisms' own chlorophyll led to the suggestion that these pigments might be of benefit in the treatment of photosensitive diseases characterized by sensitivity to visible light. How these observations were developed into the use of carotenoids for the relief of photosensitivity, and the description of the treatment of the first few patients of Mathews-Roth et al.[1-3] is given in detail elsewhere. In this chapter, the use, by Mathews-Roth et al. and others, of carotenoid pigments in the treatment of human photosensitivity, will be discussed, as will recent findings of the pigments' protective action against certain kinds of cancer. The clinical pharmacology of the carotenoid pigments shall also be discussed briefly. The reader is referred to the books by Karrer and Jucker,[4] Goodwin,[5] and Isler[6] for a detailed discussion of the chemistry of the carotenoid pigments.

II. β-CAROTENE THERAPY IN ERYTHROPOIETIC PROTOPORPHYRIA

From 1969 to 1975, the collaborative group of Mathews-Roth et al.[7,8] treated 133 patients suffering from EPP with β-carotene, using a standard protocol adhered to by all participating physicians. In July 1975, the U.S. Food and Drug Administration approved the use of β-carotene for the treatment of EPP and the authors terminated the collaborative study at that time. During the course of the study, they found the following starting dosage schedule of β-carotene to be the most satisfactory, and they still recommend it for use: 1 to 4 years of age, 60 to 80 mg/day; 5 to 8 years, 90 to 120 mg/day; 9 to 12 years, 120 to 150 mg/day; 13 to 15 years, 150 to 180 mg/day; and 16 years and older, 180 mg/day. The average dose for the patient's age should be administered for 4 to 6 weeks, and the patient should be instructed not to increase sun exposure for either 4 weeks, or until some yellow discoloration of the skin, especially of the palms of the hands, is noted. Then, exposure can be increased cautiously and gradually until the patient determines the limits of exposure to light that can be tolerated without the development of symptoms. If the degree of protection is not sufficient, the daily dose of β-carotene should be increased by 30 to 60 mg/day for children under 16, and up to a total of 300 mg/day for those over 16. If after 3 months of therapy at these higher doses (blood carotenoid levels should reach at least 800 μg/dℓ) no significant increase in tolerance to sunlight exposure has occurred, it can be assumed that β-carotene therapy will not be effective for that patient, and the medication should be discontinued.

To determine the efficacy of the treatment with β-carotene, patients were given detailed questionnaires each time they visited the physician for physical examinations and blood tests as described in the protocol. These visits occurred several times before therapy, and at 4 to 6 week intervals during therapy. The questionnaires recorded the length of time the patient could tolerate exposure to sunlight without the development of symptoms, and from the data obtained before and during therapy, a protection index was calculated.[7,8]

The authors found that 84% of the patients increased by a factor of three or more their ability to tolerate sunlight exposure without the development of symptoms. On the average, it took between 1 and 2 months for the patients who received benefit from β-carotene therapy to notice increased tolerance to sun exposure. The majority are now engaging in outdoor activities which they were unable to do before therapy started, many stating that they spend more time in the sun since taking β-carotene. This seemed to be especially gratifying to the children, who, prior to β-carotene therapy, were unable to play outdoors to any great extent; now the majority can spend hours outside with their friends.

Many patients stated that they were able to develop a sun tan for the first time in their lives. Indeed, the results of tests with artificial light sources indicate that patients with EPP react normally to sunburn radiation (290 to 320 nm), when tested with that radiation alone. However, because of the presence of light of 380 nm and above in natural sunlight, which light causes their symptoms, patients with EPP have never been able to tolerate sun exposure long enough to develop a tan. It was the impression of several of the patients that the acquisition of the tan, plus the β-carotene, added to their protection from the effects of the sun.

The majority of the patients noted that when they are taking β-carotene, those reactions from the sun that do occur are less severe in intensity and duration than before therapy. The patients not only have increased the time that they can remain in the sun symptom free, but they also do not develop significant numbers of cutaneous lesions during their increased exposure time.

The authors found that the patients' blood and stool porphyrin levels were not affected by the ingestion of large amounts of β-carotene. Thus, treatment with β-carotene ameliorates photosensitivity in EPP, but has no effect on the biochemical lesion in this disease.

Because of the difficulties in the subjective evaluation of therapeutic effect and of setting up a controlled double-blind study in the particular case of β-carotene and EPP, the authors decided to use phototesting with polychromatic light from 380 to 560 nm (as opposed to monochromatic light) as an objective measure of clinical improvement. Using a xenon arc lamp under the conditions they have developed,[7] Mathews-Roth and co-workers have not yet found a patient with EPP who did not develop immediate erythema when exposed to this light. A few patients develop, in addition to erythema, wheals, itching, and burning, but the erythema is the reproducible phenomenon. They determined the average for each patient of three phototests done before treatment with β-carotene was begun, and the average of those phototests done during treatment, after blood carotenoid levels had reached at least 400 μg/dℓ. The number of tests was large, and clearly established the reliability and reproducibility of the tests.[7,9] The authors found that tolerance to xenon arc light exposure increased in those patients who reported benefit from β-carotene therapy, but no increased tolerance to this radiation has been found in patients reporting no improvement.[7,9] Other workers have also noted increased tolerance to polychromatic xenon arc light after treatment with β-carotene.[10,11] Thus, phototesting with polychromatic xenon arc light can serve as an objective method of determining improvement in tolerance to light.[9]

It should be noted that Mathews-Roth et al. did not find a definite relationship between the ability to tolerate a given amount of xenon arc light and the ability to tolerate a given amount of sunlight. Response to artificial light shows a qualitative but not a quantitative relationship to response to sunlight.

Mathews-Roth et al. have found 21 other studies reporting increases in tolerance to sunlight in over three quarters of patients with EPP treated with high doses of β-carotene.[10-30] One study reporting little or no improvement has also appeared:[31] however, these workers used a much lower dosage of β-carotene than recommended. Later, some of the patients from the unsuccessfully treated group were given higher doses of β-carotene by another investigator, and the patients noted some increased tolerance to sun while taking the higher doses.[32] These results emphasize the importance, as mentioned earlier, of individualizing dosage to each patient, and increasing the dose until the patient reports some improvement. The problem of individualizing treatment is important to the conduct of a controlled trial and makes double-blinded designs almost impossible. At a minimum, such trials should use a dose of β-carotene large enough to produce amelioration of symptoms in the majority of patients (a minimum period of 3 months' treatment at doses giving blood levels of at least 800 μg/dℓ — for adults at least 180 mg/day should be given).

From the results of Mathews-Roth et al., and those of the other workers mentioned earlier, it can be concluded that β-carotene, when administered in sufficiently high doses, can be an effective therapy for ameliorating photosensitivity in most patients with EPP.

III. β-CAROTENE THERAPY IN OTHER PHOTOSENSITIVITY DISEASES

Since β-carotene seemed to be effective in preventing photosensitivity in EPP, it seemed logical to see if it could prevent photosensitivity in other photosensitivity diseases. Several workers have investigated this possibility.

Six groups of workers have reported some success in treating patients suffering from congenital porphyria (Gunther's disease) with β-carotene.[33-38] Mathews-Roth et al.[39] also have treated with β-carotene two children having this disease, and have noted some improvement. New lesions were significantly decreased in number and severity, and the patients have been able somewhat to increase their sun exposure.

Wennersten and Swanbeck[11,40] treated eight patients with polymorphous light eruption (PMLE) with β-carotene and found an increase in the minimal erythema dose (tested with an appropriately filtered xenon arc lamp) of all the patients after treatment. Jansen[41] treated 66 patients, 55 of whom improved as judged by patients' questionnaires and physicians' examinations. Nordlund et al.[42] found that a topical sunscreen had to be used in conjunction with oral β-carotene to get an effective therapeutic result in a severe case of PMLE.

During the course of the collaborative study, Mathews-Roth et al.[8] treated 27 patients with PMLE with high doses of β-carotene. Nine of these patients increased their ability to tolerate exposure to sunlight by more than a factor of three. Only two of these nine patients reported marked improvement in their photosensitivity with the use of β-carotene alone — the other seven had to use a sunscreen in addition to the β-carotene. (None of the 27 patients with PMLE had received any benefit from the use of nonopaque sunscreens alone.) The remaining 18 patients reported no improvement with the use of both oral β-carotene and the topical sunscreen.[8]

Thune[43] administered β-carotene to 17 patients with PMLE and found "a significant increase in minimal erythema dose for unfiltered (xenon) light at the 5% level of probability". However, only five of the 17 patients had a protection index of more than three with respect to sunlight exposure, a much lower success rate than Wennersten and Swanbeck, and Jansen, and more in line with the findings of Mathews-Roth et al. He also could not show a direct correlation between the protection index obtained for artificial light and that obtained for sunlight exposure, as they had been unable to do for EPP. He also found, as they have, that some patients benefited from the combined therapy of oral β-carotene and topical sunscreen.

Parrish et al.[44] found a similar low rate of improvement (six out of 19) in patients with PMLE treated with β-carotene, and Haeger-Aronsen et al.[24] reported improvement in 15 out of 18 patients treated. Suhonen and Plosila[45] found good to excellent results in 36 out of 50 patients with PMLE in a double-blind randomized trial, and Corbett et al.[46] found very little effect of either β-carotene or chloroquin in a controlled trial of PMLE patients.

Porphyria cutanea tarda is successfully treated by phlebotomy, but this treatment is occasionally contraindicated. Mathews-Roth et al.[8] treated one patient with this condition with β-carotene in the hope of decreasing the appearance of new cutaneous lesions, which in this condition are porphyrin-induced photoinjury. However, examination of the patient's skin and the patient's own report after 2 months of β-carotene therapy showed no appreciable decrease in the number of new lesions formed.

Kobza et al.[47] reported that β-carotene alone had no effect in the treatment of two cases of actinic reticuloid and three cases of solar urticaria, whereas Aratari et al.[48] obtained good results in one patient with actinic reticuloid using a combination of β-carotene and canthaxanthin. Mathews-Roth et al.[8] treated six cases of solar urticaria, three cases of hydroa aestivale, and two cases of actinic reticuloid with high doses of β-carotene. Only one patient with solar urticaria and one patient with hydroa aestivale reported noticeable relief. Both stated that they obtained improvement without the use of a topical sunscreen.

Bickers et al.[49] noted that two patients with hydroa vacciniforme developed no new lesions when taking β-carotene. Husquinet et al.[50] have administered β-carotene to one patient with porphyria variegata, who developed no lesions during the two summers the drug was ingested. Muhlbauer et al.[51] treated two patients with this condition and reported cutaneous remissions for periods exceeding 2 years.

Carotenoids were also reported beneficial in preventing the photosensitivity associated with a case of mucinosis follicularis triggered by light exposure[52] and associated with quinidine therapy.[53]

In summary, β-carotene treatment may be of some use in congenital porphyria, if given in high doses starting when the patients are very young. However, carotene treatment seems of limited use in PMLE; Mathews-Roth et al. would recommend it only after other treatment modalities have failed. At this time it is difficult to ascertain the use of carotenoid therapy in the other conditions listed here — further studies are needed. However, since carotenoids have little or no toxicity, it is probably worthwhile to try a therapeutic course of these pigments, if other treatments do not relieve photosensitivity.

Two independent observations suggested that perhaps carotenemia might have an effect on a normal individual's response to sunlight. Bendes[54] had observed that the presence of carotenemia in children undergoing heliotherapy for tuberculosis prevented sunburn, and Sandler[55] found that carotenemia facilitated tanning of the skin. These observations, plus the success of Mathews-Roth et al. with β-carotene in the treatment of EPP, led these authors to conduct a controlled trial to determine if the administration of high doses of β-carotene would alter the normal fair-skinned individual's response to sunlight. It was found that high doses of β-carotene had a small but statistically significant effect in increasing the minimal erythema dose for eliciting erythema produced by natural sunlight. However, the observed effects were too small to recommend the use of β-carotene as a protective agent for sunburn.[56] It was also found that the men in the group taking carotene in this study developed more pigmentation (tanning) than did the men in the placebo group, thus confirming the findings of Bendes[54] and Sandler.[55]

Anyan[57] reported that a young girl who had been carotenemic while hypothyroid had not been able to tan while in this condition. Once she became euthyroid after appropriate treatment, her carotenemia disappeared and the ability to tan returned. These findings are not necessarily at variance with those of Mathews-Roth et al.; perhaps, as Anyan suggests, marked, long-standing carotenodermia may protect the skin to such a degree that tanning may become hard to induce. In addition, one does not know the effects of hypothyroidism itself on tanning. Unfortunately, serum carotenoid levels were not given. Also, it is difficult to compare the degree of sun exposure of this youngster to the degree of sun exposure of the patients in the study of Mathews-Roth et al. As was mentioned earlier, many of their EPP patients report tanning for the first time because of their new-found ability to stay out in the sun. Some actually sunbathe on the beach. None have reported not being able to tan. These observations would all support their findings and those of Bendes[54] and Sandler[55] that tanning increases and burning decreases in carotenemic individuals. However, more work needs to be done to elucidate the extent of the ability of carotenoids to alter the reaction of the skin to sunburn.

IV. CAROTENOIDS AND CANCER PREVENTION

A. Animal Studies

Over the past 30 years, reports have appeared in the scientific literature of the antitumor effects of vitamin A in animal models. This work has been extensively reviewed[58-61] and thus will not be discussed here. Instead, the concentration will be on describing the work concerning β-carotene.

In view of the success of vitamin A in suppressing tumor development, it is not surprising that investigations were done to see if β-carotene, the precursor of vitamin A, had an antitumor effect. In 1973, Dorogokupla et al.[62] induced subcutaneous tumors in rats by injecting 9,10-dimethyl-1,2-benzanthracene (DMBA), and skin tumors in mice by topical application of DMBA. The authors found in both cases that those animals fed a diet supplemented with "unlimited amounts of red carrots" developed tumors at a slower rate than did the animals receiving the unsupplemented diet. In that year also, Epstein[63] reported inducing skin tumors by exposing hairless mice to UVB radiation (290 to 320 nm) and found that those mice which received injections of a solution of β-carotene beadlets (Roche) developed tumors at a slower rate than did mice injected with a solution of placebo beadlets. Dorogokupla et al.[62] proposed that the effect of β-carotene on tumor formation was due to the effect of vitamin A, which was formed from the administered β-carotene.

To determine whether or not the carotenoid molecule itself has an antitumor action, Mathews-Roth[64,65] decided to study the effects on tumor formation of two other carotenoids which have no vitamin A activity: canthaxanthin (4,4′-diketo β-carotene), which cannot be converted to vitamin A by mammals, and which has been found to be effective in preventing photosensitization both in vitro and in vivo, and phytoene (7,8,11,12,7′,8′,11′,12′-octahydrolycopene), which has been found to have protective activity against sunburn. Mathews-Roth and associates examined the effects of these two pigments and of β-carotene on the development in hairless mice of skin tumors induced by three different methods: (1) UVB irradiation three times a week, (2) two applications of DMBA followed by twice-weekly applications of croton oil, and (3) one application of DMBA followed by UVB irradiation twice a week. She found that administration of β-carotene, canthaxanthin, or phytoene resulted in a statistically significant delay in the appearance of skin tumors induced by UVB radiation. She also found that β-carotene and canthaxanthin had some effect in delaying the appearance of subsequent tumors, when the administration of these pigments was started after one tumor had developed.[66] However, in the case of DMBA-croton oil tumor induction, only β-carotene was effective in delaying tumor development — canthaxanthin and phytoene had no such effect. With DMBA application and low-dose UVB irradiation, both β-carotene and canthaxanthin administration caused a delay in tumor development, but phytoene had no effect. In each case, when the administration of a given pigment caused a delay in tumor development, fewer tumors per animal were found in the pigment-treated mice, as compared to the placebo-treated animals.[64,65]

These experiments suggest that carotenoid pigments, irrespective of their vitamin A activity, can prevent or slow down the growth of skin tumors induced by UVB alone. These experiments also suggest that the mechanism of tumor induction by DMBA-croton oil application must be different from that of UVB exposure. As to tumors produced by DMBA and low-dose UVB radiation, canthaxanthin may not be able to affect initiation by DMBA, but may have an effect on promotion or cocarcinogenesis by UVB. Phytoene, because of molecular differences between it and either β-carotene or canthaxanthin, may not be able to exert enough effect on the development of tumors induced by an agent other than UVB. There are suggestions in the literature that excited species such as free radicals and singlet oxygen may be involved in the development of skin tumors induced by UVB. It is interesting to note that β-carotene and canthaxanthin are extremely efficient quenchers of such species; phytoene is less so.[67]

The work of Mathews-Roth has been confirmed and extended by Santamaria et al.[68] These workers have reported that β-carotene and canthaxanthin administered orally to Swiss albino mice will delay the development of tumors induced by the application of benzo(*a*)pyrene alone (BP) or BP plus UVA radiation (320 to 400 nm). They have also found that these pigments will delay the development of tumors in hairless mice induced by the administration of 8-methoxy psoralen and UVA radiation.

Seifter et al.[69-71] have reported a series of experiments in which they have found that either vitamin A or β-carotene inhibits the size and/or development of transplanted tumors in mice. These workers also suggest that β-carotene and vitamin A administration prevent the involution of the thymus gland, which occurs when transplanted tumors take hold in the host. They have recently reported that administration of β-carotene or vitamin A can decrease the incidence of DMBA-induced tumors in rats.[72] The kinds of tumors were not specified. One concern with their studies is that the authors added β-carotene to the diet of the animals as an alcoholic solution of crystalline β-carotene. It is difficult to determine whether the added β-carotene and vitamin A survived oxidation and contributed significantly to the blood and tissue level of these substances.

In recent experiments, Mathews-Roth[65,66,198] has found that β-carotene or canthaxanthin at a dose of 100 mg/kg of diet, a bit higher than the 90 mg/kg of diet used by Seifter et al., was not protective against tumors induced by UVB irradiation. However, a dose level of 3.5 mg carotenoid per 5 g of diet, similar to the 2.5 mg/5 g of diet (plus a bit more from gavaging) used by Santamaria et al. was protective. In all of her experiments Mathews-Roth used β-carotene beadlets (Roche) as the source of her pigments. The beadlet formulation protects the pigments against oxidation. Indeed, she could detect high levels of carotenoids in the skin of the treated animals, and analysis of the supplemented foods contained the specified amounts of pigments. Santamaria et al. also used Roche pigment beadlets as the source of carotenoids, as well as freshly prepared solutions of carotenoids in oil for supplemental gavage administration. In this latter case also, little deterioration of the pigments could have occurred.

The experiments described to this point have been concerned with oral administration of carotenoids before and/or during tumor induction. Shamberger[73] reported the effect of applying topical β-carotene to haired mice undergoing tumor induction with DMBA and croton oil or resin application. He found an increase of tumor formation in the animals which received β-carotene applications. It is difficult to know what happened to the β-carotene under these conditions. Mathews-Roth et al.[7] have found that β-carotene applied topically to human skin bleaches quickly. Shamberger's finding is also somewhat reminiscent of similar findings with topically applied vitamin A, in which some doses applied topically cause excess tumor formation and other dose levels protect.

β-Carotene, canthaxanthin, and phytoene are all lipid-soluble carotenoids. Gainer et al.[74] have been investigating the water-soluble carotenoid, crocetin, and found that this pigment, prepared by them from an extract of saffron, offered protection against DMBA-croton oil induced tumors in haired mice. Mathews-Roth[75] repeated their experiments, but used instead hairless mice and crystalline crocetin produced synthetically (Roche). The route of administration, vehicle, and dose were those used by Gainer et al. She found only a small effect in delaying tumor development when tumors were induced with DMBA-croton oil applications. She found no effect in preventing the development of skin tumors induced by UVB alone.[75] The same amounts of radiation were used in these experiments as were used in the experiments mentioned earlier in which β-carotene, canthaxanthin, and phytoene had a protective effect against skin tumor formation. It is not totally clear why crocetin is not as effective as the other pigments. Perhaps the dose used was incorrect. Also, the molecular structure of crocetin differs from the other pigments, and the water solubility may affect where the pigments localize in the cell.

The several experiments reported here would suggest that the carotenoid molecule itself has some ability to prevent or delay the development of skin tumors induced by UVB radiation, and in some cases tumors induced by a chemical carcinogen (DMBA or BP) and promoted by UVB or UVA. Certainly β-carotene should have some effect on the inhibition of tumors in organs other than skin which have been shown to be inhibited by vitamin A. Carefully controlled experiments, where adequate dose levels of β-carotene are given, and carotenoid and vitamin A levels in the serum and target organs are measured, need to be performed to determine the antitumor spectrum of not only β-carotene but also other carotenoid pigments.

B. Human Studies

The reports of animal studies suggesting that vitamin A has antitumor activity have led to studies to determine whether vitamin A, or its precursor pigment in foodstuffs, β-carotene, have any such activity in human tumors.

The author has found 32 reports in the literature of epidemiological studies concerned with the effect of diet, particularly the ingestion of carotenoid-containing vegetables, on the incidence of cancer.[76-107] All but four[90,94,96,99] have suggested that there is an inverse relationship between the ingestion of carotenoid-containing vegetables and the incidence of cancer. Most of the studies have been retrospective or prospective dietary studies. However, 12 studies analyzed the serum of cancer victims and control populations.[96-107] Peto et al.[108] have reviewed many of the studies, so no attempt will be made to do that here.

Admittedly, it is risky to base conclusions of drug efficacy on dietary-intake studies, because recall of specific food intake may be incomplete. Even studies examining serum levels of the compound of interest may be suspect in that control populations may not have been chosen properly. In addition, in dietary studies, there is always the possibility that the actual active compound may not be the component upon which interest is focused, but rather another component of the foods containing it. Even negative studies of either type do not necessarily mean that the compound of interest has no effect in preventing cancer. The compound may only be effective in certain types of cancer not yet tested, or in subgroups of a population (such as people with low-normal or deficient blood levels of the compound).

Certainly the best way to determine whether a particular substance is effective in preventing human cancer is to administer it prophylactically to a large enough number of people at high risk to develop the particular kind of cancer to be studied, the study population being randomized into a group receiving the substance under study and a group receiving placebo. Indeed, the National Cancer Institute of the U.S. National Institutes of Health is presently funding six such studies in which carotenoids are administered to people at high risk for developing certain kinds of cancer.[199] In each case, the study population is divided into groups receiving carotenoid and groups receiving placebo.

All of the six placebo-control studies are due to run for several years, and may have to be extended beyond their originally planned periods, to assure that sufficient data have been obtained. Thus it will take quite a few more years before it is definitely established whether or not β-carotene and/or vitamin A have a significant anticancer effect in man. A prime requirement besides efficacy in cancer prevention for a prophylactic anticancer agent to be used in large numbers of people is that it must have little or no toxicity. As will be discussed later, β-carotene has proved to be very safe. Thus, if it is found from studies in man such as the ones presently in progress, that β-carotene is effective in preventing the development of one or more kinds of cancer, this pigment will be an extremely good candidate for use as a cancer-preventive agent in large populations. We eagerly await the results of the ongoing clinical studies described here, and others which may also be commenced in the near future.

V. SOME ASPECTS OF THE CLINICAL PHARMACOLOGY OF CAROTENOIDS

A. Role of β-Carotene in Human Nutrition

In spite of the current availability and use of synthetic vitamin A either in foods or in vitamin supplements, carotenoid vitamin A precursor-containing foods are still a very significant source of this vitamin in the diets of humans, particularly in developing countries. Diets in various parts of the world vary greatly in their content of vitamin A, either available directly in meat, fish, eggs, or dairy foods, or as the various provitamins A in vegetable sources. β-Carotene is the most important vitamin A precursor, mainly because of its prevalence in plants widely consumed by man and livestock, and because it is the provitamin A with the greatest activity.[109] In addition to availability in the diet, several factors such as severe dietary deficiency in protein; the amount, nature, and dispersion of dietary fat; and the nitrate content of the food may influence the absorption and utilization of β-carotene and other provitamins.[109-114] The transformation of β-carotene to vitamin A (retinol) in the animal (plants do not contain vitamin A as such) occurs by cleavage of the carotene molecule at the central double bond[115,116] or by cleavage of the conjugated chain adjacent to one β-ionone ring followed by sequential oxidative removal of fragments consisting of two to five carbon atoms.[117,118]

The activity of the enzyme carotene dioxygenase isolated from rat and human intestinal mucosa and also from liver seems to suggest that the cleavage of the molecule at the central double bond is the more important mechanism for metabolism of carotene in the mammal.[118,119]

Many studies have shown that the main point of conversion of ingested carotenoids is the intestinal mucosa. Conversion also occurs to a fairly significant degree in the liver. It was suggested that lung might also be a site of conversion, but data from heart-lung preparations have not confirmed this.[120] Greenberg et al.[121-123] suggested that sebaceous glands may convert β-carotene to vitamin A.

Traditionally, it has been thought that only plants and bacteria can synthesize carotenoids from simple precursors;[5] recently some evidence has been presented that bovine corpus luteum tissue could synthesize β-carotene from acetate in vitro.[124] Further confirmation of this finding must be obtained before the present view of no carotenoid synthesis in animals can be abandoned.

The manner in which carotenoids are excreted is not clear. Hashimoto[125] claims that none is excreted in the urine. Hess and Myers[126] found that urine got yellower within 45 min after giving concentrated carrot extract. Unfortunately, they did not chemically extract the urine to determine the presence of carotenoids. Almond and Logan[127] made the same observation and attempted to chemically extract the urine. They extracted a yellow pigment, but this was found not to be β-carotene. Mathews-Roth et al. also looked for carotenoids in the urine of EPP patients taking carotene, and also found that the yellow pigment contained in the urine did not seem to be a carotenoid on spectrophotometric or chromatographic analysis. More work certainly needs to be done on the question of the excretion of carotenoids.

B. Toxicity Studies

The results of studies conducted by Mathews-Roth et al.[7,8,56] on the patients with EPP treated with β-carotene and on the normal volunteers given similar doses of β-carotene for a 10-week period in their study on carotene and sunburn radiation have confirmed reports in the literature that the ingestion of large amounts of pure β-carotene has not led to toxic side effects.[128,129] Some of their patients have been taking β-carotene for 11 years, with no appearance to date of toxicity. They detected no abnormalities in blood sugar or blood urea nitrogen in either EPP patients or normal volunteers, nor have the other workers mentioned earlier who are using β-carotene in the treatment of EPP reported any toxicity. All except one patient with EPP in our study had normal total bilirubin and serum glutamate-oxaloacetate transaminase levels. This one patient had fairly consistently elevated levels of these parameters before as well as during therapy, but never complained of or gave a history of liver disease. He has not responded to carotene therapy, and has sustained high levels of blood and stool porphyrins. Although serious liver disease is seen in some cases of EPP, this patient is the only one in our group to demonstrate significant liver-function test abnormalities. Another patient has had his gall bladder removed, but his liver function tests were normal.

The complete blood count was not affected by carotene intake. Many of the patients with EPP had a mild anemia.[130] This was not worsened or improved by carotene intake. Leucopenia has been reported in patients who ingested large amounts of carrots.[131,132] This phenomenon was found to reverse itself when the patient stops eating the abnormal amount of carrots. Mathews-Roth et al. found no evidence of leucopenia in any of their patients or volunteers. The absence of this finding in their study strengthens the suggestion of earlier workers that the leucopenia was probably due to substances other than β-carotene present in the carrots.[128,129] Indeed, they reported that the administration of pure β-carotene did not induce leucopenia in a subject who had experienced leucopenia after eating large amounts of carrots.[128,129]

Methemoglobinemia has recently been shown to result from the consumption of large amounts of carrot juice.[133] This is another example of the toxicity of large amounts of vegetables and their products, and a reason why Mathews-Roth et al. do not encourage photosensitive patients to become carotenemic by ingesting large amounts of these foods.

None of the patients of Mathews-Roth et al. were found to develop abnormally high levels of serum vitamin A. This confirms what has been repeatedly reported in the literature, that individuals ingesting large amounts of β-carotene do not develop hypervitaminosis A.[128,129,134]

Kobza and collaborators[47] report that one patient treated with β-carotene reported an increase in hay fever attacks and another reported swelling of the tongue and face. Mathews-Roth et al. found no such reactions in their patients. However, it should be kept in mind that an occasional patient could be allergic to some of the other components of the beadlet preparation (gelatin, antioxidants, etc.). Perhaps such patients should be given crystalline β-carotene in vegetable oil, to determine if indeed the reaction was caused by beadlet components. If the adverse reaction disappers, then the patient can continue to take the carotene in oil, if ingestion of carotene has benefited his photosensitivity. Some patients report transient diarrhea which usually disappears either spontaneously or with lowering of the dose. Thus, it would seem that the only significant "side effect" from the ingestion of large amounts of β-carotene is carotenodermia, the yellowish coloration of the skin observed in many people who ingest large amounts of the pigment. As is well known, this coloration disappears several weeks after cessation of carotenoid intake.

Carotenoids have been found in human milk,[135,136] and there are at least two reported instances of breast-fed infants becoming carotenemic on their carotenemic mothers' milk.[127,137] It also seems that carotenoids can cross the placental barrier. DeBuys[138] reports that two children of a carotenemic mother were born carotenemic with no apparent side effects. Claussen and McCoord[139] suggest that 1 mg of carotenoid per kg fetus per day crosses the placenta. Claussen[140] reports the presence of some carotenoid in the cord blood at birth. After birth, the amount of carotenoid present is dependent on the child's diet.

We have studied the livers of two patients who had been taking β-carotene for photosensitivity. There was no evidence of any accumulation of large amounts of either β-carotene or vitamin A in the liver by chemical analysis, or evidence of morphological alterations attributable to carotenoids or vitamin A accumulation on light-microscopic examination.[138]

C. Carotenemia and Carotenodermia

Hess and Myers[126] noted that people who consumed large amounts of carrots and other vegetables often developed a yellow color of their skin; they used the term "carotenemia" for this condition. The phenomenon of yellow skin color was probably first noted by Baeltz,[142] who called it aurantiasis cutis; he did not, however, make the connection between a diet abnormally high in carotenoid-containing foods and the yellow skin color. Apparently Hess and Myers were the first to do so. In actuality, carotenodermia is the more accurate term for the phenomenon of yellow skin coloration by the ingestion of large amounts of carotenoid pigments; carotenemia literally means carotene in the blood. Carotenemia has been thoroughly reviewed in the past,[132,143,144] therefore, only highlights of this condition will be noted here.

Mammals can be divided into three groups according to their ability to accumulate carotenoids after oral administration in blood and tissues as carotenoids (not vitamin A): (1) those which can accumulate a mixture of carotenoids; (2) those which accumulate predominantly carotenes, as opposed to xanthophylls; and (3) those which cannot accumulate any carotenoids.[5] Man, the fox, the fitchert, and possibly the badger and roedeer are examples of the first group; cattle, horses, dogs, sheep, buffalo, deer, antelope, caribou, and hedgehogs are examples of the second; and goats, hares, and the common laboratory animals, mice, rats, rabbits, and guinea pigs are examples of the third group. Castle[145] studied a strain of rabbit which carried a recessive gene for "yellow fat", or the ability to accumulate carotenoids. However, Mathews-Roth et al.[65,66,146] have found that if laboratory animals are given large doses of carotenoids in their diet, they can accumulate the pigments in their blood, skin, and other organs.

In man, the most common cause of carotenemia and carotenodermia is overindulgence in carotenoid-containing vegetables and fruits. The predominant pigment found in the serum is the predominant pigment of the vegetable eaten. Usually this is β-carotene. There has been a report of high levels of lycopene, the principal pigment of the tomato, in a patient whose dietary indiscretion was tomato juice.[147] There have been a few cases reported in which carotenemia and carotenodermia developed in individuals whose vegetable intake is not abnormal. It seems that these patients have a genetic inability to split carotenoids into vitamin A to the same degree as the normal individual.[144,148-150] These patients have low to normal serum vitamin A levels, and may or may not show symptoms of vitamin A deficiency.

The development of carotenodermia seems to be quite variable; different individuals on essentially the same diet can vary in the degree of carotenodermia they develop.[132] Factors which seem to influence the development of carotenodermia, besides the amount of carotenoid-containing foods ingested, are the amount, nature, and dispersion of lipid in the diet; the digestability of the carotenoid-containing food; and the general adequacy of nutrition and level of protein in the diet.[109] It has been found that the ingestion of mineral oil definitely decreases the absorption of carotenoids (and for that matter also, that of vitamin A). The time of development of carotenemia and carotenodermia from the onset of increased carotenoid intake varies widely, depending on the amounts of and kind of carotenoid-containing foods ingested. In the studies of Mathews-Roth et al. of EPP patients and normal volunteers taking pure β-carotene and on otherwise normal diets, they found that development of carotenodermia occurred between 1½ to 3 months, but the development of serum carotene levels above normal values occurred within 1 month. Those taking the higher doses from the beginning (180 mg/day) developed carotenemia and carotenodermia somewhat earlier. Some development of carotenoid tolerance, manifested by decrease in blood levels of carotene after several months of therapy has been noted by some workers.[151] Mathews-Roth et al. have not found any conclusive evidence of this kind of phenomenon in their patients or subjects, but they have noted that some individuals, even on fairly high doses, never achieve as high a level as do others on the same dose. This may indicate the variation in different individuals' ability to absorb carotenoids.

Clinical observation of patients with carotenodermia has long demonstrated that this coloration is most obvious in those areas of the body with a prominent amount of horny layer, such as the palms of the hands, soles of the feet, knees, and other areas of skin that are subject to mechanical irritation. Gandy[152,153] also suggested that in addition to areas subject to mechanical irritation, carotenoids are deposited in greater amounts in areas subject to irritation by physical factors, including light. The observation that light may play a part in regulating the intensity of carotenodermia was suggested by Klose.[154] He observed that carotenodermia was confined to areas exposed to light in infants fed carrots, and suggested that this might be due to the influence of light. He found that in a group of infants given the same food, carotenodermia developed only in those infants whose cribs were kept near the window. Interestingly, the findings of Edwards and Duntley[155] on the distribution of carotene in the skin of the whole body fairly well parallels those areas of the body exposed to light. These findings are interesting in view of the protective function of carotenoids. They might suggest a function of the carotenoid pigments normally present in skin; i.e., a certain level of protective pigments is always present to "balance" the potential photosensitizing pigments such as cytochromes and other heme pigments which are also always present. Thus, in conditions of increased light exposure, the body might increase the deposition of carotenoids in light-exposed areas; if there was sufficient carotene present in the diet, this would be seen as some carotenodermia in the light-exposed areas, explaining the findings of Klose. It is known that visible light can penetrate through the epidermis into the dermal layer.[156] As skin contains potential photosensitizers as was pointed out earlier, it would indeed seem important that exposed skin have some kind of protective mechanism to guard against photosensitization. Mathews-Roth suggests that this protective mechanism is the normally occurring carotenoid pigments.

It is intriguing to think that there might be some increase of carotenoid pigments in light-exposed areas in view of the protective function of the carotenoids. However, other than the observations of Klose,[154] there are no data to suggest that light really does play a role in increasing the deposition of carotenoids in the skin of man. Interestingly, in certain bacteria and fungi, light indeed does play a role in the deposition of carotenoids.[157] The possibility remains that carotenoids are more prevalent in light-exposed areas simply because there is more horny layer in these areas. In addition, those areas of the body exposed to light are also usually exposed to mechanical irritation, resulting in increased formation of horny layer, so it is hard to say which factor could be more important. It would seem that certain physical and probably also anatomical factors influence the development of the horny layer. Because of the affinity of carotenoids for the horny layer, the pigments are predominant in those areas of marked horny-layer deposit. If the diet is high in carotenoids, carotenodermia becomes most marked in these areas. Grof et al.[158] suggest a histamine-mediated mechanism to explain the deposit of carotenoids in these areas. Mathews-Roth suggests that perhaps such a mechanism is not necessary since the carotenoids naturally accumulate in skin cells. When physical irritation occurs there is a more rapid division of cells and resultant thickening of the horny layer, and thus accumulation of carotenoids in the area, the degree again being related to the amount of carotenoid intake. However, the Grof suggestion is appealing if there were some mechanism in the skin for the regulation of the level of photoprotector; then some mechanism would be necessary to get more carotenoids into the area susceptible to photosensitization. Clearly, more work is needed to study the factors involved in the development of carotenodermia.

Although the presence of carotenoids in skin has been established by visual and reflectance spectrophotometric observation, until recently no one had chemically isolated carotenoids from skin. Dohi and Ono[159] had observed yellow material in the horny layer of an unstained section of skin, and Grof[158] has extracted carotenoids from the horny layer scraped from the soles and knees of a carotenemic patient. For the chemical isolation of carotenoids from human skin, specimens of abdominal skin were obtained from noncarotenemic individuals at autopsy, and whole skin, epidermis, and dermis were analyzed for the presence of carotenoids.[160] Mathews-Roth et al. found that the epidermis contained more carotenoids than did the dermis. Recently, Vahlquist et al.[161] have confirmed these findings. Some authors have proposed that β-carotene, which they thought to be present in sebum, is excreted onto the surface of the skin, and rubbed into the horny layer.[132] However, Greenberg et al.[122] could not demonstrate carotenoids in sebum, but did find vitamin A. It should be remembered that carotenoids oxidize rapidly when applied to the surface of the skin. In addition, the findings of Mathews-Roth et al. that carotene is present in both layers of the skin, and also the fact that abdominal skin does not have a prominent horny layer, but does contain carotenoids, would tend to militate against this mechanism, if indeed it exists, playing a prominent part in the accumulation of carotenoids in skin. It is more likely that carotenoids are carried to all the skin cells via the circulation and extracellular fluid, and the pigments enter the cells and bind to membranes. A precedent for this entry into cells and binding to membranes is the finding that carotenoids are present in the membranes of liver-cell mitochondria.[162] Further studies of the subcellular anatomy of skin are necessary to pinpoint exactly the localization of the carotenoid pigments in the cellular organelles of the various layers.

Another area of the body in which a localized deposit of carotenoids has been found is in a pre-existing arcus senilis of the cornea. Giorgio et al.[163] reported the case of a man in whose corneas were found golden-yellow rings, at first thought to be Kayser-Fleischer rings. On more careful examination, however, these rings did not have the characteristics of Kayser-Fleischer rings, nor did the patient have elevated serum copper and ceruloplasmin, or disease of the liver or basal ganglia, features pathognomonic of Wilson's disease, in which Kayser-Fleischer rings occur exclusively. A detailed history elicited from the patient revealed that he ate a diet prominent in vegetables, and in addition consumed daily 1 to 2 qt of carrot juice. The serum carotene level at the time of examination was 900 μg%. Neither Mathews-Roth et al. nor any of their collaborating physicians have seen any patient or normal subject with a carotenoid-stained arcus senilis, but it is conceivable that as the number of patients receiving high doses of carotenoids for the treatment of photosensitivity increases, some patients may develop this phenomenon, or smaller amounts of carotenoid deposits in the cornea. Hence the physician should be aware of the existence of this phenomenon.

Another ophthalmological finding was noted in patients receiving high doses of β-carotene for EPP by deVries-de Mol et al.[164] In a study of 39 patients, they noted that treatment with β-carotene had no effect on the patients' visual acuity, dark adaptation, and color discrimination; all of these parameters being within the normal range. They found no yellow color in the lens. No mention was made of any pigmentation in the cornea. They found that most patients seemed to have "grains" in the macular region of the retina, which the authors considered not to be pathological. More recently, glistening particles have been found in the macular regions of some patients who took more than the recommended dose of canthaxanthin, which is used as an artificial sun tanning agent in some countries.[165] These granules did not seem to affect vision. Poh-Fitzpatrick and Barbera[166] examined the eyes of patients with EPP treated with β-carotene and could find no granules of any kind in the macular region.

D. Carotenoid Metabolism in Various Disease States

Levels of carotenoids in blood and skin are influenced by several disease conditions. The blood carotene level may be markedly decreased in malabsorption diseases such as steatorrhea and sprue. Workers have found that although the actual level of carotenoids in blood is not an absolute diagnostic parameter for diseases of malabsorption, the rate at which carotene is absorbed can be used as an indication of the presence of these diseases.[167-169] Hence, a carotene loading test has been developed for use as a diagnostic test for malabsorption diseases.

Studies have been conducted on carotenoid levels in various infectious diseases. Carotenodermia has been reported in polio[170] and malaria,[132] and lowered carotene absorption (or intake?) in several fevers, pneumonia, sepsis, and grippe.[5] Claussen[140] suggested that high levels of carotene might prevent infections, after having observed that infants with high carotene levels develop less severe upper respiratory infections.

High carotene levels have also been reported in castrated males[171] and in retarded children.[172] In the latter it seems that the particulate size of the food was an important matter — the smaller the particles, the more likely the children were to develop carotenodermia. These children probably do not chew their food well, so large particle food may pass rather unchanged through the digestive tract.

Jehgers[143] suggested that carotenemia may be present in liver disease. However, Ditlefsen and Stoa[173] found decreased carotene levels in alcoholics (perhaps this may be somewhat related to diet), but patients with cirrhosis and chronic hepatitis had either normal or only moderately decreased carotene levels, with much individual variation. Blood carotene levels were found to be below normal in patients with obstructive jaundice, acute hepatitis and cholecystitis, and secondary carcinoma of the liver. Adlersberg et al.[174] found that carotenoid absorption was markedly decreased in severe hepatic damage. Interestingly, Newberne et al.[175] found that the administration of carotenoids to rats protected against liver damage induced by aflatoxin.

Patients with conditions associated with hyperlipemia have in many cases been found to have concomitant carotenemia, and in some cases though not all, carotenodermia. Examples are nephrosis, diabetes, and hypothyroidism.[132,139,143] It is not clear, however, whether the hyperlipemia is the important factor in the increased carotenoid levels. In nephrosis there is some question about decreased excretion of carotenoids, but since, as noted earlier, there is disagreement as to whether carotenoids are excreted in the urine, this point is still not settled. Interestingly and for reasons unknown, patients with nephrosis do not often develop carotenodermia.

Carotenemia and carotenodermia have been observed for many years in diabetics, and the question has long been asked as to whether this phenomenon is due to some role of insulin on carotenoid metabolism. Animal experiments suggested such a role, as it was found that in alloxan diabetic rats there was markedly less conversion of β-carotene to vitamin A in the isolated intestinal loop.[176] Ralli et al.[177,178] suggested that diabetics had an impaired ability to convert β-carotene to vitamin A in the liver. However, since that time, studies of carotene tolerance curves in diabetics and normals reveal no difference between them in absorption and conversion of carotene to vitamin A.[179,180] This latter finding would suggest that there is no direct role of insulin on carotenoid metabolism. The carotenemia and carotenodermia which result are probably due to the diet containing ample amounts of vegetables and fruits which most diabetics follow.[5] It is interesting to note that on the other hand, vegetarians do not often develop carotenodermia. This is because their diet in actuality does not contain an overabundance of carotenoid-containing vegetables, as these are low in protein. The vegetarians increase the concentration of high-protein vegetables, such as various seeds and beans, in their diet. These are not very high in carotenoid content.

Carotenemia and carotenodermia are often noted in cases of hypothyroidism. For many years it was suggested that thyroid hormone had some effect on the absorption of carotenoids and their conversion to vitamin A. Studies by Walton et al.[181] suggest that absorption of carotenoids is unimpaired in hypothyroidism, but that the mechanism for the disposal of the carotenoids once they are taken into the body as such (''intrinsic'' carotenoids — those carotenoids not split into vitamin A by the intestine, but absorbed into the circulation) was interfered with. They found that the amounts of low-density lipoproteins (Sf 3-9) responsible for carriage of carotenoids in serum are significantly higher in patients with hypothyroidism than in euthyroid patients, and also that this fraction is overloaded with respect to carotene as compared to the concentration of carotene in the Sf 3-9 fraction in euthyroid blood. Absorption and utilization of pre-formed vitamin A seems to be within normal limits. They suggest that serum lipoprotein levels may be influenced by the level of protein metabolism, which is under the influence of thyroid hormone. Conversion of high Sf to low Sf classes, however, does not seem to be affected by this hormone. These authors suggest that the conversion of absorbed and circulating carotenoids to vitamin A might be dependent on complete catabolism of the Sf 3-9 lipoprotein fraction. Thus it would seem that once carotenoids are absorbed and bound to lipoprotein, there is a defect in the disposal of the carotenoids and their carrier lipoprotein, and as a result, a decrease in the conversion of some of this intrinsic carotene into vitamin A, probably in organs other than the intestinal wall. Clearly, careful carotenoid absorption and conversion studies, with concurrent lipoprotein metabolic studies on a significantly large number of patients, are needed to establish if indeed the effect of thyroid hormone is limited to the fate of the intrinsic carotenoids, and has no effect on the absorption of carotene and the conversion of some of this absorbed carotene to vitamin A in the intestinal wall.

Several workers have reported menstrual disorders and hypervitaminosis A in women ingesting large amounts of carotenoid-containing food.[182-184] Some of the women had anorexia nervosa. Several authors have reported carotenemia in anorexia nervosa, which condition also results in menstrual disorders.[185-192] Pops and

Schwabe[182] suggest that there might be decreased catabolism of β-lipoproteins in anorexia nervosa, as there is in hypothyroidism, as was found by Walton. This may indeed be true, as lipoprotein metabolism is so closely linked to protein metabolism, which may be altered in anorexia nervosa. More investigation of lipoprotein metabolism in anorexia nervosa is certainly in order. In all of these cases, one cannot rule out the effects of other components of the ingested vegetables in causing some of the "symptoms" of anorexia such as tiredness, nervousness, and depression, occasionally attributed to carotenemia due to excessive vegetable ingestion, and also found in anorexia nervosa. In the patients of Mathews-Roth et al.[7,8] taking pure β-carotene, no such symptoms have developed, and the women patients between 17 and 45 developed no menstrual abnormalities. The unusual finding in the patients of Page[182] and Robboy et al.[188] is that of abnormally high serum vitamin A. Two recent studies also reported elevated vitamin A in women overindulging in carotenoid-containing vegetables.[193,194] All previously published reports indicate that this does not happen if ingested vegetables are the only source of carotenoids. It is not clear whether or not these patients were taking any supplemental vitamins. In studies of this kind it is important to ascertain (and report) whether or not the patients are taking any supplementary vitamins containing even small amounts of vitamin A. In addition, one must be sure that appropriate corrections are made in the vitamin A assays for the cross-reacting carotene, the latter being especially important when there is carotenemia. The recently developed methods of vitamin A analysis using high-pressure liquid chromatography (HPLC) eliminate problems of cross reactions. A recent study of vitamin A levels in subjects taking high doses of β-carotene found no elevations of vitamin A.[195] Although the studies of Mathews-Roth et al. used spectrophotometric methods, they also did not find abnormal levels of vitamin A; they used appropriate corrections for carotenoid cross reaction.

Mathews-Roth and Gulbrandsen[196] also examined the carriage of carotenoids in the lipoprotein fractions of normal individuals taking carotene and of normal individuals on a regular diet. They found that the amount of carotene in each lipoprotein fraction had increased in the serum from carotenemic individuals, but that the relative increase is significant only in the low- and high-density lipoprotein fractions. There was no significant difference between the cholesterol levels of each fraction of the carotenemic individuals' and noncarotenemic individuals' serums, indicating that there is no increase in serum β-lipoprotein in carotenemia. They also found, as Krinsky et al.[197] had reported, that the majority of the pigments were in the β-lipoprotein fraction of both the carotenemic and noncarotenemic volunteers.

From the results of their studies and from the data found in the literature, it would seem that there is no marked direct toxicity from the ingestion of large amounts of pure β-carotene. On the other hand, those individuals developing carotenemia and carotenodermia by means of increased intake of carotenoid-containing foods are at some risk for the development of some hematological abnormality. They also may encounter other difficulties because of the usually accompanying inadequate diet, and occasionally because of the misdiagnosis of their "jaundice" by a physician unaware of their dietary indiscretions, and unfamiliar with carotenemia.

Because of the potential dangers associated with excessive ingestion of carotenoid-containing foods, Mathews-Roth et al. do not recommend that photosensitive patients obtain relief from their photosensitivity by increasing the intake of such foods. Pure β-carotene, either as the beadlet preparation or as crystalline β-carotene for those allergic to beadlet components, is indeed the preparation to be recommended.

VI. SUMMARY

In this chapter evidence has been presented to show that carotenoid pigments can exert a protective action against photosensitivity in man as well as in animal and plant systems. From the evidence available to date, it would seem that administration of high levels of carotenoids is effective for the amelioration of photosensitivity in the majority of patients with EPP and does not lead to toxic side effects. Larger numbers of patients with other forms of photosensitivity (PMLE, congenital porphyria, etc.) must be studied before it definitely can be determined if carotenoids are effective in ameliorating photosensitivity in these conditions.

REFERENCES

1. Mathews-Roth, M. M., Carotenoid pigments and protection against photosensitization: how studies in bacteria suggested a treatment for a human disease, *Perspect. Biol. Med.*, 28, 127, 1984.
2. Mathews-Roth, M. M., Pathak, M. A., Fitzpatrick, T. B., Harber, L. C., and Kass, E. H., Beta-carotene as a photoprotective agent in erythropoietic protoporphyria, *New Engl. J. Med.*, 282, 1231, 1970.
3. Mathews-Roth, M. M., Pathak, M. A., Fitzpatrick, T. B., Harber, L. C., and Kass, E. H., Beta-carotene as a photoprotective agent in erythropoietic protoporphyria, *Trans. Assoc. Am. Phys.*, 83, 176, 1970.
4. Karrer, P. and Jucker, E., *Carotenoids*, transl. by E. A. Braude, Elsevier, Amsterdam, 1950.
5. Goodwin, T. W., *The Comparative Biochemistry of the Carotenoids*, Chapman & Hall, London, 1952.
6. Isler, O., *Carotenoids*, Birkhauser Verlag, Basel, 1971.
7. Mathews-Roth, M. M., Pathak, M. A., Fitzpatrick, T. B., Harber, L. C., and Kass, E. H., Beta-carotene as an oral photoprotective agent in erythropoietic protoporphyria, *JAMA*, 228, 1004, 1974.
8. Mathews-Roth, M. M., Pathak, M. A., Fitzpatrick, T. B., Harber, L. C., and Kass, E. H., Beta-carotene therapy for erythropoietic protoporphyria and other photosensitivity diseases, *Arch. Dermatol.*, 113, 1229, 1977.
9. Mathews-Roth, M. M., Kass, E. H., Fitzpatrick, T. B., Pathak, M. A., and Harber, L. C., Phototesting as an objective measure of improvement in erythropoietic protoporphyria, *Arch. Dermatol.*, 115, 1381, 1979.
10. Krook, G. and Haeger-Aronson, B., Erythropoietic protoporphyria and its treatment with beta-carotene, *Acta Derm. Venereol.*, 54, 39, 1974.
11. Wennersten, G. and Swanbeck, G., Treatment of light sensitivity with carotenoids, *Acta Derm. Venereol.*, 54, 491, 1974.
12. Anderson, C. R., Erythropoietic protoporphyria, *Arch. Dermatol.*, 106, 412, 1972.
13. Beckert, E. and Metz, J., Zur beta-Karotin-Behandlung der erythropoetische Protoporphyrie, *Hautzarzt*, 25, 195, 1974.
14. Chapel, T. A., Stewart, R. H., and Webster, S. B., Erythropoietic protoporphyria — report of a case successfully treated with beta-carotene, *Arch. Dermatol.*, 105, 572, 1972.
15. Gschnait, F. and Wolff, K., Die erythropoetische Protoporphyrie, *Hautzarzt*, 25, 72, 1974.
16. Hopsu-Havu, V. K., Terho, P. E., and Hollmen, T., Erythropoietic protoporphyria — the first case in Finland, *Ann. Clin. Res.*, 5, 181, 1973.
17. Lewis, M. B., Effect of beta-carotene on serum vitamin A levels in erythropoietic protoporphyria, *Aust. J. Dermatol.*, 13, 75, 1972.
18. Moynahan, E. J. and Leppard, B., Erythropoietic protoporphyria, *Proc. R. Soc. Med.*, 66, 1085, 1973.
19. Wurker, I., Protoporphyrinamische Lichtdermatose bei vitium cordis, *Z. Haut. Geschlechtskr.*, 47, 1006, 1072.
20. Ippen, H., Goerz, G., and Botsch, C., Photoprotection in porphyrias and the role of phlebotomy, in *Sunlight and Man*, Pathak, M. A., Ed., University of Tokyo Press, Tokyo, 1974, 658.
21. Wiskemann, A., Topical and systemic photoprotection against porphyrins, in *Sunlight and Man*, Pathak, M. A., Ed., University of Tokyo Press, Tokyo, 1974, 669.

22. Baart de la Faille, H., Suurmond, D., Went, L. N., van Steveninck, J., and Schothorst, A. A., Beta-carotene as a treatment for photohypersensitivity due to erythropoietic protoporphyria, *Dermatologica,* 145, 389, 1972.
23. Zaynoun, S. T., Hunter, J. A. A., Darby, F. J., Zarembski, P., Johnson, B. E., and Frain-Bell, W., The treatment of erythropoietic protoporphyria: experience with beta-carotene, *Br. J. Dermatol.,* 97, 663, 1977.
24. Haeger-Aronsen, B., Krook, G., and Abdulla, M., Oral carotenoids for photohypersensitivity in patients with erythropoietic protoporphyria, polymorphous light eruption and lupus erythematodes discoides, *Int. J. Dermatol.,* 18, 73, 1979.
25. Eales, L., The effects of canthaxanthin on the photo-cutaneous manifestations of porphyria, *S. Afr. Med. J.,* 54, 1050, 1978.
26. Keller, J. and Hornstein, O. P. Erythropoietische Protoporphyrie, *Dtsch. Med. Wochenschr.,* 108, 862, 1983.
27. Fergusen, J., Addo, H. J., and Cochrane, T., *Br. J. Dermatol.,* 107, (Suppl. 22), 61, 1982.
28. Meigel, W., Schaeg, G., and Goerz, G., Atypisch lokalisierte bleibende Hautveranderungen bei der erythropoieteschen Protoporphyrie (EPP) — histologische und electronmikroscopische Befunde, *Z. Hautkr.,* 56, 489, 1980.
29. MacDonald, D. M., Germain, D., and Perrot, H., The histopathology and ultrastructure of liver disease in erythropoietic protoporphyria, *Br. J. Dermatol.,* 104, 7, 1981.
30. Thomson, K., Schmidt, H., and Fischer, A., Beta-carotene in erythropoietic protoporphyria — 5 years' experience, *Dermatologica,* 159, 82, 1979.
31. Corbett, M. F., Herxheimer, A., Magnus, I. A., Ramsay, C. A., and Kobza-Black, A., The long-term treatment with beta-carotene in erythropoietic protoporphyria: a controlled trial, *Br. J. Dermatol.,* 97, 655, 1977.
32. Shafrir, A., Comments on a case of erythropoietic protoporphyria, *Proc. R. Soc. Med.,* 70, 574, 1977.
33. Seip, M., Thune, P. O., and Eriksen, L., Treatment of photosensitivity in congenital erythropoietic prophyria with beta-carotene, *Acta Derm Venereol.,* 54, 239, 1974.
34. Sneddon, I. B., Congenital prophyria, *Proc. R. Soc. Med.,* 67, 593, 1974.
35. Baart de la Faille, H. and Remme, J. J., Congenital prophyria (Gunther), *Br. J. Dermatol.,* 101, 224, 1979.
36. Storck, H. and Kaufmann, J., Kongenitale erythropoietische Porphyrie, morbus Gunther, *Dermatologica,* 157, 323, 1978.
37. Maleville, J., Babin, P., Mollard, S., Martin, C., Nordmann, Y., and Guillet, G., Prophyrie erythropoietique congenitale de Gunther et carotenoides. Essai therapeutique de quatres ans, *Ann. Dermatol. Venereol. (Paris),* 109, 883, 1982.
38. Jung, E. C., Porphyria erythropoietica congenita Gunther, *Dtsch. Med. Wochenschr.,* 102, 277, 1977.
39. Mathews-Roth, M. M., Haining, R. G., and Kinney, T. R., Beta-carotene treatment of congenital porphyria, *Am. J. Dis. Child.,* 131, 366, 1977.
40. Swanbeck, G. and Wennersten, G., Treatment of polymorphous light eruption with beta-carotene, *Acta Derm. Venereol.,* 52, 462, 1972.
41. Jansen, C. T., Beta-carotene treatment of polymorphous light eruption, *Dermatologica,* 149, 363, 1974.
42. Nordlund, J. J., Klaus, S. N., Mathews-Roth, M. M., and Pathak, M. A., New therapy for polymorphous light eruption, *Arch. Dermatol.,* 108, 710, 1973.
43. Thune, P., Chronic polymorphic light eruption: particular wave bands and the effect of carotene therapy, *Acta Derm. Venereol.,* 56, 127, 1976.
44. Parrish, J. A., Levine, M. J., Morison, W. L., Gonzalez, E., and Fitzpatrick, T. B., Comparison of PUVA and beta-carotene in the treatment of polymorphous light eruption. *Br. J. Dermatol.,* 100, 187, 1979.
45. Suhonen, R. and Plosila, M., The effect of beta-carotene in combination with canthaxanthin, Ro 8-8427 (Phenoro) in the treatment of polymorphous light eruption, *Dermatologica,* 163, 172, 1981.
46. Corbett, M. F., Hawk, J. L. M., Herxheimer, A., and Magnus, I. A., Controlled therapeutic trials in polymorphous light eruption, *Br. J. Dermatol.,* 107, 571, 1982.
47. Kobza, A., Ramsay, C. A., and Magnus, I. A., Oral beta-carotene therapy in actinic reticuloid and solar urticaria, *Br. J. Dermatol.,* 88, 157, 1973.
48. Aratari, E., Virno, G., Desirello, G., and Nazzari, G., Associazione di beta-carotene e cantaxantina nel trattamento del reticuloideattinico, *Acta Vitaminol. Enzymol.,* 4, 319, 1982.
49. Bickers, D. R., Demar, L. K., DeLeo, V., Poh-Fitzpatrick, M. B., Aronberg, J. M., and Harber, L. C., Hydroa vaccineforme, *Arch. Dermatol.,* 114, 1193, 1978.
50. Husquinet, H., Noirfalise, A., and Parent, M.-Th., Porphyria variegata, *J. Genet. Hum.,* 26, 367, 1978.

51. Muhlbauer, J. E., Pathak, M. A., Tischler, P. V., and Fitzpatrick, T. B., Variegate porphyria in New England, *JAMA*, 247, 3095, 1982.
52. Lagerholm, B. and Wennersten, G., Mucinosis follicularis provoked by light exposure, *Acta Derm. Venereol.*, 59, 153, 1979.
53. Fisher, D. A., Quinidine photosensitivity, *Arch. Dermatol.*, 120, 298, 1984.
54. Bendes, J. H., Heliotherapy in tuberculosis, *Minn. Med.*, 9, 112, 1926.
55. Sandler, A. S., Carotene in prophylactic pediatrics, *Arch. Pediatr.*, 52, 391, 1935.
56. Mathews-Roth, M. M., Pathak, M. A., Parrish, J., Fitzpatrick, T. B., Kass, E. H., Toda, K., and Clemans, W., A clinical trial of the effects of oral beta-carotene on the responses of human skin to solar radiation, *J. Invest. Dermatol.*, 59, 349, 1972.
57. Anyan, W. P., Carotenemia, *Arch. Dermatol.*, 105, 130, 1972.
58. Sporn, M. B., Dunlop, M. M., Newton, D. L., and Smith, J. M., Prevention of chemical carcinogenesis by vitamin A and its synthetic analogs (retinoids), *Fed. Proc.*, 35, 1332, 1976.
59. Sporn, M. and Newton, D., Chemoprevention of cancer with retinoids *Fed. Proc.*, 38, 2528, 1979.
60. Boutwell, R., *Selected Abstracts on Vitamin A in Cancer Biology*, International Cancer Research Data Bank, National Cancer Institute, Bethesda, Md., 1979.
61. Bolag, W. and Matter, A., From vitamin A to retinoids in experimental and clinical oncology: achievements, failures and outlook, *Ann. N.Y. Acad. Sci.*, 359, 9, 1981.
62. Dorogokupla, A. C., Troitzkaia, E. G., Adilgireieva, L. K., Postolnikov, S. L., and Chekrygina, Z. P., Effect of carotene on the development of induced tumors, *Zdravookhr. Kaz.*, 10, 32, 1973.
63. Epstein, J. H., Effects of beta-carotene on UV-induced cancer formation in the hairless mouse skin, *Photochem. Photobiol.*, 25, 211, 1973.
64. Mathews-Roth, M. M., Carotenoid pigments as anti-tumor agents, in *Current Chemotherapy and Infectious Diseases*, Nelson, J. D. and Grassi, C., Eds., American Society for Microbiology, Washington, D.C., 1980, 1503.
65. Mathews-Roth, M. M., Antitumor activity of beta-carotene, canthaxanthin and phytoene, *Oncology*, 39, 33, 1982.
66. Mathews-Roth, M. M., Carotenoid pigment administration and delay in development of UV-B-induced tumors, *Photochem. Photobiol.*, 37, 509, 1983.
67. Mathews-Roth, M. M., Wilson, T., Fujimori, E., and Krinsky, N. I., Carotenoid chromophore length and protection against photosensitization, *Photochem. Photobiol.*, 19, 217, 1974.
68. Santamaria, L., Bianchi, A., Arnabaldi, A., Andreoni, L., and Bermond, P., Dietary carotenoids block photocarcinogenic enhancement by benzo(a)pyrene and inhibit its carcinogenesis in the dark, *Experientia*, 39, 1043, 1983.
69. Seifter, E., Rettura, G., Padawer, J., Strattford, F., Goodwin, P., and Levinson, S. M., Regression of C3HBA mouse tumor due to X-ray therapy combined with supplementary beta-carotene or vitamin A, *J. Natl. Cancer Inst.*, 71, 409, 1981.
70. Seifter, E., Rettura, G., Padawer, J., and Levinson, S. M., Moloney murine sarcoma virus tumors in CBA/J mice: chemopreventive and chemotherapeutic actions of supplemental beta-carotene, *J. Natl. Cancer. Inst.*, 68, 835, 1982.
71. Rettura, G., Strattford, F., and Seifter, E., Prophylactic and therapeutic actions of supplemental beta-carotene in mice inoculated with C3HBA adenocarcinoma cells: lack of therapeutic action of supplemental ascorbic acid, *J. Natl. Cancer Inst.*, 69, 73, 1982.
72. Seifter, E., Rettura, G., and Levinson, S. M., Supplemental beta-carotene (BC): prophylactic action against 7,12 dimethylbenz(a)-anthracene (DMBA) carcinogenesis, *Fed. Proc.*, 43, 662, 1984.
73. Shamberger, R. J., Inhibitory effect of vitamin A on carcinogenesis, *J. Natl. Cancer Inst.*, 47, 667, 1971.
74. Gainer, J. L., Wallis, D. A., and Jones, J. R., The effect of crocetin on skin papillomas and Rous sarcoma, *Oncology*, 11, 222, 1976.
75. Mathews-Roth, M. M., Effect of crocetin on experimental skin tumors in hairless mice, *Oncology*, 39, 362, 1982.
76. Shekelle, R. B., Liu, S., Raynor, W. J., Lepper, A. H., Maliza, M., and Rossof, A. H., Dietary vitamin A and risk of cancer in the Western Electric study, *Lancet*, 2, 1185, 1981.
77. Mettlin, C. and Graham, S., Dietary risk factors in human bladder cancer, *Am. J. Epidemiol.*, 110, 255, 1979.
78. Phillips, R. L., Role of life-style and dietary habits in risk of cancer among Seventh-Day Adventists, *Cancer Res.*, 35, 3513, 1975.
79. Bjelke, E., Dietary vitamin A and human lung cancer, *Int. J. Cancer*, 15, 561, 1975.
80. Hirayama, T., Diet and cancer, *Nutr. Cancer*, 1, 67, 1979.
81. MacLennan, R., DaCosta, J., Day, N. L., Law, C. H., Ng, Y. K., and Shanmugarainan, K., Risk factors for lung cancer in Singapore Chinese. A population with high female incidence rates. *Int. J. Cancer*, 20, 854, 1977.

82. Gregor, A., Lee, P. M., and Roe, F. J. C., Comparison of dietary history of lung cancer cases and controls, with specific reference to vitamin A, *Nutr. Cancer,* 2, 93, 1980.
83. Graham, S., Mettlin, C., Marshall, J., Priore, R., and Rzepka, T., Diet, tobacco and alcohol in the epidemiology of cancer of the larynx, *Am. J. Epidemiol.,* 112, 422, 1980.
84. Mettlin, C., Graham, S., Priore, R., Marshall, J., and Swanson, M. M., Diet and cancer of the esophagus, *Nutr. Cancer,* 2, 143, 1981.
85. Cook-Mozaffani, P., Esophageal cancer, *Br. J. Cancer,* 39, 293, 1979.
86. Bjelke, E., Case-control study of cancer of the colon, rectum and stomach, in *Oncology 1970 — Proceedings of the Tenth International Cancer Congress,* Vol. 5, Clark, R. L., Cumley, R. W., McCay J. E., and Copeland, M. M., Eds., Year Book Medical Publishing, Chicago, 1971, 320.
87. Paymaster, J. C., Sanghri, L. D., and Gangadharan, R., Cancer in the gastrointestinal tract in western India, *Cancer,* 21, 279, 1968.
88. Haenszel, W., Kurihara, M., and Segi, M., Stomach cancer among Japanese in Hawaii, *J. Natl. Cancer Inst.,* 49, 969, 1972.
89. Ziegler, R., Morris, L. E., Blout, W. J., Pattern, L. M., Hoover, R., and Fraumeri, J., Esophageal cancer among black men in Washington, D.C. II. Role of nutrition, *J. Natl. Cancer Inst.,* 67, 1199, 1981.
90. Graham, S., Haughey, B., Marshall, J., Price, R., Byers, T., and Rzepka, T., Diet in the epidemiology of cancer of the prostate gland, *J. Natl. Cancer Inst.,* 70, 687, 1983.
91. Stich, H. F., Rosin, M. D., and Vallegera, M. O., Reduction with vitamin A and beta-carotene administration of the proportion of micronucleated buccal mucosal cells in Asian betel nut and tobacco chewers, *Lancet,* 2, 1204, 1984.
92. Correa, P., Cuello, C., Fajardo, L. F., Haenszel, W., Bolanos, O., and deRamirez, B., Diet and gastric cancer: nutrition survey in a high-risk area, *J. Natl. Cancer Inst.,* 70, 673, 1983.
93. Stahelein, H. B., Buess, E., Rosel, F., Widmer, L. K., and Brubacher, G., Vitamin A, cardiovascular risk factors and mortality, *Lancet,* 1, 394, 1982.
94. Smith, P. G. and Jick, H., Cancer among users of preparations containing vitamin A, *Cancer,* 42, 808, 1978.
95. Graham, S., Dayal, H., Swanson, M., Mittelman, A., and Wilkinson, G., Diet in the epidemiology of cancer of the colon and rectum, *J. Natl. Cancer Inst.,* 61, 709, 1978.
96. Willet, W. C., Polk, B. F., Underwood, B. A., Stampfer, M. J., Pressel, S., Rosner, B., Taylor, J. W., Schneider, K., and Hames, C. G., Relation of serum vitamins A and E and carotenoids to the risk of cancer, *New Engl. J. Med.,* 310, 430, 1984.
97. Basu, T. K., Donaldson, D., Jenner, M., Williams, D C., and Sakula, A., Plasma vitamin A in patients with bronchial carcinoma, *Br. J. Cancer,* 33, 119, 1976.
98. Atukorala, S., Basu, T. K., Dickerson, J. W. T., Donaldson, D., and Sakula, A., Vitamin A, zinc and lung cancer, *Br. J. Cancer,* 40, 927, 1979.
99. Cohen, M. H., Primack, A., Broder, L. E., and Williams, L. R., Vitamin A serum levels and dietary vitamin A intake in lung cancer patients, *Cancer Lett.,* 4, 51, 1977.
100. Ibrahim, K., Jafary, N. A., and Zuberi, S. J., Plasma vitamin A and carotene levels in squamous cell carcinoma of oral cavity and oro-pharynx, *Clin. Oncol.,* 3, 203, 1977.
101. Wald, N., Idle, M., Boreham, J., and Bailey, A., Low serum vitamin A and subsequent risk of cancer, *Lancet,* 2, 813, 1980.
102. Kark, J. D., Smith, A. H., Switzer, B. R., and Hams, C. G., Serum vitamin A (retinol) and cancer incidence in Evans County, Georgia, *J. Natl. Cancer Inst.,* 66, 7, 1981.
103. Capel, I. D. and Williams, D. C., Relation between zinc and vitamin A in cancer patients, *I.C.R.S. Med. Sci.,* 7, 361, 1979.
104. Abels, J. C., Gorham, A. T., Park, G. T., and Rhoads, C. P., Metabolic studies in patients with cancer of the gastrointestinal tract. I. Plasma vitamin A levels in patients with malignant neoplastic disease, particularly of the gastrointestinal tract, *J. Clin. Invest.,* 20, 749, 1941.
105. Clifford, P., Carcinogens in the nose and throat: nasopharyngeal carcinoma in Kenya, *Proc. R. Soc. Med.,* 65, 682, 1972.
106. Wahi, P. N., Kehar, U., and Lahin, B., Factors influencing oral and oropharyngeal cancer in India, *Br. J. Cancer,* 19, 642, 1965.
107. Wahi, P. N., Bodhke, R. R., Aroza, S., and Srivastava, M. C., Serum vitamin A studies in leukoplakia and cancer of the oral cavity, *Indian J. Pathol. Bacteriol.,* 5, 10, 1962.
108. Peto, R., Doll, R., Buckley, J. E., and Sporn, M. B., Can dietary beta-carotene materially reduce human cancer rates?, *Nature (London),* 290, 201, 1981.
109. Thompson, S. Y., Occurrence, distribution and absorption of provitamins A, *Proc. Nutr. Soc.,* 24, 136, 1965.
110. Arroyave, G., Wilson, D., Mendez, J., Behar, M., and Scrimshaw, N., Serum and liver vitamin A and lipids in children with severe protein malnutrition, *Am. J. Clin. Nutr.,* 9, 180, 1961.

111. Berger, S., Recheigl, M., Loosli, J. K., and Williams, H. Y., Protein quality and carotenoid utilization, *J. Nutr.*, 77, 174, 1962.
112. Deshmukh, D. S. and Ganguly, J., Effect of dietary protein content on the internal conversion of beta-carotene to vitamin A in rats, *Indian J. Biochem.*, 1, 204, 1964.
113. Jagannathan, S. N. and Patwardhan, V. N., Dietary protein in vitamin A metabolism. I. Influence of level of dietary protein on the utilization of orally fed preformed vitamin A and beta-carotene in the rat, *Indian J. Med. Res.*, 48, 775, 1960.
114. Wood, R. D., Effect of adding nitrate and nitrite to drinking water on the utilization of carotene by growing swine, *J. Anim. Sci.*, 26, 510, 1967.
115. Goodman, D. S. and Huang, H. S., Biosynthesis of vitamin A with rat intestinal enzymes, *Science*, 149, 879, 1966.
116. Olson, J. A., The absorption of beta-carotene and its conversion into vitamin A, *Am. J. Clin. Nutr.*, 9, 1, 1961.
117. Glover, J. and Redfern, E. R., The mechanism of the transformation of beta-carotene into vitamin A in vivo, *Biochem. J.*, 58, xv, 1954.
118. Glover, J., The conversion of beta-carotene into vitamin A, *Vitam. Horm.*, 18, 371, 1960.
119. Goodman, D. S., Metabolism of beta-carotene and vitamin A, *Prog. Biochem. Pharmacol.*, 3, 487, 1967.
120. Olson, J. A., Biosynthesis and metabolism of carotenoids and related molecules, *J. Lipid Res.*, 5, 281, 1964.
121. Cornbleet, T. and Greenberg, R., Conversion of carotene to vitamin A by sebaceous glands, *Arch. Dermatol.*, 76, 431, 1957.
122. Greenberg, R., Cornbleet, T., and Jeffay, A. L., Accumlation and excretion of vitamin A-like fluorescent material by sebaceous glands after the oral feeding of various carotenoids, *J. Invest. Dermatol.*, 32, 599, 1959.
123. Greenberg, R., Cornbleet, T., and Demovsky, R., Conversion of carotene to vitamin A by sebaceous glands, *Arch. Dermatol.*, 76, 17, 1957.
124. Austern, B. M. and Gawienowski, A. M., In vitro biosynthesis of beta-carotene by bovine corpus luteum tissue, *Lipids*, 4, 227, 1969.
125. Hashimoto, H., Carotenoid pigmentation of the skin resulting from a vegetarian diet, *JAMA*, 78, 1111, 1922.
126. Hess, A. F. and Myers, V. C., Carotenemia: a new clinical picture, *JAMA*, 73, 1743, 1919.
127. Almond, S. and Logan, R. F. L., Carotinaemia, *Br. Med. J.*, 2, 235, 1942.
128. Niemann, C. and Klein Obbink, H. J., Biochemistry and pathology of hypervitaminosis A, *Vitam. Horm.*, 12, 69, 1954.
129. Ostwald, R. and Briggs, G. M., Toxicity of the vitamins, in *Toxicants Occurring Naturally in Foods*, edited by the Food Protection Commission, National Science Foundation, Washington, D. C., 1967, 183.
130. Mathews-Roth, M. M., Anemia in erythropoietic protoporphyria, *JAMA*, 230, 824, 1974.
131. Gjerlow, J., Granulocytopeni som folge au karotinemi, *Tidsskr. Nor. Legeforen.*, 86, 33, 1966.
132. Josephs, H. W., Hypervitaminosis A and carotenemia, *Am. J. Dis. Child.*, 67, 33, 1944.
133. Keating, J. P., Lell, M. E., Strauss, A. W., Zarkowsky, H., and Smith, G. E., Infantile methemoglobinemia caused by carrot juice, *New Engl. J. Med.*, 288, 824, 1973.
134. Hoch, H., Effect of prolonged administration of carotene in the form of vegetables on the serum carotene and vitamin A levels in man, *Biochem. J.*, 37, 430, 1943.
135. Kon, S. and Mawson, M. E., Carotenoids in breast milk, Special Rep. Ser. No. 269, Medical Research Council, London, 1950.
136. Lesher, M., Brody, J. K., Williams, H. H., and May, I. G., Human milk studies, *Am. J. Dis. Child.*, 70, 182, 1945.
137. Thomson, M. L., Carotenemia in a suckling, *Arch. Dis. Child*, 18, 112, 1973.
138. DeBuys, L. R., Some clinical experiences with carotenemia, *Trans. Miss. State Med. Assoc.*, p. 133, 1923.
139. Claussen, S. W. and McCoord, A. B., The carotenoids and vitamin A of the blood, *J. Pediatr.*, 13, 635, 1938.
140. Claussen, S. W., Carotenemia and resistance to infection, *Trans. Am. Pediatr. Soc.*, 43, 27, 1931.
141. Mathews-Roth, M. M., Abraham, A. A., and Gabuzda, G., Beta-carotene content of certain organs from two patients receiving high doses of beta-carotene, *Clin. Chem.*, 22, 922, 1976.
142. Baeltz, I., as cited in Miyake I., Über Aurantiasis cutis Baeltz, *Arch. Dermatol. Syphilol.*, 147, 184, 1924.
143. Jehgers, H., Skin changes of nutritional origin, *N. Engl. J. Med.*, 228, 678, 1943.
144. Cohen, L., Observations on carotenemia, *Ann. Intern. Med.*, 48, 219, 1958.
145. Castle, W. E., Linkage relations of yellow fat in rabbits, *Proc. Natl. Acad. Sci. U.S.A.*, 19, 947, 1933.

146. Mathews-Roth, M. M., Hummel, D., and Crean, C., The carotenoid content of various organs of animals fed large amounts of beta-carotene, *Nutr. Rep. Int.*, 16, 419, 1977.
147. Reich, P., Shwachman, H., and Craig, J. M., Lycopenemia — a variant of carotenemia, *N. Engl. J. Med.*, 262, 263, 1968.
148. McLaren, D. S. and Zekian, B., Failure of enzymic cleavage of beta-carotene, *Am. J. Dis. Child.*, 121, 278, 1971.
149. Molholm-Hansen, J., Xanthosis cutis og hyperkarotenoidaemi, *Nord. Med.*, 22, 385, 1965.
150. Sharvill, D. E., Familial hypercarotenemia and hypovitaminosis A, *Proc. R. Soc. Med.*, 63, 605, 1970.
151. Urbach, C., Effect of individual vitamins A, C, E and carotene administered at high levels on their concentration in the blood, *Exp. Med. Surg.*, 10, 7, 1952.
152. Gandy, D. T., Carotenemia and carotenodermia, *South. Med. J.*, 8, 444, 1935.
153. Gandy, D. T., Carotenemia and carotenodermia, *Med. Rec. Ann. (Houston)*, 47, 274, 1964.
154. Klose, E., Hautverfarbung bei Sauglingen und Kleinkindern infolge der Nahrung, *Munch. Med. Wochenschr.*, 66, 419, 1919.
155. Edwards, E. A. and Duntley, S. Q., Pigments and color of living human skin, *Am. J. Anat.*, 63, 1, 1939.
156. Everett, M. A., Yeargers, E., Sayre, R. M., and Olson, R. L., Penetration of epidermis by ultraviolet light, *Photochem. Photobiol.*, 5, 533, 1966.
157. Batra, P., Mechanism of light-induced carotenoid synthesis in non-photosynthetic plants, in *Photophysiology*, Giese, A. C., Ed., Academic Press, New York, 1971, 47.
158. Grof, P., Bodzay, J., and Kovaks, A., Carotenodermia, *Acta Med. Acad. Sci. Hung.*, 24, 129, 1967.
159. Dohi, D. and Ono, T., as cited in Hashimoto, H., Carotenoid pigmetation of the skin resulting from a vegetarian diet, *JAMA*, 78, 1111, 1922.
160. Lee, R., Mathews-Roth, M. M., Pathak, M. A., and Parrish, J. A., The detection of carotenoid pigments in human skin, *J. Invest. Dermatol.*, 64, 175, 1974.
161. Vahlquist, A., Lee, J. B., Michaelson, G., and Rollman, O., Vitamin A in human skin. II. Concentrations of carotene, retinol and dehydroretinol in various components of normal skin, *J. Invest. Dermatol.*, 79, 94, 1982.
162. Green, D. E. and Hatefi, Y., The mitochondrion and biochemical machines, *Science*, 133, 13, 1961.
163. Giorgio, A. L., Cartwright, G. E., and Wintrobe, M. M., Pseudo Kayser-Fleischer rings, *Arch. Intern. Med.*, 113, 817, 1964.
164. deVries-de Mol, E. C., Went, L. M., and Volker-Dieben H. J., Farnsworth-Munsell 100-hue results in a series of patients with long-standing therapeutic carotenemia, *Med. Probl. Ophthalmol.*, 13, 349, 1974.
165. Cortin, P., Corriveau, L. A., Rousseau, A. P., Tardif, Y., Malenfant, M., and Biurdreault, G., Maculopathie en paillettes d'or, *Can. J. Ophthalmol.*, 17, 103, 1982.
166. Poh-Fitzpatrick, M. B. and Barbera, L. G., Absence of crystalline retinopathy after long-term therapy with beta-carotene, *J. Am. Acad. Dermatol.*, 11, 111, 1984.
167. Feldman, E. and Adlersberg, D., Response to carotenoid loading in malabsorption states, *Am. J. Med. Sci.*, 238, 733, 1959.
168. Kasper, H. and Hospach, R., Der diagnostische der Wert der Vitamin A und Carotin-bestimmung in Serum bei Maldigestion und Malabsorption, *Dtsch. Med. Wochenschr.*, 99, 198, 1974.
169. Onstad, G. R., Carotene absorption — a screening test for steatorrhea, *JAMA*, 221, 677, 1972.
170. Linn, H. W., Xanthosis cutis in poliomyelitis, *Med. J. Aust.* 1, 581, 1954.
171. Edwards, E. A., Hamilton, J. B., Duntly, S. Q., and Hubut, G., Cutaneous vascular and pigmentary changes in castrate and eunuchoid men, *Endocrinology*, 28, 119, 1941.
172. Patel, H., Dunn, H. G., Tischler, B., McBurney, A. K., and Hach, E., Carotenemia in mentally retarded children — incidence and etiology, *Can. Med. Assoc. J.*, 108, 848, 1973.
173. Ditlefsen, E.-M. L. and Stoa, K. F., Vitamin A and carotenoid levels in blood serum with special reference to values observed in disease of the skin, *Scand. J. Clin. Lab. Invest.*, 6, 210, 1954.
174. Adlersberg, D., Kann, S., Maurer, A. P., Newerly, K., Winternitz, W., and Sobotka, H., Studies on serum carotene in man, *Am. J. Dig. Dis.*, 16, 333, 1949.
175. Newberne, P., Chan, W.-C., and Rogers, A., Influence of light, riboflavin and carotene on the response of rats to the acute toxicity of aflatoxin and monocrotaline, *Toxicol. Appl. Pharmacol.*, 28, 200, 1974.
176. Rosenberg, A. and Sobel, A. E., In vitro conversion of carotene to vitamin A in alloxan diabetes, *Arch. Biochem. Biophys.*, 44, 326, 1953.
177. Pariente, A., Present, C., and Ralli, E. P., A case of carotenemia and diabetes mellitus, with necropsy report and analysis of liver for carotene, vitamin A, total fat and cholesterol, *Am. J. Med. Sci.*, 192, 365, 1936.

178. Ralli, E. P., Brandaleone, H., and Mandelbaum, T., Studies on the effect of the administration of carotene and vitamin A in patients with diabetes mellitus. I. Effect of the oral administration of carotene on the blood carotene and cholesterol of diabetic and normal individuals, *J. Lab. Clin. Med.,* 20, 1266, 1934—1935.
179. Murrill, W. A. Horton, P. B., Leiberman, E., and Newburgh, L. H., Vitamin A and carotene. II. Vitamin A and carotene metabolism in diabetics and normals, *J. Clin. Invest.,* 20, 395, 1941.
180. Ramachandran, K., Beta-carotene metabolism in diabetes, *Indian J. Med. Res.,* 61, 1831, 1973.
181. Walton, K. W., Campbell, D. A., and Tonks, D., Significance of alterations in serum lipids in thyroid dysfunction. I. The relation between serum lipoproteins, carotenoids and vitamin A in hypothyroidism and thyrotoxicosis, *Clin. Sci.,* 29, 199, 1965.
182. Page, S. W., Golden ovaries, *Aust. N. Z. J. Obstet. Gynecol.,* 11, 32, 1971.
183. Kemmann, E., Pasquale, S. A., and Kaff, R. S., Amenorrhea associated with carotenemia, *JAMA,* 249, 926, 1983.
184. Bilmoria, S., Keczkes, K., Williamson, D., and Rowell, N. R., Hypercarotenemia in weight watchers, *Clin. Exp. Dermatol.,* 4, 331, 1979.
185. Crisp, A. H. and Stonehil, E., Hypercarotencmia as a symptom of weight phobia, *Postgrad. Med. J.,* 43, 721, 1967.
186. Dally, P. J., Carotenemia occurring in a case of anorexia nervosa, *Br. Med. J.,* 2, 1333, 1959.
187. Pops, M. A. and Schwabe, A. D., Hypercarotenemia in anorexia nervosa, *JAMA,* 205, 121, 1968.
188. Robboy, M. S., Sato, A. S., and Schwabe, A. D., The hypercarotenemia in anorexia nervosa: a comparison of vitamin A and carotene levels in various forms of menstrual dysfunction and cachexia, *Am. J. Clin. Nutr.,* 27, 362, 1974.
189. Silverman, J. A., Anorexia nervosa: clinical observations in a successful treatment plan, *J. Pediatr.,* 84, 68, 1974.
190. Bhanji, S. and Mattingly, D., Anorexia nervosa. Some observations on dieters and vomiters, cholesterol and carotene, *Br. J. Psychiatr.,* 39, 238, 1981.
191. Wells, J. D., Dieting and carotenemia — sometimes a cause and effect, *J. Ark. Med. Sci.,* 80, 396, 1984.
192. Frumar, A. M., Meldrum, D. R., and Judd, H. L., Hypercarotenemia in hypothalamic amenorrhea, *Fertil. Steril.,* 32, 261, 1979.
193. Monk, B. E., Metabolic carotenemia, *Br. J. Dermatol.,* 106, 485, 1982.
194. Mathus-Vliegan, E. M. H. and Tytgat, G. M. J., Biological fate of the vitamin A transporting complex after exercise and dietary intake, *Digestion,* 27, 116, 1983.
195. Willett, W. C., Stampfer, M. J., Underwood, B. A., Taylor, J. O., and Hennekens, C. H., Vitamins A, E and carotene: effects of supplementation on their plasma levels, *Am. J. Clin. Nutr.,* 38, 559, 1983.
196. Mathews-Roth, M. M. and Gulbrandsen, C. L., Transport of beta-carotene in serum of individuals with carotenemia, *Clin. Chem.,* 20, 1578, 1974.
197. Krinsky, N. I., Cornwell, D. G., and Oncley, J. L., Transport of vitamin A and carotene in human plasma, *Arch. Biochem. Biophys.,* 73, 233, 1958.
198. Mathews-Roth, M. M., Unpublished observations.
199. Sestili, M. A., Personal communication.

Chapter 4

PHOTOCHEMOTHERAPY FOR PSORIASIS USING THE FUROCOUMARINS

Michael K. Reusch and Philip C. Hanawalt

TABLE OF CONTENTS

I. INTRODUCTION

Psoriasis is a benign skin disease characterized by epidermal hyperplasia, with a prevalence of about 1 to 3% in the general population. Encompassing a wide range of clinical patterns, from predominantly inflammatory pustular lesions to chronic, stable plaques, and varying from localized persistent lesions to total body involvement, psoriasis ranges in consequences from a minor nuisance to severe incapacitation for patients. Because of its possible early onset, its lifelong recurring character, its unpredictable course, and the lack of more than palliative treatments, psoriasis constitutes one of the major challenges in dermatology.[1]

Photochemotherapy using photosensitizing psoralen derivatives and UVA (usually referred to by its acronym PUVA) has become one of the most effective and widely used therapeutic protocols for psoriasis during the past decade. Introduced in 1974 as a systemic mode of treatment for severe, widespread psoriasis, it soon gained worldwide acceptance due to its high effectiveness and patient compliance[2] — over 100,000 patients have been treated with PUVA so far. Its usefulness in an astounding variety of skin diseases, but especially psoriasis, has made PUVA an almost indispensable tool in everyday dermatological practice and it has also spurred a dramatic general interest in photomedicine in recent years. It has concomitantly led to basic research that has considerably furthered our understanding of the photobiology of DNA and cellular responses to photoadducts in the genetic material.

However, in spite of its widespread use, its long history in folk medicine, and the spectacular recent advances in photobiology, PUVA shares the fate of most other skin therapeutics: the precise mechanism of action in vivo is not yet known. It also shares several other features with most therapeutics for chronic skin conditions: it is only a palliative treatment, and because of the chronicity of its application, long-term adverse side effects deserve special attention and concern.

In the following, the authors will briefly explore the historical roots of the PUVA regimen and recapitulate the current views on its mode of action — focusing mainly on interactions with DNA — before reviewing its therapeutic use in psoriasis. The variety of different treatment protocols will be discussed in some detail. The possible hazardous side effects of PUVA will be considered, and potential future improvement of this therapy, based upon our current understanding of its mode of action, will be presented.

II. HISTORICAL BACKGROUND AND CHEMISTRY OF FUROCOUMARINS

Furocoumarins (or furanocoumarins) are a group of tricyclic compounds that occur naturally in a large number of plants.[3] If coumarin is fused with furan so that the oxygens are oriented in the same direction, the unsubstituted heterocyclic linear compound is called psoralen. Substituted derivates of psoralen such as 8- and 5-methoxypsoralen and 4,5′,8-trimethylpsoralen are of principal interest in dermatology. A large number of other naturally occurring and newly synthesized compounds, some with angular fusion between furan and coumarin (such as angelicin) are currently being investigated but have not yet gained significant use in therapy.

The photosensitizing furocoumarins elicit two very prominent reactions in the skin: (1) phototoxic dermatitis of varying degree, and (2) hyperpigmentation.

Although both reactions bear resemblance to the erythema and hyperpigmentation arising after exposure of skin to UV irradiation, they differ considerably when examined more closely. The time course and steepness of the erythema is very different from that elicited by UVB. The PUVA-induced erythema appears later (36 to 96 hr) and lasts longer than does the UVB-induced erythema. Also, the phototoxic response to dose increments is much steeper for PUVA than for UVB. Therefore, PUVA has a much narrower therapeutic window, and dose increments have to be carefully tailored. Finally, the hyperpigmentation following PUVA is darker and lasts longer compared to the UVB-induced tanning response.

Phytophotodermatitis due to contact with plants containing psoralens and subsequent sunlight exposure has long been common knowledge in traditional medicine. The use of photo-sensitizing psoralens and sunlight for medical purposes has been documented for a period of 3400 years.[4] In the Atharva Veda, dating back to 1400 B.C. in India, the use of psoralens containing seeds, and sunlight for treatment of leukoderma (presumably vitiligo, although sometimes confused with leukoderma in leprosy) is described.[4] Similar reports from Egypt date to about 1200 A.D.[4]

In 1938, the Swiss scientist Kuske[5] was the first to identify furocoumarins as being responsible for the erythema and hyperpigmentation arising after skin contact with certain plants. He furthermore localized the action spectrum in the UVA range and reached some conclusions regarding active sites in the molecule.[5,6] Psoralens became subject to intensive research in the 1940s in Egypt, where the extract of the weed Ammi majus Linn had been traditionally used in combination with sunlight to cure vitiligo. In 1947, Fahmy and Abu-Shady[7] isolated and crystalized from that plant compounds that included 8-methoxypsoralen (8-MOP), 5-methoxypsoralen (5-MOP), and 8-isoamylin-oxypsoralen. The Egyptian physician El Mofty[8] then began clinical trials using these purified compounds for treatment of vitiligo.

Later, Lerner et al.[9] in the U.S. confirmed the remarkable effectiveness of psoralens plus sunlight in treatment of vitiligo and began routine clinical application. When in 1974 Parrish et al.[2] reported on the remarkable efficacy of orally administered 8-MOP and subsequent UVA exposure for the treatment of severe, widespread psoriasis, a dramatic interest in PUVA was spurred that swiftly led to its worldwide use in a variety of skin conditions, but especially psoriasis. The Harvard group of Fitzpatrick, Pathak, and Parrish as well as the Italian group of Musajo, Rodighiero, Dall'Acqua, and Bordin have played a major role in furthering the scientific and clinical exploration of PUVA.

III. MOLECULAR MECHANISMS OF ACTION

The effects of nonionizing radiation on living matter are initiated through the absorption of photons by chromophores. Depending upon the nature of the chromophore, the absorption spectrum can have strong interactions anywhere in the range between the UV and the IR regions. The principal molecular structures involved in absorbing photons in this energy range are usually unsaturated polycyclic compounds. Many such compounds are endogenous in biological systems, but additionally they may be incorporated systemically or topically as certain dyes (such as eosin, methylene blue, etc.), antibiotics, or furocoumarins. All of these involve the same principle: the absorption of the energy of a photon of the appropriate wavelength yields a short-lived excited (singlet) state of the compound. Many exogenous sensitizers also require oxygen to which the energy is transferred or as an intermediate in the ensuing reaction. Oxygen radicals in turn are involved in the photooxidation of various target molecules. This so-called photodynamic action was described first in 1908 and is responsible for the photoeffects of substances such as hematoporphyrin and methylene blue.[10] In contrast to most other photosensitizing drugs, known psoralens do not require oxygen for the lethal photosensitization of bacteria,[11] and therefore do not constitute a photodynamic action phenomenon for that biological effect. In 1965, it was shown that interaction with DNA bases is involved in the psoralen-mediated photosensitization.[12,13] In addition to their prominent therapeutic use for dermatological conditions, psoralens have therefore become important probes for understanding chromatin structure as well as DNA damage and its repair in a variety of cellular systems.

It is now well established that the psoralens can enter the nuclei of cells and intercalate into the DNA, forming weak complexes by van der Waal's interactions between the stacked planes of the bases. Upon irradiation with UVA, the molecule enters a short-lived excited singlet state, which can undergo spin conversion by intersystem crossing to the longer-lived triplet state. Psoralens in both the singlet and the triplet states can form covalent bonds with the 5,6-carbons of the pyrimidine bases. Either the 3,4-(pyrone-) double bond or the 4′,5′-(furyl-) bond or both sites in suitable furocoumarins can react in this way, although in general, the furan side of the molecule is much more reactive. Cyclobutane rings are formed with the pyrimidine bases, predominantly thymine, but also cytosine. For 8-MOP, the 4′,5′-monoadduct is formed by far more frequently than is the 3,4-adduct. The 4′,5′-adduct can absorb another photon and, if appropriate pyrimidines are in the plane above or below the intercalated drug, can form another covalent bond with a thymine or cytosine in the opposite DNA strand, generating an interstrand cross-link through the resulting bi(functional) adduct.

The monoadducts, diadducts, and unreacted psoralen have different absorption spectra. The psoralen absorption extends to 400 nm, while the 4′,5′-monoduct absorbs only below 380 nm, and the 3,4-monoadduct below 330 nm.[14,15] Thus, irradiation in the 380 to 400-nm range generates mostly monoadducts, and some of these can then be converted to cross-links by subsequent irradiation at wavelengths below 380 nm. By controlling the wavelength and by giving a second irradiation (after washing out unbound psoralen), it is possible to reproducibly and quantitatively control the yield of different psoralen adducts, thereby determining their respective biological effects. The 8-MOP monoadducts are repairable lesions quite similar in their effects to those of the intrastrand pyrimidine dimers produced in DNA by short (254-nm) wavelength UV radiation. On the other hand, the cross-link poses a more serious problem; it not only blocks progression of DNA polymerases along either strand, but also prevents the essential process of separation of parental DNA strands. Intuitively, one would predict that the cross-link should be a more highly lethal type of damage than is the monoadduct. There is some evidence that in cultured human cells the interstrand cross-link is a less-repairable lesion than are the monoadducts.[16] Thus, cross-links persist in the cellular DNA and may have longer-lasting inhibitory effects upon DNA replication and cellular proliferation. Studies in a model bacteria system have suggested that 8-MOP monoadducts are highly mutagenic while the cross-link represents the major lethal lesion.[17] The remarkable effectiveness of 8-MOP in photochemotherapy for psoriasis may be primarily due to cross-links, while the mutagenic/carcinogenic potential could be related more to the monoadducts. Such a hypothesis, originally suggested by Hanawalt and co-workers,[18,19] remains to be thoroughly tested, but it forms the theoretical basis for the therapeutic protocol termed "split-dose PUVA".[20]

Whereas the cyclobutyl pyrimidine dimers generated by UV irradiation are formed randomly throughout chromatin,[21] the furocoumarin adducts seem to be largely, if not completely, restricted to specific "linker" regions in the nucleosome-structured DNA.[22,23] The DNA is wrapped around the histone-rich nucleosome "cores" about every 160 bp.[24] Most of the base pairs seem thereby to be protected against interaction with many "bulky" adducts, and only the regions in between (thus called linker regions) are accessible for interaction with furocoumarins.[24,25] Within these linker regions, the potential locations for cross-links are further restricted by the nucleotide sequence, since the 4′,5′-adduct of 8-MOP is preferentially formed in DNA sequences rich in A-T base pairs[26,27] and there must always be an appropriately positioned pyrimidine in the other strand for cross-linkage to be possible.

Psoralen adducts have been used as probes to analyze the repairability of bulky lesions in cultured mammalian cells. In one recent study, the localization of psoralens (and angelicin) to linker regions was utilized to show that nucleosome core structures do not necessarily return to their original sites along the DNA sequence following the repair of bulky adducts.[28] In another study, it was shown that DNA adducts of the psoralen hydroxymethyltrioxylen are very poorly repaired, relative to those in the genome overall, in a highly repetitive sequence in primate heterochromatin known as α-DNA.[29] Further, analysis of the efficiency for forming the cross-links from the monoadduct in α-DNA suggested that the DNA helix was in a different configuration than the usual β-form in α-chromatin.[30] Thus, the unique photochemical properties of the psoralens can be used to study structural features of chromatin and DNA in different regions of the cellular genome. These features may be very important since it recently has been shown that damaged DNA may be much more efficiently repaired in some regions (e.g., active genes) of the mammalian genome than in others.[31] The risk of oncogenic transformation of cells may depend upon the distribution of genomic damage in relation to the locations of particular target regions such as "protooncogenes". That risk may also depend upon the repairability of psoralen adducts in the relevant genomic regions.[32] Clearly, there is much to be learned about the basic interaction of psoralens with DNA and the processing of psoralen adducts before meaningful risk assessments can be projected. In the meantime, we must rely upon the results of various carcinogen test systems, animal models, and the follow-up experience with patients who have received PUVA therapy.

Although the early work of Oginsky et al.[11] suggested that PUVA action does not require oxygen, more recent studies have shown that photochemically excited furocoumarins do generate oxygen radicals.[33] Since singlet oxygen is a very reactive species that can damage DNA and other molecules, that might also play an important role in biological effects, even though it is produced only in low quantities.[34] Psoralens furthermore have been shown to form covalent bonds with a variety of proteins through an autooxidization step,[35] the biological importance of which will be discussed later.

It is important to appreciate that PUVA in the skin elicits a very complex series of molecular events. Direct conjugation occurs between psoralens and DNA, RNA, and proteins, while the generation of singlet oxygen may lead additionally to cell-membrane damage with consequent release of prostaglandins and other mediators of inflammation. Our understanding of the relevant mechanisms is complicated by the fact that the bifunctional psoralens are both therapeutically effective and also strong generators of singlet oxygen.[36] Conversely, the monofunctional psoralens are generally less effective in therapy and their ability to generate singlet oxygen is weak or absent. We therefore must sort through a complex set of actions on several affected systems in order to assess the overall effects of PUVA on the skin. The most prominent actions are not necessarily the most relevant ones for any particular biological end point.

Nevertheless, monoadducts and especially biadducts to DNA appear to be the molecular events primarily responsible for the inhibition of DNA synthesis and cell proliferation in cultured cell systems. Cells in log growth phase that have been treated with large doses of PUVA demonstrate hyperchromatic enlarged nuclei or are multinucleated.[37,38] They are obviously still synthesizing DNA but at attenuated rates and without adequate mitoses for normal chromosome segregation and cell division. Whereas the effect of monoadducts on cytotoxicity and growth inhibition seems to be rather small compared to that of the cross-links, they have a distinct impact on DNA synthesis.[39] However, lethal photosensitization and inhibition of mitosis appear to be more related to the presence of cross-links than monoadducts.[37,39] In fact, cell killing was shown to be markedly enhanced when a second UVA dose was given to maximize conversion of monoadducts to cross-links in Chinese hamster cells containing photoadducts of 4,5′,8-trimethylpsoralen. This experiment, reported by Ben-Hur and Elkind,[39] was really the prototype for the split-dose PUVA approach alluded to earlier. A similar study using 8-MOP applied to cultured epidermal keratinocytes gave the same qualitative picture when the split-dose protocol was applied.[20] For cultured mammalian cells, psoralen cross-links represent a more lethal adduct than do the monoadducts.

IV. POSSIBLE MODES OF ACTION IN PSORIASIS

In the following, we will consider more broadly the possible therapeutic mechanism(s) of PUVA and also discuss how the analysis of PUVA effects has impacted upon our understanding of psoriasis.

A. Keratinocytes

The changes in proliferation and maturation of keratinocytes in psoriasis have quite naturally led to focus on the epidermis as the prime target for any kind of treatment. Whether these epidermal changes are of primary or secondary nature is not yet known. The antiproliferative effect of PUVA in a variety of cellular systems and the demonstration that PUVA inhibits the synthesis of DNA *in situ* in the epidermis seemed to provide sufficient explanation for the beneficial therapeutical effects of this treatment for psoriasis.[40,41] PUVA therapy appears to be similar in principle to other treatment modalities useful in psoriasis such as UVB, anthralin, coal tar, methotrexate, or hydroxyurea in that all of these can be regarded as potent inhibitors of DNA replication. Most of them also produce lesions in DNA and would appear to act as tumor initiators and/or promoters in various test systems. Some, such as UVB, are clearly established carcinogens for humans.

The antiproliferative action of PUVA as an exclusive explanation for therapeutical efficacy may be questioned on molecular as well as clinical grounds. Although epidermal hyperplasia is prominent in some clinical patterns of psoriasis, in others such as pustular psoriasis with its preponderance of inflammatory changes, PUVA has also been shown effective.[42] Several treatment modalities other than PUVA (including methotrexate and topical corticosteroids) recently have been shown to affect a variety of activities other than DNA replication such as mediatiors of inflammation or chemotaxis.[43,45] Although high PUVA doses result in detectable DNA adducts of psoralens in human and animal epidermis, such adducts have not been detected in the epidermis at therapeutically relevant doses.[45,47] This of course does not rule out the existence of such lesions, but it does set an upper limit upon their effective number. If one assumes that DNA adducts are formed in the human epidermis during PUVA therapy, then they may be in lower concentrations than those required for the inhibition of cell proliferation in cell cultures. It is therefore conceivable that the mechanism of action is not simply a straightforward inhibition of DNA synthesis in keratinocytes.

Keratinocytes in clinically symptomless as well as in the involved psoriatic epidermis seem to be in a metastable condition as regards their proliferative control. In the symptomless skin of psoriatic patients, small focuses of keratinocytes have been shown to have abnormally high labeling indexes or mitotic indexes[48,49] without further evidence of a psoriatic process. Trauma, in the most generalized sense, can initiate the active psoriatic process in symptomless skin in many of these patients. That response is termed the Koebner phenomenon and can be modeled as an abnormal response to a wound-healing process in the skin. It is conceivable that low levels of DNA photoadducts, by engaging the cell in DNA repair, might interfere with gene expression in the misdirected cells and provide a means to redirect the keratinocyte to the normal maturation pattern. Although there is no evidence to support this hypothesis directly, it is of interest that several treatment modalities (PUVA, methotrexate, dithranol, coal tar plus UVB) have been shown to result in the reappearance of the stratum granulosum before any significant drop in DNA labeling or mitotic indexes is evident.[48,50]

Alternatively, the interaction with DNA could be negligibly small and irrelevant, while photoconjugation of 8-MOP to proteins (enzymes) or damage to membranes via singlet oxygen could be the significant therapeutic reaction. Since phospholipase A_2, a key enzyme for the generation of both prostaglandins and leukotrienes, appears to be enhanced in psoriatic skin, an interaction on that level could be plausible.[51]

However, in the absence of direct evidence for significant actions in the epidermis other than the DNA-dependent ones hypothesized, Reusch and Hanawalt favor DNA as the relevant target for PUVA. Indirectly supporting this view is the fact that the DNA cross-linking psoralens are in fact more effective therapeutically than are the angelicins.

B. The Immune System

Although the immune system is evidently involved in psoriasis, it is not yet known whether autoantibodies against stratum corneum, changes in number and distribution of Langerhans cells, and the dermal/epidermal lymphohistiocytic infiltrate are a casual or secondary event in this disease.

Recently, changes in Langerhans cells (LC) and subtypes of epidermal T lymphocytes were shown to be consistently related to the evolution and resolution of plaques in guttate psoriasis.[52] The same study found T lymphocytes residing in the uninvolved psoriatic skin. This could imply a pathophysiological involvement of the immune system.

1. Langerhans Cells

In psoriatic lesions, LC undergo very specific changes in localization and numbers.[53] The number of LC is reduced and they are located mostly in clusters close to the papillae.[53] HLA-DR$^+$, dendritic cells are also found in the dermis in close proximity to activated T lymphocytes.[54] To what extent these cells play a role in the psoriatic process remains to be elucidated.

LC react very sensitively to a variety of treatment modalities effective for psoriasis. PUVA, UVB, or topical corticosteroids deplete the epidermis to a large extent of HLA-DR$^+$ dendritic cells.[55-57] During early PUVA therapy, cell membrane markers such as HLA-DR or ATPase disappear.[57] Within a few weeks, about 90% of the LC then disappear from the epidermis, but reappear about 3 weeks after discontinuation of PUVA therapy.[57] In contrast, therapy using etretinate increases the amount of LC continuously during treatment.[53] A variety of treatment modalities useful in psoriasis have been shown in vitro to inhibit the functional capacity of LC.[58] The mixed epidermal-lymphocyte reaction is inhibited by PUVA, UVB, corticosteroids, anthralin, methotrexate, or hyperthermia.[58] This might explain the inhibition of the induction of allergic reaction or delayed hypersensitivity reaction in animals and humans.[59,60]

So far, it is not known whether the changes of LC in psoriasis play any significant pathophysiological role. LC have been implicated to be involved in control and regulation of epidermal proliferation and maturation, although no direct evidence exists to support this hypothesis.[61] Whether PUVA or other treatment modalities can influence the psoriatic process through specific action on these cells remains to be demonstrated.

2. Lymphocytes

A large fraction of the infiltrate in psoriasis is comprised of T lymphocytes. Some of these lymphocytes have been shown to be activated and to be in close proximity to dermal, dendritic, HLA-DR+ cells.[54] Subsets of T lymphocytes in the dermis appear to remain constant throughout the process, and the ratio helper/inducer vs. cytotoxic/suppressor cells is similar to that found in circulating blood.[62] In the epidermis, conflicting reports demonstrate prevalence of T cells of either subtype.[52,63] In one study, evolution of guttate psoriasis was closely correlated with an influx of activated T helper cells, whereas resolution showed a prevalence of activated T suppressor cells.[52]

PUVA has been shown to affect lymphocytes passing through the vasculature of the dermis.[64-66] Proliferation of lymphocytes after stimulation with mitogens such as concanavalin A or phytohemagglutinin was decreased for doses that could reach the blood vessels.[64-66] This effect appears to be transient, however. Overall, the WBC counts remain unperturbed.[67] T cells of the T helper/inducer subtype seem to be affected preferentially.[67] Possibly, the drop in counts of this subtype, accompanied by a compensatory increase in Null cells with overall unaffected WBC counts, could be due to a deficiency in expression of cell surface markers.[67] Whether PUVA affects lymphocytes, possibly one subtype preferentially, in the psoriatic lesion, and whether this might be a way to beneficially affect the psoriatic process, remains to be clarified.

Interestingly, interferon lately has been related to PUVA and psoriasis. In suction blister fluid of psoriatic patients, interferon could be demonstrated.[68] Another study showed that whereas healthy volunteers did not have any detectable levels of interferon in blood, psoriatic patients did.[69] Furthermore, the levels increased significantly during PUVA treatment.[69] Since Morhenn and co-workers[70] found that recombinant γ-interferon inhibits growth of cultured keratinocytes, it is conceivable that the lymphocytic infiltrate exerts some control over the proliferation and maturation of keratinocytes in psoriasis.

C. Polymorphonuclear Neutrophils

Exocytosis of polymorphonuclear neutrophils (PMN) is among the first events during evolution of a psoriatic lesion.[71] Their number is closely correlated with the mitotic activity of the epidermis.[72] Removal of "polys" from blood by leukapheresis can result in the clearing of otherwise recalcitrant psoriasis.[73] A number of treatment modalities such as topical corticosteroids, methotrexate, or anthralin have been recently shown to interfere with chemotaxis of PMN.[43,44,74] A recent report could not demonstrate any inhibition of polymorphonuclear leukocyte functions with therapeutical doses of PUVA; only excessive ones rendered some effect.[75] It nevertheless remains conceivable that mediators related to the PMN could be affected, since exocytosis of PMN is one of the first phenomena to subside during PUVA therapy.

D. Vasculature

The capillaries in psoriasis exhibit some special, prominent features. The endothelial cells of the efferent, venular limb undergo rapid proliferation resulting in the tortuosity and dilation of the capillaries, which are "venularized" and thus fenestrated and highly permeable.[48] In growing psoriatic lesions, vascular changes are present well beyond the margin of the clinically visible plaques and epidermal changes.[76] It is interesting to note that PUVA as well as UVB first affect the vasculature, transforming the venularized capillaries back into normally appearing ones 1 week before any changes in labeling indexes of keratinocytes are noticed.[48] These changes of the vasculature go along with reappearance of the stratum granulosum.[48] Whether PUVA achieves its beneficial effect on psoriasis through the vasculature of the papillary body remains speculative.

E. Other Cells

Mast cells have been described to be the first to arrive in an early psoriatic lesion.[77]

Whether they play a significant role and whether PUVA affects them is not yet known. Macrophages constitute a larger fraction of the infiltrating cells in psoriasis, but hardly anything is known as to their possible pathogenetic role and potentially beneficial response to PUVA treatment.

F. Mediators

Recent advances in the understanding of the dynamic interactions of the various systems comprising the psoriatic plaque have led to growing interest in the mediators interconnecting them.[43] Photoconjugation with proteins, inhibition of their production due to PUVA effects on DNA or cell membranes, could very well interrupt the psoriatic process on this level. Enzymes such as phospholipase A_2 could constitute a key target.[51]

V. CLINICAL EFFECTIVENESS OF PUVA

A. Treatment Protocols

The first experimental results on the use of topical 8-MOP and UVA for psoriasis (and parapsoriasis) were reported in 1972 by Mortazawi,[78] and evidence of the effectiveness was presented at the 6th International Congress for Photobiology in Bochum in 1972.[79] The introduction of oral 8-MOP and UVA for severe psoriasis by Parrish et al.[2] in 1974 provided a powerful, new approach to psoralen photochemotherapy. This approach almost immediately gained worldwide acceptance, and its effectiveness was subsequently proved in a large number of studies all over the world.[80-88] Three large multicenter cooperative studies (comprising more than 4000 patients) performed in the U.S. and in Europe confirmed the extraordinary effectiveness of PUVA and its short-term safety.[42,89,90] Two distinctively different protocols evolved. In both cases, the dose of the 8-MOP is held constant (0.6 to 0.8 mg/kg body weight is administered orally 2 hr prior to irradiation) whereas the dose of UVA is varied. The "Boston" protocol which has been specified and improved during the long-term studies usually employs two or three weekly UVA exposures 2 hr after ingestion of 8-MOP.[89,90] Initial UVA doses are selected by reference to skin types, and the dose is then increased with every treatment. After about 25 treatments, nearly 90% of the patients are clear of psoriatic plaques, with a cumulative UVA dose of about 250 J/cm^2.[89]

The European PUVA study (EPS) employs a more frequent irradiation protocol, four sessions per week, with dose increments every third treatment.[42] Initial doses are determined either by phototoxicity testing or by skin type. The initial doses are generally lower than the ones employed in the U.S. multicenter study. This protocol results in clearing of psoriasis in 90% of the patients with nearly 20 treatments resulting in a cumulative UVA dose of about 100 J/cm^2.[42]

It is quite evident that relatively minor changes in protocol can result in greatly increased effectiveness of PUVA, which reduces the cumulative dose as well as the time span for clearing. The reason for the effectiveness of various modifications remains unclear. Referring to the discussion earlier, one might speculate that it is advantageous to frequently engage the keratinocyte in processes that might allow it to regain its appropriate maturation pathway. At the molecular level, the conversion of some persisting DNA monoadducts into cross links might be an explanation. The concomitant hyperpigmentation that appears after about 1 week could also be a disadvantage when a slower, more cautious protocol is followed, since the doses need to be increased further to achieve penetration.

B. Indications and Contraindications

Essentially all forms of psoriasis respond favorably to PUVA, but special adjustments have to be made for the erythrodermic form, which should be treated like skin type I. As will be discussed later in more detail, PUVA generally should be restricted to elderly patients with widespread, severe forms of psoriasis. Relative contraindications for its application are previous treatment with arsenic or ionizing radiation, extensive solar damage, history of nonmelanoma skin cancer, (family) history of melanoma, phemphigus, and phemphigoid. Absolute contraindications are xeroderma pigmentosum, lupus erythematosus, pregnancy and lactation, prophyria cutanea tarda, and porphyria erythropoietica.

C. Combination Treatment

The effectiveness of PUVA can often be significantly enhanced by combining it with other treatment modalities, such as etretinate, topical steroid, methotrexate, dithranol, or UVB. Especially recalcitrant psoriasis, which does not respond to any single modality, can sometimes be effectively treated by appropriate combinations. In some of those cases, the use of methotrexate and PUVA appears to be of special value.[91] The use of UVB, etretinate, topical corticosteroids, or methotrexate can result in a reduction of cumulative UVA doses by a factor of two to three and a significant reduction in number of treatments.[91-96] In one study, complete clearing could then be achieved with 12 exposures and a total cumulative dose of only 40 J/cm^2.[94]

D. Maintenance Treatment

Especially with severe forms of psoriasis, the palliative responses and the consequently occurring relapses underline the need for the long-term or permanent control of the disease. Maintenance PUVA has therefore been investigated in a number of studies. Whereas Melski et al.[89] and Roenigk et al.[90] found that maintenance significantly reduces relapses, the EPS did not confirm these results.[42] Since the long-term risks of PUVA seem to be related to the total cumulative dose,[97] the endorsement of such continuing UVA and psoralen exposure must be carefully evaluated. Wolff[88] finds that maintenance treatment during the first 2 or 3 months after clearing is a helpful compromise. In this patient group, about 70% were still clear after 9 months[88] as opposed to about 40% in the EPS without maintenance treatment.

VI. SIDE EFFECTS OF PUVA THERAPY

Some of the more significant short- and long-term side effects of PUVA therapy will now be reviewed. Almost every type of treatment entails some unwanted hazards, because few available treatments are specific enough for the disease in question. This is especially true for the medications used in dermatology since the action mechanisms of only a few drugs are understood. PUVA is no exception. Since PUVA is used in quite differing ways among various centers (treatment protocols, type and bioavailability of furocoumarin, type of lamp [admixture of UVB], combination with other modalities, maintenance treatments, etc.) it is not surprising to learn that the experiences in different centers are not necessarily in accord. The following therefore represents a selection from a large amount of data on side effects, but the emphasis admittedly reflects the authors' "bias".

A. Short-Term Side Effects

Short-term side effects are not unexpected during PUVA therapy.[42,89,90] The predominant complaints include erythema, GI problems (mainly nausea), pruritus, headache and dizziness, and the Koebner reaction. One or more of these sequelae occur in about 30% of the patients during treatment.[42,89]

Due to the narrow therapeutic window of PUVA (very steep erythema response reaction) and the long latency of the PUVA erythema, (48 to 72 hr) phototoxic reactions ranging from moderate erythema to blistering constitute the most frequent side effect. Careful evaluation of the patient's tanning history, photosensitivity testing, if indicated, and individually tailored, carefully metered dose increments limit these problems substantially. Erythema is a far more frequent side effect when the more aggressive protocol of the EPS is employed,[42] whereas the two or three times weekly protocols better allow for adjustment to individual phototoxic responses.[87,88] Erythemas are usually mild in nature and do not constitute a major problem. They generally can be handled by omitting the next scheduled treatment.

Many patients experience mild pruritus during treatment and this can be helped with lubricants. The deep burning itch which sometimes arises may require interruption of the treatment until it ceases.

PUVA can sometimes cause a Koebner phenomenon as can other treatment modalities useful in psoriasis. This is best handled by continuing therapy but with reduced doses.

Nausea, another fairly common problem, can usually be handled with dietary adjustments. Sometimes medication is required.

Sometimes the use of 5-MOP instead of 8-MOP presents a valuable alternative. 5-MOP has been shown to be as effective as 8-MOP without leading to erythema, pruritus, and nausea when applied in therapeutic doses.[98]

During treatment of a large number of patients,[42,89,90] no significant changes in laboratory parameters were reported. In general, PUVA treatment in the short term has proved to be a very safe treatment with high patient compliance. The reported side effects, occurring in less than half of the patients, are usually so mild in nature that only rarely do they require discontinuation of therapy.

B. Long-Term Side Effects

PUVA has been closely studied for its long-term side effects ever since its introduction. In fact, a much more thorough analysis of such sequelae has been carried out on PUVA than for any other type of treatment, including UVB. Extensive follow-up studies and many more case reports have revealed a number of long-term risks associated with PUVA.

Among the long-term risks are the following: ocular hazards; pigmentary changes including the induction of the so-called "PUVA-lentigines"; alterations in the immune system; premature aging of the skin with the induction of possibly precancerous lesions, such as epidermal dystrophy or bowenoid lesions; and, most important of all, the induction of malignancies.

1. Ocular Hazards

Adequate eye protection during and after PUVA treatment is of great importance, since UVA alone and the combination of UVA and psoralens can be potentially hazardous to the eye. The human ocular lens is exposed to radiation above 300 nm, and the near UV seems to be related to cataractogenesis.[99] 8-MOP enters the human lens and is still present 12 hr following oral ingestion.[100] It can be photoactivated and bind to lens proteins leading to permanent retention and lens opacities.[100] In a large variety of animals, cataract formation was observed after exposure to chronic PUVA at doses resulting in severe phototoxicity and thus much higher than the ones used for treatment.[101,102] For one animal strain, it has been shown that doses which allow for tolerable acute and chronic cutaneous photosensitization do not result in cataract formation.[103] However, it would be premature to conclude that in dose ranges used for treatment, PUVA would be safe without eye protection. Thus, suitable glasses should be worn at least during the days psoralens are ingested.

Two small follow-up studies showed no evidence of cataracts in 28 patients as long as 25 years after their exposure to PUVA for treatment of vitiligo, despite the lack of any preventive measures.[104,105] Though there are few anecdotal reports on cataracts after PUVA, the experience with large numbers of patients does not indicate any particular risk if the necessary precautions are taken. Using adequate eye protection, PUVA seems to be reasonably safe, and any ocular damage is most likely due to carelessness and/or lack of patient compliance.

2. Pigmentary Changes

PUVA induces a variety of pigmentary changes. The long-lasting diffuse hyperpigmentation of the skin is due to a stimulation of the melanocytes and concomitant, increased transfer of melanosomes to the keratinocytes.[106] Quite frequently, the melanosomes change from the usual aggregated pattern (in Caucasians) to a single, dispersed pattern as encountered in blacks.[106,107] Occasionally, melanocytes are observed in groups, and double-nucleated melanocytes may be present,[107] which may indicate induced proliferative activity. Overall, the number of melanocytes appears to be increased, sometimes showing alterations as swollen, irregularly enlarged nuclei, abnormal mitochondria, a swollen and distorted endoplasmatic reticulum, and large lysosomes.[107,108] The changes reportedly persist up to 15 months after discontinuation of PUVA therapy.[107]

Another frequent feature in long-term PUVA-treated patients is the so-called PUVA-lentigines or PUVA-freckles. They can be observed in nonsun-exposed as well as in sun-exposed areas.[109-111] After several years of PUVA treatment, lentigines of varying degree have been found in 33 to 53% of patients.[109,111] These freckle-like lesions are differently interpreted. Whereas one study[112] considers them to be similar to simple freckles, others have shown them to be mostly typical lentigines presenting with a certain increase in epidermal thickness, slight elongation of rete ridges, and a more or less pronounced budding associated with an increased number of melanocytes of normal configuration.[109,111] Ros et al.[111] report in 10 out of 250 patients a very profuse disseminated lentiginosis. In the same study[111] six patients (2.4%) developed isolated, clinically dark and irregular lentiginous lesions which on biopsy showed heavy epidermal pigmentation, extensive linear melanocytic interretal proliferation, sometimes a pagetoid spread of melanocytes, and nuclear polymorphism. Atypical melanocytes in PUVA-lentigines were sometimes found in Rhodes et al.[109] study as well. PUVA-lentigines appear to be correlated with the number of treatments,[109-111] an older age at starting PUVA, and male sex, and negatively associated with skin types V and VI.[109] So far there is no indication that PUVA-lentigines disappear after discontinuation of therapy.[109]

A peculiar mottling of skin (hyper- and hypopigmentation) has been reported in 13 out of 572 patients at sites of previous overdosage.[113] Hyperpigmented areas showed a dramatic increase in melanocytes and heavy pigmentation, whereas in the hypopigmented areas melanocytes and epidermal melanin were considerably decreased.[113] The melanocytes were large and presented with large hyperchromatic and irregular nuclei.[113] These changes were to a large extent accompanied by disturbed epidermal architecture, some pleomorphism of epidermal cells, subepidermal homogenization, and reduction of elastic fibers.[113]

The changes introduced in melanocytes alert researchers to the possible carcinogenic potential of the treatment. Although especially the "atypical lentigines" raise suspicions, there seem, despite its wide use for vitiligo in the 1950s and 1960s, to be no PUVA-related melanomas. The few melanomas reported so far, and the two melanoma *in situ* reported by Marx et al.[114] do not seem to be caused by PUVA treatment and do not represent any unusual number given for this large treatment sample. Nevertheless, any suspicious pigmentary change deserves full and immediate attention and should be removed and studied histopathologically.

3. The Immune System

PUVA obviously affects two major components of the immune system: the LC and the lymphocytes passing through the dermal blood vessels. The inhibition of proliferative response of lymphocytes to a number of mitogens,[65-67] as well as the inhibition of the induction of allergic or delayed hypersensitivity reactions, leads to concern that the immune system might be adversely affected.[59,60] In one study, after 200 exposures to PUVA, patients showed decreased response to mitogens such as PHA or ConA.[115] Two out of three patients with marked reduction of T helper/inducer cells had squamous cell carcinomas of the skin.[69] It has been suggested that PUVA might act as a pseudopromoter for tumorigenesis by reducing or inhibiting immunosurveillance in skin.[60] Malignant foci of cells in the skin would thus not be recognized and eliminated. Increased risk for squamous cell carcinomas is well established in otherwise immunosuppressed patients.[116] It remains to be elucidated whether prolonged exposure to PUVA actually can facilitate the growth of malignancies due to an effect on the immunosurveillance.

4. Premature Skin Aging

Sunlight is known to cause premature skin aging and premalignant lesions after long-term exposure. The changes can be easily appreciated in certain professionally sun-exposed people as fishermen and farmers. The aged areas and tumors elicited are therefore located in sun-exposed regions of the skin. PUVA causes concern for at least three reasons: (1) the UVA radiation per se cannot be neglected and could be a great detriment to skin structures, even without enhancement through the specific sensitizing step, (2) the overall feature of the PUVA effect seems to resemble sunlight exposure and therefore deserves at least similar attention, and (3) knowing that PUVA can cause specific damage to the genome of the cells, carcinogenic side effects must be a major concern.

Reports so far indicate that the epidermal as well as the dermal component of the skin is affected during long-term exposure of PUVA.[113-117] There is nevertheless a considerable difference of opinion about the extent and the nature of this damage and possible predispositions required to acquire it.

In at least some of the PUVA-treated patients, and obviously dose-dependent, the skin of these patients resembles what we know as aging: telangiectasias, pigmentary changes, dryness, and wrinkling.

a. Epidermal Changes

Shortly after initiation of PUVA treatment, several reports appeared indicating that keratinocytes in the epidermis can be adversely affected. Abel et al.[117] coined the expression "epidermal dystrophy" referring to little foci of keratinocytes which may have variable size, may show atypical mitoses or hyperchromatic, enlarged nuclei, sometimes giant cell formation, or multiple nuclei, and a generally disturbed orientation of the cells, indicating disturbed maturation. These changes were confirmed by Niemi et al.[108] and by Hoffmann et al.[118] The latter called them bowenoid changes of the epidermis. Abel et al. found these changes in about 50% of all patients in clinically normal, nonsun-exposed PUVA-treated skin, whereas Gschnait et al.[113] reported comparable changes only in areas of previous overdosage and not in otherwise "healthy" skin. These studies certainly present various difficulties: (1) as stated, the foci are scattered through the epidermis and can be easily missed. An intensive study searching for such foci has not been done on healthy individuals and especially extensively pretreated psoriatic patients; (2) there are differences in the light sources used and Abel's as well as Niemi's contains a significant amount of UVB (2%); and (3) the latter studies employ a treatment schedule with two irradiations per week and considerably higher cumulative doses.

It seems that these epidermal foci are a dose-dependent phenomenon and may well be less prevalent in the low-dose treatment schedules. They also seem to be at least partially reversible, and the possible role of UVB has to be ascertained. To what extent these changes could possibly be the source of actinic keratoses or even squamous cell carcinomas remains to be evaluated.

b. Dermal Changes

Changes of the dermis have been reported in a number of follow-up studies during and after chronic PUVA treatment. As has already been pointed out, a separate effect of the UVA irradiation alone has to be considered, since about 45% of the incident UVA irradiation reaches the dermis[119] and presumably constitutes the major factor in skin aging due to sunlight exposure. The clinical changes mentioned earlier, such as wrinkling, dryness of skin, and laxity of cutis reflect mostly alterations of the dermal component rather than an epidermal one. Histologic studies report on papillary changes such as increased fibrosis, an eosinophilic homogenization, and reduction of elastic fibers at the dermo-epidermal junction.[106,113] These changes have been closely correlated to the effects of PUVA overdosage in one large study.[113] Recently, premature skin aging has been evaluated in the 1380 patients that are enrolled in the NIH cancer follow-up study.[120] In these patients, moderate to severe skin aging could be found in 11% of the patients after 5 years of PUVA and usually did not pose any serious problems.

5. PUVA and Skin Malignancies

The most important concern about chronic exposure to PUVA is the possibly carcinogenic side effect of this treatment. Several aspects of PUVA therapy raise this concern. UV irradiation, especially of the UVB region, appears to be the most important exogenous factor for the induction of nonmelanoma skin cancer in man.[139] Since PUVA in many respects resembles the effects of UVB radiation on skin, concern about its possible long-term hazards is certainly appropriate. Additionally, the UVA range has been neglected for quite a while as to its biological risks, but there are indications that it might play a role in photocarcinogenesis.[121] This is of specific importance since high cumulative UVA doses will be acquired over a prolonged period of PUVA exposure. More importantly, the interactions of the psoralen adducts with the cellular DNA, and the biological impact of those lesions or their repair products poses serious questions. Furthermore, the potential reduction or impairment of immunosurveillance due to PUVA might inhibit recognition and elimination of malignant cell clones and facilitate growth of malignant tumors.

PUVA adducts to DNA resemble in some ways bulky adducts to DNA such as alkylating agents, which are used for chemotherapy and are also known to cause malignancies in human beings. As mentioned earlier (see Section III) 8-MOP monoadducts to DNA are quite similar in their effects to those of the UV-induced intrastrand pyrimidine dimers. Their repair seems to have at least one step in common with the excision-repair system for pyrimidine dimers.[16,122] Repair of monoadducts has been demonstrated in a number of cellular systems. In DNA saturated with UVC-induced pyrimidine dimers, PUVA monoadducts do not elicit any additional repair response.[20] Cross-links pose a much more complicated problem for the cell. They appear to be the more lethal lesion and there is evidence that they are repaired less than are the monoadducts.[16] Since they block the separation of complementary DNA strands, repair of cross-links has to entail mechanisms in addition to the excision-repair system that is employed for removal of pyrimidine dimers and possibly for monoadducts. These mechanisms almost certainly would include postreplicational exchanges of DNA sequences of homologous chromosomes. Sister-chromatid exchange has been shown to be elevated in keratinocytes in vitro for doses used in therapy.[122] Sister-chromatid exchange is not caused by cross-links exclusively, but by monofunctional adducts as well, with about half the efficiency of cross-links.[124] Therefore, presence of sister-chromatid exchange cannot be taken as a decisive measure for either type of DNA lesion. Enhanced levels of sister-chromatid exchange are very sensitive measures for DNA damage and can be used as an indicator system for the presence of mutagenic and possibly carcinogenic substances. However, sister-chromatid exchange is not enhanced in keratinocytes in vivo following photochemotherapy.[125]

No repair activity could be measured in human epidermis following PUVA treatment as judged by unscheduled DNA synthesis when the UVB admixture of the light source had been filtered out.[126] This could indicate that either no repair takes place, or it is extremely slow, or rather small amounts of adducts or none are present in epidermis.

Even if an efficient repair mechanism exists, this does not necessarily mean that there is no danger of carcinogenesis. The existence of the very efficient repair system for pyrimidine dimers does not prevent the induction of skin cancer due to prolonged exposure to UV irradiation, although deficiency in that system in xeroderma pigmentosum, a hereditary autosomal recessive disease, places its victims at an extremely great risk for sunlight-induced cancer. The repair event itself, which requires nicking of the damaged DNA strand, could be responsible for some enhancement of DNA sequence transpositions, which are known to result in activation of some oncogenes.

In a variety of cellular systems such as bacteria, yeasts, and mammalian cells, PUVA has been shown to be mutagenic.[115,127,128] Certainly these results cannot be simply translated into "cancer", but they must raise caution.

Already in the 1950s it was shown that PUVA can cause skin cancer in various animal strains.[129-131] The results of animal experiments conducted remain at least partly inconclusive. PUVA could be shown to induce skin carcinomas only in some mouse strains but not in others, which indicates an endogenous susceptibility of those animals.[132] Furthermore, some of the older studies used either sunlight or light sources with significant admixture of UVB radiation so that the cancer induced could result at least partly, if not exclusively, from the known carcinogenic UV spectrum. The route of application also plays a role, since intraperitoneal and topical application of 8-MOP seems to cause cancer, yet orally administered 8-MOP seems to be far less carcinogenic. Additionally, the applicability of animal results to humans has to be weighed carefully since animal results have been generated with usually far higher doses and chronic long-term exposure which cannot be easily equated with the therapeutic use of PUVA.

Since the laboratory investigations are not yet conclusive, the follow-up studies during and after clinical treatment are of even greater importance. Fortunately, PUVA has been accompanied with a great deal of scrutiny rarely employed for other treatment modalities useful in psoriasis. This careful follow-up is of extreme benefit for the evaluation of PUVA, but it also somewhat distorts the appreciation of risks related to other therapeutic regimens less extensively monitored.

Relatively soon after the initiation of PUVA therapy, case reports on the development of skin tumors following treatment were published.[118,133] These reports were substantiated with the largest to date follow-up study whose first intermediate results showed a significant increase in skin cancer.[134] After an average of 2.1 years of follow-up of 1373 patients who were enlisted in the original 16 center PUVA study, Stern et al. reported a 2.6 times higher overall rate of nonmelanoma skin cancer than expected in an age, sex, and geographically matched control population. These carcinomas were mainly confined to certain risk groups — patients with a history of ionizing radiation therapy, previous history of skin cancer, and skin types I and II. The ratio of basal cell carcinoma (BCC) to squamous cell carcinoma (SCC) was found to be inverted. Many of the SCC were located in previously nonsun-exposed skin. After an average follow-up of 5.7 years[97] of the same patient cohort, the data confirmed the trend observed in the earlier study. PUVA appears to act as a tumor promoter for people in special risk groups. Pretreatment with tar and extensive UVB irradiation appear to be additional risk factors. In this study,[97] PUVA does not seem to substantially increase the number of BCC, but rather has an effect on the tumorigenesis of SCC. More importantly, some of the tumors occur in individuals not included in the thus far identified risk groups. These malignancies appear related to high PUVA exposure, and their existence could be accounted for as an independent, complete carcinogenic effect of PUVA.

In another large study on 631 patients after 4 years of PUVA, Roenigk and Caro[135] could not confirm the inverted ratio of BCC and SCC. The overall risk to develop skin cancer was much less alarming in their patient sample. Ionizing radiation was confirmed to be a major risk factor.[135] Several European studies either failed to demonstrate any increase in cancer rates,[111,136,142] or tumors were confined to patients in special risk groups, who had been pretreated with other known carcinogens.[137] Arsenic and possibly methotrexate and topical nitrogen mustard appear to be additional risk factors.[137,142] The inversion of the BCC/SCC ratio has been confirmed so far only in one smaller patient group.[138]

The studies on association of skin carcinomas with PUVA therapy published so far have led to rather vivid discussions. It appears quite clear that there exist vastly different experiences with PUVA in various centers. The number of variables affecting possible tumorigenesis is fairly difficult to assess, but different light sources with varying degrees of UVB, various protocols, maintenance therapy, selection and pretreatment of patients, and many more factors could be mentioned.

Nevertheless, it appears clear that carcinomas can arise in certain predisposed risk groups and that there are indications of a "complete" carcinogenic effect for very high doses of PUVA. Whether these findings are relevant for even more cautious use of this treatment, as will be discussed later, remains to be elucidated. Meticulous follow-up of patients exposed to PUVA remains to be the most important task for the time being. PUVA as a possibly complete carcinogen cannot be expected to be fully elucidated until two or three decades have passed by, but that in itself is a concern.

VII. CRITICAL APPRAISAL OF THE CURRENT STATUS OF PUVA THERAPY AND CONCLUDING REMARKS

As one looks back at a decade of successful PUVA treatment for psoriasis, some assessment of its future application is appropriate. A relapsing, frequently lifelong disease such as psoriasis requires specific concerns as to its treatment. To date, none of the treatments available is more than palliative and the disease inevitably flares up again. Therapy is therefore chronically applied, possibly over decades, and long-term health threats are of major concern. On the other hand, severe, widespread psoriasis creates a major problem for the patient that is not just limited to his skin, but deeply encroaches upon and disables his personal and professional life. Alcohol and drug abuse as well as divorce rates and even suicide rates among psoriatic patients substantiate these conflicts. Treatment for psoriasis therefore has to be highly effective, safe, as little time consuming and burdening as possible, and of high patient compliance to restore a normal life. None of the therapeutic regimes at hand meets all the criteria indicated. The use of one or another modality has therefore to be weighed carefully for each individual patient with due consideration for his medical and personal needs.

PUVA has been proved to be an extremely effective treatment for severe, widespread psoriasis, with high patient compliance. Whereas short-term safety creates no problem, the potential long-term hazards, especially carcinogenicity, are of major concern. Today one is cautioned by past experiences of the indifferent use of arsenic, ionizing radiation, or corticosteroids. To weigh these concerns properly, several points regarding PUVA must be raised.

It seems quite certain that higher doses of PUVA can promote cancer in specifically predisposed patients (pretreatment with ionizing radiation, arsenic, possibly UVB, tar and methotrexate, history of nonmelanoma skin cancer, skin types I and II). A carcinogenic potential of its own as suggested by the latest Stern group study[97] still remains to be ascertained beyond doubt and is not likely to be proved before two or three decades have passed. Whereas the nonmelanoma skin cancers arisen so far appear to be manageable, the further biological potential of PUVA lentigines, especially, deserves scrupulous scrutiny.

The evaluation of PUVA hazards is hampered by various variables such as differences in protocol between various centers, doses required for clearing, admixture of significant amounts of UVB in some fluorescent lamps used, bioavailability of the psoralen medication, and others. Additionally, as has been pointed out in many discussions, the use of adequate control groups with which to compare the carcinogenic risk is another matter of dispute. Moreover, PUVA has been followed up with a scrutiny that has hardly ever been applied for other treatment modalities useful in psoriasis. The mere absence of data concerning large populations being treated with high energy UVB protocols over longer periods of time cannot allow for any statements regarding its long-term safety. In addition, severe cases of psoriasis usually have been extensively treated with a variety of therapeutic modalities including a number of known carcinogens, not even accounting for possible self-treatment of those patients with extensive sunlight exposure. To adequately account for PUVA long-term side effects, all these and more aspects have to be considered.

For the time being, PUVA represents one of the most powerful armaments in dermatology and every effort must be taken to lessen its long-term side effects. It seems advisable to reduce cumulative doses to the lowest extent possible. The EPS protocol appears to represent a regimen which not only considerably reduces total cumulative dosage, but also distinctly decreases the time span necessary for total clearing, thereby saving valuable time for the patient.

PUVA should generally be restricted to the elderly patient with severe, widespread psoriasis. Patients falling into specific risk groups require special consideration. Since PUVA damage appears to be at least partly reversible, maintenance therapy, in the authors' opinion, should be applied only in exceptional cases. Protocols that taper out over a period of 2 months after clearing appear to be a good alternative for prolonging the patient's remission.

Two further measures should help to increase long-term safety of PUVA. First, the use of combination therapy in psoriasis (PUVA plus etretinate, topical corticosteroids, methotrexate, anthralin, or UVB) can reduce cumulative doses for clearing to as little as 40 to 60 J/cm^2 and also cut down the patients' clearing time. It needs to be ascertained, nevertheless, that despite the reduction of cumulative UVA doses for total clearing, combination treatment does not otherwise increase the risk of development of skin malignancies. In this respect, the combination of PUVA with treatment modalities that employ other than DNA interactions might be advisable. Retinoids, a new class of drugs that has recently attracted a lot of interest for its potential use in prevention of cancer, may in this respect turn out to be the most valuable combination.[140] More data are necessary to assess such a beneficial effect in combination PUVA therapy. The combination of PUVA and methotrexate proves especially valuable in otherwise recalcitrant psoriasis. Second, PUVA (in combination) should then be restricted to severe flare-ups only, whereas intermittently, other modalities (such as anthralin) should be used to treat lesser relapses. Since relapses after PUVA are often of minor nature, the use of topical therapeutics appears, to be even more indicated. Rotating through different treatment modalities over a period of years before starting PUVA again could also serve as a measure to reduce the risk of one specific modality. Henceforth, in the authors' opinion, it is possible to reduce the total cumulative UVA dose of a long-term treated psoriatic patient to less than 1000 to 1500 J/cm^2 in a lifetime using the presently available therapeutic modalities. The biological impact of PUVA used in such a fashion should be considerably safer than the accumulation of thousands of Joules within a few years. As the great physician Galen taught "sola dosis facit venenum" (only the dose makes the venom) the authors' present opinion is that there is a dose range within which PUVA is reasonably safe.

Although PUVA has spurred a dramatic interest in photomedicine over the last decade, the in vivo responsible action mechanism(s) are still not known. The identification of those could lead to a more effective and safer treatment. The question of whether DNA interaction is a necessary prerequisite for effectiveness in vivo, and if so which type of interaction is mainly responsible for PUVA's beneficial effects, has to be solved. If cross-links can be proved to be the important lesion and the less carcinogenic one, improved protocols (e.g., split-dose PUVA[20] and choice of certain psoralen derivatives could help to further improve the treatment. Alternatively, it is still possible that new furocoumarins that form new potent monofunctional adducts will be shown to achieve the same goal.[141] More importantly, the understanding of repair of psoralen monoadducts and cross-links in DNA in vivo and their relationship to oncogenic transformation might help to operationally separate the beneficial and adverse effects, and lead to development of a new generation of photosensitizing agents having more target-specific properties without adverse side effects.

With the advent of PUVA in psoriasis therapy, a powerful new tool in treatment has been created that under certain precautions can be regarded as reasonably safe. It has concomitantly spurred dramatic interest in photomedicine and vastly increased our understanding of complex photochemical interactions in the skin. The unique accessibility of the skin to light (plus photosensitizing agents) will in the future quite predictably lead to new highly effective protocols for photochemotherapy.

ACKNOWLEDGMENT

Michael K. Reusch was supported by a fellowship of the Deutsche Forschungsgemeinschaft. The authors' research on DNA repair in cultured human epidermal keratinocytes has been supported by grant AG 02006 from the National Institute on Aging, U.S. Department of Health and Human Services, USPHS.

REFERENCES

1. Farber, E. M., Abel, E. A., and Charuworn, A., Recent advances in the treatment of psoriasis, *J. Am. Acad. Dermatol.*, 8, 311, 1983.
2. Parrish, J. A., Fitzpatrick, T. B., Tananbaum, L., and Pathak, M. A., Photochemotherapy of psoriasis with oral methoxsalen and long wave ultraviolet light, *N. Engl. J. Med.*, 291, 1207, 1974.
3. Kavli, G. and Volden, G., Phytophotodermatitis, *Photodermatology*, 1, 65, 1984.
4. Fitzpatrick, T. B. and Pathak, M. A., Historical aspect of methoxsalen and other furocoumarins, *J. Invest. Dermatol.*, 32, 229, 1959.

5. Kuske, H., Experimentelle Untersuchungen zur Photosensibilisierung der Haut durch pflanzliche Wirkstoffe. I. Mitteilung. Lichtsensibilisierung als Ursache verschiedener phytogener Dermatosen, *Arch. Dermatol. Syphilol.,* 178, 112, 1939.
6. Kuske, H., Perkutane Photosensibilisierung durch pflanzliche Wirkstoffe, *Dermatologica,* 82, 273, 1940.
7. Fahmy, I.R. and Abu-Shady, H., Ammi majus Linn: pharmacognostical study and isolation of a crystalline constituent, ammoidin, *Q. J. Pharmacol.,* 20, 281, 1947.
8. El Mofty, A. M., A preliminary clinical report on the treatment of leukoderma with Ammi majus Linn, *J. R. Egypt. Med. Assoc.,* 31, 651, 1948.
9. Lerner, A. B., Denton, C. R., and Fitzpatrick, T. B., Clinical and experimental studies with 8-MOP in vitiligo, *J. Invest. Dermatol.,* 20, 299, 1953.
10. Reitz, A., Untersuchungen mit photodynamischen Stoffen (Photobiologischen Sensibilisatoren), *Centr. Bakteriol. Parasitenk.,* 45, 270, 1908.
11. Oginsky, E., Green, G. S., Griffith, D. G., and Fowlks, W. L., Lethal photosensitization with 8-methoxypsoralen to long wave length ultraviolet radiation, *J. Bacteriol.,* 78, 821, 1959.
12. Musajo, L., Rodighiero, G., Colombo, G., Torlone, V., and Dall'Acqua, F., Photosensitizing furocoumarins: interaction with DNA and photoactivation of DNA containing viruses, *Experientia (Basel),* 21, 22, 1965.
13. Musajo, L., Bordin, F., Caporale, G., Marciani, S., and Rigatti, G., Photoreactions at 3654 A between pyrimidin bases and skin-photosenzitizing furocoumarins, *Photochem. Photobiol.,* 6, 711, 1967.
14. Ashwood-Smith, M. J., Poulton, G. A., Barker, M., and Mildenberger, M., 5-methoxypsoralen, an ingredient in several suntan preparations, has lethal mutagenic and clastogenic properties, *Nature (London),* 285, 407, 1980.
15. Averbeck, D., Chandra, P., and Biswas, R. P., Structural specificity in the lethal and mutagenic activity of furocoumarins in yeast cells, *Radiat. Environ. Biophys.,* 12, 241, 1975.
16. Kaye, J., Smith, C. A., and Hanawalt, P. C., DNA repair in human cells containing photoadducts of 8-methoxypsoralen or angelicin, *Cancer Res.,* 40, 698, 1980.
17. Bridges, B. A., Mottershead, R. P., and Knowles, A., Mutation induction and killing of *Escherichia coli* by DNA adducts and crosslinks: a photobiological study with 8-methoxypsoralen, *Chem. Biol. Interact,* 27, 221, 1979.
18. Hanawalt, P. C., Liu, S. C., and Parsons, S., DNA repair responses in human skin cells, *J. Invest. Dermatol.,* 99, 86, 1981.
19. Hanawalt, P. C., Kaye, J., Smith, C. A., and Zolan, M., Cellular responses to psoralen adducts in DNA, in *Psoralens in Cosmetics and Dermatology,* Pergamon Press, Elsmford, N.Y., 1982, 133.
20. Hanawalt, P. C., Calvin, N., and Liu, S. C., DNA damage and carcinogenesis: split-dose PUVA to minimize risk in photochemotherapy for psoriasis, in *Psoriasis — Proceedings of the third Int. Symp.,* Farber, E. M. and Cox, A. J., Eds., Grune & Stratton, New York, 1981, 163.
21. Niggli, H. J. and Cerutti, P. A., Nucleosomal distribution of thymine photodimers following far- and near-ultraviolet light, *Biochem. Biophys. Res. Commun.,* 105, 1215, 1982.
22. Cech, T. and Pardue, M. L., Cross-linking of DNA with trimethylpsoralen is a probe for chromatin structure, *Cell,* 11, 631, 1977.
23. Wiesehahn, G. P., Hyde, J. E., and Hearst, J. E., The photoaddition of trimethylpsoralen to *Drosophila melanogaster* nuclei: a probe for chromatin substructure, *Biochemistry,* 16, 925, 1977.
24. Gurley, L. R., Tobey, R. A., Walters, R. A., Hildebrand, C. E., Hohmann, P. G., D'Anna, J. A., Barham, S. S., and Deaven, L. L., Histone phosphorylation and chromatin structure in synchronized mammalian cells, in *Cell Cycle Regulation,* 1st ed., Jeter, J. R., Cameron, I. L., Padilla, J. M., and Zimmermann, A. M., Eds., Academic Press, New York, 1978, 37.
25. McCarty, K. S. and McCarty, K. S., Jr., Some aspects of chromatin structure and cell-cycle-related postsynthetic modifications, in *Cell Cycle Regulation,* 1st ed., Jeter, J. R., Cameron, I. L., Padilla, G. M., and Zimmermann, A. M., Eds., Academic Press, New York, 1978, 9.
26. Dall'Acqua, F., Vedaldi, D., and Recher, M., The photoreaction between furocoumarins and various DNA with different base compositions, *Photochem. Photobiol.,* 27, 33, 1978.
27. Dall'Acqua, F., Vevaldi, D., Bordin, F., and Rodighiero, G., New studies on the interaction between 8-methoxypsoralen and DNA in vitro, *J. Invest. Dermatol.,* 73, 191, 1979.
28. Zolan, M. E., Smith, C. A., Calvin, N. M., and Hanawalt, P. C., Rearrangement of mammalian chromatin structure following excision repair, *Nature (London),* 299, 462, 1982.
29. Zolan, M. E., Cortopassi, G. A., Smith, C. A., and Hanawalt, P. C., Deficient repair of chemical adducts in alpha-DNA of monkey cells, *Cell,* 28, 613, 1982.
30. Zolan, M. E., Smith, C. A., and Hanawalt, P. C., Formation and repair of furocoumarin adducts in alpha-DNA and bulk DNA of monkey cells, *Biochemistry,* 23, 63, 1984.

31. Bohr, V. A., Smith, C. A., Okumoto, D. S., and Hanawalt, P. C., DNA repair in an active gene: removal of dimers from the DHFR gene of CHO cells is much more efficient than in the genome overall, *Cell,* 40, 359, 1985.
32. Hanawalt, P. C., Intragenomic heterogeneity in DNA damage processing: potential implications for risk assessment in *Mechanisms of DNA Damage and Repair,* Simic, M., Grossman, L., and Upton, A., Eds., Plenum Press, New York, 1986, 489.
33. Poppe, W. and Grossweiner, L. I., Photodynamic sensitization by 8-methoxypsoralen via the singlet oxygen mechanism, *Photochem. Photobiol.,* 22, 217, 1975.
34. Dall'Acqua, F., Marciani, M. S., Zambon, F., and Rodighiero, G., Kinetic analysis of the photoreaction (365 nm) between psoralen and DNA, *Photochem. Photobiol.,* 29, 489, 1979.
35. Yoshikawa, K., Mori, N., Sakakibara, S., Mizuno, N., and Song, P. S., Photo-conjugation of 8-methoxypsoralen with proteins, *Photochem. Photobiol.,* 29, 1127, 1979.
36. Pathak, M. and Prakash, C. J., The nature and molecular basis of cutaneous photosensitivity reactions to psoralens and coal tar, *J. Invest. Dermatol.,* 80, 66s, 1983.
37. Pohl, J. and Christophers, E., Photoinactivation and recovery in skin fibroblasts after formation of mono- and bifunctional adducts by furocoumarins-plus-UVA, *J. Invest. Dermatol.,* 75, 306, 1980.
38. Omar, A., Wiesmann, U. N., Krebs, A., Polyploidization and hemiploidization induced by PUVA in vivo, *Dermatologica,* 159, 195, 1979.
39. Ben-Hur, E. and Elkind, M. M., Psoralen plus near ultraviolet light inactivation of cultured Chinese hamster cells and its relation to DNA cross-links, *Mutat. Res.,* 18, 315, 1973.
40. Epstein, J. and Fukuyama, K., A study of 8-methoxypsoralen (8-MOP) induced phototoxic effects on mammalian epidermal macromolecular synthesis in vivo, *J. Invest. Dermatol.,* 54, 325, 1975.
41. Walter, J. F., Vorhees, J. J., Kelsey, W. H., and Duell, E. A., Psoralen plus black light inhibits epidermal DNA synthesis, *Arch. Dermatol.,* 107, 861, 1973.
42. Henseler, T., Wolff, K., Hoenigsmann, H., and Christophers, E., Oral 8-methoxypsoralen photochemotherapy of psoriasis. The European PUVA study: a cooperative study among 18 European centers, *Lancet,* 1, 853, 1981.
43. Kragballe, K. and Voorheese, V. J., Arachidonic acid and leukotrienes in dermatology. Editorial, *J. Invest. Dermatol.,* 81, 293, 1983.
44. Walzdorfer, U., Christophers, E., and Schroeder, J. M., Methotrexate inhibits polymorphonuclear leukocyte chemotaxis in psoriasis, *Br. J. Dermatol.,* 108, 451, 1983.
45. Cech, T., Pathak, M. A., and Biswas, R. K., An electron microscopic study of the photochemical cross-linking of DNA in guinea pig epidermis by psoralen derivatives, *Biochim. Biophys. Acta,* 562, 342, 1979.
46. Lerche, D., Sondergaard, J., Wadskos, S., Leick, V., and Bohr, V., DNA interstrand cross-links visualized by electron microscopy in PUVA-treated psoriasis, *Acta Derm,. Venereol.,* 59, 15, 1979.
47. Zarebska, Z., Jarzabek-Chorzelska, M., Rzesa, G., Glinski, W., Pawinska, M., Chorzelski, T., and Jablonska, S., Detection of DNA-psoralen photoadducts in situ, *Photochem. Photobiol.,* 39, 307, 1984.
48. Braverman, I. M. and Sibley, J., Role of the microcirculation in the treatment and pathogenesis of psoriasis, *J. Invest. Dermatol.,* 78, 12, 1982.
49. Fry, L. and McMinn, R. M. H., The action of chemotherapeutic agents on psoriatic epidermis, *Br. J. Dermatol.,* 80, 373, 1968.
50. Goldberg, L. H., Cox, A. J., and Abel, E. A., The mitotic index in psoriatic plaques and their response to PUVA therapy, *Br. J. Dermatol.,* 102, 401, 1980.
51. Forster, S., Ilderton, E., Summerly, R., and Yardley, H. J., The level of phospholipase A2 activity is raised in uninvolved epidermis of psoriasis, *Br. J. Dermatol.,* 108, 103, 1983.
52. Baker, B. S., Swain, A. F., Fry, L., and Valdimarsson, H., Epidermal T lymphocytes and HLA-DR expression in epidermis, *Br. J. Dermatol.,* 110, 555, 1984.
53. Haftek, M., Faure, M., Schmitt, D., and Thivolet, J., Langerhans cells in skin from patients with psoriasis: quantitative and qualitative study of T6 and HLA-DR antigen-expressing cells and changes with aromatic retinoid administration, *J. Invest. Dermatol.,* 81, 10, 1983.
54. Bos, J. D., Hulsebosch, H. J., Krieg, S. R., Bakker, P. M., and Cormane, R. H., Immunocompetent cells in psoriasis. In situ immunophenotyping by monoclonal antibodies, *Arch. Dermatol. Res.,* 275, 181, 1983.
55. Aberer, W., Schuler, G., Stingl, G., Hoenigsmann, H., and Wolff, K., Effects of UV light on epidermal Langerhans cells, *J. Invest. Dermatol.,* 74 (Abstr.), 458, 1980.
56. Nordlund, J. J., Ackles, A. E., and Lerner, A. B., The effects of ultraviolet light and certain drugs on Ia-bearing Langerhans cells in murine epidermis, *Cell Immunol.,* 60, 50, 1981.
57. Friedmann, P. S., Ford, G., Roos, J., and Diffey, B. L., Reappearance of epidermal Langerhans cells after PUVA therapy, *Br. J. Dermatol.,* 109, 301, 1983.

58. Morhenn, V. B., Orenberg, E. K., Kaplan, J., Pfendt, E., Terrell, C., and Engleman, A. G., Inhibition of a Langerhans cell-mediated immune response by treatment modalities useful in psoriasis, *J. Invest. Dermatol.,* 81, 23, 1983.
59. Morison, W. L., et al., The influence of ultraviolet radiation on allergic contact dermatitis in the guinea pig. II. Psoralen/UVA radiation, *Br. J. Dermatol.,* 104, 165, 1981.
60. Strauss, G. H., Bridges, B. A., Greaves, M., Hall-Smith, P., Price, M., and Vella-Briffa, A., Inhibition of delayed contact hypersensitivity reaction in skin (DNCB test) by 8-methoxypsoralen photochemotherapy. Possible basis for pseudo-promoting action in skin carcinogenesis, *Lancet,* 2, 556, 1980.
61. Potten, C. S. and Allen, T. D., A model implicating the Langerhans cell in keratinocyte proliferative control, *Differentiation,* 5, 43, 1976.
62. Kaudewitz, P., Braun-Falco, O., Kind, P., Galosi, A., Rieber, P., and Riethmueller, G., Distribution of T-cell subsets as defined by monoclonal antibodies in skin lesions of psoriasis vulgaris, *Arch. Dermatol. Res.,* 276, 33, 1984.
63. Hammar, H., Gu, S. H., Johanneson, A., Sundkvist, K. G., and Biberfeld, P., Subpopulations of mononuclear cells in microscopic lesions of psoriatic patients. Selective accumulation of suppressor/cytotoxic T cells in epidermis during the evolution of the lesion, *J. Invest. Dermatol.,* 83, 416, 1984.
64. Krueger, J. P., Christophers, E., and Schlaak, M., Dose-effect of 8-methoxypsoralen and UVA on cultured human lymphocytes, *Br. J. Dermatol.,* 97, 141, 1978.
65. Kraemer, K. H., Waters, H. L., Ellingson, O. L., and Tarone, R. E., Psoralen plus ultraviolet radiation-induced inhibition of DNA synthesis and viability in human lymphoid cells in vitro, *Photochem. Photobiol.,* 30, 263, 1979.
66. Scherer, R., Kern, B., and Braun-Falco, O., UVA-induced inhibition of proliferation of PHA-stimulated lymphocytes from humans treated with 8-methoxypsoralens, *Br. J. Dermatol.,* 97, 519, 1977.
67. Moscicki, R. A., Morison, W. L., Parrish, J. A., Bloch, K. J., and Colvin, R. B., Reduction of the fraction of circulating helper-inducer T cells identified by monoclonal antibodies in psoriatic patients treated with long-term psoralen/ultraviolet-A radiation (PUVA), *J. Invest. Dermatol.,* 79, 205, 1982.
68. Bjerke, J. R., Livden, J. K., Degre, M., and Matre, R., Interferon in suction blister fluid from psoriatic lesions, *Br. J. Dermatol.,* 108, 295, 1983.
69. Diezel, W., Waschke, S. R., and Soennichsen, N., Detection of interferon in the sera of patients with psoriasis, and its enhancement by PUVA treatment, *Br. J. Dermatol.,* 109, 549, 1983.
70. Morhenn, V. B., Nickoloff, B. J., Merigan, T. C., and Basham, T. Y., The effect of gamma interferon on cultured keratinocytes, *J. Invest. Dermatol.,* 82, 410, 1984.
71. Jablosnka, S., Autoimmunity in psoriasis. II. Immunohistologic studies on various forms of psoriasis and the Koebner phenomenon, in *Psoriasis: Proceedings of the Second International Symposium,* Farber, E. M. and Cox, A. J., Eds., Yorke Medical, New York, 1977, 73.
72. Cox, A. J., Polymorphonuclear leukocytes in psoriasis, in *Psoriasis — Proc. 3rd Int. Symp.,* Farber, E. M. and Cox, A. J., Eds., Grune & Stratton, New York, 1981, 233.
73. Glinski, W., Jablonska, S. et al., *Arch. Dermatol. Res.,* 265, 339, 1979.
74. Schroeder, J. M., Koosfeld, U., and Christophers, E., Multifunction and inhibition in non-stimulated and chemotactic factor stimulated human neutrophils, *J. Invest. Dermatol.,* in press.
75. Bredberg, A. and Forsgren, A., Effects of in vitro PUVA on human leukocyte function, *Br. J. Dermatol.,* 111, 159, 1984.
76. Van de Kerkhof, P. C. M., Van Rennes, H., De Grood, R., Bauer, F. W., and Mier, P. D., Metabolic changes at the margin of the spreading psoriatic lesion, *Br. J. Dermatol.,* 108, 647, 1983.
77. Brody, I., Mast cell degranulation in the evolution of acute eruptive guttate psoriasis vulgaris, *J. Invest. Dermatol.,* 82, 460, 1984.
78. Mortazawi, S. A. M., Meladine mit UVA bei Vitiligo, Psoriasis, Parapsoriasis und Akne vulgaris, *Dermatol. Monatsschr.,* 158, 908, 1972.
79. Tronnier, H. and Schuele, D., First results of therapy with longwave UV after photosensibilisation of skin, in 6th Int. Congr. Photobiology, Bochum, West Germany, 1972.
80. Hofmann, C., Plewig, G., and Braun-Falco, O., Klinische Erfahrungen mit der 8-Methoxypsoralen-UVA-Therapie (Photochemotherapie) bei Psoriasis, *Hautarzt,* 27, 588, 1976.
81. Lakshmipathi, T., Gould, P. W., Mackenzie, L. A., Johnson, B. E., and Frain-Bell, W., Photochemotherapy in the treatment of psoriasis, *Br. J. Dermatol.,* 96, 587, 1977.
82. Mortazawi, S. A. M. and Oberste-Lehn, G., Lichtsensibilisatoren und ihre therapeutischen Faehigkeiten. Erste vorlaeufige Mitteilungen, *Z. Haut. Geschlachtkr.,* 48, 1, 1973.
83. Swanbeck, G., Thyresso-Hoek, M., Bredberg, A., and Lambert, B., Treatment of psoriasis with oral psoralens and long-wave ultra-violet light: therapeutic results and cytogenetic hazards, *Acta Derm. Venereol.,* 55, 367, 1975.
84. Thivolet, J., Ortonne, J. P., and Chouvet, B., La photochimiotherapie du psoriasis: technique et resultats, *Lyon Med.,* 237, 87, 1977.

85. Weissmann, K., Howitz, J., and Bro-Joergensen, A., Treatment of resistant psoriasis with oral 8-methoxypsoralen and long-wave ultraviolet light (PUVA), *Acta Derm. Venereol.,* 57, 73, 1977.
86. Wolff, K., Hoenigsmann, H., Gschnait, F., and Konrad, K., Photochemotherapie bei Psoriasis: klinische Erfahrungen bei 152 Patienten, *Dtsch. Med. Wochenschr.,* 100, 2471, 1975.
87. Wolff, K., Fitzpatrick, T. B., Parrish, J. A., Gschnait, F., Gilchrest, B., Hoenigsmann, H., Pathak, M. A., and Tanenbaum, L., Photochemotherapy with orally administered methoxsalen, *Arch. Dermatol.,* 112, 943, 1976.
88. Wolff, K., PUVA 1979, Klinik und Praxis, in *Fortschritte der praktischen Dermatologie und Venerologie,* Vol. 9, Braun-Falco, O. and Wolff, H. H., Eds., Springer-Verlag, Berlin, 1979.
89. Melski, J. M., Tanenbaum, L., Parrish, J. A., Fitzpatrick, T. B., Bleich, H. L., and 28 participating investigators, Oral methoxsalen photochemotherapy for the treatment of psoriasis: a cooperative clinical trial, *J. Invest. Dermatol.,* 68, 328, 1977.
90. Roenigk, H. H., Jr. et al., Photochemotherapy for psoriasis, *Arch. Dermatol.,* 115, 576, 1979.
91. Morison, W. L., Momtaz, K., Parrish, J. A., and Fitzpatrick, T. B., Combined methotrexate-PUVA therapy in treatment of psoriasis, *J. Am. Acad. Dermatol.,* 6, 46, 1982.
92. Morison, W. L., Parrish, J. A., and Fitzpatrick, T. B., Controlled study of PUVA and adjunctive topical therapy in the management of psoriasis, *Br. J. Dermatol.,* 98, 125, 1978.
93. Fritsch, P. O., Hoenigsmann, H., Jaschke, E., and Wolff, K., Augmentation of oral methoxsalen-photochemotherapy with an oral retinoic acid derivative, *J. Invest. Dermatol.,* 70, 178, 1978.
94. Heidbreder, G. and Christophers, E., Therapy of psoriasis with retinoid plus PUVA: clinical and histologic data, *Arch. Dermatol. Res.,* 264, 331, 1979.
95. Lauharanta, J., Juvakoski, T., and Lassus, A., A clinical valuation of the effects of an aromatic retinoid (Tigason), combination of retinoid and PUVA, and PUVA alone in severe psoriasis, *Br. J. Dermatol.,* 104, 325, 1981.
96. Schmoll, M., Henseler, T., and Christophers, E., Evaluation of PUVA, topical corticosteroids and the combination of both in the treatment of psoriasis, *Br. J. Dermatol.,* 99, 693, 1978.
97. Stern, R. S., Laird, N., Melski, J., Parrish, J. A., Fitzpatrick, T. B., and Bleich, H., Cutaneous squamous-cell carcinoma in patients treated with PUVA, *N. Engl. J. Med.,* 310, 1156, 1984.
98. Hoenigsmann, H., Jaschke, E., Gschnait, F., Brenner, W., Fritsch, P., and Wolff, K., 5-methoxy-psoralen (bergapten) in photochemotherapy of psoriasis, *Br. J. Dermatol.,* 101, 369, 1979.
99. Lerman, S., Ultraviolet radiation and human cataractogenesis, *Invest. Ophthalmol. Vis. Sci.,* 18 (Suppl.), 128, 1979.
100. Lerman, S., Megaw, J., and Willis, I., Potential ocular complications from PUVA therapy and their prevention, *J. Invest. Dermatol.,* 74, 197, 1980.
101. Cloud, T. M., Hakim, R., and Griffin, A. C., Photosensitization of the eye with methoxsalen. II. Chronic effects, *Arch. Ophthalmol.,* 66, 689, 1961.
102. Freeman, R. G. and Troll, D., Photosensitization of the eye by 8-methoxypsoralen, *J. Invest. Dermatol.,* 53, 449, 1969.
103. Parrish, J. A., Chylack, L. T., Jr., Woehler, M. E., Cheng, H. M., Pathak, M. A., Morison, W. L., Krugler, J., and Nelson, W. F., Dermatological and ocular examinations in rabbits chronically photosensitized with methoxsalen, *J. Invest. Dermatol.,* 73, 256, 1979.
104. Baeck, O., Hollstroem, E., Liden, S., and Throburn, W., Absence of cataract ten years after treatment with 8-methoxypsoralen, *Acta Derm. Venereol.,* 60, 79, 1980.
105. El-Mofty, A. M. and El-Mofty, A., Retrospective ocular study of patients receiving oral 8-methoxypsoralen and solar irradiation for the treatment of vitiligo, *Ann. Opthalmol.,* 946, 1979.
106. Hashimoto, K., Kohda, H., Kumakiri, M., Blender, S., and Willis, I., Psoralen-UVA-treated psoriatic lesions. Ultrastructural changes, *Arch. Dermatol.,* 114, 711, 1977.
107. Zelickson, A. S., Mottaz, J. H., and Muller, S. A., Melanocyte changes following PUVA therapy, *J. Am. Acad. Dermatol.,* 1, 422, 1979.
108. Niemi, K. M., Niemi, A., and Kanerva, L., Morphologic changes in epidermis of PUVA-treated patients with psoriasis with or without a history of arsenic treatment, *Arch. Dermatol.,* 119, 904, 1982.
109. Rhodes, A. R., Stern, R. S., and Melski, J. W., The PUVA lentigo: an analysis of predisposing factors, *J. Invest. Dermatol.,* 81, 459, 1983.
110. Kanerva, L., Niemi, K. M., and Lauharanta, J., A semiquantitative light and electron microscopic analysis of histopathologic changes in photochemotherapy-induced freckles, *Arch. Dermatol. Res.,* 276, 2, 1984.
111. Ros, A. M., Wennersten, G., and Lagerholm, B., Long-term photochemotherapy for psoriasis: a histopathological and clinical follow-up study with special emphasis on tumor incidence and behavior of pigmented lesions, *Acta Derm. Venereol.,* 63, 215, 1983.
112. Bleehen, S. S., Freckles induced by PUVA treatment, *Br. J. Dermatol.,* 20, 99, 1978.

113. Gschnait, F., Wolff, K., Hoenigsmann, H., Stingl, G., Brenner, W., Jaschke, E., and Konrad, K., Long-term photochemotherapy: histopathological and immunofluorescence observations in 243 patients, *Br. J. Dermatol.*, 103, 11, 1980.
114. Marx, J. L., Auerbach, R., Possick, P., Myrow, R., Gladstein, A. H., and Kopf, A. W., Malignant melanoma *in situ* in two patients treated with psoralens and ultraviolet A, *J. Am. Acad. Dermatol.*, 9, 904, 1983.
115. Morison, W. L. et al., Abnormal lymphocyte function following long-term PUVA therapy for psoriasis, *Br. J. Dermatol.*, 108, 445, 1983.
116. Walder, B. K. et al., Skin cancer and immunosuppression, *Lancet*, 2, 1282, 1971.
117. Abel, E. A., Cox, A. J., and Farber, E. M., Epidermal dystrophy and actinic keratoses in psoriasis patients following oral psoralen photochemotherapy (PUVA), *J. Am. Acad. Dermatol.*, 7, 333, 1982.
118. Hoffmann, C., Plewig, G., and Braun-Falco, O., Bowenoid lesions, Bowen's disease and keratoacanthomas in long-term PUVA-treated patients, *Br. J. Dermatol.*, 101, 685, 1979.
119. Everett, M. A., Yeargers, E., Sayre, R. M., and Olson, R. L., Penetration of epidermis by ultraviolet rays, *Photochem. Photobiol.*, 5, 533, 1966.
120. Stern, R. S., Parrish, J. A., Fitzpatrick, T. B., and Bleich, H. L., Actinic degeneration in association with long-term use of PUVA, *J. Invest. Dermatol.*, 84, 135, 1985.
121. Van Weelden, H., de Gruijil, F. R., and van der Leun, J. C., Tumors induced by UV-A in mice, *Photochem. Photobiol.*, 37, 879, 1983.
122. Coppey, J., Averbeck, D., and Moreno, G., Herpes virus production in monkey kidney and human cells treated with angelicin or 8-methoxypsoralen plus 365 nm light, *Photochem. Photobiol.*, 29, 797, 1979.
123. West, M. R., Johansen, M., and Faed, M. J. W., Sister chromatid exchange frequency in human epidermal cells in culture treated with 8-methoxypsoralen and long-wave UV radiation, *J. Invest. Dermatol.*, 78, 67, 1982.
124. Cassel, D. M. and Latt, S. A., *Exp. Cell Res.*, 129, 15, 1980.
125. Wolff-Schreiner, E. C., Carter, D. M., Schwarzacher, H. G., and Wolff, K., Sister chromatid exchanges in photochemotherapy, *J. Invest. Dermatol.*, 69, 387, 1977.
126. Bishop, S., DNA repair synthesis in human skin exposed to ultraviolet radiation used in PUVA (psoralen and UV-A) therapy for psoriasis, *Br. J. Dermatol.*, 101, 399, 1979.
127. Igali, S. and Gazso, L., Photosensitizing lethal and mutagenic action of 8-methoxypsoralen and near UV light in *Eschericia coli*, *Stud. Biophys.*, 29, 45, 1971.
128. Arlett, C. F., Mutagenesis in cultured mammalian cells, *Stud. Biophys.*, 36, 139, 1973.
129. Griffin, A. C., Hakim, R. E., and Knox, B., The wave length effect upon erythemal and carcinogenic responses in psoralen treated mice, *J. Invest. Dermatol.*, 31, 289, 1958.
130. Griffin, A. C., Methoxsalen in ultraviolet carcinogenesis in mouse, *J. Invest. Dermatol.*, 32, 376, 1959.
131. Pathak, M. A., Daniels, F., Hopkins, C. E., and Fitzpatrick, T. B., Ultraviolet carcinogenesis in albino and pigmented mice receiving furocoumarins: psoralen and 8-methoxypsoralen, *Nature (London)*, 183, 728, 1959.
132. Grube, D. D., Ley, R. D., and Fry, R. J. M., Photosensitizing effect of 8-methoxypsoralen on the skin of hairless mice. II. Strain and spectral differences for tumorigenesis, *Photochem. Photobiol.*, 25, 269, 1977.
133. Tam, D. W., Scott, van E. J., and Urbach, F., Bowen's disease and squamous cell carcinoma. Occurrence in a patient with psoriasis after topical, systemic, and PUVA therapy, *Arch. Dermatol.*, 115, 203, 1979.
134. Stern, R. S., Thibodeau, L. A., Kleinerman, R. A., Parrish, J. A., and Fitzpatrick, T. B., Risk of cutaneous carcinoma in patients treated with oral methoxsalen photochemotherapy for psoriasis, *N. Engl. J. Med.*, 300, 809, 1979.
135. Roenigk, H. H. and Caro, W. A., Skin cancer in the PUVA-48 cooperative study, *J. Am. Acad. Dermatol.*, 4, 319, 1981.
136. Lassus, A., Reunala, T., Idaenpaeae-Heikkilae, J., Juvakoski, T., and Salo, O., PUVA treatment and skin cancer: a follow-up study, *Acta Derm. Venereol.*, 61, 141, 1981.
137. Lindskov, A., Skin carcinomas and treatment with photochemotherapy (PUVA), *Acta Derm. Venereol.*, 63, 223, 1983.
138. Lobel, E., Paver, K., King, R., Le Guay, J., Poyzer, K., and Wargon, O., Relationship of skin cancer to PUVA therapy in Australia, *Aust. J. Dermatol.*, 22, 100, 1981.
139. Scotto, J., Kopf, A. W., and Urbach, F., Non-melanoma skin cancer among Caucasians in four areas of the United States, *Cancer*, 34, 1333, 1974.
140. Moon, R. C. and Itri, L. M., Retinoids and cancer, in *The Retinoids*, Vol. 2, Sporn, M. B., Roberts, A. B. and Goodman, D. S., Eds., Academic Press, New York, 1984, 327.

141. Averbeck, D., Dubertrat, L., Bisagni, E., Moron, J., Papadopoulo, D., Nocentini, S., Blais, J., and Zajdel, F., Photobiological and phototherapeutic properties of new monofunctional pyridopsoralens, *J. Invest. Dermatol.*, 80, 306, 1983.
142. Hoenigsmann, H., Wolff, K., Gschnait, F., Brenner, W., and Jaschke, E., Keratoses and nonmelanoma skin tumors in long-term photochemotherapy (PUVA), *J. Am. Acad. Dermatol.*, 3, 406, 1980.

Chapter 5

PHOTOCHEMOTHERAPY OF VARIOUS SKIN DISORDERS

Jeffrey D. Bernhard and Barbara A. Gilchrest

TABLE OF CONTENTS

I. RATIONALE FOR PHOTOCHEMOTHERAPY IN NONPSORIATIC DISORDERS

Since its introduction for the treatment of psoriasis in 1974, oral methoxsalen photochemotherapy (PUVA) has become an established dermatologic treatment modality.[1,2] Given its dramatic effectiveness in severe and/or recalcitrant cases of psoriasis, it is not surprising that PUVA has been used to treat a variety of other cutaneous disorders. Table 1 lists diseases that have been reported to respond to PUVA, sometimes in a single patient or small series of patients, and occasionally in well-designed clinical trials. It must be noted, however, that to date the U.S. Food and Drug Administration has approved PUVA only for the treatment of psoriasis and vitiligo.

The mechanism of action for PUVA in psoriasis is not known, but a variety of effects are probably at play. At the molecular level, 8-methoxypsoralen forms photoadducts with DNA upon irradiation with UVA, cross-linking the DNA strands, and presumably interferes with cellular replication. This antiproliferative effect provides a rationale for the use of PUVA in psoriasis and other hyperproliferative disorders. PUVA has also been documented to have a number of immunologic and anti-inflammatory effects,[3] and overall, the therapy results in a transient and dose-related downward modulation of the immune response. These phenomena may also contribute to the beneficial effect of PUVA in psoriasis, and provide a rationale for the treatment of other diseases in which immunologic and inflammatory mechanisms are apparently involved, such as alopecia areata, lichen planus, and eczematous dermatoses.

Finally, psoralen plus UVA radiation leads to an increased melanocyte population as well as increased melanin production by individual cells, providing a rationale for the treatment of pigmentary disorders such as vitiligo, in which a putative immunologic aberration may be altered by PUVA as well.

II. DISORDERS RESPONSIVE TO PUVA

A. Cutaneous T-Cell Lymphoma (CTCL)

PUVA therapy of mycosis fungoides, more recently and appropriately renamed CTCL, was first reported in 1976.[4] In this report, all nine patients with biopsy-confirmed disease (five with extensive plaques, three with erythroderma, and one with plaques and tumors) responded favorably, with complete clearing in four patients. Follow-up of this group and 2 additional patients revealed good to excellent control in 10 of the 11 for periods ranging from 8 months to 4.5 years,[5] with fewer side effects than anticipated for a comparable group of CTCL patients managed with conventional therapies such as oral prednisone (for exfoliative erythroderma), topical nitrogen mustard, or electron-beam irradiation. Subsequent reports[6-9] have confirmed the initial impression that PUVA is an attractive treatment option for CTCL.

The universal published experience has been that PUVA is highly effective at clearing eczematous and plaque-stage lesions and moderately effective at clearing early tumor-stage lesions.[4-9] In the only available long-term study,[9] 19 patients were followed 6 to 7 years and an additional 25 patients were followed for a shorter period after beginning PUVA. All those with eczematous and/or plaque-stage CTCL cleared; and after a mean follow-up of 44 months, approximately half remained free of disease. Some patients remained in continuous remission for up to 6.5 years after a single PUVA course. However, of the seven patients with early tumor-stage CTCL who cleared initially with PUVA therapy, all had relapses and only three survived 6 years despite additional conventional treatment.

Reported treatment protocols are identical to those for psoriasis.[4,5] Thick tumors and lesions in relatively shielded areas such as the scalp or axillary vault are predictably

Table 1
CUTANEOUS DISORDERS REPORTED TO RESPOND TO PHOTOCHEMOTHERAPY

Psoriasis	Certain photodermatoses
Cutaneous T-cell lymphoma	Solar urticaria
Parapsoriasis	Polymorphic light eruption
Vitiligo	Actinic reticuloid
Atopic eczema	Photosensitive eczema
Alopecia areata	Persistent light reactor
Lichen planus	Urticaria pigmentosa
Palmar-plantar dermatoses	Graft vs. host reaction
Pityriasis lichenoides	Others: see Table 2

difficult to eradicate and indefinite maintenance appears necessary to prevent relapse. Side effects of PUVA therapy experienced by CTCL patients are those also noted by psoriatics: mild nausea, pruritus, and localized erythema are common. In very elderly or debilitated patients or those with skin damage resulting from prior fluorinated steroid usage or X-irradiation, these side effects are sometimes poorly tolerated, however. One curious apparent side effect of PUVA treatment for CTCL, observed but not reported in one of the original patients[4] and subsequently reported in two patients at another center,[10] is depigmentation clinically and histologically indistinguishable from vitiligo, although such depigmentation may also occur spontaneously in CTCL.[11]

B. Parapsoriasis

Due to the lack of a universally accepted definition for parapsoriasis, discussion of this uncommon disorder and of reported therapeutic responses is difficult.[12] For the most part, the term "parapsoriasis" is used to denote a number of chronic papulosquamous disorders that may be very pruritic and that tend to respond poorly to topical or systemic treatment of any sort. Furthermore, certain variants, such as large plaque parapsoriasis, are generally accepted as precursors to mycosis fungoides (CTCL). The risk of malignant transformation and resistance to conventional treatment have made photochemotherapy a welcome treatment option for parapsoriasis.

The largest reported experiences with PUVA treatment of parapsoriasis are those from the Mayo Clinic[13] and Lyon, France,[14] although both of these series include some patients with pityriasis lichenoides, considered separately later. For the most part, treatment was administered three times weekly 2 hr after oral ingestion of 30 to 50 mg methoxsalen (based on body weight) according to protocols essentially the same as those used in the treatment of psoriasis. In one report,[15] three patients with parapsoriasis en plaques were treated with UVA after topical application of trioxsalen solution.

In the Mayo Clinic series, each of the seven patients with parapsoriasis en plaques experienced complete clearing. One patient with digitate parapsoriasis experienced considerable improvement before discontinuing treatment because he felt that the methoxsalen exacerbated his irritable bowel syndrome. One patient with atrophic large-plaque parapsoriasis developed non-Hodgkin's lymphoma before treatment with PUVA; after systemic chemotherapy, PUVA cleared more than 80% of the persistent skin lesions.[13] From their experience, these authors conclude that PUVA may be the treatment of choice for most forms of parapsoriasis because it consistently led to improvement in their patients. They further suggest that PUVA treatment for certain forms of parapsoriasis be recognized as a major advance not only for clearing recalcitrant skin lesions but also for possibly preventing evolution into mycosis fungoides. Along with symptomatic and cosmetic improvement, this potential benefit — which remains conjectural — must be weighed against the well-known side effects of PUVA.[16]

The observation that PUVA may induce long-lasting disease-free intervals if used in the early stages of mycosis fungoides[9] provides a particular incentive for its use in potentially "pre-mycotic" variants of parapsoriasis. Honigsman et al.[9] emphasize that possible long-term hazards related to frequent PUVA treatments are less relevant for patients with CTCL than for patients with benign conditions such as psoriasis, a perspective also applicable to premycotic variants of parapsoriasis, with the caveat that accurate diagnosis and close long-term follow-up of such patients is absolutely essential.

C. Vitiligo

Vitiligo is an acquired amelanosis of the skin.[17] Although its precise pathogenesis is not known, autoimmune factors are thought to be important, and it is noteworthy that vitiligo may be associated with other autoimmune disorders, such as pernicious anemia, Addison's disease, and thyroid disease. Especially when it occurs on areas such as the face, vitiligo can be a significant cosmetic liability, and in some cases a psychosocial disaster. The ancient Egyptians found that a topical application of an extract from Ammi majus, a weed that grows along the Nile, followed by exposure to sunlight, was successful in repigmenting vitiliginous skin. What is new in the current photochemotherapeutic approach to this disorder, however, is a substitute for the sun: high-intensity UVA light sources.[18]

Egyptian workers identified psoralen compounds as the active ingredients from Ammi majus in the 1940s and proposed that they be used to treat vitiligo patients.[19] In contrast to the other disorders discussed in this chapter, 8-methoxypsoralen (8-MOP or methoxsalen) and 4,5′,8-trimethylpsoralen (TMP or trioxsalen) are both effective in the treatment of vitiligo. Furthermore, TMP is frequently used when sunlight is utilized as the light source, since TMP is much less erythemogenic than 8-MOP[20] and therefore safer to use in uncontrolled situations such as outdoor self-treatment.

PUVA treatment of vitiligo is an arduous undertaking. Results vary widely according to the affected areas of skin, physician interest and sophistication, and patient motivation and compliance. Patients who want to undertake photochemotherapy of vitiligo must understand that hundreds of treatments over a period of 12 to 24 months or more may be required to achieve results. They must not expect complete resolution. The face and neck respond to treatment best, the hands and feet are most resistant. In widespread vitiligo, complete confluent regimentation of all sites is unusual.[18] In the largest recent study on photochemotherapy of vitiligo with oral methoxsalen and high-intensity UVA light sources, PUVA led to total repigmentation in only half of 22 patients with focal vitiligo.[21] Among 42 patients treated in Boston with oral TMP and artificial UVA,[22] only one third achieved at least 80% repigmentation; of those receiving fewer than 200 treatments, the proportion was one fourth. Response rates were substantially better for the face and neck alone, however, with 60% of those with 50 or more PUVA exposures achieving this degree of repigmentation.

Either 8-MOP, or TMP, or both together may be utilized in PUVA treatment of vitiligo. Mosher et al.[22] compared these agents in a study of outdoor photochemotherapy (PUVA-Sol) involving 365 patients in India treated over 2 years. They concluded that the most effective single agent was 8-MOP, and that 0.3 mg/kg, half the standard dose for treatment of psoriasis, gave optimal results. However, because TMP is less phototoxic and therefore safer, it is preferred over 8-MOP by these investigators when outdoor sunlight rather than controlled indoor irradiation is utilized for PUVA. For either drug, eye protection with appropriate UV-blocking glasses is essential. Shielding of lips, nipples, genitalia, and indeed any uninvolved areas of the body is desirable. Initial exposures should be with TMP 0.6 to 0.9 mg/kg and 1 to 2 J/cm^2 UVA or roughly 5 to 10 min of summer sun. Treatments are taken 2 or 3 times per week, never 2 days in a row, and increased by 0.5 to 1 J/cm^2 (or about 5 min of sunlight) until a maximum of 8 to 12 J/cm^2 or 45 min of sun is achieved. The dose is increased very carefully, with explicit instructions not to proceed if more than minimal erythema is observed. This is a critical point: severe phototoxicity may precipitate an isomorphic response with development of new areas of vitiligo. The physician is strongly urged to become familiar with detailed instructions provided by Mosher et al.[22] or by others experienced with this treatment.[23-27]

The first sign of a good clinical response is perifollicular repigmentation. If there is no response after 25 TMP treatments, the authors recommend a switch to 8-MOP 0.3 mg/kg with a decrease in UVA dose to 2 J/cm^2. 8-MOP may be increased to 0.6 mg/kg after 25 further treatments if there is still no repigmentation. The last resort is to utilize 8-MOP 0.3 mg/kg and TMP 0.6 mg/kg in combination. Absence of pigmentation after another 25 treatments constitutes a failure of PUVA therapy. Since relapses are most likely in lesions that have not completely repigmented, it is important not to discontinue treatment too early if the patient is a responder. Patients undertaking this treatment for vitiligo should ideally plan to continue either outdoor or indoor PUVA through the entire year.

Side effects of oral methoxsalen photochemotherapy for vitiligo are similar to those observed in PUVA treatment for other disorders, except that symptomatic erythema — which is to be avoided — may occur more often. Among 139 Finnish patients treated with 8-MOP and artificial UVA, side effects included erythema in 40%, nausea in 33%, and pruritus in 12%.[21]

Finally, topical photochemotherapy for vitiligo is an effective modality that deserves careful consideration if the vitiligo lesions are small. Children under 12 are generally not treated with oral 8-MOP, and topical 8-MOP may be considered in those cases in which the child, the parents, and the physician concur that treatment benefits are likely to outweigh the inconvenience and risks. Because topical psoralen compounds are extremely phototoxic for vitiliginous skin, this treatment is usually conducted under close medical supervision to prevent severe burns. Great precautions must be taken to prevent accidental overexposure.

In a recent study of topical photochemotherapy in 73 vitiligo patients, 7 (9%) had 100% repigmentation, 26 (36%) had 50% or greater repigmentation, 29 (40%) had some degree of repigmentation, and 11 (15%) had no repigmentation.[28] These investigators found that 0.1% topical 8-MOP was as effective as higher concentrations and caused fewer side effects. As with oral photochemotherapy, the face was the most responsive area; dorsal hands and feet, digits, palms, and soles were most recalcitrant.[28] The solution was applied 1 hr prior to UVA exposure once a week. Initial exposures were 0.25 J/cm^2, increased by 0.25 J/cm^2 weekly until moderate asymptomatic erythema was achieved. The maximum UVA dose was 11 J/cm^2. The authors conclude that topical photochemotherapy can be a safe and effective therapy in children and adults with limited involvement, but that oral psoralen is the therapy of choice for extensive disease.

D. Atopic Dermatitis

PUVA led to temporary clearing of atopic dermatitis in 15 patients reported by Morison and co-workers[29] and in 9 of 10 patients reported by Soppi and co-workers.[30] Treatment details are not given in the second study, but in the first, treatment was administered three times per week according to a protocol in which the initial dose was determined by the patient's skin type, as for psoriasis, starting with 1.5 to 3.5 J/cm^2, and increasing by 0.5 to 1.0 J/cm^2 provided there was no significant erythema present from the previous treatment.[29] Oral methoxsalen (0.6 mg/kg body weight) was given 2 hr before exposure to UVA radiation. Final UVA dose at clearing was 7 to 24 J/cm^2

(mean 13 J/cm²) after 50 to 202 days (mean 101) of treatment. In ten patients, the first sign of improvement was a decrease in itching; in five the first sign of improvement was a decrease in the severity and extent of eczema. Although improvement was noted after 4 to 21 treatments, progress was slow, and clearing required 18 to 59 (mean 36) treatments. In this study and in subsequent clinical experience, the number of treatments required to achieve improvement or clearing, the duration of treatment, and maintenance requirements are all greater in atopic eczema than in psoriasis. Furthermore, relapse of eczema occurs rapidly if adequate maintenance is not given, with 13 of 15 patients in the original series[29] requiring exposures more often than once per week and only 1 patient disease-free for 3 months without maintenance. Despite frequent maintenance treatments, relapse is common. Even 3 to 5 years after clearing, most patients still receiving therapy appear to require treatment every other week.[2]

The mechanisms of the beneficial effect of PUVA in atopic eczema are unknown. Although local and/or systemic immunologic alterations are believed to play a role, as yet unrecognized physical or chemical changes in the skin may be important as well.

As in UV phototherapy for other diseases, precipitation of recurrent herpes simplex may occur during PUVA treatment for atopic dermatitis, but is of special concern because of the risk of viral dissemination in atopic individuals. This possibility should be carefully watched for, so that diagnosis and therapeutic interventions can be made early. In addition to the usual concerns and side effects, increased "dryness" of the skin and recurrent mild folliculitis are common,[2] but respond to emollients and systemic antibiotics, respectively.

Successful PUVA treatment for atopic dermatitis can be extremely rewarding, but it is not easy. It offers an effective option for a condition that can be psychologically and physically devastating and for which other options are rather limited. However, PUVA requires not only a highly motivated patient, but an experienced physician-phototherapist with adequate time and motivation as well.

Certain other types of eczema, such as seborrheic eczema and contact dermatitis, may also respond to PUVA,[2] although there are no published treatment series.

E. Alopecia Areata

Alopecia areata (AA) is a common, distinctive form of hair loss in which circumscribed areas of near-total or total baldness usually appear rather suddenly.[31] Although the precise etiology remains unknown, immunological abnormalities are thought to play an important role. There is no uniformly acceptable or successful treatment for AA. Relapses are common, and severe cases can be psychologically devastating.

A number of centers have reported encouraging, albeit preliminary results for the use of photochemotherapy in the treatment of AA with good to excellent regrowth of hair in 60 to 70% of patients.[32-38] Psoralen has been administered topically (1% 8-MOP ointment or 0.1% 8-MOP solution) or orally (usually 0.6 mg/kg 8-MOP) and followed by UVA irradiation 1 or 2 hr later. UVA dosage has been increased gradually with each treatment as tolerated on a two or three times per week schedule with maximum UVA doses not exceeding those routinely regarded as the upper limits for psoriasis treatment, as determined by skin type and clinical response. In one study, the maximum dose per session was 11 J/cm² and the average was 8 to 8.5 J/cm², with appearance of vellus hair around the 30th session and complete regrowth usually requiring 50 to 80 sessions.[37]

Since regrowth of eyelashes which were shielded by goggles has been observed, and the shielding effect of regrown hairs has not had a deleterious effect on subsequent regrowth in some patients,[37] the proposition that PUVA acts via systemic rather than merely local immunomodulation appears tenable.

In one study, good treatment results, obtained in 11 of 17 patients, did not appear to depend on the age of onset, extent, or duration of AA,[37] while in another, patients more than 20 years old at the first attack fared better than those with earlier onset.[38] Among the 43 of 76 patients with good to excellent results in the latter study, relapse occurred in approximately half after follow-up periods as long as 68 months after completion of treatment.[38] Many, but not all, patients with disease recurrence responded to retreatment.[37,38]

More experience and prospective studies will be required to determine the prognostic factors and optimum methods for PUVA treatment of AA. In any individual patient, the overall risks must be weighed against the possible benefits, with consideration given to alternative methods of treatment and to the relapsing nature of the illness. For the moment, the findings are encouraging and suggest that PUVA belongs on the rather short list of treatments appropriate for selected cases of AA.

F. Lichen Planus

Lichen planus is a distinctive cutaneous disorder that consists of flat-topped, angular, violaceous papules that may be intensely itchy and widespread. The rationale for PUVA therapy is provided by the characteristic epidermal changes as well as band-like infiltrate of predominantly T lymphocytes in the upper dermis, at a depth accessible to UVA radiation. Furthermore, cells of the same lineage have previously been shown to be PUVA-responsive in lesions of cutaneous T-cell lymphoma.

Ortonne and co-workers[39] first reported clinical improvement in six of seven patients with widespread lichen planus treated with PUVA. The cumulative UVA dose required for clearing was 150 ± 39 J/cm^2, corresponding to an average of 31 treatment sessions over 10 weeks. Lesions on mucous membranes and on nonexposed skin areas persisted, thereby eliminating the possibility that improvements were due to spontaneous remission. Clinical remission correlated with diminution or disappearance of the superficial dermal infiltrate.

The effectiveness of PUVA treatment for lichen planus was recently confirmed by Gonzalez and co-workers[40] in a bilateral comparison study. Five of ten patients cleared completely once both sides were treated and required no maintenance treatment during follow-up periods of up to 4 years. A mean of 27 treatments was required to clear one side, with a cumulative dose 256 J/cm^2. Treatments were administered three times a week following the PUVA protocol used to treat psoriasis at the same institution. Among those who cleared completely, a mean of 6 weeks was required. Two of the ten patients developed an exacerbation of their disease that required hospitalization for symptomatic treatment, and three patients achieved partial improvement.

On the basis of the available data, PUVA would appear to be an effective, alternative therapy for some cases of generalized, symptomatic lichen planus. It must be used with caution by experienced phototherapists however, not only because of the usual inherent risks, but because of the risk of photoexacerbation in some cases.

G. Palmar and Plantar Dermatoses

A number of chronic dermatoses affect the palms and/or soles. Some of these, including certain varieties of eczema and psoriasis, can be extremely difficult to treat. By shielding unaffected areas, it is possible to localize photochemotherapy to the palms and/or soles, an approach reported to be extremely effective for localized plaque-type psoriasis, pustular psoriasis, endogenous eczema, and persistent palmar plantar pustulosis.[41-45] PUVA has also been successful in treating idiopathic hyperkeratotic dermatoses of the palms,[46] dyshidrotic eczema (pompholyx),[47] and even allergic contact dermatitis of the hands.[48] The use of bilateral comparison treatment protocols (PUVA vs. no treatment) in some studies has established effectiveness beyond question,[42,45,47] and response rates are excellent.

Although topical application of methoxsalen for localized PUVA treatment of such disorders has been successful in the hands of some experienced investigators, the risk of uneven light penetration or variable drug absorption causing burns has led to a general preference for oral administration in most centers. In one study in which topical PUVA was compared to oral PUVA, five patients who had cleared with oral psoralen on one side failed to clear with topical psoralen on the other.[45] In most instances, treatment protocols are similar to those used in treating psoriasis. Depending upon the condition being treated and the individual patient, maintenance treatments appear to be required by the majority of patients, a reflection of the inherent chronicity and recalcitrance of these disorders. However, some prolonged remissions following successful treatment have been observed.[42]

H. Photodermatoses

PUVA has been used in the prophylaxis of some photodermatoses, such as solar urticaria and polymorphic light eruption, and in the treatment of certain others, such as persistent light reactivity and actinic reticuloid. Its mechanism of action in each disorder may be different, and has not been established in any. However, proposed explanations include induced photoprotection through melanogenesis and stratum corneum thickening ("hardening"); and induction of "tolerance" either by mediator depletion or immunologic alterations. Since the action spectrum for most photodermatoses includes UVA, PUVA treatment in these disorders must be undertaken with extreme caution under careful daily observation, and must usually start with doses lower than those used in the treatment of other disorders. Whenever possible, the action spectrum and sensitivity for each individual patient should be determined before starting treatment. In some instances, as in the prophylactic use of PUVA for polymorphic light eruption, manifestations of the disease may be provoked during the course of treatment. For example, in one study[49] two of ten patients developed exacerbation of their polymorphic light eruption (PMLE). These exacerbations were treated with a tapering course of systemic corticosteroids while PUVA treatment was continued.

The most common photodermatosis, PMLE, is the one in which the largest experience with PUVA treatment has been gained.[49,50] Treatments are carried out two or three times per week, generally starting with a UVA dose of 1.5 to 4.5 J/cm^2 and increasing by 0.5 to 1.5 J/cm^2 at each treatment unless PMLE symptoms or phototoxicity are provoked. The usual dose of 8-MOP (0.6 mg/kg) is used. Photoprotection may be achieved after as few as six to ten treatments[50] or 4 to 12 weeks.[49] Increased solar tolerance may last for 6 to 8 weeks or longer; regular sunlight exposure may be required to maintain tolerance. In one study,[49] only two of six patients who responded to PUVA also responded to treatment with β-carotene when these treatments were compared in consecutive years. Overall, the response rate to PUVA appears to be 90%, while β-carotene is apparently effective in only about a third of patients with PMLE.

PUVA is also helpful in the prophylactic treatment of solar urticaria. PUVA therapy was successful in six of six patients treated three times per week by Parrish and co-workers[51] and in one patient treated four times per week by Holzle and co-workers.[52] This success is not surprising, in that other studies have shown "densisitization" by means of repeated UV exposures produces a marked decrease in the urticarial reaction. Patients not photosensitive to UVA have been treated according to usual protocols with starting doses based upon skin type, while patients sensitive to UVA were treated initially with 80% of their experimentally determined minimal urtication dose, with incremental dosage increases of 0.5 J/cm^2 as tolerated. When treatment produced an urticarial reaction, the UVA dose was held constant for subsequent treatments until no further reactions occurred. Once solar tolerance is achieved, it must be maintained by either regular sunlight exposure or maintenance treatments. In some cases, UVA alone without psoralen may be effective.[53]

PUVA treatment of the more unusual photodermatoses such as actinic reticuloid or persistent light reaction has been attempted in only a few cases. Details can be found in the original reports.[52,54,55]

It is important to exclude the diagnosis of lupus erythematosus before using PUVA to treat any disorder, and this is even more critical before using PUVA to treat photodermatoses, with which lupus photosensitivity may be confused. Another special concern is the exquisite sensitivity to UV and sometimes visible radiation often manifested by patients with photodermatoses. Whenever possible, action spectra should be determined to guide treatment and to assess the therapeutic response. For some patients, the degree of photosensitivity precludes PUVA therapy even in the face of concomitant treatment with such agents as hydroxychloroquine and prednisone.[56]

I. Pityriasis Lichenoides

Pityriasis lichenoides is a distinctive but uncommon dermatosis of unknown cause. It occurs in acute, chronic, and probably intermediate forms, and is characterized by discrete, erythematous, scaling papules that may become hemorrhagic and crusted, and may ulcerate. Aside from pruritus and the psychological effect of a generalized cutaneous eruption, varioliform (smallpox-like) scars can occur and may be cosmetically disfiguring. No uniformly safe and effective treatment has been available; and PUVA has now been successful in managing a number of long-standing cases resistant to other forms of treatment such as antibiotics, gold injections, topical and systemic steroids, and methotrexate. To date, PUVA treatment of a total of at least 19 patients with various forms of pityriasis lichenoides has been reported, with results ranging from marked improvement to total clearing without recurrence,[14,57-61] although the disorder may also recur after PUVA therapy is discontinued.[62] For the most part, treatment has been carried out twice weekly according to protocols used to treat psoriasis, and a strong argument can be made[61] that PUVA should be considered before the potentially more toxic alternatives listed earlier.

Finally, it is worth noting that UVB phototherapy has also been used successfully to treat pityriasis lichenoides.[63] Whether PUVA has significant advantages over UVB for this disorder remains to be determined. The average number of treatments required for clearing, and optimal maintenance treatment schedules are as yet unknown for both UVB and PUVA.

J. Urticarial Disorders

1. Urticaria Pigmentosa

Using the PUVA protocol they employ to treat psoriasis, Christophers and co-workers[64] reported extremely encouraging results in ten patients with urticaria pigmentosa (UP). Although macular hyperpigmentation persisted over cutaneous mast-cell infiltrates, PUVA treatment prevented whealing and dermographism following physical trauma, the clinical hallmark of this disease, within 27 treatments. In one patient, migraine headaches disappeared, coincident with a decrease in serum histamine concentration from elevated to normal, but symptoms recurred after PUVA was stopped for 8 weeks. Decreased urinary histamine metabolites and histologically detectable dermal mast cells correlated with clinical improvement in three additional patients treated by Granerus et al.[65] Changes in cutaneous mast-cell number, ultrastructure, and histamine content have been inconsistent in two other UP patient groups with clinical improvement following PUVA therapy.[66,67]

Among other possibilities, it has been suggested that PUVA decreases symptoms and whealing in urticaria pigmentosa by "stabilizing" mast-cell membranes, by chronically

Table 2
UNCOMMON DISEASES SUCCESSFULLY TREATED BY PHOTOCHEMOTHERAPY IN SMALL NUMBERS OF PATIENTS

Disease	Ref.
Lymphomatoid papulosis	77
Transient acantholytic dermatosis	81
Prurigo nodularis	78
Pustulosis vegetans[a]	79
Scleromyxoederma	80
Graft-vs.-host reaction	85
Idiopathic pruritus	82
Granuloma faciale[a]	83
Keratosis lichenoides chronica	84

[a] Topical PUVA was utilized.

depleting histamine, by influencing vascular responsiveness to histamine, and by altering cutaneous nerves or their responses. The degree of relief and required frequency of maintenance treatments appear to vary widely from patient to patient.

2. Physical Urticarias

PUVA treatment of solar urticaria was discussed earlier. Given the effectiveness of PUVA in solar urticaria and in urticaria pigmentosa, it appeared sensible to try this therapy for symptomatic dermographism. However, in a recent uncontrolled trial, only 5 of 14 patients reported improvement of pruritus sufficient to justify having undertaken the treatment, and in none was there a significant reduction in whealing.[68] Why PUVA should be quite effective in some mast cell-mediated disorders and not in others remains unclear.

K. Miscellaneous

Individual patients with a number of uncommon diseases have been treated by photochemotherapy (Table 2), but at this point it is either premature or impossible to determine the ultimate role that PUVA may play in their treatment.

III. DISORDERS UNRESPONSIVE TO PUVA OR EXACERBATED BY PUVA

As stated by Morison,[2] "there is always the possibility that the treatment might be effective if used in a different way", and negative results must be interpreted with this fact in mind. Furthermore, few investigators publish negative results, so the following list cannot be considered complete. However, diseases that evidently do not respond to PUVA treatment include: acne vulgaris, pityriasis rubra pilaris, herpes simplex, warts, scleroderma, and keloids.[2] Lupus erythematosus, and probably pemphigus vulgaris and bullous pemphigoid are exacerbated by photochemotherapy.[2] Any photosensitivity disease may be exacerbated by photochemotherapy.

IV. ORAL VS. TOPICAL PUVA

In most of the disorders discussed earlier, oral methoxsalen photochemotherapy is utilized primarily. Topical application of methoxsalen followed by exposure to UVA (topical PUVA) has also been effective in some of these.[69] Topical PUVA makes especially good sense in other localized disorders, in order to avoid unnecessary risks of oral PUVA therapy such as cataract formation, phototoxicity, and photocarcinogenesis. The latter two risks are not eliminated with topical PUVA, however, and some investigators and patients find oral treatment to be more convenient and predictable. There are no adequate studies of relative effectiveness for the two forms of photochemotherapy.

V. USE OF UVA OR UVB ALONE

Some PUVA-responsive dermatoses may respond to UVA alone. Psoriasis, for example, responds to large doses of UVA, but light sources capable of delivering such large doses (20 to 300 J/cm^2) to large surface areas within a practical time span are not presently available.[70] An exception is solar urticaria, which may respond to practical doses of UVA alone.[53]

Many PUVA-responsive dermatosis also respond to UVB phototherapy,[71] although response rates are usually higher for PUVA, especially among "resistant" patients.

VI. PUVA COMBINATIONS

The use of PUVA in combination with other modalities, such as UVB, methotrexate, and retinoids has been recently reported for the treatment of psoriasis.[72-74] It is reasonable to expect that such adjunctive therapies may also be found helpful in some of the diseases discussed earlier.

VII. RISKS AND BENEFITS

PUVA therapy is not without risks.[2,16,75,76] When faced with a patient whose skin disease may give rise to physical discomfort, psychological distress, or disfigurement, the physician must consider these risks as part of the overall risk-benefit equation in every individual case.

VIII. CONCLUSION AND FUTURE DIRECTIONS

An exciting new development in the management of disorders caused by abnormal lymphocytes (such as Sezary syndrome) is extracorporeal photophoresis, in which blood containing 8-MOP is passed from one arm vein through a UVA exposure system and then back to the patient through another arm vein. The potential for this procedure, which is still experimental, is enormous, and may be applicable to the treatment of a variety of disorders ranging from leukemia to autoimmune disorders to organ transplant rejection.

Photochemotherapy is an important treatment option for an increasing number of skin diseases. For some, it appears to be the treatment of choice. In the future, modifications designed to decrease risks and side effects and to increase effectiveness will probably expand its role.

61. Powell, F. C. and Muller, S. A., Psoralens and ultraviolet A therapy of pityriasis lichenoides, *J. Am. Acad. Dermatol.*, 10, 59, 1984.
62. Gonzalez, E., Editorial comment, in *1984 Yearbook of Dermatology*, Sober, A. J. and Fitzpatrick, T. B., Eds., Year Book Medical Publishing, Chicago, 1984, 263.
63. Levine, M. J., Phototherapy of pityriasis lichenoides, *Arch. Dermatol.*, 119, 378, 1983.
64. Christophers, E., Honigsmann, H., Wolff, K., and Langner, A., PUVA-treatment of urticaria pigmentosa, *Br. J. Dermatol.*, 98, 701, 1978.
65. Granerus, G., Roupe, G., and Swanbeck, G., Decreased urinary histamine metabolite after successful PUVA treatment of urticaria pigmentosa, *J. Invest. Dermatol.*, 76, 1, 1981.
66. Vella Briffa, D., Eady, R. A. J., James, M. P., Gatti, S., and Bleehen, S. S., Photochemotherapy (PUVA) in the treatment of urticaria pigmentosa, *Br. J. Dermatol.*, 109, 67, 1983.
67. Kolde, G., Frosch, P. J., and Czarnetzki, B. M., Response of cutaneous mast cells to PUVA in patients with urticaria pigmentosa: histomorphometric, ultrastructural, and biochemical investigations, *J. Invest. Dermatol.*, 83, 175, 1984.
68. Logan, R. A., O'Brien, T. J., and Greaves, M. W., The effect of photochemotherapy (PUVA) on symptomatic dermatographism, *Br. J. Dermatol.*, 111(Suppl. 26), 41, 1984.
69. Abel, E. A., Goldberg, L. H., and Farber, E. M., Treatment of palmoplantar psoriasis with topical methoxsalen plus long-wave ultraviolet light, *Arch. Dermatol.*, 116, 1257, 1980.
70. Parrish, J. A., Treatment of psoriasis with long-wave ultraviolet light, *Arch. Dermatol.*, 113, 1525, 1972.
71. Anderson, T. F., Waldinger, T. P., and Voorhees, J. J., UV-B phototherapy: an overview, *Arch. Dermatol.*, 120, 1502, 1984.
72. Morison, W. L., Momtaz, K., Parrish, J. A., and Fitzpatrick, T. B., Combined methotrexate-PUVA therapy in the treatment of psoriasis, *J. Am. Acad. Dermatol.*, 6, 46, 1982.
73. Momtaz-T, K. and Parrish, J. A., Combination of psoralens and ultraviolet A and ultraviolet B in the treatment of psoriasis vulgaris: a bilateral comparison study, *J. Am. Acad. Dermatol.*, 10, 481, 1984.
74. Fritsch, P. O., Honigsmann, H., Jaschke, E., and Wolff, K., Augmentation of oral-methoxsalen-photochemotherapy with an oral retinoid acid derivative, *J. Invest. Dermatol.*, 70, 178, 1978.
75. Stuttgen, G., The risk of photochemotherapy, *Int. J. Dermatol.*, 21, 198, 1982.
76. Stern, R. S., Laird, N., Melski, J., Parrish, J. A., Fitzpatrick, T. B., and Bleich, H. L., Cutaneous squamous cell carcinoma in patients treated with PUVA, *N. Engl. J. Med.*, 310, 1156, 1984.
77. Lange Wantzin, G. and Thomsen, K., PUVA-treatment in lymphomatoid papulosis, *Br. J. Dermatol.*, 107, 687, 1982.
78. Vaatainen, N., Hannuksela, M., and Karvonen, J., Local photochemotherapy in nodular prurigo, *Acta Derm. Venereol.*, 59, 544, 1979.
79. Tagami, H., Iwatsuki, K., Shirahama, S., and Yamada, M., Pustulosis vegetans, *Arch. Dermatol.*, 120, 1355, 1984.
80. Farr, P. M. and Ive, F. A., PUVA treatment of scleromyxoedema, *Br. J. Dermatol.*, 110, 347, 1984.
81. Paul, B. S. and Arndt, K. A., Response of transient acantholytic dermatosis to photochemotherapy, *Arch. Dermatol.*, 120, 121, 1984.
82. Farmer, E. R., Pruritus, in *Current Therapy in Dermatology 1985—1986*, Provost, T. T. and Farmer, E. R., Eds., B. C. Decker, Philadelphia, 1985, 253.
83. Hudson, L. D., Granuloma faciale: treatment with topical psoralen and UVA, *J. Am. Acad. Dermatol.*, 8, 559, 1983.
84. Baran, R., Panizzon, R., and Goldberg, L., The nails in keratosis lichenoides chronica, *Arch. Dermatol.*, 120, 1471, 1984.
85. Hymes, S. R., Morison, W. L., Farmer, E. R., Walters, L. L., Tutschka, P. J., and Santos, G. W., Methoxsalen and ultraviolet A radiation in treatment of chronic cutaneous graft-versus-host reaction, *J. Am. Acad. Dermatol.*, 12, 30, 1985.

ADDITIONAL REFERENCES OF INTEREST ADDED IN PROOF

Randle, H. W. and Sander, H. M., Treatment of generalized lichen nitidus with PUVA, *Int. J. Dermatol.*, 5, 330, 1985.

Yoshiike, T., Manabe, M., Negi, M., and Ogawa, H., Ichthyosis linearis circumflexa: morphologic and biochemical studies, *Br. J. Dermatol.*, 112, 277, 1985.

van den Hoogenband, H. M., van den Berg, W. H. H. W., and van Diggelen, M. W., PUVA therapy in the treatment of skin lesions of the hypereosinophilic syndrome, *Arch. Dermatol.*, 121, 450, 1985.

Robinson, H. N., Morison, W. L., and Hood, A. F., Thiazide diuretic therapy and chronic photosensitivity, *Arch. Dermatol.*, 121, 522, 1985.

Hindson, C., Spiro, J., and Downey, A., PUVA therapy of chronic actinic dermatitis, *Br. J. Dermatol.*, 113, 157, 1985.
Lindberg, L., Larko, O., and Roupe, G., Successful PUVA-treatment for musk ambrette-induced persistent light reaction, *Photodermatology*, 3, 111, 1986.
Kalimo, K., Lammintausta, K., Viander, M., and Jansen, C. T., PUVA treatment of dermatitis herpetiformis, *Photodermatology*, 3, 54, 1986.
Zaynoun, S., Jaber, L. A. A., and Kurban, A. K., Oral methoxsalen photochemotherapy of extensive pityriasis alba. Preliminary report, *J. Am. Acad. Dermatol.*, 15, 61, 1986.
Kaudewitz, P., Przybilla, B., Schmoeckel, C., and Gollhausen, R., Cutaneous lesions in histiocytosis X: successful treatment with PUVA, *J. Invest. Dermatol.*, 86, 324, 1986.
Honigsmann, H., Wolff, K., Fitzpatrick, T. B., Pathak, M. A., and Parrish, J. A., Oral photochemotherapy with psoralens and UVA (PUVA): principles and practice, in *Dermatology in General Medicine*, 3rd ed., Fitzpatrick, T. B., et al., Eds., McGraw-Hill, New York, 1987, 1533.

Chapter 6

PHOTODYNAMIC THERAPY OF CANCER

Brian C. Wilson and W. Patrick Jeeves

TABLE OF CONTENTS

I. INTRODUCTION

The concept of photodynamic therapy (PDT) for the treatment of patients with malignant tumors dates from 1903, when Jesionek and Tappeiner[1] used eosin and sunlight to treat skin cancer. The first photodynamic studies using hematoporphyrin (HP) were carried out by Hausmann[2] several years later, both in vitro and in vivo. Auler and Banzer,[3] and Figge et al.[4] in the 1940s demonstrated the affinity of various porphyrins, including HP, for malignant tissues, by studying the fluorescence produced when the tissues were exposed to UV light. During the 20 years that followed, the interest in HP for clinical oncology concentrated mainly on the detection and localization of malignant tumors by in vivo fluorescence.

In attempting to improve the in vivo localization and photodynamic properties of HP, Lipson and Baldes[5] in 1960 developed hematoporphyrin derivative (HPD) a complex mixture of porphyrin species. The superior tumor-localizing and fluorescence properties of HPD were demonstrated clinically by Lipson et al.[6-9] in esophageal, tracheobronchial, and cervical malignancies, and subsequently in larger numbers of patients with cervical cancer by Gray et al.,[10] and in a range of squamous cell and adenocarcinomas by Gregorie et al.[11] in 1968.

The first detailed clinical report of PDT with HPD was published in 1976 by Kelly and Snell,[12] who treated a patient with recurrent superficial bladder carcinoma. In the previous year, Dougherty et al.[13] and Kelly et al.[14] had shown that PDT with HPD was capable of eradicating, i.e., "curing", malignant tumors in animals. The clinical development of PDT was then systematically pursued by Dougherty's group, who published their first clinical studies in 1977[15] and 1978.[16] By the end of 1982, $\sim$1400 patients worldwide had received PDT with HPD, for many tumor sites, including skin, lung, bladder, head and neck, brain, upper and lower GI tract, female reproductive tract, and eye.

Treatment by PDT is a two-step procedure. HPD is first administered, usually intravenously, and, after a delay of typically 1 to 3 days to allow for accumulation of HPD in the tumor, the target tissue is irradiated with visible light, usually at 630 nm in the red part of the spectrum. The light may be delivered either to the surface of the tumor, or interstitially via optical fibers. The latter also enables treatment of deep-lying tumors which are accessible endoscopically. The volume of tissue treated is determined effectively by the penetration of the light in the tissue, and so as currently practiced,

PDT is suitable mainly for relatively small localized solid tumors. It should be noted that this form of treatment with HPD has in the past decade undergone several name changes, from photochemo-therapy (PCT), to photoradiation therapy (PRT), to photodynamic therapy (PDT).

In the next section, the authors will survey the clinical techniques and light-delivery systems which have been developed for clinical PDT, and, following this, the published clinical data will be reviewed site by site. The science underlying the photoirradiation of HPD-sensitized tumors will then be discussed, starting with the physics of light dosimetry in tissue, and continuing with the chemistry and photochemistry of HPD, and the in vitro and in vivo photobiology of this photosensitizer. At the end, the authors shall return to the subject of in vivo HPD fluorescence for tumor detection and localization, and will discuss the potential use of quantitative fluorescence measurements for determining the concentration of HPD in tumor and adjacent normal tissues prior to light irradiation.

II. CLINICAL TECHNIQUES AND INSTRUMENTATION

A. Hematoporphyrin Derivative (HPD) Administration

For clinical use, HPD is normally injected intravenously in isotonic saline at a concentration of $\sim$5 mg/mℓ. The dose of HPD which has been found optimum for most patients is in the range 2 to 5 mg/kg body weight, with light irradiation normally at 1 to 3 days later. The clearance rate of HPD from skin is usually very slow, so that patients may remain photosensitive for 1 to 2 months after treatment. The degree of skin sensitivity is such that exposure to sunlight or bright artificial lighting must be avoided, and mild to severe burns have been reported in patients who have not taken appropriate precautions. Recently, a "purified" form of HPD was introduced which may reduce the problem of skin photosensitivity without reducing the tumorcidal effectiveness.[17] The topical application of HPD in a suitable carrier preparation has also been proposed in order to eliminate skin photosensitization, but only preliminary results have been published to date.[18,19] Apart from skin photosensitivity, no other serious side effects have been reported from the intravenous use of HPD, although it may occasionally cause spiking fever,[20] or an inflammatory reaction at the injection site if extravasation occurs during injection.[21,22]

B. Light Sources and Delivery Systems

There are three main requirements for the light sources used for PDT:

1. Wavelength(s) which match the activation spectrum of HPD, and with sufficient penetration in tissue to treat the whole tumor volume
2. Sufficient power and power density to deliver a tumorcidal light dose in an acceptable time
3. A means of directing the light energy to the tumor site with a spatial distribution which matches the size and shape of the tumor

The absorption spectrum of HPD is shown in Figure 1. The peak at around 625 to 635 nm is the least photoactive, but nevertheless this wavelength is almost universally employed, since this disadvantage is more than offset by the rapid rise in light penetration which occurs in most tissues in the red part of the spectrum above 580 to 600 nm, and which enables effective irradiation of a range of human tumors.

For surface irradiation, incident light exposures as high as a few hundred Joules per square centimeter may be required. Thus, for clinically acceptable irradiation times of, say, 15 min, the corresponding power density required may be a few hundred milliwatts

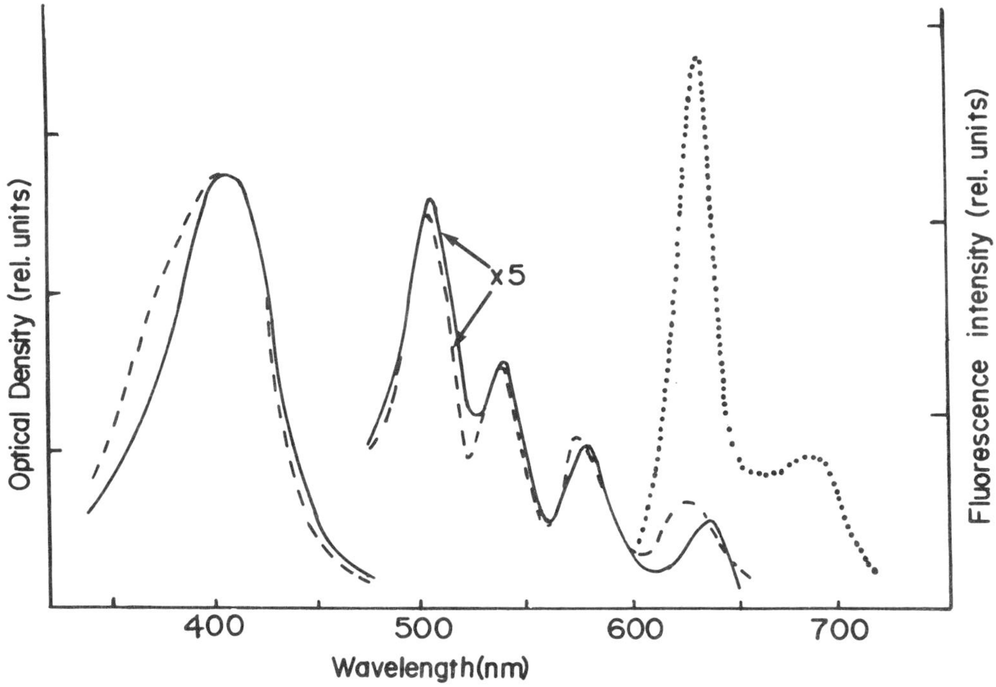

FIGURE 1. Optical spectra of HPD in mammalian cells. (-----) Absorption; (——) Fluorescence excitation for emission at 630 nm (lower 4 peaks) or at 690 nm (630 nm excitation peak); (.....) Fluorescence emission for excitation at 400 nm. Absorption and excitation spectra for the longer wavelength bands are scaled by a factor of five relative to the Soret (400 nm) band. The photodynamic action spectrum for HPD in vitro (not shown) has the same general features as the fluorescence excitation spectrum.[234,246-248]

per square centimeter in the wavelength band 625 to 635 nm. For treatment of cutaneous or subcutaneous tumors, a variety of "conventional", wavelength-filtered lamps have been used.[23-29] However, only accessible superficial tumors may be treated with these sources, and, because of the limited penetration of light through tissue (~1 to 2 cm effective depth for red light), surface irradiation may not be suitable for thick lesions or for tumors of complex shape. Both limitations may be partly overcome by using optical fibers to guide the light into or onto the tumor, and a wide range of irradiation geometries are made possible with these. Quartz fibers of core diameters 100 to 600 μm, which are virtually lossless for visible light, are commonly used. These are robust, and flexible enough to be passed through endoscopes. They can also be inserted directly into tumors, for interstitial irradiation; several fibers may inserted at different points in the tumor, either simultaneously or sequentially, to distribute the light throughout larger tumor volumes.

The output from a single fiber with a flat, cut end is a cone of light directed forward. The output tip of these fibers may be modified to give, for example, cylindrically or spherically isotropic light distributions.[30,31] Cylindrical fiber tips are particularly suitable for irradiating intralumenal lesions, or they may be used interstitially to increase the treatment volume. Examples of the use of spherical geometries are irradiation of the entire surface of the bladder wall, or of the residual tumor bed after surgical excision of malignant brain tumors. For these applications, spherically isotropic irradiance also can be produced with a cut-end fiber by filling the body cavity with a light-scattering medium such as a lipid-aqueous colloid. Optical fibers may also be combined with (micro)lenses to modify the light-beam profiles, and several other forms of light-pattern modification have been used to improve the light delivery to specific body sites, as will be discussed in the next section.

With optical fiber delivery, a laser source is needed in order to couple sufficient optical power into the fiber (beam-splitting optics then can be used to divide the laser power among several fibers simultaneously). For many tumor sites, where large areas or tissue volumes are to be irradiated, several watts of 630-nm light may be needed to avoid excessive treatment times. The equipment most commonly used for this is a dye laser optically pumped by an argon ion laser. This can be tuned to 630 nm with an argon-to-dye conversion efficiency of $\sim$20%, so that the ion laser power needed is typically 5 to 20 W. Such laser systems are now commercially available as clinical instruments.[32] The gold vapor laser, operating at 627.8 nm is a viable alternative.[33] It is cheaper, simpler, and more reliable than dye lasers, but cannot be tuned to different wavelengths and so may not be useful with other potential photosensitizers.

In reviewing the clinical data to be presented later, the authors will use, where possible, "630-nm equivalent" light exposure levels, in order to take into account the spectra of different light sources, particularly the filtered lamps used for skin tumors.

C. Treatment Protocols

In a number of clinical studies, both multiple HPD injections and multiple light exposures have been used to reduce the tumor bulk in stages, to retreat after recurrence, to reduce local complications due to rapid tumor tissue necrosis, and to minimize undue reaction of the overlying skin from external beam irradiation of subcutaneous tumors. Spiking fever has been reported in patients receiving PDT due to the sudden burden of a large volume of dead tissue, and the treatment can also cause local edema and necrotic debris, which may cause problems from obstruction in certain sites.[27] Appropriate fractionation may help reduce these complications, although fractionation protocols have not yet been systematically developed. At this time, there seem to be no contraindications to repeating PDT, either as a fractionated regime, or for retreatment. In addition, previous surgery, radiation therapy*, or chemotherapy** do not generally preclude the use of PDT. Indeed, this is one of the major advantages of the technique, and many of the clinical studies to date have been in patients who have previously failed on some or all of the other therapies.

III. CLINICAL STUDIES

In this section, the protocols and results of published clinical studies of PDT shall be reviewed. Previous reviews have been presented by Dougherty et al. in 1982[27] and 1984.[34] The latter concentrated on the treatment of skin, bladder, and lung tumors, where the greatest clinical experience has been accumulated. Even for these sites, however, no randomized clinical trials have been published, and follow-up times are rarely more than 1 year, so that there are as yet no systematic patient-survival data. As with any new treatment modality, the majority of patients treated by PDT have been advanced cases who were either ineligible for standard treatments, or who had failed to respond to or whose symptoms had recurred after such treatments. For example, in the 113 lung cancer cases reviewed by Dougherty[34] in 1984, only 14 were early-stage disease, in whom the clinical response was very much higher than in the 99 cases of advanced disease. Also, since there are many variables in the delivery of PDT (HPD dose,

* The sensitivity of normal skin to PDT is higher in areas exposed to therapeutic X-rays concurrently with, or within a few months prior to PDT.[20]

** An exception is in patients who have previously had Adriamycin® and radiation therapy, in whom flaring of the skin in the previous radiation fields upon HPD administration has been reported.[29]

light dose, time between HPD and light, light delivery technique, etc.), and since the optimum treatment protocols for each tumor site and type were initially unknown, the clinical studies have themselves provided much of the information on how PDT should be given. Thus, there have been substantial changes in clinical techniques in the last 5 years as treatments improved, so that only restricted intercomparisons of the available patient data can be made.

Given this, the most useful way to assess the effectiveness of PDT as a local treatment modality has been to examine the immediate tumor response and the recurrence rates at the treatment site. Generally, responses have been classified as: "complete response" (CR) if there was disappearance of the tumor as assessed by clinical examination, endoscopy, radiography, histology, or cytology; "significant response" (SR) or "partial response" (PR) if there was greater than 50 to 60% reduction in tumor size at, typically, 1 month after treatment; and "no response" (NR) otherwise. Because of the short time for which most clinical studies have been in place, and the limited patient numbers, local recurrence rates as a function of time after treatment generally are not analyzed in the clinical reports.

A. Skin

The results of PDT in more than 200 patients with 2000 primary and metastatic cutaneous and subcutaneous lesions have been reported since 1980, including metastatic breast cancer, malignant melanoma, basal and squamous cell carcinomas, Kaposi's sarcoma, mycosis fungoides, and Bowen's disease. Considerable success has been achieved in treating metastatic lesions from breast cancer, particularly those recurrent on the chest wall. In many cases, PDT has given local control after failure on surgery, chest-wall radiation therapy, or systemic chemotherapy. Response rates have also been encouraging in basal cell carcinoma and lightly pigmented malignant melanoma. In general, the best results have been obtained in patients with early-stage disease before the tumors have become widespread or ulcerated. With surface treatments, normal skin tolerance has been found to be quite variable, and strongly dependent on the HPD dose, the time delay, and the light exposure, with shorter time delay, or higher drug dose requiring reduction of the light dose. The high light doses needed for heavily pigmented tumors such as larger malignant melanoma nodules have made these lesions difficult to treat through overlying skin.[23,29,34] For subcutaneous lesions, skin damage can be reduced or eliminated by using interstitial fiber irradiation, which also overcomes some of the restrictions due to limited light penetration.

1. Metastatic Breast Cancer

The first reports of PDT in skin cancer were in 1977/78 by Dougherty et al.[15,16] for 113 malignant lesions in 25 patients. This included 85 breast carcinoma secondaries in 5 patients, for which complete response was obtained in 88%. The treatment produced initial erythema and edema, followed by tumor necrosis, usually within 24 hr. This group's experience with skin cancer was reviewed in 1981[26] and 1983,[35] and included 35 patients with metastatic breast carcinoma, both multiple small lesions and some single large lesions. All but one lesion showed short-term response, but clinical benefit was less clear due to the poor condition of most of the patients at the time of treatment. The best results were seen with superficial tumors, and although large, widespread cutaneous lesions did respond, pain due to extensive tumor necrosis often led to treatment being terminated. Considerable variability was noted in the skin response of different patients to the same treatment (2.5 mg/kg HPD, <72 J/cm^2 at 3 to 5 days).

Forbes et al.[24,36] treated 16 patients with breast carcinoma recurrent to the skin, using 2.5 to 3.5 mg/kg HPD with light irradiation 3 to 4 days later for 40 to 250 J/cm^2. Tumor necrosis was seen in all cases. Skin necrosis was a problem in earlier patients, but, as Dougherty also found, this could be avoided by adjustment of the treatment parameters. For the 16 breast cancer patients, the treatment was considered beneficial in 7. Wile et al.[37-40] reported a CR rate of 54% and PR rate of 29% in 225 sites of dermal recurrences of breast cancer in 21 patients treated 3 days after injection of 3 mg/kg HPD. Surface exposures less than 8 J/cm^2 were found ineffective, while those in excess of 25 J/cm^2 caused unacceptable damage to irradiated skin or normal tissue. Patients showing the greatest clinical benefit appeared to be those with large involved areas of recurrent breast cancer penetrating the epidermis, in contrast to the findings of Dougherty et al.[26,35] and Kennedy.[23,29] Fourteen resected cases of breast cancer with 71 cutaneous and 16 subcutaneous metastatic skin lesions have been treated by Konaka and Ono,[41] either by surface or interstitial irradiation. Complete remission was achieved in 40% of lesions, and significant remission also in 40%. Light exposures ranged from 5 to 43 J/cm^2, given from 1 to 7 days after 2.5 to 5 mg/kg HPD. At the same doses, greater therapeutic effect was found if the time interval between HPD administration and light was 4 days or more rather than 1 to 3 days.

2. Malignant Melanoma

In malignant melanoma, the CR rate in 50 patients from several studies reviewed by Dougherty[34] was ~50%, with PR in 30%. The response rate was found by Kennedy[23,29] to be very dependent on the degree of pigmentation, with large, heavily pigmented lesions responding poorly because of limited light penetration.

3. Basal and Squamous Cell Carcinoma

For basal cell carcinoma (BCC), 38 lesions treated by Kennedy[23,29] all responded completely with little or no damage to normal skin, and with excellent cosmetic results. Similar observations were made by Dougherty et al.[26,35] No recurrences had been observed in Kennedy's patients at the time of reporting, with the longest follow-up being 35 months. Tse et al.[42] have reported an 82% CR in 40 recurrent or de novo lesions in 3 patients with nevoid BCC syndrome, with a recurrence rate of 11% at 12 to 14 months. In 8 cases of squamous cell carcinoma (SCC) in Kennedy's study[23,29] the results were poor, with the tumor sensitivity apparently not much greater than that of normal skin. This greater sensitivity of BCC compared with SCC has been confirmed by other studies, using both HPD[43-46] and HP[28] as the photosensitizer, although some complete responses have been obtained even for SCC. Wile et al.[37-40] obtained a CR rate of only 1/16 and a partial response rate of 8/16 in 9 patients with SCC metastases of the head and neck. However, these tumors tended to be large and to spread to adjacent areas after treatment.

4. Miscellaneous

For other tumors, two cases of extensive Bowen's disease were treated successfully and with good cosmetic results by Tokuda.[47] Five cases of mycosis fungoides have been reported[43,48] with one CR and three PR. For Kaposi's sarcoma,[28,29] seven patients in one study[28] had CR rates ~80% for multiple lesions. Limited case studies have been reported by Zhonge et al.[49] and Bandieramonte et al.[50] and 12 patients with various cutaneous and subcutaneous tumors were treated with HP by Tomio et al.,[51] with 3 cases of BCC showing CR.

B. Lung

In PDT for cancer of the tracheobronchial tree, the usual light-delivery technique has been to use an optical fiber placed in a rigid or, more commonly, a flexible fiberoptic bronchoscope. This allows superficial irradiation of the tumor, insertion of the fiber into the tumor mass, or intralumenal treatment with cylindrical irradiation. A study of 73 patients was reported from Tokyo Medical College in 1984,[52] with various subsets of the data presented previously.[53-58] The lesions were treated either superficially, or for protruding tumors, interstitially using a fiberoptic bronchoscope under local anesthesia. Light exposures were in the range 120 to 240 J/cm², and irradiation was performed 48 to 72 hr after i.v. injection of 2.5 to 5 mg/kg HPD. In vivo HPD fluorescence studies were also carried out in these tumors (see later section). In early-stage disease the aim was curative, and in later-stage disease the aims were either palliative, such as to reduce bronchial obstruction and thus improve performance status, or for the purpose of extending indications for surgery and reducing the extent of resection in subsequent surgery.

In four cases of atypical squamous cell metaplasia, positive fluorescence was seen in three severely atypical lesions, but not in one mildly atypical case.[55] In all three severe cases, the atypia disappeared on brushing cytology; follow-up was 30 months in one case. In the 13 cases of early-stage central-type lung cancer, all of which were SCC, CR was obtained in 77%, and SR in the remaining 23%. In addition, among the complete responders were eight patients who were inoperable or refused surgery, of whom six were disease-free at 14 to 44 months after PDT. In five patients who had resection following PDT, there were two CR and three SR histologically; failure to achieve CR was attributed to inadequate light irradiation. In advanced cases, opening of the bronchial lumen was achieved in 60%, and "performance status was improved in most cases". In 53 foci in 45 nonearly patients,[55] the response rates were CR = 17%, SR = 45%, PR = 38%. In addition, four out of five inoperable cases of superficial infiltration in the trachea, carina, and main bronchi were made operable by PDT, and in four out of six candidates for pneumonectomy, it was possible to reduce the extent of resection to lobectomy. Surgery was performed at 3 to 4 weeks after PDT.

The effects of PDT on responding lung tumors were also described in these studies.[55] Edema was observed after 10 min of irradiation. A few hours later, a highly viscous plasma protein exudate appeared on the tumor surface, and at 1 day the surface was covered with necrosis. Regeneration of normal tissue depended on the extent of PDT damage, and for early-stage lesions was complete in 3 to 5 weeks. In the 53 advanced cases, obstructive pneumonia occurred in three. Bronchial pulmonary fistulas occurred in two cases, and particular care was required in advanced cancer with invasion to the pulmonary artery. Daily "clean-up" bronchoscopy was used for several days following treatment in order to remove necrotic tissue and exudate.

Doiron et al.[59-61] used PDT for palliation of extensive and large obstructing endobronchial tumors in 22 patients. The treatment protocol was standardized from previous experience in 40 patients. Irradiation was by fiberbronchoscopy under local anesthetic, using either cut-end fibers or cylindrical fibers, both for surface and interstitial irradiation. Prescribed light exposures were as follows: surface irradiation 100 J/cm², cylindrical fiber insertion 200 J/cm over 0.5 to 1.5 cm length, cut-end fiber insertion 200 J. These levels are generally less than those used in the Tokyo study. HPD was given at 3 mg/kg, 72 hr prior to irradiation. Clean-up bronchoscopy was carried out 2 to 3 days after treatment. Of 19 tumors causing 40 to 90% obstruction, full opening was achieved in all, with no visible residual tumor on bronchoscopy (the longest follow-up was 1 year). One carcinoma *in situ* (CIS) also responded completely. Histologies of these favorable cases were mainly SCC or adenocarcinoma. HPD fluorescence of the tumor was observed in all cases. There was little bronchial inflammation, and no bleeding from the fiber insertions. A comparable series of 17 patients with progressive obstructing lesions of the trachea or main-stem bronchus was treated for palliation by Vincent et al.[62,63] All patients had failed on standard therapies. A variety of diffusing fiber tips were used, as well as direct implantation of the cut-end fibers. The HPD dose was 2.5 to 3 mg/kg, with irradiation to 60 to 300 J/cm² 3 days later.

This treatment produced no measurable response in four patients, tumor surface necrosis in six, and >50% reduction in tumor volume in seven. Relief of obstruction was achieved in several patients for up to 4 months. Autopsy showed that the effective treatment depth was around 2 cm. Significant complications arose due to secretions and necrotic debris at the treatment site. Regular clean-up bronchoscopy was not used.

Studies with smaller numbers of patients have been reported by several other groups. Cortese et al.[64,65] treated 10 patients (11 lesions) using techniques and doses similar to those in Tokyo, but with a "microlens" fiber tip to increase the treatment area with surface illumination. CR (at 2 to 14 months) were obtained in four small superficial SCC and PR in four SCC and one tracheal cylindroma. There was concern over the possible development of bronchopleural fistula, or hemorrhage in treating large lesions. It was also suggested that hyperthermia may have been a factor in some of the treatments at higher power levels. Seven cases of advanced obstructing endobronchial cancer were treated by McCaughan et al.[44,45] with good palliation and significant improvement in performance status in most patients. One PR in an advanced tumor, and a complete response at 8 months in one early-stage lesion have been reported by Spinelli et al.[66] Forbes et al.[36] reported no benefit in six of seven cases of carcinoma of the bronchus; the failures were attributed to technical inadequacies. In the seventh, successful case (recurrent adenocarcinoma), very large HPD and light doses were used. Finally, Li et al.[67] have presented preliminary results of PDT in 21 patients with endobronchial lesions.

C. Bladder

Clinical results for PDT of superficial bladder cancer have suggested a significant role in the management of patients with diffuse CIS and noninvasive transitional cell carcinoma (TCC). T2-stage tumors with shallow invasion of the muscular layers within the depth of penetration of the light may also be suitable candidates. The advantages of PDT in these early bladder cancers are its high selectivity due to preferential uptake of HPD, the relatively noninvasive and nontraumatic nature of the treatment, and the ability to treat regardless of prior therapy and to repeat the treatment at short intervals. Results have been reported from 5 groups for over 40 patients with focal or multicentric disease.

The typical treatment protocol has been 2 to 5 mg/kg HPD, irradiation at 48 to 72 hr, irradiation time 5 to 30 min, and light exposure 50 to 360 J/cm^2 at 80 to 400 m W/cm^2. An argon/dye laser coupled to an optical fiber passed through a cystoscope gives adequate regional surface irradiation. Treatment is given uder local anesthetic, with the bladder filled with saline or oxygen.[68,69] To treat the whole bladder wall, Hisazumi et al.[70] developed a sideways-scattering fiber tip; however, the large surface area of the bladder resulted in an exposure of only 10 J/cm^2 for a treatment time around 1 hr. Jocham et al.[71] used a light-scattering medium to fill the bladder for the same purpose in studies with an animal tumor model.

A consistent pattern of tumor and normal bladder mucosa response to PDT has been reported. Within 10 to 20 min after treatment, there is marked surface edema and exudative reaction. By 1 day, tumor necrosis is seen, continuing for 1 to 2 weeks. During the 4 to 7-day period after treatment, hyperemia may be observed in the mucosa surrounding the irradiated area. Sloughing of tumor tissue, and regeneration of apparently normal mucosa follows in 4 to 10 weeks. In the initial stages, symptoms of pain, hematuria, and discharge of necrotic material in the urine have been reported. However, there have been no cases of bladder-wall perforation or bleeding, and skin photosensitivity has been the only significant complication.

In seven evaluable cases of TCC *in situ* resistant to conventional therapies, Benson et al.[68,69] achieved a 100% CR rate. Of these, two lesions subsequently recurred in the treated area and were retreated. In one case of invasive carcinoma, good control of bleeding was maintained for 30 months. Good results have also been reported from 20 patients treated in Tokyo,[72-75] with 55% CR for follow-up of 12 to 34 months; a number of these responding cases had had previous radiation therapy. Xian-wen et al.[43] obtained an early CR rate of 83% in six patients with TCC, and Zhonge et al.[49] obtained one CR and one PR for bladder cancer (histology not specified). It should be noted that, in all these, specific involved areas of the bladder were irradiated, not the whole bladder wall. However, studies of HPD fluorescence[69,73,75,76] have shown that HPD does not concentrate significantly in the normal bladder wall. Thus, the tumor responses should be similar for regional and for whole-bladder treatments for the same tumorcidal light dose.

Finally, Hisazumi et al.[70,77-79] have analyzed the response rates in 46 small and large superficial lesions in 9 patients as a function of light exposure and exposure rate. Tumors were irradiated at 48 to 72 hr after 2 to 3.2 mg/kg HPD. For tumors less than 2 cm in diameter, the CR rate increased from 2/10 to 10/12 to 14/20 (20%, 83%, and 70%, respectively) as the exposure rate increased from 150 to 200 to 300 mW/cm^2. According to in vitro experiments in dog bladder,[68,69,79] it is unlikely that this variation in response is due to thermal effects, at least up to 200 W/cm^2. CR rates as a function of total light exposure for these small tumors were 4/12 (33%) for <50 J/cm^2, 15/21 (71%) for 50 to 100 J/cm^2, and 7/9 (78%) for 100 to 250 J/cm^2. By contrast, even at the higher light doses, the response was poor for lesions greater than 2 cm (CR = 0/4, PR = 1/4).

D. Head and Neck

The largest study of PDT in head and neck tumors is from Irvine, Calif., where 21 patients with persistent or locally recurrent primary tumors (9 tongue, 3 nasopharynx, 9 miscellaneous, including 19 SCC), and 10 patients with soft-tissue regional metastases have been treated since 1981.[37-40,80] All patients had failed on conventional therapies: patients who had received radiation therapy within the previous 3 months were excluded. The treatment protocol was 3 mg/kg HPD, treated at 72 hr. External surface irradiation was used for the primary and recurrent tumors, except for three nasopharyngeal and one vocal chord lesion which were treated endoscopically. For surface irradiation, the exposures were 17 to 91 J/cm^2 per treatment, at 25 to 250 mW/cm^2. Eight patients required a second or third course of treatment. In the 21 primary tumors, there were 6 complete responses (29%) and 11 partial responses (52%), with some patients followed for over 1 year. Of these, there were nine cases of tongue cancer, with CR = 2/9 and PR = 6/9. The two complete responders had not recurred at 12 and 18 months. Two complete and three partial responses were obtained in the ten more advanced metastatic lesions; two large exophytic tumors treated interstitially did not respond.

Taketa and Imakiire[81] treated six T1-T3 SCC: two laryngeal, two oropharyngeal, and two tongue. The treatment protocol was 3 mg/kg HPD, with surface irradiation 48 hr later via fiberoptic endoscopy to 34 to 390 J/cm^2 at 200 to 500 mW/cm^2. Tumor necrosis was observed in all cases, with no deleterious effects on surrounding normal tissue. Caution was required in treating around the larynx to prevent edematous obstruction. Treatment was limited by insufficient light penetration, a problem which might be overcome using interstitial irradiation. Indications for cure were felt to be in accessible tumors where there is limited invasion in deep or submucosal areas, as for example, in CIS of the palate. In other studies, Kennedy[29] has treated four cases of SCC in the mucous membranes of the mouth and nose, with only limited success, and Xian-wen et al.[43] reported the following response rates: SCC CR = 7/12, PR = 5/12; osteogenic sarcoma of maxilla PR = 2/2; mucoepidermoid cancer of buccal mucosa

PR = 1/1. Finally, McCaughan[45] has reported two patients treated for head and neck cancer metastases.

E. Gynecologic

The first case report of PDT in gynecological malignancy was by Soma et al.[82] in 1982. Subsequently,[83] they reported four cases of vaginal carcinoma (one primary SCC, and three recurrences of adenocarcinoma), five cases of CIS of the uterine cervix, three cases of vaginal or cervical recurrence of cancer of other organs, and three cases of cancer of the vulva. Response rates were CR = 4/5 and SR = 1/5 for cancer of vagina and vulva; CR = 3/5 and SR = 2/5 for CIS of cervix; and CR = 2/3 for metastatic vaginal lesions. Reported follow-up was 2 years in one primary SCC of vagina, and 18 months for one primary vulvar lesion. There were no complications apart from photosensitivity. The treatment protocol was 2.5 mg/kg HPD, with fiber irradiation through a colposcope at 100 to 400 mW/cm^2 for exposures of 88 to 346 J/cm^2. In 1982, Ward et al.[84] reported five cases of gynecologic malignancy treated superficially or interstitially. Some tumor response was seen in all cases, and in two patients with recurrent vaginal lesions, local control was obtained at 10 and 12 months, with good healing of the tissue.

Recently, Rettenmaier et al.[18,19] reported treating six patients with nine tumors recurrent to either vagina or skin, using the relatively low light exposures of 20 to 40 J/cm^2 applied superficially to both the tumor and 2 to 4 cm of surrounding, clinically uninvolved tissue, at 72 hr after 3 mg/kg HPD. The patients had all failed on previous conventional therapies. Sites of recurrence included vulva, neovagina, vaginal apex, and perineum. CR was obtained in two cases of peritoneal cancer, while three vaginal recurrences of cervical origin showed PR. To reduce the problems of generalized skin photosensitivity, Rettenmaier and co-workers have done a preliminary study using topical application of HPD preparations in three patients with resistant premalignant lesions of the genital tract. No toxicity was observed. However, although a PR was obtained in one patient, further development of the carrier is needed in order to increase the photosensitizer penetration in tissue. A total of seven gynecologic cases have been reported from four other groups. One CR was achieved by Dahlman et al.[38] in a recurrent CIS of vulva. Xien-wen et al.[43] obtained a CR in one case of malignant papillary cancer of vagina, and PR in three cases of SCC of the vaginal cervix.

F. Brain

The photosensitivity of malignant glioma to HP in vitro and in vivo was shown initially by Diamond et al.[85,86] and Signorelli et al.[87] Wise and Taxdall[88] showed that HP was taken up by such tumors, but did not cross the normal intact blood-brain barrier. Perria et al.[89,90] in 1980 treated nine patients using HPD (five stage II-IV astrocytoma, three glioblastoma, one sarcoma). The technique was to irradiate the residual tumor bed during craniotomy after surgical removal of the bulk tumor tissue. HPD was given i.v. 24 to 96 hr before surgery at 2.5 mg/kg. The prescribed light exposure was 9 J/cm^2 using a He-Ne laser (633 nm). HPD fluorescence was observed in all tumors prior to irradiation. Although survival was not improved in this study, PDT-induced tumor necrosis was clearly shown. Preferential concentration of HPD in human gliomas compared with normal brain tissue was demonstrated by Wharen et al.[91] at the Mayo Clinic, using a fluorescence assay on tissue from three patients, who also underwent intraoperative PDT without ill effects at 24 or 48 hr after injection of 5 mg/kg HPD. The ratio of HPD concentration in tumor vs. normal brain at 24 hr was 6:1 in one case, and 17:1 in a second case. In the third patient, treated at 48 hr, the concentration in tumor was substantially lower. Preferential concentration of radiocopper-labeled HPD in some malignant and benign brain tumors at 24 hr has also been found in a preliminary study in six patients by Jeeves et al.[92]

A phase-I feasibility study of PDT has been carried out by Laws et al.[93] at the Mayo Clinic in five patients with recurrent glioblastomas. Irradiation was via an optical fiber inserted through a burr hole into the center of the tumor under stereotaxic control, delivering 900 J at 300 mW, 48 to 72 hr after injection of 5 mg/kg HPD. The study confirmed the safety of PDT administered in this way, and the presence of HPD fluorescence in the tumor tissue. Decrease in tumor size was noted in some cases. The volume of tissue treated was estimated at 5 to 10 cm^3, so that multiple fiber placement, or other light-dispersing methods, would be required to treat larger tumor volumes. (In a commentary on a study of normal brain toxicity in the rat by Cheng et al.,[94] Laws and Wharen reviewed briefly their clinical experience in 22 advanced brain tumors, and stated that the radius of cell kill with current fiberoptic techniques is 8 mm. This is less than the 1.5-cm depth of necrosis found by Perria[90] using surface irradiation.) With power levels of 300 to 400 mW through a single 400-μm fiber, Laws et al.[93] noted temperatures of around 45°C at 5 mm from the fiber tip, so that hyperthermia would have been a contributing factor in some portion of the treated volume.

Recently, McCulloch et al.[95] reported a series of 16 malignant brain tumor patients (4 metastatic tumors, 3 gliomas, 9 glioblastomas) treated intraoperatively by PDT in addition to conventional treatments. Treatment was by surface irradiation either with a fiberoptic probe coupled to an argon/dye laser (280 to 460 mW at 630 nm), or a gold vapor laser (1.5 W at 627.8 nm), or with a specially designed 1000 W filtered incandescent lamp which could be inserted into the resection cavity (24 W in 620 to 720 nm band). The lamp and gold vapor lasers allowed substantial reduction of the treatment time, which was 90 to 120 min with the argon/dye laser. Total light exposures of 100 to 150 J/cm^2 (1260 to 2520 J total delivered energy) were used, irradiation being at 48 hr after injection of 5 mg/kg HPD. Besides skin photosensitivity, McCulloch et al.[95] noted some complications due to brain edema. In the group of four metastatic lesions, two patients, in whom PDT was the only form of localized therapy, were reported clinically disease-free at 10 and 16 months. No conclusions could be drawn from the three cases of primary glioma because of the heterogeneity of response expected by conventional therapy. Changes in the light source used made results of the nine glioblastomas difficult to assess; however, it was of note that two of these patients were also continuing clinically disease-free at 35 and 42 months, the expected survival for standard treatments at 2 years being less than 10% in this group.

Although not statistically significant, these various results have encouraged other groups to initiate clinical trials of PDT in malignant brain tumors. It should be noted that to date there is no direct evidence in patients that HPD localizes in tumor cells preferentially compared with normal brain cells, or that the former are more photosensitive. Further, although in a gliosarcoma brain tumor model in the rat, Boggan et al.[96] found significant preferential uptake in tumor, only 33 to 44% of the tumor tissue fluoresced, and in treating with light, a patchy coagulation necrosis was seen, possibly corresponding to the nonuniform HPD distribution. The differential tumorcidal effect of PDT following surgical resection of accessible tumors thus remains to be established. However, even if there is only limited differential photodynamic effect between tumor and brain tissues, PDT may be preferable to extending surgical resection, since it may not damage brain tissue lying beyond the treatment depth. It may also be of value for inoperable lesions, using optical fiber light-delivery techniques.

G. GI Tract

1. Esophagus

The largest reported study of PDT in early-stage esophageal disease is from Tokyo

Medical College,[52,97,98] using endoscopic fiber irradiation under local anesthesia at 48 hr or more after 2.5 to 5 mg/kg HPD. All cases were SCC. Complete remission was achieved in three of four cases where irradiation was as prescribed (120 to 240 J/cm^2), with significant remission in the fourth patient. In two of these CRs which were not subsequently resected, the patients were disease-free at 15 and 25 months. Okuda et al.[99] in Osaka also achieved a complete remission at 14 months follow-up in an early-stage SCC treated at 90 J/cm^2.

In advanced, obstructive esophageal cancer in seven patients, significant palliation was achieved by McCaughan et al.[44-46] using interstitial or intralumenal fiber irradiation at 3 to 6 days after 3 to 5 mg/kg HPD (light exposures 120 to 1380 J, or 15 to 180 J/cm over 2 to 3 cm lengths, respectively). Multiple areas were treated, and several patients were treated more than once. The corresponding results in advanced cancer from Tokyo[97,98] were SR = 2/5, PR = 3/5 for light exposures up to 360 J/cm^2. A PR was achieved by Xian-wen et al.[43] in one case of SCC.

2. *Stomach*

The use of PDT in the stomach has been technically difficult, in that achieving appropriate light delivery has been limited by light penetration and by obstructing gastric folds.[52] Improvements are to be expected through the use of specially tailored fiber optics. In early-stage disease, the largest group is that from Tokyo, using the same techniques and ranges of light and HPD dose as for esophagus.[52,98,100] In 14 cases, the response rates were CR = 57% and PR = 43%. In three complete responders in which the tumors were not subsequently resected, the patients were disease-free at 16 to 29 months. In other studies, Okuda et al.[99] and Mimura et al.[101] treated seven patients with early gastric cancer by fiber-endoscopic irradiation, with a 71% CR rate, followed to 3 to 13 months. Their recommended treatment regime was 90 to 110 J/cm^2 at 48 to 72 hr after 3 mg/kg HPD. One early gastric cancer patient was treated by Spinelli et al.[66] with CR reported at 9 months. Lastly, Ozaki et al.[102] used an N_2/dye laser system (630 nm) in five cases of early gastric cancer, with complete disappearance of the lesions in four patients.

In advanced gastric cancer, PR was obtained in the Tokyo series for 9 of 11 cases (82%), and SR in 2. Inadequate light penetration probably contributed to the limited efficacy. Using 100 to 300 J/cm^2 at 48 to 72 hr after 2.5 to 5 mg/kg HPD, Zhonge et al.[49] achieved only one CR in ten patients, and PR in eight. However, stage of disease was not specified for each patient here. In five cases studied, all showed visible HPD fluorescence at time of treatment.

3. *Colorectal*

Spinelli et al.[66] treated three cases of advanced rectal cancer, with a 40 to 70% reduction in tumor mass. Two colorectal adenocarcinomas were treated by Douglass et al.[103] along with one esophageal and four stomach tumors, with variable tumor response. This group also studied GI damage from PDT in baboons (edema, bleeding, and perforation), and recommended that at 72 hr following 5 mg/kg HPD, the exposure should be kept below about 140 J/cm^2, increasing to 300 J/cm^2 at 96 hr. The patient studies were done at 72 hr at $\sim$110 to 140 J/cm^2.

H. Eye

PDT has been used clinically for both intraocular and orbital melanoma. Preclinical studies with amelanotic melanoma in the anterior chamber of the rabbit eye have demonstrated the tumorcidal effect of PDT, but also that normal ocular structures are also sensitive above certain HPD and light doses.[104-109] HP photosensitivity in vitro[110] and HPD uptake[111] and photodynamic response[112,113] in vivo also have been demonstrated in human retinoblastoma heterotransplants in nude mice, making this tumor type a potential candidate for clinical PDT.

In clinical studies, Bruce[114-116] treated 11 patients with malignant choroidal melanoma. Irradiation was either transcorneal, transsscleral, or both, at 300 to 6800 J/cm^2, 72 hr after 2.5 mg/kg HPD. Tumor sizes ranged from 1 to 6 mm deep and up to 12 × 10 mm in area. The immediate effects, observed within minutes of irradiation, were blanching of the tumor, with edema of the overlying retina and pinpoint intraretinal hemorrhage. Such changes were not seen in adjacent uninvolved retina. Fluorescein angiography after treatment showed dramatic and persistent modification of the vascular supply to the tumors. Exudative retinal detachment occurred in 9 of the 11 cases; in all but 3 of these the detachment resolved within 6 weeks. Other transient clinical changes were observed. Reduction in tumor mass was seen in all cases which had been followed for at least 5 months, the shrinkage in volume taking several weeks or months for completion, leaving finally a chorioretinal scar. Treatment was also successful in one metastatic choroidal tumor reported by Watanabe and Tobari.[117]

Tse et al.[42] have treated three patients with ocular tumors (three choroidal, two ciliary body, one iris), and one patient with an orbital recurrence of malignant melanoma. Transcorneal light irradiation was given in fractionated sessions between 24 and 144 hr after injection of 3 to 5 mg/kg HPD, at 32 to 720 J/cm^2 per fraction for a total of 64 to 2160 J/cm^2. All tumors showed some clinical response, although histologically all had residual tumor at enucleation between 3 days and 1 year after treatment. Most of the complications were transient, although retinal detachment was exacerbated in one case, and neovascular glaucoma developed in two eyes with anterior segment lesions.[118] The latter were possibly attributable to diffuse microvascular injury and subsequent anterior segment ischemia.

If irradiation of tumors in different parts of the eye can be achieved without causing unacceptable normal-tissue damage, the previously mentioned studies suggest that PDT may be a viable alternative to enucleation. Other applications of PDT, such as treatment of iris neovascularization, also show promise.[119]

IV. LIGHT DOSIMETRY

In PDT as currently used clinically, the light exposures and exposure rates have been arrived at largely empirically, and generally the loss of energy from the target volume and the spatial distribution of absorbed light energy have not been quantitatively determined for specific tumor sites or in individual patients.[120,121] For surface irradiation, part of the incident light is diffusely reflected (back-scattered) through the surface, and with interstitial fiber irradiation there are coupling losses between the fiber and the tissue. These losses may be substantial, and are strongly tissue and wavelength dependent. For a given light exposure, the effective treatment depth or radius depends primarily on the attenuation coefficient of the tissue at the wavelength used, and on the irradiation and tissue geometries, but the shape of the light distribution can be affected substantially by the relative values of light absorption and scattering also. Accurate light dosimetry for PDT requires detailed knowledge of the total light energy absorbed in the target tissue, and of how this energy is spatially distributed. This, in turn, requires both data on the optical properties of target tissues (absorption and scattering coefficients, angular distribution of scatter), and methods of calculating the spatial distribution of absorbed light, given the optical properties and the irradiation and tissue geometries.

In addition, a number of studies have demonstrated that tissue heating may result from light power densities of the order of those used in clinical PDT. This can cause acute thermal damage, or, at lower power densities, cytotoxic hyperthermia. The physical aspects of this will be discussed here also, since they are directly dependent on the spatial distribution characteristics of the light.

A. Propagation of Light in Tissue

Visible light is absorbed in tissue by endogenous chromophores such as hemoglobin and melanin, and by HPD itself. There may be subsequent fluorescent re-emission at longer wavelengths. It is also scattered, either isotropically or anisotropically, without change in wavelength. Blood, tissue pigmentation, and the microscopic and macroscopic structure of the tissue all affect the relative importance of scattering and absorption. At the concentrations normally found in target tissues during PDT ($\sim\mu g/g$), the absorption of light by HPD is relatively small, but may modify the spatial distribution of light.

Since the distribution of light in tissue is a complex function of the optical properties of the tissue, and of the tissue and irradiation geometries, analytic description is possible only in very simplified situations.[120] One analytic approach has been to consider the two extreme cases of "absorption-dominated" and "scattering-dominated" attenuation.[122-127] For the former, the absorption of light may be combined analytically with the light-source geometry to evaluate light distributions in homogeneous, boundary-free media. For the scattering-dominated case, diffusion theory may be applied and again, various light-source geometries can be incorporated. Svaasand and Ellingsen[124] found diffusion theory agreed well with their experimental data on light attenuation in human brain tissue in vitro. McKenzie[125] used it to compare the distribution of effective treatment depths for external beam and implanted fiber irradiation geometries, and Marynissen and Star[127] used it to model the influence of the light beam diameter in surface irradiation. An alternative, nonanalytic approach has been to use Monte Carlo techniques, in which the propagation of photons though tissue is simulated by computer. This, in principle, allows any combination of absorption and scattering characteristics, and any source and tissue geometry to be modeled. To date it has been used by Wilson and Adam[128] to demonstrate the differences in "isodose" distributions for external beam and fiber irradiation in tissues with different absorption-to-scattering ratios.

In order to study light propagation experimentally under precisely controlled conditions, and to test model predictions, various artificial "phantoms" have been developed which simulate some of the optical properties of tissue. These include highly scattering polyvinylacetate colloidal suspensions to test diffusion theory results,[127] and agar/blood phantoms to examine the contribution of blood to light attenuation.[129]

B. Optical Properties of Tissue

Until recently, there were very few measurements of the absorption and scattering coefficients of light in tissue, with the exception of skin (see, for example, Reference 130). Spectrophotometry has been used to measure the transmission spectra of mammalian tissue samples, the most extensive data being those of Preuss et al.[131-133] In 1978, Dougherty et al.[16] introduced the interstitial use of optical fibers coupled to photodetectors to map the spatial distributions of light in tissue under clinically relevant irradiation conditions. Svaasand and co-workers[122-124] subsequently measured the attenuation coefficients of various animal tissues in vitro and in vivo at several discrete wavelengths between 405 and 630 nm. In all tissues, the attenuation was strongly wavelength dependent, with the same qualitative features as found by spectrophotometry, namely high attenuation at ~425 nm (Soret band) and ~550 nm, a rapid decrease in attenuation (i.e., increase in penetration) from ~580 to 600 nm, and a slowly decreasing or roughly constant attenuation at longer wavelengths. The absolute value of the attenuation coefficients varied markedly between tissues, and in some instances, at different locations within a given tissue. The measurements also showed that the attenuation may be substantially different from in vivo to in vitro. Detailed studies by Wilson et al.[134] of the in vivo to post-mortem attenuation changes in various animal model tissues confirmed these findings, and quantified the role played by blood in different regions of the visible spectrum.

At 630 nm, the penetration depths (1/effective attenuation coefficient) of tissues which are not excessively pigmented, are typically ~1 to 5 mm. At, say, 2, 4, and 6 penetration depths for surface irradiation, the light flux is then ~10, 1, and 0.1% of the incident value, respectively. Thus the depth at which the irradiance is, for example, 1% of the surface exposure would range from <5 mm to >2 cm, depending on the tissue. This wide range of uncertainty in the light doses and treatment depths, which is reflected in many of the clinical and experimental in vivo studies of PDT, demonstrates the need to develop methods to measure the attenuation in individual target tissues.

Evaluation of the detailed spatial distributions of light for different irradiation techniques requires, in addition, a knowledge of the relative contributions of absorption and scatter to the attenuation, and of the angular distribution, or degree of isotropy, of scatter. There have been very few experiments to date to separate the absorption and scattering properties of tissues. Svaasand and Ellingsen[124] showed that scattering (by myelin) is the dominant mechanism in light attenuation by human brain tissue in vitro throughout much of the visible wavelength range, and in vivo data[134] support this above 600 nm. Scattering is a significant mechanism in light attenuation at 630 nm in muscle also,[122,126,127,135] and it seems likely that except for very heavily pigmented tissues such as some melanomas, this will be a general characteristic of most tissues at this wavelength. However, there is also evidence that the relative contribution of absorption and scatter to attenuation depends strongly on the wavelength of the light, with hemoglobin-related absorption being significant below 600 nm.[129,134] Marynissen and Star[127] concluded that scattering in bulk muscle tissue is effectively isotropic, but on a microscopic scale Wilksch et al.[135] found evidence for anisotropy. It should be noted finally that high scattering results in high diffuse reflectance and also affects the "coupling" of light into tissue from interstitial fibers, so that in addition to the spatial distribution of light, the total energy absorbed in target tissue depends on the scattering-to-absorption ratio.

Although there are very few optical data for tumor tissue,[13,122,123,126,136] it seems likely that the shape of the attenuation spectra will not be qualitatively different from that of many nonpigmented normal animal tissues, and that at 630 nm, the penetration depths will be typically in the order of a few millimeters. Thus, the qualitative characteristics of light distributions may be studied using data on appropriate normal tissues. However, the exact attenuation coefficient value, and the relative contribution of absorption and scatter are expected to vary from tumor to tumor, so that accurate light dosimetry for clinical PDT must await development of clinically acceptable methods of measuring the optical properties of individual target tissues. Finally, it is possible that the absorption and/or scattering coefficients of tumor tissue may change during extended PDT irradiation, since it has been shown that the blood flow in HPD-sensitized tumors is markedly and rapidly affected by light irradiation.[137,138]

C. Thermal Effects of Light Irradiation

Although the use of lasers, or lamps with adequate wavelength filtering, produces no IR heating of tissue, it has been demonstrated that at high enough power levels the absorption of visible light by the tissue can produce significant temperature rise. Extensive modeling of this effect has been carried out by Svaasand and co-workers.[139-141] The absolute temperature rise, and its spatial distribution, depend on the light-source geometry, on the optical and thermal properties of the tissue (including blood flow), and on the applied power density and exposure time. In practice, two different thermal effects must be considered. First, near the tip of interstitial cut-end fibers the power density may be extremely high at apparently modest power levels. This can cause blood clotting and charring of the tissue, which in turn reduces the transmission of the light in the tissue, and may destroy the fiber tip and result in failure of the treatment.[126] This extreme situation has been reported clinically above 300 mW in the lung.[65] Care must also be taken if this type of treatment is given in an oxygen atmosphere.[60]

The second, less extreme but clinically more important thermal effect may arise in any irradiation technique. For tissue temperatures above about 42°C, cell kill can be produced by hyperthermia, either alone or in synergism with PDT. Such temperatures can be reached at power densities which have been used clinically. To study this effect, there recently have been a number of in vivo measurements in animal models of the tissue temperature rise using different irradiation conditions at 630 nm. For example, Kinsey et al.[142] found that in murine tumors irradiated interstitially, the temperature at 5 mm from a cut-end fiber tip rose rapidly within the first few minutes of exposure, thereafter remaining constant at a temperature which depended on the light power, namely 3°C above basal temperature at 50 mW, 7°C at 100 mW, and 20°C at 200 mW. In an analogous experiment in a rat tumor model, Svaasand et al.[140] obtained similarly shaped temperature-time profiles, but a maximum temperature rise at 6 to 7 mm from the fiber of only 10°C/W. This was nevertheless still sufficient to produce a hyperthermic contribution to cell kill at several millimeters from the source fiber at powers above 200 mW. In mouse brain using surface irradiation, Rounds et al.[143] found a maximum temperature rise of 1.5°C/100 mW/cm^2 (up to 400 mW/cm^2).

The hyperthermia produced by light absorption during PDT may be tumorcidal in itself, and it has been suggested that this may be clinically advantageous.[144,145] In addition, the heating may act in synergism with the photodynamic effect to enhance the tumor kill. This may occur at temperatures which are sublethal for hyperthermia acting alone. Hyperthermia-PDT synergism has been demonstrated both in vitro[146] and in vivo.[147,148] It may be possible to exploit this clinically by producing hyperthermia during PDT by means other than the PDT light itself, such as, for example, conventional R. F. or microwave heating, or by using IR lasers.[144]

V. HPD CHEMISTRY AND PHOTOCHEMISTRY

The first preparation of HPD was reported in 1960, and resulted from an attempt to increase the solubility of HP, which had been previously found to have tumor-localizing properties.[5] Treatment of HP with sulfuric/acetic acid, followed by alkali solubilization, led to a product whose tumor localization and photodynamic properties were enhanced over those of the parent compound.[6] However, although the cytotoxic potential of photoactivated porphyrins had been recognized for many decades,[149] it was not until 1972 that the use of HP in the photosensitized destruction of tumors was described,[85] and it was only in 1976 that the use of HPD in patients for the treatment of localized malignancies was first reported in detail.[12] Subsequently, considerable effort has been directed toward ascertaining the chemical nature of HPD, to identifying its photoactive component(s), and to understanding the biophysical and photochemical properties responsible for its tumor-localizing and photodynamic behavior.

A. HPD Preparation and Analysis

The preparation of HPD for clinical use is a two-step procedure. First, treatment of hematoporphyrin dihydrochloride with acetic and sulfuric acids results in the formation of a complex mixture of acetylation and dehydration products (see Figure 2a). Thin layer (TLC) and high pressure liquid chromatography (HPLC) reveals the pres-

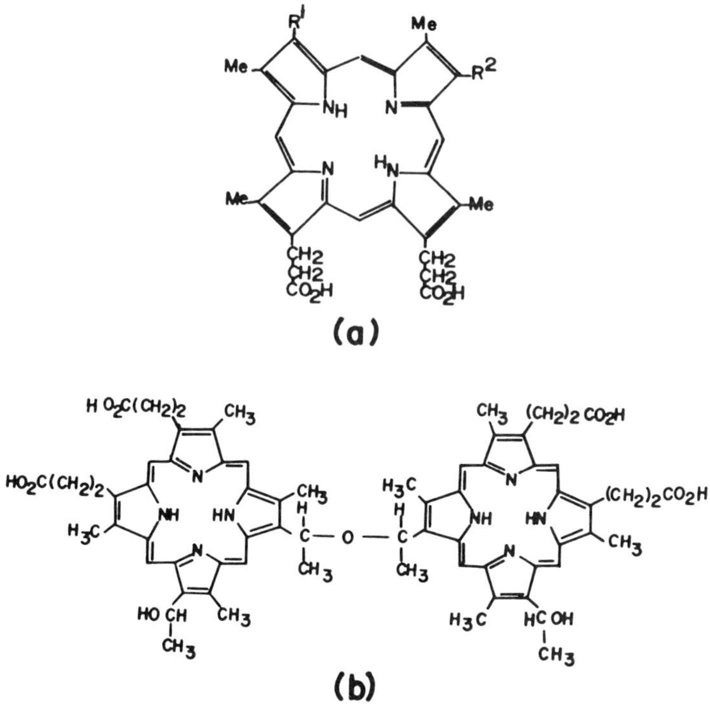

FIGURE 2. (a) Structure of HPD porphyrins. The substituents R^1 and R^2 determine the nine porphyrin species comprising HPD.[151-154] (b) Ether-linked porphyrin structure of DHE.[166,167]

ence of HP mono- and diacetates, monoacetates of hydroxyethylvinyldeuteroporphyrin (HVD), protoporphyrin (PP), and other porphyrin species.[150-157] This intermediate product is then hydrolyzed in NaOH in the second step to form the complex porphyrin mixture[35,153,154] which is used clinically (commercially known as Photofrin I^{TM}). Chromatographic analysis of this material has shown that while in large part it consists of HP, HVD, and PP, its tumor-localizing capacity resides in a separate, minor (15 to 20%) component.[35,158-165] The porphyrin species responsible for the biological activity of HPD has been identified by Dougherty and co-workers[166,167] as "dihematoporphyrin ether" (DHE), i.e., an HP dimer joined by an ether linkage (see Figure 2b), as had also been suggested by Berenbaum et al.[153]

Separation of HPD by gel-exclusion chromatography has also shown that the biological activity resides mostly in the fastest eluting component (HPD-A, or Photofrin II^{TM}).[35,165,168] TLC and HPLC of this fraction suggest that in aqueous solution, the active component(s) are present as aggregates of porphyrin species which are more hydrophobic than either HP or HVD. The average molecular weight has been determined as ~21,500 daltons.[169,170]

The ability of porphyrins to aggregate in aqueous solution, and the dependence of this on the porphyrin concentration, and on the pH and ionic strength of the medium, has been established by spectrophotometry and spectrofluorimetry.[171-178] Dilute aqueous solutions of porphyrins are characterized by (1) a light absorption spectrum having a very strong (Soret) band at a wavelength of ~395 nm, and further bands in the 500 to 650-nm region, and (2) a fluorescence emission spectrum peaking at around 615 and 675 nm (for excitation by the Soret wavelength). Dimerization and higher aggregation shift both the Soret absorption band and the fluorescence maximums toward the blue, and decrease the fluorescence yield. The high aggregation state of the

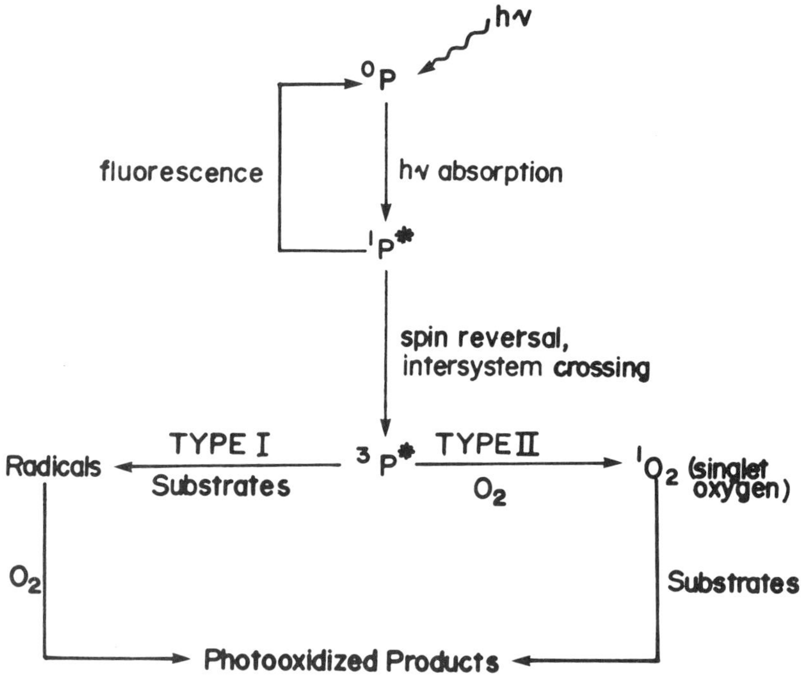

FIGURE 3. Schema of type I and type II photooxidation pathways. °P = porphyrin ground state; ¹P*, ³P* = porphyrin singlet and triplet excited states.

biologically active HPD component has been confirmed by numerous studies of its absorption and fluorescence spectra.[161,179-190] In Photofrin II^{TM}, the Soret band is shifted to 365 nm,[187,188] and the fluorescence quantum yield is much lower than for other HPD components.[160,161,183,187-189]

Photofrin II^{TM}, which presumably consists essentially of DHE, is particularly strongly and stably aggregated in aqueous solution.[35,158,169,170,185,191] However, the changes in the absorption and fluorescence spectra observed on the aging of HPD after preparation imply changes in aggregation status similar to those observed when HPD becomes bound to intracellular structures.[187,192] The addition of organic solvents or ionic surfactants to Photofrin II^{TM} induces red shifting of both the Soret absorption band and the fluorescence maximums, as well as enhancement of the fluorescence yield, suggesting disaggregation.[183,188] These changes have also been seen in the binding of DHE to human serum (HSA) or serum albumin.[181,182,189,193-195] The photophysical properties of DHE, either bound to HSA or in appropriate solvent conditions, have led to the concept that DHE may sometimes exist as a folded-over, covalent dimer.[191,193] This would have particular impact o. its localization properties.

B. HPD Photoactivation

While the characteristic fluorescence emitted by prophyrins results from radiative decay of the porphyrin excited singlet state, it is the lowest excited triplet state that acts as the reactive intermediary in porphyrin-sensitized photoreactions in biological systems.[149,196-199] Further, as illustrated in Figure 3, dye-mediated photooxidations can be produced via either a type I process (electron transfer to O_2 and/or substrate, with generation of superoxide O_2^- and/or other radical species), or a type II process (energy transfer to triplet ground state O_2 to yield reactive singlet oxygen $^1O_2$). A large number of studies have helped elucidate the ways in which the porphyrin triplet yield, and the competition between type I and type II processes, depend on the porphyrin structure and aggregation, and on the properties of the reaction medium.[200,201]

Recently, the photophysical properties of a variety of porphyrins, in particular HP and a number of its derivatives, including HPD, have been examined using nanosecond laser flash photolysis and pulse radiolysis. It was found that for most porphyrins, the triplet state is formed with a quantum efficiency greater than 0.8,[202,203] although weak free-radical formation was also observed.[203] While the triplet yield of HP in a variety of aqueous, organic, and detergent-micelle systems was also close to unity,[203-205] the low triplet-singlet extinction coefficients detected under some conditions indicated extensive HP aggregation.[204,205] Further, Reddi et al.[206-208] observed that the triplet-state quantum yield of HP increases as the porphyrin monomer-dimer equilibrium shifts toward monomer in different media. For Photofrin II^{TM}, little or no triplet state has been detected in aqueous solution, in which there is strong aggregation.[183,205] In 90% methanol, on the other hand, the triplet state has a lifetime and triplet-triplet absorption spectrum very similar to that of HP.[183,209,210]

In aerated media, the triplet states of a range of porphyrins appear to be efficiently quenched by oxygen, and the quenching constants are generally relatively insensitive to the solvent type or environment.[183,203-208] The quenching rate and yield of $^1O_2$ depend on the dissolved oxygen concentration,[205,211] and measurements of the quantum yields of $^1O_2$ and superoxide radical formation show that the former is the major product.[206-208] Indeed, although superoxide production has been implicated in some studies as an important mediator of porphyrin photocytotoxicity,[212,213] recent work has indicated that this is not a major factor in causing biological damage in PDT.[214]

Various studies have been carried out to determine the relative efficiency of the type I and type II photoreaction pathways for different porphyrin structures in aqueous and detergent-micelle systems. Thus, kinetic spectrofluorimetry has been used to evaluate the effects, on the HP-sensitized photooxidation of L-tryptophan and tryptamine, of different quenching or enhancing agents of the possible reactive intermediates.[215-218] This work has shown that while the $^1O_2$ mechanism appears to be dominant, both pathways can contribute to photooxidation.[215,216] This finding is confirmed by some,[219] but not all,[220] spin trapping and electron spin resonance (ESR) studies.

In the presence of oxygen, the polarity of the reaction medium becomes a major determinant of the relative contribution of the type-II pathway,[215-217,221,222] mainly by affecting the singlet oxygen quantum yield and lifetime.[206-208,216,217] Grossweiner et al.[180,223,224] found that the mechanism of liposomal lysis by HP and HPD changed from type II at low porphyrin concentration to a type-I (anoxic) process at high concentration, where porphyrin aggregation is increased. Indeed, in aqueous solution the singlet oxygen yield of the biologically active component of HPD is very low,[160,182,189,191,225] and Photofrin II^{TM} (or DHE) is relatively ineffective compared to HP as mediators of L-tryptophan photooxidation.[182,183,225] On the other hand, the quantum efficiency of $^1O_2$ production was close to unity when HPD-A was incorporated into liposomes,[182] and the Photofrin II^{TM}-sensitized photooxidation of L-tryptophan appears to be most efficient under conditions which promote disaggregation.[183] It is possible that the aggregates present in Photofrin I^{TM} and II^{TM} are of a special nature,[225] and that the potent photodynamic activity of DHE derives from its disaggregation and/or its conformational change upon intracellular localization.[191,226]

In summary,the biological activity of HPD (Photofrin I^{TM}) resides mainly in one component (Photofrin II^{TM}), which appears to be composed largely of DHE, and which in aqueous solution, comprises particularly stable, highly aggregated, hydrophobic porphyrins. These aggregates are characterized by small fluorescence yield, low triplet-state formation, and minimal photooxidizing efficiency. However, certain environmental conditions, including those relating to the intracellular localization of

HPD, lead to disaggregation and/or conformational changes of DHE. This enhances the fluorescence yield, and increases the efficiency of triplet-state formation, as well as promoting an effective photodynamic activity that is mediated predominantly via a type-II (singlet oxygen) pathway.

VI. IN VITRO PHOTOBIOLOGY

Although a considerable amount of information is available on porphyrin photochemistry in aqueous solutions and more complex heterogeneous media, the identity of the particular photochemical lesion(s) or critical subcellular sites responsible for the photodynamic killing of mammalian cells by HPD has not yet been established. Study of the mechanisms of HPD-photosensitized cytotoxicity is complicated by the differential ability of different porphyrin species (1) to penetrate the cell membrane, (2) to bind to particular intracellular sites, and (3) once localized, to generate reactive photochemical intermediates. Understanding of the sequence of events occurring in HPD-light inactivation of cells has been obtained through studies in a variety of in vitro models.

A. Uptake and Intracellular Localization of HPD Components

The relative hydrophobicity of the components of HPD appears to be a major factor affecting the rate of uptake by cells,[227-230] with the phototoxicity depending on the uptake of the most hydrophobic components.[158-160,165,168,231-234] While HP is an efficient photosensitizer in cell-free systems,[235-237] its relative hydrophilicity prevents it from readily crossing the cell membrane,[165,231] and its photocytotoxicity appears to be due to preferential uptake of hydrophobic contaminants.[35,158,231,238] Uptake of HPD components by cells is characteristically biphasic, with an initial rapid rise up to ~6 hr, and a subsequent plateau extending to at least 24 hr.[239,240] Incubation of cells with HPD for short times results in accumulation of relatively "loosely bound" phototoxic components, a large fraction of which can be easily washed off by serum-containing culture medium.[163,165,231,240]

The more "tightly bound", nondiffusible portion that is retained increases with increasing incubation time,[163,165,231,240] and a significantly greater degree of photosensitization is achieved for the tightly bound than for the loosely bound HPD, for the same intracellular concentration.[163,240-242] It has been hypothesized that once bound to cells, hydrophobic HPD components transform to porphyrin species that are retained in the cells.[165,230,231,233] The possibility that HPD may undergo conformational or aggregational changes as intracellular localization takes place is also suggested by studies of HPD absorption and fluorescence spectra,[191,241-248] time-resolved fluorescence microscopy,[192,249-253] and fluorescence lifetime and polarization measurements.[252-254] In particular, although the biologically active components of Photofrin I^{TM} and II^{TM} exist as unique stable porphyrin aggregates in solution,[35,225] at least partial disaggregation/monomerization is needed for cellular photoinactivation,[247,248] the quantum yield of which increases with increasing HPD incubation time prior to light irradiation.[240-242,246] A number of fluorescence microscopy studies have shown that a large fraction of cell-bound HPD is associated with cell membranes.[191,241,242,246,255] Furthermore, Kessel and Cheng[191] have recently found that whereas S-180 cells briefly incubated with DHE exhibit a weak diffuse fluorescence at the cell periphery, the pattern of fluorescence after 24 hr incubation shows localization in intracellular vesicles. This is accompanied by changes in the fluorescence emission spectra indicative of a migration of DHE from hydrophobic loci, presumably cell-membrane components, to more hydrophilic intracellular regions. Together with evidence that the relative lipid solubility of porphyrins is an important determinant of subcellular localization and resultant photodynamic toxicity,[174,227,236,256-263] these results suggest that DHE photosensitizes at cellular loci which vary with the incubation time prior to irradiation.[191] Indeed, consistent with this concept are the results of Moan and co-workers[234,264,265] that the sites of HPD photodamage depend to a large extent on the time of exposure of cells to HPD prior to irradiation.

Andreoni and co-workers[266,267] have reported selective cytotoxicity of HPD in transformed vs. normal rat thyroid cells in culture that correlates well with the significantly higher HPD uptake in the former.[267] However, such preferential uptake of porphyrins by malignant cells has not to date been confirmed by other investigators,[234,240,268,269] and it also appears that HPD uptake by different tumor cell lines does not correlate with their oncogenic potential.[234,270] Attempts have been made to improve porphyrin uptake, retention, and sensitization in tumors using two different carrier systems, namely liposomes and monoclonal antibodies (mAb). Reddi et al.[184,271-274] examined the uptake of liposome-bound porphyrins by normal and neoplastic rat hepatocytes and obtained higher rates and levels of uptake, and longer retention times, than by using the same porphyrins in aqueous solution. In addition, this enhancement was more pronounced in the tumor cells than in the normal cells, and resulted in a greatly increased tumor cell photosensitivity.[273,274] At the same time, subcellular localization studies on HeLa cells revealed differences in the intracellular distribution of various liposome-encapsulated porphyrins, including HPD, and it was suggested that the efficiency and site of photodamage might be modulated by the use of appropriate porphyrin carriers.[274] Some progress has been made by Mew et al.[275,276] in conjugating HP with mAb against mouse rhabdomyosarcoma. The conjugation did not intefere with either the antigen specificity of the antibody, or with the photodynamic activity of the HP. Similar findings have been made recently using HP covalently bound to mAb against a membrane antigen specific for human acute myelogenous leukemia cells.[277]

B. Subcellular Sites of HPD Photodamage

A number of studies have shown that the subcellular loci of HPD photodamage depend to a large extent on the length of time that cells are exposed to the HPD prior to irradiation,[191,234,264,265,278] as well as on the relative lipid solubility of the components.[174,227,236,256-261] At the same time, there is evidence that cell membranes are the major sites of photosensitization, and that at least for relatively short HPD incubation times, the plasma membrane is a particularly sensitive target.[191,264,279,280] Structural changes have been observed in the plasma membrane of cells treated by HPD components and light,[234,255,281-288] and inhibition of membrane transport[277,234,256,289-295] and enzyme systems,[288] disruption of permeability barriers,[227,234,256,291-294] and cell lysis[27,159,234,240,256,295] have also all been detected. Photoirradiation of porphyrin-incubated cells can result in membrane protein cross-linking[257] and lipid peroxidation.[296] The relative contribution of these two processes to the structural and functional alterations in plasma membrane due to protoporphyrin photoinactivation has been studied using erythrocyte membranes as a model system,[258,296-301] and plasma membrane effects have also been studied in fibroblasts,[302] reticulocytes,[288] and erythroleukemic cells.[282] Grossweiner and co-workers[179-182,223,224,303] used liposomes as a membrane model system to show that lipid peroxidation may be involved in HPD-sensitized phototocytotoxicity and cell photolysis. It is likely that at least some membrane photodamage is repairable,[159,234,255,264] and Christensen et al.[159,264,304] have found that HPD photoinactivation of cells at low light and porphyrin doses can occur in the absence of cell lysis, with cells becoming irreversibly blocked in mitosis. The contribution of plasma membrane damage to cell killing may, therefore, depend on the experimental conditions,[159,264] and ultrastructural changes have also been observed in a wide variety of subcellular structures, including mitochondria,[255,286,292,305,306] endoplasmic reticulum,[286,294,305] lysosomes,[286] and cell nucleus.[255,285-287]

Indeed, mitochrondria may be an important site of HPD photodamage. Fluorescence microscopy,[243-245,255,269] and subcellular fractionation studies[252-254,307] suggest that HPD components do localize in mitochondria. Gibson et al.[308-311] have reported HPD photosensitization of a number of mitochondrial membrane-associated respiratory enzymes. The photodamage appears to involve the mitochondrial membrane proteins themselves, rather than a disruption of membrane structural integrity,[311] and is mediated by a singlet oxygen pathway.[214] Salet et al.[312-314] have also found that both HP and HPD photoactivation can affect mitochondrial respiration via an oxygen-dependent pathway.[314] The ability of different porphyrins to photoinactivate mitochondrial respiratory enzymes depends on their relative lipid solubility.[258-260,315] In addition to effects on mitochondria and lysosomes,[316-318] HP and HPD have been shown to cause photodestruction of microsomal cytochrome P-450 and monooxygenase sytems.[319,320]

Concerning nuclear effects, subcellular fractionation[252-254,307] and fluorescence microscopy[243-245,255,269,291] suggest that intracellular HPD associated with the cell nucleus is essentially confined to the nuclear membrane or perinuclear region. Various types of DNA and chromosome damage have been reported for HP and HPD photosensitization, both in model systems and in cells. The guanine moiety is probably most susceptible to photooxidation and subsequent cross-linking to amino acids.[321,322] Single- and double-strand breaks have been detected in cell-free systems,[321,323] and the ability of DNA to act as a template for RNA synthesis is also impaired by porphyrin photosensitization.[324] Single-strand breaks have been detected in CHO and V79 Chinese hamster cells,[325-327] and in L929 cells[291] treated with HPD + light, as well as in HP-photosensitized human NHIK cells.[328-330] On the other hand, photosensitization produces fewer strand breaks than does an equitoxic dose of X-rays,[325,326,329,330] so that DNA breaks may not be a significant contributor to HPD photolethality. However, while most authors report that photodynamically induced strand breaks are repaired as efficiently as those caused by X-rays, Blazek and Hariharan[327] found a lower fractional rate of repair with PDT than for equivalent levels of X-ray breaks in CHO-K1 cells. For the two types of treatment, they also detected differences in the sensitivity of repair to inhibitors of enzymes involved in nucleotide excision repair, suggesting that the chemical nature of strand breaks by PDT and X-rays is different. DNA-protein cross links are also known to be produced in cellular DNA by HPD + light.[291,327] In CHO-K1 cells, these seem to be repaired with kinetics similar to the kinetics followed in recovery of HPD-photosensitized inhibition of DNA synthesis in human cells.[265,327] Whatever the lesion(s) responsible, HPD photodamage produces chromosome breaks in nuclear membrane-associated chromosomal regions,[331] as well as sister chromatid exchanges,[159,264,307,330,332] which are, however, produced at a much lower frequency than by an equitoxic dose of X-rays.[330] The mutagenic potential of PDT has been examined by Gomer and co-workers,[307,332-334] who found that while X-rays and UV light are very effective in producing 6-thioguanine resistant CHO cell mutants, HPD + light does not increase the level of mutagenesis significantly above background.

Although various subcellular sites of HPD photodamage have been identified in mammalian cells, the critical cellular targets responsible for cell inactivation have yet to be established. Thus, a number of studies have been done using bacterial and yeast cells as models with simpler structural organization. Bertolini et al.[335-337] examined HP photoinactivation of various Gram positive and Gram negative bacteria of different cell wall composition, concluding that for cell binding of porphyrins, cell-wall composition, and cell physiologic state may all play a role. The ability of porphyrins to penetrate the cell wall and bind to the plasma membrane has also been shown to be important for bacterial photosensitization by HPD. A variety of yeasts are likewise photoinactivated by HP or PP.[336-340] For these two porphyrins, the difference in binding to yeast cell membrane as a function of incubation time,[340] and the differential dependence of photoinactivation on a singlet oxygen pathway[338,340] indicate differences in subcellular localization. Furthermore, Ito and Ito[341] found that the presence of ethanol in the incubation medium enhanced the photosensitization for both porphyrins, with a concomitant increase in membrane damage, and in the frequency of genetic damage. They concluded that changes in HP and PP solubility, resulting from ethanol-enhanced porphyrin disaggregation, increased migration to, and subsequent photosensitization of, internal subcellular sites.

C. Factors Affecting Cell Inactivation Kinetics and Repair

In vitro survival curves have been determined by several groups for photoirradiation of cultured human cells photosensitized by HP,[283,330,342-346] and for HPD photosensitization of human,[110,158,159,168,234,240-242,248,264,265,270,287,347-350] and nonhuman[27,264,270,295,326,347,348,350-354] cells. These are characterized by a shoulder region, followed by exponential killing, suggesting that repair of photodamage can take place at low doses of light and/or of porphyrins. Comparison of survival curves for different porphyrin-incubation and light-exposure conditions has enabled a number of factors affecting HPD photocytotoxicity to be identified, as follows.

Incubation time — Moan and co-workers[158,159,168,234] have shown that the most hydrophobic and aggregated component of HPD has the greatest photoinactivation capacity. The efficiency of HPD photoinactivation increases with the length of HPD incubation, as the intracellular components of HPD undergo tighter binding to subcellular structures, accompanied by possible disaggregation and/or conformational changes.[234,240-242,246,248] Comparison of the action spectra for HPD and Photofrin II^{TM} in human NHIK-3025 cells with the absorption and fluorescence-excitation spectra in the same cells[234,247,248] led to the conclusion that while both aggregated and monomeric porphyrins are bound to the cells, only the nonaggregated fraction contributes significantly to photodynamic inactivation. Further, for a long incubation time (18 hr), the action spectra were closely similar to the fluorescence-excitation spectra of protein-bound porphyrins.

Wavelength — In experiments using three broad-band light sources spanning the visible wavelength spectrum, Kinsey et al.[348] found that the short-wavelength radiation corresponding to the cell-bound Soret band around 400 nm was 12 to 30 times more efficient in cell inactivation than red light (560 to 720 nm), with blue-green light (440 to 600 nm) being intermediate in effect. Moan and Christensen[289] also found the Soret band more effective than wavelengths around 546 nm in HP-sensitized killing of human cells, whereas Ebert et al.[355] found the Soret band less effective than the absorption band around 500 nm. Since most clinical and in vivo studies have used the HPD absorption band around 620 to 640 nm, Gomer et al.[307,334,352] have examined the action spectrum of HPD in detail in this region, and obtained maximum phototoxicity of cell-bound HPD at 630 to 632.5 nm. In the search for more potent photosensitizers in the wavelength region above 600 nm, two groups have recently reported a superior in vitro photodynamic efficiency with bonellin, a chlorin with a greater fraction of its absorption spectrum in the red region.[356,357] The tumor-localizing properties of bonellin are not yet known.

Light exposure rate — Dougherty et al.[354] found that while the efficiency of HPD phototoxicity in murine tumor cells increased with increasing intracellular HPD concentration, varying the power density of the light (milliwatts per square centimeter) twofold had no effect on cell kill. More recently, Gomer et al.[307,334] detected no significant variation in HPD photodynamic effectiveness when equal "doses" of red light

(635 nm) were delivered to CHO cells at rates between 0.5 and 60 mW/cm². However, Anderson and co-workers[349] reported a twofold increase in efficiency of MEWO cell killing by HPD + 635-nm light as the dose rate was increased from 40 to 160 mW/cm². No temperature changes were detected during irradiation, and the effect was attributed to a partial repair process. At the same time, a fivefold *decrease* in efficiency was reported by the same group for 405-nm light as the dose rate was increased from 20 to 160 mW/cm². This was attributed to a possible irreversible photodecomposition of porphyrin at the higher power densities. While Andreoni et al.[266] found that pulsed UV laser irradiation was more efficient than continuous-wave red laser light in HPD photolethality of cultured rat cells, no difference in Raji cell photolysis was detected by Cowled et al.[294] when these cells were irradiated with red light delivered either in pulsed or continuous form.

Cellular oxygenation — The evidence that singlet oxygen is the primary mediator of HPD phototoxicity in mammalian cells derives from studies both in cells[283,353,358] and in liposomes,[179,180,223] as well as work in homogeneous solutions and organized media.[216,219] However, contributions from free-radical pathways cannot be ruled out.[212,213,216,219,223,358-361] Two groups have also examined PDT with HPD using multicell spheroids as a model for solid tumors containing hypoxic cell fractions.[92,239,362] Christensen et al.[362] reported that the survival of NHIK-3025 cells in PDT-treated spheroids was significantly higher than in single-cell monolayer.

Cell cycle — This last group has also determined that the HP photoinactivation of human cells depends on the cell-cycle phase during light irradiation.[343,344] Cells were most sensitive during S-phase, as shown by reduction in the shoulder of the cell survival curves, suggesting a decreased capacity for repair of the photodamage sustained during the period of DNA synthesis. Similar cell-cycle dependence was found in a variety of human, Chinese hamster, and mouse cell lines treated with HPD and light.[347] (This study also showed no intrinsic differences in PDT sensitivity between normal C3H/$10T_{1/2}$ cells and their transformed counterparts.) Gomer and Smith,[350,351] on the other hand, did not detect any significant variation in CHO cell survival as a function of cell cycle phase, although they did observe a dose-dependent delay in cell division prior to subsequent recovery. More recently, Gomer et al.[307] reported a cell-line dependent potentiation of PDT damage by adding caffeine to the posttreatment incubation medium, suggesting that at least some cell types can effect repair of PDT damage analogous to the repair of potentially lethal damage observed after X-ray exposure. Bellnier et al.[363] have also measured a light-dose dependent delay in population growth for human bladder carcinoma cells containing tightly bound, nondiffusible HPD components. Confirmation that these cells could recover from sublethal PDT damage was obtained in classical split-dose experiments, which showed that a period of 10 hr was needed to achieve full recovery (i.e., regeneration of the shoulder on the survival curves). Repair of sublethal HPD photodamage has been demonstrated in this way also for Chinese hamster V-79 cells,[364] and for NHIK-3025 cells.[365]

Hyperthermia — Christensen et al.[146,365] have examined the effect of combining HPD photoirradiation of cells with hyperthermia at 41, 42.5, and 45°C, and found that PDT and hyperthermia acted synergistically. The enhancement of PDT photocytotoxicity by hyperthermia was maximal when hyperthermia was given after PDT treatment. The effect of PDT plus hyperthermia was to reduce the survival curve shoulder observed with PDT alone, without altering the exponential slope of the curves, thus implicating hyperthermia in inhibition of repair of sublethal PDT damage. Very similar results have been obtained for cultured EMT-6 cells by Mang and Dougherty.[366]

VII. IN VIVO PHOTOBIOLOGY

The success of PDT depends on the uptake and preferential retention in tumor vs.

adjacent normal tissues of adequate amounts of photoactive HPD components, i.e., of the most hydrophobic and highly aggregated porphyrin species. However, relatively little is known of the mechanisms of HPD accumulation, localization, and photosensitization of tumor cells in vivo. The in vitro studies mentioned earlier have been done in either cell-free or cultured-cell model systems in order to avoid many of the uncontrollable host factors inherent in in vivo experiments. These in vitro data have helped substantially in understanding the factors affecting direct photocytotoxicity. However, the extent to which this information can be applied in vivo remains to be established, since it is likely that inter- and intratissue distribution of biologically active HPD components, as well as their selective tumor tissue photosensitization, will depend on the transport of HPD from the injection site to the tumor tissue, and on morphologic and functional differences between malignant and normal tissues.

A. HPD Pharmacokinetics and Localization

The binding of HPD components to serum proteins may be a significant factor in the kinetics of HPD localization in vivo. Serum proteins are known to be involved in porphyrin transport in the bloodstream,[367-373] and HPD porphyrins bind to both serum proteins[189,193,195,374] and lipoproteins.[274,375] In particular, Jori et al.[274] found a long-lived, strongly bound HP-lipoprotein complex in the sera of patients receiving HP photodynamic therapy. When bound to serum protein, both HP and HPD retain their photooxidizing efficiency,[346,376,377] although serum and serum albumin inhibit porphyrin uptake by cultured cells[163,234,283,351,378] and liposomes.[181,224] El-Far and Pimstone[379] proposed that the relative tumor-localizing properties of porphyrins in vivo derives in part from their relative strength of binding to tumor tissue compared to plasma proteins, and suggested that preloading of serum with a porphyrin of high serum protein binding affinity, such as PP, would enhance the tumor uptake of more hydrophilic porphyrins. However, this hypothesis has been questioned.[193]

Using a tumor-bearing rat model, Gervais et al.[380] studied the tissue, plasma, and bile pharmacokinetics of HP-diacetate (HPDA), which is known to have selective tumor-localizing capacity.[150,153,381] These experiments identified an important contribution by biliary excretion and enterohepatic reabsorption to HPDA plasma kinetics, which showed an elimination half-life of 30 min. However, after an initial slow clearance, all other tissues retained detectable levels of HPDA for 4 days or more. The tumor uptake was intermediate between the relatively low levels in lung, heart, and muscle, and the high uptake in liver and kidney. Houle and Besner[382] similarly measured the uptake of HP in tumor-bearing rats. The peak concentration was highest in liver, with the levels in tumor being significantly greater than in lung, muscle, spleen, and small intestine.

Jori and co-workers[383-385] have measured the uptake, clearance, and subcellular distribution of HP in selected tissues of normal and hepatoma-bearing rats. When HP in aqueous solution was administered i.v., clearance from tumor began 24 hr later, but the HP was nonetheless retained at higher levels and for longer times in tumor than in liver or kidney, with the intracellular HP predominantly associated with the membrane fraction. On the other hand, if HP or its more hydrophobic dimethylester was incorporated into liposomes prior to injection, then significantly greater tumor-to-liver uptake ratios and longer tumor retention times were achieved,[272,273,386] due presumably in part to protection of the porphyrins from serum protein binding. Similar liposome-mediated enhancement of tumor uptake of HPD has been observed in tumor-bearing mice, and also appears to offer the possibility of promoting selective tumor uptake of otherwise non-localizing porphyrins.[274,387,388] Mew et al.[275,276] have also used porphyrin conjugation to tumor-specific monoclonal antibodies in vivo in parallel with in vitro studies of this method, to enhance HP uptake in malignant cells.

To measure the in vivo uptake and biodistribution of HPD, a number of studies have used radiolabeled material in order to avoid some of the potential problems arising in the use of either chemical extraction of porphyrins and/or HPD fluorescence techniques. Dougherty et al.[27,389] used both ^{14}C and tritium-labeled HPD in two different transplanted mouse tumor models. At all times following i.p. injection (10 mg/kg), the levels of ^{14}C-HPD and ^{3}H-HPD were very similar in all tissues studied, with the peak levels in tumor being higher than those in skin, muscle, and lung, but lower than those in liver, kidney, or spleen. Maximum concentration was reached in all tissues at 2 to 4 hr, and also in blood, for which the half-life in serum was 3 hr compared to around 25 hr in humans. The levels of radiolabeled HPD, however, did not correlate well with gross visualization of fluorescence in the tissues. Subsequently, Bugelski et al.[390] used autoradiography with ^{3}H-HPD to assess the spatial and temporal inter- and intratissue distributions in tumor-bearing mice. In normal tissues, HPD localization changed from a uniform distribution in stroma and parenchyma at early times (<3 hr) to retention of HPD in stromal and reticuloendothelial components, with elimination from parenchymal cells, at later times, as serum levels dropped. In tumor tissue, a fivefold greater concentration was maintained for up to 48 hr in stroma, pseudocapsule, and necrotic regions compared to that in tumor cells, with high grain densities at later times observed only in stromal cells believed to be macrophages. Fluorescence microscopy of tumor tissue showed essentially the same pattern. In this regard, Berns et al.[269] detected HPD fluorescence which was significantly higher and longer retained in the nonproliferating necrotic and inflammatory regions than in the proliferating tissue of a human SCC.

The differences between tumor and normal tissue HPD pharmacokinetics observed in the mouse by Bugelski et al.[390] together with the similarities in biodistibution of ^{125}I-labeled albumin and ^{3}H-HPD, was taken to imply that the uptake, retention, and distribution of HPD in tumors derive, at least in part, from the peculiar physiology of tumor tissue. Thus, the increased vascular permeability to plasma proteins, and compromised lymphatic drainage may cause accumulation of serum proteins in the interstitial fluid, and nonspecific binding to stromal components.[390-392] This hypothesis was reaffirmed by Evensen et al.,[162] who compared the uptake and clearance of the tumor-localizing isotope ^{67}Ga with ^{3}H-HPD in mice bearing Lewis lung carcinoma. The tumor to normal tissue ratios compared well with those found by Gomer and Dougherty.[389] However, the nonidentity of ^{125} I-albumin and ^{3}H-HPD tumor to normal tissue uptake led Evensen et al. to suggest that other factors may play a role in tumor localization of HPD, such as trapping in tumor macrophages, or pH-favored passive diffusion into tumor cells.

In other biodistribution studies, Sabben et al.[393] measured the time course of ^{3}H-HPD uptake in rats bearing digestive tract tumors, and found also that the serum half-life was about 3 hr, with peak uptake in all tissues at around 4 hr. Autoradiography revealed uptake in liver Kuppfer cells, and in macrophages in lung, spleen, and small intestine. Gomer and co-workers[111,396] have examined the kinetics of ^{3}H-HPD localization in ocular tumors and normal tissues. Higher levels of ^{3}H-HPD were found in athymic nude mouse eyes heterotransplanted with human retinoblastoma compared with normal control eyes.[111] Similarly, in rabbit eyes transplanted with amelanotic melanoma in the anterior chamber, tumor levels of ^{3}H-HPD were higher than in normal ocular structures. The HPD initially taken up in iris, aqueous humor, and choroid-retinal epithelium were eventually cleared from these tissues, and there was no significant uptake in cornea, lens, retina, or sclera.[394]

Care must be taken in interpreting the results of radiolabeled-HPD biodistribution studies, however, since they are based on the premise that the level of radioactivity measured truly represents the photosensitizer concentration in the tissues. In particular, whereas it was concluded from some studies that minimal exchange of tritium from ^{3}H-HPD occurs in vivo,[390,391,394] other work with tritiated porphyrins suggests that this may not be the case,[395] and further, that aqueous solutions of tritiated HPD are subject to accumulation of radioactive degradation contaminants.[396] ^{14}C-HPD is a better "gold standard" in these respects, but unfortunately is not readily available. Jeeves et al.[92,239] have reported preliminary in vivo kinetics studies in tumor-bearing hamsters and rats using HPD labeled with copper-64, a positron emitter suitable for noninvasive nuclear imaging. The γ-emitter ^{67}Cu would be an ideal label for conventional nuclear scintigraphy if the biodistribution characteristics of radiocopper-labeled HPD are shown to correctly represent the pattern of HPD uptake and clearance in normal and tumor tissues. The potential use of other radiolabeled porphyrins,[397,398] and the structurally related phthalocyanines[399] for tumor scanning and/or biodistribution studies have also been examined by other groups.

B. In Vivo HPD Photosensitization

Light irradiation of HPD-sensitized malignant and normal tissues results in a coagulation necrosis that often shows a sharp boundary between necrosed and viable tissue, with the effective depth of the necrotic region depending on the treatment conditions.[13,16,20,27,112,136,153,154,400–406] The extent of tissue destruction presumably depends in a complex way on a variety of selectable and inherent factors, whose effects have been little quantified: HPD dose administered, time delay between HPD administration and light exposure (which affects the tissue and intracellular HPD distribution), wavelength, energy density, spatial distribution of the applied light, tissue oxygenation, cellular and tissue sensitivity to HPD photodamage, repair processes, etc.

1. Wavelength Dependence

As previously discussed, in clinical PDT, red light is selected because of its relatively high penetration in tissue compared with shorter wavelengths at which HPD is more photoactive. Pimstone and Gandhi[403] used an assay of in vivo rat liver necrosis to define the optimal band of red light for PDT, and found that the depth of tissue necrosis was maximum, and essentially constant, between 620 and 650 nm. However, in vivo observations by Weishaupt and Dougherty[407] in mouse skin and SMT-F mammary carcinoma indicated that 630 to 632.5 nm was the most efficient wavelength region, in agreement with the in vitro results of Gomer et al.[352] While Bellnier et al.[408] found pulsed red laser light to be relatively ineffective in control of transplanted murine bladder tumors, Cowled et al.[294] found it no less effective than continuous-wave irradiation in the destruction of Lewis lung tumors in mice.

2. Oxygen Dependence

As is the case in vitro, a number of studies indicate that oxygen is required for in vivo porphyrin photosensitization, and that type-II processes constitute the dominant mode of action of HPD. Leg clamping prevents acute normal skin phototoxicity in HPD- and Photofrin II^{TM}-injected mice,[409] and carotenoids administered to collidine-porphyric mice provide some protection from porphyrin photosensitization.[358] It has also been found that simply clamping the blood flow to experimental tumors prevents successful PDT.[307,410] These results, and experiments in vitro with multicelluar spheroids containing hypoxic cell populations,[362] suggest that hypoxic cells in vivo are less sensitive to direct phototoxicity than their well-oxygenated counterparts.

3. Tumor Vascularization

Reduced oxygenation of tumor cells in vivo may not, however, in itself necessarily limit the effectiveness of PDT in solid tumors, since recent evidence suggests that damage to the tumor vasculature may be a primary lethal event. It is known that regression of experimental tumors can be achieved by local clamping of blood supply alone.[411] Henderson and colleagues[412,413] have observed that the clonogenic potential of cells taken from murine tumors after PDT decreased with the time delay between irradiation and removal of cells for assay, implying a delayed expression of damage. The rate of cell death was similar to that observed following cessation of blood circulation postmortem.[412] Benedict et al.[112] reported that whereas well-vascularized transplanted tumors in mice showed significant tumor destruction after PDT, poorly vascularized tumors were relatively insensitive to treatment. Castellani et al.[414] were the first to document sequential changes in the microcirculation of animal tissues treated by PDT, and Star et al.[137] directly observed disruption and eventual cessation of microcirculation in rat mammary tumors growing in "sandwich" observation chambers in vivo. Selman and co-workers[138,415,416] used radioactively labeled microspheres to demonstrate an 86% decrease in blood flow in rat urothelial tumors following PDT. This decrease occurred within 10 min of completion of irradiation, was maintained for at least 24 hr, and was correlated with significant tumor regression. A number of other studies of microvascular disruption in normal and tumor tissues following PDT have been reported.[107-109,307,394,417,418] The hypothesis that PDT is tumorcidal in vivo primarily as a consequence of vascular effects is consistent with the preferential localization of HPD in the vascular stroma of tumors, and with the observation that PDT damage can be detected in tumor cells adjacent to the microvasculature earlier than in distant cells.[27,390]

4. Hyperthermia

Henderson and Dougherty[412] have noted the similarities between the mechanisms of tumor destruction by PDT and by hyperthermia, which is also thought to be mediated in part by vascular occlusion.[419] As previously discussed, high power densities of light in PDT can produce hyperthermia which may contribute significantly to tumor cell killing,[17,136,142,145,405,406,408,420] and the additive and synergistic interactions of photodynamic and hyperthermic effects of PDT have been extensively modeled by Svaasand et al.[126,139-141] Localized microwave-induced hyperthermia of between 40.5 and 44.5°C can potentiate tumor destruction by PDT, as has been shown in two different animal tumor systems.[17,147,148,413] In the SMT-F mammary carcinoma in mice, Waldow et al.[147,148] determined that potentiation was dependent on the sequence of the treatments, the time interval between treatments, and the temperature. No detectable potentiation occurred below 40.5°C; maximal potentiation occurred when the tumor was heated to 44.5°C for 30 min immediately before or after PDT irradiation. With hyperthermia following PDT, the potentiation decreased with increasing time interval up to several hours, and except if given just prior to irradiation, essentially only additive effects occurred if the heat was applied first.

C. Action of DHE In Vivo

Finally, in addition to its dependence on the various treatment and tissue parameters, success of PDT in vivo depends specifically on the pharmacokinetics of the active HPD component, DHE. Studies on the mode of action of DHE in vivo have recently been initiated. It has been shown that DHE aggregate size is an important determinant of PDT efficacy. Aggregates less than 10,000 daltons were found by Dougherty et al.[421] to exhibit little photoactivity in a standard in vivo SMT-F tumor assay. The large aggregates are thought to be taken up by a variety of phagocytic (including vascular endothelial) cells, and may become trapped in the interstitial fluid of tumors which have inadequate lymphatic drainage. The high lipophylicity of DHE aggregates enables them to enter tumor cells by diffusion through the plasma membrane, possibly facilitated by the reduced pH of the tumor environment. Upon entry into cells, subsequent disaggregation and subcellular localization in (plasma and mitochondrial) membranes occurs, as is required for efficient photosensitization[247,248,421] and fluorescence.[191,421] The relative tumorcidal efficiency of DHE and HPD have been compared by Dougherty et al.,[422] who found DHE to be as effective as HPD in experimental tumors at about half the dose. Thus, it may be expected that the use of DHE (Photofrin II^{TM}) in clinical PDT will reduce the complications due to skin photosensitivity without significantly compromising the tumorcidal efficacy.

VIII. HPD FLUORESCENCE FOR TUMOR DETECTION AND LOCALIZATION, AND FOR HPD QUANTITATION IN VIVO

In this section, the authors will review the use of HPD fluorescence for the detection and localization of malignant tumors, and the potential use of fluorescence measurements in vivo to quantify the HPD concentration in tissues. As mentioned in the Introduction, studies prior to the development of HPD in 1960 had shown that many porphyrins, including HP, localized in malignant tissues, and that their presence could be demonstrated by fluorescence. The improved tumor-localization characteristics of HPD, and the high sensitivity for tumor detection by visualization of the salmon-red fluorescence of HPD were established in a number of clinical studies in the early 1960s (see 1972 review by Sanderson et al.[423]).

Typical HPD fluorescence excitation and emission spectra are shown in Figure 1. Maximum fluorescence yield is obtained by exciting with violet light in the Soret band around 400 nm, either using a filtered arc lamp or a krypton laser. The resulting fluorescence may be viewed, or detected photoelectronically, through secondary filters which reduce the scattered violet light and the tissue autofluorescence. Most commonly, the 630-nm emission peak, or the band between 600 and 700 nm, has been used in clinical studies. As discussed previously, the fluorescence spectra depend on the biophysical and biochemical properties of the environment, and on the complexing of HPD in tissue. The degree of porphyrin aggregation is also important, since efficient fluorescence requires monomerization, so that the in vivo localization may affect the fluorescence yield.

The absolute intensity of the detected fluorescence in vivo also depends on the intensity of the exciting light; on the absorption, scattering, and reflectance characteristics of the target tissue; on the light source and detector geometries; on the tissue geometry; and on the degree of fluorescence quenching. The depth of tissue sampled is limited by the penetration of the exciting light, and is typically 100 to 200 μm at 400-nm wavelength. The theoretical lowest concentration of HPD detectable by fluorescence is set by the level of autofluorescence from the tissue, and by the consequent ability to separate the HPD and autofluorescence spectra. These numerous complex factors have not, however, prevented successful clinical application of HPD fluorescence detection in a number of tumor sites, although they may ultimately limit the ability to quantify absolute HPD levels in target tissues.

Prior to 1974, all clinical HPD fluorescence studies were done by simply illuminating the tissue with 400-nm light, and visually assessing the degree of red fluorescence after appropriate filtering to reduce the effects of background light. Although endoscopic illumination and viewing were introduced at an early stage, there was only limited development of the instrumentation in this period to try to improve the sensitivity or to reduce the subjectivity of the technique. However, in 1974 Krohn et al.[424] modified an indirect opthalmoscope to permit quantitative, objective determination of HPD fluorescence in a study of choroidal malignant melanoma transplanted in the posterior chamber of the rabbit eye. An optical beam splitter divided the fluorescent light between the viewing system and a photomultiplier detector. Thus, an area of interest could be brought into the field of view and the 680-nm fluorescence intensity recorded simultaneously. With a threshold set at 1.5 for the tumor to background fluorescence ratio, a 94% sensitivity was achieved at 24 and 48 hr after injection of 10 mg/kg HPD.

Photoelectronic detection of fluorescence during standard flexible fiberoptic endoscopy was developed at the Mayo Clinic in the late 1970s.[425,426] This was done by delivering pulses of 405-nm light from a filtered arc lamp and detecting the 600 to 700-nm fluorescence via fiberoptic light guides separate from the main endoscopic viewing bundle. The detected signal was converted to an audible tone, whose frequency increased with the intensity of the fluorescence. The clinical assessment of this instrument was reported in a series of 10 patients with 13 radiographically occult bronchogenic lesions.[427-429] Positive fluorescence was detected in all six CIS cases examined, which were thought to be otherwise undetectable. In two cases where the fluorescence was falsely negative, the lesions were beyond the reach of the bronchoscope. Benson et al.[430] also used this instrument cystoscopically in four patients with TCC *in situ* of the urinary bladder 2 hr after HPD injection, and reported a complete correlation with fluorescence subsequently observed in resected tissue. There was no HPD fluorescence in normal bladder wall. A study using a similar fluorescence-detection technique has been reported by Lin et al.[431] in 13 patients using intravesical instillation of HPD in the bladder, with negative results.

Although the Mayo Clinic system had high sensitivity, and allowed unimpaired normal endoscopic examination, the lack of a visual image of fluorescence, and the difficulty in discriminating against autofluorescence were limitations. An alternative approach was taken at University of California, and the instrument developed was assessed clinically at three centers.[432-441] This method used a fiberoptic bronchoscope with the coherent imaging bundle connected to a high-gain image intensifier. The tissue could be illuminated by white light for normal endoscopic viewing, or (not simultaneously) by violet light to excite fluorescence. The violet light was delivered via a separate optical fiber placed through the instrument channel of the bronchoscope. Fluorescence imaging was done in the 580 to 700-nm band. In the early stages of development, a filtered arc lamp was used as the 400-nm source, but this was later replaced by a krypton ion laser producing 407 to 415-nm light. With this system, a 100-μm thick liquid phantom containing HPD at a concentration of 1 μg/g could be reliably visualized. In the clinical studies with the instrument (at various stages of its development), King et al.[437,438] reported positive fluorescence in nine out of ten patients with malignant bronchogenic carcinoma 48 to 72 hr after 1.5 to 2.5 mg/kg HPD; two of four cases of benign atypia were also detected. Similar findings were made in four cases of malignant disease by Doiron et al.[435] In a series of 38 patients, Balchum et al.[439] obtained a sensitivity of 97% and a specificity of 75% for the detection and localization of malignant bronchogenic lesions; only one CIS was missed. Examination was carried out under local anesthetic at 2 hr after i.v. injection of 2 mg/kg HPD.

A similar approach to endoscopic visualization of HPD fluorescence was followed at Tokyo Medical College,[442,443] and the technique was evaluated for detection of lung cancer in patients[54,55,443-445] and in a dog model,[444-447] and in patients with suspected bladder,[73,75] early gastric,[100,448] and esophageal tumors.[97] A further technical refinement used in some studies was spectrographic analysis of the fluorescence, in which the 630 and 690-nm HPD fluorescence emission peaks could be identified against the broad autofluorescence background, which overlaps the 630-nm peak. In 36 cases of human lung cancer examined at 48 hr after 2.5 to 4 mg/kg HPD, a true positive rate of 92% was obtained, and 3 of 4 cases of atypical squamous metaplasia were also positive. False negative results were attributed to obscuring of the fluorescence due to surface bleeding, superficial necrosis, or overlying mucosa. The studies in dogs showed very similar results for all stages of induced carcinoma and metaplasia. Fluorescence was not seen in normal bronchial mucosa. Using the same protocols cystoscopically,[73,75] fluorescence was positive in all 20 cases of primary and recurrent malignant bladder tumors. In one case of gastric cancer,[100] fluorescence yields for tumor, areas of erosion, and normal mucosa were found to be in the ratio 12:4:1 at 48 hr after HPD injection, with spectral analysis able to discriminate between HPD and autofluorescence. A subsequent study[448] in 15 patients, including 9 with early-stage disease, yielded positive fluorescence in all cases.

A. HPD Quantitation In Vivo

Lack of quantitative information on the concentration of photoactive HPD in the tumor tissue during PDT irradiation is a major limiting factor in the further refinement of clinical techniques. Three noninvasive, or minimally invasive, techniques to enable in vivo HPD quantitation have been proposed to date, namely (1) measurement of HPD fluorescence, (2) positron emission tomography (PET) or single photon emission computed tomography (SPECT) using radiolabeled HPD, and (3) reflectance spectrophotometry. Preliminary results have been published for ^{64}Cu-HPD PET scanning,[92] and reflectance spectrophotometry is also under active investigation in the authors' laboratory. Here, only the first technique, for which some detailed studies have been published, shall be considered.

The factors affecting the intensity of fluorescent light emitted when tissue containing a given concentration of HPD is illuminated with a known intensity of short wavelength light were mentioned earlier[432,436,449,450]namely

1. The fluorescence efficiency (number of fluorescent photons emitted per excitation photon absorbed) is dependent on the biochemical and biophysical environment.
2. Since fluorescence is not directly involved in photosensitization, and since the disaggregation and consequent localizing properties of HPD are complex and not fully understood, the fluorescence may not necessarily be strongly correlated with the phototoxicity in vivo.
3. The excitation and emitted photons are absorbed, scattered, and reflected by the tumor tissue, and by overlying tissue such as skin.
4. The number of exciting photons reaching the tissue, and the fraction of emitted photons subsequently detected, depends on the tissue, source, and detector geometries.
5. Due to the small penetration depth of 400-nm light, only the HPD in the surface layers contributes to the fluorescence, and this may not be representative of the distribution of HPD in the much larger depth of tissue irradiated by 630-nm light.

None of these problems is insurmountable, at least in principle, although they may each limit the accuracy of determining the HPD concentration. Solution of the first two problems will require further study of the localization and photosensitization characteristics of HPD and its component porphyrins in vivo. Problems 3 and 4 may be tackled by modeling the spatial distributions of exciting and emitted photons, using experimental data on the optical properties of tissues at these wavelengths, as exemplified by a recent study of HPD fluorescence detection in subcutaneous tumors.[451,452] Problem 5 may be circumvented by using 630-nm light to excite fluorescence, and detecting the emitted 690-nm photons, as proposed by Potter and Mang.[450] These authors have developed an instrument for "in vivo fluorimetry" using optical fibers to deliver the 630-nm light and to detect the 690-nm fluorescence, and have used it to measure HPD uptake in skin and cutaneous tumors. Similar approaches may be applicable endoscopically, and interstitial placement of the fibers may permit quantitation of fluorescence, and hence of local HPD concentration, within the tumor tissue.

ACKNOWLEDGMENTS

The authors wish to thank the National Cancer Institute of Canada and the Medical Research Council of Canada for continuing support of their research program in photodynamic therapy.

REFERENCES

1. Jesionek, A. and Tappeiner, V. H., Zur Behandlung der Hautcarcinomi mit Fluorescierenden Stoffen, *Muench. Med. Wochenschr.*, 47, 2042, 1903.
2. Hausmann, W., Die sensibilisierende Wirkung des Hamatoporphyrins, *Biochem. Z.*, 30, 276, 1911.
3. Auler, H. and Banzer, G., Untersuchengen über die Rolle der Porphyrine bei geschwulstkranken Menschen und Tieren, *Z. Krebsforsch.*, 53, 65, 1942.
4. Figge, F. H. J., Weiland, G. S., and Manganiello, L. O. J., Cancer detection and therapy. Affinity of neoplastic, embryonic, and traumatized tissue for porphyrins and metalloporphyrins, *Proc. Soc. Exp. Biol. Med.*, 68, 640, 1948.
5. Lipson, R. L. and Baldes, E. J., The photodynamic properties of a particular hematoporphyrin derivative, *Arch. Dermatol.*, 82, 508, 1960.
6. Lipson, R. L. and Baldes, E. J., The use of a derivative of hematoporphyrin in tumor detection, *J. Natl. Cancer Inst.*, 26, 1, 1961.
7. Lipson, R. L. and Baldes, E. J., Hematoporphyrin derivative: a new aid for endoscopic detection of malignant disease, *J. Thorac. Cardiovasc. Surg.*, 42, 623, 1961.
8. Lipson, R. L., Baldes, E. J., and Olsen, A. M., Further evaluation of the use of hematoporphyrin derivative as a new aid for endoscopic detection of malignant disease, *Dis. Chest*, 46, 676, 1964.
9. Lipson, R. L., Pratt, J. H., Baldes, E. J., and Dockerty, M. B., Hematoporphyrin derivative for detection of cervical cancer, *Obstet. Gynecol.*, 24, 78, 1964.
10. Grey, M. J., Lipson, R. L., Mack, J. V. S., Parker, L., and Rombyn, D., Use of hematoporphyrin derivative in detection and management of cervical cancer, *Am. J. Obstet. Gynecol.*, 9, 766, 1967.
11. Gregorie, H. B., Horger, E. O., Ward, J. L., Green, J. F., Richards, T., Robertson, H. C., and Stevenson, T. B., Hematoporphyrin-derivative fluorescence in malignant neoplasms, *Ann. Surg.*, 167, 820, 1968.
12. Kelly, J. F. and Snell, M. E., Hematoporphyrin derivative: a possible aid in the diagnosis and therapy of cancer of the bladder, *J. Urol.*, 115, 150, 1976.
13. Dougherty, T. J., Grindley, G. B., Fiel, R., Weishaupt, K. R., and Boyle, D. G., Photoradiation therapy, II. Cure of animal tumors with hematoporphyrin and light, *J. Natl. Cancer Inst.*, 55, 115, 1975.
14. Kelly, J. F., Snell, M. E., and Berenbaum, M. C., Photodynamic destruction of human bladder carcinoma, *Br. J. Cancer*, 31, 237, 1975.
15. Dougherty, T., Boyle, D., Weishaupt, K., Gomer, C., Borcicky, D., Kaufman, J., Goldfarb, A., and Grindley, G., Phototherapy of human tumors, in *Research in Photobiology*, Castellani, A., Ed., Plenum Press, New York, 1977, 735.
16. Dougherty, T. J., Kaufman, J. E., Goldfarb, A., Weishaupt, K. R., Boyle, D., and Mittleman, A., Photoradiation therapy for the treatment of malignant tumors; *Cancer Res.*, 38, 2628, 1978.
17. Dougherty, T. J., An overview of the status of photoradiation therapy, in *Porphyrin Localization and Treatment of Tumors*, Doiron, D. R. and Gomer, C. J., Eds., Alan R. Liss, New York, 1984, 75.
18. Rettenmaier, M. A., Berman, M. L., DiSaia, P. J., Burns, R. G., and Berns, M. W., Photoradiation of gynecologic malignancies, *Gynecol. Oncol.*, 17, 200, 1984.
19. Rettenmaier, M. A., Berman, M. L., DiSaia, R., Burns, G., Weinstein, G. D., McCulloch, J. L., and Berns, M. W., Gynecologic uses of photoradiation therapy, in *Porphyrin Localization and Treatment of Tumors*, Doiron, D. R. and Gomer, C. J., Eds., Alan R. Liss, New York, 1984, 767.
20. Dougherty, T. J., Lawrence, G., Kaufman, J. H., Boyle, D., Weishaupt, K. R., and Goldfarb, A., Photoradiation in the treatment of recurrent breast carcinoma, *J. Natl. Cancer Inst.*, 62, 231, 1979.

21. Kyriazis, G. A., Balin, H., and Lipson, R. L., Hematoporphyrin derivative fluorescence test colposcopy and colpophotography in the diagnosis of atypical metaplasia, dysplasia, and carcinoma *in situ* of the cervix uteri, *Am. J. Obstet. Gynecol.,* 117, 375, 1973.
22. King, E. G., Doiron, D. R., Man, G., Profio, A. E., and Huth, G., Hematoporphyrin derivative as a tumor marker in the detection and localization of pulmonary malignancy, in *Early Detection and Localization of Lung Tumors in High Risk Groups,* Bland, P. R., Ed., Springer-Verlag, Berlin, 1982, 90.
23. Kennedy, J. C. and Oswald, K., Hematoporphyrin derivative photoradiation therapy in theory and in practice, in *Porphyrins in Tumor Phototherapy,* Andreoni, A. and Cubeddu, R., Eds., Plenum Press, New York, 1984, 365.
24. Forbes, I. J., Cowled, P. A., Leong, A. S.-Y., Ward, A. D., Black, R. B., Blake, A. J., and Jacka, F. J., Phototherapy of humans tumors using hematoporphyrin derivative, *Med. J. Aust.,* 2, 489, 1980.
25. Jacka, F. and Blake, A. J., A lamp for cancer phototherapy, *Aust. J. Phys.,* 36, 221, 1983.
26. Dougherty, T. J., Photoradiation therapy for cutaneous and subcutaneous malignancies, *J. Invest. Dermatol.,* 77, 122, 1981.
27. Dougherty, T. J., Weishaupt, K. R., and Boyle, D. G., Photosensitizers, in *Cancer: Principles and Practice of Oncology,* DeVita, V. T., Hellman, S., and Rosenberg, S. A., Eds., Lippincott, Philadephia, 1982, 1836.
28. Zorat, P. J., Tomio, L., Corti, L., Calzavara, F., Scotti, G., Jori, G., Reddi, E., and Mandoliti, G., Hematoporphyrin phototherapy of malignant tumors, in *Porphyrins in Tumor Phototherapy,* Andreoni, A. and Cubeddu, R., Eds., Plenum Press, New York, 1984, 381.
29. Kennedy, J., HPD photoradiation therapy for cancer at Kingston and Hamilton, in *Porphyrin Photosensitization,* Kessel, D. and Dougherty, T. J., Eds., Plenum Press, New York, 1983, 53.
30. Doiron, D. R., Fountain, S. J., Gomer, C. J., and Keller, G. S., Delivery systems for photodynamic therapy in Proc. Porphyrin Photosensitization Workshop, Philadelphia, July, 1984. Some of these proceedings are published in *Methods in Porphyrin Photosensitization,* Kessel, D., Ed., Plenum Press, New York, 1985.
31. Russo, V., Righini, G. C., Sottini, S., and Trigari, S., Lens-ended fibers for medical applications: a new fabrication technique, *Appl. Optics,* 23, 3277, 1984.
32. Moore, J., An optimized laser system for the evaluation of HPD therapy, in *Porphyrins in Tumor Phototherapy,* Andreoni, A. and Cubeddu, R., Eds., Plenum Press, New York, 1984, 293.
33. Forbes, I. J., Photochemotherapy of human cancers with hematoporphyrin derivative, *Med. J. Aust.,* 194, 1984.
34. Dougherty, T. J., Photoradiation therapy, *Urology,* Suppl. 23, 61, 1984.
35. Dougherty, T. J., Boyle, D. G., Weishaupt, K. R., Henderson, B. A., Potter, W. R., Bellnier, D. A., and Wityk, K. E., Photoradiation therapy — clinical and drug advances, in *Porphyrin Photosensitization,* Kessel, D. and Dougherty, T. J., Eds., Plenum Press, New York, 1983, 3.
36. Forbes, I. J., Ward, A. D., Jacka, F. J., Blake, A. J., Swincer, A. G., Wilksch, P. A., Cowled, P. A., and Lee See, K., Multidisciplinary approach to phototherapy of human tumors, in *Porphyrin Localization and Treatment of Tumors,* Doiron, D. R. and Gomer, C. J., Eds., Alan R. Liss, New York, 1984, 693.
37. Wile, A. G., Dahlman, A., Burns, R. G., and Berns, M. W., Laser photoradiation therapy of cancer following hematoporphyrin sensitization, *Lasers Surg. Med.,* 2, 163, 1982.
38. Dahlman, A., Wile, A. G., Burns, R. G., Mason, G. R., Johnson, F. M., and Berns, M. W., Laser photoradiation therapy of cancer, *Cancer Res.,* 43, 430, 1983.
39. Wile, A. G., Dahlman, A., Burns, R. G., Mason, G. R., Johnson, F. M., and Berns, M. W., Laser photoradiation therapy of recurrent human breast cancer and cancer of the head and neck, in *Porphyrin Photosensitization,* Kessel, D. and Dougherty, T. J., Eds., Plenum Press, New York, 1983, 47.
40. Wile, A. G., Coffey, J., Nahabedian, M. Y., Baghdassarian, R., Mason, G. R., and Berns, M. W., Laser photoradiation therapy of cancer, an update of the experience at the University of California, *Lasers Surg. Med.,* 4, 5, 1984. Vol. 4, pp. 1-133 of this journal is reprinted as *Hematoporphyrin Derivative Photoradiation Therapy of Cancer,* Berns, M. W., Ed., Alan R. Liss, New York, 1984.
41. Konaka, C. and Ono, J., Skin metasteses from breast cancer, in *Lasers and Hematoporphyrin Derivative in Cancer,* Hayata, Y. and Dougherty, T. J., Eds., Igaka-Shoin, Tokyo, 1983, 110.
42. Tse, D. T., Dutton, J. J., Weingeist, T. A., Hermsen, V. M., and Kersten, R. C., Hematoporphyrin photoradiation therapy for intraocular and orbital malignant melanoma, *Arch Opthalmol.,* 102, 833, 1984.
43. Xian-wen, H., Xin-min, S., Jian-guo, X., Xian-jun, F., Yu-hai, Z., Qi-chu, M., Hong, S., Shi-lin, X., and Ru-gang, Z., Clinical use of hematoporphyrin derivative in malignant tumors, *Chin. Med. J.,* 96, 754, 1983.

44. McCaughan, J. S., Guy, J. T., Hawley, P., Hicks, W., Inglis, W., Laufman, L., May, E., Nims, T. A., and Sherman, R., Hematoporphyrin derivative and photoradiation therapy of malignant tumors, *Lasers Surg. Med.,* 3, 199, 1983.
45. McCaughan, J. S., Photoradiation of malignant tumors presensitized with hematoporphyrin derivative, in *Porphyrin Localization and Treatment of Tumors,* Doiron, D. R. and Gomer, C. J., Eds., Alan R. Liss, New York, 1984, 805.
46. McCaughan, J. S., Hick, W., Laufman, L., May, E., and Roach, R., Palliation of esophageal malignancy with photoradiation therapy, *Cancer,* 54, 2905, 1984.
47. Tokuda, Y., Primary skin cancer, in *Lasers and Hematoporphyrin Derivative in Cancer,* Hayata, Y. and Dougherty, T. J., Eds., Igaka-Shoin, Tokyo, 1983, 88.
48. Dougherty, T. J., Boyle, D. G., and Weishaupt, K. R., Photoradiation therapy of human tumors, in *The Science of Photomedicine,* Regan, J. D. and Parrish, J. A., Eds., Plenum Press, New York, 1982, 625.
49. Zhonge, G., Yingcai, H., and Zangmin, W., Evaluation of photoradiation therapy (PRT) in 20 cases of cancers, in *Porphyrins in Tumor Phototherapy,* Andreoni, A. and Cubeddu, R., Eds., Plenum Press, New York, 1984, 375.
50. Bandieramonte, G., Marchesini, R., Zunino, F., Melloni, E., Andreola, S., Andreoni, C., Di Pietro, S., Spinelli, P., Fava, G., and Emanuelli, H., Hematoporphyrin derivative and phototherapy in extensive basal-cell carcinoma of the dorsal skin, in *Porphyrins in Tumor Phototherapy,* Andreoni, A. and Cubeddu, R., Eds., Plenum Press, New York, 1984, 389.
51. Tomio, L., Calzavara, F., Zorat, P. L., Corti, L., Polico, C., Reddi, E., Jori, J., and Mandoliti, G., Photoradiation therapy for cutaneous and subcutaneous malignant tumors using hematoporphyrin, in *Porphyrin Localization and Treatment of Tumors,* Doiron, D. R. and Gomer, C. J., Eds., Alan R. Liss, New York, 1984, 829.
52. Hayata, Y., Kato, H., Konaka, C., Ono, J., Saito, M., Takahashi, H., and Tomono, T., Photoradiation therapy in early stage cancers, cases of the lung, esophagus, and stomach, in *Porphyrins in Tumor Phototherapy,* Andreoni, A. and Cubeddu, R., Eds., Plenum Press, New York, 1984, 405.
53. Hayata, Y., Kato, H., Konaka, C., Ono, J., and Takizawa, N., Hematoporphyrin derivative and laser photoradiation in the treatment of lung cancer, *Chest,* 81, 269, 1982.
54. Hayata, Y., Kato, H., Konaka, C., Aida, M., Ono, J., and Nishimaya, K., Hematoporphyrin derivative and photoradiation for tumor localization and treatment of lung cancer, in *Lung Cancer 1982,* Ishikawa, S., Hayata, Y., and Suemasu, K., Eds., Exerpta Medica, Amsterdam, 1982, 55.
55. Kato, H.,Lung Cancer, in *Lasers and Hematoporphyrin Derivative in Cancer,* Hayata, Y. and Dougherty, T. J., Eds., Igaka-Shoin, Tokyo, 1983, 39.
56. Kato, H., Konaka, C., Ono, J., Matsushima, Y., Nishimiya, K., Lay, J., Sawa, H., Shinohara, H., Saito, T., Kinoshita, K., Tomono, T., Aida, M., and Hayata, Y., Effectiveness of HPD and radiation therapy in lung cancer, in *Porphyrin Photosensitization,* Kessel, D. and Dougherty, T. J., Eds., Plenum Press, New York, 1983, 23.
57. Hayata, Y., Kato, H., Amemiya, R., and Ono, J., Indications of photoradiation therapy in early stage lung cancer on the basis of post-PRT histologic findings, in *Porphyrin Localization and Treatment of Tumors,* Doiron, D. R. and Gomer, C. J., Eds., Alan R. Liss, New York, 1984, 747.
58. Hayata, Y., Kato, H., Konaka, C., Ono, J., Amemiya, R., Kinoshita, K., Sakai, H., and Yamada, R., Hematoporphyrin derivative and photoradiation therapy in early stage lung cancer, *Lasers Surg. Med.,* 4, 39, 1984.
59. Doiron, D. R. and Balchum, O. J., Hematoporphyrin derivative photoradiation therapy of endobronchial lung cancer, in *Porphyrins in Tumor Phototherapy* Andreoni, A. and Cubeddu, R., Eds., Plenum Press, New York, 1984, 395.
60. Balchum, O. J., Doiron, D. R., and Huth, G. C., HPD photodynamic therapy for obstructing lung cancer, in *Porphyrin Localization and Treatment of Tumors,* Doiron, D. R. and Gomer, C. J., Eds., Alan R. Liss, New York, 1984, 727.
61. Balchum, O. J., Doiron, D. R., and Huth, G. C., Photoradiation therapy of endobronchial lung cancers employing the photodynamic action of hematoporphyrin derivative, *Lasers Surg. Med.,* 4, 13, 1984.
62. Vincent, R. G., Dougherty, T. J., Rao, U., and Doiron, D. R., Hematoporphyrin derivative in the diagnosis and treatment of lung cancer, in *Porphyrin Photosensitization,* Kessel, D. and Dougherty, T. J., Eds., Plenum Press, New York, 1983, 41.
63. Vincent, R., Dougherty, T., Rao, U., Boyle, D., and Potter, W., Photoradiation therapy in the treatment of advanced carcinoma of the trachea and bronchus, in *Porphyrin Localization and Treatment of Tumors,* Doiron, D. R. and Gomer, C. J., Eds., Alan R. Liss, New York, 1984, 759.
64. Cortese, D. A. and Kinsey, J. H., Hematoporphyrin derivative phototherapy for local treatment of cancer of the tracheobronchial tree, *Ann. Otol. Rhinol. Laryngol.,* 91, 652, 1982.

65. Cortese, D. A. and Kinsey, J. H., Endoscopic management of lung cancer with hematoporphyrin derivative phototherapy, *Mayo Clin. Proc.*, 57, 543, 1982.
66. Spinelli, P., Andreola, S., Marchesini, R., Melloni, E., Mirabile, E., Pizzetti, P., and Zunino, F., Endoscopic HPD-laser photoradiation therapy (PRT) of cancer, in *Porphyrins in Tumor Phototherapy,* Andreoni, A. and Cubeddu, R., Eds., Plenum Press, New York, 1984, 423.
67. Li, J.-H., Chen, Y.-P., Zhao, S.-D., Zhang, L.-T., and Song, S.-Z., Application of hematoporphyrin derivative and laser induced photodynamic reaction in the treatment of lung cancer: a preliminary report on 21 cases, *Lasers Surg. Med.*, 4, 31, 1984.
68. Benson, R. C., Kinsey, J. H., Cortese, D. A., Farrow, G. M., and Utz, D. C., Treatment of transitional cell carcinoma of the bladder with hematoporphyrin derivative phototherapy, *J. Urol.*, 130, 685, 1983.
69. Benson, R., The use of hematoporphyrin derivative (HPD) in the localization and treatment of transitional cell carcinoma (TCC) of the bladder, in *Porphyrin Localization and Treatment of Tumors,* Doiron, D. R. and Gomer, C. J., Eds., Alan R. Liss, New York, 1984, 795.
70. Hisazumi, H., Miyoshi, N., Naito, K., and Misaki, T., Whole bladder wall photoradiation therapy for carcinoma *in situ* of the bladder; a preliminary report, *J. Urol.*, 131, 884, 1984.
71. Jocham, D., Staehler, G., Chaussey, E., Unsold, E., Hammer, C., Lohrs, U., and Gorish, W., Integral dye-laser irradiation of photosensitized bladder tumors with the aid of a light-scattering medium, in *Porphyrin Localization and Treatment of Tumors,* Doiron, D. R. and Gomer, C. J., Eds., Alan R. Liss, New York, 1984, 249.
72. Ohi, T. and Tsuchiya, A., Superficial bladder cancers, in *Lasers and Hematoporphyrin Derivative in Cancer,* Hayata, Y. and Dougherty, T. J., Eds., Igaka-Shoin, Tokyo, 1983, 79.
73. Tsuchiya, A., Obara, N., Miwa, M., Ohi, T., Kato, H., and Hayata, Y., Hematoporphyrin derivative and laser photoradiation in the diagnosis and treatment of bladder cancer, *J. Urol.*, 130, 79, 1983.
74. Ohi, T., Kato, H., Tsuchiya, A., Obara, N., Imamura, K., Aizawa, K., Konaka, C., Ono, J., Kawate, N., Yoneyama, K., Saito, M., Takahashi, H., Tsukimura, S., Shinohara, H., Lay, J., and Hayata, Y., Photoradiation therapy with hematoporphyrin derivative and an argon dye laser of bladder cancer, in *Porphyrins in Tumor Phototherapy,* Andreoni, A. and Cubeddu, R., Eds., Plenum Press, New York, 1984, 439.
75. Obara, N., Matsumoto, T., Tsuchiya, A., Ohi, T., Kato, H., Hayata, Y., Aizawa, K., and Sassa, H., Hematoporphyrin derivative (HPD) and laser photoradiation in the diagnosis and treatment of bladder cancer, *J. Jpn. Soc. Laser Med.*, 4, 109, 1984.
76. Benson, R. C., Farrow, G. M., Kinsey, J. H., Cortese, D. A., Zincke, H., and Utz, D. C., Detection and localization of *in situ* carcinoma of the bladder with hematoporphyrin derivative, *Mayo Clin. Proc.*, 57, 548, 1982.
77. Hisazumi, H., Misaki, T., and Miyoshi, N., Photoradiation therapy of bladder tumors, *J. Urol.*, 130, 685, 1983.
78. Hisazumi, H., Misaka, T., and Miyoshi, N., Photodynamic therapy of bladder tumors, in *Lasers and Hematoporphyrin Derivative in Cancer,* Hayata, Y. and Dougherty, T. J., Eds., Igaka-Shoin, Tokyo, 1983, 85.
79. Misaki, T., Hisazumi, H., and Miyoshi, N., Photoradiation therapy of bladder tumors, in *Porphyrin Localization and Treatment of Tumors,* Doiron, D. R. and Gomer, C. J., Eds., Alan R. Liss, New York, 1984, 785.
80. Wile, A. G., Novotny, J., Mason, R. G., Passy, V., and Berns, M. W., Photoradiation therapy of head and neck cancer, *Am. J. Clin. Oncol.*, 6, 39, 1984.
81. Taketa, C. and Imakiire, M., Cancer of the ear, nose, and throat, in *Lasers and Hematoporphyrin Derivative in Cancer,* Hayata, Y. and Dougherty, T. J., Eds., Igaka-Shoin, Tokyo, 1983, 70.
82. Soma, H., Akiya, K., Nutahara, S., Kato, H., and Hayata, Y., Treatment of vaginal carcinoma with laser photoirradiation following administration of hematoporphyrin derivative, *Ann. Chir. Gynaecol.*, 71, 133, 1982.
83. Soma, H. and Nutahara, S., Cancer of the female genitalia, in *Lasers and Hematoporphyrin Derivative in Cancer,* Hayata, Y. and Dougherty, T. J., Eds., Igaka-Shoin, Tokyo, 1983, 97.
84. Ward, B. G., Forbes, I. J., Cowled, P. A., McEvoy, M. M., and Cox, L. W., The treatment of vaginal recurrences of gynecologic malignancy with phototherapy following hematoporphyrin derivative pretreatment, *Am. J. Obstet Gynecol.*, 141, 356, 1982.
85. Diamond, I., Granelli, S. G., McDonach, A. F., Neilsen, S., Wilson, C. B., and Jaenicke, R., Photodynamic therapy of malignant tumors, *Lancet,* 2, 1175, 1972.
86. Granelli, S. G., Diamond, I., McDonach, A. F., Wilson, C. B., and Neilsen, S. L., Photochemotherapy of glioma cells by visible light and hematoporphyrin, *Cancer Res.*, 35, 2567, 1975.
87. Signorelli, C. D., Ammirati, M., and Tajana, G., Photochemotherapy of human glioma cells in culture by hematoporphyrin and visible light, *Acta Neurol.*, 33, 105, 1978.
88. Wise, B. L. and Taxdall, D. R., Studies of the blood-brain barrier utilizing hematoporphyrin, *Brain Res.*, 4, 387, 1967.

89. Perria, C., Capuzzo, T., Cavagnaro, G., Datti, R., Francaviglia, N., Rivano, C., and Tercero, V. E, First attempts at the photodynamic treatment of human gliomas, *J. Neurosurg. Sci.*, 24, 119, 1980.
90. Perria, C., Photodynamic therapy of human gliomas by hematoporphyrin and He-Ne laser, *IRCS Med. Sci.*, 9, 57, 1981.
91. Wharen, R. E., Anderson, R. E., and Laws, E. R., Quantitation of hematoporphyrin derivative in human gliomas, experimental central nervous system tumors, and normal tissues, *Neurosurgery*, 12, 446, 1983.
92. Jeeves, W. P., Wilson, B. C., Firnau, G., and Brown, K., Studies of HPD and radiolabelled HPD in vivo and in vitro, in *Methods in Porphyrin Photosensitization*, Kesel, D., Ed., Plenum Press, New York, 1985, 51.
93. Laws, E. R., Cortese, D. A., Kinsey, J. H., Eagan, R. T., and Anderson, R. E., Photoradiation in the treatment of malignant brain tumors: a phase I (feasibility) study, *Neurosurgery*, 9, 672, 1981.
94. Cheng, M.-K., McKean, J., Boisvert, D., Tulip, J., and Mielke, B. W., Effects of photoradiation therapy on normal rat brain, *Neurosurgery*, 15, 804, 1984.
95. McCulloch, G. A. J., Forbes, I. J., Lee See, K., Cowled, P. A., Jacka, F. J., and Ward, A. D., Phototherapy in malignant brain tumors, in *Porphyrin Localization and Treatment of Tumors*, Doiron, D. R. and Gomer, C. J., Eds., Alan R. Liss, New York, 1984, 709.
96. Boggan, J. E., Walter, R., Edwards, M. S. B., Borcich, J. K., Davis, R. L., Koonce, M., and Berns, M. W., Distribution of hematoporphyrin derivative in the rat 9L gliosarcoma brain tumor analyzed by digital video fluorescence microscopy, *J. Neurosurg.*, 61, 1113, 1984.
97. Aida, M. and Hirashima, T., Cancer of the esophagus, in *Lasers and Hematoporphyrin Derivative in Cancer*, Hayata, Y. and Dougherty, T. J., Eds., Igaka-Shoin, Tokyo, 1983, 57.
98. Kinoshita, K., Noguchi, M., Sakai, H., Takahashi, H., Tomono, T., Saito, M., Otawa, M., Shinohara, H., Yoneyama, K., Kawate, N., Lay, J., Ono, J., Konaka, C., Aida, M., Kato, H., Kawaguchi, M., Aizawa, K., and Hayata, Y., HPD and laser for esophageal and gastric cancer, *J. Jpn. Soc. Laser Med.*, 4, 111, 1984.
99. Okuda, S., Mimura, S., Otani, T., Ichii, M., and Tatsuta, M., Experimental and clinical studies on HPD-photoradiation therapy for upper gastrointestinal cancer, in *Porphyrins in Tumor Phototherapy*, Andreoni, A. and Cubeddu, R., Eds., Plenum Press, New York, 1984, 413.
100. Aida, M. and Kawaguchi, M., Gastric cancer, in *Lasers and Hematoporphyrin Derivative in Cancer*, Hayata, Y. and Dougherty, T. J., Eds., Igaka-Shoin, Tokyo, 1983, 65.
101. Mimura, S., Ichii, M., Tatsuta, M., Otani, T., Imanishi, K., and Okuda, S., The indication of HPD photoradiation therapy to gastric cancer, *J. Jpn. Soc. Laser Med.*, 4, 83, 1984.
102. Ozaki, M., Miyagawa, H., Takechi, K., Okuda, J., and Ida, K., Treatment of gastric cancer with HPD administration and N2 dye laser photoradiation, *J. Jpn. Soc. Laser Med.*, 4, 113, 1984.
103. Douglass, H. O., Nava, H. R., Weishaupt, K. R., Boyle, D., Sugerman, M. G., Halpern, E., and Dougherty, T. J., Intra-abdominal applications of hematoporphyrin photoradiation therapy, in *Porphyrin Photosensitization*, Kessel, D. and Dougherty, T. J., Eds., Plenum Press, New York, 1983, 15.
104. Szirth, B. C., Gomer, C. J., Doiron, D. R., Lingua, R. W., McNamara, W. L., and Murphree, A. L., Photographic documentation of hematoporphyrin derivative (HPD) and fluorescein angiography in experimental ocular tumors, *J. Opthal. Photogr.*, 6, 26, 1983.
105. Liu, L. H. S. and Ni, C., Hematoporphyrin phototherapy for experimental intraocular malignant melanoma, *Arch. Opthalmol.*, 101, 901, 1983.
106. Sery, T. W. and Dougherty, T. J., Photoradiation of rabbit ocular malignant melanoma sensitized with hematoporphyrin derivative, *Curr. Eye Res.*, 3, 519, 1984.
107. Gomer, C. J., Doiron, D. R., White, L., Jester, J. V., Dunn, S., Szirth, B. C., Razan, N. J., and Murphree, A. L., Hematoporphyrin derivative photoradiation induced damage to normal and tumor tissue of the pigmented rabbit eye, *Curr. Eye Res.*, 3, 229, 1984.
108. Gomer, C. J., Murphree, A. L., Doiron, D. R., Szirth, B. C., and Razan, N. J., Preclinical examination of ocular photoradiation therapy, in *Porphyrins in Tumor Phototherapy*, Andreoni, A. and Cubeddu, R., Eds., Plenum Press, New York, 1984, 447.
109. Gomer, C. J., Doiron, D. R., Jester, J. V., Szirth, B. C., and Murphree, A. L., Hematoporphyrin derivative photoradiation therapy for the treatment of intraocular tumors: examination of acute normal tissue toxicity, *Cancer Res.*, 43, 721, 1983.
110. Sery, T. W., Photodynamic killing of retinoblastoma cells with hematoporphyrin and light, *Cancer Res.*, 39, 96, 1979.
111. Gomer, C. J., Rucker, N., Mark, C., Benedict, W. F., and Murphree, A. L., Tissue distribution of 3H-hematoporphyrin derivative in athymic nude mice hetero-transplanted with human retinoblastoma, *Invest. Opthalmol. Vis. Sci.*, 22, 118, 1982.
112. Benedict, W. F., Lingua, R. W., and Doiron, D. R., Tumor regression of human retinoblastoma in the nude mouse following photoradiation therapy: a preliminary report, *Med. Pediatr. Oncol.*, 8, 397, 1980.

113. Murphree, A. L., Benedict, W. F., and Lingua, R. W., Hematoporphyrin phototherapy in the treatment of retinoblastoma, *Invest. Opthalmol. Vis. Sci.*, 20(Suppl.), 162, 1981.
114. Bruce, R. A., Photoradiation for choroidal malignant melanoma, in *Porphyrins in Tumor Phototherapy*, Andreoni, A. and Cubeddu, R., Eds., Plenum Press, New York, 1984, 455.
115. Bruce, R. A., Photoradiation of choroidal malignant melanoma, in *Porphyrin Localization and Treatment of Tumors*, Doiron, D. R. and Gomer, C. J., Eds., Alan R. Liss, New York, 1984, 777.
116. Bruce, R. A., Evaluation of hematoporphyrin photoradiation therapy to treat choroidal melanomas, *Lasers Surg. Med.*, 4, 59, 1984.
117. Watanabe, H. and Tobari, I., Argon laser photoradiation with hematoporphyrin derivative in the treatment of metastatic choroidal tumors, *J. Jpn. Soc. Laser Med.*, 4, 259, 1984.
118. Lewis, R. A., Tse, D. T., Phelps, C. D., and Weingeist, T. A., Neovascular glaucoma after photoradiation therapy for uveal melanoma, *Arch. Opthalmol.*, 102, 839, 1984.
119. Packer, A. J., Tse, D. T., Xun-Qing, G., and Hayreh, S. S., Hematoporphyrin photoradiation therapy for iris neovascularization, *Arch. Opthalmol.*, 102, 1193, 1984.
120. Profio, A. E. and Doiron, D. R., Dosimetry considerations in phototherapy, *Med. Phys.*, 8, 190, 1981.
121. Berns, M. W., Berns, G. S., Coffey, J., and Wile, A. G., Exposure (dose) tables for hematoporphyrin derivative photoradiation therapy, *Lasers Surg. Med.*, 4, 107, 1984.
122. Svaasand, L. O., Doiron, D. R., and Profio, A. E., Light Distribution in Tissue During Photoradiation Therapy, University of Southern California Rep. MISG 900-02, USC, Los Angeles, 1981.
123. Doiron, D. R., Svaasand, L. O., and Profio, A. E., Light dosimetry in tissue: application to photoradiation therapy, in *Porphyrin Photosensitization*, Kessel, D. and Dougherty, T. J., Eds., Plenum Press, New York, 1983, 63.
134. Svaasand, L. O., Ellingsen, R., Optical properties of human brain, *Photochem. Photobiol.*, 38, 293, 1983.
125. McKenzie, A. L., How may external and interstitial illumination be compared in laser photoradiation therapy?, *Phys. Med. Biol.*, 30, 455, 1985.
126. Svaasand, L. O.,Optical dosimetry for direct and interstitial photoradiation therapy of malignant tumors, in *Porphyrin Localization and Treatment of Tumors*, Doiron, D. R. and Gomer, C. J., Eds., Alan R. Liss, New York, 1984, 91.
127. Marynissen, J. P. A. and Star, W. M., Phantom measurements for light dosimetry using isotropic and small aperture detectors, in *Porphyrin Localization and Treatment of Tumors*, Doiron, D. R. and Gomer, C. J., Eds., Alan R. Liss, New York, 1984, 133.
128. Wilson, B. C. and Adam, G., A Monte Carlo model for the absorption and flux distributions of light in tissue, *Med. Phys.*, 10, 824, 1983.
129. Wilson, B. C., Jeeves, W. P., Lowe, D. M., and Adam, G., Light propagation in animal tissues in the wavelength range 375-825 nanometers, in *Porphyrin Localization and Treatment of Tumors*, Doiron, D. R. and Gomer, C. J., Eds., Alan R. Liss, New York, 1984, 115.
130. Anderson, R. R. and Parrish, J. A., Optical properties of human skin, in *The Science of Photomedicine*, Regan, J. D. and Parrish, J. A., Eds., Plenum Press, New York, 1982, 147.
131. Preuss, L. E., Bolin, F. P., and Cain, B. W., Tissue as a medium for laser light transport — implications for photoradiation therapy, *SPIE*, 357, 77, 1982.
132. Preuss, L. E., Bolin, F. P., and Cain, B. W., A comment on spectral transmittance in mammalian skeletal muscle, *Photochem. Photobiol.*, 37, 113, 1983.
133. Bolin, F. P., Preuss, L. E., and Cain, B. W., A comparison of spectral transmittance for several mammalian tissues: effects at PRT frequencies, in *Porphyrin Localization and Treatment of Tumors*, Doiron, D. R. and Gomer, C. J., Eds., Alan R. Liss, New York, 1984, 211.
134. Wilson, B. C., Jeeves, W. P., and Lowe, D. M., In vivo and post mortem measurements of the attenuation spectra of light in mammalian tissues, *Photochem. Photobiol.*, 42, 153, 1985.
135. Wilksch, P. A., Jacka, F., and Blake, A. J., Studies of light propagation through tissue, in *Porphyrin Localization and Treatment of Tumors*, Doiron, D. R. and Gomer, C. J., Eds., Alan R. Liss, New York, 1984, 149.
136. Ellingsen, R., Svaasand, L. O., and Odegaard, A., Photoradiation therapy of Lewis lung tumors in mice with low optical dose rate and high HPD dose, in *Porphyrin Localization and Treatment of Tumors*, Doiron, D. R. and Gomer, C. J., Eds., Alan R. Liss, New York, 1984, 647.
137. Star, W. M., Marynissen, J. P. A., van der Berg-Blok, A. E., and Reinhold, H. S., Destructive effect of photoradiation on the microcirculation of a rat mammary tumor growing in "sandwich" observation chambers, in *Porphyrin Localization and Treatment of Tumors*, Doiron, D. R. and Gomer, C. J., Eds., Alan R. Liss, New York, 1984, 637.
138. Selman, S. H., Kreimer-Birnbaum, M., Klaunig, J. E., Goldblatt, P. J., Keck, R. W., and Britton, S. L., Blood flow in transplanted bladder tumors treated with hematoporphyrin derivative and light, *Cancer Res.*, 44, 1924, 1984.

139. Svaasand, L. O. and Doiron, D. R., On the Probability for Simultaneous Action of Hyperthermia and Photodynamic Effects During Photoradiation Therapy of Malignant Tumors, University of Southern California Rep. MISG 900-05, USC, Los Angeles, 1982.
140. Svaasand, L. O., Doiron, D. R., and Dougherty, T. J., Temperature rise during photoradiation therapy of malignant tumors, *Med. Phys.,* 10, 10, 1983.
141. Svaasand, L. O. and Doiron, D. R., Thermal distribution during photoradiation therapy, in *Porphyrin Photosensitization,* Kessel, D. and Dougherty, T. J., Eds., Plenum Press, New York, 1983, 77.
142. Kinsey, J. H., Cortese, D. A., and Neel, H. B., Thermal considerations in murine tumor killing using hematoporphyrin derivative phototherapy, *Cancer Res.,* 43, 1562, 1983.
143. Rounds, D. E., Doiron, D. R., Jacques, D. B., Shelden, C. H., and Olsen, R. S., Phototoxicity of brain tissue in hematoporphyrin derivative treated mice, in *Porphyrin Localization and Treatment of Tumors,* Doiron, D. R. and Gomer, C. J., Eds., Alan R. Liss, New York, 1984, 613.
144. Dougherty, T. J., Light dosimetry and thermal considerations, in *Lasers and Hematoporphyrin Derivative in Cancer,* Hayata, Y. and Dougherty, T. J., Eds., Igaka-Shoin, Tokyo, 1983, 35.
145. Berns, M. W., Coffey, J., and Wile, A. G., Laser photoradiation therapy of cancer; a possible role of hyperthermia, *Lasers Surg. Med.,* 4, 87, 1984.
146. Christensen, T., Wahl, A., and Smedshammer, L., Effects of hematoporphyrin derivative and light in combination with hyperthermia on cells in culture, *Br. J. Cancer,* 50, 85, 1984.
147. Waldow, S. M., Henderson, B. W., and Dougherty, T. J., Enhanced tumor control following sequential treatments of photodynamic therapy (PDT) and localized microwave hyperthermia in vivo, *Lasers Surg. Med.,* 4, 79, 1984.
148. Waldow, S. M. and Dougherty, T. J., Interaction of hyperthermia and photoradiation therapy, *Radiat. Res.,* 97, 380, 1984.
149. Spikes, J. D., Porphyrins and related compounds as photodynamic sensitizers, *Ann. N. Y. Acad. Sci.,* 244, 496, 1975.
150. Clezy, P. S., Hai, T. T., Henderson, R. W., and Thuc, L. V., The chemistry of pyrrolic compounds. XLV. Hematoporphyrin derivative: hematoporphyrin diacetate as the main product of the reaction of hematoporphyrin with a mixture of acetic and sulfuric acids, *Aust. J. Chem.,* 33, 585, 1980.
151. Bonnett, R., Ridge, R. J., Scourides, P. A., and Berenbaum, M. C., Hematoporphyrin derivative, *J. Chem. Soc. Chem. Commun.,* 24, 1198, 1980.
152. Bonnett, R., Ridge, R. J., and Scourides, P. A., On the nature of hematoporphyrin derivative, *J. Chem. Soc. Perkin Trans.,* 1, 3135, 1981.
153. Berenbaum, M. C., Bonnett, R., and Scourides, P. A., In vivo biological activity of the components of hematoporphyrin derivative, *Br. J. Cancer,* 45, 571, 1982.
154. Bonnett, R. and Berenbaum, M. C., HPD — a study of its components and their properties, in *Porphyrin Photosensitization,* Kessel, D. and Dougherty, T. J., Eds., Plenum Press, New York, 1983, 241.
155. Cadby, P. A., Dimitriades, E., Grant, H. G., and Ward, A. D., Separation and analysis of hematoporphyrin derivative components by high-performance liquid chromatography, *J. Chromatogr.,* 231, 273, 1982.
156. Cadby, P. A., Dimitriades, E., Grant, H. G., and Ward, A. D., The analysis of hematoporphryin derivative, in *Porphyrin Photosensitization,* Kessel, D. and Dougherty, T. J., Eds., Plenum Press, New York, 1983, 251.
157. Brault, D., Bizet, C., and Delgado, O., On the purification of hematoporphyrin IX and its acetylated derivatives, in *Porphyrins in Tumor Phototherapy,* Andreoni, A. and Cubeddu, R., Eds., Plenun Press, New York, 1984, 45.
158. Moan, J., Christensen, T., and Sommer, S., The main photosensitizing components of hematoporphryin derivative, *Cancer Lett.,* 15, 161, 1982.
159. Christensen, T., Moan, J., McGhie, J. B., Waksvik, H., and Stigum, H., Studies of HPD: chemical composition and in vitro photosensitization, in *Porphyrin Photosensitization,* Kessel, D. and Dougherty, T. J., Eds., Plenum Press, New York, 1983, 151.
160. Moan, J., Sandberg, S., Christensen, T., and Elander, S., Hematoporphyrin derivative: chemical composition, photochemical, and photosensitizing properties, in *Porphyrin Photosensitization,* Kessel, D. and Dougherty, T. J., Eds., Plenum Press, New York, 1983, 165.
161. Sommer, S., Moan, J., Christensen, T., and Evensen, J. F., A chromatographic study of hematoporphyrin derivative, in *Porphyrins in Tumor Phototherapy,* Andreoni, A. and Cubeddu, R., Eds., Plenum Press, New York, 1984, 81.
162. Evensen, J. F., Moan, J., Hindar, A., and Sommer, S., Tissue distribution of 3H-hematoporphyrin derivative and its main components, 67 Ga, and 131 I-albumin in mice bearing Lewis lung carcinoma, in *Porphyrin Localization and Treatment of Tumors,* Doiron, D. R. and Gomer, C. J., Eds., Alan R. Liss, New York, 1984, 541.

163. Hilf, R., Leakey, P. B., Sollott, S. J., and Gibson, S. L., Photodynamic inactivation of R3230 AC mammary carcinoma in vitro with hematoporphyrin derivative: effects of dose, time, and serum on uptake and phototoxicity, *Photochem. Photobiol.,* 37, 633, 1983.
164. Swincer, A. G., Trenerry, V. C., and Ward, A. D., The analysis and some chemistry of hematoporphyrin derivative, in *Porphyrin Localization and Treatment of Tumors,* Doiron, D. R. and Gomer, C. J., Eds., Alan R. Liss, New York, 1984, 285.
165. Kessel, D. and Chou, T. C., Tumor-localizing components of the porphyrin preparation hematoporphyrin derivative, *Cancer Res.,* 43, 1994, 1983.
166. Dougherty, T. J., Potter, W. R., and Weishaupt, K. R., The structure of the active component of hematoporphyrin derivative, in *Porphyrins in Tumor Phototherapy,* Andreoni, A. and Cubeddu, R., Eds., Plenum Press, New York, 1984, 23.
167. Dougherty, T. J., Potter, W. R., and Weishaupt, K. R., The structure of the active component of hematoporphyrin derivative, in *Porphyrin Localization and Treatment of Tumors,* Doiron, D. R. and Gomer, C. J., Eds., Alan R. Liss, New York, 1984, 301.
168. Moan, J., McGhie, J. B., and Christensen, T., Hematoporphyrin derivative: photosensitizing efficiency and cellular uptake of its components, *Photobiochem. Photobiophys.,* 4, 337, 1982.
169. Swincer, A. G., Ward, A. D., and Howlett, G. J., Size and aggregation of porphyrins, in Proc. Porphyrin Photosensitization Workshop, Philadelphia, July 1984.
170. Ward, A. D. and Swincer, A. G., The structure of protoporhyrin and of the aggregate fraction of hematoporphyrin derivative in solution, in *Methods of Porphyrin Photosensitization,* Kessel, D., Ed., Plenum Press, New York, 1985, 267.
171. Brown, S. B. and Shillcock, M., Equilibrium and kinetic studies of the aggregation of porphyrins in aqueous solution, *Biochem. J.,* 153, 279, 1976.
172. Brown, S. B., Hazikonstantinou, H., and Herrier, D. G., The structure of porphyrins and haems in aqueous solution, *Int. J. Biochem.,* 12, 701, 1980.
173. Brown, S. B., Aggregation of porphyrins and photosensitization, in Proc. Porphyrin Photosensitization Workshop, Philadelphia, July 1984.
174. Kessel, D. and Rossi, E., Determinants of porphyrin-sensitized photooxidation characterized by fluorescence and absorption spectra, *Photochem. Photobiol.,* 35, 37, 1982.
175. Karns, G. A., Gallagher, W. A., and Elliott, W. B., Dimerization constants of water-soluble porphyrins in aqueous alkali, *Bioorg. Chem.,* 8, 69, 1979.
176. Margalit, R., Shaklai, N., and Cohen, S., Fluorometric studies on the dimerization equilibrium of protoporphyrin IX and its haemoderivative, *Biochem. J.,* 209, 547, 1983.
177. Margalit, R. and Cohen, S., Studies on hematoporphyrin and hematoporphyrin derivative equilibria in heterogeneous systems: porphyrin-liposome binding and porphyrin aqueous dimerization, *Biochem. Biophys. Acta,* 736, 163, 1983.
178. Margalit, R., Cohen, S., and Rotenberg, M., Studies on the solvophobic nature of porphyrin self aggregation vs. the lipophylic nature of porphyrin-membrane binding, in *Methods in Porphyrin Photosensitization,* Kessel, D., Ed., Plenum Press, New York, 1985, 303.
179. Goyal, G. C., Blum, A., and Grossweiner, L. I., Photosensitization of liposomal membranes by hematoporphyrin derivative, *Cancer Res.,* 43, 5826, 1983.
180. Grossweiner, L. I. and Goyal, G. C., Photosensitized lysis of liposomes by hematoporphyrin derivative, *Photochem. Photobiol.,* 37, 529, 1983.
181. Grossweiner, L. I., Membrane photosensitization by hematoporphyrin and hematoporphyrin derivative, in *Porphyrin Localization and Treatment of Tumors,* Doiron, D. R. and Gomer, C. J., Eds., Alan R. Liss, New York, 1984, 391.
182. Grossweiner, L. I., Blum, A., and Goyal, G. C., Photophysics and photochemistry of hematoporphyrin, hematoporphyrin derivative and uroporphyrin I, in *Methods in Porphyrin Photosensitization,* Kessel D., Ed., Plenum Press, New York, 1985, 181.
183. Poletti, A., Murgia, S. M., Pasqua, A., Redi, E., and Jori, G., Photophysical and photosensitizing properties of Photofrin II, in *Porphyrins in Tumor Phototherapy,* Andreoni, A. and Cubeddu, R., Eds., Plenum Press, New York, 1984, 37.
184. Reddi, E., Garbo, G. M., and Jori, G., Fluorescence studies on the interaction of porphyrins with dipalmitoyl-phosphatid liposomes, in Proc. Porphyrin Photosensitization Workshop, Philadelphia, July 1984.
185. Andreoni, A., De Silvestri, S., Laporta, P., Jori, G., and Reddi, E., Hematoporphyrin derivative: experimental evidence for aggregated species, *Chem. Phys. Lett.,* 88, 33, 1982.
186. Andreoni, A., Cubeddu, R., De Silvestri, S., Jori, G., Laporta, R., and Reddi, E., Time-resolved fluorescence studies of hematoporphyrin in different solvent systems, *Z. Naturforsch,* 38c, 83, 1983.
187. Andreoni, A. and Cubeddu, R., Fluorescence properties of HPD and its components, in *Porphyrins in Tumor Phototherapy,* Andreoni, A. and Cubeddu, R., Eds., Plenum Press, New York, 1984, 11.
188. Andreoni, A. and Cubeddu, R., Excited-state properties of Photofrin II in different solvents, Proc. Porphyrin Photosensitization Workshop, Philadelphia, July 1984.

189. Moan, J. and Sommer, S., Fluorescence and absorption properties of the components of hematoporphyrin derivative, *Photobiochem. Photobiophys.*, 3, 93, 1981.
190. Hanzlik, C. A., Nordlund, T. M., Knox, W. H., Gibson, S. L., and Hilf, R., Picosecond fluorescence of hematoporphyrin derivative and its components in aqueous solution, in *Porphyrin Localization and Treatment of Tumors,* Doiron, D. R. and Gomer, C. J., Eds., Alan R. Liss, New York, 1984, 201.
191. Kessel, D. and Cheng, M. L., On the preparation and properties of dihematoporphyrin ether, the tumor-localizing component of HPD, *Photochem. Photobiol.*, 41, 277, 1985.
192. Bottiroli, G., Doccio, F., Freitas, I., Ramponi, R., and Sacchi, C. A., Hematoporphyrin derivative: fluorometric studies in solution and cells, in *Porphyrins in Tumor Phototherapy,* Andreoni, A. and Cubeddu, R., Eds., Plenum Press, New York, 1984, 125.
193. Grossweiner, L. I. and Goyal, G. C., Binding of hematoporphyrin derivative to human serum albumin, *Photochem. Photobiol.*, 40, 1, 1984.
194. Pottier, R., Laplante, J. P., and Kennedy, J., Mechanism of binding of hematoporphyrin derivative to human blood plasma in a spectroscopic study, in 9th Int. Congr. Photobiology and 12th Annu. Meet. Am. Soc. Photobiology, Philadelphia, 1984.
195. Moan, J., Rimington, C., Evensen, J. F., and Western, A., A binding of porphyrins to serum proteins, in *Methods in Porphyrin Photosensitization,* Kessel, D., Ed., Plenum Press, New York, 1985, 193.
196. Spikes, J. D., Photodynamic reactions in photomedicine, in *The Science of Photomedicine,* Regan, J. D. and Parrish, J. A., Eds., Plenum Press, New York, 1982, 113.
197. Jori, G. and Spikes, J. D., The photobiochemistry of porphyrins, in *Topics in Photomedicine,* Smith, K. C., Ed., Plenum Press, New York, 1983, 183.
198. Foote, C. S., Photosensitized oxidation and singlet oxygen: consequences in biological systems, in *Free Radicals in Biology,* Vol. 2, Pryor, W. A., Ed., Academic Press, New York, 1976, II 85.
199. Foote, C. S., Mechanisms of photooxygenation, in *Porphyrin Localization and Treatment of Tumors,* Doiron, D. R. and Gomer, C. J., Eds., Alan R. Liss, New York, 1984, 3.
200. Spikes, J. D., A preliminary comparison of the photosensitizing properties of porphyrins in aqueous solution and liposomal systems, in *Porphyrin Photosensitization,* Kessel, D. and Dougherty, T. J., Eds., Plenum Press, New York, 1983, 181.
201. Spikes, J. D., Photosensitizing properties of porphyrins in model cell systems, in *Porphyrins in Tumor Phototherapy,* Andreoni, A. and Cubeddu, R., Eds., Plenum Press, New York, 1984, 51.
202. Bonnett, R., Charalambides, A. A., Land, E. J., Sinclair, R. S., Tait, D., and Truscott, T. G., Triplet states of porphyrin esters, *J. Chem. Soc. Faraday Trans. 1,* 76, 852, 1980.
203. Bonnett, R., Lambert, C., Land, E. J., Scourides, P. A., Sinclair, R. S., and Truscott, T. G., The triplet and radical species of hematoporphyrin and some of its derivatives, *Photochem. Photobiol.*, 38, 1, 1983.
204. Brookfield, R., Craw, M., Lambert, C. R., Land, E. J., Redmond, R., Sinclair, R. S., and Truscott, T. G., Excited state properties of hematoporphyrin, in *Porphyrins in Tumor Phototherapy,* Andreoni, A. and Cubeddu, R., Eds., Plenum Press, New York, 1984, 3.
205. Redmond, R., Land, E. J., and Truscott, T. G., Aggregation effects on the photophysical properties of porphyrins in relation to mechanisms involved in photodynamic therapy, in *Methods in Porphyrin Photosensitization,* Kessel, D., Ed., Plenum Press, New York, 1985, 293.
206. Reddi, E., Jori, G., Rodgers, M. A. J., and Spikes, J., Flash photolysis studies of hemato- and coproporphyrins in homogeneous and microheterogeneous aqueous dispersions, *Photochem. Photobiol.*, 38, 639, 1983.
207. Reddi, E., Jori, G., and Rodgers, M. A. J., Flash photolysis studies of hemato- and copro-porphyrins in homogeneous and microheterogeneous aqueous dispersions, *Stud. Biophys.*, 94, 13, 1983.
208. Reddi, E., Rodgers, M. A. J., and Jori, G., Photophysical properties of hemato- and copro-porphyrins in homogeneous and microheterogeneous media, in *Porphyrin Localization and Treatment of Tumors,* Doiron, D. R. and Gomer, C. J., Eds., Alan R. Liss, New York, 1984, 373.
209. Murgia, S. M., Pasqua, A., and Poletti, A., Laser flash photolysis studies on hematoporphyrin IX in aqueous and micellar systems, *Med. Biol. Environ.*, 10, 278, 1982.
210. Murgia, S. M., Pasqua, A., and Poletti, A., Laser photolysis study of the hematoporphyrin IX-L-tryptophan system in solvent mixtures at different polarity, *Chem. Phys. Lett.*, in press.
211. Parker, J. G. and Stanbro, W. D., Dependence of photosensitized singlet oxygen production on porphyrin structure and solvent, in *Porphyrin Localization and Treatment of Tumors,* Doiron, D. R. and Gomer, C. J., Eds., Alan R. Liss, New York, 1984, 259.
212. Buettner, G. R. and Oberley, L. W., The apparent production of superoxide and hydroxyl radicals by hematoporphyrin and light as seen by spin-trapping, *FEBS Lett.*, 121, 161, 1980.
213. Peterson, D. A., McKelvey, S., and Edmondson, P. R., A hypothesis for the molecular mechanism of tumor killing by porphyrins and light, *Med. Hypoth.*, 7, 201, 1981.

214. Gibson, S. L., Cohen, H. J., and Hilf, R., Evidence against the production of superoxide by photoirradiation of hematoporphyrin derivative, *Photochem. Photobiol.*, 40, 441, 1984.
215. Sconfienza, C., Van de Vorst, A., and Jori, G., Type I and Type II mechanisms in the photooxidation of L-tryptophan and tryptamine sensitized by hematoporphyrin in the presence and absence of sodium dodecyl sulphate micelles, *Photochem. Photobiol.*, 31, 351, 1980.
216. Rossi, E., Van de Vorst, A., and Jori, G., Competition between the singlet oxygen and electron transfer mechanisms in the porphyrin-sensitized photooxidation of L-tryptophan and tryptamine in aqueous micellar dispersions, *Photochem. Photobiol.*, 34, 447, 1981.
217. Jori, G., Reddi, E., Tomio, L., and Calzavara, F., Factors governing the mechanism and efficiency of porphyrin-sensitized photooxidations in homogeneous solutions and organized media, in *Porphyrin Photosensitization*, Kessel, D. and Dougherty, T. J., Eds., Plenum Press, New York, 1983, 193.
218. Reddi, E., Rodgers, M. A. J., Spikes, J. D., and Jori, G., The effect of medium polarity on the hematoporphyrin-sensitized photooxidation of L-tryptophan, *Photochem. Photobiol.*, 40, 415, 1984.
219. Cannistraro, S., Jori, G., and Van de Vorst, A., Quantum yield of electron transfer and of singlet oxygen production by porphyrins: an ESR study, *Photobiochem. Photobiophys.*, 3, 353, 1982.
220. Mossoba, M. M., Rosenthal, I., Carmichael, A. J., and Riesz, P., Photochemistry of porphyrins as studied by spin trapping and electron spin resonance, *Photochem. Photobiol.*, 39, 731, 1984.
221. Cox, G. S., Whitten, D. G., and Giannotti, C., Interaction of porphyrin and metalloporpyrin excited states with molecular oxygen. Energy transfer versus electron transfer quenching mechanisms in photooxidations, *Chem. Phys. Lett.*, 67, 511, 1979.
222. Cox, G. S. and Whitten, D. G., Excited state interactions of protoporphyrin IX and related porphyrins with molecular oxygen in solutions and in organized assemblies, in *Porphyrin Photosensitization*, Kessel, D. and Dougherty, T. J., Eds., Plenum Press, New York, 1983, 279.
223. Grossweiner, L. I., Patel, A. S., and Grossweiner, J. B., Type I and Type II mechanisms in the photosensitized lysis of phosphatidylcholinic liposomes by hematoporphyrin, *Photochem. Photobiol.*, 36, 159, 1982.
224. Goyal, G. C., Blum, A., and Grossweiner, L. I., Photosensitization of liposomal membranes by hematoporphyrin derivative, *Cancer Res.*, 43, 5826, 1983.
225. Moan, J., The photochemical yield of singlet oxygen from porphyrins in different states of aggregation, *Photochem. Photobiol.*, 39, 445, 1984.
226. Dougherty, T. J., Wityk, K., Potter, W. R., and Weishaupt, K. R., Action of DHE in PDT, Proc. Porphyrin Photosensitization Workshop, Philadelphia, July 1984.
227. Kessel, D., Effects of photoactivated porphyrins at the cell surface of leukemia L1210 cells, *Biochemistry*, 16, 3443, 1977.
228. Kessel, D. and Kohn, K. I., Transport and binding of mesoporphyrin IX by leukemia L1210 cell, *Cancer Res.*, 40, 303, 1980.
229. Kessel, D., Transport and binding of hematoporphyrin derivative and related porphyrins by murine leukemia L1210 cells, *Cancer Res.*, 41, 1318, 1981.
230. Kessel, D., Determinants of hematoporphyrin-catalyzed photosensitization, *Photochem. Photobiol.*, 36, 99, 1982.
231. Kessel, D. and Chou, T. H., Porphyrin localization phenomena, in *Porphyrin Photosensitization*, Kessel, D. and Dougherty, T. J., Eds., Plenum Press, New York, 1983, 115.
232. Kessel, D., Chemical and biochemical determinants of porphyrin localization, in *Porphyrin Localization and Treatment of Tumors*, Doiron, D. R. and Gomer, C. J., Eds., Alan R. Liss, New York, 1984, 405.
233. Kessel, D., Components of hematoporphyrin derivative and their tumor-localizing capacity, *Cancer Res.*, 42, 1703, 1982.
234. Moan, J., Christensen, J., and Jacobsen, P. B., Porphyrin-sensitized photoinactivation of cells in vitro, in *Porphyrin Localization and Treatment of Tumors*, Doiron, D. R. and Gomer, C. J., Eds., Alan R. Liss, New York, 1984, 419.
235. Cuzzo, G., Gennari, G., Jori, G., and Spikes, J. D., The effect of chemical structure on the photosensitizing efficiency of porphyrins, *Photochem. Photobiol.*, 25, 389, 1977.
236. Jori, G., Reddi, E., Rossi, E., Cozzani, I., Tomio, L., Zorat, P. L., Pizzi, G. B., and Calzavara, F., Porphyrin-sensitized photoreactions and their use in cancer therapy, *Med. Biol. Environ.*, 8, 141, 1980.
237. Tomio, L., Reddi, E., Jori, G., Zorat, P., Pizzi, G. B., and Calzavara, F., Hematoporphyrin as a sensitizer in tumor phototherapy, *Springer Ser. Opt. Sci.*, 22, 76, 1980.
238. Dougherty, T. J., Hematoporphyrin as a photosensitizer of tumors, *Photochem. Photobiol.*, 38, 377, 1983.
239. Firnau, G., Maass, D., Wilson, B. C., and Jeeves, W. P., ^{64}Cu labelling of hematoporphyrin derivative for non-invasive measurements of tumor uptake, in *Porphyrin Localization and Treatment of Tumors*, Doiron, D. R. and Gomer, C. J., Eds., Alan R. Liss, New York, 1984, 629.

240. Henderson, B. A., Bellnier, D. A., Ziring, B., and Dougherty, T. J., Aspects of the cellular uptake and retention of hematoporphyrin derivative and their correlation with the biological response to PRT in vitro, in *Porphyrin Photosensitization,* Kessel, D. and Dougherty, T. J., Eds., Plenum Press, New York, 1983, 129.
241. Bellnier, D. A. and Lin, C.-W., Photodynamic killing of cultured bladder cells with hematoporphyrin derivative and light, in *Porphyrin Localization and Treatment of Tumors,* Doiron, D. R. and Gomer, C. J., Eds., Alan R. Liss, New York, 1984, 361.
242. Bellnier, D. A. and Lin, C.-W., Photodynamic destruction of cultured human bladder cells by hematoporphyrin derivative: effects of porphyrin molecular aggregation, *Photobiochem. Photobiophys.,* 6, 357, 1983.
243. Berns, M. W., Dahlman, A., Johnson, F. M., Burns, R., Sperling, D., Guiltiman, M., Siemens, A., Walter, R., Wright, W., Hammer-Wilson, M., and Wile, A., In vitro cellular effects of hematoporphyrin derivative, *Cancer Res.,* 42, 2325, 1982.
244. Berns, M. W., Wile, A. Dahlman, A., Johnson, F. M., Burns, R., Sperling, D., Guiltiman, M., Siemens, A., Walter, R., and Hammer-Wilson, M., Cellular uptake, excretion, and localization of hematoporphyrin derivative (HPD), in *Porphyrin Photosensitization,* Kessel, D. and Dougherty, T. J., Eds., Plenum Press, New York, 1983, 139.
245. Berns, M. W., Wilson, M., Rentzepis, P., Burns, R., and Wile, A., Cell biology and hematoporphyrin derivative (HPD), *Lasers Surg. Med.,* 2, 261, 1983.
246. Moan, J., Fluorescence of porphyrins in cells, in *Porphyrins in Tumor Phototherapy,* Andreoni, A. and Cubeddu, R., Eds., Plenum Press, New York, 1984, 109.
247. Moan, J. and Sommer, S., Action spectra for hematoporphyrin derivative and Photofrin II with respect to sensitization of human cells in vitro to photoinactivation, *Photochem. Photobiol.,* 40, 631, 1984.
248. Moan, J., Hovic, B., and Sommer, S., A device to determine fluence-response curves for photoinactivation of cells in vitro, *Photobiochem. Photobiophys.,* 8, 11, 1984.
249. Docchio, F., Ramponi, R., Sacchi, C. A., Bottiroli, G., and Freitas, I., Time-resolved fluorescence microscopy of hematoporphyrin derivative in cells, *Lasers Surg. Med.,* 2, 21, 1982.
250. Bottiroli, G., Freitas, I., Docchio, F., Ramponi, R., and Sacchi, C. A., Towards a better understanding of the mechanism of action of hematoporphyrin derivative at the cellular level, in Proc. 13th Cancer Congr., Seattle, 1982.
251. Schneckenburger, H., Unsold, E., Weisheimer, W., and Jocham, D., Time-resolved laser fluorescence and photobleaching of single cells after photosensitization with hematoporphyrin derivative (HPD), in *Porphyrins in Tumor Phototherapy,* Andreoni, A. and Cubeddu, R., Eds., Plenum Press, New York, 1984, 137.
252. Hisazumi, H., Miyoshi, N., Ueki, O., and Nakajima, K., Cellular binding of hematoporphyrin derivative in human bladder cancer cell lines: KK-47 and T24, in *Porphyrin Localization and Treatment of Tumors,* Doiron, D. R. and Gomer, C. J., Eds., Alan R. Liss, New York, 1984, 443.
253. Hisazumi, H., Miyoshi, N., Ueki, O., Nishimo, A., and Nakajima, K., Cellular uptake of hematoporphyrin derivative in KK-47 bladder cancer cells, *Urol. Res.,* 12, 143, 1984.
254. Miyoshi, N., Hisazumi, H., Ueki, O., and Nakajima, K., Cellular binding of hematoporphyrin derivative in human bladder cancer cell lines: KK-47, *Photochem. Photobiol.,* 39, 359, 1984.
255. Moan, J., Johannessen, P. V., Christensen, T., Espevik, T., and McGhie, J. B., Porphyrin sensitized photoinactivation of human cells in vitro, *Am. J. Pathol.,* 109, 184, 1982.
256. Kohn, K. and Kessel, D., On the mode of cytotoxic action of photoactivated porphyrins, *Biochem. Pharmacol.,* 28, 2465, 1979.
257. Brun, A., Hovding, G., and Romslo, I., Protoporphyrin-induced photohemolysis: differences related to the subcellular distribution of protoporphyrin in erythropoietic protoporphyria and when added to normal red cells, *Int. J. Biochem.,* 13, 225, 1981.
258. Sandberg, S. and Romslo, I., Porphyrin-induced photodamage at the cellular and subcellular level as related to the solubility of the porphyrin, *Chin. Chem. Acta,* 109, 193, 1981.
259. Sandberg, S. and Romslo, I., Phototoxicity of protoporphyrin as related to its subcellular localization in mice livers after short-term feeding with griseofulvin, *Biochem. J.,* 198, 67, 1981.
260. Sandberg, S., Romslo, I., Hovding, G., and Biorndal, T., Porphyrin-induced photodamage as related to the subcellular localization of the porphyrins, *Acta Derm. Venereol. Suppl.,* 100, 75, 1982.
261. Sandberg, S. and Romslo, I., Subcellular localization and phototoxicity of porphyrins produced in the livers of griseofulvin fed mice and hexachlorobenzene fed rats, Proc. Porphyrin Photosensitization Workshop. Philadelphia, July 1984.
262. Emiliani, C. and Delmelle, M., The lipid solubility of porphyrins modulates their phototoxicity in membrane models, *Photochem. Photobiol.,* 37, 487, 1983.
263. Emiliani, C., Delmelle, M., Cannistraro, S., and Van de Vorst, A., Solubility of hematoporphyrin and photodynamic damages in liposomal systems: optical and electron spin resonance studies, *Photobiochem. Photobiophys.,* 5, 119, 1983.

264. Christensen, T., Sandquist, T., Feren, K., Waksvik, H., and Moan, J., Retention and photodynamic effects of hematoporphyrin derivative in cells after prolonged cultivation in the presence of porphyrin, *Br. J. Cancer,* 48, 35, 1983.
265. Moan, J., McGhie, J. M., and Jacobsen, P. B., Photodynamic effects on cells in vitro exposed to hematoporphyrin derivative and light, *Photochem. Photobiol.,* 37, 599, 1983.
266. Andreoni, A., Cubeddu, R., De Silvestri, S., Laporta, P., Ambesi-Impiombato, F. S., Esposito, M., Mastrocinque, M., and Tramontano, D., Effects of laser irradiation on hematoporphyrin-treated normal and transformed thyroid cells in culture, *Cancer Res.,* 43, 2076, 1983.
267. Ambesi-Impiombato, F. S., Andreoni, A., Cubeddu, R., Esposito, M., Mastrocinque, M., and Tramontano, D., Effects of HPD and laser on transformed and corresponding normal cultured cells: differential cytotoxicity as an in vitro model for tumor photochemotherapy, in *Porphyrins in Tumor Phototherapy,* Andreoni, A. and Cubeddu, R., Eds., Plenum Press, New York, 1984, 143.
268. Chang, C. and Dougherty, T. J., Photoradiation therapy: kinetics and thermodynamics of porphyrin uptake and loss in normal and malignant cells in culture, *Radiat. Res.,* 74, 498, 1978.
269. Berns, M. W., Hammer-Wilson, M., Walter, R. J., Wright, W., Chow, M.-H., Nahabedian, M., and Wile, A., Uptake and localization of HPD and "active fractions" in tissue culture and in serially-biopsied human tumors, in *Porphyrin Localization and Treatment of Tumors,* Doiron, D. R. and Gomer, C. J., Eds., Alan R. Liss, New York, 1984, 501.
270. Moan, J., Steen, H. B., Feren, K., and Christensen, T., Uptake of hematoporphyrin derivative and sensitized photoinactivation of C3H cells with different oncogenic potential, *Cancer Lett.,* 14, 291, 1981.
271. Jori, G., Tomio, L., Reddi, E., Rossi, E., Cirto, L., Zorat, P. L., and Calzavara, F., Preferential delivery of liposome-incorporated porphyrins to neoplastic cells in tumor-bearing rats, *Br. J. Cancer,* 48, 307, 1983.
272. Jori, G., Cozzani, I., Reddi, E., Tomio, L., Mandoliti, G., and Malvadi, G., In vitro and in vivo studies on the subcellular localization of hematoporphyrin and its dimethylester with normal and malignant cells, in *Porphyrin Localization and Treatment of Tumors,* Doiron, D. R. and Gomer, C. J., Eds., Alan R. Liss, New York, 1984, 471.
273. Cozzani, I., Jori, G., Reddi, E., Tomio, L., Zorat, P. L., Sicuro, T., and Malvadi, G., Interaction of free and liposome bound porphyrins with normal and malignant cells: biochemical and photosensitization studies in vitro and in vivo, in *Porphyrins in Tumor Phototherapy,* Andreoni, A. and Cubeddu, R., Eds., Plenum Press, New York, 1984, 157.
274. Jori, G., Reddi, E., Salvato, B., Tomio, L., and Cozzani, I., Liposomes and lipoproteins as efficient carriers of porphyrins to tumor cells in vitro and in vivo, in Proc. Porphyrin Photosensitization Workshop, Philadelphia, July 1984.
275. Mew, D., Wat, C.-K., Towers, G. H. N., and Levy, J. G., Photoimmunotherapy: treatment of animal tumors with tumor-specific monoclonal antibody-hematoporphyrin conjugates, *J. Immunol.,* 130, 1473, 1983.
276. Wat, C.-K., Mew, D., Levy, J. G., and Towers, G. H. N., Photosensitizer-protein conjugate: potential use as photoimmunotherapeutic agents, in *Porphyrin Localization and Treatment of Tumors,* Doiron, D. R. and Gomer, C. J., Eds., Alan R. Liss, New York, 1984, 351.
277. Wat, C.-K., Mew, D., Levy, J. G., and Towers, G. H. N., In vitro killing of human leukemic cells by hematoporphyrin CAMAL conjugate, in Proc. Porphyrin Photosensitization Workshop, Philadelphia, July 1984.
278. Fritsch, P., Gschnait, F., Honigsmann, H., and Wolff, K., Protective action of beta-carotene against lethal photosensitization of fibroblasts in vitro, *Br. J. Dermatol.,* 94, 263, 1976.
279. Girotti, A. W., Mechanisms of photosensitization, *Photochem. Photobiol.,* 38, 745, 1983.
280. Kessel, D., Hematoporphyrin and HPD: photophysics, photochemistry, and phototherapy, *Photochem. Photobiol.,* 39, 851, 1984.
281. Malik, Z. and Djaldetti, M., Destruction of erythroleukemic, myelocytic leukemic, and Burkitt lymphoma cells by photoactivated protoporphyrin, *Int. J. Cancer,* 26, 495, 1980.
282. Malik, Z., Lejbkowicz, F., and Salzberg, S., Cholesterol impregnation into erythroleukemic cell membrane induces resistance to hematoporphyrin photodynamic effect, in *Porphyrins in Tumor Phototherapy,* Andreoni, A. and Cubeddu, R., Eds., Plenum Press, New York, 1984, 185.
283. Moan, J., Pettersen, E. O., and Christensen, T., The mechanism of photodynamic inactivation of human cells in vitro in the presence of hematoporphyrin, *Br. J. Cancer,* 39, 398, 1979.
284. Volden, G., Christensen, T., and Moan, J., Photodynamic membrane damage of hematoporphyrin-treated NH1K 3025 cells in vitro, *Photobiochem. Photobiophys.,* 3, 105, 1981.
285. Pan, Q. Q., Photodynamic damage of human cancer cell line and normal diploid cell line in vitro in the presence of hematoporphyrin derivative, *Chung Hua I Hsueh Tsa Chin.,* 63, 330, 1983.
286. Ting-I, C. and Zhonge-he, G., Photodynamic effect of the He-Ne laser with HDP on the ultrastructure of rhabdomyosarcoma cell, in *Porphyrins in Tumor Phototherapy,* Andreoni, A. and Cubeddu, R., Eds., Plenum Press, New York, 1984, 193.

287. Bellnier, D. A. and Lin, C.-W., Giant cell formation in bladder tumor cells following hematoporphyrin derivative-sensitized photoirradiation, *Photochem. Photobiol.*, 39, 425, 1984.
288. Breitbart, H. and Malik, Z., The effects of photoactivated protoporphyrin on reticulocyte membranes, intracellular activities, and hemoglobin precipitation, *Photochem. Photobiol.*, 35, 365, 1982.
289. Moan, J. and Christensen, T., Photodynamic effects on human cells exposed to light in the presence of hematoporphyrin. Localization of the active dye, *Cancer Lett.*, 11, 209, 1981.
290. Dubbelman, T. M. A. R. and Van Steveninck, J., Hematoporphyrin derivative induced photodynamic effects on transport processes across the membrane of L-929 cells, in Proc. 10th Annu. Meet. Am. Soc. Photobiology, 1982, 151.
291. Dubbelman, T. M. A. R., Lienhouts, K., and Van Steveninck, J., Photodynamic inactivation of L929 cells after treatment with hematoporphyrin derivative, in *Porphyrins in Tumor Phototherapy*, Andreoni, A. and Cubeddu, R., Eds., Plenum Press, New York, 1984, 167.
292. Latham, P. S. and Bloomer, J. R., Protoporphyrin-induced photodamage: studies using cultured skin fibroblasts, *Photochem. Photobiol.*, 37, 553, 1983.
293. Creekmore, S. P. and Zaharko, D. S., Modification of chemotherapeutic effects on L1210 cells using hematoporphyrin and light, *Cancer Res.*, 43, 5252, 1983.
294. Cowled, P. A., Grace, J. R., and Forbes, I. J., Comparison of the efficacy of pulsed and continuous-wave red laser light in induction of photocytotoxicity by hematoporphyrin derivative, *Photochem. Photobiol.*, 39, 115, 1984.
295. Bellnier, D. A. and Dougherty, T. J., Membrane lysis in Chinese hamster ovary cells treated with hematoporphyrin derivative plus light, *Photochem. Photobiol.*, 36, 43, 1982.
296. Goldstein, B. D. and Harbor, L. C., Erythropoietic protoporphyria: lipid peroxidation and red cell membrane damage associated with photohemolysis, *J. Clin. Invest.*, 51, 892, 1972.
297. Fleischer, A. S., Harber, L. C., Cook, J. S., and Baer, R. L., Mechanism of in vitro photohemolysis in erythropoietic protoporphyria, *J. Invest. Dermatol.*, 46, 505, 1966.
298. Schothorst, A. A., Van Steveninck, J., West, L. N., and Surmond, D., Protoporphyrin-induced photohemolysis in protoporphyria and in normal red blood cells, *Clin. Chem. Acta*, 28, 41, 1979.
299. Lamola, A. A. and Doleiden, F. H., Cross-linking of membrane proteins and protoporphyrin-sensitized photohemolysis, *Photochem. Photobiol.*, 31, 597, 1980.
300. Girotti, A. W. and Deziel, M. R., Photodynamic action of protoporphyrin on resealed erythrocyte membranes: mechanisms of release of trapped markers, in *Porphyrin Photosensitization*, Kessel, D. and Dougherty, T. J., Eds., Plenum Press, New York, 1983, 213.
301. Van Steveninck, J., Dubbelman, T. M. A. R., and Verweij, H., Photodynamic membrane damage, in *Porphyrin Photosensitization*, Kessel, D. and Dougherty, T. J., Eds., Plenum Press, New York, 1983, 227.
302. Wakulchik, S. D., Schiltz, J. R., and Bickers, D. R., Photolysis of protoporphyrin-treated human fibroblasts in vitro: studies on the mechanism, *J. Lab. Clin. Med.*, 96, 158, 1980.
303. Grossweiner, L. I. and Grossweiner, J. B., Hydrodynamic effects in the photosensitized lysis of liposomes, *Photochem. Photobiol.*, 35, 583, 1982.
304. Christensen, T., Multiplication of human NHIK-3025 cells exposed to porphyrins in combination with light, *Br. J. Cancer*, 44, 433, 1981.
305. Kato, H., Konaka, C., Ono, J., Kawate, N., Yoneyama, K., Nishimiya, K., Shinohara, H., Saito, M., Sakai, H., Tomono, T., Aizawa, K., Yoshihama, I., and Hayata, Y., Cytomorphological changes of HPD and photodynamic therapy, in Proc. Porphyrin Photosensitization Workshop, Philadelphia, July 1984.
306. Coppola, A., Viggiani, E., Salzarula, L., and Rasile, G., Ultrastructural changes in lymphoma cells treated with hematoporphyrin and light, *Am. J. Pathol.*, 99, 175, 1980.
307. Gomer, C. J., Rucker, N., Razum, N. J., Pregerson, K., and Murphree, A. L., Cell and tissue effects related to porphyrin photoradiation therapy, in Proc. Porphyrin Photosensitization Workshop, Piladelphia, July 1984.
308. Gibson, S. L. and Hilf, R., Photosensitization of mitochondrial cytochrome c oxidase by hematoporphyrin derivative and related porphyrins in vitro and in vivo, *Cancer Res.*, 43, 4191, 1983.
309. Hilf, R., Smail, D. B., Murant, R. S., Leakey, P. B., and Gibson, S. L., Hematoporphyrin derivative induced photosensitivity of mitochondrial succinate dehydrogenase and selected cytosolic enzymes of R3230AC mammary adenocarcinoma of rats, *Cancer Res.*, 44, 1483, 1984.
310. Gibson, S. L., Leakey, P. B., Crute, J., and Hilf, R., Photosensitization of mitochrondrial cytochrome c oxidase by hematoporphyrin derivative (HPD) in vitro and in vivo, in *Porphyrin Localization and Treatment of Tumors*, Doiron, D. R. and Gomer, C. J., Eds., Alan R. Liss, New York, 1984, 323.
311. Hilf, R., Perlin, D. S., Murant, R. S., and Gibson, S. L., Photosensitization of mitochrondria by hematoporphyrin derivative (HPD): effects on ATPase-mediated proton transport and membrane integrity, in Proc. Porphyrin Photosensitization Workshop, Philadelphia, July 1984.

312. Salet, C. and Moreno, G., Photodynamic effects of hematoporphyrin on respiration and calcium uptake in isolated mitochondria, *Int. J. Radiat. Biol.*, 39, 227, 1981.
313. Salet, C., Moreno, G., and Vinzens, F., Photodynamic effects induced by furocoumarins on a membrane system. Comparison with hematoporphyrin, *Photochem. Photobiol.*, 36, 291, 1982.
314. Salet, C., Moreno, G., Vever-Bizet, C., and Brault, D., Anoxic photodamage in the presence of porphyrins: evidence for the lack of effects on mitochondrial membranes, *Photochem. Photobiol.*, 40, 145, 1984.
315. Sandberg, S. and Romslo, I., Porphyrin-sensitized photodynamic damage of isolated rat liver mitochondria, *Biochim. Biophys. Acta*, 593, 187, 1980.
316. Slater, T. F. and Riley, P. A., Photosensitization and lysosomal damage, *Nature (London)*, 209, 193, 1966.
317. Torinuki, W., Miura, T., and Seiji, M., Lysosome destruction and lipoperoxidase formation due to active oxygen generated from hematoporphyrin and UV irradiation, *Br. J. Dermatol.*, 102, 17, 1980.
318. Christensen, T., Volden, G., Moan, J., and Sandquist, T., Release of lysosomal enzymes and lactate dehydrogenase due to hematoporphyrin derivative and light irradiation of NHIK-3025 cells in vitro, *Ann. Clin. Res.*, 14, 46, 1982.
319. Bickers, D. R., Dixit, R., and Mukhtar, H., Hematoporphyrin photosensitization of epidermal microsomes results in destruction of cytochrome P-450 and in decreased monooxygenase activities and heme content, *Biochem. Biophys. Res. Commun.*, 108, 1032, 1982.
320. Dixit, R., Mukhtar, H., and Bickers, D. R., Destruction of microsomal cytochrome P-450 by reactive oxygen species generated during photosensitization of hematoporphyrin derivative, *Photochem. Photobiol.*, 37, 173, 1983.
321. Gutter, B., Speck, W. T., and Rosenkrantz, H. S., The photodynamic modification of DNA by hematoporphyrin, *Biochim. Biophys. Acta*, 475, 307, 1977.
322. Dubbelman, T. M. A. R., Van Steveninck, A. L., and Van Steveninck, J., Hematoporphyrin-induced photooxidation and photodynamic crosslinking of nucleic acids and their constituents, *Biochim. Biophys. Acta*, 719, 47, 1982.
323. Fiel, R. J., Datta-Gupta, N., Mask, E. H., and Howard, J. C., Induction of DNA damage by porphyrin photosensitizers, *Cancer Res.*, 41, 3543, 1981.
324. Musser, D. A., Datta-Gupta, N., and Fiel, R. J., Inhibition of DNA dependent RNA synthesis by porphyrin photosensitizers, *Biochem. Biophys. Res. Commun.*, 97, 918, 1980.
325. Gomer, C. J., Weishaupt, K. R., and Dougherty, T. J., The effects of hematoporphyrin derivative and visible red light on V-79 cells, *Radiat. Res.*, 74, 586, 1978.
326. Gomer, C. J., DNA damage and repair in CHO cells following hematoporphyrin photoradiation, *Cancer Lett.*, 11, 161, 1980.
327. Blazek, E. R. and Hariharan, P. V., Alkaline elution studies of hematoporphyrin derivative photosensitized DNA damage and repair in Chinese hamster ovary cells, *Photochem. Photobiol.*, 40, 5, 1984.
328. Boye, E. and Moan, J., The photodynamic effect of hematoporphyrin on DNA, *Photochem. Photobiol.*, 31, 223, 1980.
329. Moan, J. and Boye, E., Photodynamic effect on DNA and cell survival of human cells sensitized by hematoporphyrin, *Photobiochem. Photobiophys.*, 2, 301, 1981.
330. Moan, J., Watsvik, H., and Christensen, T., DNA single strand breaks and sister chromatid exchange induced by treatment with hematoporphyrin and light or X-rays in human NHIK 3025 cells, *Cancer Res.*, 40, 2915, 1980.
331. Evensen, J. F. and Moan, J., Photodynamic action and chromosomal damage: a comparison of hematoporphyrin derivative (HPD) and light with X-irradiation, *Br. J. Cancer*, 45, 456, 1982.
332. Gomer, C. J., Rucker, N., Banerjee, A., and Benedict, W. F., Comparison of mutagenicity and induction of sister chromatid exchange in Chinese hamster cells exposed to hematoporphyrin derivative photoradiation, ionizing radiation, or ultraviolet radiation, *Cancer Res.*, 43, 2622, 1983.
333. Gomer, C. J., Strniste, G. F., and Chen, D. J., Cell inactivation by hematoporphyrin derivative and red light, *Proc. Am. Soc. Photobiol.*, 8, 104, 1980.
334. Gomer, C. J., Doiron, D. R., Dunn, S., Rucker, N., Razum, N. J., and Fountain, S. W., Examination of action spectrum, dose rate, and mutagenic properties of hematoporphyrin derivative photoradiation therapy, in *Porphyrin Localization and Treatment of Tumors*, Doiron, D. R. and Gomer, C. J., Eds., Alan R. Liss, New York, 1984, 459.
335. Bertoloni, G., Dall'Acqua, M., Vazzoler, M., Salvato, B., and Jori, G., Photosensitizing action of hematoporphyrin on some bacterial strains, *Med. Biol. Environ.*, 10, 238, 1982.
336. Bertoloni, G., Salvato, B., Dall'Acqua, M., Vazzoler, M., and Jori, G., Hematoporphyrin-sensitized photoinactivation of *Streptococcans faecalis*, *Photochem. Photobiol.*, 39, 811, 1984.
337. Bertolini, G., Dall'Acqua, M., Vazzoler, M., Salvato, B., and Jori, G., Bacterial and yeast cells as models for studying hematoporphyrin photosensitization, in *Porphyrins in Tumor Phototherapy*, Andreoni, A. and Cubeddu, R., Eds., Plenum Press, New York, 1984, 177.

338. Stenstrom, A. G. K., Moan, J., Brunborg, G., and Elkund, T., Photodynamic inactivation of yeast cells sensitized by hematoporphyrin, *Photochem. Photobiol.,* 32, 349, 1980.
339. Ito, T., Photodynamic action of hematoporphyrin on yeast cells — a kinetic approach, *Photochem. Photobiol.,* 34, 521, 1981.
340. Ito, T., Mode of photodynamic action with special reference to the sensitizer-cell interaction: porphyrins and membrane damage, *Stud. Biophys.,* 94, 1, 1983.
341. Ito, T. and Ito, A., Enhancement of porphyrin photosensitization of yeast cells by ethanol, *Photochem. Photobiol.,* 40, 429, 1984.
342. Moan, J. Christensen, T., Photodynamic inactivation of cancer cells in vitro. Effect of irradiation temperature and dose fractionation, *Cancer Lett.,* 6, 331, 1979.
343. Christensen, T. and Moan, J., Photodynamic inactivation of synchronized human cells in vitro in the presence of hematoporphyrin, *Cancer Res.,* 39, 3735, 1979.
344. Christensen, T., Moan, J., Wibe, E., and Oftenbro, R., Photodynamic effect of hematoporphyrin throughout the cell cycle of the human cell line NHIK-3025 cultivated in vitro, *Br. J. Cancer,* 39, 64, 1979.
345. Moan, J., Smedshammer, L., and Christensen, T., Photodynamic effects on human cells exposed to light in the presence of hematoporphyrin: pH effects, *Cancer Lett.,* 9, 327, 1980.,
346. Moan, J. and Christensen, T., Cellular uptake and photodynamic effect of hematoporphyrin, *Photobiochem. Photobiophys.,* 2, 291, 1981.
347. Christensen, T., Feren, K., Moan, J., and Pettersen, E., Photodynamic effects of hematoporphyrin derivative on synchronized and asynchronous cells of different origin, *Br. J. Cancer,* 44, 717, 1981.
348. Kinsey, J. H., Cortese, D. A., Moses, H. C., Ryan, R. J., and Branum, E. L., Photodynamic effect of hematoporphyrin derivative as a function of optical spectrum and incident energy density, *Cancer Res.,* 41, 5020, 1981.
349. Anderson, R. E., Wharen, R. E., Jones, C. A., and Laws, E. R., Parameters of hematoporphyrin derivative cell killing efficiency: decomposition of HPD at high power densities, in *Porphyrin Localization and Treatment of Tumors,* Doiron, D. R. and Gomer, C. J., Eds., Alan R. Liss, New York, 1984, 483.
350. Gomer, C. J., Cellular studies pertaining to hematoporphyrin photoradiation therapy, *Radiat. Res.,* 83, 374, 1980.
351. Gomer, C. J. and Smith, D. M., Photoinactivation of Chinese hamster cells by hematoporphyrin derivative and red light, *Photochem. Photobiol.,* 32, 341, 1980.
352. Gomer, C. J., Doiron, D. R., Rucker, N., Razum, N. J., and Fountain, S. W., Action spectrum (620-640 nm) for hematoporphyrin derivative induced cell killing, *Photochem. Photobiol.,* 39, 365, 1984.
353. Weishaupt, K. R., Gomer, C. J., and Dougherty, T. J., Identification of singlet oxygen as the cytotoxic agent in photodestruction of a murine tumor, *Cancer Res.,* 36, 2326, 1976.
354. Dougherty, T. J., Gomer, C. J., and Weishaupt, K. R.,Energetics and efficiency of photoinactivation of murine tumor cells containing hematoporphyrin, *Cancer Res.,* 36, 2330, 1976.
355. Ebert, P. S., Smith, P. D., Bonner, R. F., Hess, R. A., Costa, J. L., and Tschudy, D. P., Effect of defined wavelength and succinyl acetone on the photoinactivation of leukemia L1210 cells in vitro by hematoporphyrin, *Photobiochem. Photobiophys.,* 6, 165, 1983.
356. Kessel, D. and Dutton, C. J., Photodynamic effects: porphyrin vs. chlorin, *Photochem. Photobiol.,* 40, 403, 1984.
357. Monfrecola, G., Martellotta, D., Galli, R., Bruno, G., Cariello, L., Zanetti, L., and Santoianni, P., Effect of He-Ne laser on human erythrocytes incubated with hematoporphyrin derivative and bonellin: comparative study, in *Porphyrins in Tumor Phototherapy,* Andreoni, A. and Cubeddu, R., Eds., Plenum Press, New York, 1984, 95.
358. Mathews-Roth, M. M., Porphyrin photosensitization and carotenoid protection in mice; in vitro and in vivo studies, *Photochem. Photobiol.,* 40, 63, 1984.
359. Cannistraro, S. and Van de Vorst, A., Photosensitization by hematoporphyrin: ESR evidence for free radical induction in unsaturated fatty acids and for singlet oxygen production, *Biochem. Biophys. Res. Commun.,* 74, 1177, 1977.
360. Felix, C. C., Reszka, K., and Sealy, R. C., Free radicals from photoreduction of hematoporphyrin in aqueous solution, *Photochem. Photobiol.,* 37, 141, 1983.
361. Hariharan, P. V., Courtney, J., and Eleczko, S., Production of hydroxyl radicals in cell systems exposed to hematoporphyrin and red light, *Int. J. Radiat. Biol.,* 37, 691, 1980.
362. Christensen, T., Moan, J., Sandquist, T., and Smedshammer, L., Multicellular spheroids as an in vitro model system for photoradiation therapy in the presence of HPD, in *Porphyrin Localization and Treatment of Tumors,* Doiron, D. R. and Gomer, C. J., Eds., Alan R. Liss, New York, 1984, 381.

363. Bellnier, D. A., Prout, G. R., and Lin, C.-W., Recovery from damage induced by hematoporphyrin derivative plus light in cultured human bladder carcinoma cells, in Proc. Porphyrin Photosensitization Workshop, Philadelphia, July 1984.
364. Weishaupt, K. R., Boyle, D. C., Mensinger, M., and Dougherty, T. J., Cellular and tissue repair of photodynamic damage, *Proc. Am. Assoc. Cancer Res.*, 20, 227, 1979.
365. Christensen, T., Smedshammer, L., Wahl, A., and Moan, J., Photodynamic effects and hyperthermia in vitro, in *Methods in Porphyrin Photosensitization*, Kessel, D., Ed., Plenum Press, New York, 1985, 69.
366. Mang, T. S. and Dougherty, T. J., The influence of hyperthermia on the photodynamic therapy survival of tumor cells in vitro, in Proc. Porphyrin Photosensitization Workshop, Philadelphia, July 1984.
367. Goldacre, R. J. and Slyven, B., On the access of blood-borne dyes to various tumor regions, *Br. J. Cancer*, 16, 306, 1962.
368. Peterson, H. I. and Appelgren, K. L., Experimental studies on the uptake and retention of labelled proteins in a rat tumor, *Eur. J. Cancer*, 9, 543, 1973.
369. Tipping, E., Kelterer, B., and Koskelo, P., The binding of porphyrins by ligandin, *Biochem. J.*, 169, 509, 1978.
370. Muller-Eberhard, U., Hemopexin, *N. Engl. J. Med.*, 283, 1090, 1970.
371. Morgan, W. T. and Muller-Eberhard, U., Interactions of porphyrins with rabbit hemopexin, *J. Biol. Chem.*, 247, 7181, 1972.
372. Muller-Eberhard, U., Porphyrin-binding proteins in serum, *Ann. N.Y. Acad. Sci.*, 244, 624, 1975.
373. Koskelo, P. and Muller-Eberhard, U., Interaction of porphyrins, *Semin. Hematol.*, 45, 221, 1979.
374. Pottier, R., Laplante, J.-P., and Kennedy, J., Mechanism of binding of hematoporphyrin derivative to human blood plasma — a spectroscopic study, in Proc. 12th Ann. Meet. Am. Soc. Photobiol., Philadelphia, July 1984.
375. Morliere, P., Reyftamnn, J. P., Goldstein, B. D., Santus, R., Dubertret, L., and Lagrange, D., Interaction of LDL with porphyrins, in Proc. 12th Annu Meet. Am. Soc. Photobiology, Philadelphia, July 1984.
376. Reddi, E., Rodgers, M. A. J., and Jori, G., Photophysical and photosensitizing properties of hematoporphyrin bound with human serum albumin, in *Porphyrin Localization and Treatment of Tumors*, Doiron, D. R. and Gomer, C. J., Eds., Alan R. Liss, New York, 1984, 373.
377. Richard, P., Blum, A., and Grossweiner, L. I., Hematoporphyrin photosensitization of serum albumin and subtilisin BPN, *Photochem. Photobiol.*, 37, 287, 1983.
378. Goyal, G. C., Blum, A., and Grossweiner, L. I., Photosensitization of liposomes and human serum albumin by hematoporphyrin derivative and uroporphyrin I, in Proc. 12th Annu Meet. Am. Soc. Photobiology, Philadelphia, July 1984.
379. El-Far, M. A. and Pimstone, N. R., Tumor localization of uroporphyrin isomers I and III and their correlation to albumin binding, *Proc. Int. Meet. Cell Biochem. Funct.*, 1, 157, 1983.
380. Gervais, P., Band, P., Moisan, R., Mailhot, S., and Besner, J. G., Tissular distribution and pharmacokinetics of hematoporphyrin diacetate on rats, in *Porphyrin Localization and Treatment of Tumors*, Doiron, D. R. and Gomer, C. J., Eds., Alan R. Liss, New York, 1984, 571.
381. Henderson, R. W., Christie, G. S., Clezy, P. S., and Lineham, J., Hematoporphyrin diacetate: a probe to distinguish malignant from normal tissue by selective fluorescence, *Br. J. Exp. Pathol.*, 61, 345, 1980.
382. Houle, J. M. and Besner, J. G., Tissular uptake of hematoporphyrin in rats, in Proc. Porphyrin Photosensitization Workshop, Philadelphia, July 1984.
383. Jori, G., Reddi, E., Tomio, L., Salvato, B., Zorat, P. L., and Calzavara, F., Time-course of hematoporphyrin distribution in selected tissues of normal rats and in ascites hepatoma, *Tumori*, 65, 43, 1979.
384. Cozzani, I., Jori, J., Reddi, E., Fortunato, A., Granati, B., Felice, M., Tomio, L., and Zorat, P., Distribution of endogenous and injected porphyrins at the subcellular level in rat hepatocytes and in ascites hepatoma, *Chem. Biol. Interact.*, 37, 67, 1981.
385. Tomio, L., Zorat, P. L., Jori, G., Reddi, E., Salvato, B., Corti, L., and Calzavara, F., Elimination pathway of hematoporphyrin from normal and tumor-bearing rats, *Tumori*, 68, 283, 1982.
386. Jori, G., Tomio, L., Reddi, E., Rossi, E., Cirto, L., Zorat, P. L., and Calzavara, F., Preferential delivery of liposome-incorporated porphyrins to neoplastic cells in tumour-bearing rats, *Br. J. Cancer*, 48, 307, 1983.
387. Grossweiner, L. I., Hill, J. H., Seter, A. J., and Harris, D. M., Liposomal delivery of HPD in a mouse tumor model, in Proc. 12th Annu. Meet. Am. Soc. Photobiology, Philadelphia, July 1984.
388. Straight, R. C. and Spikes, J. D., Effect of liposomes on tumor tissue accumulation of fluorescent porphyrins, in Proc. Porphyrin Photosensitization Workshop, Philadelphia, July 1984.
389. Gomer, C. J. and Dougherty, T. J., Determination of 3H and 14C hematoporphyrin derivative distribution in malignant and normal tissue, *Cancer Res.*, 39, 146, 1979.

390. Bugelski, P. J., Porter, C. W., and Dougherty, T. J., Autoradiographic distribution of hematoporphyrin derivative in normal and tumor tissue, *Cancer Res.*, 41, 4606, 1981.
391. Musser, D. A., Wagner, J. M., Weber, F. J., and Datta-Gupta, N., The binding of tumor localizing porphyrins to a fibrin matrix and their effects following photoirradiation, *Res. Commun. Chem. Pathol. Pharmacol.*, 28, 505, 1980.
392. Musser, D. A., Wagner, J. M., and Datta-Gupta, N., The interaction of tumor localizing porphyrins with collagen and elastin, *Res. Commun. Chem. Pathol Pharmacol.*, 36, 251, 1982.
393. Sabben, G., Bosshard, A., and Lambert, R., Time dependence of 3H hematoporphyrin derivative distribution in the digestive tract in the rat, in *Porphyrins in Tumor Phototherapy*, Andreoni, A. and Cubeddu, R., Eds., Plenum Press, New York, 1984, 243.
394. Gomer, C. J., Szirth, B. C., Doiron, D. R., Jester, J. V., Lingua, R. W., Mark, C., Benedict, W. F., and Murphree, A. L., Preclinical evaluation of hematoporphyrin derivative for the treatment of intraocular tumors: a preliminary report, in *Porphyrin Photosensitization*, Kessel, D. and Dougherty, T. J., Eds., Plenum Press, New York, 1983, 109.
395. Winkelman, J., Slater, G., and Grossman, J., The concentration in tumor and other tissues of parenterally administered tritium- and 14C-labeled tetraphenylporphinesulfate, *Cancer Res.*, 27, 2060, 1967.
396. Gomer, C. J. and Little, F. M., Documentation of radioactive contaminants in tritium labelled hematoporphyrin derivative, in *Porphyrin Localization and Treatment of Tumors*, Doiron, D. R. and Gomer, C. J., Eds., Alan R. Liss, New York, 1984, 591.
397. Thaller, R. A., Lyster, D. M., and Dolphin, D., Potential use of radiolabelled porphyrins for tumor scanning, in *Porphyrin Photosensitization*, Kessel, D. and Dougherty, T. J., Eds., Plenum Press, New York, 1983, 265.
398. Wang, T. S. T., Fawwas, R. A., and Tomashefsky, P., Metalloporphyrin derivatives: structure-localization properties, in *Radiopharmaceutical Structure-Activity Relationships*, Spencer, R. P., Ed., Grune & Stratton, New York, 1981, 225.
399. Van Lier, J. E., Ali, H., and Rousseau, J., Phthalocyanines labelled with gamma emitting radionuclides as possible tumor scanning agents, in *Porphyrin Localization and Treatment of Tumors*, Doiron, D. R. and Gomer, C. J., Eds., Alan R. Liss, New York, 1984, 315.
400. Bonnett, R., Berenbaum, M. C., and Kaur, H.,Chemical and biological studies on hematoporphyrin derivative: an unexpected photosensitization in brain, in *Porphyrins in Tumor Phototherapy*, Andreoni, A. and Cubeddu, R., Eds., Plenum Press, New York, 1984, 67.
401. Dougherty, T. J., Thoma, R. E., Boyle, D. G., and Weishaupt, K. R., Interstitial photoradiation therapy for primary solid tumors in pet cats and dogs, *Cancer Res.*, 41, 404, 1981.
402. Bellnier, D. A., Prout, G. R., and Lin, C.-W., Treatment of experimental bladder carcinomas with hematoporphyrin derivative plus 514.5 nm argon laser light, in Proc. Porphyrin Photosensitization Workshop, Philadelphia, July 1984.
403. Pimstone, N. R. and Gandhi, S. N., Optimal photodynamic band of red light in hematoporphyrin derivative (HPD) photoradiation therapy of cancer, in *Porphyrin Localization and Treatment of Tumors*, Doiron, D. R. and Gomer, C. J., Eds., Alan R. Liss, New York, 1984, 673.
404. Pezzotta, S., Spanu, G., Andreoni, A., Cubeddu, R., and Giordana, M. T., Experimental ENU induced brain tumors with HPD and dye laser light, in *Porphyrins in Tumor Phototherapy*, Andreoni, A. and Cubeddu, R., Eds., Plenum Press, New York, 1984, 227.
405. Coleridge-Smith, P. D., Brown, S. G., Mills, T. N., Hobbs, K. E. F., and Salmon, P. R., Quantitative study of photodynamic therapy in the rat, in Proc. Porphyrin Photosensitization Workshop, Philadelphia, July 1984.
406. Ellingsen, R., Svaasand, L. O., and Odegaard, A., Photoradiation therapy (PRT) of Lewis lung carcinoma in B_6D_2 mice; dosimetry considerations, in *Porphyrins in Tumor Phototherapy*, Andreoni, A. and Cubeddu, R., Eds., Plenum Press, New York, 1984, 213.
407. Weishaupt, K. R. and Dougherty, T. J., Photoradiation therapy: optimum wavelength for in vivo activation of hematoporphyrin derivative with laser irradiation, *Radiat. Res.*, 74, 586, 1978.
408. Bellnier, D. A., Lin, C.-W., Parrish, J. A., and Mock, P. C., Hematoporphyrin derivative and pulse laser photoradiation, in *Porphyrin Localization and Treatment of Tumors*, Doiron, D. R. and Gomer, C. J., Eds., Alan R. Liss, New York, 1984, 533.
409. Gomer, C. J. and Razum, N. J., Acute skin response in albino mice following porphyrin photosensitization under oxic and anoxic conditions, *Photochem. Photobiol.*, 40, 435, 1984.
410. Gomer, C. J., Doiron, D. R., Dunn, S., Rucker, N., Razum, N., and Fountain, S., Determinants of hematoporphyrin derivative photosensitization: action spectrum, dose rate, and oxygen effects, *Photochem. Photobiol.*, 37S, S91, 1983.
411. Denekamp, J., Hill, S. A., and Hobson, B., Vascular occlusion and tumor cell death, *Eur. J. Cancer Clin. Oncol.*, 19, 271, 1983.

412. Henderson, B. W. and Dougherty, T. J., Studies on the mechanism of tumor destruction by photoradiation therapy, in *Porphyrin Localization and Treatment of Tumors,* Doiron, D. R. and Gomer, C. J., Eds., Alan R. Liss, New York, 1984, 601.
413. Henderson, B. W., Waldow, S. M., and Dougherty, T. J., Tumor control and tumor cell survival kinetics following in vivo treatments with either photodynamic therapy, hyperthermia, or combined modalities, in Proc. Porphyrin Photosensitization Workshop, Philadelphia, July 1984.
414. Castellani, A., Pace, G. P., and Concioli, M., Photodynamic effect of hematoporphyrin on blood microcirculation, *J. Pathol. Bacteriol.,* 86, 99, 1963.
415. Selman, S. H., Keck, R., Klaunig, J. E., Kreimer-Birnbaum, M., Goldblatt, P. J., and Britton, S. L., Acute blood flow changes in transplantable FANFT-induced urothelial tumors treated with hematoporphyrin derivative and light, *Surg. Forum,* 34, 676, 1983.
416. Kreimer-Birnbaum, M., Klaunig, J. E., Keck, R., Goldblatt, P. J., Britton, S. L., and Selman, S. H., Studies with hematoporphyrin derivative in transplantable urothelial tumors, in *Porphyrins in Tumor Phototherapy,* Andreoni, A. and Cubeddu, R., Eds., Plenum Press, New York, 1984, 235.
417. Bicher, H. I., Hetzel, F. W., Vaupel, P., and Sandhu, T. S., Microcirculation modifications by localized microwave hyperthermia and hematoporphyrin phototherapy, *Bibl. Anat.,* 20, 628, 1981.
418. Hetzel, F. W. and Farmer, H., Dose effect relationships in a mouse mammary tumor, in *Porphyrin Localization and Treatment of Tumors,* Dorion, D. R. and Gomer, C. J., Eds., Alan R. Liss, New York, 1984, 583.
419. Kang, M. S., Song, C. W., and Levitt, S. H., Role of vascular function in response of tumors in vivo to hyperthermia, *Cancer Res.,* 40, 1130, 1980.
420. Marchesini, R., Zunino, F., Melloni, E., Pezzoni, G., Savi, G., Locati, L., Bandieramonte, G., Spinelli, P., Emanuelli, H., and Fava, G., Study of irradiation parameters in HPD phototherapy of MS-2 tumor model, in *Porphyrins in Tumor Phototherapy,* Andreoni, A. and Cubeddu, R., Eds., Plenum Press, New York, 1984, 221.
421. Dougherty, T. J., Wityk, K., Potter, W. R., and Weishaupt, K. R., Action of DHE in PDT, in Proc. Porphyrin Photosensitization Workshop, Philadelphia, July 1984.
422. Dougherty, T. J., Malone, P. B., Weishaupt, K. R., Sullivan, L. A., Hartley, E. L., Boyle, D. G., Potter, W. R., Takita, H., and Regal, A.-M., Efficacy and toxicity of HPD vs. DHE, in Proc. Porphyrin Photosensitization Workshop, Philadelphia, July 1984.
423. Sanderson, D. R., Fontana, R. S., Lipson, R. L., and Baldes, E. J., Hematoporphyrin as a diagnostic tool, *Cancer,* 30, 1368, 1972.
424. Krohn, D. L., Jacobs, R., and Morris, D.A. , Diagnosis of model choroidal malignant melanoma by hematoporphyrin derivative fluorescence in rabbits, *Invest. Opthalmol.,* 13, 244, 1974.
425. Kinsey, J. H., Cortese, D. A., and Sanderson, D. R., Detection of hematoporphyrin fluorescence during fiberoptic bronchoscopy to localize early bronchogenic carcinoma, *Mayo Clin. Proc.,* 53, 594, 1978.
426. Kinsey, J. H. and Cortese, D. A., Endoscopic system for simultaneous visual examination and electronic detection of fluorescence, *Rev. Sci. Instrum.,* 51, 1403, 1980.
427. Cortese, D. A., Kinsey, J. H., Woolner, L. B., Payne, W. S., Sanderson, D. R., and Fontana, R. S., Clinical application of a new endoscopic technique for detection of *in situ* bronchial carcinoma, *Mayo Clin. Proc.,* 54, 635, 1979.
428. Cortese, D. A. and Kinsey, J. H., Hematoporphyrin derivative fluorescence for lung cancer localization, *Semin. Resp. Med.,* 3, 37, 1981.
429. Cortese, D. A., Kinsey, J. H., Woolner, L. B., Sanderson, D. R., and Fontana, R. S., Hematoporphyrin derivative in the detection and localization of radiographically occult lung cancer, *Am. Rev. Resp. Dis.,* 126, 1087, 1982.
430. Benson, R. C., Farrow, G. M., Kinsey, J. H., Cortese, D. A., Zincke, H., and Utz, D. C., Detection and localization of *in situ* carcinoma of the bladder with hematoporphyrin derivative, *Mayo Clin. Proc.,* 57, 548, 1982.
431. Lin, C.-W., Bellnier, D. A., Fujime, M., and Prout, G. R., HPD photodetection of bladder carcinoma, in *Porphyrin Localization and Treatment of Tumors,* Doiron, D. R. and Gomer, C. J., Eds., Alan R. Liss, New York, 1984, 187.
432. Profio, A. E. and Doiron, D. R., A feasibility study of the use of fluorescence bronchoscopy for localization of small lung tumors, *Phys. Med. Biol.,* 22, 949, 1977.
433. Profio, A. E., Doiron, D. R., and Huth, G. C., Fluorescence bronchoscope for lung tumor localization, *IEEE Trans. Nucl. Sci.* 24, 521, 1977.
434. Profio, A. E., Doiron, D. R., and King, E. G., Laser fluorescence bronchoscope for localization of occult lung tumors, *Med. Phys.,* 6, 523, 1979.
435. Doiron, D. R., Profio, A. E., Vincent, R. G., and Dougherty, T. J., Fluorescence bronchoscopy for detection of lung cancer, *Chest,* 76, 37, 1979.
436. Doiron, D., Fluorescence Bronchoscopy for the Early Localization of Lung Cancer, Ph.D. thesis, University of California, Santa Barbara, 1982.

437. King, E. G., Man, G., le Richie, J., Amy, R., Profio, A. E., and Doiron, D. R., Fluorescence bronchoscopy in the localization of bronchogenic carcinoma, *Cancer,* 49, 777, 1982.
438. King, E. G., Doiron, D. R., Man, G., Profio, A. E., and Huth, G., Hematoporphyrin derivative as a tumor marker in the detection and localization of pulmonary malignancy, *Recent Results Cancer Res.,* 82, 90, 1982.
439. Balchum, O. J., Doiron, D. R., Profio, A. E., and Huth, G. C., Fluorescence bronchoscopy for localizing early bronchial cancer and carcinoma in situ, *Recent Results Cancer Res.,* 82, 97, 1982.
440. Profio, A. E., Doiron, D. R., Balchum, O. J., and Huth, G. C., Fluorescence bronchoscopy for localization of carcinoma *in situ, Med. Phys.,* 10, 35, 1983.
441. Profio, A. E., Carvlin, M. J., Sarnaik, J., and Wudl, L. R., Fluorescence of hematoporphyrin-derivative for detection and characterization of tumors, in *Porphyrins in Tumor Phototherapy,* Andreoni, A. and Cubeddu, R. Eds., Plenum Press, New York, 1984, 321.
442. Hayata, Y., Laser equipment and technique, in *Lasers and Hematoporphyrin Derivative in Cancer,* Hayata, Y. and Dougherty, T. J., Eds., Igaka-Shoin, Tokyo, 1983, 11.
443. Aizawa, K., Kato, H., Ono, J., Konaka, C., Kawate, N., Yoneyama, K., Otawa, M., Shinohara, H., Saito, M., Takahashi, H., Nakamera, H., Yamada, T., Sakai, H., and Hayata, Y., A new diagnostic system for malignant tumors using hematoporphyrin derivative, laser photoradiation, and a spectroscope, in *Porphyrin Localization and Treatment of Tumors,* Doiron, D. R. and Gomer, C. J., Eds., Alan R. Liss, New York, 1984, 227.
444. Hayata, Y., Kato, H., Konaka, C., Ono, J., Matsushima, Y., Yoneyama, K., and Nishimiya, K., Fiberoptic bronchoscopic laser photoradiation for tumor localization in lung cancer, *Chest,* 82, 10, 1982.
445. Hayata, Y., Kato, H., Ono, J., Matsushima, Y., Hayashi, N., Saito, T., and Kawate, N., Fluorescence fiberoptic bronchoscopy in the diagnosis of early stage lung cancer, *Recent Results Cancer Res.,* 82, 121, 1982.
446. Hayata, Y., Kato, H., Konaka, C., Hayashi, N., Tahara, M., Saito, T., and Ono, J., Fiberoptic bronchoscopic photoradiation in experimentally induced canine lung cancer, *Cancer,* 51, 50, 1983.
447. Kawauchi, T., Basic studies in experimentally induced lung cancer, in *Lasers and Hematoporphyrin Derivative in Cancer,* Hayata, Y. and Dougherty, T. J., Eds., Igaka-Shoin, Tokyo, 1983, 25.
448. Aizawa, K., Endoscopic detection of hematoporphyrin derivative fluorescence, in *Lasers and Hematoporphyrin Derivative in Cancer,* Hayata, Y. and Dougherty, T. J., Eds., Igaka-Shoin, Tokyo, 1983, 21.
449. Profio, A. E. and Sarnaik, J., Fluorescence of HPD for tumor detection and dosimetry in photoradiation therapy, in *Porphyrin Localization and Treatment of Tumors,* Doiron, D. R. and Gomer, C. J., Eds., Alan R. Liss, New York, 1984, 163.
450. Potter, W. R. and Mang, T. S., Photofrin II levels by in vivo fluorescence photometry, in *Porphyrin Localization and Treatment of Tumors,* Doiron, D. R. and Gomer, C. J., Eds., Alan R. Liss, New York, 1984, 177.
451. van der Putten, W. J. M. and van Gemert, M. J. C., Hematoporphyrin derivative fluorescence in vitro and in an animal model, *Phys. Med. Biol.,* 28, 633, 1983.
452. van der Putten, W. J. M. and van Gemert, M. J. C., A modelling approach to the detection of subcutaneous tumors by hematoporphyrin derivative fluorescence, *Phys. Med. Biol.,* 28, 639, 1983.

Chapter 7

PHOTOIMMUNOTHERAPY

Julia G. Levy and Daphne Mew

TABLE OF CONTENTS

I. INTRODUCTION

The preliminary work which is discussed in this chapter involves utilization of a number of concepts, basic scientific information, and technologies which are already established but which have not been, until recently, used in combination.

A. Phototherapy

It is well established that a number of porphyrins are preferentially taken up and retained by neoplastic cells or lesions in comparison to normal cells or tissue.[1-4] This characteristic of some photochemicals has been used to advantage as a diagnostic tool to locate and delineate tumors in man.[5-8] Photoactivation of porphyrins at the site of tumor lesions has the dual advantages that the resultant fluorescence permits localization of malignant tissue, and that the activation process may deliver a cytotoxic effect to those cells for which the porphyrins have affinity. Lethal photoactivation involves the production of singlet oxygen.[9-12] Although it has been shown that singlet oxygen can have a toxic effect when it reacts with subcellular organelles, proteins, and nucleic acids,[13-16] the main target site for cytotoxicity is thought to be the cell membrane.[10]

There have been a number of studies in which hematoporphyrin (HP) has been used as an anticancer agent,[17,18] but more recently hematoporphyrin derivative (HPD — prepared by acetylation and hydrolysis of HP dehydrochloride) has been found to be a more effective agent in a number of systems.[19-25]

B. Serotherapy

The idea that antibodies with specificity for tumor-associated antigens (TAA) could have a significant role in the treatment and/or localization of tumor masses is not new. Such antibodies, conjugated to either cytotoxic drugs or radioisotopes would provide the specificity for the delivery of such probes to the desired site. Attractive as this concept is, attempts to apply it successfully have been disappointingly few.

There are rationales, particularly since the advent of monoclonal antibodies (Mab), which justify the use of antibodies with specificity for TAA alone as means of cancer therapy in certain circumstances. Antibodies with specificity for membrane-bound TAA could bind to tumor cell membranes and thus lead to tumor cell lysis by complement mediation or tumor cell killing by antibody-dependent cellular cytotoxicity (ADCC). Binding of antibodies to the surface of tumor cells may also sensitize the cells to destruction by reticuloendothelial cells. Alternately, it has been suggested that circulating antitumor antibodies may modulate the host's immune response to TAA by eliminating freely circulating TAA, and by so doing, permit host cellular mechanisms to target on tumor cells rather than being blocked by soluble antigens. While these approaches might appear feasible on a theoretical basis, serotherapy for cancer has had only marginal success.[26-28]

The development of hybridoma technology[29] and the concomitant potential for the production of essentially unlimited quantities of Mab of defined isotype and specificity has added a new impetus to the concept of serotherapy. It was hoped that the use of Mab would allow for more critical assessment of the therapeutic potential for antitumor serotherapy. In a review on such studies in animal models,[30] it was concluded that although Mab were more effective therapeutically than conventional antisera, there were clearly significant limitations regarding the potential clinical applications of monoclonal serotherapy.

Murine Mab have been applied clinically in the treatment of a number of leukemia and lymphoma patients.[31-33] The most consistent finding in these clinical trials was a dramatic but transient drop in the number of circulating leukemia or lymphoma cells. In instances where skin lesions constituted symptoms of the disease, these too would undergo almost immediate regression. Unfortunately, these effects usually lasted only a few days. Occasionally, longer remissions have been noted using this type of serotherapy.[32,33] Of course, another complication arising from the use of nonhuman Mab in repeated dosages is that patients so treated may develop their own antibodies to the Mab, thereby neutralizing its effect.

C. Magic Bullets (Immunotoxins)

There is very little disagreement regarding the conclusion that antibodies (monoclonal or conventional) on their own, are not going to revolutionize either diagnosis or therapy of human cancer, although they may have defined application in serological diagnosis and prognosis. However, the concept of the "magic bullet", as proposed by Paul Ehrlich at the turn of the century, still pertains. He referred to these hypothetical agents as "bodies which possess a particular affinity for a certain organ . . . as a carrier by which to bring therapeutically active groups to the organ in question".[34] It is clear that antibodies, and particularly Mab, have the appropriate characteristic to fulfill the criterion as the "bodies which possess affinity for a certain organ". Thus it is not surprising that since the advent of hybridoma technology, a great deal of work has been done by many investigators on the production of magic bullets which can be used as tailor-made homing devices for either the localization or treatment of tumors by conjugating either radioisotopes or cytotoxic agents to the appropriate Mab. In such studies, it has been possible to demonstrate unequivocally that Mab can act as carriers to designated targets either in vitro or in vivo, thus bestowing the required characteristic of specificity on the experimental conjugates. Toxic agents such as protein toxins (e.g., ricin-A chain, diphtheria toxin, abrin, *P. aeruginosa* exotoxin A, *Clostridium perfringens* phospholipase C, bovine pancreatic ribonuclease) or chemotherapeutic drugs (e.g., daunomycin, methotrexate, chlorambucil) when linked to Mab displayed appreciable levels of cytotoxicity for antigen-expressing cells.[35-47] Mab conjugated to a variety of radioisotopes (e.g., ^{131}I, ^{125}I, ^{111}In, ^{99m}Tc) have also been shown to display preferential target localization in vivo, thus demonstrating their potential use in imaging.[48-50]

Most experimental work which has been published on immunotoxins involving Mab and cytotoxic agents has shown clearly that when such reagents are tested in vitro on selected cell lines, they demonstrate both the discriminatory specificity and toxicity required. However, their ability to function under in vivo conditions is more questionable. One of the major problems encountered by researchers attempting to use magic bullets (and exemplified most clearly by those using radiolabeled Mab for imaging) is that these reagents, once injected, tend to accumulate rapidly in cells of the reticuloendothelial system because these cells are either actively phagocytic or possess Fc receptors to which Mab adhere. This leads to an appreciable build up of the conjugated antibodies in a variety of tissues other than those for which the reagent was targeted. Major areas for nonspecific accumulation of labeled antibodies include the liver, spleen, lungs, and kidneys.[51] Levels of materials in these organs do not diminish significantly for more than 48 hr. Problems arising from this kind of nonspecific accumulation of labeled Mab are obvious. If the Mab is conjugated to a cytotoxic agent, significant numbers of cells, other than the desired target, will be adversely affected by the toxin. When radioactively labeled Mab are used for imaging purposes, nonspecific accumulation of labeled Mab may obscure detection of the desired target unless it is located in an area peripheral to the soft organs of the body. Preliminary imaging studies carried out in the authors' laboratory indicated that favorable ratios of accumulation of isotope (in this case ^{99m}Tc) in tumor tissue vs. nonspecific tissue begin to appear after 48 hr, a time at which the activity of ^{99m}Tc is greatly reduced.[52] Other problems associated with magic bullets as such include the requirements for definitive information on both the way in which the cytotoxic reagent acts (i.e., membrane or intracellular), whether the conjugate, once located on the tumor cell surface is endocytosed, sloughed off, or remains on the cell surface, or whether levels of free-circulating TAA will inhibit activity of the immunotoxin at the tumor site.

II. PHOTOIMMUNOTHERAPY

Work in the authors' laboratory, during the past 3 years has been directed toward the development of a new kind of magic bullet: Mab to TAA conjugated to either HP, HPD, or, in the future, other photochemicals. The work has been carried out at all times in collaboration with Drs. Towers and Wat (Department of Botany, University of British Columbia). There are a number of reasons why this type of conjugate, once conceived, appeared to offer a number of obvious advantages over other immunotoxins currently being investigated. A magic bullet of this sort is essentially innocuous unless the photochemicals are activated by light. Thus, tumor-bearing animals injected with anti-TAA Mab conjugated to HP or HPD, could be maintained under dark conditions until the optimal time at which accumulation of the Mab at the tumor site was maximal, and nonspecifically accumulated Mab in other organs and tissues had turned over. Studies have been carried out to investigate the feasibility of using HP-conjugated Mab as an immunotoxin, both in vivo and in vitro, and the following section summarizes the results of these experiments.

A. Photoimmunotherapy in an Animal Model

The animal model used in this study involved a transplantable methylcholanthrene-induced rhabdomyosarcoma (M-1) which is syngeneic in DBA/2 mice. The tumor line and its maintenance have been described in detail elsewhere.[53] Hybridomas which secrete Mab with specificity for this tumor were generated by fusing splenocytes from M-1-immune DBA/2 mice with NS-1 cells by standard procedures.[54] Hybridomas secreting Mab with specificity for M-1 tumor membrane extracts were selected by the enzyme-linked immunosorbent assay (ELISA).[55] One such Mab (termed anti-M-1) was selected for further studies. It was cloned, purified, and subjected to specificity testing according to standard procedures.[55]

The anti-M-1 Mab and an irrelevant control Mab were conjugated to HP 2 HCl (Sigma) by a standard neutral EDCl procedure as described.[55] Fluorescence quantification on conjugated materials showed that in each case, between 40 and 60 HP molecules were associated per Ig molecule, and that most of these were covalently bound. Studies on the activity of both the Mab and the HP following conjugation showed that both reagents had retained their specific biological activities as assayed by ELISA and light-activated hemolysis (Figures 1 and 2).

Initially, experiments were carried out to determine whether anti-M-1 HP conjugates were capable of killing M-1 tumor cells in vitro with greater selectivity and specificity than either HP or the Mab alone. Test cultures were set up on ^{51}Cr labeled M-1 cells ($2 \times 10^6/m\ell$) in RPM1 1640 medium to which were added anti-M-1 HP, anti-M-1 alone, HP alone, the irrelevant Mab-HP conjugate, or PBS. In all appropriate cases, Mab concentration was 10 $\mu g/10^7$ cells and HP (either alone or conjugated) was at 2 $\mu g/10^7$ cells. Treated cells were incubated for 1 hr at 37°C in a humidified incubator at 6% CO_2, washed three times in dark conditions, resuspended in a final volume of 2.0 $m\ell$ RPM1, and exposed to incandescent light (power intensity was 22.5 mW/cm^2) at 30°C for 2 hr. All tubes were incubated for a further 6 hr at 37°C, centrifuged, and supernatants were evaluated for ^{51}Cr release or for viable cell numbers. The results of one such experiment are shown (Figure 3) and clearly demonstrate that only those cells treated with the anti-M-1 HP conjugate demonstrated a cytotoxic effect. Clearly HP

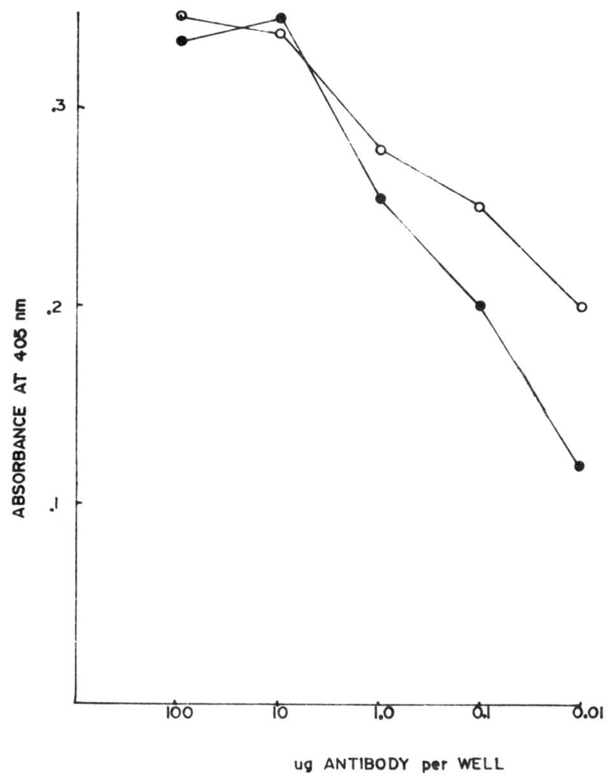

FIGURE 1. ELISA titration of the Mab anti-M-1 and the anti-M-1-HP conjugate on M-1 antigen. The immunological activity of the anti-M-1-HP conjugate (O) and anti-M-1 (•) were compared on ELISA plates coated with 25 μg/mℓ of M-1 tumor membrane extracts. Each point represents the average value obtained from quadruplicate samples in which standard deviations were <7.0%.

alone or HP conjugated to another Mab, under these conditions, were not toxic to the M-1 cell line. Specificity controls were run, in which it was shown that no such effect was seen with the anti-M-1 HP conjugate when it was tested with either normal DBA/2 splenocytes or another DBA/2 tumor line (data not shown), thus establishing the specificity of the reaction.

Subsequent experiments were carried out to determine whether the anti-M-1 HP conjugate could be used in vivo to target and destroy developing tumors in mice. A series of experiments were carried out which involved a number of identical procedures. DBA/2 mice were shaved on their backs and injected subcutaneously about 1 cm above the tail with a known number of M-1 cells (numbers varied between 5×10^4 and 2×10^5 on different experiments). The anti-M-1 HP conjugates were administered intravenously to test animals (2.0 mg Mab: 0.268 mg HP per kilogram body weight). Depending on the experiment, mice were either exposed to light immediately after receiving HP conjugates or left in the dark for various lengths of time. The light intensity used was 22.5 mW/cm^2 as measured by a YSI Kettering model 65 radiometer. The light source was positioned 24 cm away from experimental animals. A standard 100 W GE 125 V light bulb emitting wavelengths between 200 and 1600 nm was used. Early experiments showed that while a small but significant delay was noted in appearance of tumors in mice treated with anti-M-1 HP in comparison to controls, no increase in survival was observed, and the effects were not impressive. The possibility was consid-

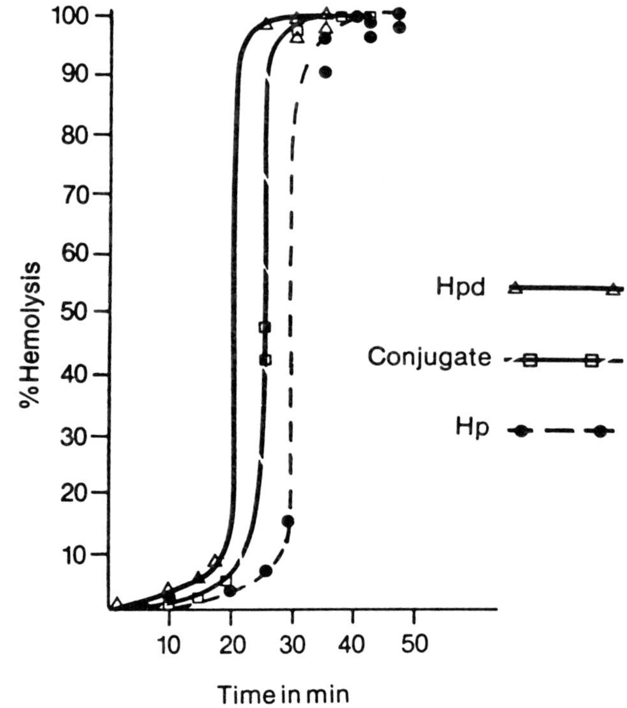

FIGURE 2. The hemolytic activity of free HP, HPD, or anti-M-1-HP conjugate. 10 μg of equivalent porphyrin concentrations of HP (•), HPD (Δ), or anti-M1-HP (□) were assayed on human RBC for hemolytic activity. Each point is the mean of triplicate samples in which SDs were <5.0%.

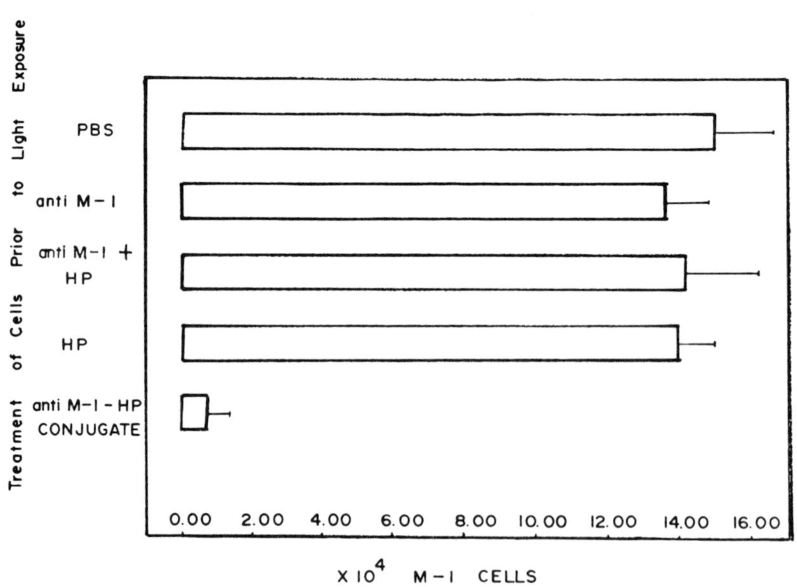

FIGURE 3. In vitro cytotoxicity of anti-M-1-HP conjugate and controls on M-1 tumor cells. Each bar represents the mean viable cell number ± standard deviation for six duplicates after conjugate and control treatment (10 μg Hp:40 μg Ab/5 × 10^5 cells) and light activation (22.5 mW/cm^2).

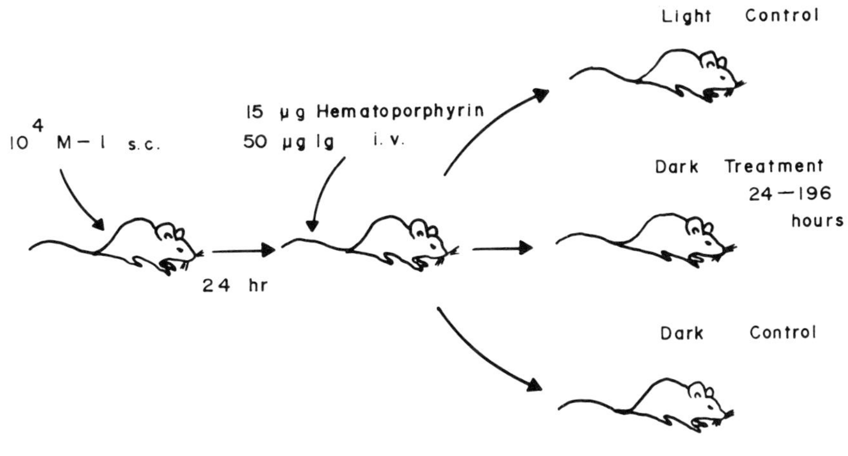

FIGURE 4. Experimental design for the in vivo effect of anti-M-1-HP in a timing experiment. Treated mice were exposed to white light (22.5 mW/cm²) for 4 hr.

Table 1
THE INFLUENCE OF TIMING OF DARK INCUBATION FOLLOWED BY LIGHT EXPOSURE OF M-1 BEARING DBA/2J MICE TREATED WITH ANTI-M-1 HP CONJUGATE ON THE TIME OF PALPABLE TUMOR INDUCTION AND SURVIVAL

No. of hr in darkness	N[a]	Mean tumor induction time ± Sx (days)	p^b	Mean survival time ± Sx (days)	p^b	% Survivors[c]
0	14	13.9 ± 1.8		31.9 ± 3.8		0
12	7	15.0 ± 3.3		32.7 ± 3.4		0
24	7	15.1 ± 3.2		33.0 ± 4.5		0
48	7	15.9 ± 3.1		34.0 ± 3.9		0
96	14	15.9 ± 3.2		35.6 ± 3.8	<0.01	7
120	7	16.3 ± 2.6	<0.025	34.7 ± 4.0		0
144	7	17.3 ± 2.3	<0.005	33.9 ± 4.2		14
168	8	16.3 ± 1.6	<0.025	36.0 ± 4.8	<0.025	25
192	14	18.7 ± 3.8	<0.005	39.7 ± 5.9	<0.005	30

Note: Drug concentrations used were 0.268 mg HP: 2.0 mg Ab/kg body weight.

[a] Number of experimental animals per group.
[b] Values were compared to the 0 time control group by student's *t* test.
[c] Survival designates those animals which remain free of tumor during the course of the experiments and for periods up to 6 months thereafter.

ered that HP activation was occurring as the conjugate circulated in the blood of the test animals since light penetration would probably be sufficient under the conditions used. Subsequent experiments involved maintenance of anti-M-1 HP treated animals in the dark for various lengths of time prior to light exposure. The experimental design is shown in Figure 4. The results (Table 1) were somewhat surprising. First, they clearly showed that anti-M-1 HP had a significant effect on the development and growth of the M-1 tumor, in that under the optimal conditions, some of the mice so treated failed to develop tumors at all. Further, it was clear that as the period during which animals were maintained in the dark increased, the mean time for development of palpable tumors increased significantly. Thus, conjugate-treated animals which were maintained under dark conditions for 120 to 192 hr prior to light treatment fared signifi-

Table 2
THE INFLUENCE OF TIMING OF DARK INCUBATION FOLLOWED BY LIGHT EXPOSURE OF M-1 BEARING DBA/2J MICE TREATED WITH HP, ANTI-M-1, OR PBS ON PALPABLE TUMOR INDUCTION

No. of hr in darkness	N[a]	Mean tumor induction time Sx (days)		
		HP	anti-M-1	PBS
0	5	8.0 ± 2.9	8.1 ± 2.0	8.1 ± 2.1
48	5	8.0 ± 2.9	N.D.	N.D.
96	10	8.2 ± 1.8	N.D.	N.D.
192	10	8.3 ± 1.8	8.2 ± 1.9	8.1 ± 2.0
>192[b]	10	7.9 ± 1.8	8.0 ± 2.9	8.2 ± 2.0

Note: Drug concentrations used were 0.268 mg HP per kilogram body weight and 2.0 mg Ab per kilogram body weight.

[a] Number of experimental animals per group.
[b] This group was maintained in the dark for the duration of the experiment.

cantly better than did other groups, and some of those in the 144 to 192 hr groups did not develop tumors (these animals were monitored over a 6-months period and were still tumor free at the end of that time). This effect was shown to be reproducible and specific for the anti-M-1 HP conjugate since control animals treated with equivalent dose levels of HP alone, anti-M-1, or PBS all developed tumors at times no different from control animals (Table 2). We also showed that light activation was critical to the tumoricidal effect, since animals treated with the conjugate and left in the dark all developed palpable tumors about 9 to 10 days after injection of M-1 cells. The specificity of the effect seen was also established since it was shown that the anti-M-1 HP conjugate had no effect on the growth of the EL4 tumor in B6D2 F animals (data not shown).

The findings in this study were surprising in a number of ways. In the first place, it was found that optimal targeting for Mab-HP conjugates to localize in tumor tissue apparently took almost 8 days. Only in those animals treated with the appropriate conjugate and maintained in the dark for this period of time prior to light activation, did significant numbers of "cures" occur. These observations were unexpected in two ways: in dark control animals, M-1 tumors became palpable on about day 9 or 10; thus the amount of tumor cell killing in "activated" animals must be appreciable; and, all previous studies on tumor cell targeting with Mab have "assumed" that optimal uptake by tumor must occur certainly within 72 hr after administration of the Mab. The result observed in these studies would indicate that this assumption may not be appropriate, at least in some systems. Further, continued examination of animals which remained tumor free after treatment showed no deleterious side effects. The tumor injection site, clearly visible due to shaving the surrounding area, showed no visible symptoms of tissue damage (erythema, edema, etc.). The actual dose levels of HP used in this study (0.268 mg/kg body weight) were significantly below (approximately 1/20 to 1/50) the dose levels used in reported studies using HP alone, and results indicate that levels used here may be below toxic side effects levels. One could thus argue that the use of Mab as targeting agents, conjugated to photochemicals such as HP may have a useful application in cancer therapy since they may enhance tumor localization of the drug while omitting unpleasant side effects.

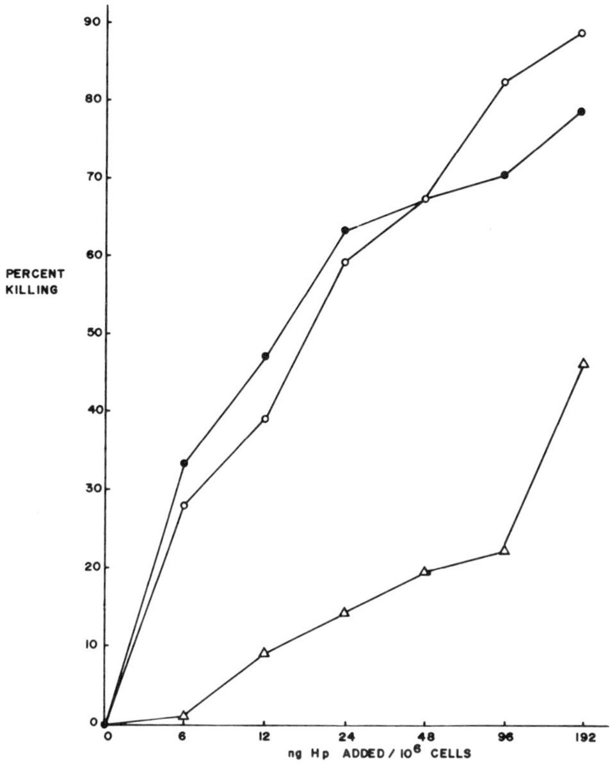

FIGURE 5. Activity of CAMAL-1 HP on WC-6. (○), positive control (HP-rabbit antihuman leukocyte serum; (•), HP-CAMAL-1; (△), negative control (HP-L1210 Mab).

B. Human TAA and Potential Use of Photoimmunotherapy

In the authors' laboratory, a leukemia-associated antigen (LAA), expressed at high frequency in or on cells of patients with either acute nonlymphocytic leukemia (ANLL) or chronic granulocytic leukemia (CGL) has been identified.[56] An Mab specific for this antigen (CAMAL) has been generated and is referred to as CAMAL-1.[57] The antigen can be identified on leukemia cells by immunofluorescence using either conventional antiserum[56] or CAMAL-1.[57] It can also be identified using indirect immunoperoxidase.[58] It was of interest to determine whether CAMAL-1 HP conjugates would be capable of delivering a lethal signal to cells expressing the CAMAL antigen. A cell line which expressed significant quantities of membrane-associated CAMAL was essentially tailor made for this study by fusing a selected murine cell line (WEHI-3B $HGPRT^-$)[59] with peripheral blood cells from a patient with CGL using standard polyethylene glycol procedures for fusion. (Fusion was performed by Vincent Lum of the authors' laboratory.) Human-mouse hybridomas which secreted the CAMAL antigen were selected by ELISA screening,[60] and one such stable cell line, WC-6, was selected for further study. Initial experiments were set up in which WC-6 cells were exposed under dark conditions to CAMAL-1 HP or equivalent levels of an irrelevant Mab-HP conjugate, HP alone, Mab alone, or HP plus Mab mixtures. Procedures were essentially those described in the previous section. Cells were washed, resuspended, and exposed to light. Viability was assessed, after treatment, by measurement of ^{51}Cr release, viable counts, or 18-hr ^3H-thymidine uptake of cultured surviving cells.[61] Representative results (Figure 5) show that the only significant levels of killing are achieved when the CAMAL-1 HP conjugate is used. This conjugate had no effect on the CA-

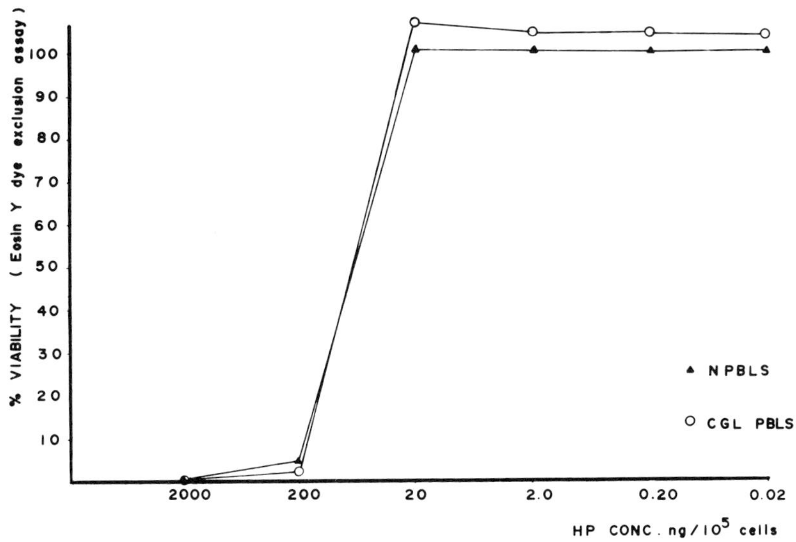

FIGURE 6. Titration of HP activity on normal and CGL peripheral blood leukocytes (PBLs). The photocytotoxic activity of HP was titrated on normal (▲) and CGL (○) PBLs. The viability of the treated cells was determined with 0.2% Eosin Y vital dye. Each point represents the mean of quadruplicate samples. The standard deviations were <5% in all cases.

MAL negative WEHI cell line nor was it effective when cells were left under dark conditions. These results established both the specificity and the sensitivity of the CAMAL-1 HP conjugate for cell killing under these conditions.

Since CAMAL-1 HP was capable of specifically reacting with and killing cell lines which express the CAMAL antigen, further experiments were carried out to determine whether such treatment would kill cells from patients with either ANLL or CGL. Ficoll-Hypaque separated cells from either bone marrow aspirates or peripheral blood from patients with myeloproliferative disease or other conditions were used. A number of HP-antibody conjugates were prepared according to previously described procedures.[55,61]

Free HP was titrated for nonspecific adherence, using the standard protocol, and it was established that levels between 2000 and 200 ng/10^5 cells/200 $\mu\ell$ were highly toxic to both normal and leukemic cells even after washing. However, levels of 20 ng/10^5 cells per 200 $\mu\ell$ had no apparent effects on viability (Figure 6). Further experiments were carried out using this dosage of HP, whether free or conjugated to Mab. Initial studies were carried out on normal peripheral blood leukocytes (PBLs). The results (Figure 7) showed that only the positive control (RαHu rabbit antiserum directed to normal human PBL antigens) caused significant loss of viability in these cells and that all other preparations had no effect. A series of experiments were subsequently carried out (the results are shown in Figures 8 and 9) in which it can be seen that while RαHu-HP kills essentially all cells in every case, CAMAL-1 HP specifically kills cells only from patients with CGL or ANLL. Other controls had no significant effect on cell viability. Serial experiments were run with two CGL patients, since their blood was available regularly over the period of time they were being treated on the cell separator. In all cases, only RαHu-HP and CAMAL-1 HP had any significant effect on viability. In these experiments, treated cells were cultured for a short time and viability was estimated by ^{3}H-thymidine uptake. A set of results is shown in Table 3, in which it can be seen that viability counts correlate well with subsequent viability of cell populations of cells as assessed by ^{3}H-thymidine uptake.

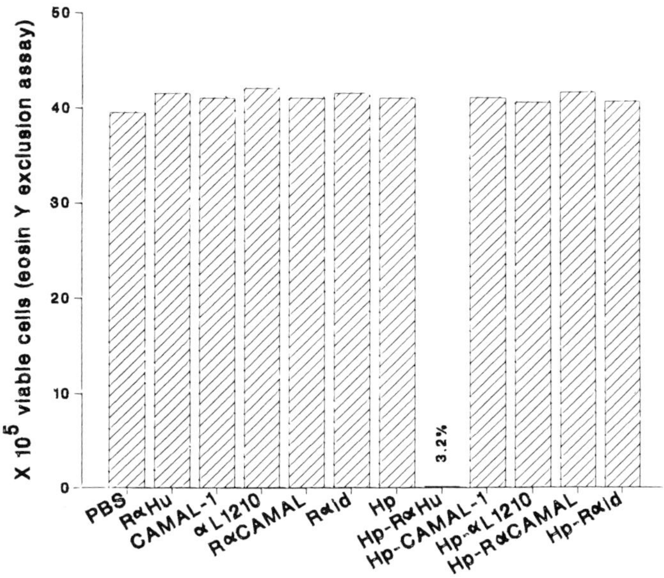

FIGURE 7. HP-antibody conjugate and control activity on normal PBLs. The phototoxic activity of different Hp-antibody conjugates and controls were tested on normal PBLs. The viability of the treated cells was assayed with 0.2% Eosin Y vital dye. Each bar represents the mean of duplicate samples. The standard deviations were <5%. The PBS control was taken as the 100% value.

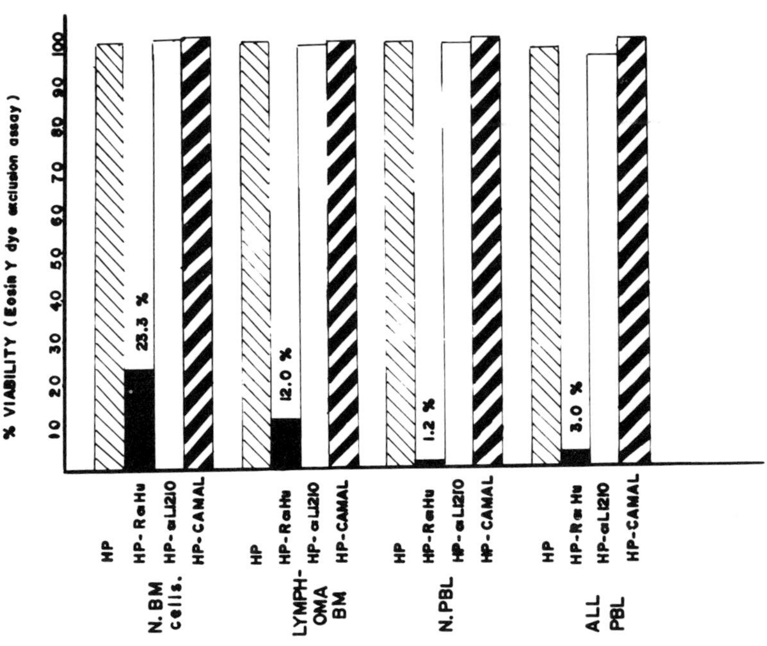

FIGURE 8. The effects of HP-antibody conjugates on normal and ANLL PBLs, normal, and lymphoma bone marrow cells. The photocytotoxic activity of HP ▨, Hp-RαHu ■, Hp-anti-L1210 □ and HP-CAMAL-1 ▨ was tested on four different irrelevant tumor or normal cell samples. Each bar is the mean of triplicate samples. Cell viability was measured with 0.2% Eosin Y vital dye. The viable cell number of the PBS control was used as the 100% value.

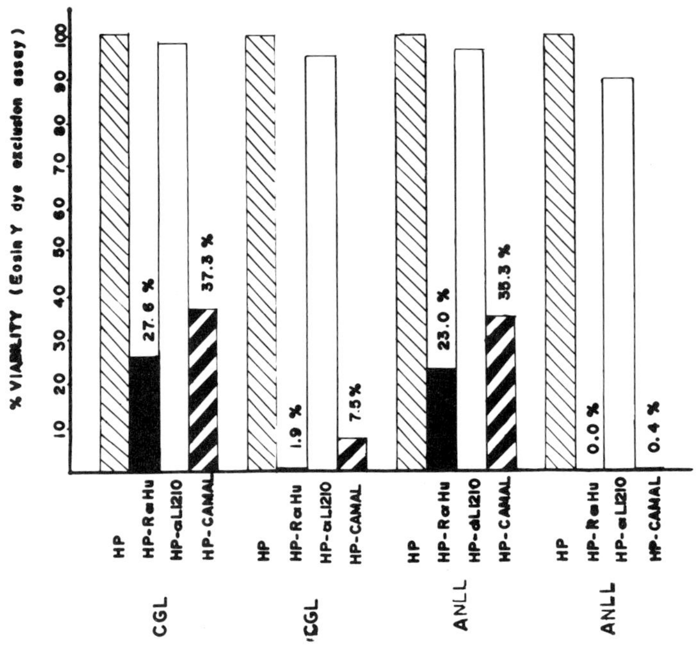

FIGURE 9. The effect of HP-antibody conjugates on CGL and ANLL samples from four different patients. The photocytotoxic activity of HP ▧, HP-RαHu ■, HP-anti-L1210 □, and HP-CAMAL-1 ▨ was tested on PBLs drawn from four leukemia patients. Cell viability was measured with 0.2% Eosin Y vital dye. Each bar is the mean of triplicate samples. The viable cell number of the PBS control was used as the 100% value.

Table 3
EFFECT OF HP-ANTIBODY CONJUGATES ON PBL SAMPLES FROM CGL PATIENT II

	No. (×10)	% viable	cpm (^{3}H)	% ^{3}H incorp.
PBS	252.90 ± 21.17	—	52,472,50 ± 5,690.53	
HP	288.67 ± 19.07	100.00	48,325.90 ± 4,601,51	92.10
HP-Rα Hu	5.53 ± 1.12	1.92	2,456.00 ± 1,267.69	4.68
Hp-CA-MAL 1	18.83 ± 0.15	7.47	7,156.83 ± 2,439.42	13.64
HP-M-1	242.67 ± 10.69	96.30	42,745.17 ± 5,665.96	81.46

While these data are admittedly preliminary, they do indicate that Mab-HP conjugates may be used as a reliable mechanism for destroying human tumor cells in vitro and possibly in vivo at lower dose levels than could be used effectively with HP or HPD alone.

III. PROSPECTS FOR PHOTOIMMUNOTHERAPY

In these present, very preliminary studies, the authors have attempted to investigate the possible advantages in the use of photochemicals conjugated to Mab as opposed to other cytotoxic agents or to photochemicals on their own. Theoretically this approach has a number of appealing features over other methods.

1. Coupling of photochemicals to Mab should increase their specificity for the desired target and therefore reduce dose requirements and possible side effects.
2. Because such a "photoimmunotoxin" is not cytotoxic until it is activated, its effects in vivo on organs which tend to accumulate Mab (spleen, lung, liver, kidneys) nonspecifically in the period immediately after their administration, should be minimal, and such material which will not ultimately find the desired target, will be eliminated during the early period of treatment.
3. Optimal time can be allowed to achieve maximum target accumulation of the Mab prior to activation with light, thus minimizing side effects and enhancing desired effects.
4. HP is a small molecule and it is apparent that a large number (40 to 60) can be linked to Mab without impairing antibody specificity.
5. The cytotoxic effect of HP can be delivered at the membrane level, thus eliminating the necessity for the Mab-HP conjugate to enter the cell.

In the murine model, reported on here, a number of possibly significant observations have been made. The authors found that the specific Mab (anti-M-1) conjugated to HP was able to kill, in vitro, M-1 tumor cells under conditions which HP on its own, HP plus anti-M-1, or an irrelevant HP-Mab conjugate could not, thus establishing (at least in this system) that the appropriate conjugate did bestow increased specificity and thus sensitivity for specific killing. Further, the in vivo studies revealed a number of interesting observations. The authors demonstrated that intravenously administered anti-M-1-HP could specifically "find" subcutaneously injected M-1 tumor cells, and that this effect was specific since effects were seen only in experimental groups so treated. This showed that the increased sensitivity demonstrated in vitro pertained in the in vivo situation. A second interesting observation made in this study related to optimal timing. It was of considerable surprise that the experimental animals left under "dark" conditions for 7 to 8 days fared better than animals which were exposed to light earlier during the experiments. These results imply that "homing" of Mab to tumor sites may take considerably longer than one would expect, but this may not be the case in all instances. It should be realized that tumor mass, in this system, becomes palpable at about day 9 and is thus small (but substantial in terms of cell numbers) on days 7 to 8. The apparent "lag" in effects may therefore be due to poor vascularization of the tumor during early phases of study rendering the nidus of tumor cells relatively inaccessible to circulating Mab until later. These possibilities are being investigated. However, this experimental flier does indicate that use of this kind of magic bullet may be of use in treatment. It is estimated that the as yet unpalpable tumor mass on day 8 of the M-1 experiments was in the range of 2 to 6×10^6 tumor cells. These experiments showed that there was sufficient antibody on the surface of these cells and sufficient light penetration under the experimental conditions used here, to eliminate essentially all these cells, since a significant number of these animals so treated (30%) remained free of tumor.

The experimental program reported on here involving human TAA was a study on the ability of another Mab, CAMAL-1, conjugated to HP to kill either a cell line expressing the antigen or cells from patients with CGL or ANLL. These studies showed unequivocally that CAMAL-1 HP showed specificity and sensitivity in both systems, which were not demonstrated by any of the control preparations. The possible applications of these findings will be discussed later.

Photochemicals, particularly HPD, are already proving to be useful tools in the treatment of some malignancies. It is possible that in conjunction with appropriate Mab, their application may be broader. It does not take a lot of imagination to speculate on the possibility that Mab-HP conjugates could be used, with appropriate laser technology, in the detection of *in situ* tumors of the lung, providing suitable Mab are developed for such tumors. In fact, HP alone has been reported to be of use for tumor localization in this disease.[6] In such a modality, it is possible that such a procedure could be used for both detection and treatment since again, photochemicals have been used alone for such treatments.[23]

While the Mab used by Levy et al. (CAMAL-1) has not yet been shown to be expressed on the malignant self-renewing cell in leukemia patients, this work is underway. If it or other appropriate Mab are capable of reacting with this population, then in vitro treatment of remission marrow cells from leukemia patients would provide a suitable method for bone marrow purging in candidates for autologous marrow transplantation.

Recent advances in the use of laser light sources have already been applied to the treatment of a variety of cancers. Using an argon laser, Rettenmaier et al.[62] and Dahlman et al.[63] have used HPD to treat cancers including gynecological, dermal breast carcinomas, squamous ENT tumors, and superclavicular nodes of colon carcinomas. The possibility exists that this type of therapy may be improved if specificity and sensitivity are enhanced with the use of Mab.

IV. OTHER APPLICATIONS IN PHOTOIMMUNOLOGY

On of the more interesting applications for the use of Mab conjugated to photochemicals such as HP may be in the definition and localization of cellular components within the cytoskeleton. It has already been demonstrated that subcellular localization of fluorescent compounds can be achieved using laser-stimulated fluorescence and fluorescence microscopy.[25] One of the more interesting areas of research utilizing this type of technology lies in the possibility that oncogene products and other subcellular components may be identified in normal or malignant cells using these sensitive procedures. Since Mab are currently available to some oncogene products such as *fps*,[64] such studies are feasible and may yield valuable information.

There is no doubt that the combination of Mab with photochemicals gives us a potentially powerful tool both at the clinically applied and at the basic investigative level.

REFERENCES

1. Figge, F. H. J., Weiland, G. S., and Manganiello, L. O. J., Cancer detection and therapy: affinity of neoplastic embryonic and traumatized regenerating tissues for porphyrins and metalloporphyrins, *Proc. Soc. Exp. Biol. Med.*, 68, 640, 1948.
2. Lipson, R., Baldes, E., and Olsen, A., The use of a derivative of hematoporphyrin in tumor detection, *J. Natl. Cancer Inst.*, 26, 1, 1961.
3. Barker, D. S., Heneerson, R. W., and Storey, E., The *in vivo* localization of porphyrins, *Br. J. Exp. Pathol.*, 51, 628, 1970.
4. Dougherty, T. J., Hematoporphyrin as a photosensitizer of tumors, *Photochem. Photobiol.*, 38, 377, 1983.
5. Sanderson, D. R., Fontana, R. S., Lipson, R. L., and Baldes, E. J., Hematoporphyrin as a diagnostic tool: a preliminary report of new techniques, *Cancer*, 30, 1368, 1972.
6. Doiron, D. R., Profio, E., Vincent, R. G., and Dougherty, T. J., Fluorescent bronchoscopy for detection of lung cancer, *Chest*, 76, 27, 1979.
7. Kelly, J. F. and Snell, M. E., Hematoporphyrin derivative: a possible aid in the diagnosis and therapy of carcinoma of the bladder, *J. Urol.*, 115, 150, 1976.
8. Kyriazis, G. A., Balin, H., and Lipson, R. L., Hematoporphyrin derivative fluorescence test colposcopy and colpophotography in the diagnosis of atypical metaplasia, desplasia, and carcinoma *in situ* of the cervix uteri, *Am. J. Obstet. Gynecol.*, 117, 375, 1973.

9. Spikes, J. D., Porphyrins and related compounds as photodynamic sensitizers, *Ann. N. Y. Acad. Sci.*, 176, 496, 1975.
10. Bellnier, D. A. and Dougherty, T. J., Membrane lysis in Chinese hamster ovary cells treated with hematoporphyrin derivative plus light, *Photochem. Photobiol.*, 36, 43, 1982.
11. Kvello Stenstrom, A. G., Moan, J., Brumborg, G., and Eklund, T., Photodynamic inactivation of yeast cells sensitized by hematoporphyrin, *Photochem. Photobiol.*, 32, 349, 1982.
12. Weishaupt, K. R., Gomer, C. J., and Dougherty, T. J., Identification of singlet oxygen as the cytotoxic agent in photo-inactivation of a murine tumor, *Cancer Res.*, 36, 2326, 1976.
13. Kohn, K. and Dessel, D., On the mode of cytotoxic action of photoactivated porphyrins, *Biochem. Pharmacol.*, 28, 2465, 1979.
14. Lamola, A. A. and Doleiden, F. H.,Cross-linking of membrane proteins and protoporphyrin-sensitized photohemolysis, *Photochem. Photobiol.*, 31, 597, 1980.
15. Canti, G., Marelli, O., Ricci, L., and Nicolin, A., Haematoporphyrin treated murine lymphocytes: *in vitro* inhibition of DNA synthesis and light-mediated in activation of cells responsible for GVHR, *Photochem. Photobiol.*, 34, 589, 1981.
16. Moan, J., McGhie, J., and Jacobsen, P. B., Photodynamic effects on cells *in vitro* exposed to hematoporphyrin derivative and light, *Photochem. Photobiol.*, 37, 599, 1983.
17. Diamond, I., Granelli, S. G., McDonagh, A. F., Nielsen, S., Wilson, C. B., and Jaenicke, R., Photodynamic therapy of malignant tumors, *Lancet*, 2, 1175, 1972.
18. Granelli, S. G., Diamond, I., McDonagh, A. F., Wilson, C. B., and Nielsen, S. L., Photochemotherapy of glioma cells by visible light and hematoporphyrin, *Cancer Res.*, 35, 2357, 1975.
19. Dougherty, T. J., Boyle, D. G., Weishaupt, K. R., Henderson, B. W., Potter, W. R., Bellnier, D. A., and Wityk, K. E., in *Porphyrin Photosensititization*, Kessel, D. and Dougherty, T. J., Eds., Plenum Press, New York, 1983, 3.
20. Moan, J., Christensen, T., and Sommer, S., The main photosensitizing components of hematoporphyrin derivative, *Cancer Lett.*, 15, 161, 1982.
21. Dougherty, T. J., Grindey, G. B., Fiel, R., Weishaupt, K. R., and Boyle, D. G., Photoradiation therapy II: cure of animal tumors with hematoporphyrin and light, *J. Natl. Cancer Inst.*, 55, 115, 1975.
22. Kaufman, J., Dougherty, T. J., Goldfarb, A., Johnson, R., Didolkar, N., Creaven, P., Weishaupt, K., and Boyle, D., Photoradiation of malignant tumors, *Proc. Am. Soc. Clin. Oncol.*, 18, 275, 1977.
23. Hayata, Y., Harubami, K., Konaka, C., Onon, J., and Takizawa, N., Photoporphyrin derivative and laser photoradiation in the treatment of lung cancer, *Chest*, 81, 269, 1977.
24. Ha, X.-W., Sun, S.-M., Xie, J.-G., Fan, X.-J., Zhang, Y.-H., Mei, Q.-C., Shen, H., and Xu, S.-L., Clinical use of hematoporphyrin derivative in malignant tumors, *Chin. Med. J.*, 96, 734, 1983.
25. Berns, M. W., Dahlman, A., Johnson, F. M., Burns, R., Sperling, D., Guiltman, M., Siemins, A., Walter, R., Wright, W., Hammer-Wilsoin, M., and Wile, A., *In vitro* cellular effects of hematoporphyrin derivative, *Cancer Res.*, 42, 2325, 1982.
26. Laszlo, J., Buckley, C. E., and Amos, D. B., Infusion of isologous immune plasma in chronic lymphocytic leukemia, *Blood*, 31, 104, 1968.
27. Fass, L., Herberman, R. B., and Ziegler, J., Evaluation of effects of remission plasma on untreated patients with Burkitt's lymphoma, *J. Natl. Cancer Inst.*, 44, 145, 1970.
28. DeCarvalho, S., Preliminary experimentation with specific immunotherapy of neoplastic disease in man. I. Immediate effects of hyperimmune equine gamma globulins, *Cancer*, 16, 306, 1963.
29. Kohler, G. and Milstein, C., Continuous cultures of fused cells secreting antibody of predefined specificity, *Nature (London)*, 256, 495, 1975.
30. Foon, K. A., Bernhard, M. I., and Oldham, R. K., Monoclonal antibody therapy: assessment of animal tumor models, *J. Biol. Resp. Modif.*, 1, 277, 1982.
31. Ritz, J. and Schlossman, S. F., Utilization of monoclonal antibodies in the treatment of leukemia and lymphoma, *Blood*, 59, 1, 1982.
32. Miller, R. A. and Levy, R., Response of cutaneous T cell lymphomas to therapy with hybridoma monoclonal antibody, *Lancet*, 2, 226, 1981.
33. Miller, R. A., Maloney, D. G., Warnicki, E. R., and Levy, R., Treatment of B cell lymphoma with monoclonal anti-idiotype antibody, *N. Engl. J. Med.*, 306, 517, 1982.
34. Moolten, F. L., Schreiber, B. M., and Zahdel, S. H., Antibodies conjugated to potent cytotoxin as specific antitumor agents, *Immunol. Rev.*, 62, 47, 1982.
35. Youle, R. J. and Neviele, D. M., Anti-thy 1.2 monoclonal antibody linked to ricin is a potent cell-type-specific toxin, *Proc. Natl. Acad. Sci. U.S.A.*, 77, 483, 1980.
36. Raso, V., Ritz, J., Basala, M., and Schlossman, S. F., Monoclonal antibody-Ricin A chain conjugate selectivity cytotoxic for cells bearing the common acute lymphoblastic leukemia antigen, *Cancer Res.*, 42, 457, 1982.
37. Seon, B. K., Specific killing of human T-leukemia cells by immunotoxins prepared with Ricin A chain and monoclonal anti-human T cell leukemia antibodies, *Cancer Res.*, 44, 259, 1984.

38. Bernhard, M. I., Foon, K. A., Deltmann, T. N., Key, M. E., Hang, K. M., Clarke, G. C., Christensen, W. L., Hayer, L. C., Hanna, M. G., and Oldham, R. K., Guinea pig line 10 hepatocarcinoma model: characterization of monoclonal antibody and *in vivo* effect of unconjugated antibody and antibody conjugated to diphtheria toxin A chain, *Cancer Res.*, 43, 4420, 1983.
39. Ross, W. C. J., Thorpe, P. E., Cumber, A. J., Edwards, D. C., Hinson, C. A., and Davies, A. J. S., Increased toxicity of diphtheria toxin for human lymphoblastoid cells following covalent linkage to anti-(human lymphocyte) globulin or its $F(ab')_2$ fragment, *Eur. J. Biochem.*, 104, 381, 1980.
40. Thorpe, P. E., Cumber, A. J., Williams, N., Edwards, D. E., Ross, W. C. J., and Davies, A. J. S., Abrogation of the non-specific toxicity of abrin conjugated to anti-lymphocyte globulin, *Clin. Exp. Immunol.*, 43, 195, 1981.
41. Thorpe, P. E., Brown, A. N. F., Ross, W. C. J., Cumber, A. J., Detre, S. I., Edwards, D. C., Davies, A. J. S., and Stirpe, F., Cytotoxicity acquired by conjugation of an Anti-Thy 1.1 monoclonal antibody and the ribosome-inactivating protein gelonin, *Eur. J. Biochem.*, 116, 447, 1981.
42. Moolten, F. L., Genetic mosaicism and the control of cancer, *Med. Hypothesis*, 2, 79, 1976.
43. Flickinger, R. A. and Trost, S. R., Cytotoxicity of antibody-phospholypase c conjugates on cultued friend leukemia cells, *Eur. J. Cancer*, 12, 159, 1976.
44. Moolten, F. L., Schreiker, B. M., and Zajdel, S. H., Antibodies cnjugated to potent cytotoxin as specific antitumor agents, *Immunol. Rev.*, 62, 47, 1982.
45. Calendi, E., Constanzi, G., Indweri, F., Lotti, G., and Zini, C., Histoimmunologic specificity of an anti-lymphoid tissue sarcoma gamma globulin bound methotrexate, *Bull. Chin. Pharmacol.*, 108, 25, 1969.
46. Ghose, T., Norwell, S. T., Guclu, A., and MacDonald, A. S., Immunochemotherapy of human malignant melanoma with chlorambucil-carrying antibody, *Br. Med. J.*, 3, 495, 1975.
47. Hurwitz, E., Levy, R., Maron, R., Wilcher, M., Arnon, R., and Sela, M., The covalent binding of daunomycin and adriamycin to antibodies with retention of both drug and antibody activities, *Cancer Res.*, 35, 1175, 1975.
48. Moshakis, V., McIlhenney, R. A. J., Raghaven, D., and Neville, A. M., Monoclonal antibodies to detect human tumors: an experimental approach, *J. Clin. Pathol.*,34, 314, 1981.
49. Ballou, B., Levine, G., Hakala, T. R., and Solter, D., Tumor location detected with radioactively labelled monoclonal antibody and external scintigraphy, *Science*, 206, 844, 1979.
50. Bernhard, M. I., Hwang, K. M., Foon, K. A., Keenan, A. M., Kessler, R. M., Frinkke, J. M., Tallam, D. H., Hanna, M. G., Peters, L., and Oldham, R. K., Localization of ^{111}In- and ^{125}I-labelled monoclonal antibody in guinea pigs bearing line 10 hepatocarcinoma tumors, *Cancer Res.*, 43, 4429, 1983.
51. Scheinberg, D. A., Strand, M., and Ganscow, O. A., Tumorimaging with radioactive metal chelates conjugated to monoclonal antibodies, *Science*, 215, 1511, 1982.
52. Mew, D., Alcorn, D., Lyster, D., Rhodes, R. D., and Levy, J. G., Imaging studies with Tc-99m labeled tumor specific monoclonal antibodies to a transplantable murine tumor, in Proc. 2nd Int. Symp. Radioimmune Imaging, 1982, 265.
53. Whitney, R. B. and Levy, J. G., Effects of sera from tumor-bearing mice on mitogen and allogeneic cell stimulation of normal lymphoid cells, *J. Natl. Cancer Inst.*, 54, 733, 1975.
54. Oi, V. T. and Herzenber, L. A., Immunoglobulin producing hybrid cell lines, in *Selected Methods in Cellular Immunology*, Mishell, B. B. and Shügi, S. M., Eds., W. H. Freeman, San Francisco, 1980, 351.
55. Mew, D., Wat, C.-K., Towers, G. H. N., and Levy, J. G., Photoimmunotherapy: treatment of animal tumors with tumor-specific monoclonal antibody-hematoporphyrin conjugates, *J. Immunol.*, 130, 1973, 1983.
56. Malcolm, A. J., Shipman, R. C., and Levy, J. G., Detection of a tumor-associated antigen on the surface of human myelogenous leukemia cells, *J. Immunol.*, 128, 2599, 1982.
57. Malcolm, A. J., Shipman, R. C., Logan, P. M., and Levy, J. G., A monoclonal antibody to myelogenous leukemia: isolation and characterization, *Exp. Hematol.*, 12, 539, 1984.
58. Logan, P. M., Malcolm, A. J., and Levy, J. G., Detection of a common leukemia associated antigen in human myelogenous leukemia using immunoperoxidase, *Diagn. Immunol.*, 2, 86, 1984.
59. Van Kessel, A. H. M. G., Tetteroo, P. A. T., von dem Borne, A. E. G. Kr., Hagemeijer, A., and Bootsma, D., Expression of human myeloid-associated surface antigens in human-mouse myeloid cell hybrids, *Proc. Natl. Acad. Sci. U.S.A.*, 80, 3748, 1983.
60. Lum, V. and Levy, J. G., Production of a human-mouse hybrid cell line which secretes a luekemia-associated antigen, *J. Natl. Cancer Inst.*, submitted for publication, 1985.
61. Mew, D., Lum, V., Wat, C.-K., Towers, G. H. N., and Levy, J. G., Killing of cell lines and cells from patients with myelogenous leukemia using a monoclonal antibody (CAMAL-1)-hematoporphyrin conjugate, *J. Natl. Cancer Inst.*, submitted for publication, 1984.
62. Rettenmaier, M. A., Berman, M. L., DiSaia, P. J., Burns, R. G., and Berns, M. W., Photoradiation therapy of gynecological malignancies, *Cancer Res.*, 43, 642, 1983.

63. Dahlman, A., Wile, A. G., Burns, R. G., Mason, G. R., Johnson, F. M., and Berns, M. W., Laser photoradiation therapy of cancer, *Cancer Res.*, 43, 430, 1983.
64. Ingman-Baker, J., Hinze, E., Levy, J. G., and Pawson, A.,Monoclonal antibodies to the transforming protein of Fujinami aviansarcoma virus discriminates between different *fps*-encoded proteins, *J. Virol.*, 35, 572, 1984.

INDEX

A

B

C

I

K

L

M

N

O

P

Q

R

S

T

U

V

W

X

Y

Photomedicine

Volume III

Editors

Ehud Ben-Hur, Ph.D.
Department of Radiobiology
Nuclear Research Center-Negev
Beer-Sheva, Israel

Ionel Rosenthal, Ph.D.
Head
Division of Food Technology
Agricultural Research Organization, Volcani Center
Bet Dagan, Israel

CRC Press, Inc.
Boca Raton, Florida

Library of Congress-in-Publication Data

Photomedicine.

Includes bibliographies and index.
1. Phototherapy. 2. Photochemotherapy. 3. Light--
Physiological effect. I. Ben-Hur, Ehud. II. Rosenthal,
Ionel. [DNLM: 1. Phototherapy. WB 480 P5745]
RM837.P48 1987 615.8'3 86-34293
ISBN-0-8493-4673-8 (set)

Direct all inquiries to CRC Press, Inc., 2000 Corporate Blvd., N.W., Boca Raton, Florida, 33431.

International Standard Book Number 0-8493-4674-6 (v. 1)
International Standard Book Number 0-8493-4675-4 (v. 2)
International Standard Book Number 0-8493-4676-2 (v. 3)
International Standard Book Number 0-8493-4673-8 (set)

Library of Congress Card Number 86-34293
Printed in the United States

INTRODUCTION

Go outside and play in the sun. It's good for you.

My mother

The use of sunlight and drugs for the treatment of skin diseases has been documented for over 3400 years; for an even longer time, the reddening, blistering, and tanning effects of sunlight have probably been known. With the discovery of lasers a new dimension was added in the study and application of light in medical therapy. Ophthalmologists adopted the laser in the clinic as a photocoagulator for the treatment of detached retina, and the use of laser as a scalpel for noncontact, noninvasive, and even subcellular surgery is at an earlier state of acceptance. In addition to surgical uses, new, promising ideas are continuing to emerge. Thus, laser can be used to diagnose and treat malignant tumors using photoradiation therapy. This renewed interest, stimulated by the mutual interplay of both scientific and technological innovations, is characterized by a multidisciplinary approach involving physicists, chemists, biochemists, and physicians.

Our objective has been to collect in these three volumes the most up-to-date assessment of our understanding of light in medicine. Since *Photomedicine* was defined as an informative guide to practical applications rather than an esoteric study of medical discipline, the level of medical rigor was reasonably relaxed.

Given limitations on length, the chapters are not intended to be all embracing reviews of the field, but rather to present an overview of key ideas and directions with the objective of delineating the most promising and exciting problems. We hope that the text is sufficiently introductory to stimulate the curiosity and interest of a neophyte, and to simultaneously provide the specialist with a rather short, but current summary of the status of this field. Most important, we hope that the volumes will further highlight this rapidly developing science and spur current and new researchers and ideas.

Ehud Ben-Hur
Ionel Rosenthal

THE EDITORS

Ehud Ben-Hur, Ph.D., was born in Israel in 1940. After graduation from the Hebrew University of Jerusalem in 1965, he went on to study biochemistry at the Technion, Israel Institute of Technology at Haifa, where he obtained his M.Sc. and doctorate degrees. He then joined the Biology Department of Brookhaven National Laboratory as Research Associate where he completed postdoctoral work on the radiobiology of cultured mammalian cells under the auspices of Dr. M. M. Elkind. Upon returning to Israel in 1973, he first joined the Department of Cellular Biochemistry at the Hebrew University and then the Nuclear Research Center-Negev, in 1975, where he is currently engaged in studies of biological effects of ionizing and nonionizing radiations.

The main thrust of his research activity in the past was related to radiation-induced damage in DNA and its repair. During the last few years he has become interested in photodynamic therapy of cancer and is actively involved with Dr. I. Rosenthal in developing new and improved photosensitizers for this purpose.

Dr. Ben-Hur is affiliated with the Department of Radiation Biology, Colorado State University. He is also affiliated with Ben-Gurion University, Beer-Sheva, Israel, where he teaches photobiology. Dr. Ben-Hur has published over 80 papers in scientific journals, is a member of the American Society for Photobiology and the Radiation Research Societies of both the U.S. and Israel, and is on the Editorial Board of the *International Journal of Radiation Biology.*

Dr. Ben-Hur is married with two children and lives most of the time in Beer-Sheva.

Ionel Rosenthal, Ph.D., received his degree in Chemical Engineering from the Polytechnic Institute in Bucharest (Romania) and Ph.D. degree from the Freinberg Graduate School of the Weizmann Institute of Science, Rehovoth, Israel. Dr. Rosenthal has had a very colorful professional career which has included Plant Engineer at "Mahteshim" Chemical Co. and Senior Scientist at the Department of Organic Chemistry at the Weizmann Institute of Science and at the Department of Organic Chemistry, Nuclear Research Center-Negev. Currently he is Principal Scientist at the Department of Food Science, Agricultural Research Organization, Bet-Dagan, and Professor in the Department of Agricultural Biochemistry at the Faculty of Agriculture of the Hebrew University, Jerusalem.

His scientific interests in organic photobiochemistry and food chemistry (and its spin-off: cookery) have resulted in more than 100 research publications in these areas.

ACKNOWLEDGMENTS

The editors wish to acknowledge their indebtedness to the authors for their ready response to our request to contribute, and for their kindness and understanding in accepting their editorial efforts.

CONTRIBUTORS

Esther Azizi, M.D.
Senior Dermatologist
Department of Dermatology
Sheba Medical Center
Tel-Hashomer, Israel

Roger W. Barnes, B.S.
Assistant to the Director
Office of Device Evaluation
U.S. Food and Drug Administration
Silver Springs, Maryland

Ehud Ben-Hur, Ph.D.
Department of Radiobiology
Nuclear Research Center-Negev
Beer-Sheva, Israel

Jeffrey D. Bernhard, M.D.
Director, Division of Dermatology
Director, Phototherapy Center
Assistant Professor
Department of Medicine
University of Massachusetts Medical Center
Worcester, Massachusetts

Homer S. Black, Ph.D.
Director
Photobiology Laboratory
Veterans Administration Medical Center
Baylor College of Medicine
Houston, Texas

Larry E. Bockstahler
Division of Life Sciences
Optical Radiation Branch
U.S. Food and Drug Administration
Rockville, Maryland

S. G. Bown
Department of Surgery
University College London
London, England

John A. S. Carruth
Senior Lecturer in Otolaryngology
Southampton University Hospitals
Southampton, England

James E. Cleaver, Ph.D.
Professor
Laboratory of Radiobiology and Environmental Health
University of California
San Francisco, California

Farrington Daniels, Jr., M.D., M.P.H.
Emeritus Professor of Medicine
The New York Hospital-Cornell Medical Center
New York, New York

John H. Epstein, M.D.
Clinical Professor of Dermatology
University of California Medical School
San Francisco, California

I. Farine, M.D.
Professor and Head
Orthopedic Department
Sheba Medical Center
Tel-Hashomer, Israel

Richard P. Felten
Department of Health and Human Services
Division of Life Sciences
U.S. Food and Drug Administration
Rockville, Maryland

Anna Flint, Ph.D.
Department of Dermatology
Sheba Medical Center
Tel-Hashomer, Israel

Orna Geyer, M.D.
Department of Ophthalmology
Tel Aviv Medical Center
Tel Aviv, Israel

Barbara A. Gilchrest, M.D.
Professor and Chairman
Department of Dermatology
Boston University School of Medicine
Boston, Massachusetts

Philip C. Hanawalt, Ph.D.
Professor and Chairman
Department of Biological Sciences
Stanford University
Stanford, California

M. Heim, M.D.
Department of Rehabilitation
Sheba Medical Center
Tel-Hashomer, Israel

Kiki B. Hellman, Ph.D.
Senior Scientist
Division of Life Sciences
U.S. Food and Drug Administration
Rockville, Maryland

Victoria M. Hitchins, Ph.D.
Division of Life Sciences
Optical Radiation Branch
U.S. Food and Drug Administration
Rockville, Maryland

H. Horoszowski, M.D.
Head
Department of Orthopedy
Sheba Medical Center
Tel-Hashomer, Israel

Elizabeth D. Jacobson, Ph.D.
Acting Director
Office of Science and Technology
Center for Devices and Radiological Health
U.S. Food and Drug Administration
Rockville, Maryland

W. Patrick Jeeves, Ph.D.
Department of Medical Physics
Ontario Cancer Treatment and Research Foundation
Hamilton, Ontario, Canada

Anthony Lamanna
Division of Life Sciences
Optical Radiation Branch
U.S. Food and Drug Administration
Rockville, Maryland

Robert J. Landry, Ph.D.
Electrooptics Branch
U.S. Food and Drug Administration
Rockville, Maryland

Moshe Lazar, M.D.
Department of Ophthalmology
Ichilov Hospital
Tel-Aviv, Israel

Sidney Lerman, Ph.D.
Professor
Department of Opthalmology
Emory University
Atlanta, Georgia

Jerome I. Levine, Ph.D.
Division of Life Sciences
Optical Radiation Branch
U.S. Food and Drug Administration
Rockville, Maryland

Julia G. Levy, Ph.D.
Professor
Department of Microbiology
University of British Columbia
Vancouver, Canada

C. David Lytle, Ph.D.
Division of Life Sciences
Optical Radiation Branch
U.S. Food and Drug Administration
Rockville, Maryland

I. A. Magnus, M.D., F.R.C.P.
Professor Emeritus
Photobiology Department
Institute of Dermatology
London, England

Micheline M. Mathews-Roth, M.D.
Associate Professor of Medicine
Channing Laboratory
Harvard Medical School
Boston, Massachusetts

Daphne Mew, Ph.D.
Faculty of Medicine
Foothill Hospital
University of Calgary
Calgary, Alberta, Canada

K. Mohan, Ph.D.
Science & Technology
U.S. Food and Drug Administration
Rockville, Maryland

Volume III

Chapter 1

THE PHTHALOCYANINES: SENSITIZERS WITH POTENTIAL FOR PHOTODYNAMIC THERAPY OF CANCER

E. Ben-Hur, I. Rosenthal, S. G. Bown, and D. Phillips

TABLE OF CONTENTS

I. INTRODUCTION

Photodynamic inactivation of biological systems by exposure to certain dyes and visible light has been studied since the beginning of this century.[1] Oxygen is required for these photochemical reactions, which proceed either by direct interaction of the electronically excited dye molecule with a cellular target, followed by reaction of the transients formed with molecular oxygen (type I), or via interaction of the dye in its excited triplet state with oxygen and generation of active oxygen species (type II).[2] Clinical application of this process was prompted by the observation that malignant tumors can retain porphyrin derivatives for longer time than adjacent normal tissues. Subsequent exposure to light of proper wavelength can cause eradication of the tumor. This modality, termed photodynamic therapy, uses primarily hematoporphyrin derivative (HPD) as photosensitizer.[3] Although the main absorbtion band of HPD is around 400 nm, for therapy the dye is usually activated by red light (λ = 630 nm) where only a minor absorption peak exists, because of the increased transparency of tissues in the red.

HPD is a complex mixture of porphyrins of a somewhat variable composition derived from hematoporphyrin by treatment with a mixture of sulfuric acid and acetic acid (1:19).[4,5] This chemical reaction leads to formation of a large number of derivatives, dependent on the reaction conditions, which have been partially characterized as the dimethyl esters, dicarboxylic acids, and dehydro derivatives. The uncertainty of the active constituents in HPD and its chemical lability, may create difficulties in ensuring reproducibility in its use.

The potential advantage of phototherapy, of total and exclusive eradication of tumors without damage to surrounding tissues even in inoperable cases such as blockage of airways, justifies the search for alternative sensitizers. Ideally such a compound should (1) show a preferential retention/affinity for malignant tumors, (2) possess a very low toxicity, and (3) exhibit an efficient photodynamic effect activated by light of wavelength longer than 600 nm.

Metallophthalocyanines are porphyrin-like compounds (Figure 1) that absorb strongly in the red (the Q band, 600 to 700 nm, $\varepsilon \sim 10^5 \ell$/mol/cm). These dyes have important technological applications as catalysts, dyes, and pigments. There has also been a considerable interest in the photophysical properties of phthalocyanines since the first description of their photoconduction and semiconductor properties,[6] and their use as laser dyes.[7] The mechanistic photochemistry of several phthalocyanines has been investigated in recent years.[8-13]

The phthalocyanines can be easily synthesized and purified, and exhibit a very unusual chemical stability. Phthalocyanine derivatives are reported to be nontoxic. Thus, salts of copper phthalocyanine sulfonic acid are practically nontoxic to various species, from protozoa to dogs, in doses up to 100 mg/kg. Furthermore, this latter dye was reported to be retained preferentially in experimentally produced intracranial neoplasms in mice.[14] Similarly, uranyl tetrasulfonate phthalocyanine was shown to accumulate in brain tumors,[15] and (^{99}Tc) tetrasulfophthalocyanine concentrated to some extent in mammary adenocarcinoma in rats.[16] These observations make phthalocyanine derivates most promising photosensitizers for use in photodynamic therapy.[17] However, in spite of their structural resemblance to the biologically important porphyrins, the photobiology of phthalocyanines barely has been studied.

This chapter describes the photobiological activity of phthalocyanines in vitro and in vivo and indicates the potential for photodynamic therapy.

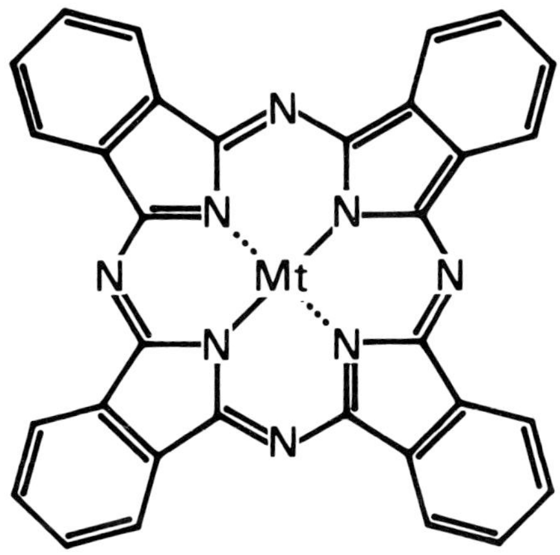

FIGURE 1. The molecular structure of metallophthalocyanine.

II. PHOTOBIOLOGY OF PHTHALOCYANINES IN VITRO

A. Photosensitized Killing of Cultured Cells

The initial observation that phthalocyanines can inactivate cultured mammalian cells following exposure to visible light was performed on Chinese hamster fibroblasts.[17] It was very early recognized that the metal atom complexed with the phthalocyanine ring was crucial for photobiological activity.[18] The screening assay of 11 different phthalocyanines indicated that aluminum phthalocyanine (AlPC) was most photobiologically active, followed by zinc derivative which was only one fifth as active under similar experimental conditions. Therefore, most of the experimental data were collected employing either AlPC or its water-soluble derivative containing sulfonic acid groups (AlPCS). Thus, exposure of Chinese hamster cells to white fluorescent light following an overnight incubation with AlPC causes exponential cell killing which is preceded by a pronounced shoulder (Figure 2). The shoulder on the survival curve was reduced, and the final slope became steeper when the concentration of AlPC in the growth medium prior to light exposure was increased. The cells became progressively more sensitive to light as a function of incubation time with the dye, reaching a maximum at about 3 hr for AlPC (Figure 3). The loss of sensitivity of the cells containing the dye, upon incubation in dye-free medium, is more rapid. This loss did not occur when incubation was carried out in medium or buffer devoid of blood serum. This is most probably due to the very low solubility of AlPC in water. In the growth medium, AlPC tightly binds to serum proteins,[19] which serve as a vehicle for its transport into and out of the cell.

The kinetics of photosensitization by AlPCS are much slower than those of AlPC, and saturation does not take place even after overnight incubation. The reverse process, loss of photosensitivity, is also slower with AlPCS as compared to AlPC.

The concentration dependence of photosensitivity (expressed as $1/F_{10}$, the reciprocal of the fluence required to reduce survival to 10%) induced by the two aluminum phthalocyanines is shown in Figure 4. While the response to both derivatives is linear over a wide range of concentrations, the slope of the AlPC curve is steeper and displaced upward, indicating a higher potency for the same concentration. While some of this difference is due to the fact that after 16-hr incubation, the effect of AlPCS has not yet reached a maximum (Figure 3), other factors might play a role. For example, their different solubilities could lead to AlPCS localization at intrinsically less-sensitive subcellular targets. The possibility that the two phthalocyanines have different photo-

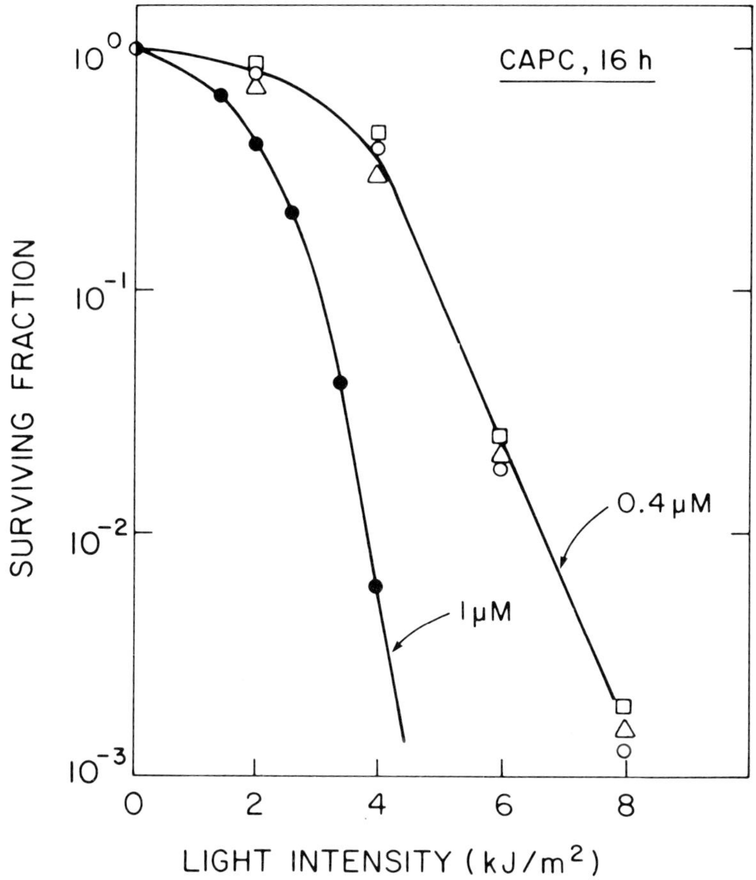

FIGURE 2. Survival of Chinese hamster cells exposed for 16 hr to 0.4 or 1 μM AlPC, as indicated, follwed by light exposure. Squares denote exposure to light in PBS containing 1 M glycerol, triangles are for PBS in D_2O, and circles are for PBS in H_2O.

reactivities is less likely since the photophysics of phthalocyanines is not expected to be affected by the substitution with sulfonic acid residues on the benzene rings.

B. Factors Affecting Photosensitization by Phthalocyanines

Cell physiology is an important factor that can determine the sensitivity of the cell to cytotoxic agents. The position of the cell in the cell cycle is crucial in this respect, and the response to UV and ionizing radiation displays a pronounced age structure (see Chapter 2). Thus, Chinese hamster cells are most sensitive to X-irradiation at the G_1/S border and are most resistant in late S phase. Conversely, the photosensitization by AlPC is equally efficient throughout the cell cycle, and the age-response function is virtually flat (Figure 5).

The external interference with the cellular processes operating, postphotoexcitation, to repair the photodamage, could result in enhanced sensitivity. In preliminary experiments, the production of DNA strand breakage was observed using the sensitive alkali elution technique. The yield of these breaks was about half of that produced by X-irradiation, at equisurvival doses. Such DNA breaks stimulate the synthesis of poly(ADP-ribose), which is involved in their repair. Inhibition of poly(ADP-ribose) polymerase, e.g., by 3-aminobenzamide (3-ABA), following exposure to ionizing ra-

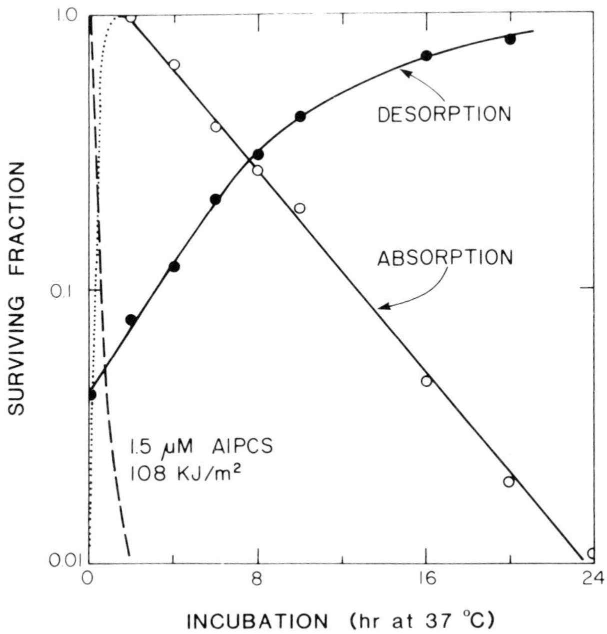

FIGURE 3. Kinetics of photosensitizaiton by phthalocyanines. Solid circles denote the development of photosensitivity as a function of time in the presence of 1.5 μM AlPCS followed by exposure to 108 kJ/m². The slashed curve is for 0.4 μM AlPC followed by 8 kJ/m². Open circles show the disappearance of photosensitivity when cells incubated for 16 hr with 1.5 μM AlPCS are rinsed and exposed to 108 kJ/m² after various incubation times in dye-free growth medium. The dotted curve is for 0.4 μM AlPC and 6 kJ/m².

diation, enhances cell killing.[20] In contrast, 3-ABA had no effect after exposure to AlPC and light (Figure 6).[21] This is consistent with the observation that the level of NAD^+, the precursor of poly(ADP-ribose) synthesis, is not depleted after photosensitization by AlPC. Presumably, the DNA damage produced is expressed as strand breaks only in the alkali conditions of the assay.

A more generalized treatment that inhibits repair processes operating on radiation damage is exposure to D_2O. Postirradiation incubation in D_2O medium enhances radiation response[22] and a similar effect occurs after AlPC photosensitization (Figure 6). It should be pointed out that the presence of D_2O during light exposure only, has no effect on the cell killing (Figure 2, see Section II.D for discussion). However, since in principle cells could repair sublethal damage during exposure that lasts for more than 1 hr at lower light fluence rate (see Chapter 2), the presence of D_2O under such conditions could increase the bioresponse by inhibiting repair processes.

C. Action Spectrum

Determination of action spectrum, i.e., the relative efficiency of various wavelengths in causing a specific bioeffect, is of fundamental importance in photobiology since it identifies the chromophore responsible for light absorption. In phototherapy, action spectrum indicates the optimal wavelength to be used for treatment. A dye laser pumped by a copper-vapor laser was employed for obtaining the action spectrum. The absorption spectrum of AlPC, in ethanolic solution, in aqueous solution after complexation with bovine serum albumin (BSA), or with yeast RNA, and the action spec-

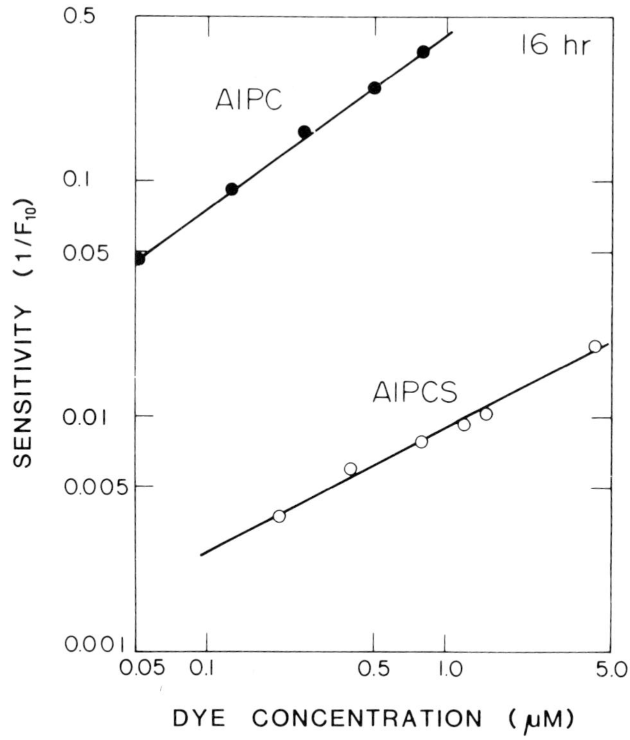

FIGURE 4. The dependence of Chinese hamster cell photosensitivity on phthalocyanine concentration. Cells were incubated for 16 hr with AlPC (•) or AlPCS (O) and the reciprocal of the fluence required to reduce survival to 10% ($1/F_{10}$) was plotted as a function of dye concentration.

trum are shown in Figure 7. It is evident that the absorption spectrum of AlPC is drastically modified after complexation with biological macromolecules. The binding constant is so high that the complex is not dissociated when run through a molecular sieve column of Sephadex G-25.[19] It is noteworthy that the action spectrum for Chinese hamster cell killing follows none of the absorption spectra of the complexes of AlPC. This would tend to rule out proteins and RNA as the cellular targets for AlPC action in the cell. Since no dye binding to DNA could be observed, and because of the dye hydrophobicity, a most likely target is the lipid constituents of cellular membranes.

Unlike white fluorescent light, survival curves resulting from laser exposure were essentially exponential. The fluence required to kill 90% of the cells was approximately tenfold lower using 680-nm laser light than a fluorescent lamp. While the average fluence rate was similar in both cases, the laser in contradistinction to fluorescent light consisted of 4000 pulses per second. Thus, the actual power density during the pulse was much higher than during exposure to fluorescent light and could drastically modify the photophysics of the sensitizer. This could be the reason for the disappearance of the shoulder from the survival curve. At any rate, the action spectrum suggests that wavelengths about 680 nm should be employed for maximum photoefficiency in sensitization reaction with AlPC.

D. Mechanism of Photosensitizing Inactivation

The first question that arises concerning the mechanism of phthalocyanine photosensitization is the involvement of molecular oxygen. Figure 8 shows that oxygen is defin-

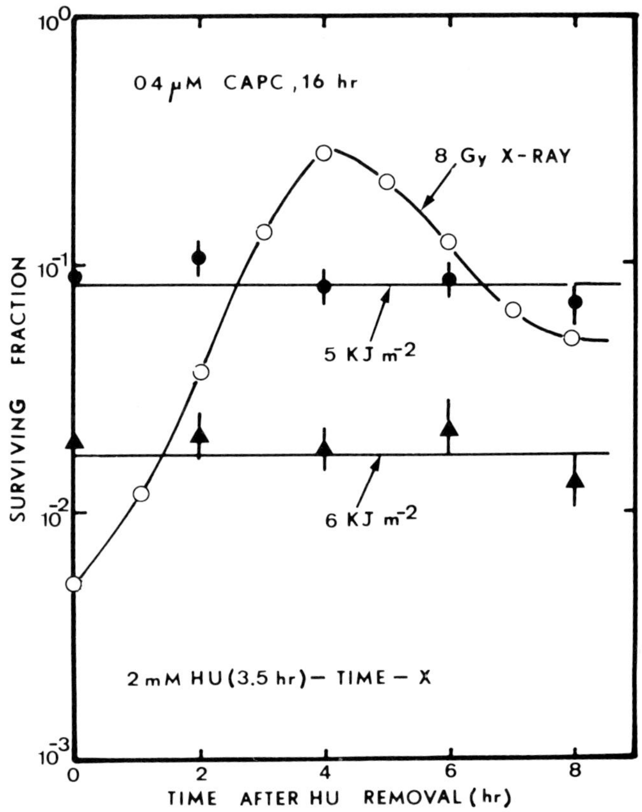

FIGURE 5. Age-response of Chinese hamster cells exposed to X-irradiation or AlPC plus light. Cells were synchronized at the G_1/S border by HU after 16 hr in 0.4 μM AlPC. The cells were exposed to 8 Gy X-rays or fluorescent light, as indicated, at various times after removal of HU. Incubation after HU removal was in growth medium containing 0.4 μM AlPC to avoid loss of photosensitivity.

itely required, although some effect can be observed in its absence.[23] Direct measurements of the residual oxygen indicated $pO_2 < 5$ ppm.

In a photodynamic process, the molecular oxygen could be activated to singlet molecular oxygen. Apparently this is not the case for AlPC, since the presence of D_2O had no effect on photosensitivity (Figure 2). In the presence of D_2O, the half-lifetime of $^1O_2$ is approximately ten times longer than in H_2O[24] and therefore type II reactions are enhanced when carried out in deuterated solvent. Also shown in Figure 2 are results of the exposure performed in 1 *M* glycerol. Under these conditions, glycerol could serve as an effective scavenger for hydroxyl radicals. The fact that glycerol did not protect against photosensitization by AlPC strongly suggests that free hydroxyl radical is not an intermediate in cell killing.

Another oxygen-reactive species which can be cytotoxic is superoxide radical anion.[25] The generation of $O_2^{\cdot-}$ has been detected in light-exposed solutions of phthalocyanines by spin trapping and electron spin resonance (ESR) spectroscopy.[18] Among the phthalocyanines tested, Ga, Li, metal-free, Al, and Zn phthalocyanine were found to generate $O_2^{\cdot-}$ with quantum yields of 4.22×10^{-4}, 1.40×10^{-4}, 8.47×10^{-5}, 2.61×10^{-5}, and 9.26×10^{-6}, respectively. Of these, AlPC was the most active biologically, followed by Zn derivative (five times less active). GaPC, the best producer of $O_2^{\cdot-}$ was completely

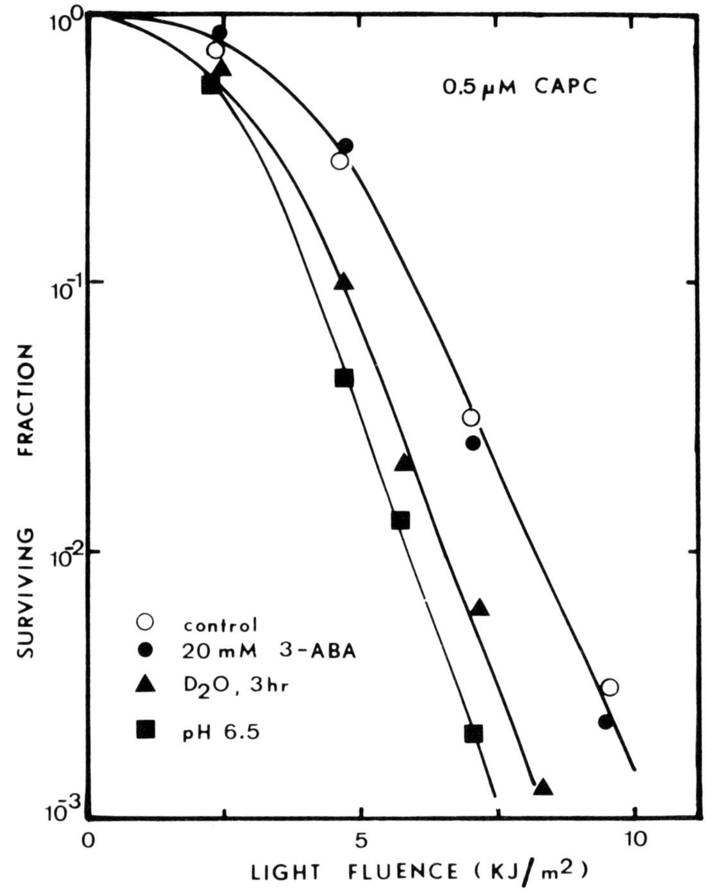

FIGURE 6. Survival of Chinese hamster cells exposed to fluorescent white light after 16-hr incubation with 0.5 μM AlPC. Following light exposure the cells were incubated for 2 to 3 hr in growth medium containing 20 mM 3-ABA or 90% D_2O, as indicated. In one case, exposure to light was in PBS at pH 6.5 instead of the usual 7.5 (■).

inactive in causing photosensitized inactivation of Chinese hamster cells. Thus, no correlation could be found between production of $O_2^{\cdot -}$ and the photobiological activity of phthalocyanines. Furthermore, in cells incubated with superoxide dismutase (SOD) before, during, and after exposure to light, under conditions that protect the cells from $O_2^{\cdot -}$ induced damage,[21] no protection against photosensitized inactivation was observed. This supports the conclusion that $O_2^{\cdot -}$ is not an intermediate in the oxygen effect during photosensitization.

These results, in conjunction with the previous elimination of the involvement of singlet molecular oxygen and hydroxyl radicals in phthalocyanine phototoxicity, favor a type I dye-sensitization mechanism. This mechanism is based on the initial reaction between excited sensitizer and specific cell targets to generate reactive transients, which further react with molecular oxygen. It also suggests a higher affinity or localization effect of the sensitizer to the cell, as compared to a mechanism mediated by relatively free active oxygen species. This hypothesis is supported by the experimental procedure employed in a cultured cells study, which requires the photoexcitation of the dye attached to the cell.[23]

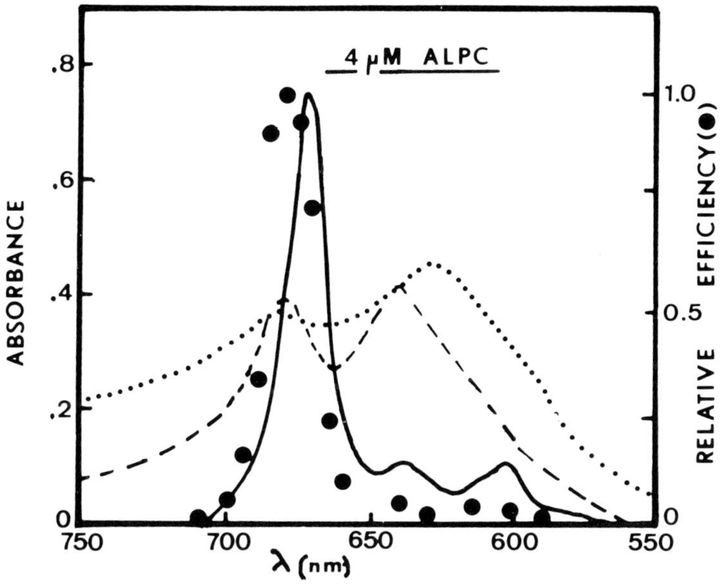

FIGURE 7. Absorption spectra of 4 μM AlPC in ethanol (——), in aqueous solution bound to 0.5% BSA (---), and in aqueous solution bound to 500 $\mu g/m\ell$ yeast RNA. Filled circles denote the relative efficiency of various wavelengths in inactivation of Chinese hamster cells, normalized to 680 nm. Efficiency was taken as $1/F_{10}$ of cells incubated for 16 hr in 0.5 μM AlPC and then exposed to monochromatic light from a dye laser pumped by a copper-vapor laser.

III. PHOTODYNAMIC THERAPY WITH PHTHALOCYANINES IN VIVO

A. General Considerations

Photodynamic therapy in vivo is an extremely complicated and poorly understood subject. Although it is well documented that sensitization with HPD and subsequent phototherapy with red light at 630 nm can produce tumor necrosis,[26] there is little evidence on which parameters determine the extent of necrosis produced and how the effects vary between normal and neoplastic tissue.[27] The primary mechanism of tissue damage appears to be occlusion of small blood vessels, and damage to other cells is secondary to this.[28] There are little data on how areas necrosed by photodynamic therapy (PDT) heal. There is good evidence that small tumors show more fluorescence than the surrounding normal tissue in which they arise, particularly in the bladder and bronchi.[29,30] Safe destruction of these lesions *in situ* may be possible, but it is not clear whether this is true selective destruction of the tumor or whether both normal and neoplastic mucosa are destroyed, but only the normal mucosa regenerates.

With this background information containing so many unanswered questions, it is difficult to compare new photosensitizing agents with the one that has been most studied, HPD. However, Ben-Hur et al. describe here preliminary experiments using AlPCS which show that phthalocyanines do indeed produce photodynamic effects in both normal and tumor tissue.

Initial pharmacodynamic studies were carried out to determine the distribution of AlPCS in normal tissues and in a transplantable fibrosarcoma at varying times after systemic administration. Normal and tumor-bearing rats (150 to 250 g Wistar and CBH) were given 5 mg/kg AlPCS by injection into a tail vein, and the animals killed at times from 1 to 72 hr later. At post-mortem, specimens of liver and tumor were

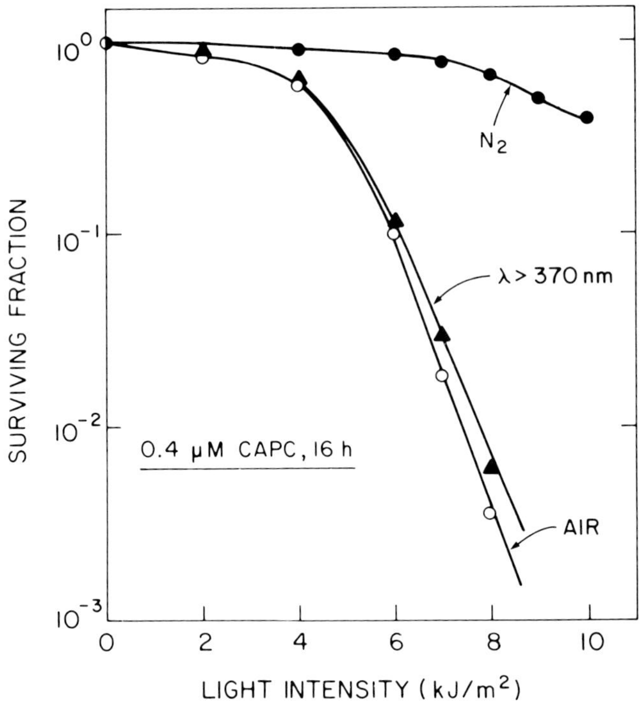

FIGURE 8. Survival of Chinese hamster cells incubated for 16 hr with 0.4 μM AlPC followed by exposure to white fluorescent light either in air or under nitrogen, as indicated. Triangles are for exposure with a UV cutoff filter to remove $\lambda < 370$ nm.

removed, weighed, homogenized in saline, and centrifuged. The level of AlPCS in the supertanant was measured in a fluorescence spectrophotometer with excitation at 610 nm and measurement of fluorescence at 680 nm. Organs from unsensitized animals had no detectable fluorescence at 680 nm. The results are shown in Figure 9. This distribution pattern is similar to that described by Rousseau et al.[16] for phthalocyanines measured by radioactive labeling and also to that described for HPD.[31] No specific tumor retention has been shown by this method for AlPCS or HPD. Ben-Hur et al. have yet to carry out detailed experiments on the localization of AlPCS within tumors or the measurements of tumor fluorescence in vivo.

Most studies on PDT in vivo look only at the effects on tumors and monitor results by the long-term cure or regrowth of the tumor. There are few publications comparing the tissue damage produced under comparable treatment conditions in tumors and the normal tissue in which the tumors arose, or indeed in any normal tissue.[27] This makes it difficult to compare the relative merits of new sensitizers such as AlPCS with HPD. The experiments described here are preliminary attempts to determine which factors control the necrosis produced in normal liver and a transplantable fibrosarcoma. Liver was used as a convenient normal organ to study, as it is relatively homogeneous and retains high concentrations of AlPCS and HPD. As the distribution curves show that peak levels of AlPCS are reached 3 to 6 hr after i.v. administration, all treatments were carried out in this time period after sensitization. AlPCS has a strong absorption peak in the region of 675 to 680 nm in the red part of the visible spectrum (Figure 7), so phototherapy was applied using light at 675 nm from an argon-pumped dye laser (using DCM as the laser dye).

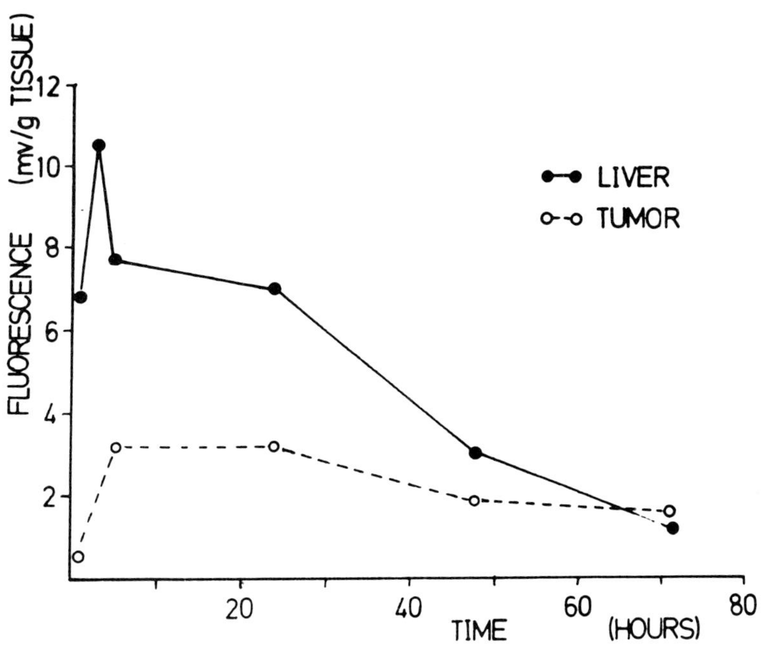

FIGURE 9. Tissue concentrations of AlPCS in liver and tumor measured by fluorescence after saline extraction.

B. Normal Liver

Initial experiments were performed on normal Wistar rats sensitized with 5 mg/kg AlPCS 3 to 6 hr prior to treatment. At laparotomy, a 200-μm fiber from the laser with a bare tip was inserted into an exposed lobe of liver at its thickest region. The laser, preset to give 100 mW at the fiber tip, was switched on for exposure times from 50 to 1500 sec, after which the fiber track was marked with a thin thread suture, the fiber removed, and the abdomen closed. The animals were killed 2 to 4 days later. A few minutes before death, the rats were injected with 1-mℓ 0.5% Evans blue as a vital stain to make identification of necrotic areas simple. At post-mortem, the livers were removed, fixed in formalin, and sliced at 1-mm intervals. The dimensions of the area of macroscopic necrosis were measured directly. Control animals were treated in an identical manner, but without prior sensitization with AlPCS. An example of the macroscopic appearance on sectioning the liver is shown in Figure 10.

Quantitative results are shown in Figure 11 in which the area of necrosis is plotted against the applied energy. Using a small-diameter fiber with a plane, bare tip applied interstitially, so the tip is in the contact with the tissue, it would be expected that there should be a thermal effect in the immediate vicinity of the fiber tip. However, it is evident from control animals that with energies of up to 150 J this never causes necrosis of more than 2.5 mm^2. In contrast, the necrosis in the sensitized animals increases with the applied energy (though not linearly) to reach a mean of 62 mm^2 at 150 J, showing that AlPCS is an effective sensitizer for producing nonthermal tissue damage in normal liver. These results are similar to those reported with hematoporphyrin,[27] although there are no such data available for HPD in normal tissue.

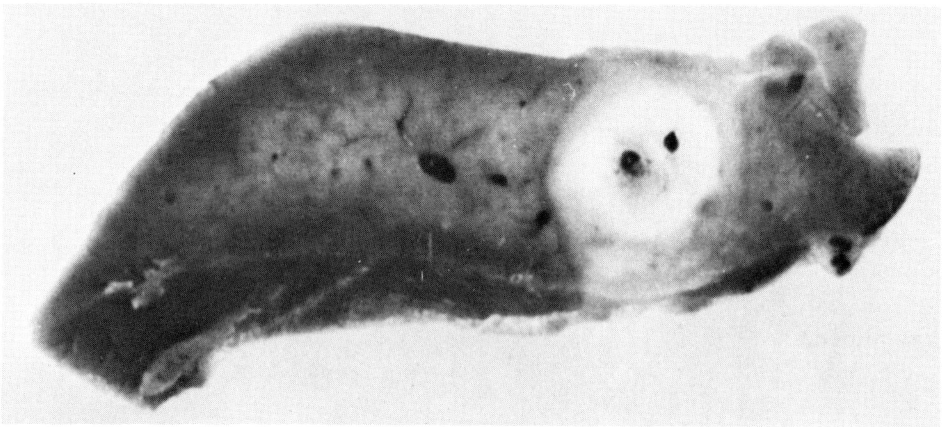

FIGURE 10. Macroscopic necrosis in liver (Evans blue given 15 min before death). The necrotic area is white.

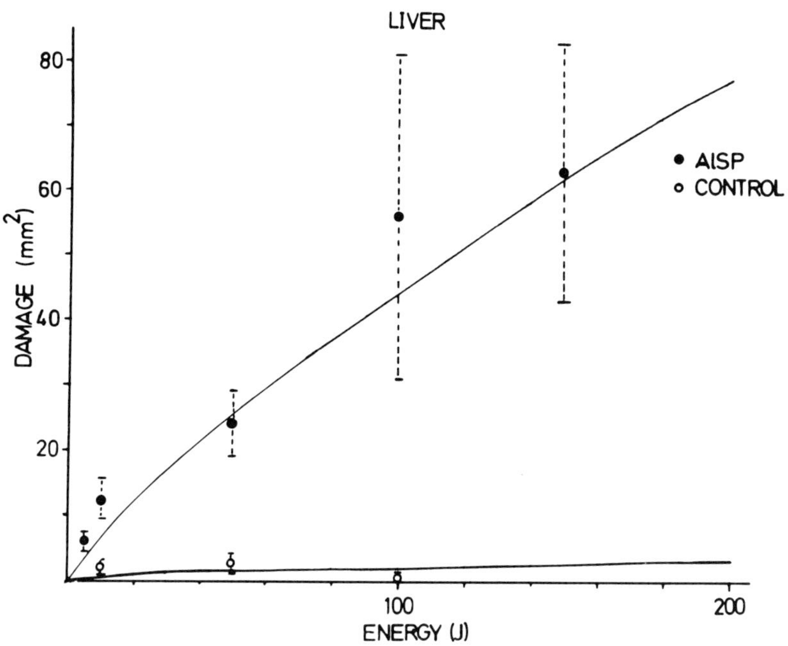

FIGURE 11. Effect of applied energy on area of necrosis produced in normal liver.

C. Tumors

The tumor model used in these experiments was a transplantable fibrosarcoma in CBH rats. As noted earlier, no selective retention of AlPCS has been demonstrated, in the same way that no selective retention of HPD was found in a similar tumor model in mice.[31] However, it is a useful model for studying photodynamic effects. Tumors were prepared for transplantation by removal immediately after killing donor rats, and homogenized in physiological saline. Transplantation was carried out by injecting 0.5 mℓ of the homogenate subcutaneously into the flank of recipient rats. Tumors grew to a size of 10 mm in 12 to 16 days and were treated at this time. Spontaneous necrosis was rare in tumors less than 10 to 12 mm in diameter. Experiments were carried out which were similar to those performed on the liver. Rats were sensitized with 5 mg/kg

AlPCS 3 to 5 hr prior to interstitial laser therapy at 100 mW for a range of exposure times. Animals were killed 2 to 4 days after treatment and the tumors prepared in the same way as with the livers described earlier. However, analysis of the results proved more difficult. Lesions in liver were roughly spherical but the areas of laser-induced necrosis in tumors were irregular in shape (Figure 12). In view of this problem, the maximum area of laser necrosis in sections taken perpendicular to the fiber track was measured, and this is shown in Figure 13, plotted against the applied energy for each tumor. From these results, it is clear that it is possible to produce nonthermal tissue necrosis in both normal liver and a transplantable fibrosarcoma with light at 675 nm following sensitization with AlPCS, and that the extent of damage depends on the light energy applied. At comparable energy levels, more damage was produced in tumors than in liver despite the lower AlPCS level. This is probably due to the deeper penetration of the light into the tumor as liver absorbs more strongly in the red, but illustrates well that the final biological effect of PDT depends on more factors than just the absolute level of AlPCS and the total applied energy. Also, an area that is close to spontaneous necrosis will probably be more vulnerable to PDT than the growing edge of the tumor. Treatment of such areas may explain the relatively large area of necrosis seen in treated tumors in unsensitized control animals, compared with the small area of necrosis produced in the liver of control animals.

There is good evidence that photodynamic damage produced in tumors using HPD primarily occludes small blood vessels rather than killing individual malignant cells.[28] Ben-Hur et al. do not have direct evidence of this yet with AlPCS, but two of their observations would be consistent with this: (1) small blood vessels are not seen in laser-necrosed areas, whereas they are seen in spontaneously necrotic zones, and (2) the necrotic areas are irregular in shape, which could be due to occlusion of some, but not all, the vessels within the tumor. It is well documented that most solid tumors stimulate intense neovascularization at their growing edges with relative ischemia at their centers,[32] so if PDT attacks vessels, the therapeutic light should be directed at the growing edges, not the center. This is supported by external illumination of tumors when a more regular area of necrosis is formed as in Figure 14. This makes it even more important that research is directed at what happens where tumors are invading normal tissue. So far, PDT research has hardly touched on this area.

In summary, current evidence suggests that PDT with AlPCS produces damage to normal liver and a transplantable fibrosarcoma which is similar to that produced using HPD. AlPCS has the advantages of being a single, stable substance of known structure and relatively easy to synthesize, with a strong absorption peak in a suitable part of the spectrum. Much work lies ahead to establish what clinical role PDT may have, and to assess the effects of different phthalocyanines, but as a group they could have a bright future as the sons and heirs of HPD.

NOTE ADDED IN PROOF

Since the initial publication of these studies suggesting the use of phthalocyanines in photodynamic therapy of tumors, there has been vivid activity in this field: Spikes, J. D., Phthalocyanines as photosensitizers in biological systems and for the photodynamic therapy of tumors, *Photochem. Photobiol.*, 43, 691, 1986. See also Selman, S. H., Kreimer-Birnbaum, M., Chaudhuri, K., Garbo, G. M., Seaman, D. A., Keck, R. W., Ben-Hur, E., and Rosenthal, I., Photodynamic treatment of transplantable bladder tumors in rodents after pretreatment with chloroaluminum tetrasulfophthalocyanine, *J. Urol.*, 136, 141, 1986.

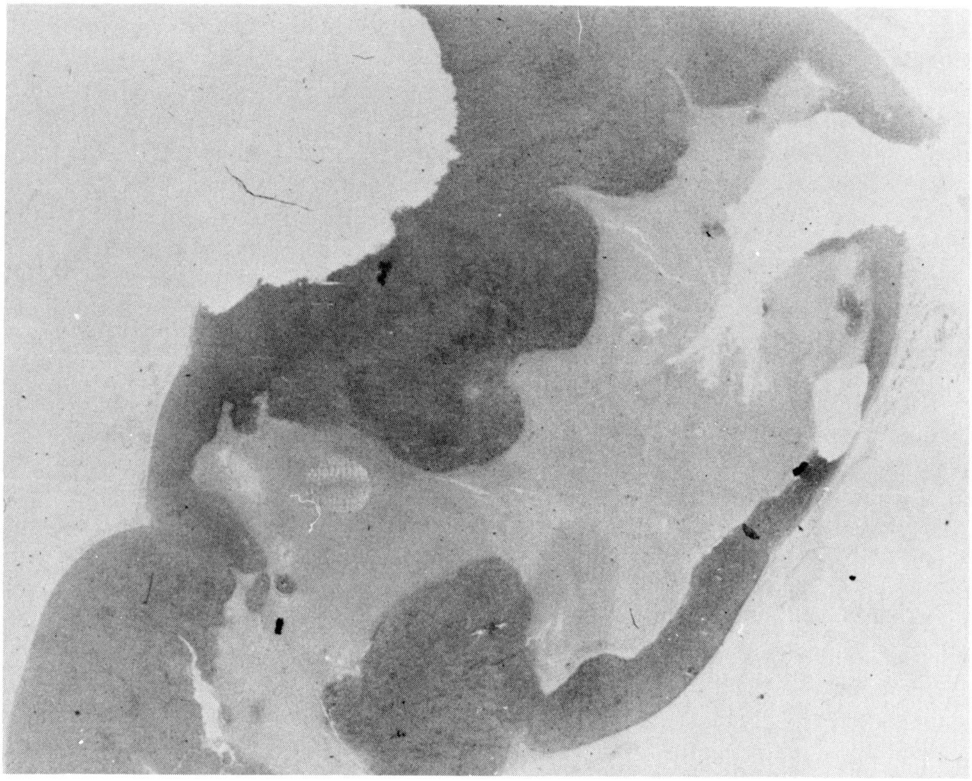

FIGURE 12. Necrosis of tumor using interstitial fiber. The necrotic area is pale. (H & E stain; magnification × 6.)

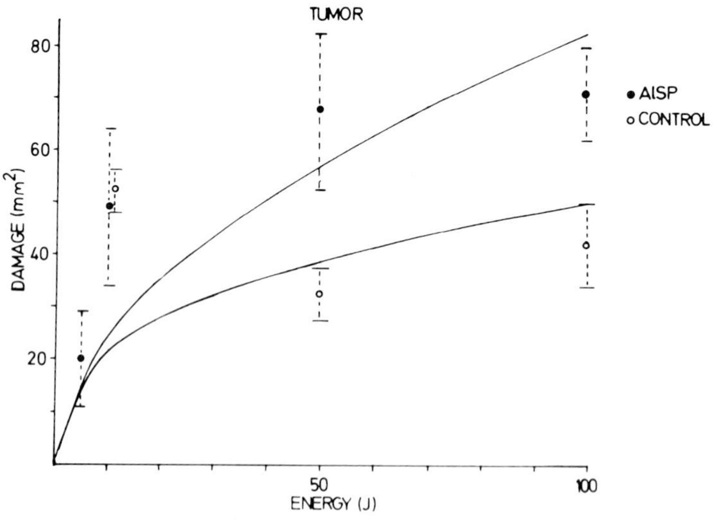

FIGURE 13. Effect of applied energy on area of necrosis produced in tumors. Regression lines on this figure and Figure 11 are the best fit to give the same mathematical function linking necrosis with applied energy in liver and tumor. However, the purpose of the figures is to illustrate the effects with and without sensitizer and not to quantify the results with precision.

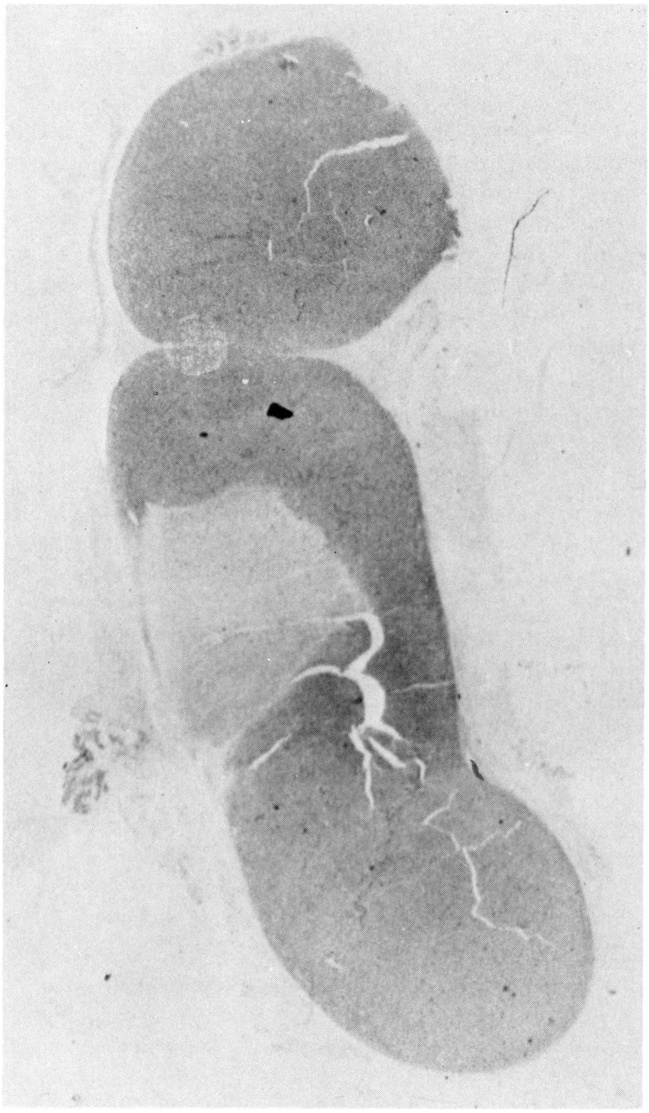

FIGURE 14. Necrosis of tumors with fiber tip held above tumor surface. The necrotic area is pale. (H & E stain; magnification × 8.)

ACKNOWLEDGMENTS

The authors thank Dr. I. McCubbin and Mr. R. Spencer at the Royal Institution and Mr. P. D. Coleridge-Smith and Miss C. J. Tralau at the Rayne Institute for their help in these experiments. Work at both establishments was funded by the Imperial Cancer Research Fund, London.

REFERENCES

1. Raab, O., Uber die Wirkung fluorescirender Stoffe auf Infusorien, *Z. Biol. (Munich)*, 39, 524, 1900.
2. Foote, C. S., Photosensitized oxidation and singlet oxygen: consequences in biological systems, in *Free Radicals in Biology*, Vol. 2, Pryor, W. A., Ed., Academic Press, New York, 1976, 85.
3. Dougherty, T. J., An overview of the status of photoradiation therapy, in *Porphyrin Localization and Treatment of Tumors*, Doiron, D. R. and Gomer, C. J., Eds., Alan R. Liss, New York, 1980.
4. Bonnett, R., Ridge, R. J., Scourides, P. A., and Berenbaum, M. C., On the nature of "Haematoporphyrin Derivative", *J. Chem. Soc. Perkin Trans. 1*, 3135, 1981.
5. Clezy, P. S., Hai, T. T., Henderson, R. W., and van Thuc, L., The chemistry of pyrrolic compounds. XLV. Hematoporphyrin derivative: hematoporphyrin diacetate as the main product of the reaction of hematoporphyrin with a mixture of acetic and sulfuric acids, *Aust. J. Chem.*, 33, 587, 1980.
6. Moser, F. H. and Thomas, A. C., *Phthalocyanine Compounds*, Reinhold, New York, 1963.
7. Kugel, R., Svirmickas, A., Katz, J. J., and Hindman, J. C., Laser properties of some phthalocyanines, *Opt. Commun.*, 23, 189, 1977.
8. Darwent, J. R., McCubbin, I., and Phillips, D., Excited singlet and triplet state electron-transfer reactions of aluminum (III) sulfonated phthalocyanine, *J. Chem. Soc. Faraday Trans. 2*, 78, 347, 1982.
9. Day, P., Hill, H. A. O., and Prince, M. G., Some reactions of cobalt phthalocyanines, *J. Chem. Soc. A*, 90, 1968.
10. Dolphin, D., James, B. R., Murray, A. J., and Thornback, J. R., Synthetic and oxidation studies of ruthenium (II) phthalocyanine complexes, *Can. J. Chem.*, 58, 1125, 1980.
11. McVie, J., Sinclair, R. S., and Thuscott, T. G., Triplet state of copper and metal-free phthalocyanines, *J. Chem. Soc. Faraday Trans. 2*, 74, 1870, 1978.
12. Muralidharan, S. and Ferrundi, G., Sequential biphotonic process in Rhodium (III) phthalocyanines, *J. Phys. Chem.*, 87, 4877, 1983.
13. Vincett, P. S., Voigt, E. M., and Rieckhoff, K. E., Phosphorescence and fluorescence of phthalocyanines, *J. Chem. Phys.*, 55, 4131, 1971.
14. Wrenn, F. R., Good, M. L., and Handler, P., The use of positron-emitting radioisotopes for the localization of brain tumors, *Science*, 113, 525, 1951.
15. Frigerio, N. A., Metal Phthalocyanines, U.S. Patent 3,027,391, 1962.
16. Rousseau, J., Autenrieth, D., and van Lier, J. E., Synthesis, tissue distribution and tumor uptake of (^{99}Tc) tetrasulfophthalocyanine, *Int. J. Appl. Radiat. Isotopes*, 34, 571, 1983.
17. Ben-Hur, E. and Rosenthal, I., The phthalocyanines: a new class of mammalian cells photosensitizers with a potential for cancer phototherapy, *Int. J. Radiat. Biol.*, 47, 145, 1985.
18. Ben-Hur, E., Carmichael, A., Riesz, P., and Rosenthal, I., Photochemical generation of superoxide radical and the cytotoxicity of phthalocyanines, *Int. J. Radiat. Biol.*, 48, 837, 1985.
19. Shimoni, M., Rosenthal, I., and Ben-Hur, E., Tritium-labeled chloroaluminum phthalocyanine, *J. Labelled Compd. Radiopharm.*, in press.
20. Ben-Hur, E., Involvement of poly(ADP-ribose) in the radiation response of mammalian cells, *Int. J. Radiat. Biol.*, 46, 659, 1984.
21. Ben-Hur, E. and Rosenthal, I., Factors affecting the photokilling of cultured Chinese hamster cells by phthalocyanines, *Radiat. Res.*, 103, 403, 1985.
22. Ben-Hur, E. and Riklis, E., Deuterium oxide enhancement of Chinese hamster cells response to gamma radiation, *Radiat. Res.*, 81, 224, 1980.
23. Ben-Hur, E. and Rosenthal, I., Photosensitized inactivation of Chinese hamster cells by phthalocyanines, *Photochem. Photobiol.*, 42, 129, 1985; 43, 615, 1986.
24. Nilsson, R. and Kearns, D. R., A remarkable deuterium effect on the rate of photosensitized oxidation of alcohol dehydrogenase and trypsin, *Photochem. Photobiol.*, 17, 65, 1973.

25. Michelson, A. M., McCord, J. M., and Fridovich, I., Eds., *Superoxide and Superoxide Dismutase,* Academic Press, London, 1977.
26. Dougherty, T. J., Kaufman, J. E., and Goldfarb, A., Photoradiation therapy for the treatment of malignant tumors, *Cancer Res.,* 38, 2628, 1978.
27. Pimstone, N. R., Horner, I. J., Shaylor-Billings, J., and Gandhi, S. N., Haematoporphyrin augumented phototherapy, dosimetric studies in experimental liver cancer in rat, *Lasers Med. Surg.,* 357, 60, 1982.
28. Henderson, B. W., Waldow, S. M., Mang, T. S., Potter, W. R., Malone, P. B., and Dougherty, T. J., Tumor destruction and kinetics of tumor cell death in two experimental mouse tumors following photodynamic therapy, *Cancer Res.,* 45 572, 1985.
29. Benson, R. C., Use of HPD in the localization and treatment of transitional cell carcinoma of the bladder, in *Porphyrin Localization and Treatment of Tumors,* Doiron, D. R. and Gomer, C. J., Eds., Alan R. Liss, New York, 1984.
30. Doiron, D. R., Profio, E., Vincent, R. G., and Dougherty, T. J., Fluorescence bronchoscopy for detection of lung cancer, *Chest,* 76, 27, 1979.
31. Gomer, C. J. and Dougherty, T. J., Determination of ^{3}H and ^{14}CHPD distribution in malignant and normal tissue, *Cancer Res.,* 39, 146, 1979.
32. Goldacre, R. J. and Sulven, B., On the access of blood borne dyes to various tumor regions, *Br. J. Cancer,* 16, 306, 1962.

Chapter 2

LASERS IN SURGERY AND MEDICINE

J. A. S. Carruth

TABLE OF CONTENTS

I. INTRODUCTION

It would be impossible to cover the subject of lasers in medicine and surgery in great depth as books have been written on it and, indeed, on the use of lasers in the individual disciplines. This chapter will attempt to provide an up-to-date review of the lasers currently used in clinical practice and their role in those disciplines not covered by other chapters in this book. Those uses which are in regular clinical practice will be distinguished from those which are being researched and evaluated.

II. THE LASER

Laser is an acronym for light amplification by stimulated emission of radiation. Although Einstein predicted that stimulated emission could be achieved, it was not until 1960 that the first laser was produced by T.H. Maiman using a synthetic ruby as the lasing medium. Since then, a very large number of laser systems have been developed and their use, in science and industry, is now so great that it has been suggested that when the age in which we live is finally named, it will be known as the "laser age" rather than the atomic or space age.

A laser is a device which produces coherent light; an intense beam of pure, essentially monochromatic, light which does not diverge and in which all the light waves are in phase and travel in the same direction.

The lasing medium, which in most medical lasers is a gas, is contained in the laser tube which has a fully reflective mirror at one end and a partially reflective mirror at the other, which allows access to the laser beam. The lasing medium is excited or pumped electrically to invert the population ratio of excited to nonexcited molecules, which normally has a great excess in the nonexcited state. An excited molecule will decay to the low-energy ground state with the release of a photon — a quantum of radiant energy — a light particle. When an incident photon strikes a molecule in the high-energy state, it stimulates it to emit an identical photon and to decay to the ground state — this is stimulated emission. The released photons are reflected back into the lasing medium from the mirrors and this optical feedback causes the "cascade effect" with a rapid build up of light energy within the tube and this coherent light can then be released through the partially reflective mirror.

The lasing medium names the laser and also determines the wavelength of the coherent light. The wavelength determines the site of absorption of the beam in the body tissues and this defines the clinical role of the laser. As each laser produces essentially a single wavelength, it has only one major clinical role; to change roles one must change lasers. However, there is significant overlap in the clinical roles of the three most commonly used lasers. The carbon dioxide laser is used primarily for cutting, but will coagulate small vessels. The argon laser is used for blood vessel coagulation, but at high-power levels it will cause thermal tissue destruction. The neodymium YAG laser can be used both to coagulate vessels and to remove tissue, and its role lies between those of the other two lasers.

III. LASER SYSTEMS

It is beyond the scope of this chapter to give the detailed specifications of all the surgical lasers which are available and, if given, these details would rapidly become obsolete as surgical lasers are being improved and developed very rapidly. A significant range of machines is available for each type of laser and the majority of lasers of a particular type have essentially similar specifications, except that argon lasers are of two powers, low power (up to 5 W) and high power (up to 20 W). The newer carbon dioxide lasers can be used in pulsed mode and neodymium YAG lasers may be continuous wave or pulsed. The power output of the dye laser is dependent on the power or input from the driving argon laser tube.

Almost all the machines are single-laser systems, but some machines incorporate two lasers and it seems possible that machines incorporating all three lasers, Argon, CO_2, and Nd YAG, may soon be available. With the clinical dye laser, the argon driving beam should be accessible and controllable for clinical use.

In the choice of a machine, the purchaser must ensure that the machine conforms with all the appropriate safety codes of his country and is sufficiently versatile for all present and anticipated future uses. There is a tendency for machines to be purchased for more than one specialist and all users must be happy with the machine, although compromises will be necessary on occasion. Another important feature which must be considered is the quality of service and back up provided by the manufacturer. Regular service visits should be provided and an extended warranty should be available for the replacement of expensive components. If machines have similar specifications and back up, then the choice will depend on the preference of the individual and will be determined by the delivery system, microscope attachment, micromanipulator, and size of machine, etc.

A. The Carbon Dioxide Laser

The carbon dioxide laser is a continuous wave (c.w.) laser which produces IR coherent light at a wavelength of 10.6 μm. This wavelength is absorbed by water and, therefore, by the soft tissues of the body which contain 70 to 90% water. The beam is absorbed by intracellular water causing it to boil and the cells to rupture, releasing cell contents into the beam at high speed, where they are carbonized and fall as carbon soot around the wound. It has been shown experimentally that there are no viable cells or cell components in the vapor and, as there is no shock impact with the c.w. laser, there is no tendency to scatter tissue or to force cells into adjacent tissues. It has also been shown experimentally that there is no increase in the spread of malignant cells via the venous or lymphatic routes.

Tissue destruction is by cell vaporization which takes place at the relatively low temperature of 100°C. As tissues conduct heat poorly, there is a very thin layer of damaged cells between the laser wound and adjacent normal tissues. The width of this zone of damage may be as little as 50 μm, and as a result, healing is rapid, not complicated by edema, and relatively pain-free compared to healing after tissue removal by conventional techniques. It has also been shown experimentally that after removal of mucosa from the oral cavity, there is a reduced contracture of the defect compared to a scalpel wound.

The CO_2 laser beam cannot, at present, be transmitted by a flexible fiber, but some rather thick and relatively inefficient fibers are now being introduced and it seems likely that better fibers will soon be available. The beam is transmitted from the laser tube via a self-supporting articulated arm with mirrors at the angles to the microscope attachment, colposcope, bronchoscope, or hand piece. The design of the articulated arm is very important as it must be self-supporting, adequately long, and fully flexible to allow a complete range of surgical procedures to be performed.

All CO_2 machines incorporate a low-powered, visible, helium neon laser aiming beam to define the beam path and target area.

When used under the operating microscope or colposcope, tissue dissection is very precise. The aiming beam is positioned in the well-illuminated, magnified target area by a micromanipulator on the microscope attachment and, with proximal control of the beam, there is no tendency for instrument shake, which is possible when operating with long instruments under the microscope. Penetration of the beam into the tissues is controlled by adjusting the power of the beam and the exposure of the tissues to the beam. For work on the larynx, power levels of 10 to 15 W are often used, but higher power levels are used in gynecology and in surgery of the mouth and tongue. However, as with the diathermy, if an incision is made with the CO_2 laser by tissue vaporization at a high power level, vessel coagulation is less good and a compromise power level giving good cutting and coagulation is often used.

The exposure is controlled by a mechanical shutter which allows various fractions of a second exposure to be selected and "continuous" when the exposure will be controlled by the foot switch.

There is evidence that if a "short, sharp" dissection technique is used with high power and short exposure, then lateral tissue damage is further reduced. The CO_2 lasers which have recently been introduced which offer a wide range of pulse widths of very short duration may prove to offer significant benefits in this respect over machines with mechanical shutters and longer pulses of energy.

In addition, there are no instruments or blood in the wound and, as the beam does not denature the tissues through which it is passing, the progress of dissection can be followed with great accuracy.

Dissection is relatively bloodless as the beam seals vessels of up to 0.5 to 1.0 mm in diameter, and larger vessels may be sealed by defocusing the beam. Lymphatic channels are also sealed, possibly reducing the spread of tumor cells by this route, and nerve endings are sealed, contributing to the lack of postoperative pain and the absence of neuroma formation.

The CO_2 laser can be used to remove tissue in one of two ways. First, after biopsy, the whole lesion may be vaporized until a bed of healthy tissue is reached. The second and better method is to use the laser as a scalpel and to excise a lesion with adequate margins. With this technique, the specimen to be removed must be kept under tension to avoid heat contracture during dissection, but the whole lesion is then available for histological assessment, which is of great importance in many lesions which may show dysplasia in one area and malignancy in another. The dissection is also more precise and the final defect no larger.

No instruments are required to deliver the light energy to the tissues, although suction is needed to aspirate the vapor produced by tissue destruction, not only to allow visual access but also to remove the vapor at source to prevent it from scalding and damaging the tissues. The absence of instruments is of vital importance in pediatric laryngology and neurosurgery, where the access may be severely limited.

B. The Argon Laser

This is also a c.w. laser, and it produces blue/green coherent light at several wavelengths, but the majority of its energy is at 0.488 μm and 0.514 μm. This blue/green light is absorbed by its complementary color red, and its absorption by hemoglobin is of great clinical importance. However, the beam will pass through clear and colorless structures without absorption and thermal damage, enabling treatment of vascular structures to be carried out through overlying, clear, normal structures. For example, disorders of the retina can be treated through, and without damage to, the clear, anterior parts of the eye, and the dermal capillaries of a port wine stain can be coagulated through the normal overlying epidermis.

When used to coagulate blood vessels directly, as in the control of bleeding from ulcers of the upper GI tract, and in the coagulation of normal vessels and vascular malformations in neurosurgery, the argon laser offers advantages over both bipolar and monopolar diathermy. A precise lesion of the vessel can be achieved by delivering a controlled dose of energy without damage to adjacent structures. There is also no mechanical contact with the vessel and, therefore, no risk of causing vessel rupture or clot dislodgement, which may occur when the diathermy is removed from the vessel.

For tissue destruction, higher power levels are needed, but destruction is relatively slow and it appears to be of value only in the removal of small and inaccessible tumors. Some early work was carried out on lesions of the tracheobronchial tree, but the argon laser was soon replaced by the Nd YAG laser in this field. The use of the argon laser to remove the bulk of small, acoustic neuromas has been described by otoneurosurgeons who also utilize this laser for microsurgical procedures on the middle ear. Tissue destruction with this laser lacks the precision of that performed by CO_2 laser, as the beam is absorbed to a greater depth in the tissues and there is some danger of spread of heat with thermal damage to adjacent structures.

The argon laser can be transmitted via a flexible fiber, and this may be coupled to a slit lamp for ophthalmology, the operating microscope for otology, a hand piece for dermatology, or to a fine, flexible fiber for work via the biopsy channel of a flexible fiberoptic endoscope. For aiming, an attenuated beam is used and, both with the slit lamp and microscope, a shutter is provided which closes when the main beam is activated to protect the eyes of the surgeon. However, if the hand piece or an unshuttered endoscope is used, eye protection must be worn by the surgeon and, with all usage, protective eye wear must be provided for all in the operating theater.

C. The Neodymium YAG Laser

The neodymium yttrium aluminum garnet (Nd YAG) laser produces low IR coherent light at a wavelength of 1.06 μm, which is deeply absorbed with no specific color or tissue absorption. At high power levels, effective thermal tissue destruction can be performed with better hemostasis than that provided by the CO_2 laser. However, the tissue removal lacks the precision of the CO_2 laser and, in tumor removal, tissue is often coagulated by laser and then removed mechanically. The deeper penetration of this laser means that there is a significant risk of causing thermal damage to adjacent tissues, even beyond the organ being treated.

At the periphery of tumor destroyed by the Nd YAG laser, there appears to be an area of killed cells which are replaced by fibrous scar tissue without loss of physical integrity, and tumor involving the wall of a hollow viscus may be treated without serious risk of perforation.

At lower power levels, the Nd YAG laser can be used to control blood vessels — either to stop hemorrhage or to coagulate vessels which may bleed in, for example, upper GI ulcers. As the absorption is not color dependent, the laser energy is not absorbed by the blood, as with the argon laser, but is absorbed by the vessel wall and perivascular tissues, enabling larger vessels to be controlled.

The Nd YAG laser beam can be transmitted down a flexible fiber and, for use in both tumor destruction and the control of hemorrhage, a special fiber must be used and the design of the fiber appears to be critical. A similar fiber is used with the argon laser for endoscopic work. The laser fiber is surrounded by a coaxial Teflon sheath down which CO_2 or N_2 is passed to keep the tip of the fiber free from blood and debris. If the tip becomes contaminated, then the blood or debris on the tip will absorb laser energy causing damage to the fiber. During a long tumor treatment session, the fiber may have to be removed several times for cleaning, and may even have to be recut. In addition, if the metal ferrule at the tip becomes hot, it will cause the Teflon to swell and make it impossible to remove the fiber from the endoscope.

The nonlinear effects of the pulsed Nd YAG laser are being investigated in the shattering of gallstones and urinary calculi by photoacoustic effects. This laser is also being researched and investigated for its use in the creation of very precise lesions within the eye.

D. The Ruby Laser

The first laser proudced by Maiman used a synthetic ruby as the lasing medium, and much of the early experimental work on the skin and on the CNS was carried out with the pulsed, ruby laser. However, it is not now widely used clinically, as the c.w. lasers are much easier to control, but it is still used in dermatology for the treatment of some nevi and blue and black tattoos.

E. The Krypton Laser

This laser is used similarly to the argon laser in the performance of retinal photocoagulation.

F. The Excimer Laser

This laser produces coherent light at a wavelength of 0.315 μm and some interesting experimental work has been carried out using this pulsed laser to destroy cutaneous, melanized cells and single melanosomes *in situ,* without damage to the cutaneous vessels.

G. The Dye Laser

This laser is different from all the others used in medicine at present in that its wavelength can be tuned over a significant range of the spectrum. The exciting energy for the dye laser is provided by an argon laser, but the power available from the dye laser is only about 25% of that of the argon tube. The beam from the argon laser is passed into the dye stream of the dye laser, and the design of this stream appears to be very important in the efficiency of the laser and its stability. A rhodamine dye is used most commonly.

Experimental work has been carried out on the selective destruction of the blood vessels in the skin with both the c.w. and pulsed dye lasers, and it appears they may prove to be better than the argon laser in this field, although clinical studies are awaited.

The main clinical role of the dye laser, at present, is the activation of hematoporphyrin derivative (HPD) in malignant tissues in photodynamic therapy (PDT). In the development of PDT, a wide range of light sources was used, but recently, it has become apparent that the best source of red light at wavelengths around 0.630 μm, which is the optimal wavelength for activation of HPD, and for tissue penetration, is the dye laser. It is possible that other sensitizers will be developed in the future which are activated by other wavelengths, and the dye laser is flexible, as the wavelength can be tuned to the exciting wavelength of any new sensitizers.

In the majority of clinical dye lasers, the exciting argon beam can be diverted from the dye laser and utilized in any of its clinical roles via a suitable fiber. This feature makes the dye laser a very versatile machine.

The beam is transmitted via a flexible fiber, and a number of tips have been designed for the end of the treatment fiber to enable various lesions to be treated. When a tumor is irradiated by a flat-cut fiber, the beam diverges and the power falls off toward the beam edge. A microlens can be created at the tip to give a flat-sided beam which will photoactivate HPD to a depth of about 1 cm, depending on the tissue. For the irradiation of a circumferential lesion, or for implantation into a large tumor, a diffusing cylinder is used at the fiber tip, and a diffusing bulb can be used to treat the inside of a hollow viscus.

H. The Gold Vapor Laser

For PDT, this pulsed laser produces much higher power levels than the dye laser at a wavelength of 0.628 μm. However, the wavelength is fixed and cannot be changed to activate other sensitizers, but the metal in this tube can be changed to copper without too much difficulty and the copper vapor laser could be used to excite a dye laser. The pulsed, red light appears to be at least as good as the c.w. red light from a dye laser in PDT.

IV. SAFETY

Total safety is of the utmost importance, as serious accidents could irreparably damage the reputation of surgical and medical lasers. The general public is well aware of the industrial and scientific uses of lasers and tends to identify medical lasers with these devices. Although medical and surgical lasers are much less powerful, all those used therapeutically are in the highest power class — Class 4 — and their use requires extreme caution.

It must be shown that safe lasers are being used in a safe manner by fully trained medical personnel on appropriate clinical conditions. It would be impossible and inappropriate to attempt to detail all the safety regulations which are in force worldwide, but in practice many of them are essentially similar.

A. The Machine

Each country will have electrical and laser safety standards and the manufacturers and importers must ensure that any machine conforms with all the appropriate regulations.

B. Safe Usage

Regulations in Great Britain and in several other countries insist on the appointment of a local laser safety officer who will advise on and supervise all aspects of safety in a laser unit. He will also produce laser safety codes for the use of each laser in each clinical situation. One of the main hazards is that the laser beam could be reflected back into the operating theater from an instrument or retractor and damage the eyes of the patient, the surgeon, or a member of the operating theater personnel. The eyes of the patient must be protected by appropriate pads firmly fixed in place, but it is essential that all other members of the operating theater personnel are provided with fully protective, laser-proof eye wear appropriate to the wavelength of the laser being used. The area in which the laser is being used must be designated a laser-controlled area and access to it limited to those who are essential to the procedure.

There are also some specific hazards to particular clinical situations, such as the danger of anesthetic-tube combustion with the CO_2 laser in laryngology, but these will be discussed briefly in the sections on the relevant disciplines.

C. Appropriate Use

Lasers should be used only when they offer clear, and scientifically proven, advantages over established, conventional techniques. The words of Dr. Leon Goldman, one of the fathers of laser surgery must never be forgotten: "If you don't need a laser don't use one".

Before using a medical or surgical laser, the specialist must have an adequate training in his discipline to know which cases are suitable for treatment by laser, and he must have an adequate training in laser techniques and laser safety.

A number of national laser associations are now arranging laser training programs in collaboration with the appropriate specialist organizations, and are offering some form of accreditation to identify those who are adequately trained.

V. CLINICAL USE OF LASERS IN MEDICINE AND SURGERY

The role of lasers in a number of disciplines is described in other chapters, and an outline will be given of the role of lasers in the disciplines not covered in these chapters.

A. Chest Medicine

The prognosis for carcinoma of the bronchus remains extremely poor, and a large number of patients die in severe distress with asphyxia from obstruction of one of the main air passages by tumor. A number of studies have been reported from several centers on the use of both the Nd YAG and CO_2 lasers in the provision of palliation for these patients, by the thermal removal of tumor and PDT has also been used to provide palliation in these cases.

As the Nd YAG laser delivery fiber can be introduced into the tracheobronchial tree, via the biopsy channel of a fiberoptic bronchoscope, it can be used and is well tolerated under local anesthetic. In theory, it can be used to remove tumor more peripherally than the CO_2 laser which must be used with a rigid instrument. However, in practice, palliation can be achieved regularly only in those cases in which the obstruction is of one of the major air passages and both lasers can be used. A significant amount of work with the Nd YAG laser in this field is now done via a rigid bronchoscope, which allows coagulated and necrotic tumor fragments to be removed with forceps, and allows better suction of blood, secretions, and vapor.

Until recently, bronchoscope attachments have not been widely available for the CO_2 laser, and relatively few cases have been performed. However, with increasing use of the CO_2 laser, via the bronchoscope, it should soon be possible to compare these two lasers directly. Tissue destruction is more rapid with the CO_2 laser, even with the somewhat reduced power at the longer operating distance, and hemostasis seems to be adequate, although it appears to be better with the Nd YAG laser.

Treatment of carcinoma of the bronchus with the Nd YAG laser has to date been only palliative and, to maintain palliation, repeated treatments may be necessary. In these cases, the ability to operate with the Nd YAG laser under local anesthetic is of great importance. Treatment for cure is possible in the rare, benign tumors of the tracheobronchial tree and in some cases of stenosis of the trachea and main bronchi but, in cases of stenosis, if the stenotic area is thick, there is significant chance of recurrence.

Photodynamic therapy is described elsewhere, but it is so important in chest medicine that mention must be made of it here. After injection, HPD is widely distributed in the body, but is then selectively retained by malignant tissues. Therefore, 3 days after an i.v. injection, the HPD will have been cleared from normal tissues but will be retained in any malignant tissues.

When a tumor containing HPD is exposed to blue/violet light, it will fluoresce, and this is utilized in the early diagnosis of preinvasive carcinoma of the bronchus. In a patient who is developing carcinoma of the bronchus, progressively more atypical cells can be found in the sputum over a long period of time, often measured in years. If these patients with abnormal cells can be identified by an adequate sputum screening program, it is often difficult to find the site of abnormal mucosa in the tracheobronchial tree with normal "white light" bronchoscopy. However, after injection of HPD, fluorescent bronchoscopy is carried out using a bronchoscope which incorporates a krypton laser as the light source and an image intensifier to identify the fluorescent areas. With this system, the patches of atypical mucosa can be identified relatively easily. Fluorescence is found in severe dysplasia, carcinoma *in situ,* and invasive carcinoma, and there is an excellent correlation between fluorescence and histology.

If the tumor containing HPD is exposed to red light, at a wavelength of 0.630 μm, produced by an argon-pumped dye laser, singlet oxygen is produced, and this highly reactive, transient state of the oxygen molecule is cytotoxic to the malignant cells, but surrounding normal tissues which do not contain HPD are undamaged. Work from both Japan and the U.S. suggests that if tumors can be identified early, and ideally in the preinvasive phase, they may be treated for cure, and a small number of cases have been reported with excellent results.

In advanced cases, palliation can be achieved by necrosing obstructing tumor under local anesthetic, and this technique is better tolerated than tumor destruction by Nd YAG laser. Follow-up bronchoscopy is regularly needed to remove necrotic tumor by suction and, if tumor invades a major vessel, tumor necrosis may lead to severe hemorrhage.

B. Cardiovascular

One of the most common causes of death in the world is myocardial infarction from coronary artery disease. The improved monitoring and medical management of patients who reach the hospital having had a ''coronary'' has resulted in a fall in mortality from 30 to 15%, but details of the medical care of the coronary patient are not relevant to this chapter.

A wide range of techniques has been used in an attempt to revascularize the heart. These techniques range from the ''Beck procedure'', in which asbestos and phenol were applied to the myocardium, through the implantation of the internal mammary artery into the muscle of the left ventricle, to coronary artery bypass graft using the saphenous vein, and finally, the internal mammary to coronary bypass graft, with a free venous or arterial graft to vessels not accessible to the internal mammary artery.

Coronary artery bypass graft is now a standard surgical procedure which gives good results and is performed more than 40,000 times each year in the U.S.

In the acute phase of myocardial infarction, attempts have been made to recannalize the obstructed vessel directly, by either an inflatable balloon catheter or by mechanical cannalization of the thrombus, using a soft-tipped, metal guide wire followed by dissolution of the clot with streptokinase.

In the management of the chronic, ischemic heart, the operation of percutaneous, transluminal, coronary angioplasty was introduced by Gruentzig in 1969. An inflatable balloon is inserted by a double catheter method into the obstructed vessel, which is then dilated. This procedure has high, immediate success rates and long-term patency.

It would appear that lasers may prove to be of value in the recannalization of the vessel acutely obstructed by thrombus and in vessels obstructed by atheromatous plaque.

A considerable amount of research work has been carried out to determine the effects of laser radiation on atheromatous plaque and on thrombus, and clinical work is just beginning. It has been shown that the CO_2, argon, and Nd YAG lasers can be used to destroy atheromatous plaque in cadaver coronary vessels, and that if the total energy delivered is less than a critical level, and if the beam if kept coaxial to the vessel, there is only a small risk of vessel perforation.

In experiments on swine made ''atheromatous'', laser lesions created in plaques of atheroma healed the same as lesions of the normal vessel. Endothelialization was rapid and there was a low thrombogenicity in small lesions.

Thrombus in cadaver vessels could be tunneled easily and, in artificially produced thrombus in the arteries of dogs, 1 cm of thrombus could be vaporized in 1 sec at a power of 1 W with the Argon laser, using a specially designed, intraluminal laser catheter. A large amount of minute bubbles were produced on clot vaporization with no evidence of debris and no distal embolization.

The use of a fiberoptic scope has enabled precise lesions to be created by Nd YAG laser under direct visual rather than fluoroscopic control. Recently attempts have been made to recannalize coronary arteries, 95% obstructed by atheroma, during open heart surgery for coronary artery bypass graft. The argon laser was used and, out of five arteries, three were successfully recannalized.

Another approach to the revascularization of the heart is based on the study of the hearts of snakes and reptiles which are vascularized by channels into the myocardium from the ventricular cavity.

In 1965, myocardial acupuncture was shown to provide some protection to the myocardium, but the channels subsequently became closed by fibrosis.

Mirhoseini et al.[6] used a high-powered CO_2 laser to create 20 channels per square centimeter through the myocardium into the left ventricular cavity of dogs, and showed that the channels remained patent and free from debris, and were able to provide significant protection for the myocardium after ligation of the coronary artery supplying the perforated area.

One clinical case has been treated in which the patient had anteroapical hypokinesia. The patient survived with a reduction in the hypokinesia.

Another application of lasers is in the nonsurgical treatment of both valvular stenosis and regurgitation. Under direct visual control, with a fiberoptic instrument, it has been shown that it is possible to perforate the valvular leaflets in artificially produced valve stenosis with relief of the stenosis and no damage to other cardiac structures. If the chordae tendineae of the mitral valve are irradiated by the argon laser in the postmortem heart, they shorten by about 23%, without a significant change in resistance to traction, and this technique may be of value in the nonsurgical treatment of some forms of mitral regurgitation.

C. Dermatology

The carbon dioxide laser may be used either to excise or to vaporize a wide range of skin lesions but with no particular advantages over conventional techniques except in multiple lesions such as warts or condylomata acuminata when the lesions can be vaporized rapidly under local or general anesthetic with clean, relatively pain-free healing and a low recurrence rate.

The argon laser is, at present, the treatment of choice for the hitherto untreatable port wine stain. This birthmark, which often occurs on the face, is thought to occur in 0.1% of the population and has a devastating effect on the life of an individual. A wide range of treatments have been used in the past, including thorium X, radioactive phosphorus, dermabrasion, cryotherapy, and excision and grafting, but no treatment has given consistently good results. Patients were told that no treatment was available and were advised to cover the mark with make-up. However, only about 30% of patients use make-up, and it is particularly difficult for men to wear the heavy make-up which is necessary.

A port wine stain consists of an abnormal network of capillaries in the outer dermis, with an overlying normal epidermis. There appears to be both an increase in number and in the dilatation of these vessels and several histological types have been described, depending on the depth of the vessels, number of vessels per unit area, and their dilatation and percentage filling with erythrocytes. Results appear to be better in those which are superficial, numerous, dilated, and well filled.

In the treatment of a port wine stain, the argon laser beam will pass through the clear epidermis with minimal absorption and thermal damage and will then be absorbed by the hemoglobin in the capillaries causing thermal damage and thrombosis. It has been shown, histologically, that the thrombosed vessel layer is replaced by colorless scar tissue over a period of months, with a marked lightening of the port wine stain, and that the epidermis returns to normal or shows minimal, atrophic changes. Treatment is not possible in those with racially pigmented skin, as the argon laser energy is absorbed by the pigment in the basal layer of the epidermis, causing more damage to that layer with an increased chance of scarring.

Results appear to be better in older patients, but, in the studies reported, few children have been treated, probably because most treatment has been carried out under local anesthetic, and this is extremely difficult in children. Work is in progress to determine the youngest age at which treatment can be carried out successfully. Results are better on dark marks than on light ones, and facial ones respond better than those on the limbs, but this is, in part, due to the fact that limb marks occupy a very large area and are, technically, very difficult to treat.

However, some who should respond well do not and vice versa, and many workers carry out a test patch to determine the response of an individual's birthmark to this form of treatment. Some use a fixed power for all patients, but perhaps a better technique is to determine the "minimal blanching power" for each patient. Under local anesthetic, using an intermittent repeated pulse of 0.2 sec, the beam is moved over the birthmark, and the power raised in increments of 0.2 W until blanching is achieved, and an area of the birthmark of about 1 cm^2 is then erased as the test patch. Blanching is achieved in the majority of patients at between 0.6 and 1.0 W. It appears certain that the incidence of scarring is directly related to the total energy delivered, and, with the minimal blanching technique, the scarring rate of less than 2% should be achieved, whereas with the higher-power techniques, 12 to 17% scarring has been reported.

The epidermis is not totally clear and suffers some reversible damage. The tested area will weep for a few days and a scab forms which separates after 10 to 14 days leaving the treated area red. The color fades until the test patch can be assessed at 4 to 6 months.

If a good result is obtained — a marked lightening in color with normal skin texture — treatment will be carried out under general or local anesthetic. If the area is still too red, a further test patch will be carried out at a higher power. If the minimal-blanching power technique has been used, a second test patch will be needed in about 20% of the cases. If scarring of the test patch occurs, either atrophic or hypertrophic, with the minimal-blanching power, then an adult may be untreatable, but a child will be retested after a few years.

In treatment, the power used varies from clinic to clinic from 0.6 W to 2.5 W or even higher. The exposure also varies and, although many use an intermittent pulse of 0.2 sec, some use continuous exposure using an "air brush" technique.

For treatment around the eyes, where goggles or pads cannot be used, protective contact lenses are available and must be used.

The zebra stripe technique has been described in which treated stripes of 5 mm were separated by similar untreated stripes to allow better healing. However, most workers simply use the beam to block out the mark with varying degrees of imprint overlap. The technique is prolonged and fatiguing, and computer-controlled robot arms are being developed to treat large areas automatically. The areas around the eyes, etc. will be treated by hand.

The majority of studies quote a good to excellent result in 60 to 80% of cases, which have a good prognosis, with a lower percentage in children and those with light marks.

Some preliminary work suggests that the CO_2 laser may be of value in the treatment of light marks. However, as the CO_2 laser must remove the epidermis before it can coagulate the abnormal capillaries, it represents simply a form of high-precision dermabrasion and scarring may occur. Further results must be awaited before this technique is widely accepted.

The argon laser may be used to treat a wide range of vascular abnormalities of the skin including spider nevi, telangiectasia, and the strawberry nevus, although in most cases the strawberry nevus should be left to resolve spontaneously. The adenomata sebaceum of tuberous sclerosis will also respond well, but few of the children have sufficient intelligence to appreciate the cosmetic improvement.

Blue and black tattoos may be treated by the ruby laser and good results are claimed. For other tattoos, the skin may be removed down to the pigment which is then vaporized. Both the CO_2 and argon lasers may be used, but there is no color-selective absorption with the argon laser and no epidermal sparing. The pigment is removed and the wound allowed to heal. The final result will depend on the area of the tattoo and the depth of the pigment. At best, a pale negative of the original tattoo will remain with offensive names and symbols blurred. Patients must be told what to expect and many will prefer a pale scar to the stigma of a tattoo.

Some preliminary studies suggest that the tuneable dye laser, both pulsed and c.w., may be better than the argon laser for the selective destruction of cutaneous blood vessels and clinical results are awaited.

D. Gastroenterology

Bleeding from ulcers of the upper GI tract still represents a severe problem with a significant mortality, which has remained relatively constant for many years, in those patients who continue to bleed and in those who rebleed.

Both the argon and Nd YAG lasers have been used, via the flexible endoscope, in a number of uncontrolled and a lesser number of controlled trials, to assess the value of laser photocoagulation in the control of the acute bleed and in the prevention of rebleeding.

Some of the preliminary results, mainly from Germany, were so impressive that it was considered by some that controlled trials were unethical. The work from Germany was carried out with the Nd YAG laser and, from the literature and in particular from a study of two consecutive trials carried out in London, first with the argon laser and then with the Nd YAG laser, it would appear that the Nd YAG laser will prove to be the better in the treatment of upper GI hemorrhage.

However, a number of controlled trials have been carried out with the argon laser. It has been shown in all that the acute bleed can be controlled by photocoagulation in the majority of patients. It has been found that the most reliable predicting sign for rebleeding is a visible vessel in the ulcer crater. In the studies from University College Hospital, London, photocoagulation of patients with a visible vessel resulted in a reduction in rebleeding and, in two of the studies, there was a marked reduction in mortality in the laser-treated, compared to the control group.

The uncontrolled trials reported by Keifharber and co-workers from Germany, indicate that with the Nd YAG laser, the acute bleed can be controlled in almost all cases, and that hemostasis was permanent in 95% of patients. This study was remarkable in that it included esophageal varices with control of hemorrhage in 92% of these. However, 30% were found to rebleed and, after the hemorrhage had been controlled by photocoagulation, elective sclerotherapy was carried out.

In controlled trials, it was found to be possible to control the acute bleed, and in two studies there was a significant reduction in the rate of rebleeding. In some studies, a reduction in the need for transfusion, acute surgery, and a reduction in mortality rates were reported.

The exact role of both lasers in this field is still to be defined, with further research to determine the optimum power and exposure levels for treatment. With more clinical studies, laser photocoagulation must play an important role in this field.

Patients with carcinoma of the esophagus which cannot be treated die with severe dysphagia, which is a most distressing symptom. Palliation can be achieved by the use of a prosthetic, rigid tube passed through the tumor to provide a relatively narrow lumen to allow the patient to swallow his own saliva and a liquid diet.

Recent work has been carried out on the palliation of these patients by relieving the obstruction with the Nd YAG laser. The procedure can be carried out under a local anesthetic and is well tolerated, although several treatment sessions may be needed to provide an adequate lumen. Significant palliation can be achieved and, if the symptoms recur, retreatment is possible. There appears to be no serious risk of perforation as, at the periphery of tumor destroyed by the Nd YAG laser, there is an area of damaged cells replaced by scar tissue without loss of physical integrity. Similarly, obstructing tumors of the stomach may be removed by this laser.

PDT has been used to provide palliation in inoperable tumors of the esophagus and, in some stomach tumors and in noninvasive carcinoma of the stomach, there appears to be a possibility of treating these for cure. With large tumors, if the whole thickness of the viscus wall is involved by tumor, or, if the tumor invades a major vessel, there is a danger of perforation or severe hemorrhage.

E. Gynecology

The incidence of the "premalignant", cervical, intraepithelial neoplasia (CIN) appears to be increasing, and the word epidemic has been used on both sides of the Atlantic. A number of etiological agents have been blamed for this increase and are related to a greater sexual freedom. The disease is also now occurring in a much younger age group in women and girls who still wish to have children. The initial diagnosis is made by cervical smear, and debate continues as to the optimum pattern of screening with this test.

Three grades of CIN are recognized, CIN I — mild dysplasia, CIN II — moderate dysplasia, and CIN III — severe dysplasia and carcinoma *in situ*. The natural history of this disease is still uncertain, in particular its progression to invasive carcinoma. However, it has been estimated that 70% of CIN I undergo spontaneous remission and a similar percentage of CIN II and III persist. Carcinoma *in situ* appears to be a definite precursor to invasive carcinoma, although its spontaneous disappearance has, on occasion, been reported.

When cases of CIN I and II are diagnosed, it is reasonable to treat any infection present and to perform colposcopy if the abnormal smear persists at the end of three months. Colposcopy will be performed at once if CIN II is diagnosed.

With a greater understanding of the natural history of CIN, it became apparent that the performance of hysterectomy was inappropriate and that conization of the cervix, which had been the treatment of choice, was also excessively radical in most cases, particularly in young women. As a result, there has been an increasing tendency to treat appropriate lesions by local destruction under the microscopic control of the operating colposcope. It must be stressed that a gynecologist must first become an expert in diagnosis of cervical disease with the colposcope before laser surgery is undertaken.

The lesion will be examined at colposcopy and a biopsy taken. If the whole lesion is visible and does not extend into the cervical canal, if endocervical curretage is negative for dysplastic cells, and if there is no evidence of invasive carcinoma on histology, then the lesion is suitable for local destruction.

Two techniques have been used in the past, electrodiathermy and cryosurgery. A general anesthetic is necessary with electrodiathermy to enable adequate destruction to be performed. Postoperative discharge is often profuse and there may be stenosis or incompetence of the cervix as late complications. Many believe that cryotherapy lacks precision and that the depth of tissue destruction cannot be predicted accurately and may be affected by cervical vascularity, leaving disease in the cervical crypts. In the postoperative period, discharge is profuse as the tissue destroyed by freezing separates and cervical stenosis may occur. However, the technique is well tolerated as an outpatient "office" procedure.

A number of studies have now been reported which show that for the removal of CIN lesions of the cervix, the CO_2 laser is superior to these other two modalities.

The lesion can be vaporized with great precision to a depth of at least 4 mm to ensure that disease in the cervical crypts is removed, and most workers remove tissue to a depth of 7 mm with a 2- to 5-mm margin around the lesion. Two methods are available to remove the correct amount of tissue. First, with experience, the depth can be judged accurately under the stereoscopic colposcope and the depth can then be confirmed with measuring rods. Second, using carefully calculated power density and exposure settings, a number of holes can be drilled in the cervix in the knowledge that they will be to the desired depth and then the cervix can be destroyed to this depth.

Hemorrhage during the procedure is minimal and can be controlled by defocusing the beam. There are no severe problems from postoperative hemorrhage and discharge is slight. In the long term, there is no significant scarring of the cervical canal and the patients can still have children.

A large majority of patients can tolerate the procedure without anesthetic, although many experience some discomfort. Local anesthetic by injection gun may be used, and a very small number of patients need a general anesthetic. However, this laser is now in wide, routine use in a large number of colposcopy clinics, and a very large number of women are treated on an outpatient basis.

Some CIN lesions disappear into the cervical canal, or no lesion is visible but endocervical curretage is positive for dysplastic cells. In these cases, there is an incidence of occult invasive carcinoma not evident until the cervical cone is examined histologically. In these cases, which require cone biopsy, there is evidence to show that biopsy can be performed better by CO_2 laser than by the conventional, "cold knife" technique.

In cases of dysplasia of the vagina and vulva which are difficult to treat by conventional techniques, excision or vaporization may be performed by CO_2 laser, and stainless steel mirrors may be used to reflect the beam into inaccessible areas. For lesions of the vulva, conventional surgery is often disfiguring and leads to severe sexual dysfunction. These lesions can also be removed by CO_2 laser without significant sequelae.

The lack of tissue reaction following CO_2 laser surgery has made it a valuable addition to the armamentarium of the surgeon performing pelvic reconstructive surgery for infertility. A number of cases have been reported in the literature with encouraging results, often in patients who already have had standard microsurgical procedures.

The CO_2 laser is of value in the treatment of condylomata accuminata, which may be very numerous and are difficult to remove by conventional techniques, but can be vaporized easily and rapidly with the laser.

F. Neurosurgery

The neurosurgeon works under severe constraints as he must perform his surgical maneuvers without either mechanical or thermal damage to vital structures, and without causing even a temporary interruption of the circulation to the neuraxis.

The use of lasers offers very considerable potential advantages to the neurosurgeon, and early reports of the use of the CO_2, Argon, and Nd YAG lasers in this surgical discipline suggest that this potential can be realized. Lasers offer true, no-touch surgery and as there is no shock impact with the c.w. lasers, there is no significant mechanical trauma, particularly as smaller access is needed with reduced retraction. There is also no significant risk of thermal damage to adjacent tissues if the CO_2 laser is used to incise or excise tissues of the CNS and if the Argon laser is used for precise vessel coagulation. However, the Nd YAG laser, which is deeply absorbed in the tissues, must be used only to debulk large tumors, or there may be damage from spread of the laser energy.

In the coagulation of blood vessels with the argon laser, a localized lesion can be created by delivering a precise dose of energy without spread of the energy beyond the lesion. There is also no mechanical contact with the vessel and no risk of vessel rupture or dislodgement of the clot which may occur when the diathermy, which must touch the vessel, is removed.

A relatively small amount of uncontrolled clinical work has been reported to date, but a considerable amount of research work has been carried out to determine the effects of the lasers at various wavelengths on the CNS and its blood supply.

Much of the early research work was carried out with the pulsed ruby laser, but it soon became apparent that this laser was too harsh and uncontrollable for clinical work. Early work with the CO_2 laser showed that precise lesions could be created in the brain with no damage to adjacent areas beyond "a fraction of a millimeter" from the laser wound. However, early work was hampered by the use of research lasers which were not suitable for surgical procedures.

With the development of clinical, surgical lasers, there was a renewed interest in their use in neurosurgery, and further research and clinical studies were begun in Japan and Austria in 1975. Research confirmed that precise atraumatic incisions could be created in the brain with the CO_2 laser with the characteristic three layers adjacent to the wound. These layers are first, the layer of carbonization, second a layer of vaccuolization and, beyond this, the third layer of edema. It was also suggested that neurones were more sensitive to laser energy than astrocytes and that healing was not normal with a delay in astrocytosis. Using a number of techniques including evoked response and fluorescence angiography, it was shown that there was no damage to the spinal cord or brain beyond the laser wound and its immediate adjacent layers of damaged tissues. It was found that where access was possible, precise lesions of the spinal cord could be created for pain relief.

In research on blood vessels, it was shown that the argon laser could be used to occlude experimental aneurysms without disturbance to the circulation of the parent vessel. It was also shown by Jain[9] that, end to end and end to side, microvascular anastamoses could be performed without sutures in the vessel wall and with a high patency rate with the Nd YAG laser. He also confirmed the potential advantages of the argon laser, described earlier in the coagulation of vessels around the CNS.

In the studies reported to date, a wide range of tumors have been removed, and it has been clearly shown in most cases that the exposure and need for retraction are reduced, as is the intraoperative blood loss. The operative time is shortened and the postoperative course is smoother than expected.

In the approach through soft tissues to the CNS, some advantages have been claimed for removal of the muscle from the vertebral column by laser in the performance of laminectomy.

In the removal of a tumor, the capsule is exposed and incised with the CO_2 laser. The center of the tumor is then exenterated, either by excision of tumor mass by CO_2 laser, or the bulk may be removed by suction in soft tumors, the Cavitron Ultrasonic aspirator, or diathermy loop. In large, hemorrhagic tumors, the Nd YAG laser may be of value in the removal of tumor with better hemostasis than that provided by the CO_2 laser, but extreme care is needed to prevent damage to adjacent structures by spread of energy through the tissues. When the tumor has been exenterated, the capsule may be drawn medially and removed or vaporized by the CO_2 laser after the brain has been protected with wet gauze.

If a meningioma invades a venous sinus, the extra sinus portion is removed and then, by application of the Nd YAG laser to the intrasinus portion, it can be coagulated and shrunk without the need for a sinusplasty.

Kelly and associates have developed an elegant stereotactic technique for the removal of deeply seated tumors. An exact model of the tumor is created by computer, and the laser beam is displayed visually for the surgeon. With the tumor in the center of the stereotactic frame, the approach is made through a silent area of the brain and a retractor introduced. The tumor is then removed by laser and the deficit created compared with preoperative scans to show the completeness of excision. Few patients have been treated to date, but the role of this technique appears certain to expand.

A number of pituitary tumors have been removed with the CO_2 laser via the transfrontal route, and the Nd YAG laser is of value in the removal of thick bone from the front of the pituitary fossa in the transsphenoidal approach.

A small number of cases have been reported in which the argon and CO_2 lasers were used to treat arteriovenous malformations. The feeding vessels are coagulated, and then the vessel divided and the malformation removed. Further bipolar coagulation of the feeding vessels may be needed if the CO_2 laser is used. Very few aneurysms have been treated by argon laser and, despite encouraging results, more clinical trials are needed.

Many surgeons must, for financial reasons, use only one laser, but Fasano has described the use of CO_2, Argon, and Nd YAG lasers in the performance of certain procedures, and in this way the optimal role for each will be defined.

It has been shown that PDT is feasible for brain tumors, and a small number of cases have been treated with encouraging preliminary results. The tumor has been removed by conventional or laser neurosurgical techniques and the tumor bed treated with PDT. Further clinical trials are awaited.

G. Otolaryngology

The carbon dioxide laser is of the very greatest value in laryngology, in particular, in children. The ability of this laser to remove tissue from the airway with precision, no bleeding or postoperative edema, and reduced postoperative contracture is unique to this modality. The lack of tissue reaction makes it possible to treat some cases of laryngeal stenosis endoscopically, without recurrence.

In the larynx, benign nodules, edema, and polypi may be treated with excellent results. To remove tiny nodules, a protected anesthetic tube is inserted and the subglottic region is protected with wet gauze. The beam is then aimed at the gauze and moved toward the cord, allowing the edge of the beam to shave the nodule from the cord. Vocal cord edema and polypi may be either excised or vaporized. To remove a lesion from the larynx requires special instruments, as the lesion must be drawn medially with forceps to keep the pedicle under tension while it is cut by the laser. Suction must be incorporated either into the forceps or into the laryngoscope to remove the vapor at source while one hand retracts the lesion and the other uses the laser micromanipulator.

The condition beyond all others which must be treated by CO_2 laser is recurrent, respiratory papillomatosis. This unpleasant condition often begins in childhood between the ages of 2 to 5 years, and the larynx becomes filled with frond-like, viral warts. After removal, these tend to recur regularly until the age of puberty when some regress but others continue into adult life. Patients require scores of operations, and there is, at present, no method of regularly preventing recurrence. The CO_2 laser reduces the morbidity of each procedure and enables the surgeon to remove all visible disease, which is not possible with conventional techniques which provoke nuisance bleeding. In the postoperative period, there is no edema and, in the long term, minimal scarring. The voice will be maintained and tracheostomy avoided in almost all cases.

In the treatment of congenital or posttraumatic stenoses of the larynx and upper trachea, the CO_2 laser can recreate the airway by vaporization of the stenotic area, and the airway can be restored in about 60% of cases by laser surgery alone. If recurrence occurs after laser excision, an indwelling "stent" will be required, which can be inserted either endoscopically or at an open operation.

Areas of leukoplakia should be excised so that the whole lesion can be studied histologically. If dysplasia or carcinoma *in situ* is diagnosed and a complete excision has been performed, then the patient may be kept under careful observation and further lesions removed as necessary.

In Great Britain, cases of early vocal cord carcinoma are treated by radiotherapy and a "cure rate" of 80 to 90% can be achieved without significant morbidity. However, work from Boston suggests that an equal cure rate can be achieved by endoscopic laser excision of these lesions. If histology shows excision to be incomplete, further laser surgery or radiotherapy will be performed. With this technique, which can be performed as often as necessary, the patient has laryngoscopy, biopsy, and treatment under one anesthetic and is spared the short-term morbidity and long-term complications of a course of radiotherapy. However, the voice may not be quite as good as that after a successful, well-given course of radiotherapy. It would seem certain that laser excision will become the treatment of choice for certain patients, particularly the young ones in centers of laser excellence.

Some work has been done on the treatment of more advanced tumors by endoscopic, laser excision, but only a few patients have been treated and further controlled studies are needed.

In advanced cases, tumor bulk may be reduced at the time of biopsy, which may contribute to the success of radiotherapy and will enable the extent of the tumor to be assessed more accurately with better staging, planning of treatment, and reporting of results. If a tumor is obstructing the airway, enough of it may be removed to relieve the obstruction, avoiding either an emergency laryngectomy or a tracheostomy with its risk of tumor seeding around the stoma.

In the performance of resection of lesions of the mouth and tongue, the CO_2 laser does not cure cancer, and the ability of the surgeon to cure malignant disease will depend on his ability to remove a tumor with histologically free margins.

Lesions of the mouth may be excised with reduced blood loss and a very low morbidity. It has been shown experimentally that there is less contraction of a wound of the buccal mucosa created by CO_2 laser than of a similar wound created by scalpel, and large areas of mucosa may be removed and left ungrafted without significant contracture.

It has been shown that muscle in the mouth epithelializes rapidly, and lesions of the tongue may be resected and the defect left unsutured and ungrafted. The tongue remnant is not tethered by sutures, and the residual muscle hypertrophies, and the tongue retains maximal, residual function. The patient is also spared the morbidity of a skin graft in the mouth.

There is an extremely low postoperative morbidity, and many patients after major tongue resections require no analgesics or mild, oral analgesics only. Nasogastric feeding is not needed and patients resume their diet on the evening after surgery. Most patients can leave the hospital on the first or second postoperative day.

In otology, the role of the argon laser is being evaluated. In the middle ear, it can be used to perform no-touch, atraumatic surgery to the ossicular chain using a 50-μm spot. A number of cases of laser stapedectomy have been performed, and a better preservation of high-tone hearing is claimed. The laser may be used also to fix a graft of temporalis fascia onto the drum remnant by spot welding it into place with tiny spots of argon laser energy.

In otoneurosurgery, the argon laser has been used to remove the center of acoustic neuromas to facilitate removal of the capsule without damage to adjacent structures.

There is a specific hazard to the use of the CO_2 laser in laryngology. If a plastic or rubber anesthetic tube carrying anesthetic gases is struck by the CO_2 beam, it will ignite and fatalities have been recorded. A number of techniques are now available to overcome this hazard, and it is essential for the surgeon and anesthetist to work as a team.

No anesthetic tube may be used and the patient ventilated by intermittent jet ventilation delivered via the laryngoscope. Some patients are not suitable for this technique, and the cords move on each injection of gas and the subglottis cannot be protected with wet gauze.

The most commonly used tube is one protected with a wrapping of aluminum foil with the cuff and subglottis protected with wet gauze, but the tube is made rough and rigid by the wrapping, and a tube must be prepared for each procedure. Laser-proof, metal tubes are now available which are safe, but they are not cuffed and have a relatively poor, internal/external diameter ratio. Whichever technique is used, total safety must be achieved.

H. PDT

Although this subject is dealt with at length in another chapter, it is of such crucial importance that mention must be made of it here. The dye laser is somewhat peripheral to the technique, being just the best method of producing activating red light, but PRT has come to be identified recently with a number of national and international laser societies.

After an i.v. injection, HPD is widely distributed in the body and is then selectively retained by malignant tissues, so that 3 days after the injection, it will have been cleared from normal tissues but will be retained by any malignant tissues. There is some retention also in the liver, spleen, and skin, and patients experience skin photosensitivity for 3 to 4 weeks. If the tumor containing HPD is exposed to blue/violet light, it will fluoresce, and this can be used to identify sites of malignancy. If it is exposed to red light at a wavelength of 0.630 μm, which is not the optimum wavelength for the activation of HPD but is chosen for tissue penetration, singlet oxygen is produced and this highly reactive, transient state of the oxygen molecule is cytotoxic to the malignant tissues, but leaves the surrounding normal tissues, which do not contain HPD, undamaged.

A number of clinical trials have been performed and reference is made to them in the relevant sections in this chapter.

I. Urology

In the treatment of carcinoma *in situ* and invasive carcinoma of the bladder, a large amount of work is carried out endoscopically using electrosurgical techniques via a cystoscope.

Several authors who have written on lasers in urology have stressed that these electrosurgical techniques are very successful in the majority of cases, and that it will be difficult to show that lasers offer advantage over them.

However, several features of bladder tumors make them unsuitable for a purely surgical approach by conventional techniques. Recurrence is common after the electrosurgical resection of a solitary tumor, and more so after the removal of multiple tumors. It has been shown that the disease is very often multifocal, with disease in apparently normal areas of the bladder, and many of the recurrences are blamed on this. In addition, it has been suggested that tumor fragments may become implanted within the bladder, as an unusually high proportion of the recurrences are in the dome of the bladder, and it is thought that the tumor fragments float up to this site.

In the treatement of a bladder tumor, the exophytic portion of the lesion must be destroyed, without the release of viable fragments, and the tumor in the substance of the bladder wall must be destroyed without perforation of the viscus, and without damage to adjacent structures.

The first effects of laser radiation on the bladder wall were reported in 1966, but clinical work had to await the development of appropriate delivery systems.

Research has been carried out to show the effects of argon, CO_2, and Nd YAG lasers on the bladder. They create the characteristic lesions. The argon laser has been used to destroy some superficial tumors, and there is apparently less sensation with this laser than with the more commonly used Nd YAG laser. A CO_2 laser cystoscope was described in 1978, but little clinical work has been reported to date.

The majority of the research and clinical work has been carried out with the Nd YAG laser, which is delivered to the bladder by glass fiber, via a narrow, rigid cystoscope, incorporating an Albarren Bridge which allows the laser light to be directed to all parts of the bladder.

Techniques have been developed largely in Germany by Rothenberger and his coworkers[12] who have introduced a transurethral, ultrasonic technique for measuring the thickness of the bladder wall. They, and other workers both in Germany and Canada, have shown that if the bladder wall is irradiated with a power of 45 W for 2 sec, there is no significant risk of bladder perforation or injury to adjacent viscera.

Rothenberger et al.[12] have described the treatment of more than 100 cases, treating small lesions by laser alone and other larger lesions are first removed by an electrosurgical technique and the base then coagulated by Nd YAG laser. Their early results show an improved survival in ''laser cases'' over those treated by electrosurgical transurethral resection alone. Similar encouraging results using the argon laser to treat superficial lesions have been reported by Smith and Dixon.

In the management of urethral stricture, recurrence is common after dilatation or electrosurgery, and the deep penetration of Nd YAG laser energy also resulted in recurrence. However, good results have been reported using the argon laser, and lack of tissue reaction following CO_2 laser tissue destruction must make it the ''laser of choice'' for this problem, when suitable delivery systems are available.

The use of the CO_2 laser in the treatment of condylomata acuminata has already been mentioned in the section on gynecology, and other local lesions of the penis, such as erythroplasia of Queyrat and balanitis xerotica obliterans can be well treated by CO_2 laser vaporization. In addition, amputation of the penis can be avoided in some cases of carcinoma by local tumor destruction using the Nd YAG laser, and some encouraging preliminary results have been reported.

It was hoped that surgery to the kidney could be performed by laser without the need to clamp the renal vessels, but this hope has not been fully realized. Some exciting research work is in progress in the photoacoustic destruction of renal stones using the pulsed Nd YAG laser.

The multifocal nature of bladder malignancy makes it most suitable for treatment by PDT. After an injection of HPD, the whole of the inside of the bladder can be treated by either filling it with a diffusing medium or by using a diffusing bulb on the end of the delivery fiber. Some excellent early results have been reported.

VI. CONCLUSIONS

We are only in the infancy of the laser age and lasers have already turned science fiction into science fact in medicine and surgery. Further exciting developments can be anticipated.

REFERENCES

1. Andrews, A. H. and Polanyi, T. G., *Microscopic and Endoscopic Surgery with the CO_2 Laser,* John Wright, Bristol, England, 1982.
2. Atsumi, K., *New Frontiers in Laser Medicine and Surgery,* Excerpta Medica, Amsterdam, 1983.
3. Fleischer, D. and Bown, S. G., Endoscopic laser therapy for upper gastro-intestinal carcinoma, in *Therapeutic Laser Endoscopy in Gasto-Intestinal Disease,* Fleischer, D., Jensen, D., and Bright-Asare, P., Eds., Martinus Nijhoff, The Hague, 1983.
4. Hetzel, M. R., Millard, F. J. C., Ayesh, R., Bridges, C. E., Nanson, E. M., Swain, C. P., and Williams, J. P., Laser treatment for carcinoma of the bronchus, *Br. Med. J.,* 286, 1983.
5. Choy, D. S., Stertzer, S., Rotterdam, H. Z., Sharrock, N., and Kaminow, I. P., Transluminal laser catheter angioplasty, *Am. J. Cardiol.,* 50, 1206, 1982.
6. Mirhoseini, M., Muckerheide, M., and Cayton, M. M., Transventricular revascularisation by laser, *Lasers Med. Surg.,* 2, 187, 1982.
7. Arndt, K. A., Noe, J. M., and Rosen, S., *Cutaneous Laser Therapy,* John Wiley & Sons, Chichester, 1983.
8. Bellina, J. H. and Bandieramonte, G., *Principles and Practice of Gynaecologic Laser Surgery,* Plenum Press, New York, 1984.
9. Jain, K. K., Lasers in neurosurgery: a review, *Lasers Med. Surg.,* 2, 217, 1983.
10. Simpson, G. T. and Shapshay, S. M., The use of lasers in otolaryngologic surgery, *Otolaryngol. Clin. N. Am.,* 16(4), 1983.
11. Andreoni, A. and Cubeddu, R., *Porphyrins in Tumor Phototherapy,* Plenum Press, New York, 1984.
12. Rothenberger, K., Pensel, J., Hofstetter, A., Keodotsch, K., and Frank, F., Transurethral laser coagulation for treatment of urinary bladder tumours, *Lasers Med. Surg.,* 2, 255, 1983.

Chapter 3

LASERS IN OPHTHALMOLOGY

Orna Geyer and Moshe Lazar

TABLE OF CONTENTS

I. INTRODUCTION

In the Middle Ages it was already known that looking at the sun during an eclipse was liable to cause retinal damage. Since then, many investigators utilized sunlight for production of lesions on the retina. Meyer-Schwickerath in 1945 used different optical instruments to concentrate the sunlight on the retina in order to produce therapeutic lesions on it. Since 1956, when the first photocoagulator was developed, light has been used in clinical practice for the treatment of eye diseases. It is therefore not surprising that ophthalmology was one of the first specialties to introduce the laser to medicine. In 1963, only 3 years after the laser was invented, H. C. Zweng performed the first laser photocoagulation on the retina in a rhesus monkey. Today, laser therapy is used in a large variety of eye diseases. Its application has expanded from the retina to the iris, the ciliary body, and recently to the lens and the cornea. Today, laser can be applied with beneficial effect to every ocular tissue.

II. ANATOMY OF THE EYE

The eye consists of a fibrovascular sphere, the walls of which are formed by three layers: cornea and sclera, uvea, and retina (Figure 1). The cornea, which is transparent, covers the anterior part of the eye. Posteriorly, it continues as a white opaque fibrous coat, the sclera. The covering of the eye has another two layers in addition to cornea and sclera, i.e., uveal tract and retina. The uveal tract consists from behind of choroid, ciliary body, and iris, all continuous with each other. The choroid is highly pigmented and is made up of blood vessels. The ciliary body has the form of a triangle, the outer side of which corresponds to the ciliary muscle and the inner side to the ciliary processes; the latter produce aqueous humor. The iris is a diaphragm with a central hole, the pupil, which regulates the amount of light reaching the retina. The iris is attached at its periphery to the ciliary body. It forms a curtain dividing the space between the cornea and the lens into the anterior and posterior chambers of the eye. On the inner side of the choroid lies the structureless membrane of Bruch. The retina, the innermost layer, is the nervous tunic of the eye. It consists essentially of the nuclei and processes of three layers of nervous elements placed one on the other and forming synapses. They are the visual photoreceptors (rods and cones), the bipolar cells, and the ganglion cells. The processes of the ganglion cells form the nerve fiber layer of the retina and further on they form the optic nerve. The point where the nerve fibers of the retina become continuous with the optic nerve is called "optic disc". Here is the site of the entry of the central artery and vein of the retina. The macula, the center of the retina, appears as a small circular area of a deeper red than the rest of the retina; it is free from blood vessels. The fovea centralis is the most important part of the retina, since most of the visual cells are concentrated in it. It is represented by an intensely bright spot in the middle of the macula (Figure 2). If there is damage to the macula, notwithstanding the fact that the retina itself is normal, the visual acuity becomes very poor. The light which enters the eye is transmitted through the ocular media to the receptor cells of the retina where it is absorbed and thereby initiates a neuroelectric response which is propagated by the optic nerve to the visual center in the occipital cortex of the brain.

The eyeball contains the vitreous, the lens, and the aqueous humor. The vitreous is a transparent, colorless, gelatinous mass which fills the posterior four fifths of the globe. The lens of the eye is a transparent, elastic, biconvex body of crystalline appearance placed between iris and vitreous. It is surrounded by a capsule which is attached to the ciliary processes by the zonules. The aqueous humor, which fills the anterior and posterior chamber of the eye, is secreted by the ciliary body into the

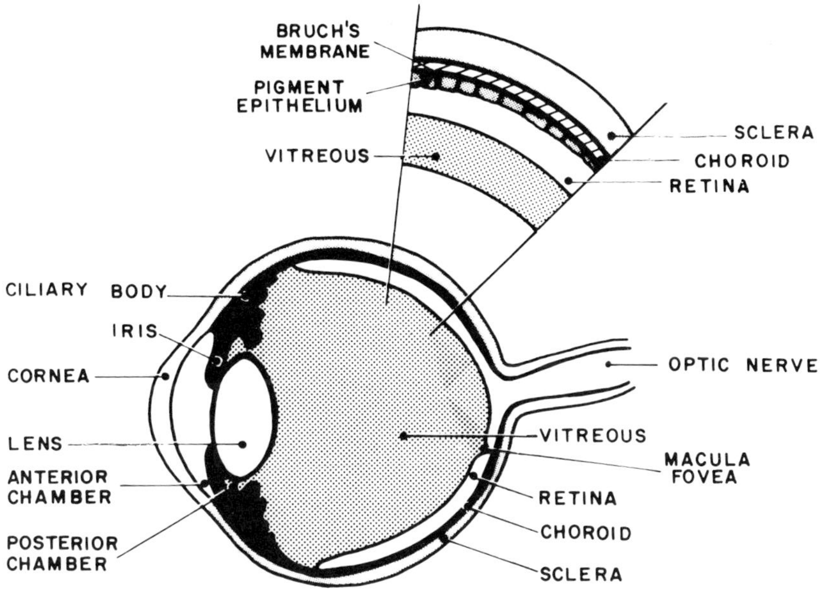

FIGURE 1. Anatomy of the eye.

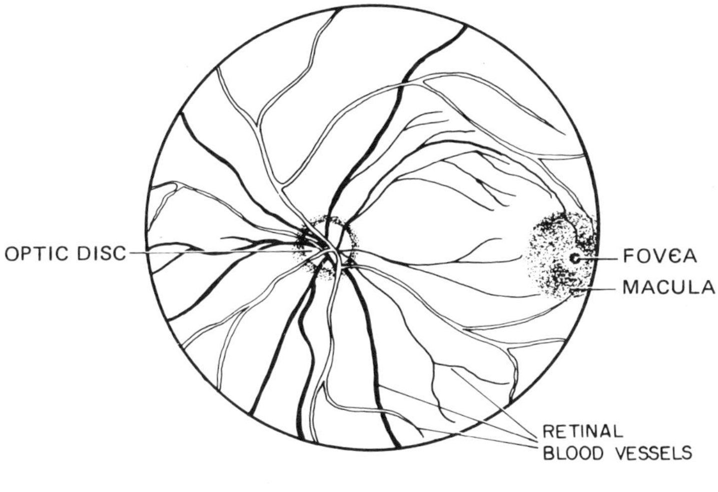

A

FIGURE 2. Anatomic regions of the posterior retina. (A) Diagram; (B) photography.

posterior chamber, from here at the anterior chamber angle through perforated collagenous trabecular sheets into the channel of Schlemm, leaving the eye into the blood. The equilibrium between the production and the drainage of the aqueous forms the intraocular pressure which normally ranges between 10 and 20 mmHg.

III. EFFECTS OF LASERS IN OCULAR TISSUES

The laser provides the eye surgeons' community with a clean, precise, and noninvasive method for the treatment of many eye diseases. Each of the many types of laser now in use in ophthalmology has its own unique mechanism of action. There are four

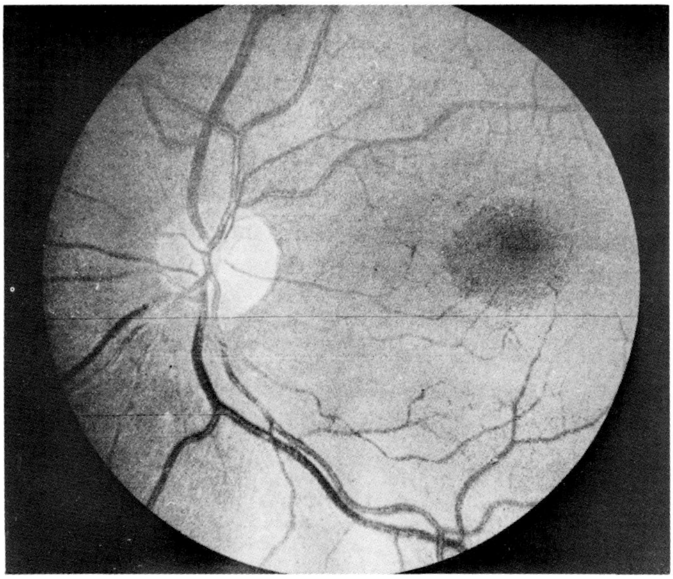

FIGURE 2B.

effects on which these lasers rely: (1) photocoagulation, (2) photodisruption, (3) ablative photodecomposition, and (4) vaporization.

A. Photocoagulation

The photothermal mechanism consists of the absorption of light by the irradiated tissue and the conversion of that light energy into heat energy. The result of this photothermal mechanism is that tissue is heated to a varying degree depending on such parameters as thermal diffusivity, depth of absorption, and rate of energy deposition. This leads to thermal damage in the form of dehydration, protein denaturation, and scarring. At the point of focus of the laser beam, tissue can even be boiled away and removed. Factors which influence the coagulative result of a given power are the amount of energy carried by a specific wavelength, the spot size, and the exposure time. For all presently available lasers working by the photocoagulation system, there is approximately a 75 to 95% transmission through the ocular media. The effects of these lasers inside the eye are dependent on absorption of the beam by the ocular pigments melanin, hemoglobin, and xanthophyl (Figure 3). Melanin is present in the retinal pigment epithelium and in the uvea. It absorbs light wavelengths of 400 to 600 nm. Hemoglobin is found in eye tissue containing blood vessels. Its maximum absorption is at blue (488 nm), green (514.5 nm), and yellow (568.2 nm) wavelengths. Xanthophyl is present only in the fovea. The maximal absorption of this pigment is in the blue region of the spectrum. Absorption of laser energy by xanthophyl should be avoided since it causes permanent damage to the fovea.

The main disadvantage of the lasers which rely on the mechanism of photocoagulation is their dependence on tissue absorption of radiation by pigments. Therefore, these lasers cannot be applied to transparent tissue in the eye. The lasers presently in use for photocoagulation are argon and krypton.

1. Argon Laser

The argon blue-green laser operates in a continuous-wave mode. Most of the argon lasers produce light that is 70% blue (488 nm) and 30% green (514.5 nm). Use of both

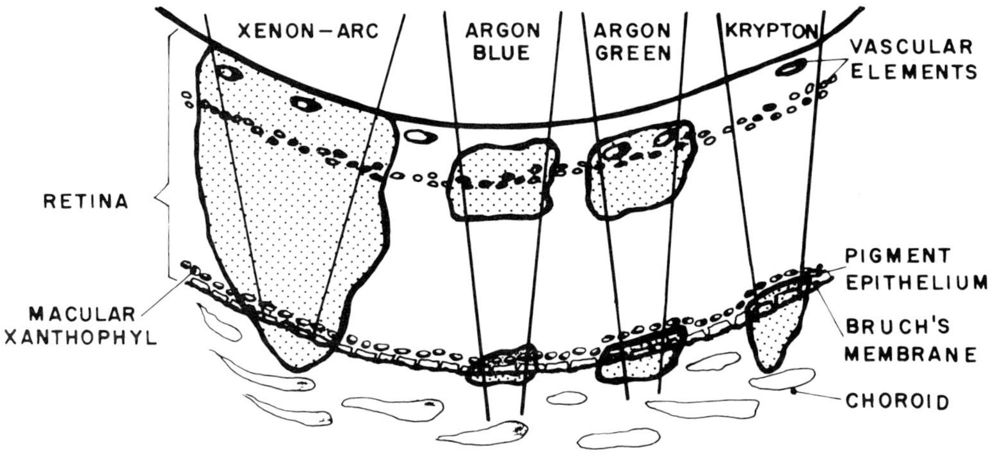

FIGURE 3. Beam retinal absorption for xenon arc, argon, and krypton lasers.

wavelengths is necessary, because high power is not available at a single wavelength. In the argon laser, the total power output ranges from 3 to 5 W. The blue argon laser is well absorbed by melanin, hemoglobin, and xanthophyl. Therefore, it has a more superficial effect in the retina.[1] As this laser is absorbed so well by ocular pigments, it can be used for photocoagulation both in the posterior and anterior segment. The green argon is not absorbed by xanthophyl. Thus this wavelength can be used for treatment of macular diseases with minimal damage to the fovea.

2. Krypton Laser

The red krypton laser beam (647.1 nm) is well absorbed by melanin, while it is not absorbed by the other two ocular pigments (xanthophyl and hemoglobin). Therefore, this laser has a deeper effect in the choroid than the Argon laser. Krypton laser is advantageously used for macular photocoagulation and for laser treatment in the presence of extensive retinal hemorrhage. The maximal power with conventional krypton laser is 1.5 W.

3. The Dye Laser

This laser was adapted a short time ago for ophthalmic use. It is able to produce any wavelength in the visible portion of the spectrum at high powers due to various dyes contained in the dye lasers. The dye laser bridges the gap between the green and red. Laser operation in the yellow and orange spectral region potentially offers great advantages in treating many eye diseases. The reason for this versatility is the rapid change in the absorption of hemoglobin with wavelengths over 600 to 650 nm ranging from 70% to less than 20%. Thus, the ophthalmologist can run up and down the absorption curve with a twist of the micrometer dial, adapting the treatment exactly as needed. Ultimately, the dye laser has been experimentally applied to photoradiation therapy of tumors. By this technique, the cancer cells are selectively destroyed by laser light. The treatment relies on a dye which is selectively absorbed by the cancer cells. Subsequent illumination by laser light produces material toxic to the dye-containing cells, leading to rapid tumor cell death.[2]

B. Photodisruption

This is a breakdown-mode surgery. The process involves focusing a very short pulse of laser energy to a small spot, producing a very high power density and thus a large electric field, resulting in optical breakdown and in the formation of plasma. The plasma produces a local high-pressure shock which expands outward as a miniature explosion. As a result, the tissue is mechanically ruptured. Breakdown-mode surgery provides a precise, noninvasive tool for microsurgery on transparent tissue. The process, based on mechanical effects produced by the laser beam, offers a mechanism of tissue dissection which does not involve thermal effects or temperature rise. Therefore, it is most closely related to mechanical mechanism of conventional surgery. The Neodymium-YAG (ythrium, aluminum, garent) laser is the only photodisruptor commercially available for ophthalmic use. This laser emits light of 1064.5 nm, i.e., in the IR range. Since light in this wavelength is not seen, its visualization is achieved via helium-neon laser beam (632.8 nm), colinear with the YAG laser beam.

The mechanism by which these short laser pulses are generated varies, and two methods are in common use. In the Q-switch technique, an electrooptical shutter in the laser cavity is used to produce a pulse of 2 to 14 nsec. In the mode-locked technique, a dye cell in the laser cavity is used to generate a pulse with a similar total duration which consists of several (6 to 10, average 7) short 30 psec spikes. The dye in the cell is opaque to low intensities of light, but becomes transparent at very high intensity. Thus it acts as a gate which lets pass only very intense, very brief pulses of light, and thereafter shuts.[3] The peak power density which can be generated by the Neodymium-YAG laser lies between 1.0 to 1.6×10^{12} W/cm^2. Its main use in ophthalmology is for the rupturing of cloudy membranes which form across the optical path of the eye blocking and scattering light, resulting in poor vision.

C. Ablative Photodecomposition

This process appears to proceed by a photomechanical mechanism. It entails the absorption of the far-uv light (wavelengths less than 200 nm) in a very thin layer (approximately 1 μm) at the surface of the irradiated organic polymer. Photoenergy is converted into electronic excitation leading to the breaking of molecular bonds without the energy degrading into heat. The long chain polymer molecules break up into similar volatile fragments which then ablate away from the surface. The useful feature of this process is that the fragments carry away almost all the energy deposited on the surface of the laser beam. This energy, unless carried away, would degrade into heat, diffuse into the surrounding material, and ultimately produce thermal damage. Thus the pulse duration of the laser must be short, so that the entire process, including energy deposition, takes place in a shorter time than the time required for energy to diffuse into the surrounding material. The laser which works by the ablative photodecomposition system is the excimer laser. The advantage of this laser is that it etches material out of the target in a specific pattern to a specific depth, leaving clearly defined edges where tissue has been removed. It has been shown that the corneal epithelium is extremely sensitive to the 193-nm light emitted by the argon-fluoride excimer laser.[4] The application of the excimer laser is still experimental. If offers interesting short-term possibilities in the area of cornea sculpting. The excimer laser has been used to remove corneal tissue, controlling the depth of the incision with great precision, without damaging the adjacent tissue.[5] This has implications for a variety of corneal surgical procedures. Laser ablation can be used to remove the sharp area of cornea to any depth as would be done in a lamellar keratectomy. Additionally, in cornea transplantation, a controlled penetrating corneal incision can be made. It also can be useful in a new surgical correction for myopia (nearsightedness) called radial keratotomy. As this procedure is currently carried out, radial grooves are cut into the outer surface of the cornea with a scalpel to flatten the radius of curvature of the cornea, thus turning myopics to normals, eliminating the need for corrective lenses. By using a properly shaped mask with the excimer laser, an ophthalmic surgeon could gently create all the required grooves in the cornea.

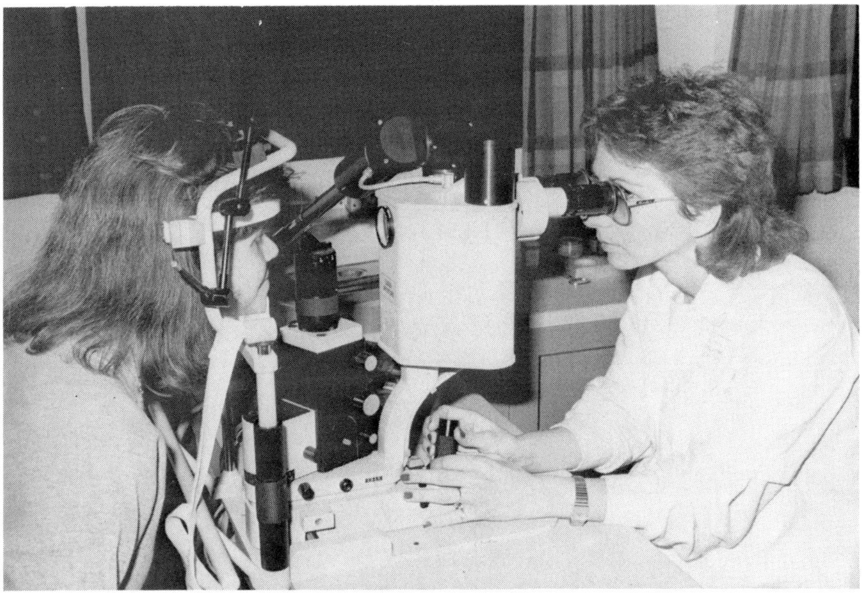

FIGURE 4. Slit lamp — a laser delivery system.

Another use of the excimer laser may be in strabismus surgery (the condition of being cross-eyed). In this operation, extraocular muscles are cut. Using the excimer laser, muscles can be removed much more gently and precisely than with knife or scissors.

The excimer laser holds much hope for future application, but for the time being, its use is still in the experimental stages.

D. Vaporization

Longer IR wavelengths (2000 to 10,000 nm), contrary to visible and near-IR wavelengths, are absorbed by water and by the water contents of tissue. This characteristic is utilized for simultaneously cutting coagulation and tissue vaporization.

The CO_2 laser (10,600 nm), the only laser commercially available in IR regions, works by vaporization. It uses a gas of carbon dioxide molecules. The output can be in a continuous wave or pulsed mode, and the power ranges up to 80 W. The CO_2 laser has been used experimentally for orbital surgery and for vaporization of lid tumors. It has also been used for filtering procedures in glaucoma[6] as well as for full-thickness ocular-wall resection for intraocular tumors.[7]

E. Laser Delivery Systems

Laser energy of the most useful lasers in ophthalmology (argon, krypton, and YAG laser) is delivered on ocular tissue transcorneally. Therefore, the slit-lamp biomicroscope is the most common method of delivery (Figure 4). The patient's head is fixed to the instrument, while the eye surgeon observes the eye through the slit lamp and manually focuses the laser beam. In the argon and krypton laser, the physician sees the laser focal spot. In the Neodymium-YAG laser, visualization of the laser focal spot is achieved via a helium-neon laser red-beam collinear with the YAG laser beam.

The laser treatment, presently widely used, is a noninvasive, easy method, painless and therefore not requiring anesthesia and not necessitating hospitalization.

Another delivery system used in the operating room during surgery is the endophotocoagulation.[8] In this case, the laser beam is delivered into the eye by fiberoptics. This

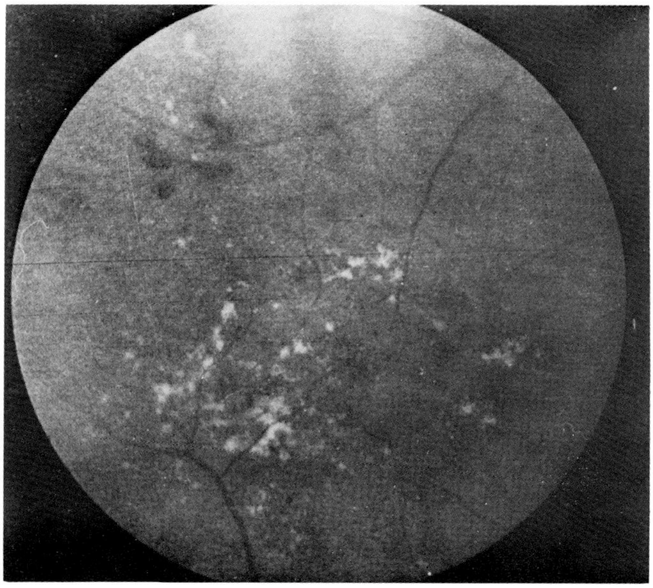

FIGURE 5. Nonproliferative diabetic retinopathy: microaneurysms, retinal hemorrhages, and exudates.

delivery system has been applied in conjunction with vitrectomy surgery for coagulating retinal holes, in pan-retinal photocoagulation, and in coagulation of bleeding sources.

IV. CLINICAL APPLICATIONS OF LASERS IN OPHTHALMOLOGY

The main conditions treated by laser are (1) retinal diseases — the most common being diabetic retinopathy, senile macular degeneration, retinal vein occlusion, and retinal tears; (2) glaucoma; and (3) secondary cataract.

A. Laser Treatment in Retinal Diseases

1. Diabetic Retinopathy

Diabetes mellitus is a complex metabolic disorder characterized by insufficient insulin production and/or inadequate utilization. Diabetes has a devastating effect on the small blood vessels and causes widespread damage to many body tissues, including the eye. Two percent of the general population suffer from diabetes.

Diabetic retinopathy is the fourth leading cause of legal blindness and the most common cause of blindness with onset in young adult life. Forty percent of all diabetics have some retinopathy. The frequency of retinopathy rises with increasing duration of diabetes. Eighty percent of diabetic patients suffer from retinopathy after 25 years.

Diabetic retinopathy is divided into two stages, i.e., the nonproliferative and the proliferative stages.

a. Nonproliferative Diabetic Retinopathy

In such patients, the funduscopic changes are confined to the retina only. These retinal changes are characterized by microaneurysms, dilatation of veins, and retinal hemorrhages. Leaking from vessels may result in retinal edema and hard and soft exudates (Figure 5). Edema in the macular area is responsible for impairment of visual acuity. Macular edema can be focal or diffuse. The first category is characterized by

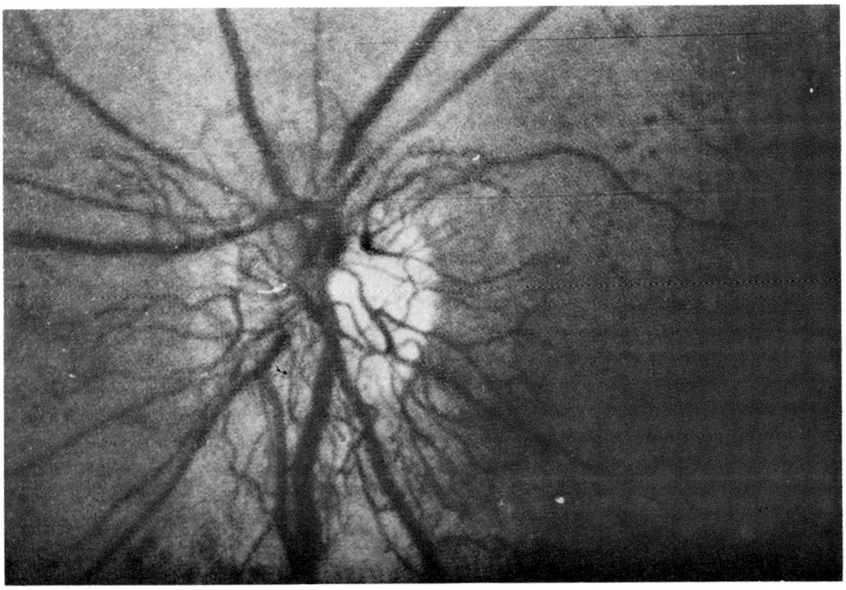

FIGURE 6. Proliferative diabetic retinopathy with large neovascular frond extending off optic disc.

areas of focal leakage from specific capillary lesions (microaneurysms, dilated capillary segments) which are often associated with hard exudate rings. Diffuse macular edema presents a diffuse leakage from extensive areas of retinal vessels around the macula and the optic disc.

b. Proliferative Diabetic Retinopathy

Five to ten percent of diabetics with retinopathy develop proliferative changes. These occur in the presence of severe microvascular abnormalities, particularly capillary nonperfusion which results in retinal ischemia. It is postulated that a vasoformative substance is released from the retina which stimulates the growth of new vessels (Figure 6). This occurs at the edge of capillary nonperfusion at the periphery of the retina and at the optic disc. Neovascularization can be limited to the surface of the retina or may invade the vitreous. These new vessels may bleed, resulting in vitreous hemorrhage. Fibrovascular proliferation in the vitreous results in vitreous constriction, retinal traction, retinal detachment, and loss of vision.

Evaluation of diabetic retinopathy and of other retinal vascular diseases is greatly aided by fluorescein angiography, which consists in ophthalmoscopy, following the i.v. injection of fluorescein (Figure 7). This special technique provides sequential evaluation of the anatomic-physiological status of the choroid and retinal vessels. In the case of diabetic retinopathy, large numbers of microaneurysms not diagnosed by ophthalmoscopic examination can be seen with fluorescein angiography. Focal areas of closure of capillaries are seen as areas of nonperfusion. The site of capillary leakage is also identified by fluorescein angiography (Figure 8).

c. Photocoagulation Treatment of Diabetic Retinopathy

The diabetic retinopathy study[9,10] conducted by the National Eye Institute in 15 clinical centers in various parts of the U.S. showed that photocoagulation therapy is beneficial in several specific stages of proliferative retinopathy, as well as in macular edema.

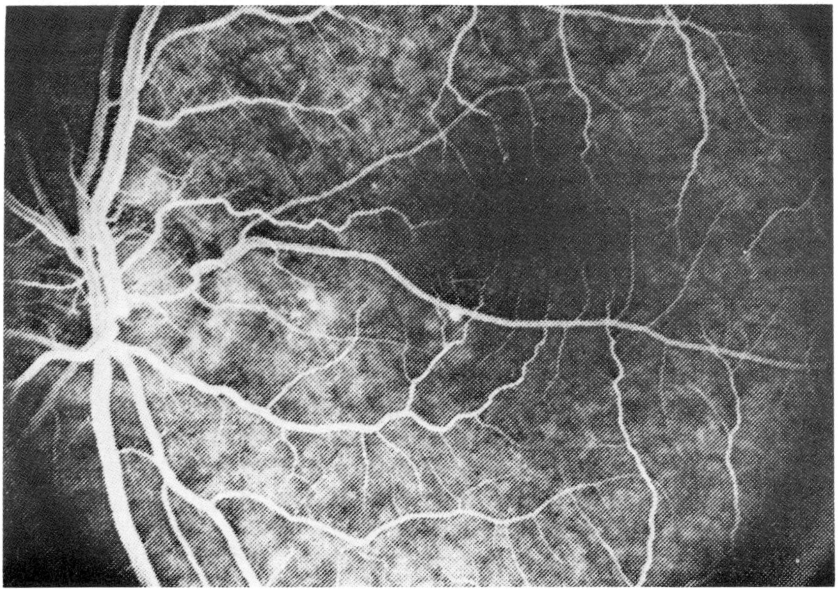

FIGURE 7. Normal fluorescein angiogram demonstrating filling of retinal blood vessels with dye.

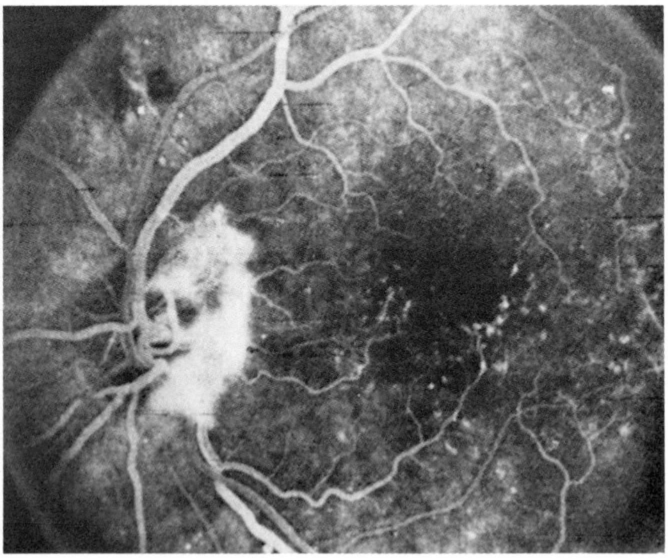

FIGURE 8. Fluorescein angiogram of early nonproliferative diabetic retinopathy showing marked leakage of from optic disc and microaneurysm formation.

The photocoagulation in diabetic retinopathy patients utilizes the argon laser, but it can also be done with the krypton laser. The technique employed is the pan-retinal photocoagulation (Figure 9). This technique consists in the placement of multiple mild retinal burns throughout the fundus, sparing the macular area and the optic disc. The burns are done with low laser beam intensity (0.6 to 0.7 W) and with spot size of 200

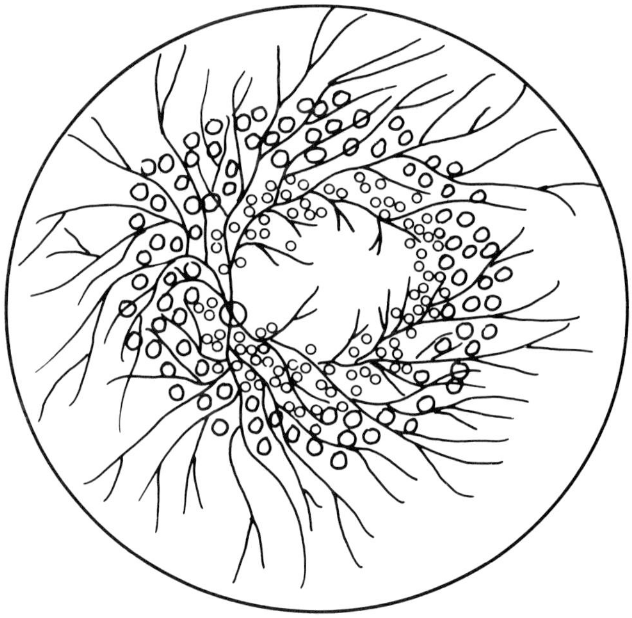

FIGURE 9. Pan-retinal photocoagulation: multiple retinal burns placed throughout the fundus sparing the macular area and optic disc.

to 500 μm. The idea of pan-retinal photocoagulation originated from the observation that proliferative diabetic retinopathy does not develop in eyes with chorio-retinitis scars. It is based on the hypothesis that by destroying a portion of hypoxic tissue, the remainding will be adequately supplied. By this technique, a large portion of the retinal tissue is destroyed, with reduction of its metabolic requirement without producing a significant central visual loss. Theoretically, the destruction of retinal hypoxic areas reduces the vasoformative factor and thus diminishes the stimulus to neovascularization which as a result will regress or be obliterated. The technique therefore affects the cause of neovascularization production; pan-retinal photocoagulation not only prevents the formation or causes regression of neovascularization, but it also creates multiple sites of chorioretinal adhesions which resist retinal traction by the fibrovascular membranes in the proliferative stage. Pan-retinal photocoagulation induces a dramatic regression of new vessels, especially of disc neovascularization, within 2 weeks of the therapy.

Although pan-retinal photocoagulation benefits patients with proliferative diabetic retinopathy, some adverse effects have been associated with this treatment. They include peripheral field loss, reduction of several lines in visual capacity, macular edema, choroidal detachment, serous retinal detachment, tractional retinal detachment, and vitreous hemorrhage.

Photocoagulation treatment of focal diabetic macular edema was proved to be effective in retaining pretreatment visual acuity, but not in improving it. The treatment consists in focal treatment of leaky aneurysms and dilated capillary segments, particularly those located in the center of hard exudate rings. After laser treatment, there is reabsorption of the hard exudate.[11] In diffuse macular edema with extensive leakage, but relatively mild retinopathy in the periphery, a significant absorption of edema was demonstrated in eyes treated by the "grid" photocoagulation technique (Figure 10).[12]

may be extracted easily in some septic loosenings. However, in a large percentage of cases, the removal of a broken artificial joint and the removal of bone cement may be extremely difficult. There is no specific way of removing the cement, and each surgeon has developed his own technique or instruments to remove all the cement and at the same time to avoid breaking the bone or weakening it by cutting a window.

The CO_2 laser scalpel is of real advantage of this type of surgery, particularly where there is a cement plug deep in the shaft of the bone. In such a case, it is then possible to sublimate the plug,[6] or to perforate the cement without any pressure on the bone, thus minimizing the risk of fractures and facilitating the extraction.

Another use of the CO_2 laser scalpel is the cutting of the polyethylene components of endoprosthesis for adapting to the shape of the joints or removing them again without weakening the bone or the interface bone cement by chiseling.

D. Bone Cutting and Bone Consolidation

Specific to orthopedics are the problems of cutting the bone and bone consolidation.[7] Bone cutting was experimentally studied on fresh animal bones moving at predetermined speeds under a fixed laser beam, with the output being the only variable. Tauber et al. determined the cut depth at high voltage as related to the time during which the beam remains on the bone. When augmenting the voltage by 80 to 120% using a hyperpulsed beam, the cut is not deeper, but narrower, by up to 50%. More precisely, the laser scalpel does not really cut the bone, but creates a separation, altering organic substances and melting the inorganic ones to a char deposit in the depth of the cut. These deposits are composed principally of crystals of hydroxyhepatites which remain amorphic in the bone and do not alter the consolidation. If the laser scalpel is activated a second time on the same cut, the energy of the beam is partially absorbed by the deposed char, which melts. This results primarily in a thermal effect and a loss of cutting property. The result is an attrition chamber in the depth of the bone cut, with a broad destroyed area surrounded by devitalized bone. Therefore the osteotomy using the laser scalpel has to be done by one pass-by only, giving the beam enough time on the bone to enhance its penetration into the medullary canal. This is achieved when the whole bone becomes lucent, due to the dephasing of the beam at the end of the cut. If the cut is not deep enough at the end of the first cut, the char in the bottom of the cut has to be removed with a thin osteotome prior to reactivation of the laser on the same cut.

Bone consolidation was also studied experimentally.[7] Tauber et al. found no significant difference with respect to the consolidation time or percentage of pseudarthrosis in bones osteomized by osteotomes, manual or motorized saws, or CO_2 laser scalpel. Microscopically, the cut edge of a bone cut by laser scalpel presents a thermal devitalized zone of approximately 2 mm. Then there are three to five layers of empty osteocytes, and without transition, normal cells. Therefore the first phase of callus formation is slower following laser osteotomy. The callus then hardens massively, and the final time of ossification is similar to classical osteotomy. The authors presume that the hydroxyhepatite crystals liberated and crystallized by the thermal effect by the laser beam could be a catalyst of osteogenesis. Nevertheless, in clinical applications, when consolidation has to be achieved, they always try to use a dynamic compressive osteosynthesis device such as an external fixator which enhances compression for days following the operation (Muller, Wagner, or Sheba type).

E. Surgery of Infective Tissues

The authors found some advantages to use of the CO_2 laser scalpel in surgery on infective tissues, such as excision of infected postoperative wounds, especially for the treatment of infected total joints and of infected pressure sores in paraplegic patients.

In these cases, they were able to perform primary closures by wide-rotation plastics with a good security.

F. Disadvantages

Contrary to these findings, in the surgery of fine structures, the laser scalpel presented no advantage to the authors. In general, a rather high complication rate made them even believe that the existing device is dangerous. For instance, in hand surgery, the proximity of nerves, vessels, and tendons obliges the use of protective techniques. While in the hip, thigh, knee, and elbow region, it is easy to utilize wet gauzes, or to infiltrate the tissues with a large quantity of saline solution around the region to be protected, in fine structure surgery only a few milliliters can be injected. This small quantity of fluid is immediately vaporized by the laser beam and the thermal diffusion creates damages incompatible with good functional results.

In arthroscopic surgery of large joints, the authors did not find any advantage in introducing the laser beam in the operative arthroscope, but they have the feeling that in conjunction with microfiberoptics, a new field of endoscopic surgery of little joints may be opened.

III. CONCLUSIONS

In conclusion, the CO_2 laser scalpel is of definite surgical importance, provided that the indications of its utilization are carefully considered. In the authors' opinion, coagulopathies and malignant tumors are an absolute indication of use of the CO_2 laser scalpel, while joint surgery in inflammatory diseases is a relative indication. There are some advantages in using the device in surgery of infective tissues. It seems that the field will gain a new impetus with the application of microfiberoptics.

REFERENCES

1. Moore, Y. H., Laser energy in orthopaedic surgery, in *Proc. Int. Congr. Orthopaedic Surgeons,* Excerpta Medica, Amsterdam, 1973, 1077.
2. Horoszowski, H., Farine, I., and Engel, J., The laser in orthopaedic surgery, in *Proc. 1st Int. Symp. Laser Surgery,* Kaplan, I., Ed., Jerusalem Academic Press, Jerusalem, 1976, 139.
3. Farine, I. and Horoszowski, H., The use of laser scalpel in orthopaedic surgery, in *Proc. 2nd Int. Symp. Laser Surgery,* Kaplan, I., Ed., Jerusalem Academic Press, Jerusalem, 1978, 351.
4. Horoszowski, H., Heim, M., Seligshon, U., and Farine, I., Use of the laser scalpel in orthopaedic surgery on hemophilic patients, in *Haemophilia,* Seligshon, U., Rimon, A., and Horoszowski, H., Eds., Castle House, London, 1981, 189.
5. Kaplan, I., Personal communication.
6. Ganel, A., Farine, D., Horoszowski, H., Shlezinger, Z., and Farine, I., Bone cement melting by a laser beam, gas analysis, *Lasers Electro-optics,* 1, 12, 1981.
7. Tauber, C., Farine, I., Gasner, S., and Horoszowski, H., Fracture healing in rabbits after osteotomy using the CO_2 laser, *Acta Orthop. Scand.,* 50, 385, 1979.

Chapter 5

DIAGNOSTIC USES OF LIGHT

A. Werner, A. Flint, and E. Azizi

TABLE OF CONTENTS

I. GENERAL INTRODUCTION

Diagnostic skills are acquired by knowledge, experience, and a keen power of observation where the integrative contribution of all our senses is needed, with vision playing the key role. Since vision is light-dependent, no diagnosis is possible without light. The direct role of light in diagnosis is based essentially on the fact that almost any biological material or its constituents can be identified by the light it absorbs or transmits, and that biological compounds may be characterized by their specific absorption spectra.

The first step to physical diagnosis is the detection of any signs which result from optical interaction of light with the patient's skin. This includes absorption of light by the skin and its chromophores (i.e., melanin, hemoglobin, bilirubin, etc.) and its reflection which is detected by the naked eye as a change in skin color. Thus, pallor, cyanosis, or jaundice are easily recognized and interpreted by the experienced physician as signs of underlying disorders. It is a fact that in dermatology the clinical diagnosis rests exclusively on the gross morphological appearance of skin lesions, their size, thickness, anatomic distribution, and pigment variations. By modifying conditions or light sources, it is possible to diagnose — still with the naked eye — additional changes in the skin and underlying tissues.

Diagnostic laboratory procedures are also often based on a characteristic color change in tissues, body secretions, microorganisms, and others, or on changes in optical density of a biochemical solution; all resulting from optical interaction between light and the test sample.

When dealing with tissue not accessible to direct inspection, such as the lung, bronchial tree, GI tract, or urinary bladder, examination is made possible by the use of flexible fiber endoscopes which transmit light to and from the test area.

Absorption of light by biological systems may lead to more complex processes followed by pathological consequences. Under these circumstances, the wavelengths and the intensity of the exciting light will determine the extent of the photochemical reaction and its resulting clinical manifestations. Thus sun exposure may lead to sunburn, to phototoxic and photoallergic skin reactions, or to the elicitation of metabolic, inherited, or idiopathic photosensitivity disorders. The diagnosis of these conditions rests not only on their characteristic clinical features, but also on the artificial induction of the abnormal response by exposure of the skin to appropriate wavelengths of the solar spectrum in a series of phototests. The procedure will enable the clinician to identify the specific wavelength responsible for the abnormal response and thereby permit him to select appropriate preventive measures.

It would be presumptuous to attempt a comprehensive description of all the uses of light in diagnosis. Due to the authors' dermatological and oncological orientation they have limited themselves to two major fields only: one concerns the diagnosis of disorders in which sensitivity to light is a prominent feature; the other focuses on the phenomenon of fluorescence as an aid to clinical and laboratory diagnostic procedures. These two major fields will be discussed in detail.

II. SKIN PHOTOTESTING

A. Introduction

In skin phototesting, UV radiation and occasionally the visible part of the electromagnetic spectrum is utilized. The lower limit of UV is set arbitrarily at a wavelength of 10 nm (10^{-8}m) and extends to 380 to 400 nm, where the visible part of the spectrum begins and spans up to 800 nm. The lower part of the UV spectrum, referred to as the vacuum UV, is of little interest. Even through emitted by the sun, none of it reaches the surface of earth, being scattered and absorbed by the constituents of the atmosphere.

The scattering of light follows Rayleigh's law and is proportional to $1/\lambda$. Obviously, scatter is a strong function of the wavelength λ. The factor $1/\lambda$ implies that scatter of UV radiation of 200 nm is 40 times larger than that of blue light ($\lambda = 400$ nm). The absorption is governed by the exponential law $I = Io \times e^{-kx}$ where Io is the intensity of the initial beam at a particular wavelength, I is the intensity of the beam after a penetration through distance x, and k is a constant which varies with wavelength.

Due to these two main processes, practically none of the radiation below 300 nm emitted by the sun reaches the surface of the earth. Paradoxically, although the sun is the greatest benefactor of all species, many of the adverse conditions discussed in this chapter are triggered by radiation coming from the sun. Wavelengths shorter than 300 nm are highly germicidal and carcinogenic.

For convenience, the UV spectrum of interest in dermatology is divided into three regions, namely: AVC (< 290 nm), UVB (290 to 320 nm), and UVA (320 to 400 nm). The laws governing scatter and absorption from of the atmosphere apply also to interaction of light with matter such as solids, liquids, and suspensions as well as to the constituents of the skin. However, due to the complex composition of the human skin, these laws are of not much use in predicting how much of the light beam is absorbed. Experiments which were conducted on excised skin are also of little interest as far as skin phototesting is concerned, though there is some evidence indicating that the epidermis transmits UVB radiation. No measurements have been carried out in vivo.

The only measurable physical entity is the reflection of the light from the skin. Similar to scatter and absorption, reflection varies with wavelength. It is as low as 5% at 300 nm, it rises to some 70% in the visible region at 500 to 700 nm, and falls again beyond that wavelength.

Transmission, scatter, and reflection complement each other to determine the color of the skin. As certain wavelengths are more intensely absorbed by the chromophores in the skin, only the remaining fraction which is selectively reflected is observed by the eye.

Many factors may determine the reflectance and absorbance, that is, the fraction which is reflected and absorbed by the skin. The size and distribution of melanocytes, the number of blood vessels in the skin, the amount of blood they carry, or the saturation of oxygen in the hemoglobin, all have an effect on the amount of light absorbed or reflected. Still, under controlled conditions, reflected light may serve as a useful tool in photodiagnosis.

B. The Minimal Erythema Dose, Diagnostic Implications

The response of the skin to UV light exposure depends on the wavelength and on the delivered dose. It also varies from subject to subject, from one area of the skin to another, and sometimes in the same subject from day to day. It is therefore difficult to define the standard response. However, for clinical purposes, the determination of the minimal erythema dose (MED) for a given UV source, in a given subject, remains a practical tool.

The MED is defined as the least exposure, at a specific wavelength, that will induce on the skin, 24 hr after irradiation, an easily perceptible, uniform redness with sharp borders.[1] It is expressed in energy units of radiation delivered per unit area (J/cm^2). It can also be presented as exposure-duration for a UV source of known irradiance, where irradiance is expressed in watts per unit area (W/cm^2) and $1\ J/cm^2 = 1\ W/cm^2$.

1. Uses of the MED

The main use of the MED is to measure, if only approximately, the skin sensitivity to UV irradiation.[2] The MED thus enables one to determine if a subject has a normal or abnormal response to UV radiation. In the latter case, the MED may be either lower or higher than normal, appear immediately or later than normal, or present itself as a morphologically abnormal reaction (i.e., papular, urticarial, etc.). This may be helpful in the diagnosis of light-associated diseases.

2. Method of Determination of the MED

a. Irradiation Equipment

As the wavelengths most effective in producing skin erythema are within the UVB range, these are usually selected for evaluation of the MED. The broad-band fluorescent sunlamps, with peak emission in this spectral range, are in common use. Selection of narrower wavebands may be obtained by the use of interference filters, or monochromators, in conjunction with high-power xenon or mercury lamps.

Artificial UV radiation sources require some time for their output to stabilize, therefore, before starting a test, a waiting period of a few minutes is required, as indicated by the lamps' manufacturer. The distance from source to skin should be kept constant during the test procedure, and preferably minimal in order to make the exposure time as short as possible (the intensity of light is a function of the inverse square law). The calibration of the source must be checked periodically.

b. Irradiation Procedure

The skin of the lower back is the site of choice for testing: it is a fairly large area, relatively flat, and usually untanned. Outside of the areas deliberately exposed, the skin of the patient should be shielded with opaque material. Care must be taken to protect the eyes of the persons exposed, as wavelengths shorter than 310 nm may produce conjunctivitis and keratitis, and lead to cataract formation.[3]

The patient is postured perpendicularly to the light source, either sitting straight in a chair, or better, stretched out prone on a table, at a fixed distance from the lamp. While the lamp distance is kept constant, the dose is varied by altering the duration of the irradiation. Five to eight 1-cm square areas of skin are irradiated with the UV source in a stepwise series of doses, the lowest dose being one third to one half the estimated MED, based on previous calibration of the UV source and on the skin type of the subject.[4]

The dose in subsequent squares should progress geometrically in increments of 20 to 40%. When the UV source provides a relatively large and uniform radiation field, the exposure series may be administered simultaneously by successive covering of corresponding windows cut out on an opaque template, which is closely attached to the test site by an adhesive tape (Figure 1).

If the source allows for only a small irradiation field, each exposure dose should be administered to a single skin patch, one at a time. The irradiated skin is examined the following day, 20 to 24 hr postirradiation. The MED is then determined, usually by visual assessment alone, under standardized light conditions.

3. Interpretation of Results

The square that received the lowest dose of UVR and meets the criteria of perceptible erythema with distinct borders corresponds to the MED.

a. The Normal MED Response

The range of normal values of the MED for selected wavelengths and wavebands compiled from finding of several studies is presented in Table 1. It illustrates the fact that with the commonly used broadband UV sources, the normal MED values may differ by a factor of two to three.

The mean MED for midday-midsummer sunlight at a latitude of 44° N is approximately 20'[5] and in Israel at a latitude of 32° N, the mean MED is 19.3 ± 5.2' for skin types 2 and 3.[6] Data on some common sources of radiation and their MED are provided in Table 2.

The MED is influenced by the testing conditions and will depend on factors such as the skin color of the subject, his ability to remain passive during testing, the spectral distribution of the radiation applied, and possible shadow effects. Moreover, the normal MED response at 300 nm may vary by 50% from one day to another in the same skin region, in the same patient.[7] Nevertheless, as long as one keeps to the definition of the MED given earlier, and keeps standard the test procedure, the use of the MED as a diagnostic measure of individual photosensitivity is probably reliable.[8]

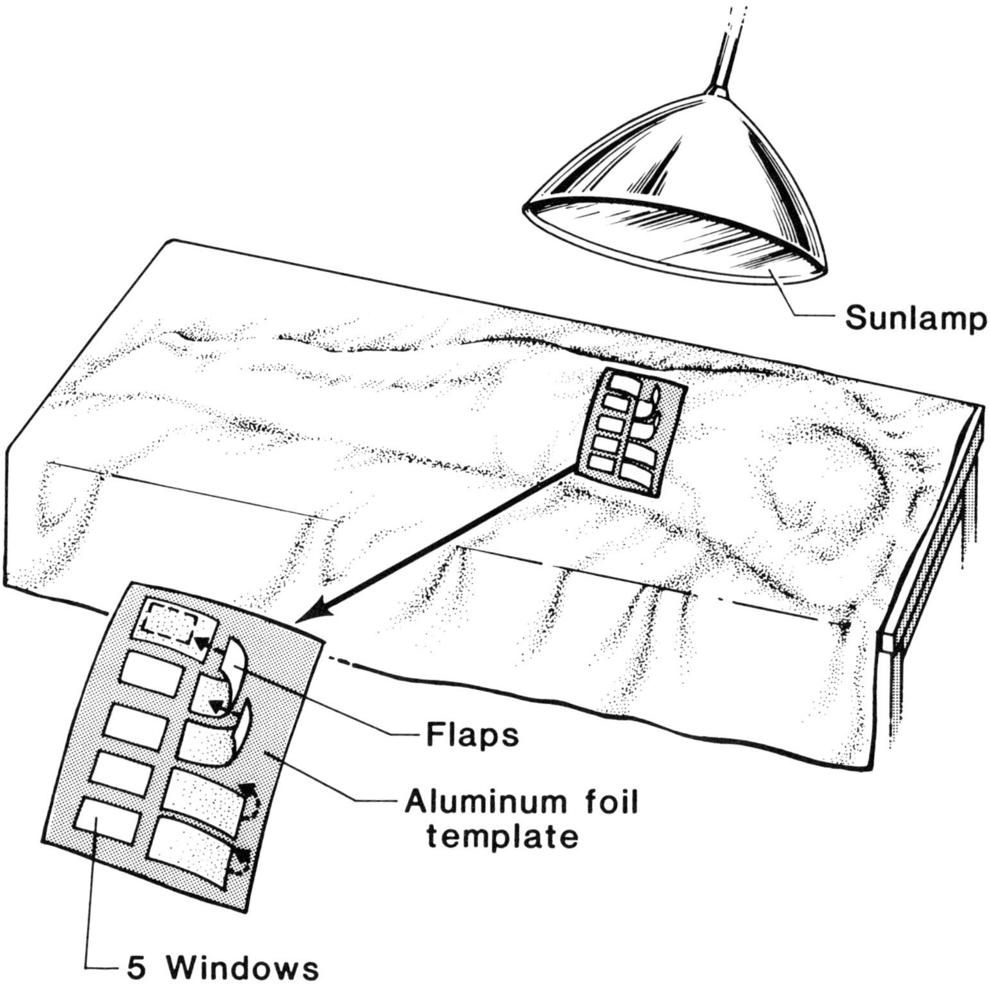

FIGURE 1. Irradiation procedure for skin phototesting.

b. Pathological MED

The MED is considered pathological (1) if it is lower or higher than normal, (2) if the response is morphologically abnormal, (3) if the response is immediate (within minutes after exposure) or if it is delayed by more than 24 hr. Pathological MED to sunlight has been found in light-associated disorders such as polymorphous light eruption, actinic reticuloid, xeroderma pigmentosum, lupus erythematosus, and solar urticaria.[9,10] They will be discussed elsewhere in this chapter.

C. The Diagnosis of Chemical Photosensitivity

In clinical medicine, chemical photosensitivity (CPHS) refers to a condition in which a specific chemical present in a tissue (most commonly a drug), will produce adverse reactions when activated by electromagnetic radiation of appropriate wavelength. Radiation alone or the agent alone will not produce any response.

CPHS can be classified as either phototoxicity or photoallergy, each having its characteristic clinicopathological features, discussed in more detail elsewhere in this book. The CPHS response is characterized (1) by its action spectrum, namely the wavelengths and radiation doses which activate the reaction; and (2) by a specific dose-response

Table 1
VALUES OF MED FOR SELECTED WAVELENGTHS AND SELECTED WAVEBANDS COMPILED FROM THE FINDINGS OF SEVERAL STUDIES[5]

	MED range (mJ/cm²)	MED rounded value (mJ/cm²)
UVC (254 nm)	6.9—12.0	10
UVB (300 nm)	11.6—25.2	20
UVB (310 nm)	205.0—238.0	200
UVB (broadband)	18.3—47.8	30
	MED range (J/cm²)	MED rounded value (J/cm²)
UVA (337.1 nm)		
Immediate	12.7—21.2	15
At 24 hr	14.8—27.5	20
UVA (broadband)		
Immediate	5—30	13
Persisting 4—6 hr	14—60	30
Persisting 24 hr	25—80	50

curve. The aim of appropriate diagnostic procedures is to determine these two parameters, one dealing with the physics of light, and the other with pharmacology.[11]

A correlation between the action spectrum of the CPHS reaction and the absorption spectrum of the suspected chemical strongly confirms the diagnosis of CPHS.[12] However, a lack of correlation may signify that the chemical was altered in the organism after exposure to light, as is the case for sulfanilamide, phenothiazides, halogenated salicylanilides, or psoralens.

For further support of the diagnosis of CPHS, elicitation of the response in patients or in experimental biological systems is necessary, using the suspected chemical followed by irradiation.

1. Phototoxicity

The majority of phototoxic agents are activated by radiation in the UVA range, with several important exceptions: sulfonamides and vinblastin are activated by UVB radiation, chlorthiazide by both UVB and UVA,[13] and some synthetic dyes such as eosin and rose bengal by visible light.[14]

The amount of energy required for the production of a phototoxic (PHOX) reaction varies with the PHOX chemical. For instance, sensitizers such as 8-methoxypsoralen (8-MOP), chlorpromazine, and coal tar require a low UVA dose (approximately 10 J/cm² or less), and therefore the diagnostic elicitation test may be performed with an artificial UVA radiation source of low output, such as a fluorescent black light. For the identification of photosensitizers requiring higher activating energy, a xenon arc, which has high output in the UVA and in the visible spectral regions, is desirable. This latter light source, because of its spectral qualities, is most suitable for screening new potential PHOX agents whose action spectra have not been previously established.[15]

2. Biological Assessment of Phototoxic Reactions

a. In Vitro Assays

Several in vitro single-cell systems, including cell cultures, RBC, fungi, and viruses have been used for this purpose. After introducing into these systems the chemical to be investigated and exposing them to irradiation, an assessment is made of the biological and chemical alterations which may result from the PHOX effect.[16] It must be

Table 2
DATA ON SOME COMMON SOURCES OF RADIATION[4]

Source	Typical source skin distance (cm)	Irradiance ($\mu W/cm^2$) UVC	UVB	UVA	VIS	MED, s	Application
Incandescent							
Single 40-W bulb	100	0	0	0.2	21	—	Illumination
Fluorescent, in planar array							
F40 T12 daylight, 5 lamps	30	0	1	10	300	—	Illumination
F40 BL (black light) 40 lamps	20	0	5	3,000	200	1,000[a]	Photochemotherapy
F40 T12 PUVA, 40 lamps	20	0	10	10,000	600	300[a]	Photochemotherapy
FS40 T12 sunlamp, 6 lamps	50	2	500	750	50	30	Phototherapy; suntan
F20 T12 special blue, 10 lamps	30	0	1	3	2,900 (425—475 nm)		Phototherapy of neonatal jaundice
Mercury discharge							
G15T8 germicidal, 3 lamps	50	200	3	5	10	30	Germicidal uses
Havonia alpine	100	30	50	130	350	90	Phototherapy
Kromayer	Contact	250	500	1,200	3,000	5	Phototherapy; phototesting
Xenon arc							
800 W	50	20	40	30	1,000	10	Research; solar simulation
Sunlight, clear sky							
Sun directly overhead, total	—	0	400	60,000	60,000	600	
Sun directly overhead, minus sky	—	0	150	3,500	50,000	1,500	
Sun 20° from overhead, total	—	0	250	4,500	50,000	1,000	

[a] Patients sensitized with psoralen.

noted, however, that most of these systems cannot serve with certainty as a predictive model for photosensitizing potential in man. Some have produced false-positive responses, while others have failed to detect various well-known PHOX agents.

Recently, Morison et al.[17] have described a method that appears to be a very reliable in vitro test for the detection of PHOX chemicals, and that seems to avoid the drawback of false-negative results. It makes use of a human lymphoid cell system, to which the potential sensitizer is introduced, which is then exposed to UVB and UVA broadband sources of radiation. The interaction between the chemical and the nonionizing radiation results in an alteration of the cell function, which does not occur either with the chemical alone, or with radiation alone. This alteration is measured by inhibition of the incorporation of tritiated thymidine into the DNA of the cells.

b. In Vivo Assays

1. Animal Assays

The PHOX reaction can be demonstrated easily in a number of animal models such as mice, rats, rabbits, and guinea pigs.[18,19] In these animals, it manifests itself as edema, erythema, and necrosis of the irradiated ears or tails.[20] By weighing the excised tail while still wet, immediately after the response, and again after drying, and comparing the weights with those of the tail of an identically treated but nonirradiated control animal, the PHOX reaction can be assessed.[21] The technique is well suited for studying the dose-response and kinetics of PHOX responses.

Animal assays have their limitations, however, and cannot replicate the actual environmental conditions in which man lives. A number of techniques have been developed, therefore, for human assays.

2. Human Assays

PHOX reactions in humans may be elicited either as a local response, following topical or intradermal administration of a potential photosensitizer, or as a more extensive response, following its systemic administration. Tests which involve only small areas of the skin are, of course, less hazardous.

For direct application of topical PHOX agents to the skin,[13] the test agents are dissolved in ethanol or methanol, in a nonirritating concentration of 2.5%, or — if insoluble — in hydrophylic ointment USP 5 $\mu\ell/cm^2$, and are applied to an area of the skin of the lumbar region, which is then covered with an occlusive dressing for 6 hr. After this time lapse, the area is irradiated with UVA and visible light, using a 150-W xenon arc lamp, equipped with a Schott W6-345 filter. The initial exposure dose, experimentally found, is 17.5 J/cm^2 UVA. If no response is obtained, higher doses may be used. In case of a negative response, the test may be repeated on scarified skin, using the same radiation dose.

It is important to examine the sites immediately after irradiation and again at 24, 48, and 72 hr later. PHOX responses may vary in time pattern; most substances will produce a peak at 24 hr, however fragrances will peak at 48 hr and psoralens at 72 hr.

For intradermal injection of photosensitizers,[13] 0.1-mℓ aliquots of the chemicals are dissolved or suspended in a saline solution adjusted to physiological pH. This solution is injected intradermally into the skin of the lower back of the tested subjects. When the action spectra of the agents are unknown, different test sites are exposed to either UVB, UVA, or visible light. Experience has shown that for these chemicals with unknown action spectra, the starting dose with UVA should be 30 J/cm^2, increasing to 40 J/cm^2 if necessary. With solar-simulating UVB, suberythemal doses (three fourths MED) are given. Larger doses would result in excessive erythema, which would obscure a PHOX reaction. Reading is done after 24 hr. This technique has the great advantage of duplicating the actual clinical situation and is relatively easy to perform. However, it fails to demonstrate the phototoxicity of tolbutamide and chlordiazepoxide. A procedure that duplicates the actual conditions patients may encounter during therapy consists of the systemic administration of the PHOX agent in therapeutic doses 2 to 3 hr prior to several hours of exposure to sunlight.[22] This is rarely done however, because of the difficulties in controlling the exact radiation dose under outdoor exposure conditions, and the hazard of severe PHOX reactions from overexposure.

The oral[23] or i.v.[24] administration of PHOX agents followed by artificial irradiation with controlled dosimetry is now reserved for establishing the threshold dose for safe response to photochemotherapeutic regimens such as PUVA treatment, and the photodiagnosis or treatment of neoplasms with hematoporphyrin derivatives (HPD).

For quantification of the PHOX response, the minimal phototoxic dose (MPD) is commonly used. The MPD can be expressed as a function of either the radiation dose or the concentration of the chemical. The minimal PHOX radiation dose (MPDr) is the least radiation dose, in J/cm^2, required to produce a PHOX reaction, when the concentration of the chemical is kept constant.[25,26] As explained earlier, the MPDr must be taken into consideration in photochemotherapy in order to avoid untoward severe phototoxic reactions and to obtain a beneficial therapeutic effect.[23]

Table 3
MINIMUM PHOTOTOXIC CONCENTRATIONS OF SOME COMMONLY USED DRUGS[27]

Drug	MPDc (%)
Demethylchlortetracycline	0.0016
Chlortetracycline	0.031
Tetracycline HCl	0.031
Chlorthiazide	0.125
Chlorpromazine	0.0004
Promethazine	0.0008
Sulfisoxizole	0.125
Sulfanilamide	0.125
Griseofulvin	0.125
Tolbutamide	0.063

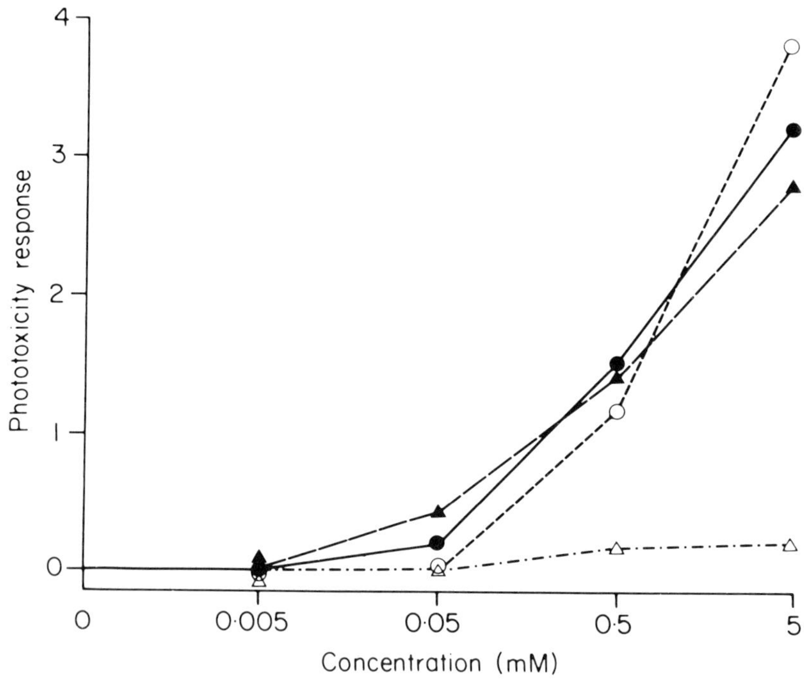

FIGURE 2. Phototoxicity assay for tar preparations on guinea pig skin.

The minimal PHOX chemical dose (MPDc) is the least drug concentration expressed in milligram percent or milligram per kilogram required to produce a PHOX reaction when the radiation dose is kept constant.[27]

The MPDc is used essentially for comparing the potency of phototoxic compounds. As an example, the comparative phototoxic concentrations for ten drugs, in human skin, are presented in Table 3. The MPDc was recorded as the least concentration required to elicit a positive response in three out of four subjects tested. Kocheaver et al.[28] have similarly established the MPDc of various tar preparations in guinea pigs (Figure 2).

Table 4
SCALE FOR GRADING ERYTHEMA AND PIGMENTATION[23]

Scale	Degree of erythema	Scale	Degree of pigmentation
0	No erythema	0	No change in basic skin color
1	Minimal perceptible erythema	1	Minimal perceptible tan
2	Light pink erythema	2	Light brown pigmentation
3	Marked erythema, red in color, but no edema	3	Moderate or medium brown pigmentation
4	Fiery red erythema with edema and tenderness	4	Dark brown pigmentation
5	Violaceous red erythema accompanied by marked edema, blistering, and tenderness	5	Intense deep brown or black pigmentation

Both MPDr and MPDc are determined at the time of maximum response, by a well-defined erythema and pigmentation in the tested skin area according to standard grading criteria[23] (Table 4).

Pigmentation readings are preferably done under diascopic pressure if erythema interferes. Variables which should be kept constant in this procedure are the method of administration or application of the chemical, the time lapse between administration and irradiation, the size of the area exposed, and the wavelength. The skin type and degree of pigmentation of the patient must also be taken into consideration. When both chemical concentration and radiation dose are kept constant, the intensity of the biological response reflects the phototoxic potential of the chemical.[29]

3. *Photoallergy*

Photoallergy, in contradistinction from phototoxicity, is essentially an immune response, in which specifically sensitized lymphocytes interact with the photoallergen. The diagnosis of this condition relies on the characteristic clinical picture and histological findings[30] as well as on the demonstration of light-dependent, cell-mediated immune response of the skin.

The in vivo tests that may provide either direct or indirect support to this diagnosis are, respectively, the photopatch test performed on patients, the induction of the photoallergic response to suspected chemicals, or the transfer of this hypersensitivity to naive animals. Also available for the further substantiation of the diagnosis of photoallergy are several in vitro assays such as a migration inhibition test and the transformation of human lymphocytes by a hapten-protein conjugate.[30,32] These, however, will not be discussed here as they do not make direct use of light.

a. *The Photopatch Test*

The photopatch test[18,33] consists of exposure of the patient's skin to the eliciting wavelengths of light 48 hr after topical application of the suspected chemicals to the same area. The photosensitizers are usually applied to symmetrical 1- to 2-cm^2 areas on the patient's back in two to four sets: two sets when only UVA is tested and four sets when UVB and visible light are also tested. The concentration of photosensitizers to be used is usually determined by trial and error. Ideally, the concentrations should be lower than the toxic or irritating dose, and below the phototoxic dose in the case of chemicals with that property. Ordinarily, a 1 to 10% solution in ethanol or acetone, or a suspension in petrolatum are used. The application sites are protected from ambient light with opaque material. The patches are removed and the sites are examined 48 hr later. Any skin change is noted.

One set of patch-test sites is then irradiated with UVA, which is the most common spectrum for eliciting photoallergic reactions, at a dose of 1 J/cm^2, which is below the MED UVA for most subjects. Since photoallergy is sometimes produced by visible light or by UVB, depending on the action spectra of the photosensitizers, other sites are correspondingly exposed to suberythemal doses of UVB and visible light. One set of sites treated by the test chemical is left unirradiated, and serves as a control. Also as control, untreated areas will be irradiated by the various wavebands.

The test areas are examined at 24 and 48 hr, and 1 week after irradiation. The diagnosis of photoallergy is made by finding, at the irradiated patch-test sites, an erythema with or without edema and vesiculation, and no response at the control sites exposed to the chemical alone or to irradiation alone.

When positive reactions of equal intensity are found in both irradiated and nonirradiated treated sites, the diagnosis is likely to be an ordinary contact dermatitis. If, however, the irradiated patch site shows a stronger reaction than the nonirradiated one, one may assume the coexistence of photocontact dermatitis with contact dermatitis.[34] The possibility of a false-positive response due to a nonspecific effect of radiation on the inflammatory reaction of an ordinary contact dermatitis should be considered as well, under the latter circumstances. When both treated and untreated irradiated sites are positive, the diagnosis of persistent light reactor has to be considered. Patients with this exceptional condition may have abnormal responses to light in the absence of the chemical on the skin.[35-37] Misinterpretations may still arise when a phototoxic response, rather than a photoallergic response, occurs, in spite of the low concentrations of photosensitizers used for photopatch testing.[38] Here, a histologic evaluation of the test sites may help in differentiating between the two responses.[30]

When all or some reactions are negative, the possibility of false-negative reactions should be considered. These can be due to either insufficient irradiation, poor penetration of the chemical into the skin, or to the possibility that the agent acted as a barrier to radiation.[34]

In some cases, a positive response to photopatch testing may be obtained with a chemical other than the primary photosensitizer, but usually related to it. This phenomenon of photo-cross-reactivity occurs, for example, between tri- and dibromosalicylanilide, or tetrachlorosalicylanilide (TCSA).[39]

These are only some examples which illustrate how difficult it is to interpret photopatch tests and how they must be viewed with a critical mind.

b. Induction of the Photoallergic Response in Experimental Animals

Historically, photoallergic contact sensitization had been experimentally induced in human volunteers by using halogenated salicylanilides and related chemicals.[37,39] However, the induction of altered reactivity states in human subjects who could eventually become persistent light reactors, is questionable from the ethical point of view.

Methods have been designed for the induction of photoallergy in various animal models, most commonly guinea pigs. The procedure is done by topical application of a high concentration of the chemical (10 to 100 times the "eliciting" concentration) to the depilated nuchal region of the animal, which is then exposed to UVA radiation. This procedure, in order to be successful, must be repeated three to ten times over a 7- to 10-day period. In order to facilitate the sensitization process, several adjuncts have been used: topical application of a detergent such as 20% sodium lauryl sulfate or cellophane stripping of the skin, which may reduce the barrier function of the skin.[40,41]

Intradermal injections of Freund's complete adjuvant[42,43] or prior exposure to UVB radiation are alternative methods.[36,44] An excellent critical review of these techniques has been recently published.[19]

The technique for the elicitation test is similar to that of the photopatch test. It is performed 2 to 3 weeks after induction, with two or three different concentrations of the chemical applied on the depilated lumbar region of the animal. The concentration of chemical required for eliciting the photoallergic response in these models varies with the substance tested, and may also depend on the method of induction. For example, in testing musk ambrette, a concentration of 5% is required when the skin-stripping technique is used for induction, but a concentration of only 0.1% will elicit a positive reaction when Freund's adjuvant is used.[42]

Another technique for induction of photoallergy, using mice as experimental models, has been recently described.[45-47] The animals in the two latter studies were first pretreated with cyclophosphamide which, by inhibiting the suppression of the response, enhances the sensitization process. Induction of photoallergy to TCSA, for example, is performed by administration of an i.p. injection of 150 mg/kg of cyclophosphamide, followed 3 to 4 days later by an application of a 1% concentration of TCSA to the shaved backs of the animals, which are then exposed to 1 mJ/cm^2 UVA while their ears are shielded from radiation. For elicitation of the response (7 days later), 1% TCSA is applied to the ears, and the animal is again exposed to 1 mJ/cm^2 UVA radiation. The photoallergic response is assessed 24 hr later by measuring the thickness of the ears.

c. *Transfer of Hypersensitivity to Naive Animals*

This has been done with TCSA using guinea pigs.[48,49] The naive animals were injected with peritoneal mononuclears from donor animals photosensitized with TCSA. The recipient animal was challenged with TCSA 24 hr later at two symmetrical sites, of which one was irradiated with UVA. Only the site that was exposed to both TCSA and UVA radiation showed a positive response. Control animals similarly treated which did not receive the sensitized cells from donors showed no reaction. Transfer of photoallergy to TCSA to mice by injecting naive animals with lymphoid cells from sensitized syngeneic mice also has been reported.[46,47]

D. The Diagnosis of Idiopathic Photodermatoses by Light Testing

Photodermatoses are a variety of metabolic, genetic, or idiopathic disorders of altered reactivity of the skin or other organs to nonionizing radiation, in which no exogenous photosensitizer can be identified as a pathogen. The role of light in the pathogenesis of these conditions is usually obvious from the clinical manifestations in affected patients. These manifestations are either an accentuation of the normal erythema response, an abnormal time course of this response, or a typical skin rash, accompanied sometimes by other systemic malfunctions. The clinical symptoms, signs, and pathogenesis of this group of diseases are described in several excellent textbooks to which the reader is referred.[4,34,50-53] In addition, each of these conditions has its characteristic action spectrum confined to either UVC, UVB, UVA, and visible regions, or to a broad spectral range (Table 5).

Confirmation of the diagnosis of a photodermatosis may be obtained first by light-testing the skin to demonstrate an altered MED and/or an abnormal erythema time course. In the next step, the provocation of typical lesions could be attempted by challenging the patient's skin with the appropriate wavelengths of electromagnetic radiation. For this purpose, one should search for those wavelengths which will produce a definite abnormal skin response with the least dose of energy, thus defining the action spectrum of the specific disorder. Final diagnosis obviously should be substantiated by

Table 5
ACTION SPECTRA FOR ABNORMAL SKIN RESPONSE TO LIGHT

Condition	UVC only	UVB only	UVA only	Visible only	Broad spectral range	Main features of abnormal response
Polymorphous light eruption	−	+	+	+	+	Persistent erythema papules
Photosensitive eczema	−	+	−	−	−	Erythematous papule
Solar urticaria	−	+	+	+	+	Immediate whealing
Actinic reticuloid	−	+	+	+	+	Reduced MED UVB; persistent erythematous papule
Xeroderma pigmentosum	+	+	+	−	+	Reduced MED; delayed erythema
Lupus erythematosus	−	+	−	−	−	Persistent erythema; reproduction of skin lesions
Pemphigus erythematosus	−	+	−	−	−	Reproduction of skin lesions
Familial benign chromic pemphigus	−	−	−	−	+	Erythematous papule
Keratosis follicularis	−	+	−	−	−	Reproduction of skin lesions
Disseminated superficial actinic porokeratosis	−	−	−	−	+	Reproduction of lesions
Porphyria	−	−	−	+	−	Erythema; burning pain and whealing may occur

histological examination of genuine and artificially induced lesions, as well as by relevant laboratory tests.

Due to the considerable variations among patients and the fact that in each particular individual the degree of sensitivity may vary from time to time, it is very difficult to formulate a standard light-test procedure. Nevertheless, the authors have attempted, in the following section, to provide some guidelines for phototesting of patients with photodermatoses, based on review of recent as well as older studies in the field.

In general, for preliminary screening of the patient's spectral sensitivity, one may choose broadband light sources such as sunlamps, UVB, or UVA lamps. Narrower wavebands of monochromatic radiation may be used for more precise determination of the action spectrum of the specific disorder. The sequence of wavelengths administered in a test series and the dose range should be selected on the basis of previous experience as discussed later, but adjusted individually to the patient under investigation. The size and anatomical location of the test site, as well as optimal timing, are important factors that may significantly affect the results of such procedures.

A negative phototest series should not by itself exclude the diagnosis of a photodermatosis, but when positive information is obtained, it is most valuable for establishing the diagnosis. This information will also be helpful for application of preventive therapeutic measures, such as the selection of suitable light-absorbing sunscreens and the induction in patients of tolerance to pathogenic wavelengths or of protective tanning.[54-58]

Table 6
SCREENING PHOTOTESTS FOR PLE[60,61,65,70]

Light source	Energy requirements for reproduction of pathological responses: Single exposure	Repeated exposures
UVB	>1—8 multiples of MED (0.04 J/cm²—5.82 J/cm²)	3—5 multiples of MED repeated 3—4 times a week (total 5—6 times)
UVA[a]	60—100 J/cm²	50—100 J/cm² repeated 4 times
Broadband, UVB, UVA, and visible	3—6 multiples of MED	

[a] High-intensity UVA light source, such as the UVASUN 5000, Mutzhas Comp FRG, suitable for the irradiation of large body sites with high doses of UVA.

1. Polymorphous Light Eruption

Polymorphous light eruption (PLE) represents an abnormal cutaneous response to light. It includes a variety of clinical forms and is subject to seasonal and geographical variations. Photobiological studies have established that PLE patients may fall into three major sensitivity groups: (1) abnormality to UVB only,[51,59-62] (2) abnormality to UVA only,[57,63-65] and (3) abnormality to a broad spectral range — UVB, UVA, and sometimes beyond into the visible.[62,63,66-72]

It is therefore suggested that for initial screening of a patient's sensitivity, broadband light sources be used (starting with either one of the previously mentioned spectral ranges), in order to determine the MED and/or the threshold dose for an abnormal cutaneous response. Optimal timing for diagnostic phototests in PLE is after the winter months prior to the patient's first exposure to the spring sun.[65] The patient should be free from recent genuine lesions, and test sites not smaller than 5 by 10 cm should be chosen, preferably in sun-exposed areas where the first manifestations of the disorder usually appear.

The MED UVB values are within normal limits in most studies of PLE.[9,73] However, in many cases abnormal skin responses can be reproduced with dosage levels lower than those required to produce minimal erythema in normal subjects.[10,51,62,72] Dose requirements for reproduction of pathological erythematous or papular responses, by either single or repeated exposures to broadband light sources, are presented in Table 6.

For more precise determination of the action spectra in PLE patients, exposure to narrower wavebands of 10-nm increments within the suspicious spectral range may be of help. Energy requirements for induction of abnormal papular responses and/or erythema in PLE patients by monochromatic radiation are presented in Table 7. Test sites must be examined at several time intervals, as the latent period between exposure and the appearance of lesions may vary from a few hours to more than 2 days (commonly 18 hr).

2. Photosensitive Eczema

The action spectrum of this condition lies almost exclusively in the UVB range in the 300-nm region. The patient's MED at this waveband may be lower than normal. An abnormal papular reaction may be obtained by administering one to three multiples of the patient's MED, from a monochromatic 300-nm radiation source.[74] Photopatch

Table 7
PLE — ACTION SPECTRA FOR INDUCTION OF ABNORMAL PAPULAR RESPONSE AND/OR ERYTHEMA[62]

Wavelength (nm)	Radiation dose range (mJ/cm²)
295	1.5—22
300	1.5—56
305	1.5—150
310	1.5—390
315	3.9—2,700
320	10—2,700
325	27—2,700
330	27—2,700
340	68—2,700
365	4,700—<10,000
400	<15,000—90,000
430	82,000—<100,000
460	<130,000
500	39,000—75,000

tests are usually negative, but ordinary patch tests to common contact allergens may be positive.[51]

3. Solar Urticaria

The urticarial response to sun exposure is relatively easy to investigate since the lesions appear within minutes after a challenge exposure. The action spectrum of the common idiopathic form of solar urticaria varies greatly: sometimes it is confined to a relatively narrow spectral range, either to UVB, UVA, or to visible light only;[75-78] sometimes it is broader than one spectral range, with sensitivity to either UVB and UVA, UVB and visible, UVA and visible, or visible and heat;[77] and sometimes patients will show an abnormal response to the full UV and visible spectrum from 250 to 500 nm.[9,75] Besides the considerable variability in spectral range among patients, the wavelengths responsible for the reaction in any given individual patient may vary over time.[76,79]

To aid in the subclassification of patients with solar urticaria, passive tranfer experiments from affected patients to normal individuals have been used. They have been successful in producing the urticarial response either with serum transfer prior to irradiation, or after irradiation.[75,80] These procedures, however, carry the risk of transferring the hepatitis B virus or other pathogens, and are now regarded as unethical.

The MED UVB in solar urticaria is essentially normal. The minimal, or threshold whealing dose is generally higher than the MED,[76] although it may be occasionally be lower.[9] In those cases showing no abnormal response to UVB, even doses of up to 10 MED failed to produce whealing.

When using broadband light sources, the presence of certain inhibitory wavelengths may sometimes interfere with the development of the abnormal response in certain cases of solar urticaria.[78] For this reason, tests should be conducted with monochromatic light also. The energy requirements for threshold responses to monochromatic wavelengths in solar urticaria are presented in Table 8.

In some cases, heat may enhance the urticarial response. In one patient, the production of whealing with relatively cold monochromatic light (from 450 to 475 nm) re-

Table 8
SOLAR URTICARIA — ACTION SPECTRA FOR INDUCTION OF IMMEDIATE ERYTHEMA AND URTICARIA[9,56,76-78]

Wavelength (nm)	Immediate erythema dose range (mJ/cm²)	Threshold urticaria dose range (mJ/cm²)
290	—	5.1×10^2
300	—	$2.7—6.0 \times 10^3$
310	1.0—2.3	4.5—100
313	5.6×10^2	—
350		2×10^3[a]
360	—	1.8×10^4
370	—	4.1×10^4
400	—	10^3[a]
410	—	$1.9 \times 10^3—2.1 \times 10^3$
425	12.5×10^3	$1.3 \times 10^3—18 \times 10^3$
450	$1.9—3.7 \times 10^3$	$3.7—26 \times 10^3$
475	—	6×10^4
500	—	$5 \times 10^3—8.6 \times 10^4$
550		10^4[a]
600		$>2.5 \times 10^4$[a]

[a] 30-nm bandwidth.

quired ten times the dose sufficient to produce the response with a relatively hot broadband light source, such as a slide projector with a spectral range from 360 to 690 nm (45×10^3 mJ/cm² as compared to 4.4×10^3 mJ/cm²).[77] Heat tests should be performed separately in suspicious cases, either by applying a warm metal (approximately 43°C) to the skin for a few seconds,[77] or by heating the skin with an electric fan for 5 to 10 min.[78]

Patients with peak reactivity in the visible (400 to 600 nm) should be differentiated from patients with porphyria cutanea tarda (PCT) or erythropoietic protoporphyria (EPP), which may also produce an urticarial response in this spectral range.

Solar urticaria can be a manifestation of anthracene phototoxicity. Photopatch tests with 10 $\mu\ell$/cm² of 0.25% anthracene solution at 360 nm, will be of help in differentiating this condition from the common idiopathic case. The threshold phototoxic dose by this procedure is $(1.0 \pm 0.6) \times 10^3$ J/cm² for immediate erythema, $(1.9 \pm 1.0) \times 10^3$ J/cm² for delayed erythema and $(2.8 \pm 1.9) \times 10^3$ J/cm² for wheal and flare.[81]

4. Actinic Reticuloid

Patients with actinic reticuloid (AR) have an abnormal reactivity over a wide spectral range, extending from 300 to 400 nm and not uncommonly to 600 nm. In the 300-nm region, the MED is greatly reduced in most cases, ranging from 0.2 to 5 mJ/cm², although occasional normal values have been reported.[9,34,73,82,83]

The abnormal response at the test site consists of an erythematous papular swelling which appears 6 to 18 hr after irradiation and may persist 1 week.

A similar marked abnormal response may be obtained after exposure of the patient's skin to long-wavelength UV at 350 or 360 nm, with doses ranging from 100 to 3000 mJ/cm².[9,34,82,83] Response to the visible, however, requires 150 to 5000 mJ/cm² in the 400 to 405-nm region[9] and up to 8000 mJ/cm² in the 500-nm region,[82] and usually produces only an erythematous reaction without swelling.

The biopsy taken from the test sites will be pathognomonic as early as 2 to 4 hr after irradiation, showing a dense lymphoma-like infiltrate in the upper dermis, with lymphocytes and eosinophiles invading the epidermis and reactive acanthosis.[83] Although photopatch tests are usually negative in AR, false-positive reactions may occur due to the marked sensitivity of the skin to UVA radiation. An abnormal in vitro sensitivity to UVA of fibroblasts and lymphocytes cultured from AR patients has recently been reported.[84,85]

5. Actinic Lichen Planus

The entire action spectrum of actinic lichen planus (ALP) has not been fully explored. The MED UVB is normal. The characteristic tiny violaceous coalescent papules typical of ALP in their morphology and histology can be induced in affected patients, by challenge with UVB radiation, starting from suberythemal doses up to 5 to 12 MED, repeatedly administered at intervals of 2 to 3 days for up to 3 to 4 weeks.[86-89] It is likely that severe erythematous reactions are necessary in order to trigger the abnormal response in this condition.

A long delay of up to 4 to 6 months in the appearance of experimentally induced lesions has been reported.[86] UVA doses of up to 3 J/cm^2 failed to produce erythema or any inflammatory reaction in ALP patients.[89]

6. Lupus Erythematosus

In lupus erythematosus (LE), both in the discoid and systemic types, the effective wavelengths for provocation of skin lesions are in the UVB range. It is easier to reproduce lesions in patients who present a definite history of photosensitivity.[9,90] The MED UVB in affected patients is generally normal.

Abnormal responses to 330 nm have also been reported at doses as low as 1.1×10^2 mJ/cm^2 to 1.2×10^3 mJ/cm^2.[9] A challenge of a single exposure of eight to ten MED followed 3 days later with five MED, is usually successful in inducing abnormal persistent erythema and edema, which may fade eventually or may evolve, within a period of 6 weeks, into scaly atrophic LE-like lesions. The biopsy of these artificially induced lesions, taken 6 to 7 weeks after exposure, reveals a typical histological picture with epidermal strophy and patchy liquefaction degeneration of the basal-cell layer.[9] In some of the patients, there is positive immunofluorescence (IMF) of the biopsy specimen from the test sites 7 to 24 weeks after exposure.[9,91,92]

Another provocation test consists of challenging the same skin site repeatedly with UVB radiation at 2- to 4-day intervals, at doses which maintain a brisk erythema (about five to ten MED). Persistent erythema and LE-like lesions may appear within 7 to 12 days, showing at times a typical histological picture.[90,93]

7. Pemphigus Erythematosus (Senear-Usher Syndrome)

The limited number of studies in pemphigus erythematosus suggests that the action spectrum of this disease lies in the UVB range, with a peak reactivity of 300 nm.[94-96] The MED UVB is generally normal. The threshold dose for induction of an abnormal response is 180 mJ/cm^2 or about two MED.[94] Blistering, in the irradiated site, may occur within 48 hr after exposure. The histology of the artificially induced lesions may show typical acantholysis as well as a positive direct IMF test.[94,97]

8. Familial Benign Chronic Pemphigus (Hailey-Hailey Disease)

In this disease, doses of UVB and UVA radiation in the range of 5 to 15 MED may produce abnormal erythema and edema after 48 hr.[98] The histology from the test site may show acantholytic vesicles. The direct IMF test is negative.

These results cannot be considered practical for accurate diagnostic purposes, but when positive they may help in the detection of unaffected heterozygote or homozygote relatives in this congenital disorder with incomplete penetration.[99]

Table 9
XERODEMA PIGMENTOSUM — ACTION SPECTRA FOR MED AT 24 AND 72 HR AFTER IRRADIATION[9]

Wavelength (nm)	Radiation dose (mJ/cm²)	
	MED, 24 hr	MED, 72 hr
250	2.3	2.3
280	3.0	2.3
293	2.3	1.1
296	1.9	1.5
300	2.5	2.0
305	7.0	2.0
310	2.5	25
320	50	25
330	100	50
340	1000	1000
350	No reaction	No reaction

9. Keratosis Follicularis (Darier's Disease)

The action spectrum for sunlight sensitivity in keratosis follicularis falls in the UVB range. Lesions typical in their morphology and histology were obtained in affected patients, with exposures for 3 min to UV radiation repeated three times daily using a Hanovia hot quartz lamp. These lesions persisted for 6 weeks and eventually vanished.[100]

10. Xeroderma Pigmentosum

The characteristic abnormality in affected patients is a delayed erythema over a wide spectral range from 250 to 340 nm, with the most effective wavelengths at 293 to 297 nm.[9,101,102] The peak intensity of the erythema for all wavelengths occurs at 72 hr. The MED for most wavelengths may be slightly lower than normal. The energy levels required to produce MED at 24 and 72 hr with various wavelengths is presented in Table 9.

11. Disseminated Superficial Actinic Porokeratosis (DSAP)

There is only one case report on the induction of characteristic skin lesions in this disease. A 275-W standard UV sunlamp was used for daily skin irradiations at a distance of 25 to 38 cm over a period of 124 days. No doses were recorded, but the exposure time was gradually increased, without causing undue discomfort. The histology of the experimentally induced lesions was typical, showing the cornoid lamella and other findings of DSAP.[103]

12. Porphyria

In PCT, EPP, and variegate porphria (VP), the characteristic action spectrum of photosensitivity, where demonstrable, corresponds with the absorption spectrum of porphyrin and lies in the visible region, most often around 400 nm, and less often at 500 to 600 nm.[53] The abnormal skin reaction produced experimentally is only an erythema in most cases. Burning pain and whealing, however, are not uncommon in PCT or EPP and must be distinguished from the nonporphyric types of solar urticaria. The threshold dose for the abnormal response may vary from 0.5 J/cm² to more than 50 J/cm² in different patients. A suitable light source for providing these energies should have a range of intensities from 10 to 107 μW/cm^2 for a wavelength of 400 to 410 nm.[73] The degree of photosensitivity in the various porphyrias depends on the type of porphyrins present, as well as their skin and plasma levels.

In patients with PCT, the threshold radiation doses at 405 nm were in the range of 1.2 to 3.8 J/cm^2, provided the plasma uroporphyrin level was higher than 10 mg%. In a group of patients with EPP, whose plasma protoporphyrin levels were lower than 4 mg%, burning pain was induced by exposure to either natural sunlight or a tungsten lamp for a period of 9 to 15 min, a dose equivalent to 26.2 to 585 J/cm^2. In the same group of EPP patients, an abnormal skin response could not be obtained on the same day by monochromatic radiation of 405 nm, except in one patient with the highest circulating level of erythrocyte protoporphyrin (1780 μg%). However, in this group, three out of six patients produced abnormal skin responses when the irradiated sites were heated and monochromatic radiation of 405 nm was used, in the dose range of 2 to 9.5 $\times$ 10^3 mJ/cm^2, as long as the RBC protoporphyrin level was close to 1000 μg% or higher.[73]

Repeated phototesting in porphyria may be useful in the evaluation of any treatment regimen. A good response to venesections, for instance, will manifest itself by a higher threshold for abnormal reactions. However, variations in the individual degree of photosensitivity will sometimes occur without any therapy.[53]

III. FLUORESCENCE

A. Introduction

The use of fluorescence in diagnosis is not a recent development. The natural fluorescence of some molecules has been observed since the turn of the century, when the examination of the skin under Wood's light became a helpful procedure for the detection and control of superficial fungus infections and for the diagnosis of various dermatological conditions. The red fluorescence of porphyrins also, for a long time, has been observed with Wood's light in the urine and feces of patients with porphyria.

In ophthalmology, sodium fluorescein has been used for staining of the anterior segment of the eye, to detect foreign bodies, corneal or conjunctival injuries, and herpes lesions, and this fluorescent dye was later found useful in the fitting of contact lenses.

However, fluorescence as a tool in diagnosis has seen a rapid development in the past three to four decades with the introduction of fluorescent markers capable of staining tissues and cells in a way that reveals structures and processes which could not be visualized before.

The phenomenon of fluorescence permits the visualization of substances which the eye cannot ordinarily observe. Some biological molecules possess natural fluorescence, while others can be easily marked with fluorescent dyes, making fluorescence a convenient diagnostic probe.

For fluorescence to occur, a molecule must absorb energy in the form of UV or visible light. According to the quantum theory, light is propagated as integral units, called quanta or photons. The energy of a photon can be expressed by $E = h\nu$, where E is the energy in joules, $h = 6.62 \times 10^{34}$ joule-sec is Planck's constant, and ν is the frequency in sec^{-1}. The wavelength of the photon can then be determined by introducing $\lambda = c/\nu$ where λ is the wavelength in centimeters and c is again a constant, representing the velocity of light in vacuum and is equal to 3×10^{10} cm/sec.

The wavelength of the absorbed light is very much dependent on the particular structure of the molecule and to a lesser extent on the environment. The molecule is characterized by a series of closely spaced energy levels. Under the process of absorption, the molecule is raised from its lowest energy state — the ground state — to an excited state of higher energy. The process can take place only when the energy of the absorbed photon equals the energy difference between the two levels.

The excited state is very short lived on the order of 10 nsec, after which the molecule returns to its normal ground state by either dissipating its excess energy as heat or by emitting a photon of fluorescence. The excited state is also highly reactive, and photochemical changes may occur. In the process of the internal rearrangement of the molecule in the excited state, a small amount of energy is lost and the emitted fluorescence is of lower energy (longer wavelength) than the light absorbed (Stokes law).

Not all the photons absorbed by a solution cause molecules to fluoresce; some molecules lose their energy through competing processes such as internal collisions or energy transfer to other molecules. The fraction of the absorbed energy which can be reemitted as fluorescence is termed "quantum yield" (Q):

$$Q = \text{number of photons emitted/number of photons absorbed}$$

In the absence of competing processes, the quantum yield of fluorescence equals 1, but very few substances approach this number. Several factors affect the quantum yield: some are internal, concerning the configuration of the molecule in its ground state and excited state and its chemical structure; others are environmental, concerning the solvent, temperature, pH, etc. Quantum yields approaching 1 have been reported to some dyes in organic solvents, some of them in use as fluorescent markers, as will be discussed in more detail later.

B. IMF

The fluorescent antibody technique was introduced in the early 1940s by Coons et al.[104,105] and developed later into a precise method.[106] It stimulated much interest in the fields of both clinical diagnosis and research, and found innumerable applications in the various medical disciplines and basic sciences.

The technique, essentially, utilizes fluorescent dyes for labeling proteins. These proteins are usually antibodies specific to a cellular, or an extracellular, antigen which is being explored. If the homologous antigen is present in the test system, fluorescence will be seen at the site of the antigen-antibody reaction when the preparation is examined with the fluorescence microscope.

Labeled antibodies have an advantage over ordinary cytochemical dyes in that they localize with more precision in tissues and cells. By identifying the site of the immunological reaction, they allow for a more dynamic interpretation of function and disease processes. When the label is a fluorescent dye, antigen-antibody reactions can be detected in extremely low concentrations. The technique of immunofluorescent staining can be completed within a few hours; it is therefore more rapid than autoradiography and even faster than most cytochemical stainings.

1. Labeling of the Antibody with a Fluorescent Marker

A large number of fluorescent dyes are in existence but only a few are suitable as markers in IMF microscopy. Essentially, an acceptable fluorescent marker should have a high fluorescent yield; it should not denature the antibody after conjugation. The conjugates should be stable on storage, and their use should be relatively simple and inexpensive.[107,108] So far, two substances have proved to conform closely enough to these criteria: fluorescein isothiocyanate and lissamine rhodamine B200.

a. Fluorescein Isothiocyanate (FITC)

This dye was first synthesized by Riggs et al.[109] and it has proved to date to be the best fluorescent marker. Its fluorescence efficiency is good with a quantum yield of

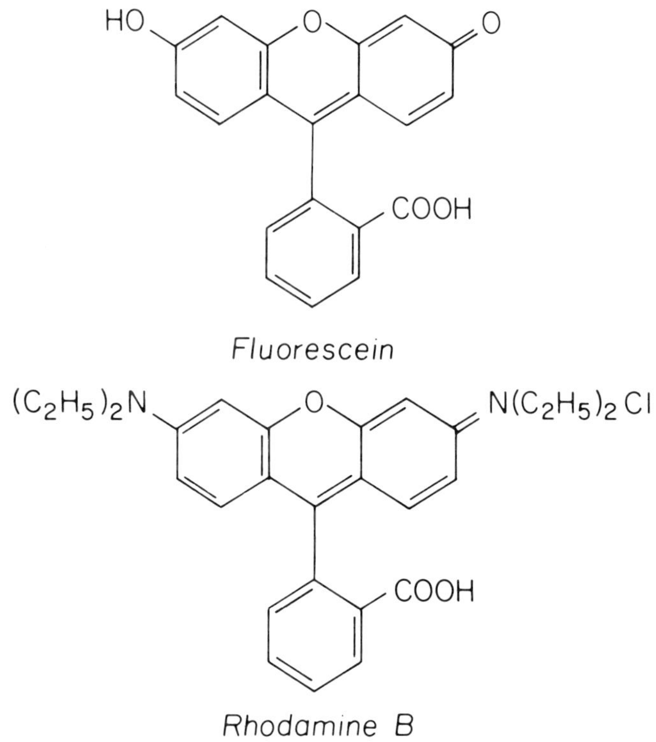

FIGURE 3. Structural formulas of the two most common fluorescent markers, FITC and RB200.

40%, and it combines well with proteins. It has yellow-greenish fluorescence. The chemical formula is shown in Figure 3 and the absorption and fluorescence spectra in Figure 4. It can be obtained commercially in a crystalline form.

FITC in aqueous solution loses fluorescence in time. The stained specimen, however, will retain its brightness for a few days, although it will fade gradually under the light of the microscope.

b. Lissamine Rhodamine B 200 (RB200)

This dye is available in powder form. It is soluble in water in which it shows an orange fluorescence. Its labeling efficiency is fairly good, although less than FITC. Its quantum yield is 25%. It will not combine directly with protein, unless first treated with phosphorus pentachloride.[110]

The fluorescence of RB200 also fades on exposure to light, but it will first change its color from orange to yellow. The yellow component does not fade as quickly as FITC, allowing for a more prolonged inspection of the specimen (more than 1 hr). FITC and RB200 can be used together for double staining of specimens.

c. Other Markers

Acridine orange described by West et al.,[111] DANS described by Weber,[112] and other fluorescent labels are more rarely used in special tracing experiments. The procedures for labeling of antibodies with fluorescent markers, and for the purification of the conjugate are described extensively in several excellent texts.[113,114]

2. Underlying Principles of Immunofluorescent Methods

There are several ways for the fluorescent staining of the antigen-antibody complex.

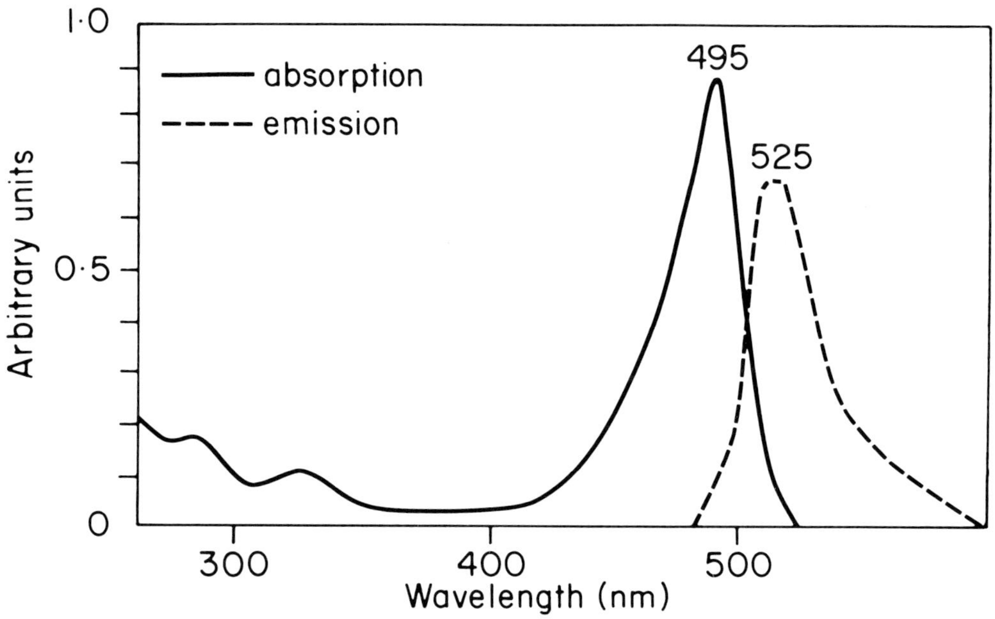

FIGURE 4. Absorption and fluorescent spectra of FITC.

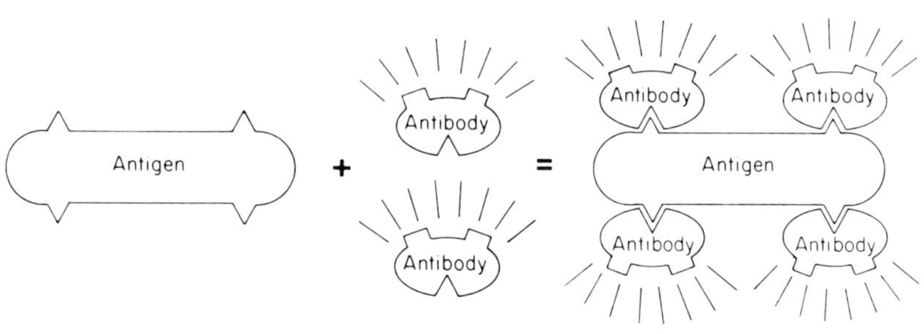

FIGURE 5. Schematic representation of direct staining reaction. Radiating lines indicate fluorescence. (From Goldman, M., *Fluorescent Antibody Methods,* Academic Press, New York, 1968. With permission.)

a. The Direct Technique

The specimen placed on a microscopic slide is exposed to a selected fluorescent-labeled antibody. The preparation is incubated, washed, dried, and mounted (usually in slightly alkaline buffered glycerin). Fluorescence will be evident at the site of the antigen-antibody reaction if the corresponding antigen is present in the tested specimen (Figure 5).

The specificity of this reaction can be verified with the following controls: (1) no fluorescence should result if the conjugate is not specific to the antigen under investigation, and (2) the fluorescent staining may be blocked by prior or simultaneous addition of unlabeled antibodies, which will occupy all antigenic sites and thus prevent them from reacting with the fluorescent antibody.

b. The Indirect Technique

This method utilizes an unlabeled antibody in an intermediate step. An antigen is first exposed to an unlabeled antibody derived from the serum of a known species

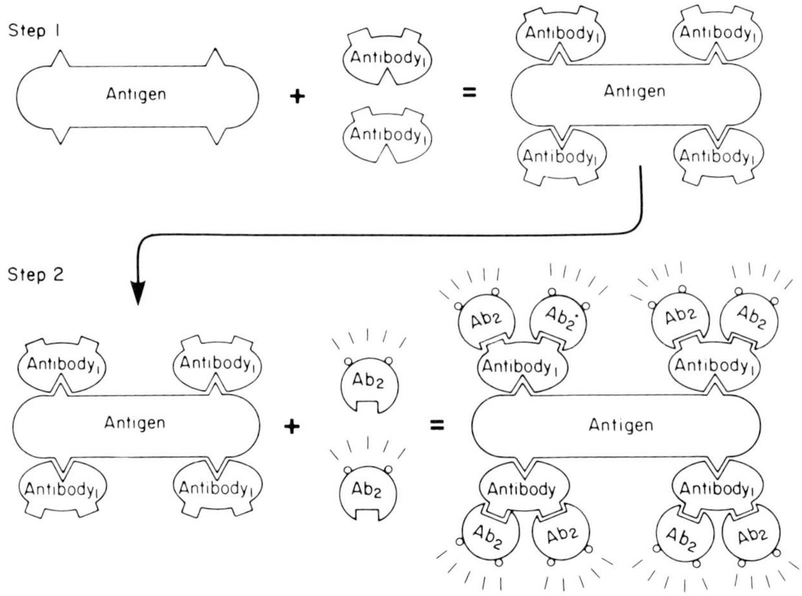

FIGURE 6. Schematic representation of antiglobulin (indirect) staining reaction. In step 1, nonfluorescent antibody$_1$ is bound to the primary antigen. In step 2, fluorescent antiglobulin (Ab_2) binds to antibody$_1$ acting as a secondary antigen. Radiating lines indicate fluorescence. (From Goldman, M., *Fluorescent Antibody Methods,* Academic Press, New York, 1968. With permission.)

"A". If an antigen-antibody reaction occurs, the complex is rendered fluorescent by adding a known labeled antibody derived from antiserum of another species "B" directed against the serum of species "A" (Figure 6). The specimen is treated and mounted as described under direct staining. This technique is suitable for the identification of either an unknown antigen or an unknown antibody. In order to detect antibodies to tissue antigens in human serum, animal tissues are used as substrates. This is made possible by the fact that antibodies to tissue antigens are not always species specific. The animal tissue is exposed to the human serum to be investigated and anti-human IgG-FITC conjugate is added. If antibodies are present in the patient's serum, the complex will fluoresce. The advantage of this method is that it is more sensitive than the direct technique, as the intermediate unlabeled antigen-antibody complex carries a greater number of sites to which the fluorescent markers can be attached. In some instances, antibodies or antigens can not be detected by the direct method but can be demonstrated by fluorescence with the indirect method. The indirect method, however, requires more complicated control for proof of specificity.[115]

c. Utilizing Unlabeled Complement in Addition to Unlabeled Antibody in the First Step

This indirect technique is based on the staining of a complement bound to an antigen-antibody complex by a fluorescent-labeled anticomplement from another species serum (Figure 7). The use of this technique allows the detection of either one of the unknown primary components attached to the complex.

The method was first described by Goldwasser and Shepard[116] who stained rickettsiae by first coating the organisms with unlabeled specific antiserum, together with unlabeled normal guinea pig complement. By adding labeled anticomplement (antiguinea pig globulin), the rickettsiae bodies were rendered fluorescent.

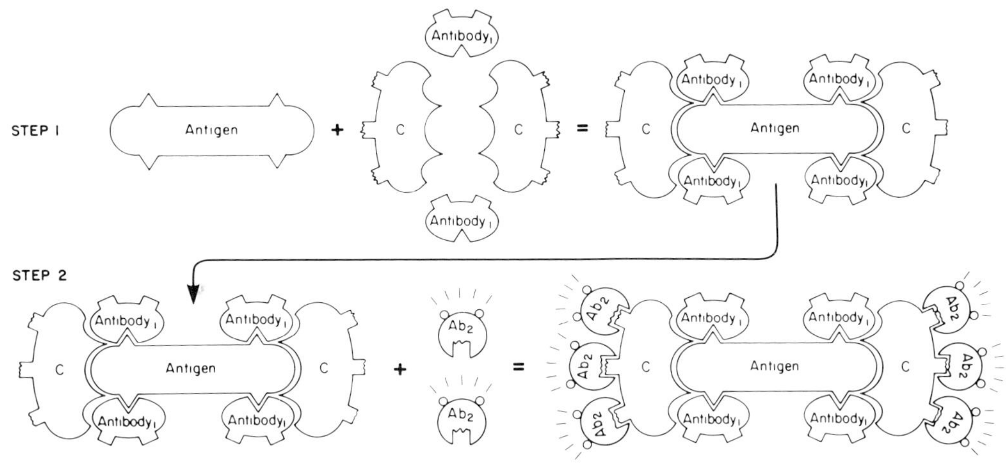

FIGURE 7. Schematic representation of anticomplement staining reaction. In step 1, nonfluorescent complement (C) becomes bound to the reaction product of the primary antigen and its nonfluorescent antibody$_I$. In step 2, fluorescent anticomplement (Ab_2) attaches to C acting as a secondary antigen. Radiating lines indicate fluorescence. (From Goldman, M., *Fluorescent Antibody Methods,* Academic Press, New York, 1968. With permission.)

This indirect method thus entails two intermediate substances. The complement is first fixed by the unlabeled antigen-antibody complex, prior to the addition of the fluorescent anticomplement, as described earlier.

The advantage here is that all antisera can be used, regardless of the animal species, and the final staining is obtained by only one single labeled anticomplement serum. Here the controls are somewhat more complicated, since the introduction of another reagent increases the potential sources of error.

d. Special Staining Procedures

Double staining — This procedure is used to localize two different antigens in the same specimen, by using a different fluorescent marker (usually FITC and RB200) for each corresponding antibody.

Combination with routine staining — It is often useful to compare the fluorescent preparation with the same preparation that has been routinely stained and examined under the light microscope. This allows cell counting, comparison of pathological changes, and better localization of the pathological process. The two procedures should be done separately, as a routinely stained preparation is not suitable for fluorescent staining and the fluorescence of an immunofluorescent preparation will be destroyed by routine staining.

3. Materials Used in the Performance of the Tests

Although the principles behind the immunofluorescent method are relatively simple, the technique requires careful preparation of the materials necessary for its performance. Specific antisera are obtained by immunizing laboratory animals with pure antigenic materials.[113,114,117] Methods for isolation and purification of immunoglobulins and bacterial or other antigens suitable for the production of specific antisera are described in the literature.[118]

The most commonly used FITC conjugates for staining tissue are antihuman IgG, IgA and IgM, C3, C4, C1q, properdin, fibrinogen, and albumin, all of which can be obtained commercially. For storage, they should be frozen at −20°C.

Some of the substrates used for studying antibodies to tissue antigens are rat liver, for the detection of antinuclear antibody; rat kidney and mouse stomach, for anti-smooth muscle antibody; rat kidney, for antimitochondrial antibody; mouse stomach, for antiparietal cell antibody; human or baboon thyroid, for cytoplasmic antithyroid antibody; rat muscle, for antiskeletal muscle antibody; and human skin, for antiepithelial antibodies.

4. Preparing Specimens for Examination with the IMF Technique

a. Serum Samples

Ten milliliters of blood are collected in a test tube without anticoagulant. Serum samples should be separated from the collected blood as soon as possible, and sent to the laboratory without freezing.

It is best not to send clotted whole blood, as hemolysis may interfere with proper staining. However, if a specimen is submitted hemolysed, the technique could still be tried, as it may still yield fair results.[119]

b. Tissue Specimens

The site of the biopsy should be carefully selected. As in diagnostic histopathology, it is often useful to submit a specimen which includes the normal margin of the observed lesion. Samples should be processed, as soon as possible, on the same day. There are several methods of fixation of specimens for a limited time period, either by using a liquid fixation,[120] or by quick-freezing the sample mounted in O.C.T. compound (Lab-Tek products) in an hexane bath.[121] By these ways, specimens can be preserved if they must be sent to a distant laboratory for processing.

Stained specimens are best examined under the microscope immediately after staining, when the fluorescence is at its brightest. They can still be examined later, however, if stored at 0 to 5°C in a box covered with polyethylene, but the fluorescence will have faded about 30% overnight and 50% after 1 week. Specimens stained with FITC and preserved in elvanol (polyvinyl alcohol) keep for a few weeks, although the intensity of the fluorescence will be diminished. Fading of preparations stained for IMF is more of a problem after their exposure to excitation light sources, and the greater the excitation efficiency, the worse it becomes. This phenomenon, which is particularly associated with FITC, has stimulated the search for antifading substances which would allow for more prolonged inspection of specimens under the fluorescence microscope, but so far no real solution has been found. It is therefore best to use a fresh specimen for photographic purposes.

Recently, a marked retardation of fading has been obtained by mounting stained preparations with glycerol containing *p*-phenylenediamine.[122] Promising results were obtained with other materials, in particular 1,4-diazobicyclo-(2,2,2)-octane (DABCO).

5. The Fluorescence Microscope

Under the fluorescence microscope, a powerful light source excites the fluorescent marker in the specimen, and the emitted fluorescence is then observed through the eyepiece, in a darkened room. A fluorescence microscope is based on principles similar to those of an ordinary light microscope. It consists of three main components: (1) a light source of uniform brightness, normally equipped with a condenser lens to focus the beam of light on the object; (2) an object holder, for holding the specimen, which can be moved in relation to the optical system; and (3) an optical system consisting of an eyepiece and a turret with a number of objective lenses of various magnifications which can be snapped into position in the light path. To this basic configuration some extra features are added for fluorescence microscopy. A set of optical filters is introduced to separate between the exciting light emitted by the light source and the fluorescence ensuing from the specimen. The filters are either interference filters designed to transmit with high efficiency isolated narrow-band wavelengths, or broadband fil-

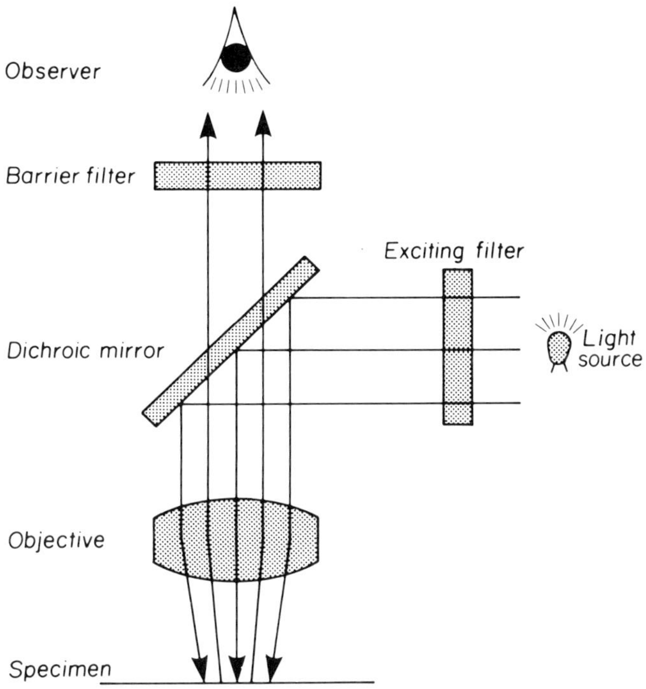

FIGURE 8. The light path of the fluorescence microscope.

ters, or cut-off filters which transmit all the light above a certain cut-off value. One set of filters, the "exciting filter", which transmits the exciting wavelengths range, is placed between the light source and the condenser lens. Another set of filters, the "barrier filter", is placed between the objective and the eyepiece, and transmits the fluorescence spectrum while suppressing the exciting light, thus preventing it from reaching the observer's eye. In fluorescence microscopy, care must be taken that optical components of the microscope, microscopic slides, and covers are made of nonfluorescing material.

Two types of microscopes are in use in fluorescence microscopy. One is the normal type, based on the transmitted-light method, where a direct line can be drawn between the light source, the specimen, and the eye of the observer. Another type, based on the reflected-light method, has an additional dichroic mirror which is placed at an angle of 45° to the light source. The mirror reflects, with high efficiency, light below a certain wavelength and directs it at the specimen, for excitation of fluorescence (Figure 8). This mirror is also transparent to light of longer wavelength, allowing the fluorescence emitted by the specimen to reach the eye of the observer. The advantage of the reflected-light method lies in the fact that the exciting light is not attenuated by passing through the specimen, and will permit the viewing of thick specimens.

The most frequently used exciting light sources in fluorescence microscopy are high-pressure mercury lamps. These lamps emit a line spectrum with certain powerful lines in the near UV and the visible, most suitable for use in fluorescence microscopy. The emitted lines, however, do not always coincide with the absorption maximums of the fluorescent marker. For instance, the absorption maximum of FITC lies at 492 nm, but at this particular wavelength and in the immediate vicinity, no spectral lines are emitted by the mercury lamp, and a compromise has to be made. Use is made, instead, of the very strong lines emitted at 405 and 438 nm which lie within the absorption curve of the fluorescent conjugate (Figure 9). For this purpose, a suitable exciting filter is chosen which transmits in this region.

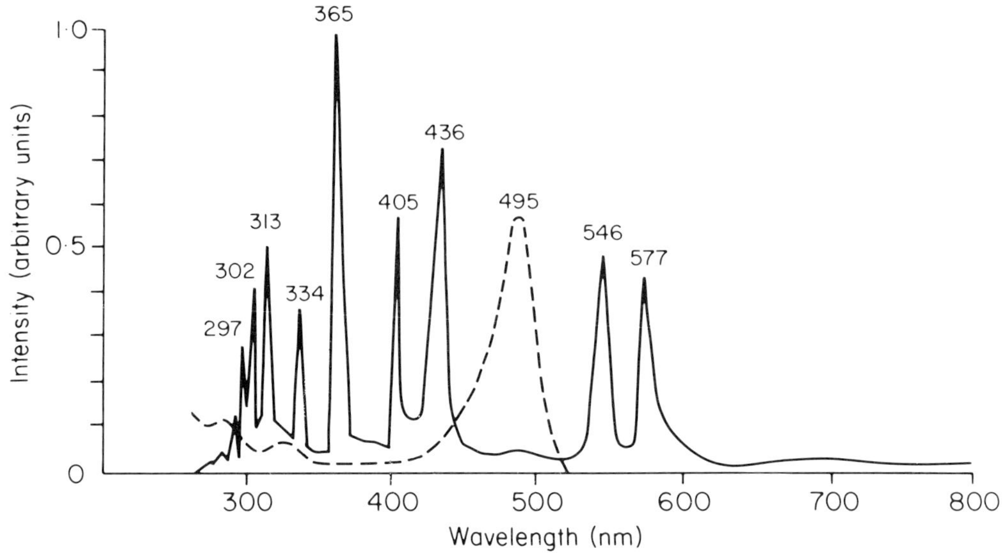

FIGURE 9. Emission spectrum of a high-pressure mercury lamp, and absorption curve of FITC superimposed on it.

The object holder is placed beyond the focal point of the objective lens, and an inverted image of the object magnified up to ×100 can be obtained. The eyepiece magnifies the image additionally (up to ×20) and focuses it onto the eye of the observer. In fluorescence microscopy, the most frequently used objective lens magnification is ×40, which is the same as for ordinary pathological slides.

The resolution of the optical system (and the microscope) may be expressed by $R = \lambda/(2n \sin\alpha)$, where R is the resolution (defined as the minimum possible distance one can distinguish between two adjacent points); λ is the wavelength of the illuminating light (λ and R are expressed in the same units, normally in nanometers); n is the refractive index of the medium between the specimen and the objective lens; and α is half the angle of acceptance of the lens (the angle of the broadest cone of light passing from a point of the specimen into the lens). $n\sin\alpha$ designates the numerical aperture NA which is normally inscribed on the objective lens. At a given wavelength, for good resolution (a small R), NA must be as large as possible. The highest value of NA encountered is 0.95, and in oil immersion objectives it is normally 1.35.

6. Diagnostic Application of IMF

IMF is used routinely in practically all branches of medicine. An extensive review of its numerous applications in diagnosis is therefore impossible in this test, where discussion will be limited to a short description of the most commonly performed tests.

a. IMF in the Diagnosis of Infectious Diseases

The detection of bacteria — Bacteria can be stained with suitable specific antisera conjugated with fluorescent markers, and rapidly identified under the fluorescence microscope. It can be done in direct preparations of clinical specimens or in brief cultures of the pathogens on selective media. The latter method is preferable as it enriches the bacterial population that is being sought and obviates the confusion often found in a direct clinical specimen where mixed flora and autofluorescence of the background may interfere with the diagnosis.

This direct fluorescent antibody technique can be used for most of the important human pathogens. It has the advantage of permitting rapid diagnosis, but it does not replace the isolation of the organisms in culture, essential for antibiotic sensitivity tests.

The indirect method is used for the detection of antibody to pathogenic organisms in the serum of patients, a major application of the fluorescence technique.

The detection of virus — Rapid diagnosis of viral disease is made possible by using the IMF technique on specimens taken directly from the infected patient. Virus inoculated in tissue culture can also be identified by IMF, and post-mortem diagnosis can be made from secretions as well as tissue specimens. The technique for collecting these specimens is extensively described.[123]

Among the viruses that can be rapidly diagnosed are the respiratory syncytial virus; influenza A; paramyxo viruses (parainfluenza, mumps); measles; picornaviruses (polio, coxsackie, echo); rabies;[124] adenoviruses (method not fully perfected yet); rubella (limited use); herpesvirus hominis; varicella-herpes zoster; cytomegalovirus; and Epstein-Barr virus.

In addition to the identification of viruses, IMF permits the detection of virus-specific immunoglobulins in the patient's serum. The immune response in viral infections is characterized by the appearance of specific IgM at first, followed by the appearance of specific IgG, the latter persisting for a long time. The demonstration of the presence of specific IgM which is transient and indicates acute or recent infection has, therefore, useful diagnostic value.[123]

b. FTA-ABS Test for Syphilis

Deacon et al.[125] introduced the fluorescent treponemal antibody (FTA) test improved later[126] into the more sensitive and more specific FTA absorption (FTA-ABS) test, which has become one of the most useful serological treponemal tests for syphilis.

The FTA test is an indirect IMF test which utilizes a suspension of *Treponema pallidum* (Nichol's strain) as the antigen. A thin film of this suspension is allowed to air dry on a microscopic slide. The test serum is applied in appropriate dilutions and the preparation is incubated at 37°C for 30 min. It is then washed to remove any free test serum, and a fluorescent antihuman globulin conjugate is added. The *T. pallidum* will fluoresce under the fluorescence microscope, if the test serum belongs to a patient with syphilis, yaws, pinta, or related treponematoses.

The FTA-ABS test excludes false-positive reactions due to antibodies to the nonpathogenic *Treponema microdentium* and the Reiter treponema antigens which are often present in normal sera. These antibodies are first removed from the test serum, by absorbing them with a water-soluble extract of Reiter treponemas, after which the test is performed as described earlier.

No serological test for syphilis (STS) is infallible. Only in untreated secondary syphilis will all STS approach a reactivity of 100%. The FTA-ABS test, however, compares well with the other STS.[127] In primary syphilis, it is positive earlier than the VDRL slide test. In syphilis of long duration, the FTA-ABS test detects positive reactions in a greater percentage of cases than the Treponema pallidum immobilization (TPI) test. In all stages of the disease, the FTA-ABS test gives positive reactions in a higher percentage of cases than either the VDRL or the TPI test.

False-positive FTA-ABS test results may occur infrequently. They are due to the presence of macroglobulins and antinuclear factors in the serum and have been reported in sera of patients with autoimmune diseases, viral infections, malignancies, and pregnancy.[128]

The FTA-ABS test is not a screening test for syphilis. It is a good confirmation test when the disease is suspected clinically and it must always be used in conjunction with other laboratory tests. The FTA-ABS test is also performed with cerebrospinal fluid, where it is helpful, but it does not replace the other laboratory procedures, such as the protein determination and the cerebrospinal fluid (CSF) cell count, in assessing the activity of neurosyphilis.[128]

Table 10
IMF FINDINGS IN BULLOUS SKIN DISEASES[130-132]

Bullous skin diseases	Direct IMF: Tissue antibodies	Direct IMF: Pattern	Direct IMF: Binding sites	Indirect IMF — circulating antibodies	Remarks
Pemphigus group	IgG, sometimes IgM, IgA complement	Intercellular	Intraepidermal	Intercellular circulating (IC) antibodies	IC circulating antibodies found in most active cases in high titers; a fall in titer indicates remission or disease control[130]
Pemphigoid	IgG complement, sometimes IgM	Linear	Epidermal basement membrane zone (BMZ)	BMZ circulating antibodies	Indirect IMF positive 75% of cases; if indirect IMF is negative, suspect concurrent internal malignancy[131-132]
Dermatitis herpetiformis (DH)	IgA complement	Granular	Tips of dermal papillae, in normal-appearing skin	—	The "IgA test" (direct IMF on biopsy tissue) is a screening test for DH
Herpes gestationis	Complement (C_3), sometimes IgG	Linear	Epidermal BMZ in perilesional skin	BMZ circulating antibodies, complement binding IgG	Immunofluorescent pattern similar to that of pemphigoid

c. Diagnosis of Bullous Skin Disease with the IMF Technique

The diagnosis of the bullous skin diseases is generally based on careful observation of their clinical and histopathological features. Often enough, however, the symptomatology is unclear and the histopathology atypical. Since each of these conditions carries a different prognosis and responds to different therapeutic regimens, it is important to differentiate among them.

IMF has become an invaluable tool in the diagnosis of the three important groups of bullous diseases: pemphigus, pemphigoid, and dermatitis herpetiformis, and in the differentiation of most conditions presenting with bullae on the skin.[121,130-132]

The direct method will bring into evidence the binding sites of tissue antibodies in the skin biopsy specimen. The indirect method, performed on the patient's serum, may reveal circulating antibodies, found in some disorders and not in others, the presence or absence of which may be of prognostic value and help in the treatment and follow up of the patient.

The IMF findings in the main bullous skin diseases are summarized in Table 10.

d. The Diagnosis of LE

Both the direct and indirect IMF techniques are useful diagnostic tools in this condition. Direct IMF will detect deposits of IgG and C3c at the basement membrane zone (BMZ) in over 90% of the cases in skin lesions of both discoid and systemic LE, and in about 60% of the cases in the apparently normal skin of systemic LE.[129] A positive band-like pattern of IMF at the BMZ is therefore highly suggestive of LE.

This pattern is also found in some other diseases, but in bullous pemphigoid, circulating anti-BMZ antibodies can also be detected in the patient's serum, which is not the case in LE. The differential diagnosis from dermatitis herpetifomis is made simple by the fact that here, the BMZ antibodies are always IgA and not IgG, as is the case in LE.

Indirect IMF will detect antibodies to nuclei (antinuclear factor), which are present in sera of patients with active systemic LE and other collagen diseases.

The test is performed by applying the patient's serum to a tissue section (here the substrate is usually rat liver) which has been first damaged by freezing and thawing, in order to expose the nuclear antigens. Antihuman IgG conjugate is then added. Fluorescence will occur within the cell nucleus if an antinuclear factor is present in the test serum.

Several patterns of fluorescence are recognized, each with its own prognostic significance.[133] These patterns can be homogenous, peripheral, nucleolar, or speckle-like, and reflect the binding of the circulating antibody to different constituents of the nuclei.

The most important antinuclear antibodies in systemic LE are those directed to the double-stranded DNA. Their presence is almost specific for systemic LE, and is usually detected and quantitated by a precipitation technique. However, systemic LE can also be identified by indirect IMF using *Crithidia luciliae* as a substrate, a protozoan that has a nucleus containing only native, double-stranded DNA. The titer of antibodies to DNA is an indication of disease activity.

C. Wood's Lamp

The lamp introduced by R. W. Wood in 1903 consists of a high-pressure mercury arc equipped with a nickel oxide phosphor filter. Because it had the property of exciting substances to fluoresce in the dark, Wood believed it to be suitable for solving a number of military problems, and made it the object of a secret communication. Nonetheless, he had not failed to mention some observations on the aspects of the skin when illuminated by it, and his discovery was soon to find its main applications in the dermatology office,[134] where it became standard equipment for the detection of fungus infections of the scalp.

Essentially, Wood's lamp is a source of UVA radiation with maximum output at 365 nm, where wavelengths shorter than 320 nm and longer than 400 nm have been cut off. Commercially available Wood's lamps are easy to handle, safe, and relatively inexpensive.

When UVA radiation is absorbed by a fluorescent substance, radiation of longer wavelengths, usually from 400 to 700 nm (visible light), is emitted. The phenomenon is best observed in a shaded room, after the eyes have adapted to the dark. The areas to be examined are brought as close as possible to the lamp, but in order to avoid potential harmful effects on the eyes, examiners and patients should not look directly into the light source.

In the dermatology office, Wood's light is used for the viewing of fluorescent substances intrinsic or extrinsic to the skin, not ordinarily visible in normal daylight. It is also most valuable for a better visualization of melanin pigmentation, the determination of its presence or absence, and the depth of its localization.

The wavelength of the fluorescence observed depends on the substance which emits it, and therefore some substances can often be readily identified: porphyrin fluoresces red; fluorescein, yellow green; pteridine, blue green; normal dermal and epidermal proteins, blue white; and tetracycline, yellow.

1. Tinea Capitis

The bright blue-green fluorescence observed with Wood's light in some forms of tinea capitis is due to pteridine,[135] a substance metabolized by fungi in the presence of keratin. Not all fungi fluoresce, but the common microsporon canis and microsporon Audouini have this property, as well as the agent of favus, *Trichophyton schoenleini.* Although topically applied greases and various drugs in ointments also sometimes fluoresce under Wood's light, they should not interfere with the diagnosis, as they never have the characteristic blue-green fluorescence of pteridine.

Fungus infections of the scalp are not as prevalent as they were several decades ago; still, Wood's lamp remains the important tool for their early detection and for their differentiation from other scalp diseases. When fungus is present, the extent of the infection is readily visible since each stub of broken infected hair fluoresces individually. The patient's contacts can be rapidly screened for fluorescence, thus allowing the detection of subclinical cases and potential carriers in families and schools. Microscopic examinations of hairs and scales, as well as cultures on Sabouraud's medium, will confirm the diagnosis and must also be done in suspicious nonfluorescent cases, for absence of fluorescence does not exclude the diagnosis of fungus infection.[136] When present, fluorescence is also extremely helpful in assessing the results of therapy: when a fungistatic antibiotic such as griseofulvin is given orally, the part of the infected hair growing out of the follicle will soon appear healthy and nonfluorescent, whereas the tip of the hair will fluoresce. Cutting off these hairs hastens the cure.

2. Erythrasma

In this condition, the coral-red fluorescence observed with Wood's light in the groins, toe webs, and sometimes in other intertriginous regions of affected patients, is due to porphyrin, metabolized by the aerobic *Corynebacterium minutissimum* in the presence of keratin,[135] or possibly in the presence of glandular secretions of the skin.[137] It is not clear whether a single porphyrin or a combination of compounds is responsible for the phenomenon. When erythrasma manifests itself by hyperkeratotic inflammatory lesions, the fluorescence can always be seen. However, in about 17% of examined cases, the intertriginous areas will fluoresce coal red in the absence of clinical disease, and in these cases *C. minutissimum* can always be cultivated.[138] The organism likely exists as a saprophyte. *C. minutissimum* will grow on soy-agar medium with 10% sheep blood incubated at 37°C. The culture will also fluoresce under Wood's light.

The characteristic fluorescence of erythrasma permits a rapid and easy differential diagnosis from other inflammatory diseases of the intertriginous areas, such as psoriasis, moniliasis, and dermatophytoses. It is best not to examine patients immediately after bathing, as porphyrin is water soluble and washes off readily from the areas.

3. Pseudomonas aeruginosa

This organism commonly invades surface wounds and stasis ulcerations, and is often responsible for invasive sepsis and death in severely burned patients.

Pathogenic forms of *P. aeruginosa* produce two pigments. The first, the blue-green pyocyanin recognized by examination of infected surfaces in regular daylight, is often obscured by the presence of eschars in wounds or by silver precipitated from the aqueous 0.5% solution of silver nitrate used in the treatment of burns. The other pigment, even more frequently found in these infections, is pyoverdin or fluorescein,[139] which fluoresces blue green under Wood's light. This fluorescence is so clear that it can be observed even when the organisms are outnumbered 300 to 1 by nonfluorescent bacteria. It is a fact that an increase of bacterial density in wounds always preceeds local or systemic signs of *P. aeruginosa* infection. Fluorescence can be detected with Wood's light before the bacterial density approaches 10^5 bacteria per square centimeter, an indication of imminent sepsis.[139] Daily examinations of severe wounds under Wood's light will therefore allow for early institution of aggressive treatment in *P. aeruginosa* infection, thus preventing septicemia.

4. Pityriasis Versicolor

This infection of the skin by Malassezia furfur is easily diagnosed when it is widespread over the trunk and intertriginous regions. The lesions are finely scaly, yellowish or brown macules and plaques, with depigmented areas. The organism is recognized on direct microscopic examination of the scales soaked in 15% potassium hydroxide solution. However, Wood's light examination emphasizes both the depigmented areas, which appear whiter than in regular daylight, and the pigmented areas which fluoresce a pale, yellowish brown. The infection often reveals itself to be more extensive than seen in normal daylight. More adequate treatment can thus be instituted, reducing the chances of relapse.

5. Diagnosis of Pigmentary Disorders

Skin fluorescence occurs only at depths to which the exciting light has penetrated. Most of the blue-white fluorescence observed with Wood's light reflects the superficial layer of the skin, namely the stratum corneum and the epidermis.[140,141] The normal white skin is strikingly different when viewed with Wood's light than with regular daylight. In the absence of visible light, only the fluorescent layers will be seen. Epidermal melanin, which strongly absorbs both visible and UV light, will appear black, as no fluorescence and no color will result in pigmented areas. Moreover, the contrast in epidermal pigment variations under these conditions is more accentuated, as no background is reflected from the dermis in the absence of deeply penetrating visible wavelengths. In fact, dermal pigmentation is less evident or completely absent under Wood's lamp. Using Wood's light, it is therefore possible to diagnose pigmentary disorders and to delineate the extent of these conditions. It is also valuable in differentiating hypopigmentation from absence of epidermal pigmentation (vitiligo). In addition, it enables the estimation of the depth of cutaneous hyperpigmentation and the differentiation between epidermal and dermal hypermelanosis.

6. Oral Diagnosis

The organic portions of the enamel and dentin are primarily responsible for the normal, light-blue fluorescence of healthy teeth under Wood's light. Dentin has a higher organic content than enamel, and fluoresces at a higher intensity.[142] Decalcification of the teeth interferes with the fluorescence, and twice as many carious teeth can be detected when Wood's lamp is used as complement to the clinical examination.[143]

Aside from the early detection of caries, Wood's lamp examination of teeth is helpful in the following conditions. Extrinsic stains show up as darkened regions of the teeth. Calculi have a yellow-orange fluorescence. Tetracycline deposits in patients who have taken the drug during the period of the tooth development can be detected by yellow fluorescence in deciduous as well as in permanent teeth. In the permanent teeth, however, the color of the fluorescence may change due to exposure to light and oxidizing agents, and it may look like a simple intrinsic stain. Plaque could be made to fluoresce brightly by staining it with fluorescein; it could be brushed off easily, helping in prophylaxis. Porphyrin deposits in the teeth of patients with congenital EPP will show red fluorescence.

7. Screening Tests for Porphyria

a. Porphyrin Screening Tests in Urine

A normal, freshly voided specimen of urine contains chromogens which have greenish fluorescence under Wood's light. Patients with PCT and erythropoietic porphyria (EP) excrete uroporphyrins which fluoresce pinkish red. Collecting a specimen of urine and examining it under Wood's light is an easy office screening test for these conditions.[144] The details are discussed elsewhere in this book.

These tests are also helpful in the detection of coproporphynuria found in cases of lead poisoning, liver disease, some chemical intoxications, and infections. Although they are extremely useful in the diagnosis of PCT and EP, they will not help in other forms of porphyria.

In acute intermittent porphyria (AIP) and porphyria variegata (PV), where porphybilinogen is excreted in the urine, the Watson-Schwartz test is commonly used.[145] It is a test which does not require the use of UV light.

In EPP, no porphyrin is present in the urine, and the previously mentioned tests are therefore useless. EPP must be diagnosed by examination of the blood.

b. Porphyrin Screening Tests for RBC Porphyrins

Chemical test for RBC porphyrins — Excessive protoporphyrins and coproporphyrins found in the blood of patients with EPP can be detected by Harber's procedure,[146] as a bright-red fluorescence seen under Wood's light. The small amounts of porphyrin present in normal blood are not detected with this test.

Fluorescence microscopy test for RBC protoporphyrins — The fluorescence microscope is used here instead of Wood's lamp. One part of anticoagulated blood is diluted in ten parts of normal saline. This specimen is examined under oil immersion with the fluorescence microscope. In cases of EPP, 5 to 30% of the erythrocytes will fluoresce. In cases of congenital EP, up to 50% will fluoresce. In lead poisoning, the fluorescence will be fainter but will affect practically all the RBC. In the performance of this test, an iodine tungsten quartz lamp is preferable to the mercury vapor lamp which makes the fluorescence fade too rapidly.[142]

Microfluorometric assay for free erythrocyte porphyrins — This is a rapid, accurate, and inexpensive quantitative test which can be performed by the laboratory on two drops of blood collected on filter paper for cases of EPP and EP.[148]

8. Miscellaneous Uses of Wood's Lamp

A number of additional microorganisms, aside from those already mentioned, produce fluorescent substances under certain circumstances. The presence of a porphyrin produced by *C. acnes* can be revealed in seborrheic individuals, by a red fluorescence of the skin follicles of the chest, the back, and the face, when viewed under Wood's light.

Other organisms in culture produce a red, fluorescent substance.[149]

These include *Pasteurella, S. pyogenes,* and *E. coli,* which have one factor in common, namely the ability to metabolize δ-aminolevulenic acid (DAL), a precursor of porphyrin, into porphyrin. Some of these organisms are incapable of degrading whole blood, but if DAL is added to the culture media, they will produce porphyrin, whose fluorescence under Wood's light will help in early diagnosis. Wood's light has been found useful in testing dyes, cosmetics, and topical medications for potential phototoxicity.[150]

D. Photodiagnosis of Malignant Neoplasms

This method of investigation raised much interest in the early 1960s. It was based on the observation that certain fluorescent substances concentrate selectively within malignant tissue and can be detected by excitation with light of appropriate wavelength. Direct illumination and endoscopic apparatus has been applied for excitation.

The first substances administered to patients were tetracycline and acridine orange.[151-153] The results were inconsistent and the validity of utilizing tetracycline fluorescence was questioned when it became apparent that it was exclusively restricted to collagen fibers, histocytes, and macrophages, and that it was actually absent in viable tumor cells.[154] Porphyrins seemed to be more promising, since it was shown that certain tetrapyrroles concentrate in neoplastic loci[155,156] and will accumulate in malignant cells as HPD.

In the intervening years, most efforts were diverted to the application of HPD to phototherapy and less to diagnosis. Recently, with the introduction of laser beams and advanced endoscopic equipment, several groups became involved in using fluorescent endoscopy in diagnosis, particularly for bronchogenic tumors.

HPD for clinical use is prepared by treating commercially available hematoporphyrin dihydrochloride with sulfuric acid in glacial acid, followed by several steps of recrystallization, a method described by Lipson et al.,[157] or a slight modification thereof. Its coral-red fluorescence is more intense than that of a whole series of synthetic pophyrins tested for hard- and soft-tissue binding.[158] The tumor-localizing component of HPD has been identified recently as dihematoporphyrin ether (DHE), a compound which is less photosensitizing than the unfractionated HPD. The use of a purified DHE can therefore minimize the troublesome adverse reactions encountered in the skin and the eyes when patients given the unfractionated HPD are unprotected from sunlight.[156]

The observed fluorescence with HPD is different when viewing viable and nonviable cells. In living tumor cells, it is confined to the cytoplasm, whereas nonviable cells show a different fluorescence resulting from the adherence of HPD to the cell membrane.[159]

It is not known why tumor cells accumulate porphyrins selectively. This property does not seem to depend on the cell type of the malignant tissue. For instance, bronchial and esophageal tumors fluoresce in a similar fashion.[160] Squamous cell carcinoma will fluoresce more intensely than adenocarcinoma.[161] This variation does not interfere with diagnosis, as the proportion of positive fluorescence has been found to be even higher in the latter type of carcinoma than in the former.

The procedure is relatively simple to perform. It consists of administering a slow i.v. injection of the drug at doses of 1.5 to 2 mg/kg of body weight. In some instances, doses as high as 4 mg/kg were applied without adverse reactions, except for increased photosensitivity to sunlight.[162] Examination is usually performed 3 hr after administration, though sometimes it is done 2 to 3 days later. The rationale for the longer time interval is based on the assumption that uptake of the drug is the same in normal and cancerous tissue, however, the clearance rate of the former is greater, thus rendering cancerous tissue to become more prominent against the background fluorescence of normal tissue. Studies on cultures substantiate this assumption, as no preferential uptake by normal vs. neoplastic cells has been observed.[163]

Originally, xenon and mercury lamps were used for excitation of HPD fluorescence in direct and endoscopic illumination. In order to separate the excitation wavelength of 405 nm, narrow-band interference filters were introduced in the light path. For viewing of fluorescence at 630 nm, wide band-pass filters were used. This method was cumbersome and inefficient, as the filters did not completely eliminate the red light from the illuminating beam. In addition, when used in endoscopy, the amount of light reaching the tumor was low because of the heavy filtration, thus rendering the detection of the tumor difficult. To improve the method, an image intensifier and audio signals were introduced.[164,165] With the introduction of a krypton ion laser, with a nearly monochromatic light (wavelength range 406.7 to 415.4 nm), less filtration was needed and a noticable improvement was achieved. An additional factor contributing to the advancement of the viewing is the improvement in endoscopic equipment. A rapid switch to a white, nonfluorescent light source facilitates accurate localization of the tumor and precise biopsy sampling.

So far, HPD has been used for photodiagnosis of neoplastic tumors in the urinary bladder,[166] the lung,[162,167] the bronchi and esophagus,[168] the oropharynx,[169] the uterine cervix,[170] and many others.[171]

Some malignant lesions of the oropharynx and uterine cervix fluoresce impressively, whereas others, similar in gross and microscopic appearance, show no fluorescence at all.[161] Reasons that could account for negative tumor fluorescence are its obstruction by necrotic debris which act as a screen, or by the presence of pigment (i.e., in malignant melanoma) or blood within the lesions, both of which strongly absorb the activating light.

Notwithstanding these limitations, the method has promising diagnostic value under certain circumstances. It is helpful in assessing the margins of known malignant tumors prior to therapy, as maximal fluorescence is seen in the advancing edge of the neoplasms.[168] It has been found useful in the detection and localization of early tumors, such as bronchial carcinoma *in situ* not detectable by either X-ray or conventional light-reflectance bronchoscopy,[164] especially in those patients with carcinoma cells in their sputum or bronchial secretions, where other imaging methods failed to reveal the tumor.[168]

Research is currently aimed at identifying prophyrins with higher affinity to tumor cells and higher fluorescence yield. Liposome-encapsulated hematoporphyrin and hematoporphyrin dimethyl ester, which accumulate preferentially in experimental tumors in rats,[172] and liposome-incorporated uroporphyrin I octamethyl ester, which associates firmly with tumor cell membranes in lymphoma-bearing mice,[173] could eventually become suitable for tumor photodiagnosis.

More sensitive optical devices with intensifying systems and video analyzers could further improve the technique.[164] They were shown to provide fluorescence imaging of bronchial tumors no larger than a few millimeters in size, which were barely visible under white-light bronchoscopy. Such a system could also reduce the frequency of false-positive readings by permitting quantitative measurements of contrast and fluorescent yield.

Whatever the future of this method, it cannot, at this time, be depended upon as a sole means for the diagnosis of malignant lesions, and further research is required.

REFERENCES

1. Willis, I. and Kligman, K. M., Aminobenzoic acid and its esters, *Arch. Dermatol.*, 102, 405, 1970.
2. Stern, W. K. and Urbach, F., The diagnostic significance of the minimal erythemal dose, *Arch. Dermatol.*, 105, 387, 1972.
3. Parrish, J. A., Anderson, R. R., Urbach, F., and Pitts, D., Eds., Effects of ultraviolet radiation on the eye, in *UVA-Biological Effects of Ultraviolet Radiation with Emphasis on Human Responses to Longwave Ultraviolet*, Plenum Press, New York, 1978, chap. 9.
4. Parrish, J. A., White, H. A. D., and Pathak, M. A., Photomedicine, in *Dermatology in General Medicine*, Fitzpatrick, T. B., Eisen, A. L., Wolff, K., Freedberg, I. M., and Austen, K. F., Eds., McGraw-Hill, New York, 1979, chap. 101.
5. Hawk, J. L. M. and Parrish, J. A., Responses of normal skin to ultraviolet radiation, in *The Science of Photomedicine*, Regan, J. D. and Parrish, J. A., Eds., Plenum Press, New York, 1982, 219.
6. Azizi, E., Kushelevsky, A. P., and Schewach-Millet, M., The efficacy of topical sunscreen preparations on the human skin, combined indoor-outdoor study, *Isr. J. Med. Sci.*, 20, 569, 1984.

7. Rottier, P. B., Untersuchungen mit einem kleinen Spectrographen zur Bestimmung der Hautempfindlichkeit gegenuber 300 und 250 mu, *Strahlentherapy,* 119, 591, 1962.
8. Magnus, I. A., Principles of investigation of photosensitivity of the skin, in *Dermatological Photobiology,* Blackwell Scientific, Oxford, 1976, chap. 9.
9. Cripps, D. J., Instrumentation and action spectra in light associated diseases, *J. Invest. Dermatol.,* 77, 20, 1981.
10. Magnus, I. A., Studies with a monochromator in the common idiopathic photodermatoses, *Br. J. Dermatol.,* 76, 245, 1964.
11. Blum, H. F., Domestic animal diseases produced by light, *J. Am. Vet. Med. Assoc.,* 93, 185, 1938.
12. Magnus, I. A., Drug and chemical photosensitization, in *Dermatological Photobiology,* Blackwell Scientific, Oxford, 1976, chap. 16.
13. Kaidbey, K. H. and Kligman, A. M., Identification of systemic phototoxic drugs by human intradermal assay, *J. Invest. Dermatol.,* 70, 272, 1978.
14. Morikawa, F., Fukuda, M., Naganuma, M., and Nakayama, Y., Phototoxic reactions to xanthene dyes induced by visible light, *J. Dermatol.,* 3, 59, 1976.
15. Kaidbey, K. H. and Kligman, A. M., Identification of topical photosensitizing agents in humans, *J. Invest. Dermatol.,* 70, 149, 1978.
16. Harber, L. C., Baer, R. L., and Bickers, D. R., Techniques of evaluating phototoxicity and photoallergy in biological systems, including man, with particular emphasis on immunologic aspects, in *Sunlight and Man,* Fitzpatrick, T. B., Pathak, M. A., Harber, L. C., Seiji, M., and Kukita, A., Eds., University of Tokyo Press, Tokyo, 1974, 515.
17. Morison, W. L., McAuliffe, D. J., Parrish, J. A., and Bloch, K. J., In vitro assays for phototoxic chemicals, *J. Invest. Dermatol.,* 78, 460, 1982.
18. Morikawa, F., Nakayama, Y., Fukuda, M., Hamano, M., Yokoyama, Y., Nagura, T., Ishihara, M., and Toda, K., Techniques for evaluation of phototoxicity and photoallergy in laboratory animals and man, in *Sunlight and Man,* Fitzpatrick, T. B., Pathak, M. A., Harber, L. C., Seiji, M., and Kukita, A., Eds., University of Tokyo Press, Tokyo, 1974, 529.
19. Harber, L. C., Current status of mammalian and human models for predicting drug photosensitivity, *J. Invest. Dermatol.,* 77, 65, 1981.
20. Ison, A. and Blank, H., Testing drug phototoxicity in mice, *J. Invest. Dermatol.,* 49, 508, 1967.
21. Ljunggren, B. and Moeller, H., Phenothiazine phototoxicity, an experimental study in chlorpromazine and its metabolites, *J. Invest. Dermatol.,* 68, 313, 1977.
22. Frost, P., Weinstein, G. D., and Gomet, E. L., Methacycline and demeclocycline in relation to sunlight, *JAMA,* 216(2), 326, 1971.
23. Wolff, K., Gschnait, F., Honigsmann, H., Konrad, K., Parrish, J. A., and Fitzpatrick, T. B., Phototesting and dosimetry for photochemotherapy, *Br. J. Dermatol.,* 96, 1, 1977.
24. Lipson, R. L., Baldes, E. J., and Olsen, A. M., Hematoporphyrin derivative: a new aid for endoscopic detection of malignant disease, *J. Thorac. Cardiovasc. Surg.,* 42, 623, 1961.
25. Tanenbaum, L., Parrish, J. A., Pathak, M. A., Anderson, R. R., and Fitzpatrick, T. B., Tar phototoxicity and phototherapy for psoriasis, *Arch. Dermtol.,* 111, 467, 1975.
26. Parrish, J. A., Anderson, R. R., Urbach, F., and Pitts, D., Eds., Adverse cutaneous reactions to UVA, in *UVA-Biological Effects of Ultraviolet Radiation with Emphasis on Human Responses to Longwave Ultraviolet,* Plenum Press, New York, 1978, 141.
27. Kligman, A. and Briet, R., Identity of phototoxic drugs by human assay, *J. Invest. Dermatol.,* 51, 90, 1968.
28. Kocheavar, I. A., Armstrong, R. B., Einbinder, J., Walther, R. R., and Harber, L. C., Coal tar phototoxicity, active compounds and action spectra, *Photochem. Photobiol.,* 36, 65, 1982.
29. Kaidbey, K. H. and Klingman, A. M., A clinical and histologic study of coal tar phototoxicity in humans, *Arch. Dermatol.,* 113, 592, 1977.
30. Ackerman, A. B., *Histologic Diagnosis of Inflammatory Skin Diseases,* Lea & Febiger, Philadelphia, 1978, 226, 500, 502.
31. Herman, P. S. and Sams, W. M., Requirement for carrier protein in salicyanilide sensitivity: the migration inhibition test in contact photoallergy, *J. Lab. Clin. Med.,* 77, 572, 1971.
32. Jung, E. G., Dummler, N., and Immich, H., Photoallergie durch 4-chlor-2-hydroxy benzesaure-*N*-butylamid, *Arch. Klin. Exp. Dermatol.,* 232, 403, 1968.
33. Morison, W. L. and Kocheavar, I. E., Photoallergy, in *Photoimmunology,* Parrish, J. A., Kripke, M. L., and Morison, W. L., Eds., Plenum Press, New York, 1983, chap. 12.
34. Magnus, I. A., Other photodermatoses, in *Dermatological Photobiology,* Blackwell Scientific, Oxford, 1976, chap. 14.
35. Freeman, R. G. and Knox, J. M., The action spectrum of photocontact dermatitis caused by halogenated salicylanilide and related compounds, *Arch. Dermatol.,* 97, 130, 1968.
36. Cripps, D. J. and Enter, T., Absorption and action spectra studies on bithionol and halogenated salicylanilide photosensitivity, *Br. J. Dermatol.,* 82, 230, 1970.

37. Willis, I. and Kligman, A. M., The mechanism of the persistent light reactor, *J. Invest. Dermatol.*, 51, 385, 1968.
38. Osmundsen, T. E., Photopatch testing, *Trans. St John's Hosp. Dermatol. Soc.*, 55, 160, 1969.
39. Willis, I. and Kligman, A. M., The mechanism of photoallergic contact dermatitis, *J. Invest. Dermatol.*, 51, 378, 1968.
40. Hozio, T., The induction of photocontact sensitivity in guinea pigs without UVB radiation, *J. Invest. Dermatol.*, 67, 591, 1976.
41. Kocheavar, I. E., Zalar, G. L., Einbinder, J., and Harber, L. C., Assay of contact photosensitivity to musk ambrette in guinea pigs, *J. Invest. Dermatol.*, 73, 144, 1979.
42. Ichikawa, H., Armstrong, R. B., and Harber, L. C., Photoallergic contact dermatitis in guinea pigs: improved induction technique using Freund's complete adjuvant, *J. Invest. Dermatol.*, 76, 498, 1981.
43. Maurer, T. H., Weirich, E. G., and Hess, R., Predictive animal testing for photocontact allergenicity, *Br. J. Dermatol.*, 103, 593, 1980.
44. Vinson, L. J. and Borselli, V. F., A guinea pig assay of the photosensitizing potential of tropical germicides, *J. Soc. Cosmet. Chem.*, 17, 123, 1966.
45. Miyachi, Y. and Takigawa, M., Mechanisms of contact photosensitivity in mice. II. Langerhans cells are required for successful induction of contact photosensitivity to TCSA, *J. Invest. Dermatol.*, 78, 363, 1982.
46. Maguire, H. C., Jr. and Kaidbey, K., Experimental photoallergic contact dermatitis: a mouse model, *J. Invest. Dermatol.*, 79, 147, 1982.
47. Granstein, R. D., Morison, W. L., and Kripke, M. L., The role of UVB radiation in the induction and elicitation of photocontact hypersensitivity to TCSA in the mouse, *J. Invest. Dermatol.*, 80, 158, 1983.
48. Harber, L. C., Targovnik, S. E., and Baer, R. L., Contact photosensitivity to halogenated salicylanilides: in man and guinea pigs, *Arch. Dermatol.*, 96, 646, 1967.
49. Harber, L. C. and Baer, R. L., Mechanisms of drug photosensitivity reactions, *Toxicol. Appl. Pharmacol.*, 3, 58, 1969.
50. Magnus, I. A., The classification and diagnosis of the photodermatoses, in *Dermatological Photobiology*, Blackwell Scientific, Oxford, 1976, chap. 12.
51. Magnus, I. A., Polymorphic light eruption and summer prurigo, in *Dermatological Photobiology*, Blackwell Scientific, Oxford, 1976, chap. 13.
52. Magnus, I. A., Solar urticaria, in *Dermatological Photobiology*, Blackwell Scientific, Oxford, 1976, chap. 15.
53. Magnus, I. A., The porphyrias, in *Dermatological Photobiology*, Blackwell Scientific, Oxford, 1976, chap. 17.
54. Morison, W. L., Momtaz, K., Mosher, D. B., and Parrish, J. A., UVB phototherapy in the prophylaxis of polymorphous light eruption, *Br. J. Dermatol.*, 106, 231, 1982.
55. Parrish, J. A., Levine, M. J., Morison, W. L., Gonzalez, E., and Fitzpatrick, T. B., Comparison of PUVA and beta-carotene in the treatment of polymorphous light eruption, *Br. J. Dermatol.*, 100, 187, 1979.
56. Diffey, B. L., Farr, P. M., and Ive, F. A., Home phototherapy of solar urticaria: a case report, *Photodermatology*, 1, 145, 1984.
57. Gschnait, F., Honignsmann, H., Brenner, W., Fritsch, P., and Wolff, K., Induction of UV light tolerance by PUVA in patients with polymorphous light eruption, *Br. J. Dermatol.*, 99, 293, 1978.
58. Ramsey, C. A., Solar urticaria treatment by inducing tolerance to artificial radiation and natural light, *Arch. Dermatol.*, 113, 1222, 1977.
59. Epstein, J. H., Polymorphous light eruptions. Wavelength dependency and energy studies, *Arch. Dermatol.*, 85, 82, 1962.
60. Epstein, J. H., Polymorphous light eruption, *Ann. Allergy*, 24, 397, 1966.
61. Epstein, J. H., Polymorphous light eruption, *J. Am. Acad. Dermatol.*, 3, 329, 1980.
62. Frain-Bell, W., Dickson, A., Herd, J., and Sturrock, I., The action spectrum in polymorphic light eruption, *Br. J. Dermatol.*, 82, 243, 1973.
63. Wiskemann, A. and Wulf, K., Utersuchungen uber den auslosenden Spektralbereich und die direkte Lichtpigmentierung bei chronischen und akuten Lichtausschlagen, *Arch. Klin. Exp. Dermatol.*, 209, 443, 1959.
64. Birt, A. R. and Davis, R. A., Hereditary polymorphic light eruption of American Indians, *Int. J. Dermatol.*, 14, 105, 1975.
65. Holzle, E., Plewig, G., Hofmann, C., and Roser-Mass, E., Polymorphous light eruption, *J. Am. Acad. Dermatol.*, 7, 111, 1982.
66. Verhagen, A. R. H. B., Light tests and pathogenic wavelengths in chronic polymorphous light dermatosis, *Dermatologica*, 133, 302, 1966.
67. Frain-Bell, W., Mackensie, L. A., and Whitham, E., Chronic polymorphic light eruption (a study of 25 cases), *Br. J. Dermatol.*, 81, 885, 1969.

68. Thune, P., Chronic polymorphic light eruption. Particular wavebands and the effect of carotene therapy, *Acta Derm. Vernereol.*, 56, 127, 1976.
69. Nordlund, J. J., Klaus, S. N., Mathews-Roth, M. M., and Pathak, M. A., New therapy for polymorphous light eruption, *Arch. Dermatol.*, 108, 710, 1973.
70. Jansen, C. T., Erythemal and pigmentary phototest reactions in polymorphous light eruptions, *Acta Derm. Vernereol.*, 59, 499, 1979.
71. Rottier, P. B., Testing of the skin for reactions to ultraviolet irradiation, *Dermatologtica*, 127, 260, 1963.
72. Rottier, P. B., Light dermatoses without abnormal reactions to single test irradiations, *Dermatologica*, 129, 286, 1964.
73. Cripps, D. J., Ultraviolet light and photosensitivity, in *Experimental and Clinical Photoimmunology*, Daynes, R. A. and Kruger, G. K., Eds., CRC Press, Boca Raton, Fla., 1983, chap. 9.
74. Ramsay, C. A. and Kobza Black, A., Photosensitive eczema, *Trans. St. John's Hosp. Dermatol. Soc.*, 59, 152, 1973.
75. Harber, L. C., Holloway, R. M., Wheatley, V. R., and Baer, R. L,, Immunologic and biophysical studies in solar urticaria, *J. Invest. Dermatol.*, 41, 439, 1963.
76. Ive, H., Lloyd, J., and Magnus, I., Action spectra in idiopathic solar urticaria, *Br. J. Dermatol.*, 77, 229, 1965.
77. Willis, I. and Epstein, J. H., Solar vs. heat induced urticaria, *Arch. Dermatol.*, 110, 389, 1974.
78. Hasei, K. and Ichihashi, M., Solar urticaria, *Arch. Dermatol.*, 118, 346, 1982.
79. Morison, W. L., Solar urticaria, in *Photoimmunology*, Parrish, J. A., Kripke, M. L., and Morison, W. L., Eds., Plenum Press, New York, 1983, chap. 11.
80. Sams, W. M., Jr., Solar urticaria, studies of active serum factor, *J. Allergy*, 45, 295, 1970.
81. Kaidbey, K. H. and Nonaka, S., Action spectrum for anthracene induced photosensitization of human skin, *Photochem. Photobiol.*, 39, 375, 1984.
82. Ive, F. A., Magnus, I. A., Warin, R. P., and Wilson Jones, E., Actinic reticuloid, a chronic dermatosis associated with severe photosensitivity and the histologic resemblence to lymphoma, *Br. J. Dermatol.*, 81, 469, 1969.
83. Frain-Bell, W., Lakshmipathi, T., Rogers, J., and Willock, J., The syndrome of chronic photosensitivity dermatitis and actinic reticuloid, *Br. J. Dermatol.*, 91, 617, 1974.
84. Gianelli, F., Botcherby, P. K., and Magnus, I. A., Sensitivity to UVA in fibroblasts from actinic reticuloid, *Photochem. Photobiol.*, 39, 645, 1984.
85. McAuliffe, D. J., Gange, R. W., and Parrish, J. A., Abnormal ultraviolet sensitivity of lymphocytes from a patient with actinic reticuloid, *Photochem. Photobiol.*, 39, 765, 1984.
86. Katzenellenbogen, I., Lichen planus actinicus (lichen planus in subtropical countries), *Dermatologica*, 124, 10, 1962.
87. Verhagen, A. R. H. B. and Koten, J. W., Lichenoid melanodermatitis, *Br. J. Dermatol.*, 101, 651, 1979.
88. Isaacson, D., Turner, M., and Elgart, M. L., Summertime actinic lichenoid eruption, *J. Am. Acad. Dermatol.*, 4, 404, 1981.
89. Van der Schroeff, J. G., Schothorst, A. A., and Kanaar, P., Induction of actinic lichen planus with artificial UV sources, *Arch. Dermatol.*, 119, 498, 1983.
90. Lester, R. S., Burnham, T. K., Fine, G., and Murray, K., Immunologic concepts of light reaction in lupus erythematosus and polymorphous light eruption, *Arch. Dermatol.*, 96, 1, 1976.
91. Burnham, T. K., Neblett, T. R., and Fine, G., The application of the fluorescent antibody technique to the investigation of lupus erythematosus and various dermatoses, *J. Invest. Dermatol.*, 41, 451, 1963.
92. Cripps, D. J. and Rankin, J., Action spectra of lupus erythematosus and experimental immunofluorescence, *Arch. Dermatol.*, 107, 563, 1973.
93. Freeman, R. G., Knox, J. H., and Owens, D. W., Cutaneous lesions of lupus erythematosus induced by monochromatic light, *Arch. Dermatol.*, 100, 677, 1969.
94. Deschamps, P., Pedailles, S., Michel, M., and Leroy, D., Photoinduction of lesions in a patient with pemphigus erythematosus, *Photodermatology*, 1, 38, 1984.
95. Jacobs, S. E., Pemphigus erythematosus and ultraviolet light, *Arch. Dermatol.*, 91, 139, 1965.
96. Cram, D. L. and Winkelmann, R. K., Ultraviolet induced acantholysis in pemphigus, *Arch. Dermatol.*, 92, 7, 1965.
97. Cram, D. L. and Fukuyama, K., Immunohistochemistry of ultraviolet induced pemphigus and pemphigoid lesions, *Arch. Dermatol.*, 106, 819, 1972.
98. Suhonen, R. and Niemi, K. M., Ultraviolet light in familial benign chronic pemphigus, *J. Cut. Pathol.*, 6, 414, 1979.
99. Cram, M. D., Muller, S. A., and Winkelmann, R. K., Ultraviolet induced acantholysis in familial pemphigus, *Arch. Dermatol.*, 96, 636, 1967.

100. Shelly, U. B., Arthur, R. P., and Pillsbury, D. M., A view of keratosis follicularis (Darier's disease) as a neoplastic process, *Arch. Dermatol.*, 80, 332, 1959.
101. Cripps, D. J., Ramsay, C. A., and Ruch, D. M., Xerodema pigmentosum: abnormal monochromatic action spectrum and autoradiographic studies, *J. Invest. Dermatol.*, 56, 281, 1970.
102. Rottier, P. B., *Proc. 1st Int. Photobiology Congr.*, H. Veenan and Sons, Amsterdam, 1954.
103. Chernosky, M. E. and Anderson, D. E., Disseminated superficial actinic porokeratosis, *Arch. Dermatol.*, 99, 401, 1969.
104. Coons, A. H., Creech, H. J., Jones, R. N., and Berliner, E., The demonstration of pneumococcal antigen in tissues by use of fluorescent antibody, *J. Immunol.*, 45, 159, 1942.
105. Coons, A. H., Creech, H. J., and Jones, R. N., Immunological properties of antibody containing a fluorescent group, *Proc. Soc. Exp. Biol. Med.*, 47, 200, 1941.
106. Coons, A. H. and Kaplan, M. H., Localisation of antigen in tissue cells. II. Improvements in a method for detection of antigen by means of fluorescent antibody, *J. Exp. Med.*, 91, 1, 1950.
107. Chadwick, C. S., McEntegart, M. G., and Nairn, R. C., Fluorescent protein tracers: a simple alternative to fluorescein, *Lancet*, 1, 412, 1958.
108. Pearse, A. G. E., *Histochemistry — Theoretical and Applied*, 3rd ed., Churchill Livingstone, London, 1968.
109. Riggs, J. L., Seiwald, R. J., Burckhalter, J. H., Downs, G. M., and Metcalf, T. G., Isothiocyanate compounds as fluorescent labeling agents for immune serum, *Am. J. Pathol.*, 34, 1081, 1958.
110. Chadwick, C. S., McEntegart, M. G., and Nairn, R. C., Fluorescent protein tracers, a trial of new fluorochromes and the development of an alternative to fluorescein, *Immunology*, 1, 315, 1958.
111. West, S. S., Hurst, R. E., and Menter, J. M., *Biopolymers*, 16, 685, 1977.
112. Weber, G., Fluorescent conjugates of ovalbumin and bovine serum albumin, *Biochem. J.*, 51, 155, 1952.
113. Kawamura, A., Jr. and Aoyama, Y., Eds., *Immunofluorescence in Medical Science*, University of Tokyo Press/Springer-Verlag, New York, 1983.
114. Nairn, R. C. and Marrack, J. R., *Fluorescent Protein Tracing*, 3rd ed., Williams & Wilkins, Baltimore, 1969.
115. Goldman, M., *Fluorescent Antibody Methods*, Academic Press, New York, 1968.
116. Goldwasser, R. A. and Shepard, C. C., Staining of complement and modification of fluorescent antibody procedures, *J. Immunol.*, 80, 122, 1958.
117. Holborow, E. J., Ed., *Standardization in Immunofluorescence*, Blackwell Scientific, Oxford, 1970.
118. Fahey, J. L. and McLaughlin, C., Preparation of antisera for 6,6, gama-globulin beta-globulin, gama-macroglobulins and for type i and ii common gama-globulin determinant, *J. Immunol.*, 91, 484, 1963.
119. Goldman, M., Personal communication, 1984.
120. Michel, B., Milner, Y., and David, K., Preservation of tissue-fixed immunoglobulins in skin-biopsies of patients with LE and bullous diseases, *J. Invest. Dermatol.*, 59, 449, 1972.
121. Black, M. M., Meyrick, T., and Bhogal, B., The value of immunofluorescence techniques in the diagnosis of bullous disorders: a review, *Clin. Exp. Dermatol.*, 8, 337, 1983.
122. Johnson, G. D., Davidson, R. S., McNamee, K. C., Goodwin, D., and Holborow, E. J., Fading of immunofluorescence during microscopy. A study of the phenomenon and its remedy, *J. Immunol. Methods*, 55, 231, 1982.
123. Gardner, P. S. and McQuillin, J., *Rapid Virus Diagnosis*, 2nd ed., Butterworths, Washington, D.C., 1980.
124. Kaplan, M. N. and Koprowski, H., Eds., *Laboratory Techniques in Rabies*, 3rd ed., monogr. Ser. No. 231, World Health Organization, Geneva, 1980.
125. Deacon, W. E., Falcone, V. H., and Harris, A., A fluorescent test for treponemal antibodies, *Proc. Soc. Exp. Biol. Med.*, 96, 477, 1957.
126. Hunter, E. F., Deacon, W. E., and Meyer, P. E., An improved FTA test for syphilis, the absorption procedure (FTA ABS), *Public Health Rep.*, 79, 410, 1964.
127. Deacon, W. E., Lucas, J. B., and Price, E. V., Fluorescent treponemal antibody absorption (FTA.ABS.) test for syphilis, *JAMA*, 198, 624, 1966.
128. Felman, Y. M. and Nikitas, J. A., Syphilis serology today, *Arch. Dermatol.*, 116, 84, 1980.
129. Rook, A., Wilkinson, D. S., and Ebling, F. J. G., *Textbook of Dermatology*, 2nd ed., Blackwell Scientific, Oxford, 1972, 1089.
130. Jablonska, S., et al., Uses of immunofluorescence test of skin and sera. Utilization of immunofluorescence in the diagnosis of bullous diseases, lupus erythematosus and certain other dermatoses, *Arch. Dermatol.*, 111, 371, 1975.
131. Person, J. R. and Rogers, R. S., Bullous and cicatricial pemphigoid clinical, histopathological and immunopathological correlations, *Mayo Clin. Proc.*, 52, 54, 1977.

132. Hodge, L., Marsden, R. A., Black, M. M., Bhogal, B., and Corbett, M. F., Bullous pemphigoid: the frequency of mucosal involvement and concurrent malignancy related to indirect immunofluorescence findings, *Br. J. Dermatol.*, 105, 65, 1981.
133. Burnham, T. K., Antinuclear antibodies: prognostic significance of nuclear immunofluorescence patterns in lupus erythematosus, *Arch. Dermatol.*, 111, 203, 1975.
134. Margarot, J. and Deveze, P., Aspect de quelques dermatoses en lumiere ultra para violette, *Bull. Soc. Sci. Med. Biol.*, 375, 1925.
135. Fitzpatrick, T. B., Arndt, K. A., Clark, W. H., Eisen, A. L., Van Scott, E. J., and Vaughan, J. H., Eds., *Dermatology in General Medicine*, McGraw-Hill, New York, 1971, pp. 1156, 1697, and 1786.
136. Levin, O. L. and Behrman, H. T., The problem of non-fluorescent ringworm of the scalp, *JAMA*, 128, 350, 1945.
137. Marples, M. J., *The Ecology of the Human Skin*, Charles C Thomas, Springfield, Ill., 1965, 675.
138. Burns, R. E., Breer, J. E., Mikhail, G., and Livingood, C. S., The significance of coral-red fluorescence of the skin, *Arch. Dermatol.*, 96, 1967, 436.
139. Ward, C. G., Clarkson, J. G., Taplin, D., and Polk, H. C., Jr., Wood's light fluorescence and pseudomonas burn wound infection, *JAMA*, 202, 1039, 1970.
140. Findlay, G. H., Blue skin, *Br. J. Dermatol.*, 83, 127, 1970.
141. Gilchrest, B. A., Fitzpatrick, T. B., Anderson, R. R., and Parrish, J. A., Localization of melanin pigmentation in the skin with Wood's lamp, *Br. J. Dermatol.*, 96, 245, 1977.
142. Hefferren, J., Cooley, R., Hall, J., Olsen, N., and Lyon, H., Use of ultraviolet illumination in oral diagnosis, *J. Am. Dent. Assoc.*, 82, 1353, 1971.
143. Benedict, H. C., The fluorescence of teeth as another method of attack on the problem of dental caries, *J. Dent. Res.*, 9(Abstr.), 274, 1929.
144. Cripps, D. J. and Peters, H. A., Fluorescing erythrocytes and porphyrin screening tests on urine blood and stool, *Arch. Dermatol.*, 96, 712, 1967.
145. Watson, C. J. and Schwartz, S., Simple test for urinary porphibilinogen, *Proc. Soc. Exp. Biol. Med.*, 47, 393, 1941.
146. Harber, L. C. and Bickers, D. R., The porphyrias. Basic science aspects, clinical diagnosis and management, in *Year Book of Dermatology*, Malkinson and Pearson, Eds., Year Book Medical Publishing, Chicago, 1975.
147. Rimington, C. and Cripps, D., Biochemical and fluorescence-microscopy screening tests for erythropoietic protoporphyria, *Lancet*, 1, 624, 1965.
148. Poh-Fitzpatrick, M. B., Piomelli, S., Young, P., Hsu, H., and Harber, L. C., Quantitative assay for erythrocyte porphyrins, *Arch. Dermatol.*, 110, 225, 1974.
149. Ghadially, F. N., Neish, W. J., and Dawkins, H. C., Mechanisms involved in the production of red fluorescence of human and experimental tumors, *J. Pathol. Bacteriol*, 85, 77, 1963.
150. Daniels, J., Jr., A simple microbiological method for demonstrating phototoxic compounds, *J. Invest. Dermatol.*, 44, 259, 1965.
151. Sandlow, L. J. and Necheles, H., Tetracycline fluorescence in detecting malignancy, *JAMA*, 189, 363, 1964.
152. Quint, R. H., Hett, J. H., and Wallace, F. J., Fluorescence cystoscopy, *J. Urol.*, 95, 208, 1966.
153. Wilbanks, G. D. and Richard, R. M., Fluorescence of cervical intraepithelial neoplasia induced by tetracycline and acridine orange, *Am. J. Obstet. Gynecol.*, 106, 726, 1970.
154. Malis, I. and Cooper, J. F., The localization of tetracycline microfluorescence in tumors of the urinary tract, *J. Urol.*, 99, 183, 1968.
155. Figge, F. H. J., Weiland, G. S., and Manganiello, L. O. J., Cancer detection and therapy: affinity of neoplastic, embryonic and traumatized tissues for porphyrins and metalloporphyrins, *Proc. Soc. Exp. Biol. Med.*, 68, 640, 1948.
156. Kessel, D., Hematoporphyrin and HPD: photophysics, photochemistry and phototherapy, *Photochem. Photobiol.*, 39, 851, 1984.
157. Lipson, R. L., Baldes, E. J., and Olsen, A. M., The use of a derivative of hematoporphyrin in tumor detection, *J. Natl. Cancer Inst.*, 26, 1, 1961.
158. Barker, D. S., Henderson, R. W., and Storey, E., The in vivo localization of porphyrin, *Br. J. Exp. Pathol.*, 51, 628, 1970.
159. Sanderson, D. R., Fontana, R. S., Lipson, R. L., and Baldes, E. J., Hematoporphyrin as a diagnostic tool: a report of new techniques, *Cancer*, 30, 1368, 1972.
160. Lipson, R. L., Baldes, E. J., and Olsen, A. M., Hematoporphyrin derivative: a new aid for endoscopic detection of malignant disease, *J. Thorac. Cardiovasc. Surg.*, 42, 623, 1961.
161. Gregorie, H. B., Horger, E. O., Ward, J. L., Green, J. F., Richards, T., Robertson, H. C., and Stevenson, T. B., Hermatoporphyrin derivative fluorescence in malignant neoplasms, *Ann. Surg.*, 167, 820, 1968.
162. Hayata, Y., Kato, H., Konara, C. et al., Fiberoptic bronchoscopic laser photoradiation for tumor localization in lung cancer, *Chest*, 82, 10, 1982.

163. Berns, M. W., Wilson, M., Rentzepis, P., Burns, R., and Wile, A., Cell biology of hematoporphyrin derivative (HPD), *Lasers. Med. Surg.*, 2, 261, 1983.
164. Profio, A. E., Doiron, D. R., Balchum, O. J., and Huth, G. C., Fluorescence bronchoscopy for localization of carcinoma in situ, *Med. Phys.*, 10, 35, 1983.
165. Profio, A. E., Doiron, D. R., and Sarnik, J., Flurometer for endoscopic diagnosis of tumors, *Med. Phys.*, 11, 516, 1984.
166. Benson, R. C., Farrow, G. M., Kinsey, J. H., Cortese, D. A., Zinke, H., and Utz, D. C., Detection and localization of in-situ carcinoma of the bladder with hematoporphyrin derivative, *Mayo Clin. Proc.*, 57, 548, 1982.
167. King, E. G., Doiron, D. R., Man, G., Profio, A. E., and Huth, G., Hematoporphyrin derivative as a tumor marker in the detection and localization of pulmonary malignancy, *Recent Results Cancer Res.*, 82, 90, 1982.
168. Lipson, R. L., Baldes, E. J., and Olsen, A. M., Further evaluation of the use of hematoporphyrin derivative as a new aid for the endoscopic detection of malignant disease, *Dis. Chest*, 46, 676, 1964.
169. Leonard, J. R. and Beck, W. L., Hematoporphyrin fluorescence: an aid in diagnosis of malignant neoplasms, *Laryngoscope*, 81, 365, 1971.
170. Gray, M. J., Lipson, R., Maeck, J. V. S., Parker, L., and Romeyn, D., Use of hematoporphyrin derivative in detection and management of cervical cancer: a preliminary report, *Am. J. Obstet. Gynecol.*, 99, 766, 1967.
171. Gregorie, H. B. and Green, J. F., Hematoporphyrin derivative fluorescence in malignant neoplasms, *J. Sci. Med. Assoc.*, 61, 157, 1965.
172. Jori, G., Tomio, L., Reddi, E., Rossi, E., Cirto, L., Zorat, P. L., and Calzavara, F., Preferential delivery of liposome-incorporated porphyrins to neoplastic cells in tumor-bearing rats, *Br. J. Cancer*, 48, 307, 1983.
173. Calzavara, F., Tomio, L., Zorat, P. L., Cirto, L., and Jori, G., The effect of chemical structure on the tumor-localizing and photosensitizing properties of porphyrins, *Photochem. Photobiol.*, 39, 705, 1984.

Chapter 6

SOURCES AND MEASUREMENTS OF OPTICAL RADIATION FOR MEDICAL APPLICATIONS

R.J. Landry, R. Barnes, and K. Mohan

TABLE OF CONTENTS

I. INTRODUCTION

The use of optical radiation in diagnostic and therapeutic medical devices has been increasing rapidly. The safety and effectiveness of many of these devices is critically linked to the nature and amount of optical radiation being emitted. This radiation needs to be characterized, sometimes with a high degree of accuracy, before one can fully evaluate the safety and effectiveness of these devices. The process of such characterization requires, among other things, knowledge of:

- The nomenclature, quantities, and units in radiometry
- The methods of measurement of optical radiation, the equipment used in such measurements, the sources of measurement error, and the overall uncertainties in the measurement
- The biological effects of the radiation being emitted
- The conditions of use of the device including the use of accessories such as filters

This chapter will outline important considerations in optical radiation measurements, which are applicable, in general, to both laser and nonlaser (noncoherent) sources, along with a description of the optical radiation emissions from selected noncoherent therapeutic medical devices. For purposes of this chapter, optical radiation includes radiation from the far UV to the far IR with wavelengths of 1 nm to 1 mm, respectively.

II. RADIOMETRY

A. Nomenclature, Quantities, and Units

The nomenclature of optical radiation is sometimes confusing because not all terms are uniquely defined. Even those that are defined are often used erroneously. For example, the UV portion of the electromagnetic spectrum is usually divided into three parts designated as UVA (315 to 380 to 400 nm), UVB (280 to 315 to 320 nm), and UVC (100 to 280 nm). (This is based on the recommendation of the Committee on Photobiology of the CIE [Commission International de l'Eclairage — the International Commission on Illumination] as found in the *International Lighting Vocabulary,* published by the CIE in 1970.) For some, the separation of the visible from the UV is taken to be at 380 nm, while for others, it is taken to be at 400 nm. In like manner, the separation between UVA and UVB may be at 315 nm or at 320 nm. The term "light", for example, refers to visible radiation as evaluated by the human eye, but is sometimes used erroneously to denote invisible radiation. An example of this is "UV light". Therefore, precise definitions of names are necessary. One way to avoid confusion is to present information about the optical radiation source in terms of spectral radiometric quantities as defined below.

The essential radiometric quantities and units of interest for the various medical applications are listed and defined in Table 1. The quantities listed in Table 1 may be measured with a radiometer (see Section II.B) and include the radiant power (P) or energy (Q) emitted by a source, which have units of watts (W) and joules (J), respectively. Irradiance (E) is the radiant power per unit area incident upon a surface, and has units of W/m^2. Radiance (L) is the radiant power emitted from a source per unit area of the source and per unit solid angle and has units of $W/sr/m^2$. The energy equivalents of irradiance (E) and radiance (L) are radiant exposure (H) and integrated radiance (tL), respectively.

For standardization purposes, the SI units listed in Table 1 should be used. However, much of the data likely to be encountered are not given in SI units but in mixed units.

Table 1
SELECTED RADIOMETRIC QUANTITIES, SYMBOLS, DEFINING EQUATIONS, AND UNITS[a]

Quantity[b]	Symbol	Defining equation	Units (abbreviations)
Radiant energy	Q		Joule (J)
Radiant power	ϕ		Watt (W)
Irradiance	E	$E = d\phi/dA$	Watt per square meter (W/m^2)
Radiant exposure	H	$H = dQ/dA$	Joule per square meter (J/m^2)
Radiance	L	$L = d^2\phi/d\Omega dA \cos\theta$	Watt per steradian per square meter ($W/sr/m^2$)
Integrated radiance		$tL = d^2Q/d\Omega dA \cos\theta$	Joule per steradian per square meter ($J/sr/m^2$)

Note: Ω is the solid angle through which flux from point source is radiated. θ is the angle between line of sight and normal to surface considered.

[a] For a more complete listing of quantities and units, see Reference 11.
[b] Quantities may be restricted to a narrow wavelength band by adding the word "spectral" and indicating the wavelength. The corresponding symbols are changed by adding the subscript λ. For example, spectral irradiance is the irradiance at a specific wavelength per unit wavelength and has the symbol E_λ.

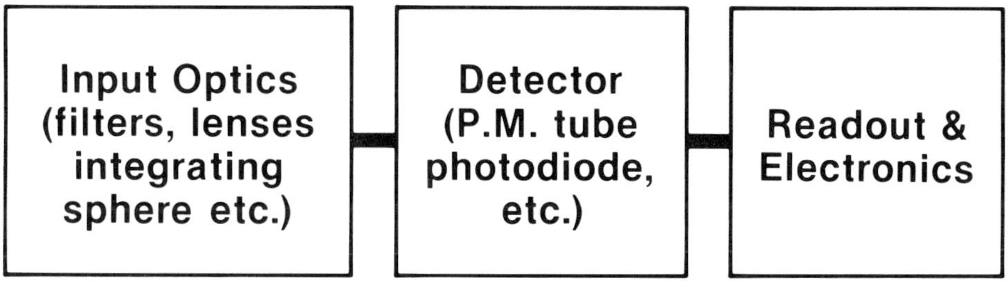

FIGURE 1. Block diagram of a radiometer.

Therefore, the data in this chapter will be given in the mixed units more likely to be encountered. Thus, for example, the units of spectral irradiance are presented in units of microwatts per square centimeter per nanometer instead of the units watts per cubic meter. (The data can be converted to SI units with the use of the relationship $1\ \mu W/cm^2/nm = 10^7\ W/m^3$.)

B. Measurement Systems and Instruments

While it is beyond the scope of this chapter to delve deeply into the complexities of measurement systems and instruments, a brief generalized description is necessary. For a more detailed discussion, see References 3 to 5.

A functional block diagram of a typical radiometer system is shown in Figure 1. The major components of a radiometer include input optics, a detector, signal-processing components (usually electronics), and the readout. The input optics are used to define the field of view and sometimes the spectral response of the instrument. The input optics, therefore, are components that the optical radiation passes through, or reflects from, before it reaches the detector. The optics usually modify the radiation that reaches the detector, and include components such as transmission filters, windows, lenses, mirrors, etc. The input optics may combine one or more of these components in order to eliminate undesired radiation at certain wavelengths which might otherwise be simultaneously transmitted to the detector with the desired wavelengths.

For measurement of the spectral quantities, dispersive optics, such as a prism or a diffraction grating monochromator, are used to disperse or spatially separate portions of the radiation into narrow wavelength bands which are then scanned (by rotating the grating or prism) into the detector. The radiation output from the monochromator is restricted to a narrow waveband with a peak output at the central wavelength selected, and a bandwidth (usually defined to be the full width at half maximum) typically between 1 and 10 nm.

An integrating sphere, which is a hollow sphere having a diffusely reflecting white finish on the inner surface, and relatively small input and output ports (usually offset by 90°), is commonly used in the input optics of a system for the measurement of spectral irradiance. Multiple diffuse (random) internal reflections result in highly uniform radiance over the entire inner surface of the sphere. A sharp-edged circular entrance aperture (port) located at the surface of the sphere provides a nearly perfect cosine angular response to the instrument. (A cosine response is one for which the relative sensitivity or response of the system varies as the cosine of the angle between the direction of the incident radiation and the perpendicular to the input aperture of the system.[2]) Such a cosine receptor allows for the quantification of irradiance on a flat surface such as the skin. It also provides for uniform illumination of the monochromator dispersing element (grating(s) or prism) regardless of the location or spatial distribution of the source(s) and the polarization of the radiation from the source(s). A cosine angular response can also be obtained with the use of a reflecting diffuser plate with a white matte finish. A more directional response with relatively more radiation transmitted in the forward direction when compared to a cosine response can be obtained with the use of a transmissive diffuser. The diffuser can be a glass or quartz (fused silica) plate with its surface(s) ground or etched so that the incident optical radiation is scattered or diffused. Such diffusers are sometimes used with radiometers. Alignment of the optical system with respect to the source to be measured is usually much less critical with a system that uses a diffuser than with one that does not.

There are a number of electronic and thermal optical radiation detectors available including silicon or vacuum photodiodes, lead sulfide or pyroelectric detectors, calorimeters (thermal), or photomultiplier (PM) tubes. Each type of detector has a characteristic response to each wavelength of radiation. For some, the "spectral response" is uniform or "flat" over a broad wavelength region. For others, there is a dramatically increased response for particular wavelength regions. For this reason, the selection of the detector is governed by the wavelength of the optical radiation to be measured. Silicon photodiodes, vacuum photodiodes, or PM tubes are normally used for the UV, visible, and near-IR regions of the spectrum. Lead sulfide, thermopiles, or germanium photodiodes are used for the IR region of the spectrum. Calorimeters and pyroelectric detectors are used for the UV, visible, and IR regions of the spectrum. Cooling the detector (typically in the range of 0 to −20°C) can enhance the signal-to-noise characteristics of some of these detectors.

The electronic signal from one of the previously mentioned detectors is amplified and displayed on a meter readout in a variety of units. The signal may also be displayed on a chart as a scan of detector signal vs. wavelength or time, or it may be converted to a digital output and processed for storage and/or output. The spectroradiometer system may be interfaced with a microprocessor or minicomputer which also may control all measurement system functions such as wavelength scan direction, speed, and wavelength range, and provide for automatic data processing.

There are several common methods of displaying spectral data.[3] Two are shown in Figure 2, in which spectral irradiance data for a common fluorescent lamp are plotted on both a logarithmic and linear scale as a function of wavelength over the range 250 to 700 nm. From Figure 2, it can be seen that an estimate of the spectral irradiance for a wide range of wavelengths less than about 350 nm can be obtained from the logarithmic plot, but not from the single linear plot with its limited dynamic range. Thus, an

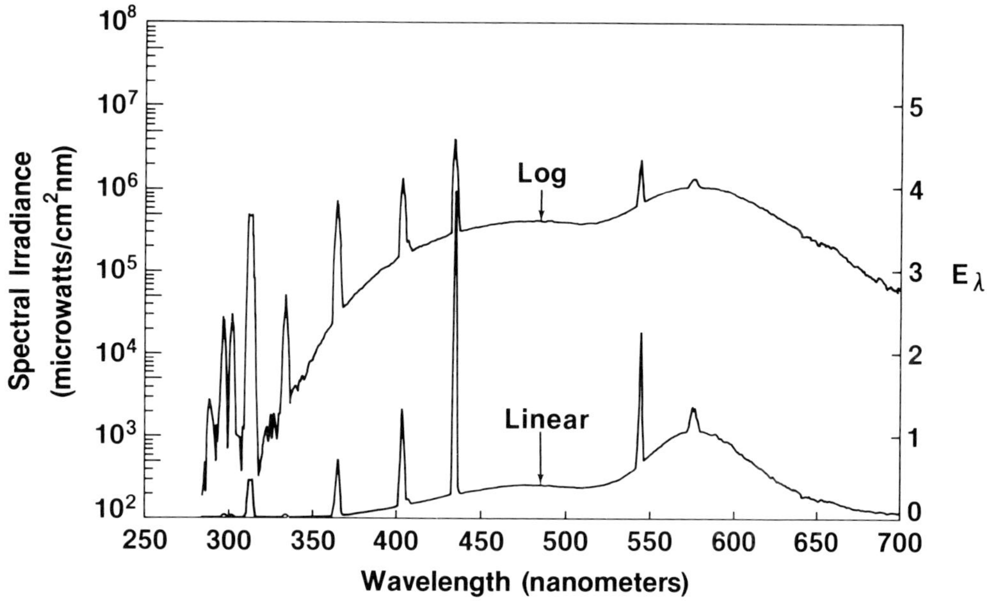

FIGURE 2. Logarithmic and linear plots of spectral irradiance for a fluorescent lamp at a distance of 2 m. The bandwidth of the spectroradiometer used to obtain the data was 1 nm.

Table 2
TYPICAL SPECTRORADIOMETER SENSITIVITY

Wavelength range (nm)	Sensitivity[a]		
	W/m^{-3}	$W/cm^2/nm$	$\mu W/cm^2/mm$
200—250	10^5	10^{-8}	10^{-2}
250—800	10^4	10^{-9}	10^{-3}
800—2500	10^7	10^{-6}	10^0
2000—2500	10^8	10^{-5}	10^1

[a] Sensitivities obtained with a 10:1 signal to noise ratio. The monochromator bandwidth of the spectroradiometer system is 2 nm, and the system utilizes a PM tube for measurement of UV and visible radiation, and a lead sulfide detector for measurement of IR radiation.

evaluation of the irradiance may require multiple displays, depending upon the region(s) of the spectrum of interest. A single linear display may be useful when all of the spectral irradiance values are of the same order of magnitude.

The minimum level of radiation that can be detected by a spectroradiometer is referred to as its sensitivity. It is usually wavelength dependent. Typical sensitivities of such systems are listed in Table 2 for various wavelength ranges in both SI units (W/m^3) and in more conventional units of ($\mu W/cm^2/nm$) and ($W/cm^2/nm$). The range of intensities over which measurements can be made with such systems can be six decades.

For accurate characterization of an optical source, it is sometimes necessary to measure the temporal characteristics of the optical radiation. The major components of a measurement system to determine the characteristics of pulsed or repetitively pulsed sources of optical radiation include input optics and a fast detector such as a photo-

Table 3
SPECTRAL CALIBRATION STANDARD LAMPS

Standard lamp type	Quantity	Wavelength range (nm)[a]	Uncertainty at specified wavelengths: Wavelength (nm)	Uncertainty (%)
Tungsten ribbon lamp	Spectral radiance	225—2400	225	2
			1000	0.7
			2400	0.6
Quartz-halogen lamp	Spectral irradiance	250—1600	250	2.6
			555—1600	1.2
Deuterium lamp	Spectral irradiance	200—350	200—350	6

[a] Primary standards of the U.S. National Bureau of Standards.

diode or pm tube, with the addition of an oscilloscope or high-speed analog to digital converted (A/D) and microcomputer as the signal readout. Care must be taken in the choice of the detector for such a system to assure that its temporal response (risetime) is faster than any temporal characteristic (e.g., rise time) of the optical radiation to be characterized. In addition, the temporal response time of the oscilloscope or sampling rate of an A/D converter must also be fast enough so that it does not distort the shape of the pulse to be characterized, or otherwise lose information about the high-frequency components of the detected signal.

C. Calibrations

In order to measure the absolute radiometric quantities (e.g., power, energy, irradiance) or the spectral radiometric quantities (e.g., spectral irradiance, spectral radiance), the radiometer or spectroradiometer must be calibrated with the use of calibration standards which are commercially available.

Electrically calibrated standard calorimeters (thermal detectors) are used to calibrate instruments which are used to measure power, energy, irradiance, or radiant exposure.[5] These calorimeters have a response that does not vary as a function of wavelength over a broad wavelength range (e.g., from 200 to 13,000 nm). Also, spectral standard lamps, which are precisely calibrated at specific wavelengths, can be used as transfer standards to calibrate spectroradiometers. The standard lamps are used as a source of known radiation output to calibrate an instrument at specified wavelengths, but an interpolation procedure must be used to obtain values at intermediate wavelengths.[7] The calibration factors obtained for the spectroradiometer are then used to determine the values for the unknown source to be measured. Standard lamps of spectral radiance and irradiance are available for this purpose and are listed in Table 3 along with the uncertainties associated with their calibration.

D. Measurement Error

There are uncertainties in any measurement. An accurate presentation of the results of measurements requires that these uncertainties be specified and, to the extent possible, estimated and presented with the data. The sources of error which give rise to the uncertainty in the data may arise from any of the components of the system (e.g., input and dispersive optics, detectors and signal-processing electronics, and readouts) used to measure the radiation, and may vary with wavelength. As noted earlier, there are also uncertainties associated with the calibration standards.

The following includes many of the common sources of error. More comprehensive discussions of sources of error in optical radiation measurements can be found elsewhere.[6,8]

1. Alignment of Optical Components

The alignment of optical components can become a very significant source of error, particularly when an instrument or measurement system has a very narrow field of view or acceptance angle (e.g., 10^{-5} sr). For instance, in the measurement of radiance or spectral radiance, a slight misalignment of a lens used to image a small area of a source (e.g., an incandescent light bulb or a tungsten filament) onto the entrance aperture of the detector or monochromator can result in a significant error in the determination of the radiance of the source. The extent of the uncertainty associated with this source of error can be estimated by making repeated measurements.

2. Contamination and Degradation of Optical Components

Dust, dirt, or other contamination on optical components such as lenses, mirrors, filters, etc. can cause scattering and absorption of optical radiation and thereby lead to errors. Some optical components simply degrade with time. For example, some thin-film interference filters degrade with time and their reflection and transmission characteristics change.

Errors associated with degradation and contamination of optical components can be discovered and quantified by periodic checks on the calibration of the measurement instrument, followed by isolation of the faulty component. Degraded components should be replaced and contaminated components should be cleaned.

3. Stray Light (Ambient)

Ambient stray light (e.g., room light), can become a significant source of error when it is within an order of magnitude of the optical radiation which is to be measured. The operator must exercise care to eliminate ambient stray light by turning off the room lights or by shielding the instrument. If this is not possible, the operator must take the ambient radiation levels into account by measuring them and subtracting their contribution from the total reading.

4. Electronic Noise

Electronic noise, like ambient stray light, can become a significant source of error when the intensity of the signal of the optical radiation to be measured is of the same order of magnitude as that of the electronic noise of the measurement system itself (sometimes referred to as "darkage"). For weak signals, the darkage can give rise to a large uncertainty in the measurement data.

5. Meter Response Time

The rate at which the meter readout of the instrument plus other signal-processing circuitry responds to the onset of radiation, or meter response time, can be a source of error if the reading is taken before the instrument has had sufficient time to arrive at its true, final equilibrium value. In this regard, it is important to account for the specific time constants for the instrument being used so that any uncertainty that may be associated with this source of error is minimized.

6. Nonlinearity of Response

This can become a significant source of error when the intensity of the radiation to be measured is very much greater or smaller than intensities at which the instrument was calibrated. In such cases, the accuracy of the reading obtained is dependent on the linearity of response of the measurement system. If the response is nonlinear, the error may be significant.

7. Stability of Source

The instability associated with the radiation source to be measured is unquestionably a source of uncertainty in measurement data. Further, the degree of the uncertainty is directly proportional to fluctuations of the radiation emissions from the source (i.e., the greater the instability, the larger the uncertainty).

8. Temperature-Dependent Response

The response of some detectors depends on the temperature. Significant errors can arise when measurements are made at an ambient temperature different from that prevailing when the instrument was calibrated. The temperature-dependent response for some detectors can be very large, on the order of 0.14 to 0.3%/°C.[8]

9. Polarization of Radiation

The electric field associated with light waves that are transmitted through an integrating sphere vibrates randomly and is not aligned in one plane. Such light is said to be randomly polarized. Light waves for which the electric field vibrates in only one plane are said to be plane polarized. In general, polarization is produced by reflection, scattering, and by transmission through special materials. Optical polarizing filters such as "Polaroid" sunglasses utilize a special material that attenuates (reduces or absorbs) polarized light that is reflected from smooth surfaces such as road or lake, or scattered from the clear blue sky by dust particles in the air. In a spectroradiometer, the amount of optical radiation transmitted through a grating monochromator is polarization dependent because the grating acts as a polarizing filter. Therefore, significant errors in measurement can result if the optical radiation to be measured is polarized or partially polarized, and if it is not depolarized with the use of an integrating sphere or reflective diffuser plate.

10. Out-of-Band Leakage

This refers to the inability of monochromators or filters to transmit radiation only in a selected wavelength band. It is not uncommon that a filter will successfully block radiation close on either side of the band of a wavelength intended to be transmitted, and yet still allow passage of radiation of wavelengths at some distance from that band (see Figure 14). Out-of-band leakage is a common source of error that can give rise to very large uncertainties (10,000%) in measurement data.[6] This source of error and means for minimizing and estimating the uncertainty associated with it are described more fully elsewhere.[9]

11. Wavelength Scale Inaccuracy

The error associated with setting the monochromator wavelength scale results in uncertainties in the spectroradiometer calibration factors that are derived from standard lamps. The magnitude of the uncertainty is proportional to the rate of change of the calibration factors as a function of wavelength (the greater the rate of change, the larger the uncertainty). The uncertainty in the calibration factors, in turn, result in a corresponding uncertainty in the data obtained from the source to be characterized. For example, a 0.1% inaccuracy in setting the wavelength scale of a laboratory-grade spectroradiometer system may result in an uncertainty of 2% for any specific wavelength.

In addition to these sources of error, the operator may fail to select proper system components or use poor measurement techniques. This source of error can be highly

Table 4
TYPICAL SPECTRORADIOMETRIC UNCERTAINTIES

Source of error	Uncertainty (%) (quartz halogen)	Uncertainty (%) (fluorescent)
Nonlinearity of response	1.0	3.0
Measurement system instability	2.7	2.7
Lamp instability	0.5	3.0
Distance measurement	0.8	0.8
Wavelength scale inaccuracies	2.0	2.0
Uncertainties in working standard	5.0	5.0
Total	6.2	7.4

variable. Some uncertainty is also associated with the transfer of the calibration factors to the instrument, since this process involves measurements with its uncertainties. The uncertainty associated with these sources of error must be estimated, accounted for, and combined so that a meaningful estimate of the total uncertainty involved in the measurement data can be obtained.

E. Estimates of Total Uncertainty

Once the uncertainty for each of the known sources of error in a measurement system has been determined or estimated, it is then possible to obtain an estimate of the total uncertainty in the measurement data. For example, the significant and known sources of error and the uncertainties associated with a laboratory-grade spectroradiometer are listed in Table 4. The uncertainties associated with other sources of error were determined to be insignificant for this system. This spectroradiometer system was used to obtain most of the data in this chapter, and is described in detail in Reference 3. (The data for Figures 17 and 25 were obtained with other instrumentation, as described in the reference cited.) In this example, and for these sources of error, the total uncertainty was taken as a combination in quadrature (the square root of the sum of the squares) of each item listed. This is the manner in which the total uncertainty associated with calibration standards is reported. Also, in this example, the measurement uncertainties associated with nonlinearity of response and lamp instability for fluorescent lamps are larger than those associated with incandescent lamps.

All too often radiometric data are presented without a discussion of measurement errors and a statement of the estimated uncertainty in the data, leaving the reader without an understanding of the variability that may be associated with the data. This is not to say that measurements must always have a small uncertainty. However, a discussion of errors and an estimate of the uncertainty in the data should always be presented with the data. Large errors in optical radiation measurements can and do occur, but the recognition of this fact need not prevent utilization of the data obtained, as long as the limits of uncertainty are expressed.

III. BIOLOGICAL EFFECTS AND RISK ANALYSIS OF OPTICAL RADIATION

Because of man's long-term familiarity with light, it is regarded by many as being "safe" or innocuous. However, this long-term familiarity does not render it harmless. Exposure to bright light sources can result in absorption of energy that can be damaging to the eyes. Most sources that produce light also produce other invisible radiations which can produce adverse effects. Some well-recognized bioeffects of optical radia-

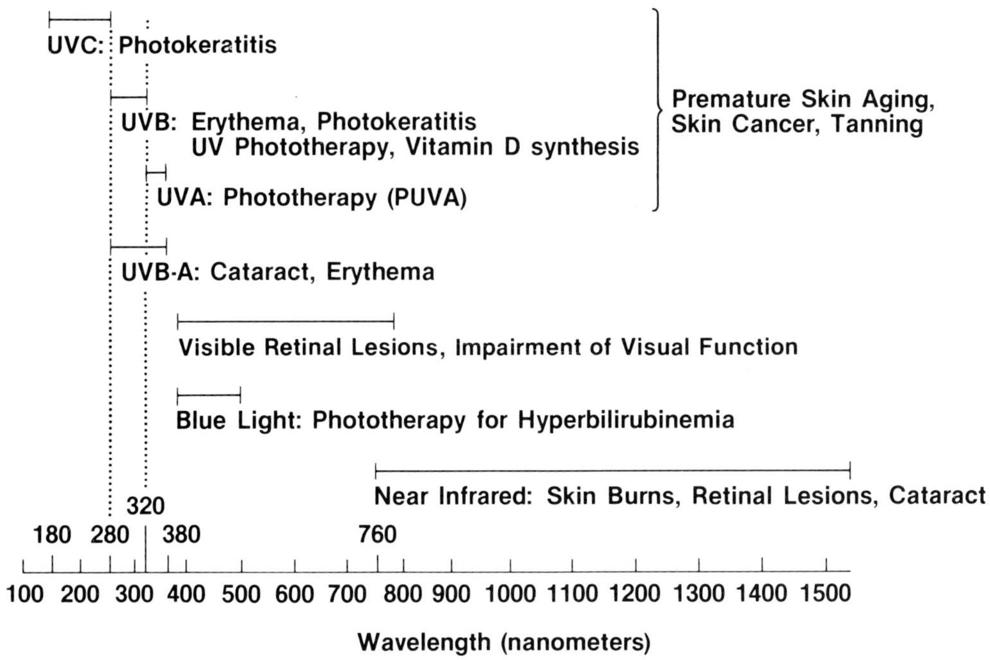

FIGURE 3. Bioeffects of optical radiation.

tion are schematically and qualitatively shown as a function of wavelength in Figure 3. They include both therapeutic and adverse effects.

Many biological effects of optical radiation are well known and are thoroughly discussed in Chapter 11. However, a summary of them is presented in Figure 3 for purposes of the following discussion of optical radiation-emitting medical devices. Using the spectral categories indicated in Figure 3, UVB and UVC radiations are known to cause photokeratitis (UV radiation induced inflammation of the cornea). UVB and UVA radiation are responsible for erythema and tanning. In addition, UV radiation can cause premature aging of the skin, skin cancer, and cataracts. It should be noted that these effects are attributable to the UV radiation irrespective of its particular origin. Terrestrial sunlight (at the surface of the earth), for example, contains both UVB and UVA radiation and poses risks to sunbathers and others who are exposed to sunlight. Visible and IR radiation also have a potential for inducing harmful biological effects.

In spite of the knowledge that radiation can have many effects, the action spectrum (i.e., that data which describe the relative ability of the radiation as a function of wavelength to produce a given biological effect) for only the acute effects of photokeratitis and erythema are well understood. The wavelength dependence of the various biological effects has been taken into account (to the extent it is known) in the various safety standards that have been developed for both lasers and nonlaser (noncoherent) optical radiation sources. A widely used safety standard for noncoherent sources is the one that has been developed by the American Conference of Governmental Industrial Hygienists (ACGIH).[10] The standard specifies threshold limit values (TLVs) to provide protection from the acute effects (photokeratitis and erythema) of UV radiation. This standard also specifies proposed TLVs to provide protection from visible and near-IR radiation damage to the retina, from IR radiation damage to the lens of the eye, and for IR whole-body thermal effects. The ACGIH TLVs do not take into account the chronic effects, nor do they account for the influence of chemicals that may affect biological sensitivity.

In all cases where an action spectrum (also called the biological function) is specified, the biologically effective radiant quantity (e.g., effective irradiance [E_{eff}]) can be determined and compared to the TLV. For example, E_{eff} is determined according to the formulation:

$$E_{eff} = \Sigma\, E_{\lambda} \cdot S_{\lambda} \cdot \Delta\lambda$$

where $E\lambda$ and $S\lambda$ represent the spectral radiometric quantity and the biological weighting function, respectively, and where $\Delta\lambda$ is the bandwidth for the spectral radiometric data. With the use of E_{eff} and after taking into account the specific use conditions, the expected exposure can be compared to the TLV for a comparative determination of the risks. The ACGIH TLVs are intended to be used as a guideline for occupational exposure and not as a fine line between safe and dangerous levels. Although the ACGIH TLVs were developed as an occupational-exposure guideline, they are sometimes used to estimate the potential hazards that may exist for nonoccupational exposures.

In most instances, a realistic risk analysis must also take into account the specific use conditions. For that reason, some caution must be exercised in extending such a risk analysis routinely to medical situations. For example, for treatment of patients, exposure to the optical radiation may have to be in excess of the ACGIH occupational exposure guideline in order to achieve a therapeutic effect. In other cases, the medical use involves periodic and chronic exposure to the emissions from sunlamps and other phototherapy equipment, in which the significant risks of interest may not relate to the acute exposure at all. In any case, occupationally exposed medical personnel should not be exposed to levels that exceed accepted exposure standards.

As a final point, it should be noted that like the adverse effects, the effectiveness of diagnostic and therapeutic effects of optical radiation is also wavelength dependent. Thus, the relative safety and effectiveness of medical devices that utilize optical radiation are wavelength dependent. This will be addressed in greater detail in connection with specific medical devices described in subsequent sections.

IV. NONCOHERENT RADIATION — SOLAR RADIATION, LAMP CHARACTERISTICS, AND OPTICAL FILTERS

A. General

Sources of noncoherent optical radiation include incandescent and quartz halogen lamps, high-pressure discharge lamps (mercury vapor, metal halide, and sodium vapor lamps), low-pressure lamps (fluorescent, germicidal, sodium vapor), compact arc lamps (xenon, mercury-xenon, etc.) and the sun. Because all of these sources of optical radiation have been or are being used in the various medical specialities, it is important to have a working knowledge of their emission characteristics. For the man-made sources, it is also helpful to have a working knowledge of their theory of operation. Man-made sources of noncoherent optical radiation generally used in medical devices can be grouped into the two broad categories of incandescent and luminescent sources. In general, incandescent sources emit a continuous spectrum, whereas luminescent sources emit mainly in discrete spectral lines. Some high-pressure discharge sources produce a continuous spectrum as well as a line spectrum. These sources are described more fully in the following paragraphs and elsewhere.[11]

B. Solar Radiation

Optical radiation from the sun has been used for the treatment of various disorders such as psoriasis and hyperbilirubinemia. The sun emits energy as if it were a hot, radiating blackbody at a color temperature of about 6500 K.[11] On a clear day, about

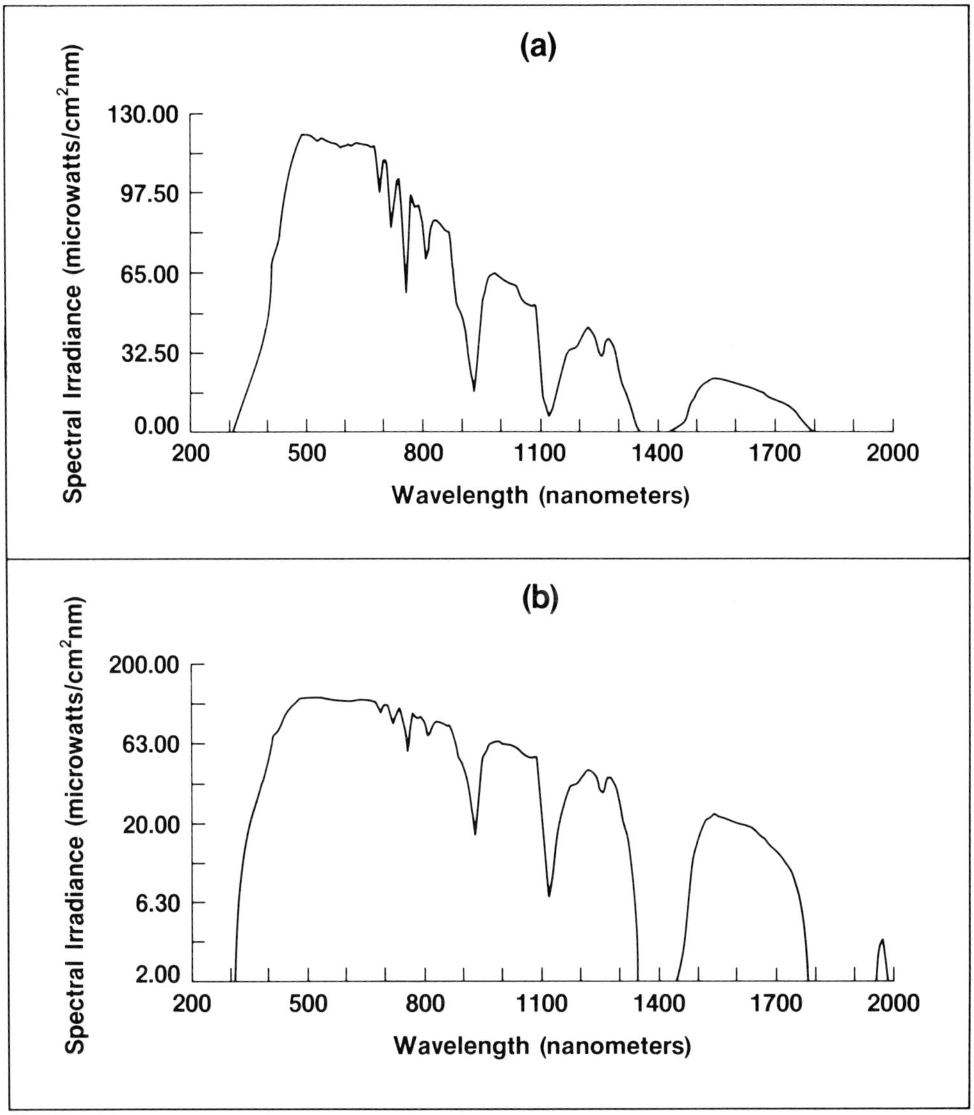

FIGURE 4. Linear plot (a) and logarithmic plot (b) of solar-terrestrial spectral irradiance at sea level. (Data obtained from American National Standard Institute (ANSI), Draft Requirements for Nonprescription Sunglasses and Fashion Eyewear, ANSI Z-80.3.)

75% of the sun's energy reaching the earth's atmosphere penetrates to the surface at the equator at sea level.[11]

The illuminance on the earth's surface by the sun may exceed 10,000 fc, whereas on a cloudy day, it can fall to less than 1000 fc. (The illuminance is the photometric analog of irradiance; it is the irradiance as evaluated by the human eye.) For comparison purposes, the illuminance in offices is typically in the range from 50 to 150 fc. The relative spectral power distribution of terrestrial solar radiation is shown in Figure 4. About half of the sun's energy reaches the earth's surface as visible radiation. The other half consists of invisible UV and IR radiation. Due to absorption by the ozone in the atmosphere, there is no measurable solar terrestrial UV radiation at wavelengths below about 290 nm.

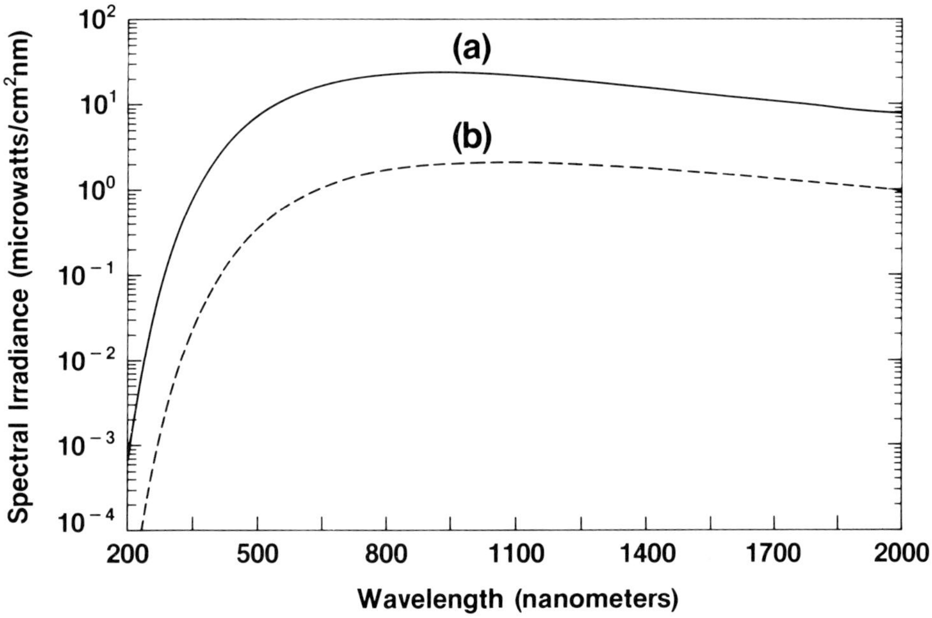

FIGURE 5. Spectral irradiance plots for (a) a 1000-W quartz halogen lamp, and (b) a 100-W frosted incandescent lamp, both at a distance of 50 cm. The bandwidth of the spectroradiometer used to obtain the data was 5 nm.

C. Incandescent Sources

Incandescent sources operating at temperatures below about 870 K emit only invisible energy in the IR.[11] A coal stove, an electric iron, and some radiant heating panels are examples of such incandescent radiating bodies. For incandescent sources operating at higher temperatures, electronic transitions in atoms and molecules result in the emission of visible radiation along with IR radiation. Examples of such sources are the heating elements on electric ranges and common "quartz heaters". Quartz heaters contain a wire within a quartz rod which, as electric current passes through it, produces a red glow and emits heat as IR radiation. The color temperature for such sources is about 1200 K. (The color temperature of a light source is that temperature of a blackbody cavity which most nearly approximates the color of the light source.) At still higher temperatures, substantial levels of visible and UV radiation are emitted. The incandescence of a tungsten filament is caused by the heating action of an electric current through the filament. Tungsten-filament lamps operating at a color temperature of about 2800 K emit substantial levels of UVA, visible, and IR radiation. Quartz-halogen tungsten lamps have a quartz glass envelope, and when operating at a temperature of about 3100 K, also emit substantial levels of UVB and UVC. It is interesting to note that most modern incandescent lamps are filled with a mixture of argon and nitrogen to increase lamp efficiency and the lifetime of the filament. Spectral irradiance plots of a 100-W incandescent lamp and a 1000-W quartz halogen lamp operating at temperatures of about 2800 K and 3200 K, respectively, are shown in Figure 5. In general, about 75 to 80% of the electrical input power to an incandescent lamp is radiated as IR energy.[11] Finally, because quartz glass is relatively transparent in the UVB and UVC regions, quartz halogen lamps that are operated at temperatures higher than 3100 K represent a potential UV radiation hazard.

D. Luminescent Sources

In luminescent sources such as tubular fluorescent lamps, an electric arc discharge is established through the mercury vapor present in the tube at low pressure (0.008 torr),[11] when the proper voltage is applied to the electrodes at the end of the tube. Although the discharge produces excitation of single-valence electrons in the vapor, which emits some visible light, it emits mostly UV radiation (primarily at 253.7 nm). This UV radiation is usually converted into visible light by a phosphor coating on the inside of the tube. Other common types of gaseous discharge lamps include the high-intensity discharge lamps such as the mercury vapor, metal halide, and high-pressure sodium vapor lamps. For these lamps, the arc discharge tube is usually contained within another glass or ceramic bulb for temperature stability purposes. The inside surface of the outer envelope may be either clear or coated with a phosphor as in most fluorescent lamps. The vapor pressure in the tube through which the discharge is established is on the order of 100,000 times greater than that of the fluorescent lamp.[11] This higher pressure results in a shift of the emitted radiation into longer wavelengths. For a high-pressure mercury vapor lamp without a phosphor coating on the inside of the outer bulb, this results in a characteristic greenish-blue light. Unlike incandescent lamps, most gas discharge lamps, including fluorescent lamps, must be operated with an electrical current-limiting device, called a ballast. While most lamps use an external electrical ballast, other "self-ballasted" lamps utilize a resistive element within the lamp to limit the current.

The spectral distribution of the optical radiation of fluorescent lamps depends upon the blend of the fluorescent chemicals in the phosphor coating (usually appearing, when not lighted, as a matte white, translucent coating) on the inside of the tube. Most of the fluorescent lamps have been developed, of course, for lighting purposes, and the industry has developed families of lamps that have common spectral characteristics. The spectral emission characteristics for some of these common fluorescent lamps are shown in Figure 6, while those for clear (uncoated with a phosphor) high-pressure discharge lamps (mercury vapor, metal halide, and high-pressure sodium vapor) are shown in Figure 7. From these two figures, it can be seen that both fluorescent lamps and the high-intensity discharge lamps emit optical radiation over a broad wavelength range from within the UV through the visible and up to about 800 nm into the IR wavelength range. Unlike incandescent lamps, these lamps emit a combination of continuous and line spectra and very little IR radiation. The primary difference between mercury-vapor and metal-halide lamps is that the latter contain various metal halides (e.g., thallium, indium, scandium, and dysprosium iodides) in addition to the mercury and argon that are present in both types of lamps. The addition of these metal halides results in a more "white" light, and higher electrical efficiency when compared to clear mercury-vapor lamps. High-pressure sodium vapor lamps have maximum emissions in the wavelength range from 550 to 650 nm. Thus, the light that is emitted from these lamps appears to be golden white in color. The spectral emission characteristics for some selected phosphor-coated mercury-vapor and metal-halide lamps are shown in Figure 8. The spectral emission characteristics of low-pressure mercury-vapor lamps that have an high silica content glass or quartz envelope, and low-pressure sodium-vapor lamps are shown in Figure 9. Both of these lamps emit the characteristic line spectra for these metal vapors. As indicated earlier, the line spectral emissions from low-pressure mercury vapor are primarily in the UV. The low-pressure mercury-vapor lamps with emissions shown in Figure 9 are generally called "germicidal" lamps because of the bactericidal action of the UV radiation emissions (primarily at 253.7 nm). The spectral emission from the low-pressure sodium lamps can be readily recognized because of its characteristic yellow color, with most of the emissions in the sodium doublet "D" lines at 598 nm.

A special type of fluorescent lamp that is used in industrial, commercial, scientific, entertainment, and certain medical and dental applications is one that utilizes a phosphor with emissions primarily in the UVA-wavelength range, with peak emissions between 350 and 370 nm. The spectral emission characteristics of two of these lamps, which are commonly referred to as "blacklight" lamps, are shown in Figure 10.

With a few exceptions, almost all of the fluorescent, blacklight, and mercury-vapor and metal-halide lamps emit some shortwave UV radiation in the 300-nm wavelength region, primarily in the principal emission lines of mercury at 297, 303, and 313 nm. At close exposure distances (within about 2 to 3 in.), the UV levels emitted at wavelengths less than 315 nm may result in UV exposures that exceed acceptable safety guidelines for some lamps. Further, because blacklight sources, by design, emit intense levels of UVA, close exposure distances (within about 6 to 12 in.) to these sources may also result in exposures that exceed safety guidelines. For these reasons, caution should be exercised in the design and use of facilities and devices that utilize these types of lamps. Of course, the levels of UV radiation emissions from germicidal lamps are intended to be high, and appropriate engineering controls or other protective measures must be taken to preclude acute injury to the eyes and skin of persons who may be exposed.

Another less common type of high-pressure gas-discharge lamp is the short-arc or compact-arc lamps. These lamps can utilize mercury, mercury-xenon, or xenon gas and are characterized by an electrode-stabilized arc which is short when compared with the diameter of the discharge tube. The arc length for these lamps, depending upon the rated power and application, can range from a fraction of a millimeter (0.3 mm) to about 12 mm.[11] The pressure in these lamps during operation ranges from 10 to 50 atm and is much higher than the 2 to 3 atm in the mercury-vapor or metal-halide lamps discussed earlier.[11] Because of the potential explosion hazard associated with these higher pressures, they must be operated in an enclosure at all times. Like the high-intensity discharge lamps, the short-arc lamps require a ballast to start the arc and to limit the current. However, unlike the high-intensity discharge lamps, the short-arc lamps are designed for DC operation, because this results in better arc stability and longer life when compared to short-arc lamps that are operated with AC. The spectral emission characteristics of a short-arc lamp in the UV and visible wavelength ranges are shown in Figure 11. The emissions from this lamp consist of a combination of both a continuous and line spectra that extends from the UV, through the visible, and well into the IR up to about 2000 nm. Because these arcs have the highest radiance of any continuously operating light source, they also represent the greatest risk of retinal injury. Furthermore, the intense UV emissions from these lamps are also capable of producing serious eye and skin injury. For these reasons, considerable care must be taken in the design and use of systems that utilize these sources.

E. Other Lamp Characteristics

It is important to be aware that different lamps have different finite working lifetimes, and that the light output of all lamps decreases monotonically with age. The light output for some quartz halogen lamps can initially increase when operated at constant current. Typical rated life data (the time, in hours, at which 50% of the lamps will still be in operation) for various types of lamps are shown in Table 5.[11]

It should be noted that the decrease (depreciation) of light output with time can be fairly large if lamps are not operated at the proper electrical design values. In this regard, it is important to be aware that the output of fluorescent lamps, which are used extensively in medical devices, varies considerably with lamp ballasting and with current load through the lamp. (The higher the current through the lamp, the greater the light output but the faster the depreciation.) For example, when operating at relatively

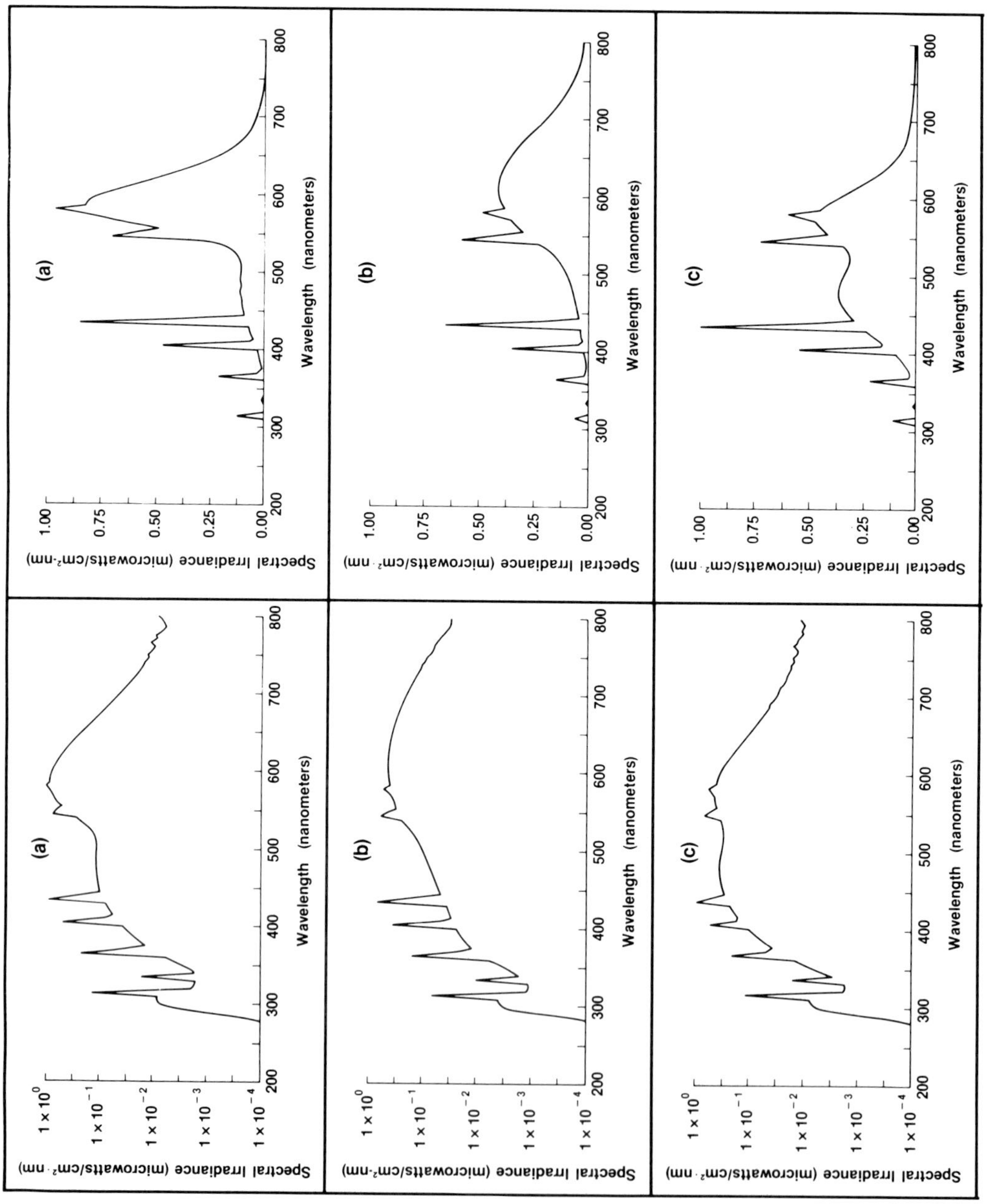
(a)
(b)
(c)
Spectral Irradiance (microwatts/cm²·nm)
1 × 10⁰
1 × 10⁻¹
1 × 10⁻²
1 × 10⁻³
1 × 10⁻⁴
1.00
0.75
0.50
0.25
0.00
200
300
400
500
600
700
800
Wavelength (nanometers)

FIGURE 6a-c.

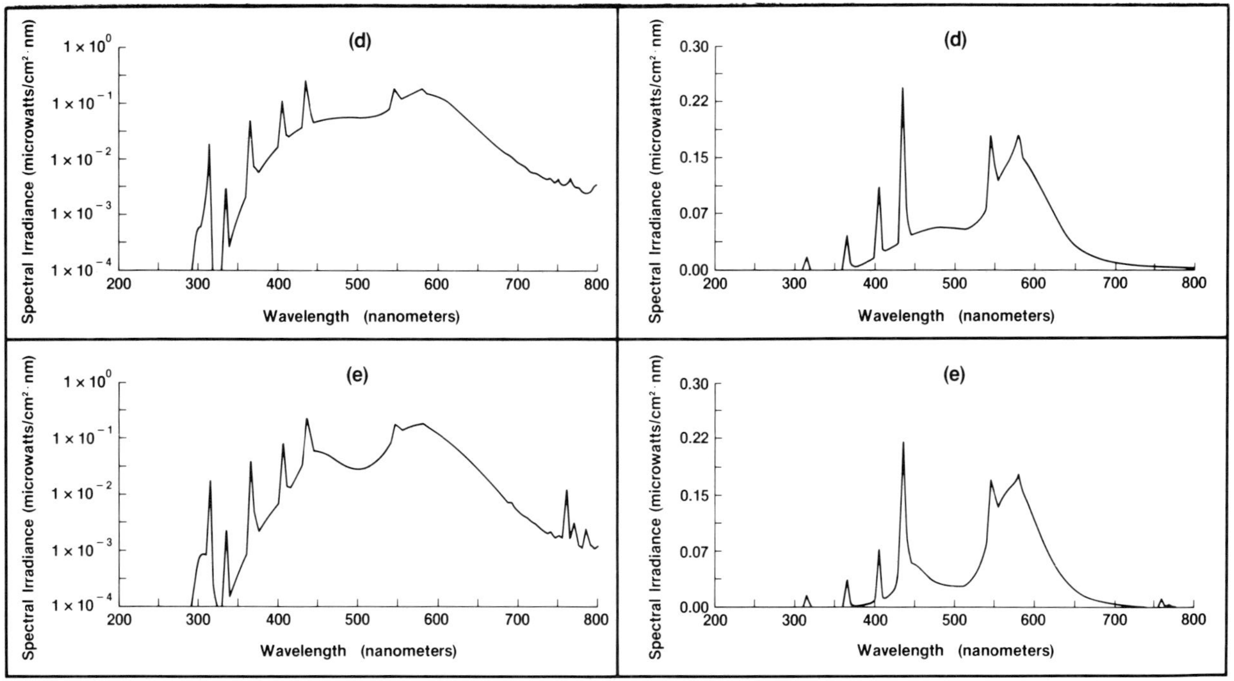

FIGURE 6. Logarithmic and linear plots of spectral irradiance for various fluorescent lamps. (a) Warm white; (b) warm white deluxe; (c) daylight; (d) cool white; (e) deluxe cool white, all at a distance of 1 m. The bandwidth of the spectroradiometer used to obtain the data was 5 nm.

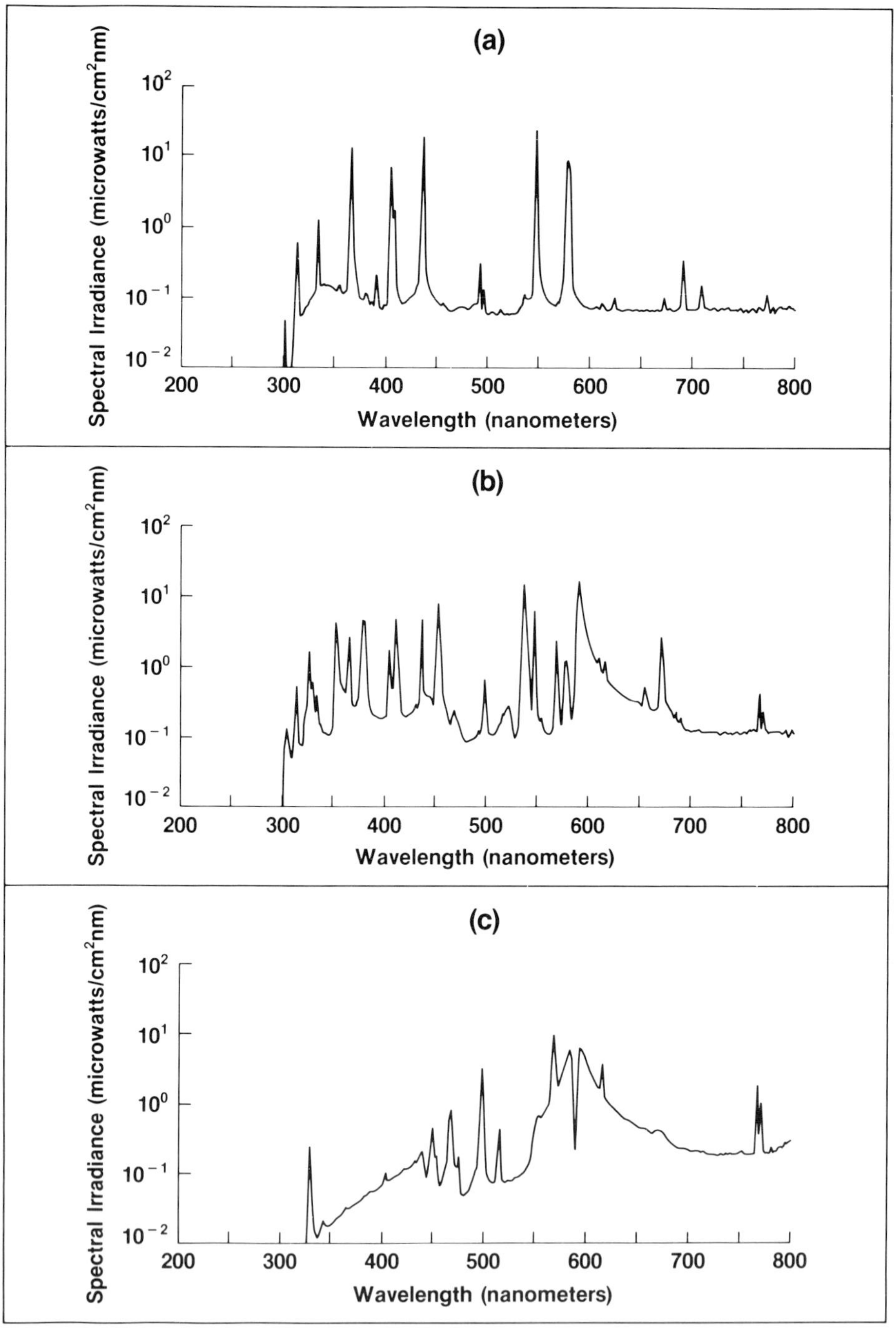

FIGURE 7. Spectral irradiance plots of uncoated high-intensity discharge lamps. (a) 400-W mercury vapor; (b) 400-W metal halide; (c) 400-W sodium vapor, all at a distance of 2 m. The bandwidth of the spectroradiometer used to obtain the data was 1 nm.

low design currents, the lamp output at 10,000 hr may be at about 90% of the initial output, whereas at relatively high design currents it may be at about 60% of the initial light output at 10,000 hr.[11] Furthermore, the lamp output of fluorescent lamps is also dependent on the bulb wall temperature. Typically, these lamps are designed to operate

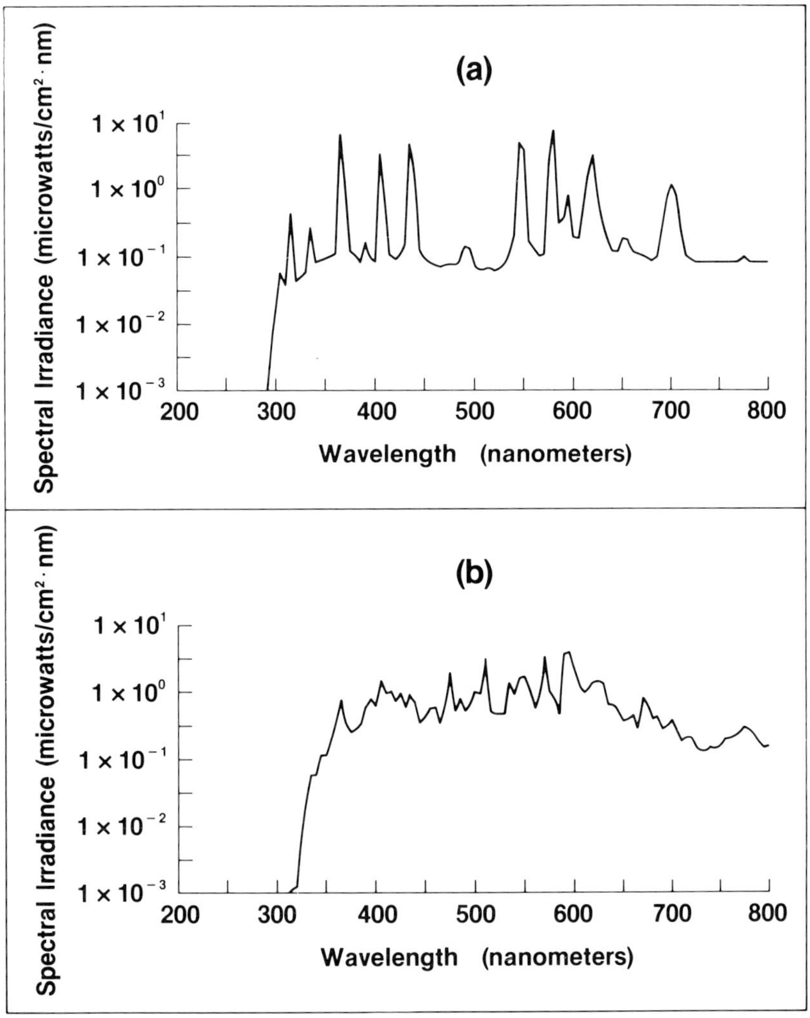

FIGURE 8. Spectral irradiance plots of phosphor-coated high-intensity discharge lamps. (a) 400-W mercury vapor; (b) 400-W metal halide, both at a distance of 2 m. The bandwidth of the spectroradiometer used to obtain the data was 5 nm.

with maximum output with a bulb temperature near 38°C.[11] When the bulb temperature is near 20 or 60°C, the output of the lamp may be reduced to about 70% of its value at the optimum bulb wall temperature. Because of these variables, measurements of the radiant output of the lamp may be the only practical way to control the exposure dose if precise intensity or dose must be maintained.

F. Optical Filters

In general, the emissions from medical devices that utilize these sources are modified through the use of optical filters. Typically, the source is located behind glass or plastic filters or filter materials to restrict or define the emissions to the desired effective wavelengths and to eliminate or minimize unnecessary radiation emissions. Commercially available transmissive/reflective type cut-off or band-pass filters can be used to accomplish this. Spectral transmittance curves for selected UV and visible wavelength cut-off filters are shown in Figure 12. Spectral transmittance curves for selected band-pass filters are shown in Figure 13 and Figure 14 for the UV and visible wavelengths.

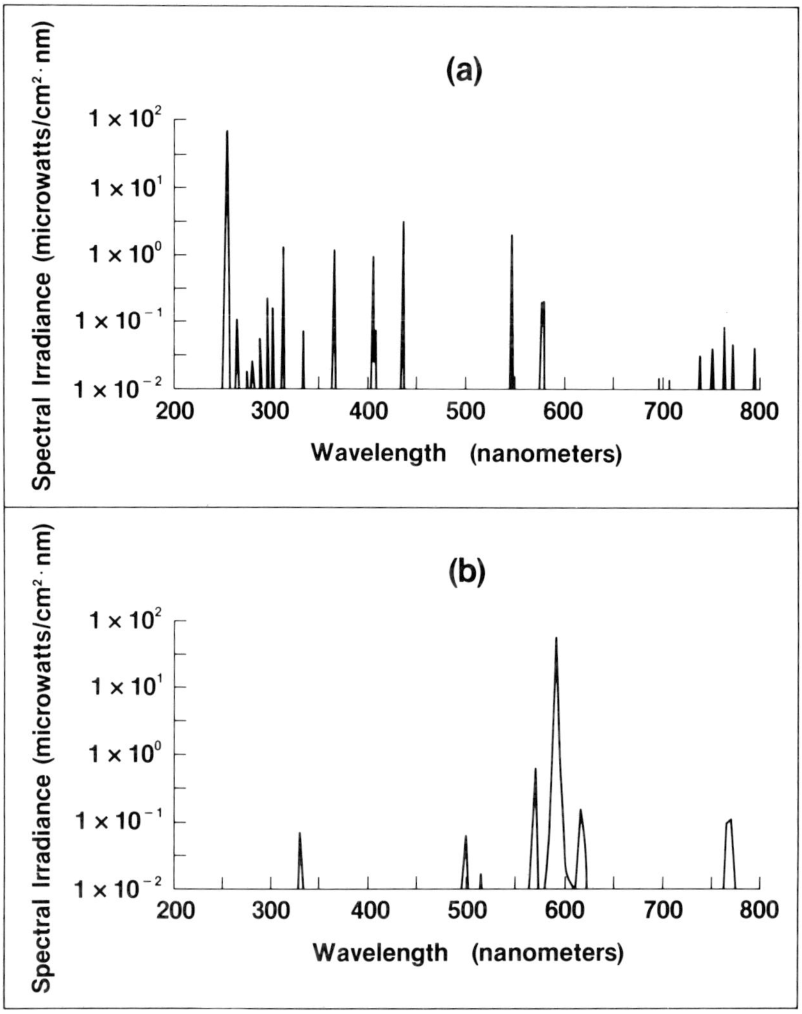

FIGURE 9. Spectral irradiance plots of low-pressure lamps. (a) 15-W mercury-vapor (germicidal) lamp at a distance of 2 m; (b) 90-W sodium lamp at a distance of 1 m. The bandwidth of the spectroradiometer used to obtain the data was 5 nm.

V. THERAPEUTIC MEDICAL DEVICES

In the following sections, specific optical radiation therapeutic medical devices are described using the terms and concepts introduced in the first four parts.

A. Phototherapy Devices for Hyperbilirubinemia

Hyperbilirubinemia (neonatal jaundice), a condition in which there is an excess of bilirubin in the blood, is a relatively common condition in newborn infants. It occurs most frequently in premature infants and in the absence of an identifiable disease. In 1958, it was reported that exposure to sunlight could reduce the level of bilirubin in the blood.[12] Presently, the use of man-made sources of light to degrade bilirubin has become widely accepted.

Typically, the nude infant is placed in an incubator under a canopy that contains a bank of fluorescent lamps as shown in the sketch of a phototherapy unit in Figure 15. A wide variety of fluorescent lamps are used in the various devices for this medical

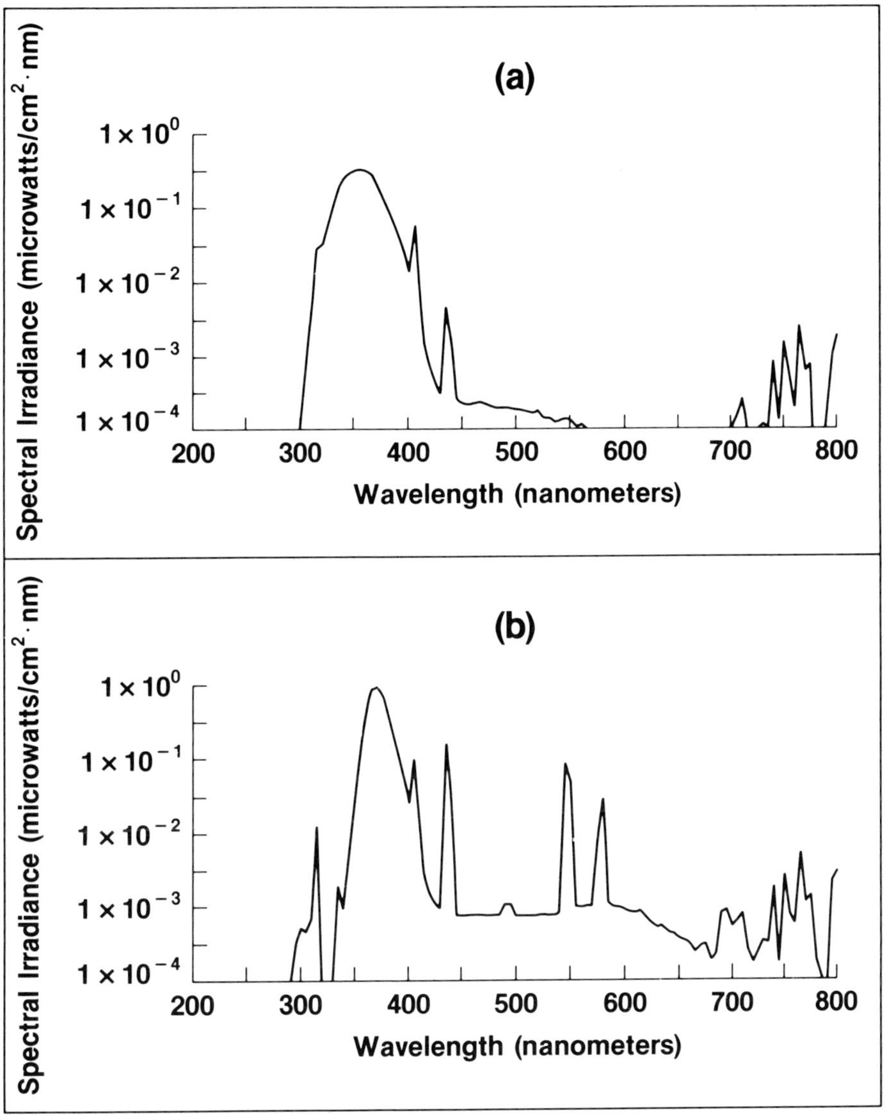

FIGURE 10. Spectral irradiance plots of blacklight lamps. (a) 15-W type BLB; (b) 15-W type BL both at a distance of 1.1 m. The bandwidth for the spectroradiometer used to obtain the data was 5 nm.

problem. Because it is known that blue light is most effective in breaking down bilirubin, some of the lamps used have phosphors that emit principally in the blue-wavelength range.[13] However, all of the lamp types used appear to be effective in breaking down bilirubin. Plots of spectral irradiance at the infant position for one commercially available unit are shown in Figure 16 with the spectral irradiance (from ambient lighting) and without the phototherapy unit at the same infant position.[14] As can be seen from this figure, there is a considerable difference in the spectral irradiance levels to which the infants undergoing phototherapy are exposed when compared to those not undergoing phototherapy. Also, as can be seen from Figure 16, there is considerable difference in the phototherapy spectral irradiance levels at the end of the incubator when compared to those at the center of the incubator at the infant position. The data for the phototherapy unit were obtained with a unit that was equipped with eight 20-

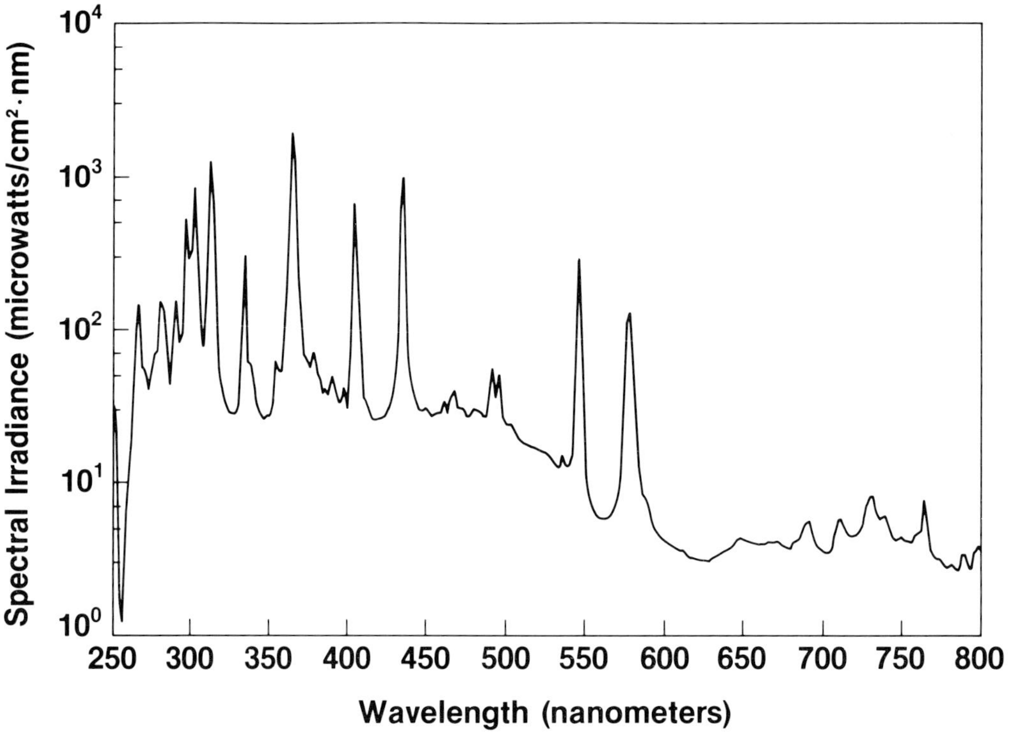

FIGURE 11. Spectral irradiance plot at a distance of 12 cm from the final lens of 1000-W xenon lamp system. The bandwidth for the spectroradiometer used to obtain the data was 5 nm.

Table 5
TYPICAL RATED LIFE FOR SELECTED LAMP TYPES

Lamp type	Rated life (hr)
Incandescent	<2,000
Fluorescent (including germicidal and blacklights)	7,500—20,000
High-intensity discharge	12,000—24,000

W "daylight" fluorescent lamps. The mean phototherapy irradiance levels at the infant position for selected wavelength bands found at eight neonatal care units that utilized this same phototherapy unit are listed in Table 6.[14] The mean irradiance levels found at the infant positions for infants not undergoing phototherapy for the same eight neonatal care units are also listed in Table 6.

For convenience purposes, some phototherapy units utilize a single incandescent-type spotlight which is mounted in a floor-standing fixture. Such a device can be easily moved around and allows for greater physical access to the infant. Unnecessary IR radiation emissions at wavelengths greater than about 750 nm, which could result in heating the infant and in additional insensible water loss, are minimized with the use of IR-absorbing filters. The spectral irradiance at a distance of 0.75 m for one commercially available device is shown in Figure 17.

For either type of phototherapy device, the total dose or infant exposure is influenced by several factors. These include the line voltage at which the phototherapy units are operated; proximity to outside windows, which can result in significant additional

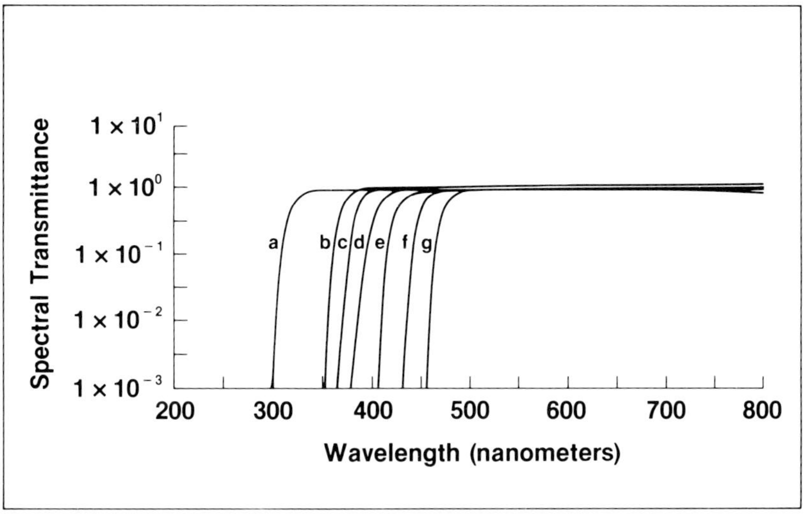

FIGURE 12. Spectral transmittance for selected UV and visible wavelength cut-off filters. (a) 320-nm cut-off; (b) 365-nm cut-off; (c) plexiglass G; (d) plexiglass UF-3; (e) 425-nm cut-off; (f) 440-nm cut-off; (g) 490-nm cut-off.

optical radiation exposure from sunlight (e.g., about 25% of phototherapy irradiance levels); weather conditions, whether sunny vs. cloudy; the use of window shades and infant covering (clothing or blankets); and the specific movement of the infant from one location to another location in the same room or nursery.[14]

In order to prevent retinal damage to the eyes of the infants undergoing phototherapy at relatively high irradiance levels of visible light, opaque coverings are placed over the eyes. In addition, almost all fluorescent lamps emit some levels of erythema-producing UVB. In order to prevent an excessive exposure, a plastic or plexiglass sheet that absorbs the UV radiation below about 350 nm is located in the canopy directly below the fluorescent lamps.

B. Sunlamps

Sunlamps are devices that emit UV radiation for the purpose of inducing skin tanning. Since the radiation they emit is designed to affect the structure or function of the body, they are medical devices. There are two basic types, which can be distinguished depending upon the UV spectral power distribution (spectral irradiance) that is utilized to produce the tanning effect. So-called "UVB" sunlamps emit UV radiation in the wavelength range between 260 and 320 nm for the tanning effect. So-called "UVA" sunlamps emit intense levels (e.g., 10 to 30 mW/cm^2) at the user exposure position in the wavelength range between 320 and 400 nm with relatively little or no measurable radiation at wavelengths less than 320 nm (less than 5%). Historically, UVB sunlamps were developed and used long before UVA sunlamps. The UVA sunlamps were only developed in the late 1970s.

Sketches of selected sunlamp sources are shown in Figure 18. Many of the sunlamp products utilize a single lamp and are table mounted or are designed to be used on an extendable tripod stand. Other sunlamp products utilize vertical rows of fluorescent sunlamps in a booth, similar to a telephone booth, with or without reflective walls. Yet others are in the form of canopied beds with a row of fluorescent lamps below a clear sheet of plastic upon which the user lies, and another row of fluorescent lamps, also behind shielding, that hangs overhead in a canopy. Typically, sunlamp products utilize either UVA or UVB type lamps, although some products utilize both types.

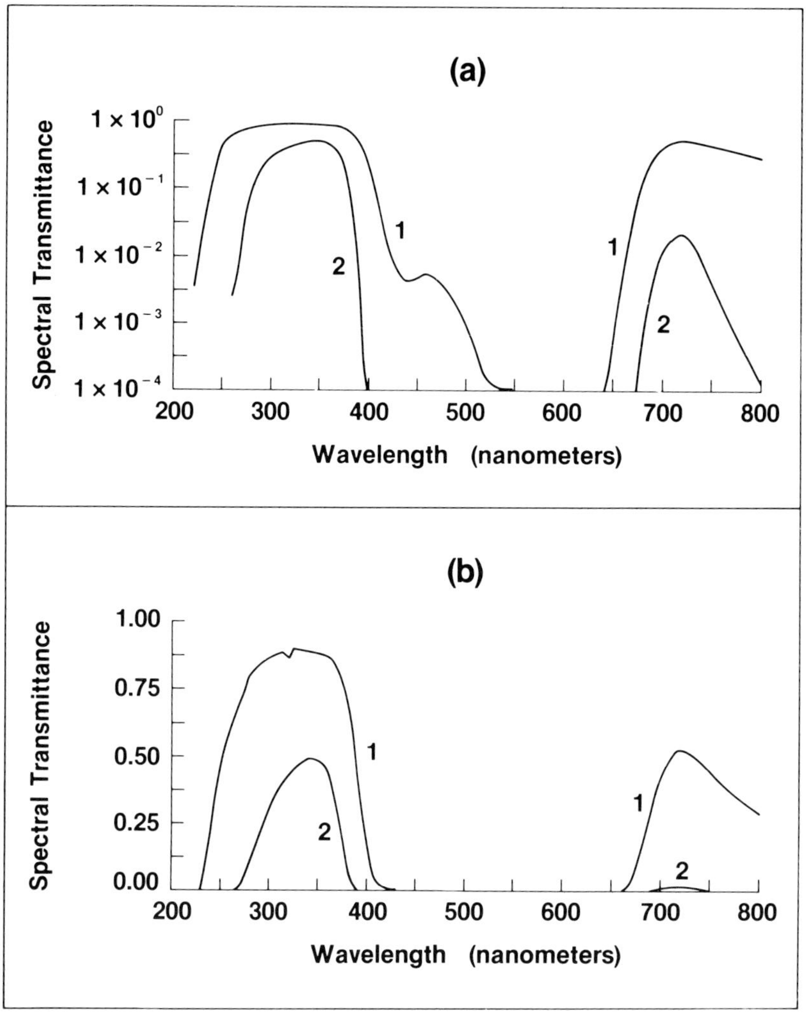

FIGURE 13. (a) Logarithmic plots and (b) linear plots of spectral transmittance for two UV radiation band-pass filters.

The UVB sunlamps can be divided into three basic types. A very common type utilizes a mercury-vapor discharge bulb with an internal "reflector" commonly referred to as an "RS" type sunlamp. The others are a fluorescent sunlamp, with a special UV-emitting phosphor commonly referred to as an "FS" type sunlamp, and a mercury-vapor discharge lamp for use in a device with an external metal reflector. The UVA sunlamps can likewise be divided into two types: a fluorescent-tube type with a special phosphor that emits UV radiation primarily in the UVA wavelength range, and a metal-halide lamp type. Special filters are used with the bare quartz UVB type lamps to limit the emission of short wavelength UV radiation for wavelengths less than 260 nm, and with the UVA type lamps to control the spectral emissions to the desired wavelength range.

The spectral emission characteristics for selected sunlamp products are shown in Figure 19. The differences in spectral power distribution for the UVA and UVB types can be seen in this figure. In no case do these products produce a spectral power distribution like that of terrestrial solar radiation (see Figure 5).

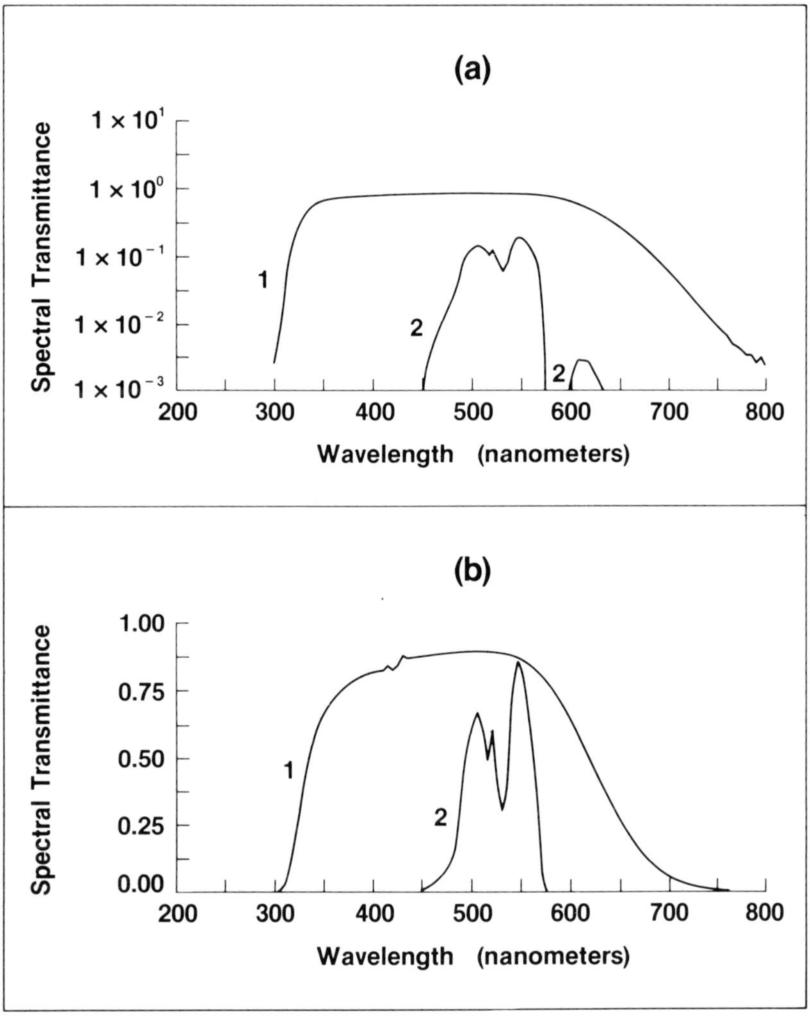

FIGURE 14. (a) Logarithmic plots and (b) linear plots of spectral transmittance for (1) a near-UV and visible band-pass filter and (2) a visible band-pass filter.

Sunlamp products, particularly the UVB types, have resulted in serious eye and skin injuries (sunburn) to many people. Although it is less likely that the UVA type sunlamp products will result in UV radiation skin burns if instructions for use are followed, UVA can penetrate deeper into the skin than UVB and may increase the risk of other health hazards such as edema and vascular-system damage. As with chronic exposure to the sun, chronic exposure to the emissions from either UVA or UVB types of sunlamps may increase the risk of premature skin aging or skin cancer. Because of the possibility of long-term skin damage, both the Food and Drug Administration (FDA) and the American Academy of Dermatology advise the public to avoid tanning for cosmetic purposes.[15] Further, in order to minimize the risk of acute eye or skin injury to persons who make the personal decision that a cosmetic tan outweighs the risks, the FDA developed and enforces a Product Performance Standard promulgated under the Radiation Control for Health and Safety Act (PL 90-602).[16] Among other requirements, this standard requires that manufacturers provide a timer to automatically shut off the lamps after a preset exposure period, protective goggles or eyewear, and adequate instructions for use. Because sunlamps are also medical devices, they are also

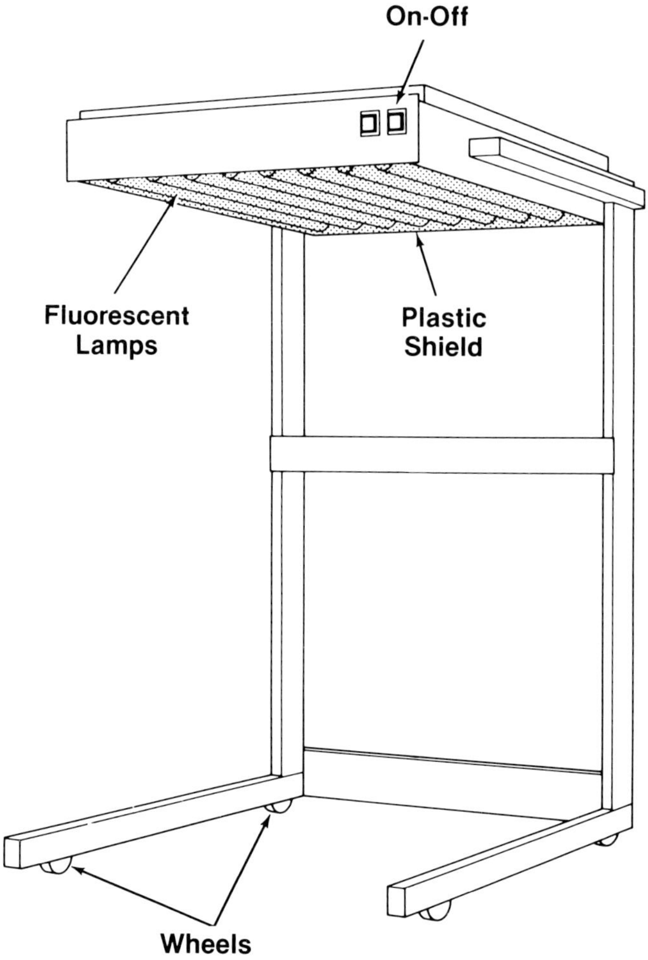

FIGURE 15. Sketch of fluorescent lamp phototherapy unit for the treatment of neonatal jaundice.

subject to the provisions of the Medical Device Amendments of 1976 to the Federal Food, Drug, and Cosmetic Act.[17]

C. Psoriasis Lamps

Psoriasis is a disfiguring and often incapacitating skin disease that affects from 1 to 3% of the world population. The distinctive psoriatic lesion is a vivid red thickening or elevation of the skin that is covered almost to its edge by a silvery scale. Phototherapy treatment for psoriasis with UVB is sometimes done following the application of crude coal tar ointment (photochemotherapy). At other times, UVB radiation is used alone.

PUVA therapy, a photochemotherapy which is used for the treatment of extensive psoriasis, involves the oral administration of a photosensitizing drug (8-methoxypsoralen) followed by exposure of the afflicted areas to high-intensity UVA radiation.

UVB type sunlamps and special blacklight UVA lamps are used for the treatment of psoriasis. For whole-body irradiation, phototherapy booths similar to the one shown in the Figure 18 sketch are used. These booths usually contain a large number of lamps closely spaced in order to achieve uniformity of exposure over the body surface.

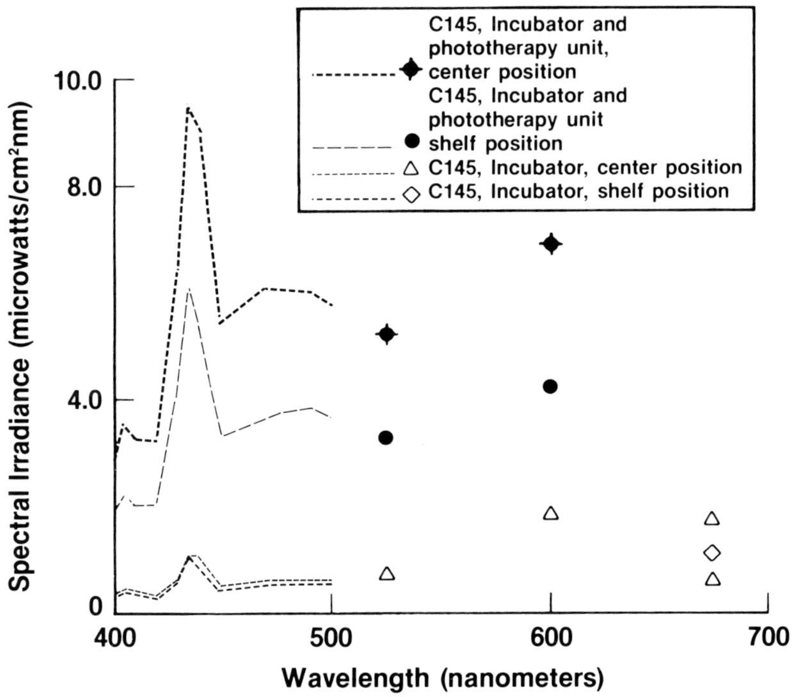

FIGURE 16. Spectral irradiance plots with and without a phototherapy unit in operation, at a center position and near one end on a shelf of an incubator (at the approximate upper surface of the infant). The bandwidth of the spectroradiometer used to obtain the data was 13.5 nm.[14]

Table 6
PHOTOTHERAPY AND AMBIENT IRRADIANCE LEVELS ($\mu W/cm^2$)

	Wavelength band		
	400—500 nm	500—675 nm	400—675 nm
Phototherapy	714 ± 97	1135 ± 154	1849 ± 248
Ambient conditions	89 ± 36	253 ± 118	342 ± 153

The spectral power distribution of a blacklight lamp used for PUVA therapy is shown in Figure 20. The spectral power distribution of these lamps is particularly important for both safety and effectiveness of the treatment and should have the spectral distribution listed in Table 7.[18]

A problem may occur if blacklight lamps that have different phosphors and different spectral power distributions are used as replacement lamps. From Table 7, it can be seen that the peak emission for the PUVA therapy lamps occurs between 350 and 360 nm and that about 98% of the energy is contained between 320 and 400 nm. Other blacklight phosphors which have a different spectral power distribution that is shifted to longer wavelengths with peak emissions near 365 nm and with a range of 340 to 400 nm are readily available commercially. A relative plot of the spectral power distribution of the emissions from two blacklight type phosphors are shown in Figure 21.

Results of animal tests indicate that if the longer-wavelength emission lamps are used as replacements for the shorter-wavelength emission lamps, the UV dose could be too

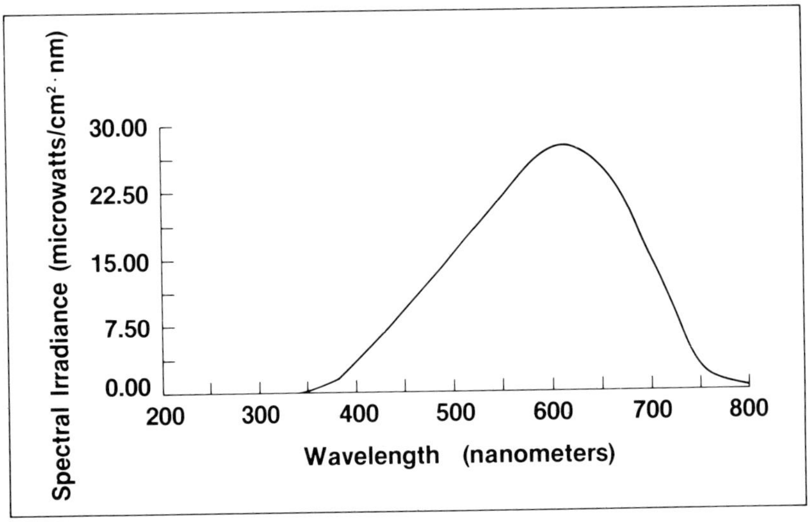

FIGURE 17. Spectral irradiance plot of an incandescent phototherapy unit at a distance of 0.75 m. The bandwidth for the spectroradiometer used to obtain the data was 5 nm.

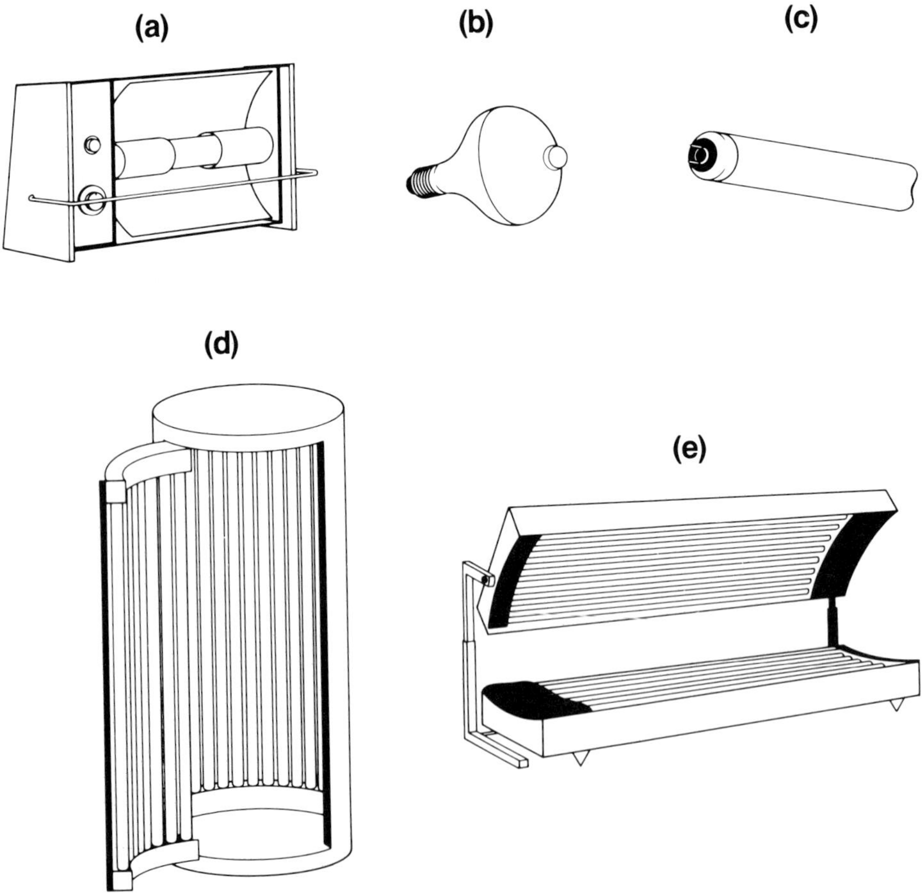

FIGURE 18. Sketches of various sunlamp types. (a) Quartz; (b) RS; (c) fluorescent; (d) booth; (e) bed.

low to achieve the desired result.[19] Conversely, if the shorter-wavelength emission lamps are used as replacements in a booth equipped with the longer-wavelength emission lamps, the patient could receive too high a UV dose and suffer from sunburn and blistering.[19] It is, therefore, most important for clinicians to select replacement lamps with the desired spectral distribution.

Finally, as with sunlamps, appropriate protective eyewear must be used with these devices, and in the case of PUVA therapy, appropriate protective skin covering and eyewear should be used following the UV exposure treatment as prescribed by the physician, since sunlight contains high intensities of similar biologically effective wavelengths.

D. Dental Photopolymerization Devices

Dental photopolymerization devices that emit intense UV radiation or blue light are used in dental restorative work for curing resins that are used as fillings, and for sealing pits and cracks in teeth. UV radiation photopolymerization devices were developed in the early 1970s, while visible blue-light emitting devices were developed nearly a decade later.

Sketches of two typical devices are shown in Figure 22. The sources for the UV radiation photopolymerization devices include medium- or high-pressure mercury-vapor lamps, quartz halogen, and fluorescent lamps. In most cases, a liquid-filled flexible "light guide", or a quartz rod on a hand-held "pistol grip" housing is used to direct the UV radiation to the surfaces of the teeth to be treated. The visible-light curing devices use quartz halogen lamps and, like the UV devices, the light is directed to the surfaces of the teeth to be treated by flexible fiberoptic light guides or by a glass rod. The relative spectral power distribution curves for a selected UV and some visible photopolymerization devices are shown in Figures 23 and 24, respectively. These devices utilize band-pass filters to restrict the optical radiation to the desired wavelength range. Total radiation outputs at the applicator tip of the different UV devices range from about 10 to 400 mW/cm^2, while those for the visible light curing devices range from about 200 to 900 mW/cm^2. The typical exposure times per application for both types of devices are in the range of about 10 through 60 sec.

Because these devices emit intense UV or blue-light radiation, instructions provided for safe use of these devices should be followed. In addition, because of concerns about long-term risks to the eyes (retina) of users of the intense visible blue-light type curing devices, especially for those dental personnel who use them extensively, it is recommended that appropriate protective eyewear that is commercially available be used.[20] The spectral transmission curves of some appropriate commercially available eyewear are shown in Figure 25.

E. Radiant Warmers

Radiant warmers and heat lamps, which emit primarily IR radiation, are used in devices intended for partial or whole-body heating. There are three basic types of devices which can be distinguished according to the different color temperatures at which they are operated. Sketches of two types are shown in Figure 26. One of these types, using an overhead canopy, which can be opaque or transparent, operates at a sufficiently low color temperature (about 800 K) that no visible radiation can be seen. Another type, also in the form of an overhead canopy, utilizes quartz heater elements that can be operated at higher color temperatures (1275 K). Elements of this type glow, and visible light is emitted as well as IR radiation. For these devices, more than 99% of the radiation emissions are in the IR range at wavelengths greater than 700 nm when being operated at maximum output. A third device is the common incandescent-type "heat lamp" that is generally operated at a color temperature close to 2500 K. Typically, a

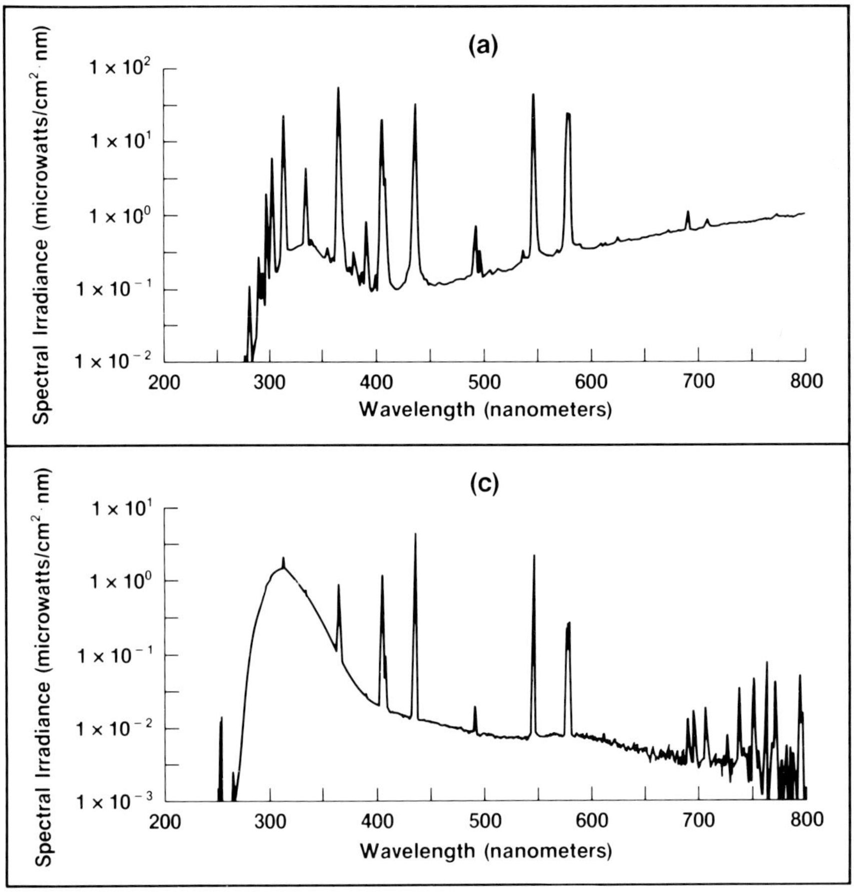

FIGURE 19a,c.

FIGURE 19. Spectral irradiance plots for various sunlamp sources. (a) A UVB quartz lamp at a distance of 1 m; (b) a UVB RS lamp at a distance of 0.75 m; (c) a UVB fluorescent lamp at a distance of 1 m; (d) a UVA fluorescent lamp at a distance of 1 m; (e) a high-intensity UVA lamp at a distance of 1 m. The bandwidth of the spectroradiometer used to obtain the data was 1 nm for a, b, c, and e, and 5 nm for d.

red-colored visible-light absorbing filter is used to minimize optical radiation emissions in the visible wavelength range. For these devices, the IR irradiance at the exposure position may be about 10 to 30 mW/cm^2. Because these levels are high for chronic exposure of the eyes, eye patches should be used.[4] In addition, a temperature sensor should be used to monitor the body temperature and control continuous exposures because of concerns for overheating with possible serious morbidity.

VI. SUMMARY

In order to properly characterize the optical radiation emissions from medical devices, it is necessary to use proper nomenclature and quantities and units. It is also necessary to be familiar with optical radiation measurement techniques, and to have an understanding of measurement instrumentation and the various sources of measurement error. In addition, in order to properly evaluate the exposure from optical radiation medical devices, it is necessary to be familiar with the bioeffects of optical

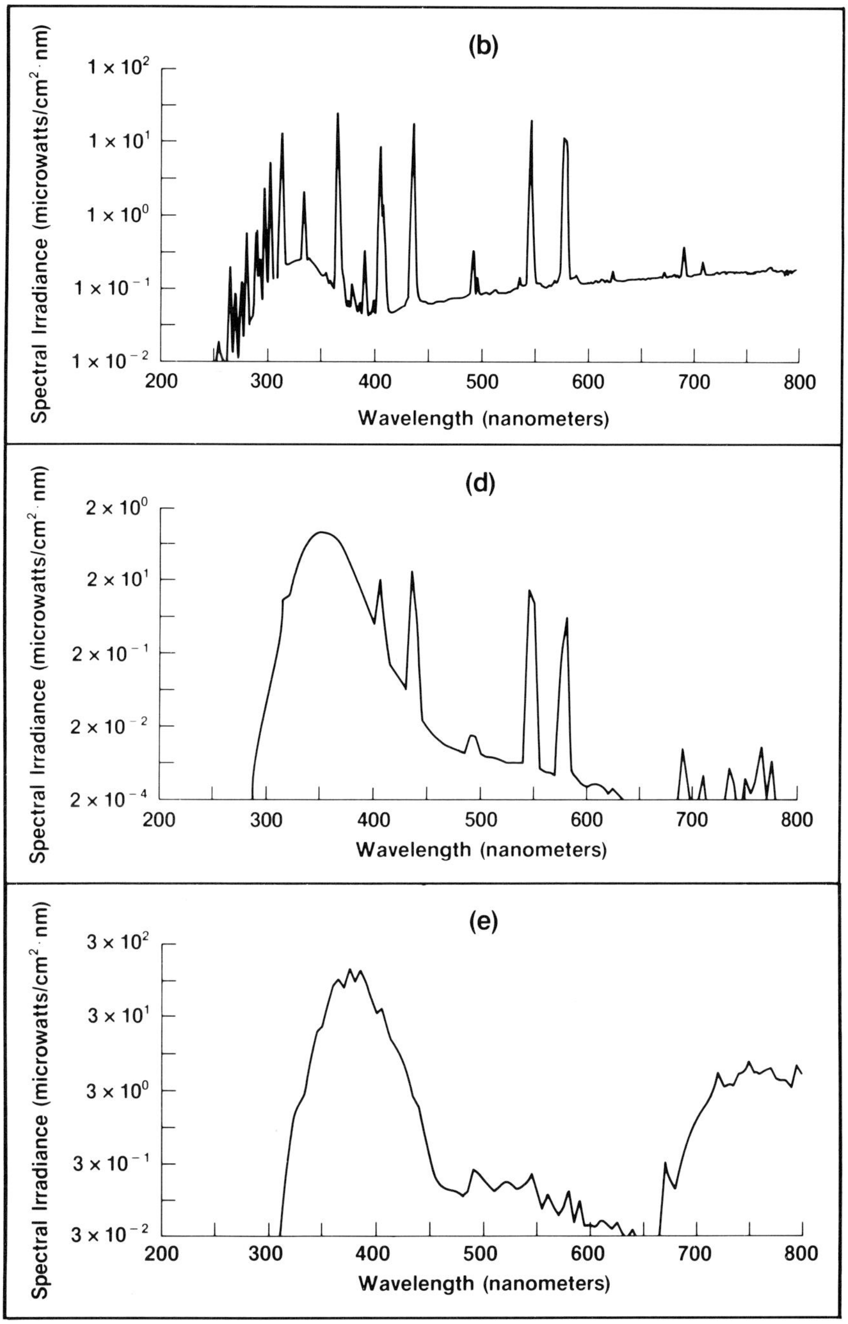

FIGURE 19b,d,e.

radiation, the operating and emission characteristics of the various man-made sources, and optical filters that can be used to modify the lamp emissions. Finally, it is necessary to be familiar with the techniques and considerations for evaluating the safety and effectiveness of the medical device, which must take into account use conditions.

The mention of commercial products or their use in connection with material reported herein is not to be construed as either an actual or implied endorsement of such products by the Food and Drug Administration.

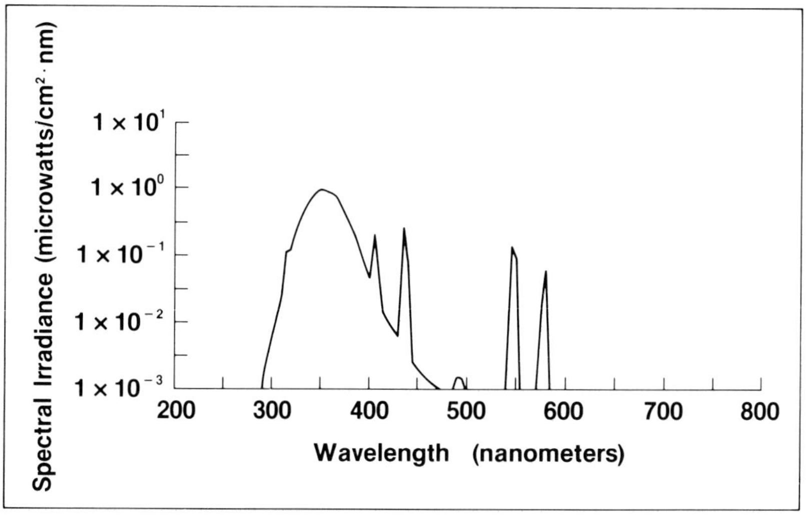

FIGURE 20. Spectral irradiance plot of a PUVA lamp at a distance of 1 m. The bandwidth of the spectroradiometer used to obtain the data was 5 nm.

Table 7
PUVA THERAPY SPECTRAL POWER DISTRIBUTION

Wavelength band (nm)	Output (as a % of total irradiance between 320—400 nm)
<310	<1
>310—320	1—3
>320—330	4—8
>330—340	11—17
>340—350	18—25
>350—360	19—28
>360—370	15—23
>370—380	8—12
>380—390	3—7
>390—400	1—3

ACKNOWLEDGMENT

The authors thank Mr. Robert James and Ms. Sharon Miller for obtaining the spectral radiometric data presented in this chapter.

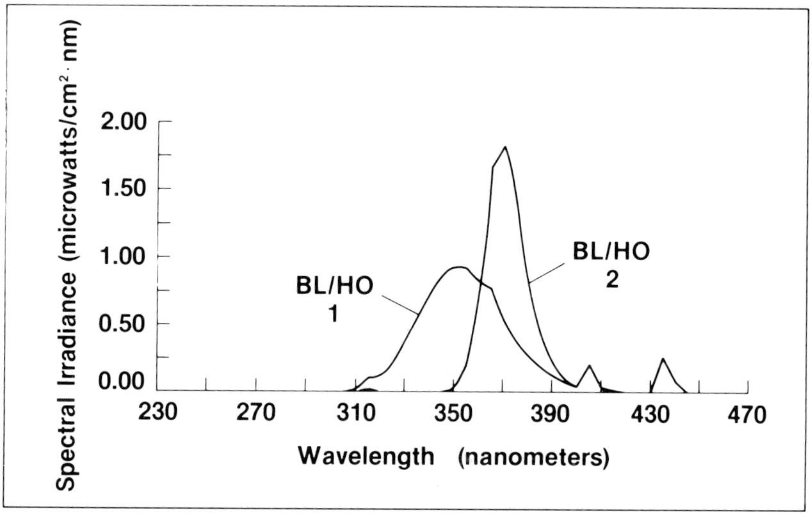

FIGURE 21. Linear spectral irradiance plots of two blacklight lamps with different phosphors at a distance of 1 m. The bandwidth for the spectroradiometer used to obtain the data was 5 nm.

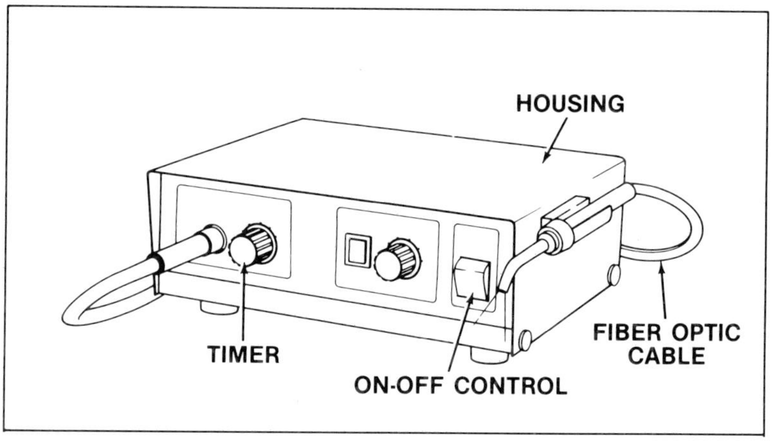

a

FIGURE 22. Sketches of dental photopolymerization devices. (a) Flexible fiber-optic visible-light curing unit; (b) a pistol-grip UVA-radiation curing type.

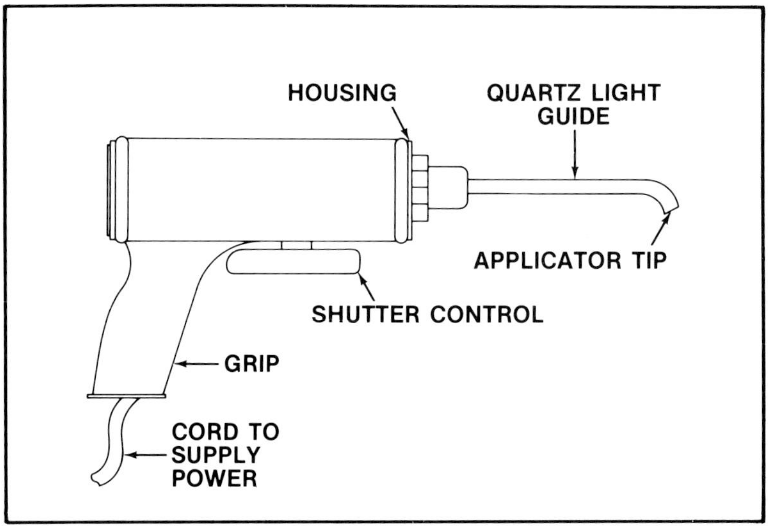

FIGURE 22B.

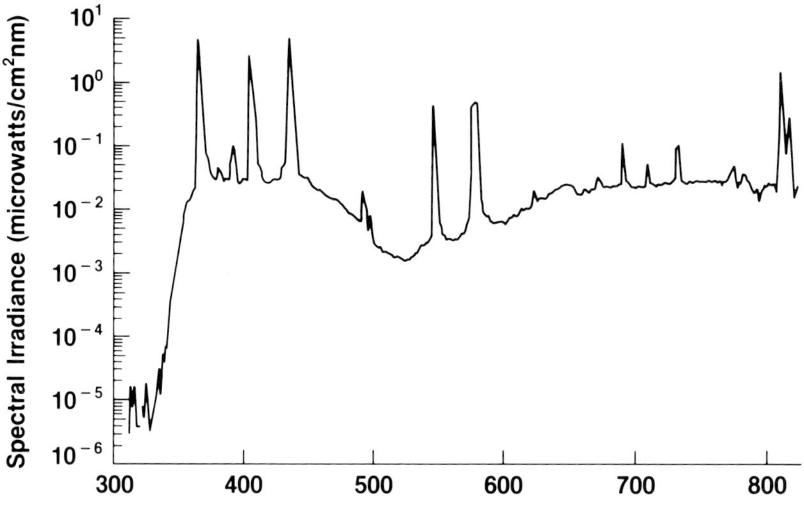

FIGURE 23. Spectral irradiance plot for a UVA curing device at a distance of 30.4 cm from the applicator tip. The bandwidth for the spectroradiometer used to obtain the data was 2 nm.

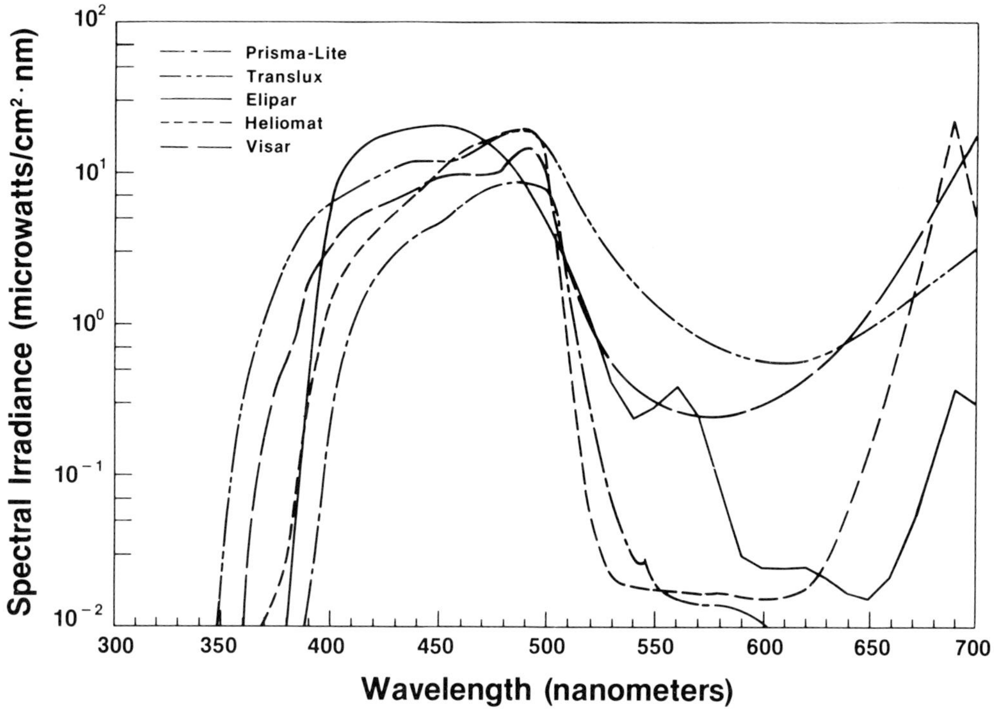

FIGURE 24. Spectral irradiance plots for various light curing devices at a distance of 10 cm from the applicator tips. The bandwidth for the spectroradiometer used to obtain the data was 2 nm.[19]

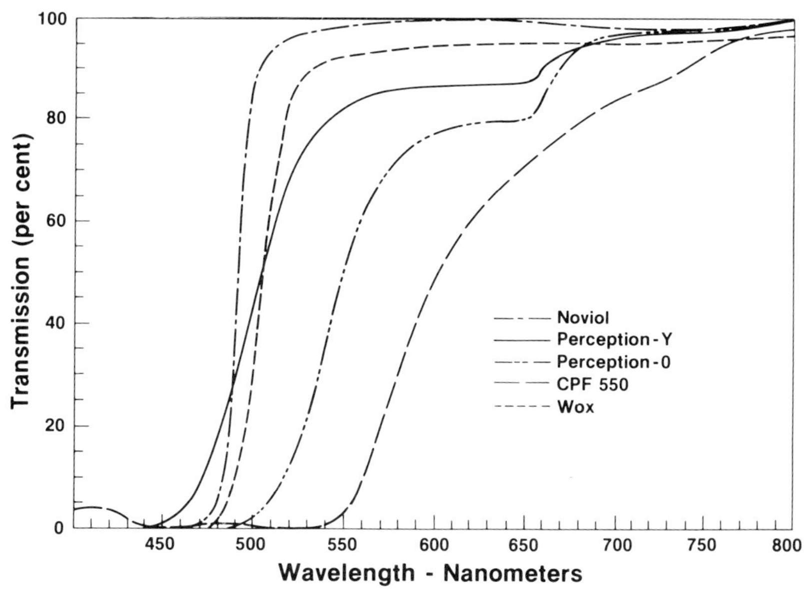

FIGURE 25. Spectral transmission of selected protective eyewear that can be used with visible-light curing devices. The eyewear tested includes: American Optical Corp., Southbridge, Mass., Noviol (safety thickness — 3 mm); Buffalo Dental Manufacturing Co., Syosset, N.Y., Guardian Perception Yellow and Perception Orange; Peepers International, Buffalo, N.Y., Wox; and Corning, Corning, N.Y., CPF 550. (Corning eyewear is a photochromic material, which darkens when activated by optical radiation. The transmission curve is for the unactivated state.)

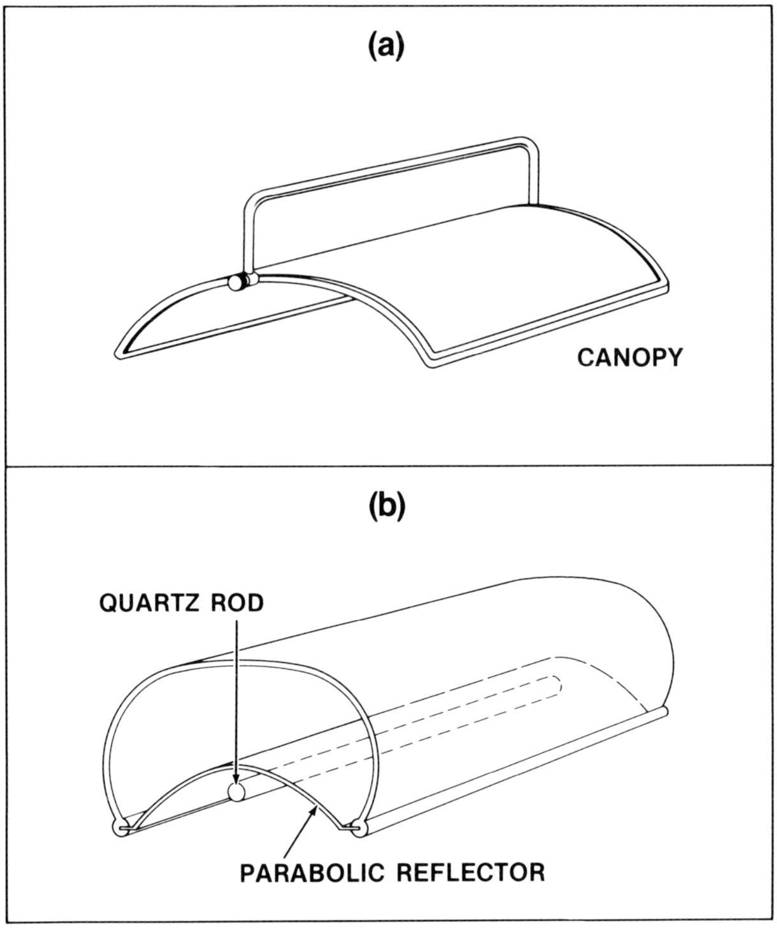

FIGURE 26. Sketches of infant radiant warmer types for: (a) a dark canopy, and (b) a quartz-rod unit.

REFERENCES

1. Nicodemus, F. E., Self-Study Manual on Optical Radiation Measurements, Part I, Concepts, Tech. Note 910-2, U.S. National Bureau of Standards, Gaithersburg, Md., 1978, chap. 4.
2. Nicodemus, F. E., Self-Study Manual on Optical Radiation Measurements, Part 1, Concepts, Tech. Note 910-2, U.S. National Bureau of Standards, Gaithersburg, Md., 1978, chap. 5.
3. Knight, W. B., Landry, R. J., and Bostrom, R. B., Recent BRH noncoherent optical radiation measurement instrumentation, in Symp. Biological Effects and Measurement of Light Sources, Publ. No. 81-8156, Food and Drug Administration, U.S. Department of Health and Human Services, Rockville, Md., 1981, 12.
4. Sliney, D. H. and Wolbarsht, M. L., *Safety with Lasers and Other Optical Sources*, Plenum Press, New York, 1980.
5. James, R. H., Optical radiation detectors and measurement devices, in Natl. Conf. Measurements of Laser Emissions for Regulatory Purposes, James, R. H., Ed., Publ. No. 76-8037, Food and Drug Administration, U.S. Department of Health, Education and Welfare, Rockville, Md., 1976, 26.
6. Kostkowski, H. J., Uncertainties in the measurement of incoherent radiation, in Symp. Biological Effects and Measurements of Light sources, Hazard, D. G., Ed., Publ. No. 77-8002, Food and Drug Administration, U.S. Department of Health, Education and Welfare, Rockville, Md., 1977, 197.

7. Saunders, R. D. and Shumaker, J. B., Optical Radiation Measurements. The NBS Scale of Spectral Irradiance, Tech. Note 594-13, U.S. National Bureau of Standards, Gaithersburg, Md., 1977, 13.
8. Landry, R. J., Sources of error in optical measurement systems, in Natl. Conf. Measurements of Laser Emissions for Regulatory Purposes, James, R. H., Ed., Publ. No. 76-8037, Food and Drug Administration, U.S. Department of Health, Education and Welfare, Rockville, Md., 1976, 40.
9. Landry, R. J. and Andersen, F. A., Optical radiation measurements: instrumentation and sources of error, *J. Natl. Cancer Inst.*, 69(1), 155, 1982.
10. Am. Conf. of Governmental Industrial Hygienists, Threshold Limit Values for Chemical Substances and Physical Agents in the Work Environment and Biological Exposure Indices with Intended Changes for 1984-85, Cincinnati, 1984.
11. Kaufman, J. E., Ed., *IES Lighting Handbook,* Ref. vol., Illuminating Engineering Society of North America, New York, 1981.
12. Cremer, R. J., Perryman, P. W., and Richards, D. W., Influence of light on the hyperbilirubinemia of infants, *Lancet,* 1, 1094, 1958.
13. Albrecht, R. M. and Roney, P. L., Phototherapy for neonatal hyperbilirubinemia: a survey of U.S. hospitals in 1974, in Symp. Biological Effects and Measurement of Light Sources, Publ. No. 81-8156, Food and Drug Administration, U.S. Department of Health and Human Services, Rockville, Md., 1981.
14. Landry, R. J., Scheidt, P. C., and Hammond, R. W., Ambient light and phototherapy conditions of eight neonatal care units: a summary report, *Pediatrics,* 75(Suppl.), 434, 1985.
15. Chissler, P. and Thompson, R., Tanning Beds are not without Drawbacks, FDA Consumer, Publ. No. 84-8228, Food and Drug Administration, U.S. Department of Health and Human Services, Washington, D.C., 1983/1984.
16. Sunlamp Products and Ultraviolet Lamps Intended for Use in Sunlamp Products, Title 21, Code of Federal Regulations, Part 1040.20, U.S. Department of Health and Human Services, Washington, D.C., 1980.
17. Medical Devices, Title 21, Code of Federal Regulations, Parts 800-895, U.S. Department of Health and Human Services, Washington, D.C., 1980.
18. Devices Used as a Radiation Source (in Combination with Psoralen Drug) in the Photochemotherapy for Psoriases, *Fed. Regist.,* 45, 8870.
19. Coles, C. A., Forbes, P. D., and Davies, R. E., Different biological effectiveness of blacklight fluorescent lamps available for therapy with psoralens plus ultraviolet A, *J. Am. Acad. Dermatol.,* 2(4), 599, 1984.
20. Ellingson, O. E., Landry, R. J., and Bostrom, R. B., An evaluation of optical radiation emissions from dental visible photopolymerization devices, in press.

Chapter 7

SAFETY MEASURES IN OPTICAL RADIATION TREATMENT

Elizabeth D. Jacobson, Larry E. Bockstahler, Richard P. Felten, Kiki B. Hellman, Victoria M. Hitchins, Anthony Lamanna, Jerome I. Levine, C. David Lytle, Stephen M. Sykes, and Morris Waxler

TABLE OF CONTENTS

8. The sensitivity of tissues to damage by optical radiation can vary from person to person. Some individuals are hypersensitive to optical radiation as a result of inherent genetic disease or chemically induced photosensitivity.
9. Some photomedical treatments utilize photosensitizing chemicals, and this introduces another set of variables capable of influencing optical radiation/tissue interactions.
10. Other potential risks associated with the treatments are unrelated to the optical radiation itself; for example, electrical shocks, electrocution, or burns resulting from electrical malfunctions or operator error.

This chapter provides information to aid in identifying potential risks associated with present and possible future clinical applications of optical radiation. Since many of the hazards can be anticipated, safety measures can be taken to prevent them or minimize their consequences. Appendix 1 lists voluntary and Federal optical radiation safety standards that contain exposure limits and guidance for safe use of optical radiation products. Appendix 2 lists potentially photosensitizing drugs and other pharmaceutical agents.

II. PHYSICAL PRINCIPLES AND MECHANISMS OF INTERACTIONS

The diverse uses of optical radiation in photomedicine for diagnostic and therapeutic procedures employ a wide range of power levels over the UV, visible, and IR wavelength regions. Knowledge of the types of interactions that occur between optical radiation and different tissues is essential before safety aspects can be considered. There are two principal modes of interaction that are likely to occur: photochemical and thermal. Both interactions are strongly wavelength dependent.

A. The Optical Spectrum

The optical spectrum is divided by the Commission International de l'Eclairage (CIE) into wavelength regions based on the biological effects associated with each region (CIE publication *International Lighting Vocabulary, 1970,* as referenced in Sliney and Wolbarsht[1]). The major divisions are wavelengths normally visible to man (visible, 400 to 760 nm), wavelengths shorter than those of the visible region (UV, 100 to 400 nm), and wavelengths longer than those of the visible region (IR, 760 to 1 million nm). The UV region is further divided into three regions: UVA, 315 to 400 nm; UVB, 280 to 315 nm; and UVC, 100 to 280 nm. Other terms used to refer to UV wavelengths include near UV (300 to 400 nm: UVA and part of UVB), far or germicidal UV (200 to 300 nm: UVC and part of UVB), and vacuum UV (wavelengths below 200 nm, absorbed by air). The IR region is similarly subdivided into three regions: IRA, 760 to 1400 nm; IRB, 1400 to 3000 nm; and IRC, 3000 to 1 million nm.

B. Chromophores

Chromophores are key to the interaction of optical radiation with tissue. The interaction of optical radiation with a particular biological tissue depends on two primary physical characteristics of the radiation: wavelength and power (exposure rate). The first law of photobiology states that radiation must be absorbed to cause an effect. The radiation may be absorbed by one or more chromophores in the tissue. The extent of absorption, which varies from one chromophore to another, is a function of wavelength. The location of chromophores within cells and tissues determines where the radiation is absorbed. Examples of important chromophores are DNA and protein (in nearly all cells) in the UV region,[2] hemoglobin (blood), rhodopsin and xanthophyll (eye) in the visible region,[3,4] and water (all cells and tissue) in the IR region beyond

1400 nm.[5] Melanin in the skin and eye is important in both the UV and visible regions.[6-8] In addition, exogenous chromophores can be introduced into tissue to enhance absorption for certain clinical procedures, e.g., 8-methoxypsoralen for treating psoriasis[9] or hematoporphyrin derivative (HPD) for treating tumors.[10,11]

C. Tissue Penetration Depth

The depth that optical radiation of a given wavelength can penetrate depends not only on absorption by the different chromophores in the tissue, but also on reflection and scattering.[6,8] The term transmission may also be used to connote penetration. Reflection occurs primarily at surfaces or tissue interfaces (both exterior and interior surfaces) and is usually higher when the incident radiation is not perpendicular to the tissue surface. In general, scattering is greater for shorter wavelengths. Furthermore, scattering varies greatly from tissue to tissue, e.g., the stratum corneum in the skin scatters all wavelengths, whereas in the eye, visible wavelengths are much less scattered.

Overall, the combined absorption and scattering properties of both the skin and eyes determine that for wavelengths less than 1400 nm (which includes the UV, visible and near-IR regions), usually the longer the wavelength, the greater the transmission (deeper penetration).[2,12] For wavelengths greater than 1400 nm (mid- and far-IR regions), absorption by water becomes important and can result in absorption very near the surface of the exposed tissue.[3]

D. Photochemical Process

Absorption of photons in the UV and visible wavelength ranges takes place via the electronic structure of molecules, usually in those having one or more ring structures.[13,14] Excited states in the molecular electronic configurations result in chemically active structures which may result in altered molecules, hence the term "photochemical effect". For short-wavelength UV radiation (<200 nm), ionization may also be a major occurrence.[13] The number of altered molecules is proportional to the number of absorbed photons, i.e., to the radiation exposure (at nonsaturating exposures). Thus, the photochemical effect is proportional to the radiation exposure, where exposure is the product of the exposure rate and the exposure time. The same photochemical effect is obtained for half the exposure time at twice the exposure rate, i.e., the exposure rate and exposure time are reciprocal.[2] Reciprocity of exposure rate and time is considered a useful test for the occurrence of a photochemical response. Inactivation of enzymes by UV radiation is a good example, where the reciprocity has been demonstrated over a 10^5-fold range of exposure rates and exposure times.[2]

Another important aspect of the photochemical effect is that while there is no threshold exposure for molecular changes, a certain accumulation of chemical damage may be necessary to elicit a given biological response (such as erythema), resulting in a threshold exposure for the biological response.

In vivo, the photochemically altered compounds are not isolated, but are part of dynamic metabolic systems. Therefore, their concentrations may change with time as a result of metabolic activity. Furthermore, as the exposure time is extended to those times comparable for metabolically induced changes, reciprocity may no longer be demonstrable. Thus, a demonstration of reciprocity not only provides evidence for photochemical effects, but also indicates that the photochemically induced changes persist for a long time compared to the exposure and assay times. Lack of reciprocity is not necessarily evidence against a photochemical mechanism, but may indeed yield valuable information about the metabolic processes involved.

Diagnostic uses of optical radiation often employ spectroscopic techniques and usually result in low exposure levels. Examples of diagnostic uses are (1) diaphanography, where the transmission of light through tissue is used instead of X-rays; (2) doppler blood-flow measurements, where the rate of blood flow is detected via reflected light pulses; and (3) flicker/EEG tests for epilepsy and other neurological dysfunctions, where a blinking light stimulates specific brain pathways via the visual system. Other uses include fluorescent dye staining, e.g., fluorescein angiography and dental plaque detection, hematoporphyrin tumor detection, and rose bengal dye detection of corneal damage, where the fluorescence of bound dye is used to locate and quantitate tissue sites which specifically bind the dye.

The use of low-powered lasers to "stimulate" the body to possibly control pain or enhance wound healing is currently under investigation. Ophthalmoscopy, where reflected light is used to detect irregularities in the retina and adjoining tissue, uses somewhat higher levels of visible radiation for diagnostic purposes. One diagnostic method, thermography, uses the body itself as the source of IR radiation.

Photochemical processes are used in several different therapeutic procedures, including uses of UV radiation for its germicidal effect, and procedures that use photosensitizing chemicals. Psoralen plus UVA radiation (PUVA) therapy is used to treat psoriasis, vitiligo, and mycosis fungoides, and HPD plus visible radiation therapy is used to destroy tumors. In addition, inadvertant injuries, e.g., erythema, skin cancer, corneal keratitis, cataracts, and retinal damage, may take place through photochemical processes (to be discussed later).

E. Thermal Processes

Absorption of photons in the IR wavelength region takes place through the molecular bonds yielding primarily molecular vibrations (molecular librations and rotations are also possible).[2,13] Molecular vibrational energy may also be obtained with UV and visible wavelengths when the energy absorbed at the electronic level is transduced directly or indirectly by the molecule to vibrational energy.[13] The vibrational energy cannot be stored in the molecule as energy is stored in a chemical bond, so that the vibrational energy is dissipated into the surroundings as thermal energy.[13] If it is desired to increase the thermal energy of a segment of tissue in order to produce a local increase of temperature, a minimum exposure rate is needed to introduce energy faster than it is dissipated to surrounding tissue. Because of heat dissipation, reciprocity of exposure time and exposure rate is not likely to hold for thermal processes. Heat dissipation depends on many factors, e.g., temperature differential between heated tissue and surrounding tissue, heat capacity of the involved tissues, and intervening fluid circulation.

Very high local temperatures can be obtained when energy is introduced much faster than it can be dissipated. Thus, lasers with very narrow, intense beams can easily burn and cut many substances. In some cases, the radiant energy is deposited in short pulses, so that it is the momentary exposure rate which is important, rather than the average exposure rate (power) in which relatively lengthy pauses between pulses are also included. For pulse lengths in the microsecond and longer range (including continuous wave emissions), the local temperature may reach the temperature at which tissue coagulates (75°C).[15] This is the basis for "thermal damage". For much shorter pulses in the nano- and picosecond range, the tissue temperature may rise quickly, vaporizing tissue which is composed primarily of water. The rapid expansion of the vapor behaves like an explosion that destroys local tissue before the heat can be transferred to cause coagulation.[15] Coagulation will occur on the edges of the vaporized area (1) because some heat will be transferred from the hotter, vaporized region, and (2) since the laser beam edge, which is not powerful enough to vaporize tissue, will still be sufficiently powerful to coagulate a thin ring of tissue surrounding the vaporized region. While the physical basis for much of the vaporizing energy deposition is thermal, the subsequent tissue damage is caused primarily by mechanical (thermoacoustic) disruption. "Structural damage" is a term used to describe this biological effect.

Ionization of tissue molecules may also occur at very high power levels because of the intense heating and electrical fields present in focused, coherent laser beams that also contribute to structural damage. Furthermore, if the pulse duration is much shorter than the time for the structural damaging process to occur, reciprocity of power and pulse duration is theoretically possible, because the total energy necessary to cause the damage is deposited before the thermal responses dissipate the energy.[16] It should also be pointed out that vaporization may lead to a "plume" of tissue-containing steam which acts as a cloud and may shield the tissue from subsequent interaction with the laser beam.[17] This plume may also contain a plasma of free electrons and ionized material.[18]

Optical radiation is used therapeutically to produce thermal effects in two basically different ways. The traditional heat lamp and a modification of it, the infant radiant warmer, operate by directly heating large areas of tissue with IR radiation. Another method of producing local thermal effects is by irradiating with narrow, intense beams of laser radiation. The latter is often used in surgery and may emit any of a number of different wavelengths of optical radiation, depending on the lasing material.

III. THE EYE

A. Ophthalmic Uses of Optical Radiation

In perhaps no other medical specialty has the use of optical radiation for diagnosis and treatment had such a long history or grown more rapidly than in ophthalmology. The use of optical radiation to observe the interior of the eye began before the invention of electric lighting, and attempts to treat disorders of the eye with optical radiation are nearly as old.[19]

1. Diagnostic Uses

Diagnostic uses of optical radiation in the eye have grown considerably. Ophthalmoscopy, retinoscopy, and slit-lamp biomicroscopy have changed radically as new sources of optical radiation, such as tungsten, xenon arc, tungsten-halogen and metal-halide lamps, and lasers, were developed. In addition, intraocular observation has become more wavelength selective to accentuate particular features of the retina.[20] Polarization also has been used.[21] Once the value of fluorescein angiography was established for the diagnosis of various diseases of the retinal pigment epithelium and retinal vasculature, new techniques were developed to study the rapid changes in the retinal fluorescein pattern using high-speed flash photography or "cine photography".[22,23] Fundus photography is used to document many clinical phenomena.[24] Vitreous photofluorometry is a new approach to measuring the degree to which the blood-ocular barrier has been breached in a diseased eye or after ocular surgery.[25] The development of microsurgery in the eye has been accompanied by newer methods, such as the use of fiberoptics to see the surgical site.

Optical radiation techniques have been developed to observe, in vivo, the attachments of the vitreous to the retina. In addition, different methods are available for the in vivo measurement of the optical density of various ocular tissues. Moreover, optical radiation is commonly used to evoke electrophysiologic responses from the retina and the brain. Sometimes these stimuli are intense, full-field, broad-spectrum flashes from xenon sources; other times they are low-level alternating displays from a video-display terminal. For other diagnostic uses, a small spot of intense optical radiation is used to evoke an electrophysiological response from a limited region of the retina. Furthermore, intense optical radiation is sometimes used to bleach the retina and study visual recovery, or as a photostress test to evoke certain phenomena. Optical radiation is used with rose bengal dye to visualize damaged corneal epithelial cells and is used in conjunction with fluorescein to measure intraocular pressure.

2. Therapeutic Uses

Meyer-Schwickerath first treated retinal detachment in 1949 by using focused sunlight to burn a hole in the retina.[19] Later, in 1954, he introduced the use of optical radiation from a xenon-arc source. In addition, he treated angiomatosis, diabetic retinopathy, perivasculitis, and retinoblastoma with this ''light-coagulation''.[19] The ability to control this kind of photocoagulation was advanced considerably by Schepens' development of the binocular microscope for use in ophthalmology.[19] By the 1960s, Zaret, Campbell, and others[19] were using the laser instead of the xenon-arc photocoagulator for the treatment of retinal detachments. Since the introduction of the argon laser for retinal-detachment surgery, the list of ocular disorders treated by laser surgery has grown to include acute and chronic glaucoma, diabetic retinal disease, retinal neovascularization, senile macular degeneration, secondary cataracts, and many other ocular disorders.[26] Indeed, the surgical laser is already a powerful addition to the ophthalmologist's treatment regimen. Modification of laser surgical parameters has been, and continues to be, a search for more effective treatment of particular disorders.

B. Safety Considerations

The eye is generally considered the organ of the body most sensitive to the destructive effects of optical radiation. Indeed, the laser performance standard promulgated by the Center for Devices and Radiological Health (CDRH), Food and Drug Administration, and the voluntary standards of the American National Standards Institute (ANSI) and the American Conference of Governmental Industrial Hygenists (ACGIH) are derived almost entirely from experimental studies on the effects of optical radiation on the eye. (See items 3, 4, and 11, Appendix 1, for information about these standards.) The threshold limit values (TLVs) for occupational exposure to nonlaser sources of UV radiation also rely primarily on eye data.[27,28] TLVs have been developed to protect the retina from thermal as well as some kinds of photochemical damage.[29] Considerations of ocular safety are, therefore, of utmost importance for all types of diagnoses and treatments utilizing optical radiation, and apply to both patient and attending personnel.[30]

The safety of lasers and other sources of optical radiation has been reviewed in detail by Sliney and Wolbarsht[1] with special emphasis on minimizing eye damage. Pitts et al.,[31] Ederer,[32] and Waxler[33] provide a detailed risk analysis of optical radiation and cataractogenesis. The relationship of optical radiation to other long-term visual problems also has been analyzed recently: visual development by Fulton et al.,[34] visual aging by Owsley et al.[35] and Marshall et al.,[36] retinal degeneration by Newsome et al.,[37] photoreceptor effects by Massof et al.[38] and drug photosensitization by Dayhaw-Barker et al.[39] Good vision depends on the transparency of the cornea and lens, and on the ability of the retina to detect photons and transmit this information to the rest of the brain. Optical radiation above certain intensities can damage any of these ocular tissues.

1. The Cornea

Most of the UV radiation incident on the eye, including all UV radiation below 280 nm, is absorbed by the cornea. The amount of absorption decreases with increasing wavelength to approximately 400 nm, where the cornea is essentially transparent. The primary photochemical effect of excessive UV absorption by the cornea is photokeratitis, which is an inflammation of the corneal epithelium. This condition is delayed in onset by several hours and, owing to the high density of sensory fibers, is extremely painful. Because of the rapid turnover of the corneal epithelium, this condition rarely lasts over two days. The action spectrum maximum for photokeratitis was determined to be 265 to 275 nm.[40] In addition, photosensitizing drugs, such as trifluoperazine, may not only reduce the threshold for photic injury to the corneal epithelium, but also shift the damage to longer wavelengths in the UV region and to the endothelium of the cornea, a potentially more dangerous consequence.[41] High-intensity exposures involving thermal or mechanical disruption are extremely hazardous and may damage underlying tissues such as the corneal stroma or Descemet's membrane, leading to a permanent clouding of the cornea. Such a condition is correctable only by corneal-transplant surgery.

2. The Crystalline Lens

Nearly all of the UVB and UVA radiation that passes through the cornea is absorbed by the lens. High exposure rates of UV radiation can damage the lens, apparently by thermal mechanisms. Lower exposure rates can produce photochemical damage to the lens, leading to cataracts. The unique metabolism of the lens, and slow repair rate and the cumulative nature of photochemical events contribute to the slow development of cataracts. Since there are no pain fibers in the lens, as there are in the cornea, the damage accumulates covertly. While the UVB region is the more hazardous wavelength region for producing lens opacities, the UVA region is increasingly recognized as playing a potentially important role in the development of senile cataracts and lens browning.[42] Human epidemiologic data on sunlight and cataracts from six continents support the experimental evidence that UV radiation is cataractogenic.[32] Intense blue optical radiation can also produce cataracts in individuals with brown crystalline lenses. Various photosensitizing drugs, such as the psoralens, can increase the risk of UV radiation-induced cataracts.[43]

While IR radiation may thermally accelerate UV radiation-induced cataractogenesis, the exposure rates necessary to produce lens damage by direct absorption of IR photons by the lens or from heat transfer following IR radiation absorption by the iris are very high. Although early epidemiologic data suggested an association between IR radiation and cataracts, recent data do not confirm this association.[31] Nevertheless, potential lens damage should be considered in any situation where the iris is dealt significant thermal or mechanical insult.

3. The Retina

a. Thermally Induced Lesions

Retinal thermal effects are mediated primarily by melanin absorption of photons in the pigment epithelium, but can also be mediated by other chromophores (e.g., hemoglobin, xanthophyll) given a sufficiently high exposure rate at the appropriate wavlengths.[1] A thermally induced lesion can be readily observed ophthalmoscopically as a small, white spot on the retina. The threshold for such a lesion is lowered with increasing spot size.

b. Photochemical Lesions

Photochemical lesions may occur in any layer of the retina, but are generally observed most prominently in the photoreceptors and pigment epithelium. These effects are not necessarily visible by routine ophthalmoscopic examination.[44] Wavelengths shorter than about 460 nm are more effective at producing photochemical retinal effects, and the threshold for photochemical retinal effects may be unaffected by spot size.[45]

The contribution to long-term visual problems of photochemical, retinal damage by visible wavelengths is suggested by some of the cumulative and delayed retinal changes observed following exposures at wavelengths less than 480 nm. The presence of photosensitizing drugs in the retina increases the risk of damage by visible wavelengths.

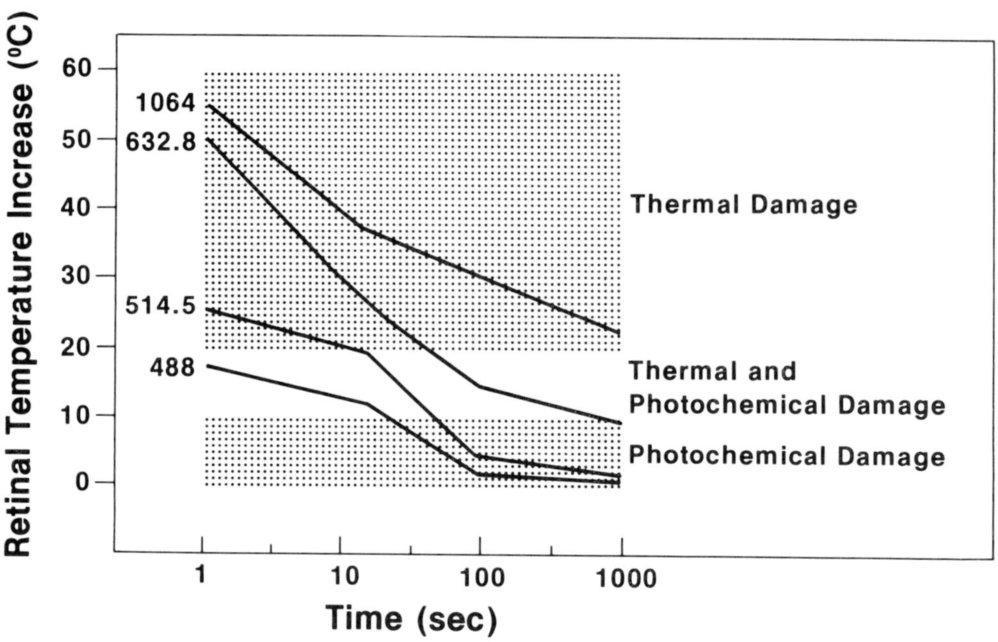

FIGURE 1. Threshold (minimum) retinal temperature rise for retinal damage with blue (488 nm), green (514.5 nm), red (632.8 nm), and IR (1064 nm) optical radiation. For brief exposures, threshold damage is entirely or largely thermal (i.e., photocoagulation, associated with a retinal temperature increase of 10 to 20°C). For lengthy exposures, threshold damage is photochemical (i.e., associated with a retinal temperature increase of less than 10°C), except for IR optical radiation which has minimal photochemical damage potential. (From Mainster, M. A., Ham, W. T., and Delori, F. C., *Am. Acad. Ophthalmol.*, 90, 927, 1983. With permission.)

Wavelengths longer than 600 nm probably produce retinal damage through thermal mechanisms, except in situations when a compound such as hematoporphyrin is introduced into the eye. Although there is no direct evidence, theroretical analysis suggests that an increased thermal load on the retina from any source will accelerate damaging photochemical processes.[46] Figure 1 shows the relationship between wavelength, time, and retinal temperature that defines the kinds of retinal damage that can occur.[47]

UV radiation of about 300 to 325 nm is hazardous to the retina even in eyes with the crystalline lens in place (phakic eyes).[40] Even though the cornea and crystalline lens absorb more than 98% of these wavelengths (300 to 325 nm), the biological effectiveness of the wavelengths transmitted (2%) is sufficiently high to damage the retina.[48] UV radiation at these and longer wavelengths is extremely hazardous to the retina in aphakes and pseudophakes not having adequate UV radiation protection incorporated into their spectacles or intraocular lens implant.

Photosensitizing drugs may increase the risk of retinal damage, and the possibility of long-term retinal degeneration exists. Recent evidence implicates UV radiation in cystoid macula edema,[132] confirming suspicions that optical radiation contributes to this disorder.[37,49]

Since the cornea is relatively transparent to the IRA radiation, and melanin in the retina has a broad absorption spectrum, IRA radition can damage the retina by thermal mechanisms. The retina also can be damaged by shock waves initiated by pico- and nanosecond pulses of high-intensity IR radiation (1060 nm).[50] This thermoacoustic event is not wavelength dependent.

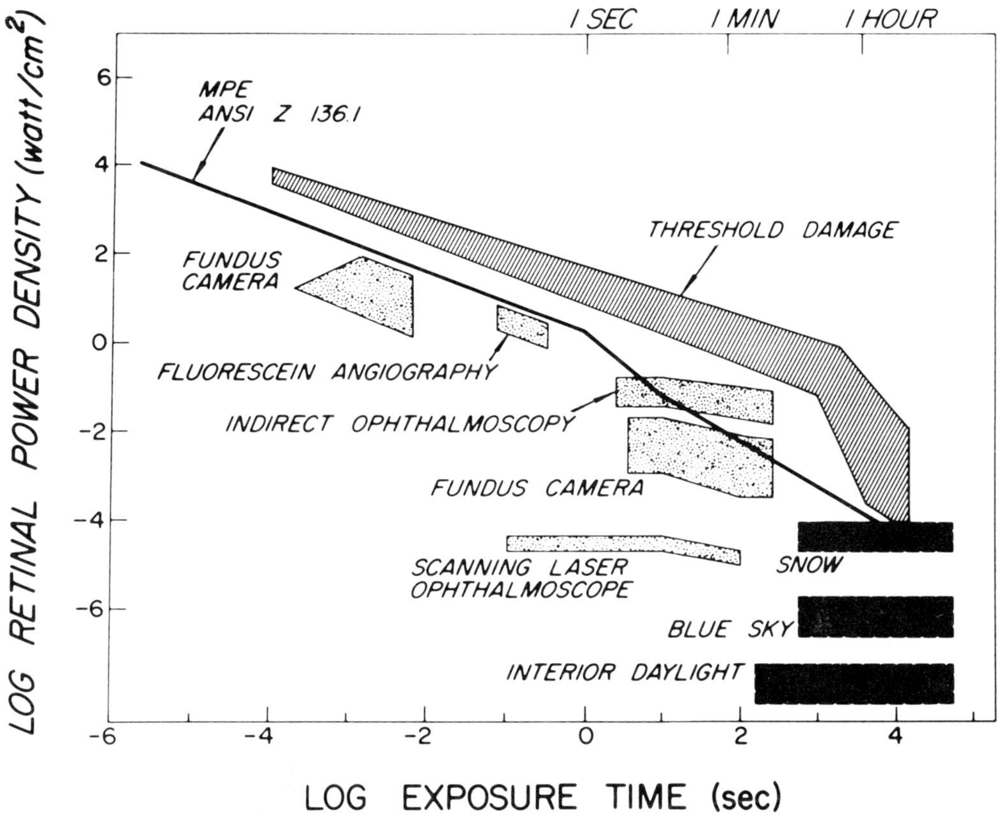

FIGURE 2. Maximum permissible exposure levels from the ANSI Z-136.1 standard (solid line) compared with clinical instrument exposure levels and experimental retinal damage thresholds. All data are scaled by appropriate spectral weighting factors. (See Mainster et al.[47] for details and further references.) (From Mainster, M. A., Ham, W. T., and Delori, F. C., *Am. Acad. Ophthalmol.*, 90, 927, 1983. With permission.)

C. Safety Measures

1. Diagnostic Devices

Mainster et al.[47] recommend that emissions of invisible radiation from ophthalmoscopes, slit lamps, and surgical microscopes be reduced below the maximum permissible limits of the ANSI Z-136.1 voluntary standard (see Appendix 1, item 3). In addition, they suggest similar limitations on blue optical radiation emissions (400 to 480 nm) (see Figure 2). Comparable limits have been proposed by others[51] and are reasonable, given both the acute effects and possible long-term consequences for visual health. Special blocking filters may be necessary for intraocular viewing of aphakes and pseudophakes, unless the filtering provided by the ophthalmic device has already taken this into account. Since the amount of visible radiation necessary to see the interior of the patient's eye has been determined,[52,53] there is no need to exceed these levels.

2. Therapeutic Devices

Safety considerations for therapeutic ophthalmic devices are more complex than those for diagnostic devices. Since most surgical lasers in ophthalmology and other specialties are Class 4, according to the Federal laser performance standard (see item 11, Appendix 1), the output beam by definition has sufficient irradiance to result in hazardous reflections. Working protective shutters or filters must be provided within the viewing optics of the instrument to protect the operator during surgery. Finally, the laser treatment area should be adequately labeled and secured to prevent unauthorized personnel from gaining access during surgery.

Depending on the particular surgical device, filters or protective eyewear may need to be worn by the staff, particularly in cases where the laser output is in the IR or UV regions. For protection of the patient's eye, ophthalmologists must select the appropriate wavelength when working on particular parts of the eye and select the appropriate safe working distances from other sensitive ocular structures.[50,54]

Perhaps the greatest risk from the neodymium-yttrium aluminum garnet (Nd-YAG) laser, which emits invisible IR radiation (1064 nm), results from improper focusing. If the laser is improperly focused within an intraocular lens, lens cracking may result. Similarly, a laser beam focused too close to an intraocular lens may result in pitting. Focusing too close to adjacent structures such as the cornea, iris, or retina can damage them. In addition, many laser treatments, including Nd:YAG posterior capsulotomy and argon trabeculoplasty, result in an increase in intraocular pressure. The risks of this side effect must be weighed against the benefits of treatment.

Protection of the eyes is also of critical importance during, and many times after, medical phototherapies. For example, patients' eyes must be protected from UV radiation during and up to 24 hr following the administration of 8-methoxypsoralen (8-MOP) for PUVA therapy.[55] A list of potential photosensitizers is contained in Appendix 2. Dayhaw-Barker et al.[39] have specifically reviewed drug-photosensitized damage in the eye.

It is currently difficult to reduce ocular safety in the clinical setting to a simple checklist of dos and don'ts. Technology and application are advancing at a rapid pace, and safety procedures pertinent to one application may not be applicable to another. Additionally, bioeffects data and adverse side effects information are limited, particularly in situations involving new applications of high-powered, pulsed lasers. The end result is that safety procedures and surgical concerns are often based on incomplete information about a rapidly changing specialty. Nevertheless, there are considerable data on the bioeffects of optical radiation on ocular tissue that can and should be used in assuring ocular safety.

IV. THE SKIN

Because of its location and large surface area, the skin is the major tissue of the body that is exposed to optical radiation from diagnostic, therapeutic, and surgical optical devices. For those devices that emit primarily UV and visible radiation, photochemical effects are important when considering hazards to the skin. In addition, intense beams of laser radiation of any wavelength can cause thermal damage to the skin.

A. Penetration

The various photobiological responses of the skin are determined by depth of penetration into and absorption by the various chromophores in different layers of the skin. At wavelengths in the visible and near-IR (IRA and IRB) regions, radiation penetrates through the stratum corneum, the derma, and into the subcutis layers of the skin. Neither shorter wavelengths in the UV region nor very long wavelengths in the far-IR region (IRC) penetrate deeply beyond the dermal layer. The outer layer, the stratum corneum, filters out most of the UV and far-IR radiation. However, near-IR wavelengths do penetrate deeply into the various layers of the skin (see Figure 3).

B. Chromophores

When the skin is exposed to UV and visible radiation from various diagnostic, therapeutic, and surgical devices, photons are absorbed by the various chromophores in

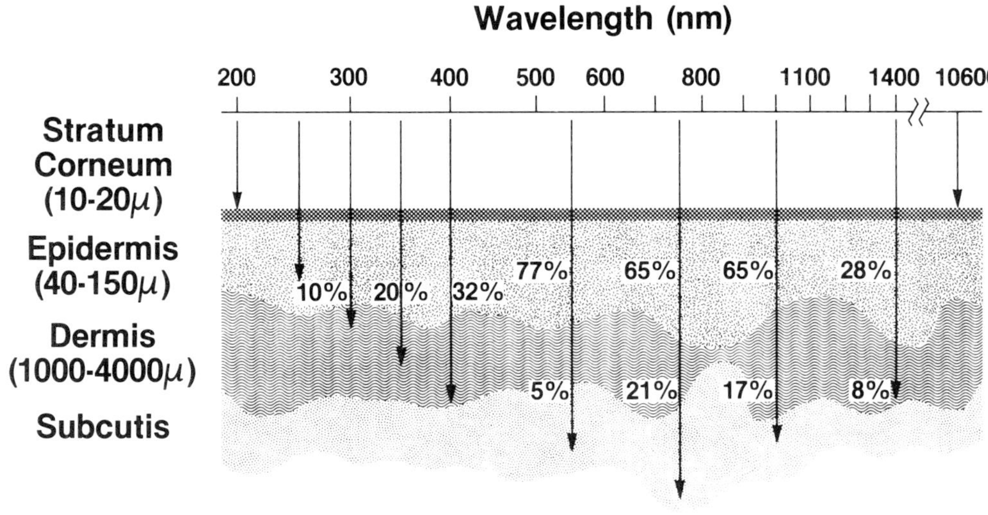

FIGURE 3. Schematic diagram showing percent penetration depth into the skin for different wavelengths.[1]

the skin. Biochemical constituents of the skin that act as chromophores include melanin, blood, carotenoids, proteins, nucleic acids, and skin surface lipids. Proteins and nucleic acids of the stratum corneum and epidermis absorb strongly at wavelengths below 290 nm and act as a primary UV radiation shield. Melanin, located in the stratum corneum and the epidermis, absorbs most strongly in the UV region and progressively less at increasing wavelengths through the visible region and into the IR region. Blood, containing hemoglobin, is located in the dermis, and primarily absorbs the shorter and middle (blue and green) visible wavelengths. Although hemoglobin can also absorb UV radiation, it is normally shielded from this radiation by the stratum corneum and epidermis. Water is the primary strong absorber in the IRC range and limits the penetration of the far-IR region to the stratum corneum.

C. Safety Considerations

1. UV Radiation Effects

The photochemical hazards of UV radiation to the skin can be acute or chronic (delayed).

a. Acute Effect: Erythema

The primary acute UV radiation-induced responses of the skin is erythema, which is characterized by transient redness. In severe cases, there is also edema, blistering, and desquamation.

Many variables can affect UV-induced erythema. Wavelength is an important factor in determining the effectiveness of inducing erythema. The erythemal action spectrum[56-59] shows that UVB and UVC wavelengths are approximately 1000 times more erythemogenic than UVA wavelengths. Maximum effectiveness occurs near 296 nm,[59] and the effectiveness decreases rapidly at a longer wavelengths. Most UVC and UVB radiation is absorbed in the stratum corneum, and erythema appears 4 to 6 hr after exposure and then fades after 12 to 36 hr. UVB radiation-induced erythema is usually followed by melanogenesis.

UVA radiation causes both immediate (within minutes of exposure) darkening of preexisting melanin as well as long-term melanogenesis. UVA radiation penetrates further into the skin than UVB radiation and can produce vascular damage in the dermis.[60] For most people who tan, UVA radiation, unlike UVB, is more melanogenic than erythemogenic,[61] i.e., the threshold exposure for melanogenesis is lower than that for erythemogenesis. Repeated exposures to UV radiation lead to accumulation of damage in the skin and the lowering of subsequent erythema and tanning threshold exposures.[60,61] Also, less energy is required to produce erythema in skin that is first hydrated in water and then exposed to UV radiation.[62-64] Other variables affecting erythema include duration of exposure, distance from the optical source, irradiance, skin type, degree of pigmentation of the skin, thickness of stratum corneum, anatomic location, use of topical or systemic sunscreens, and photosensitivity.

b. Delayed Effects

Prolonged exposures to UV radiation can lead to the long-term, delayed effects of photoaging and cancer.

1. Photoaging

Photoaging or actinic damage is primarily characterized by wrinkling and yellowing of the skin. In addition, there may be dryness, loss of elasticity, hyper and hypopigmentation, and cracking of the skin. Since melanin acts as a UV-radiation screen, Caucasians are more susceptible to actinic damage than dark-skinned individuals. UVA, UVB, and IR radiation are all probably partially responsible for actinic damage.[65,66]

2. Carcinogenesis

The risk of developing skin cancer is influenced by several factors including individual genetic makeup, exposure to UV radiation, and other factors. Fair-skinned individuals who tan poorly are at greater risk of developing skin cancer than darkly pigmented individuals. In animals, wavelengths between 280 and 315 nm are most effective in inducing tumor formation, and time-to-tumor development decreases with increasing UV-radiation exposure.[67] Recently, UVA radiation has also been shown to be photocarcinogenic in mice.[68,69] However, it has been estimated that UVA radiation is about 1600 times less carcinogenic than UVB radiation.[70]

Persons with genetic diseases such as albinism and xeroderma pigmentosum also have high incidences of skin neoplasms. Photodermatoses will be described in Section VII.

2. Visible and IR Radiation Effects

a. Low Power Levels: Little Effect

The wavelength dependence of two primarily UV-radiation photochemical effects, erythema and tanning, continues into the short-wavelength portion of the visible region, although much higher levels of visible radiation are required (e.g., minimal erythemal dose [MED] of 2.7×10^6 J/m^2 for 405 nm, compared to 2.1×10^2 J/m^2 for 296 nm).[59] For skin not photosensitized by exogenous chemicals or not sensitive through genetic disease, other photochemical effects of visible or IR radiation are unlikely. Furthermore, at low power levels, when there is no increase in temperature, no effects have been clearly documented. The controversial laser biostimulation procedures for pain control and wound healing fall into this category.

b. High Power Levels: Tissue Destruction

At very high instantaneous power levels of irradiation, heating can occur at all wavelengths, resulting in tissue destruction by thermal and thermoacoustic structural disruption (see Section II). Thermal injuries to large areas of the skin can lead to serious loss of body fluids, secondary infections, and toxemia. In the medical setting, only surgical lasers are likely to be hazardous in this manner. Lasers in dermatology include those used for blanching port wine stains and tatoos, where photochemical fading and thermal destruction of vascular and pigmented lesions are used, and for removing blemishes such as tumors and warts, where thermal effects are primary. In these uses, any incidental exposure is hazardous.

At high power levels of IR radiation, the exposure rate becomes important when considering hazards to the skin. For large areas of skin or whole-body irradiation, even relatively low exposures to IR radiation may be intolerable, because the heat dissipation mechanisms of the body become overloaded. For a 2-cm^2 patch of skin and 0.5 sec exposure, a first-degree burn is obtained with 12 W/cm^2 for broad wavelength radiation from a carbon arc (mostly visible and IR), and 34 W/cm^2 yields a third-degree burn.[1] On the other hand, the tolerance limit for exposure of a larger area of skin (12 × 12 cm) to IRA and IRB radiation is 0.17 W/cm^2.[71] To put these values in perspective, an IR infant warmer can expose an infant to 10 to 30 mW/cm^2 of whole-body irradiation.[1]

D. Safety Measures

There are several safety measures that should be considered to prevent hazards from photochemical effects occurring as a result of therapeutic procedures. The total exposure delivered at a specific treatment session or over a series of treatments should be kept to a minimum. This is especially true for UV-radiation exposure. It is important to protect the skin with clothing, surgical drapes, or topical sunscreens during treatment. Reflected and scattered radiation from surfaces such as glass or metal, e.g., medical instruments, could also be of consequence to the patient and medical personnel. The wavelength and intensity of UV radiation and the distribution of radiation delivered to a patient should be monitored. Many fluorescent sunlamps used for phototherapy have greater UV-radiation output in the center than at either ends of the bulbs.[72] Lamps should be arranged so that areas of the body that require more UV radiation receive more.

Topical sunscreen preparations protect the skin by either absorbing or reflecting UV radiation. There are three major categories of sunscreens: p-aminobenzoic acid (PABA) and its derivatives, non-PABA compounds (e.g., benzophenones, cinnamates), and physical (opaque) sunscreens (e.g., titanium dioxide, zinc oxide). The PABA sunscreens are most effective at protecting the skin from UVB radiation, while opaque sunscreens can protect from the effects of UVA and visible radiation. For a detailed description of sunscreens, refer to the proposed rule: "Sunscreen Drug Products for Over-The-Counter Human Use" (Federal Register, cited in Appendix 1, item 13).

Use of appropriate sunscreens is particularly important for persons who have received psoralens or other photosensitizing compounds such as HPD for therapeutic purposes. Photosensitizing dyes or drugs may persist in the patient for days to even months after treatment, and the patient could suffer severe erythema when exposed to UV or visible radiation. These persons should also wear protective clothing and eyewear.

IR emitters of the heat-lamp type generally are used to irradiate large areas and can overheat and dessicate by overloading the heat-dissipation systems of the patient. Prevention of heat overloading is accomplished most easily by reducing the intensity. For healthy, alert individuals, the pain threshold is reached before tissue destruction can occur, so that normally, burn hazard can be minimized. However, for infants, some invalid, elderly persons, and anesthetized patients, the normal avoidance protective reactions are not adequate. These individuals require special attention to avoid injury. In addition, direct contact with the lamp can also result in severe burns and must be avoided by both patients and medical personnel.

When using a laser, the wavelength, pulse duration, and energy output are all important factors to consider when trying to avoid unnecessary damage (e.g., permanent scarring) to the skin being treated and to the adjacent normal skin. In choosing the appropriate wavelength for treatment, the physician should consider the absorption characteristics of the lesion or the pigments in the skin. For example, the argon laser is useful in treating port wine stains, since the blue-green radiation is selectively absorbed by the oxygenated hemoglobin in the superficial dermal ectatic vessels.

V. OTHER TISSUES

A. Laser Surgery

The internal tissues of the body are generally well protected from optical radiation hazards. It is only when they are exposed during surgery, or when UV-transmitting fiberoptics are inserted into a body cavity through an endoscope, that adverse effects can occur. Presently, the single most probable source of such adverse effects is the transfer of thermal energy from the site of laser surgery to nontarget, sensitive, normal tissue. An additional problem unique to laser surgery is the potential hazard resulting from reflected laser energy. Reflection poses safety concerns for the patient, the surgeon, and assisting medical personnel. Radiation from surgical lamps, microscopes, and other equipment may also become a problem for tissue exposed during surgery.

B. Photochemotherapy

Several chemical compounds are currently being used in conjunction with exposure to optical radiation in so-called photochemotherapy. Possible adverse effects from radiation and/or from accumulation of photosensitizing chemicals may develop as new optical radiation sources and photosensitizing compounds enter the physician's repertoire.

C. Safety Considerations

1. UV Radiation Effects

The UV region of the optical radiation spectrum presents unique hazards to patients and/or attending staff depending on the specific wavelength absorption characteristic of the tissues being exposed. It is important to prevent unnecessary exposure to UV radiation in those instances where no benefit is derived from exposure. Sources of inadvertant exposure to UV radiation include germicidal lamps used in surgical suites, laboratories, or clinics for the purpose of maintaining sterile conditions. Although the individual exposure per unit time might be small, the cumulative effect could be large. A more obvious UV radiation hazard is that posed by UV radiation-emitting laser devices, presently under development for medical applications. Certain UV wavelengths[73] may be ideal for treating calcified heart valves and removing calcified plaque from veins and arteries. Excimer lasers emitting UV radiation at 193 nm have been used to remove single layers of corneal epithelial cells.[74]

2. Thermal Effects: Tissue Damage

The generation of heat as a by-product of optical radiation treatment always poses a potential danger to tissue, whether the heat source is a laser or a surgical lamp. However, heating of tissue is the generalized biological effect desired in most laser surgeries. The specific end point obtained, such as cutting, vaporizing, or coagulating, is directly related to the amount of thermal energy applied. Transfer of thermal energy from the specific site of interest to normal, noninvolved tissues results in potential thermal damage in these normal tissues. This becomes an especially important concern when surgery is performed near structures vital to normal health, such as in the brain or near the spinal cord. Care must also be taken during laser surgery to prevent the transfer of thermal energy to nearby vascular systems to prevent vessel rupture and hemorrhage, or coagulation and cessation of blood flow. Obviously, appropriate interlocks and filters must be used on laser equipment to prevent accidental exposure to the operating room personnel. Surgical use of CO_2 or Nd:YAG lasers, which emit radiation in the invisible IR region, increases the potential for accidental exposure.

It is worth pointing out that industrial and laboratory laser accidents sometimes occur to experienced laser users who momentarily disregard the usual safety procedures.[1]

3. Photochemical Effects: Photosensitization Reactions

There is growing interest in the medical community in the use of specific chemical compounds that are activated by optical radiation. A recent example is the use of HPD activated by 630-nm optical radiation to treat tumors.[75] Since optical radiation-sensitive compounds may remain sequestered in body tissues for extended periods posttreatment, exposure of these ''sensitized'' patients to any intense radiation of the appropriate wavelength may result in extreme photosensitization reactions (see Section VII).

D. Safety Measures

1. General

Protection from unnecessary UV emitted from surgical lamps and germicidal and therapeutic UV sources, can be provided by (1) use of appropriate filters, (2) limiting access to these sources, and (3) shielding nontarget tissues, e.g., the eyes of patients receiving skin phototherapy. Also, protective eyewear and clothing should be worn by medical personnel.

2. Laser Surgery

A guide to the use of lasers in the operating room has recently been published.[76] Safety with high-intensity lasers is especially important, because serious damage can occur in a very short time, in fact, even with a single pulse. Specific laser parameters, particularly wavelength and power level, must be carefully chosen, so that the appropriate chromophores are affected. For example, a CO_2 laser is ideal for most surgical cutting where highly vascular sites are not involved; however, for highly vascular areas, the Nd:YAG laser may be preferred because of its good coagulation properties and greater depth of penetration. Only the amount of energy absolutely necessary to achieve the desired result should be used. The laser must always be aimed accurately, and it should be activated only when all other safety measures have been assured, e.g., when no surgical instruments intervene between the laser and the target tissue. Finally, the thermal load to adjacent, normal tissue can be reduced by packing moist pads around the site to absorb heat as well as stray laser energy. In addition, specially designed laser-absorbing platforms can be used to absorb stray laser energy.[77]

During laser surgery, there is always the possibility that radiation will be reflected from the exposure site to the patient or the medical staff. The degree of hazard varies with the wavelength of radiation reflected and the site of reflected-energy absorption.[77] Recent research has shown that the reflected wavelength can be significantly different from the incident wavelength.[77] Although most of these data were obtained using very powerful lasers, similar reflection at lower energies cannot be ruled out.

In general, patient protection from reflected energy is relatively easy to manage, since all tissues, other than target tissues, can be covered by absorbing materials. Nonreflecting surgical instruments, absorbing platforms, and moist, absorbing pads can be used. Protection of the medical staff is somewhat more involved. Exposures of the hands can generally be avoided by wearing gloves and avoiding the reflected beam. Exposures of the eyes are more complicated, especially for the surgeon. If procedures are being performed using viewing optics, such as a microscope or other optical viewing instrument, appropriate filters can be placed in the viewing optics to absorb reflected energy. During other laser surgical procedures, protective eyewear should be worn also by attending staff. Choosing the proper eyewear is important so that hazardous wavelengths are removed, yet sufficient optical radiation passes through to enable the surgery to be performed in a safe and effective manner.[77]

An additional hazard to the patient is that of esophageal/tracheal fires that can occur during certain types of laser surgery.[78-80] When a patient is kept on a respirator during tracheal/bronchial/esophageal/oral surgery, care must be taken that the intubation tube is protected from laser exposure and that the tip of the intubation tube is not too close to the surgical site. Reducing oxygen concentration; choosing proper intubation tube material; wrapping the tube with reflective, nonabsorbing tape; and using moist pads around the surgical area will all reduce this hazard. Engineering controls such as emission indicators, warning labels and signs, and key control for switches are important and are described in Sliney and Wolbarsht.[1]

VI. THE IMMUNE SYSTEM

A. UV Radiation Effects

The immune system is a network of tissues, cells, and soluble chemical mediators that form the host's defense against harmful internal and external environmental agents or antigens. Studies over the last 15 years, described in review articles,[81-89] have documented that UV radiation affects the immune system. Most of these studies were conducted in the UVB wavelength range. Both localized effects, i.e., effects on immunocompetent cells in skin, and systemic effects were observed.

B. Safety Considerations

1. Alteration of Lymphocyte Function

The skin acts as a barrier between the host and external environment and, as such, contains elements important for host defense against external toxic agents, including UV radiation. The immunocompetent cells of skin consist of Langerhans cells and keratinocytes in the epidermis, and lymphocytes in the dermis. In addition to exposure of epidermal cells, UV and visible radiation also result in exposure of blood and circulating components of the immune system present in dermal vascular tissue. It has been estimated that an equivalent of the entire blood volume may pass through the skin two or more times during the time it takes to achieve a mild sunburn in a normal individual.[88] UV radiation reaching blood vessels could alter the function of circulating lymphocytes, most likely through photochemical processes.

2. Carcinogenesis

Experimental animal studies have shown that UV radiation-induced effects on the immune system may be important in UV radiation-induced carcinogenesis.[90-92] UVB radiation has been shown to impair the host's ability to recognize and reject UV-induced tumors. This tolerance to UV-induced tumors is due to the generation of suppressor T-lymphocytes.[93-95] These T-lymphocytes are specific for syngeneic UV radiation-induced tumors, since they reduce the immune response only against these tumors; the rejection of skin grafts and chemically induced syngeneic tumors is not affected by these T-lymphocyte suppressor cells.[95-97]

DeFabo and Noonan[98] have determined the action spectrum in the mouse for UV radiation-induced immune suppression; maximum suppression occurred between 260 and 270 nm. It was suggested that urocanic acid may be the photoreceptor for the UV radiation-induced immunological effects observed.

Although exposure to UV radiation affects the functional processes of many different immunocomponent cells,[99-101] the precise mechanism for the UV radiation-induced immunological effects observed in the mouse tumor model studies is not completely understood. However, these studies have shown that UV radiation modifies the immune system so that the host becomes tolerant to the growing tumors, and further support the concept that UV radiation-induced localized and systemic effects in the immune system may be important in UV radiation-induced skin cancer. It is not known whether this occurs in humans. However, studies showing that patients undergoing immunosuppressive therapies have an increased risk of cancer suggest this possibility.[102-107]

It has been shown that PUVA treatment induces skin cancer in mice[108,109] as well as in psoriatic patients.[110] The work of Kripke et al.[111] suggests that different immune responses are involved in PUVA- and UVB radiation-induced tumorigenesis. Unlike UVB radiation-induced tumors, PUVA produces tumors that grow progressively in both normal and immunosuppressed animals

3. Contact Hypersensitivity

Contact hypersensitivity, an immunologically mediated event, is also affected by UV radiation. UV radiation-induced suppression of allergic contact dermatitis in mice and guinea pigs is associated with suppressor lymphocytes specific for the contact sensitizer.[99,112,113] UVA radiation in conjunction with photosensitizing agents, such as psoralens (PUVA), also reduces contact hypersensitivity in animals[113-115] and humans.[116]

4. Role of the Aging Process

Changes in immune mechanisms occur as part of the normal aging process.[117] These changes begin at the time of sexual maturation and are most evident in humans after age 50. Moreover, studies have shown that aging is an important factor in UV radiation-induced effects on immune processes.[118] This suggests that older individuals may be more at risk from photoimmunologically mediated health effects.

C. Safety Measures

Safety measures pertinent to the immune system are similar to those for irradiation of skin (see Section IV.D).

VII. PHOTOSENSITIVITY

A. Genetic and Acquired Disorders and Safety Measures

Some individuals have inherited or acquired disorders that cause them to be photosensitive. Inherited genetic disorders which show photosensitivity include xeroderma pigmentosum (XP), Bloom's syndrome, Cockayne's syndrome, and Rothmund-Thomson syndrome. The better-characterized disorders, including means to help protect patients, are now briefly reviewed. For additional clinical information, consult medical texts on dermatology or photosensitivity diseases.[119-121] The clinician treating photosensitive patients should know which wavelengths cause or exacerbate the patient's symptoms and then carefully consider the risks before administering phototherapy.

XP, Bloom's, and Cockayne's syndromes are rare autosomal recessive disorders characterized by acute sensitivity to sunlight and a predisposition to cancer, predominantly on sun-exposed areas of the body. XP patients must be protected from exposure to UV radiation with protective clothing and sunscreens such as titanium dioxide compound. None of these patients should receive PUVA treatment.

The list of disorders that are caused by or are exacerbated by optical radiation includes porphyrias, polymorphic light eruption (PMLE), actinic reticuloid disorder, solar urticaria, lupus erythematosus, and disorders of tryptophan metabolism such as pellagra and Hartnup syndrome. Several of these disorders are discussed later.

All porphyrias are probably due to a defect in porphyrin synthesis. An interaction between optical radiation and the excess circulating porphyrins in the skin leads to erythema and edema. The action spectrum for skin photosensitivity in a patient with erythropoietic protoporphyria showed a maximum response at 400 nm.[122] Sunscreens employing a physical blocking agent such as zinc oxide or titanium dioxide should be used.

PMLE is an acquired skin disorder characterized by a delayed, abnormal response to optical radiation. While 290 to 320-nm radiation is the most effective in inducing PMLE in patients, even UVA and visible radiation can induce this response. After each exposure to sunlight and subsequent attack of PMLE, patients have greater tolerance to sunlight that may be due to tanning and a thickening of the stratum corneum. For further protection, patients should use broad-spectrum sunscreens or topical corticosteroids.

Actinic reticuloid patients are sensitive to long-wave UV radiation (320 to 400 nm) and shorter-wavelength visible radiation. Therefore, only incandescent lamps are acceptable sources of optical radiation for these patients.

Approximately one third to one half of all patients with lupus erythematosus are sensitive to UVB radiation; these patients should avoid UVB radiation.

B. Induced Photosensitivity

1. Chemically Induced Photosensitivity

Certain drugs and photosensitizing chemicals present in some foods, plants, cosmetics, and soaps can induce hypersensitivity to optical radiation in many individuals. Common photosensitizing agents are listed in Appendix 2. These agents can sensitize the skin and eyes of individuals to low energies of optical radiation over wide ranges of wavelengths. For example, therapeutic doses of 8-MOP (0.4 to 0.6 mg/kg) sensitized the tissues of psoriasis patients to UVA radiation exposures of only 2 to 5 J/cm^2.[9]

a. Safety Considerations

Potential hazards of chemically induced photosensitization include acute effects (skin erythema, edema, burns, and damage to the cornea and retina) and chronic effects (premature skin aging, skin cancer, and cataract formation). Photosensitization can occur by either (1) phototoxic reactions that result from absorption of radiation energy by chemical photosensitizers and the subsequent transfer of this energy to molecules leading to their chemical alteration, or (2) photoallergic responses caused by photochemical changes in the structure of the photosensitizers so that they are recognized by the body as foreign molecules, leading to the production of antibodies.[123] Photoallergic reactions of the cornea and conjunctiva have been described.[124] However, there are a number of compounds that can produce intraocular photosensitized damage (e.g., psoralens, phenothiazines).[39] Intraocular photoallergic reactions apparently have not been reported. In general, phototoxic reactions are more common than photoallergic reactions and can be either acute or chronic. The acute response of the skin is similar to the sunburn response and is generally characterized by erythema and edema followed by hyperpigmentation and desquamation. The end point of chronic skin damage can be cutaneous cancer formation.[125] The end point of chronic photosensitized eye damage can be visual impairment due to cataract formation or retinal damage.

b. Safety Measures

It is important that phototherapy patients be warned that chemical photosensitizers can increase their sensitivity to optical radiation. Patients should be screened by questioning before treatment as to whether they are taking medication that is known to be potentially photosensitizing, or if they believe that they are especially sensitive to sunlight or other optical radiation.

2. Phototherapies Based on Drug-Induced Photosensitivity

Phototherapies that involve the treatment of patients with both a photosensitizing drug and optical radiation have been developed. Examples include PUVA photochemotherapy[9] and photoradiation therapy of tumors with HPD and optical radiation.[11] Continued development of phototherapies that combine treatment of patients with optical radiation and photosensitizing chemicals is anticipated.

a. Safety Considerations

Each therapy that involves the exposure of patients to a photosensitizing drug and optical radiation has specific safety considerations and associated risks. However, these therapies also share several general safety concerns as described below.

b. Safety Measures

1. Photosensitizing Drug Treatment

Patients should be carefully screened before treatment. Patients who are unusually sensitive to optical radiation or who are allergic to the photosensitizing drug should not be permitted to undergo phototherapy treatment. It should also be determined that they are not taking other potentially photosensitizing medications. Photosensitizing drugs should be administered to patients only by physicians or properly trained personnel. It is important for both safety and effectiveness that the dose of photosensitizing drug be properly controlled. This dose will depend on previous phototherapy, previous optical or other radiation exposure, tissue sensitivity to optical radiation, and the usual factors of body weight, age, etc.

2. Optical Radiation Treatment

The exposure of each patient to optical radiation is critical and depends on the type of phototherapy, sensitivity of the target tissue to optical radiation, previous phototherapy or other radiation-exposure history, and other factors. In general, if the radiation exposure is too low, the treatment will be ineffective. If the exposure is too high, tissue burns can occur. The margin between safe/effective levels of tissue exposure to therapeutic optical radiation and hazardous levels can be small.[126]

The time when optical radiation exposure is begun following initiation of drug treatment is also critical. The patient's tissues are usually irradiated during the period when the photosensitizing drug has reached the optimal concentration level in the target tissues (time of maximum effectiveness).

Unnecessary optical radiation exposures should be avoided. If the action spectrum for a phototherapeutic procedure is known, then a radiation source should be selected that emits only wavelengths that are necessary for the intended biological response. Proper and adequate radiation measuring techniques should be used to ensure that correct exposures are being delivered to the target tissues (see Chapter 10).

Another aspect of the optical radiation treatment for a given required exposure is the selection of exposure rate and exposure time. The range should be known or determined where reciprocity between exposure rate and exposure time holds for the desired biological effect. Within this range, an exposure time should be selected when the drug level remains optimal in the target tissue(s). A further consideration is possible patient discomfort; generally, patient discomfort is reduced with shorter exposure times.

3. Other Concerns

Nontarget tissues containing photosensitizing drugs should be protected from exposure to radiation. For example, appropriate eye protection from wavelengths below 400 nm is necessary for skin phototherapies to reduce the risk of eye damage.

Some photosensitizing drugs remain active in body tissues of patients for times ranging from several hours to weeks following phototherapy. Protection of these tissues from additional exposure to optical radiation specific for the particular photosensitizer is necessary until the concentration of photosensitizing drug is reduced to a safe level. Additional safety concerns associated with psoralen photochemotherapy have been described.[126]

VIII. NONRADIATION CONCERNS IN PHOTOMEDICINE

A. Electrical Hazards

Health care facilities are frequently a jungle of electrical wires and equipment. The National Fire Protection Association (NFPA) has written into the National Electrical Code, Article 517,[127] which deals specifically with electrical construction and installation safety criteria for permanently fixed or installed equipment in health care facilities. Portable or moveable health care equipment installation and electrical safety criteria information is also available.[128]

All new equipment should be tested upon arrival by a person experienced in electronics for proper voltage, wattage, and grounding before it is put into use. Accidents can occur if medical personnel using the equipment are unaware of proper voltage and grounding requirements. For example, grounding pins can be removed or adapters used in ungrounded outlets, leaving the user with a false sense of security. Cords should be inspected at both ends for loose or defective fittings, and should be examined for breaks where they enter the equipment housing. Whenever possible, two conductor cords should be replaced with three conductors, using one of the conductors as a grounding wire. Wall outlets should be of the grounding type. Wet areas should have the added protection of a ground-fault interrupter device. All electrical equipment should be connected to a common ground system. Before putting new devices or equipment into use, the contacting surfaces of all electrical terminals should be secured.

Lasers present particular problems of safety that are well described.[1] The high-powered laser has a power supply and is probably water cooled. The cooling lines to the laser should be inspected for leaks and also pressure tested for possible rupture. The power supply should be well grounded and shielded. The entire system should have a common grounding system. If it becomes necessary to repair the power supply, one should ensure that it is properly discharged to prevent a potential electrocution hazard. Repairs should be made only by a qualified expert.

If the various electrical systems in use are carrying different voltages, it is prudent to label or code each of the wires or cables with the voltage the conductor is carrying. When replacement of wires is necessary, one should replace them with wires of the proper size and power rating. This can be determined from either the manufacturer or reliable service personnel.

B. Fire Hazards

Another potential hazard associated with the clinical use of optical radiation treatment is fire, and a common cause of fire is poor housekeeping. Transformers, power supplies, and electronic equipment produce heat and require cooling. Highly flammable items such as paper supplies, laboratory supplies, janitorial waxes, and cleaners should not be stored in these equipment areas. Although low-voltage equipment areas are considered safe by some individuals, fires have been caused by low-voltage equipment failures. Low-voltage cables need to be inspected as carefully as high-voltage cables. Short circuits in either can cause fires.

C. Other Concerns

1. Lamp Handling and Placement

Many modern optical radiation-emitting devices now come equipped with high-intensity discharge lamps, which can be dangerous to install or change, because they are filled with gas at high pressure. When these lamps are installed, care must be taken to keep the lamps absolutely clean. It is recommended that the lamps be cleaned with alcohol and that they be handled with lintless gloves (to avoid producing spots of charred lint on the hot lamp). In addition, one should wear protection for the eyes and face.

In phototherapy clinics, care should be taken to ensure that lamps are properly secured in the lamp sockets. High-intensity discharge lamps produce temperatures which can cause severe burns on contact. Care should be taken to allow lamps to cool before opening the housing and handling or replacing lamps. These lamps should be shielded against breakage and body contact. Fluorescent lamps are easily broken when dropped or when pressure is applied to the lamps. Some of the fluorescent lamp phosphors can be dangerous if inhaled.

2. Patient Stress

Another important safety consideration frequently overlooked is patient stress. For example, some phototherapy procedures require that the patient stand immobile in unfamiliar exposure chambers, perhaps with dark or opaque eye shields, for a period of time during treatment. Under such conditions, some patients have fainted or even experienced cardiac stress.[126] Phototherapy exposure chambers should be adequately ventilated to control temperature, contain a hand rail, and have a window to allow observation of the patient during irradiation. Also, in case of any emergency, it should be easy for a patient to open the door and leave the chamber.

3. Errors

Errors by health professionals are always a possibility and can be a cause of serious accidents during optical radiation therapy. There are many opportunities for mistakes to occur. Examples include giving patients accidental overexposures of radiation, neglecting eye protection during treatment for skin disease, using the wrong radiation source(s), damaging nontarget tissues, etc. Therefore, it is essential that at least one individual in the phototherapy clinic be in charge of safety. This person should receive proper optical radiation safety training and be responsible for educating other personnel on safety matters (see Section IX on developing a safety program). It is also recommended that information on general laboratory[129] and hospital safety[130] be kept readily available.

IX. ELEMENTS OF A SAFETY PROGRAM

There are at least six essential elements in the development of a safety program: selection of the medical procedure; selection of the medical device; provision of a safe environment; training in the use of the device; maintenance of equipment, including dosimetry and calibration; and medical surveillance.

A. Selection of the Medical Procedure

The selection of medical procedure sets the stage for subsequent requirements of a safety program. Where there are alternative procedures, the safest and most effective

"Documentation of the Threshold Limit Values", 4th ed., 1980
Comment: This companion reference document for the TLV booklet contains the rationale and primary references for both the adopted and proposed TLVs. The ACGIH voluntary laser standard is presented on p. 458 of this edition.

5. "A Recommended Standard for Occupational Exposure to Ultraviolet Radiation"
U.S. DHEW, Public Health Service
Center for Disease Control
National Institute for Occupational Safety and Health (NIOSH)
Comment: A criteria document prepared for protection against UV radiation in accordance with specific sections of the Occupational Safety and Health Act and transmitted to the Occupational Safety and Health Administration, U.S. Department of Labor for review and consideration in its standard development process. Available from Superintendent of Documents, Government Printing Office, Washington, D.C. 20402. Stock No. 017-033-00012.
6. Air Force Occupational Safety and Health (AFOSH)
Standard 161-10, "Health Hazards Control for Laser Radiation",
Department of the Air Force, Washington, D.C. 20330
Comment: Safety manual for use of lasers in the Air Force.
7. Laser Institute of America
5151 Monroe Street
Toledo, Ohio 43623
Comment: Offers numerous courses for the physician and operating staff on lasers used in medicine and surgery.
8. "Laser Safety in Surgery and Medicine"
LASE, INC.
8150 Corporate Park Drive, Suite 222
Cincinnati, Ohio 45242
"Laser Safety Training Manual"
Rockwell Associates, Inc.
Seminar Management
P.O. Box 43018
Cincinnati, Ohio 45243
Published by Rockwell Associates, Inc. 7754 Camargo Road P.O. Box 43010 Cincinnati, Ohio 45243.
9. Public Law 90-602
October 18, 1968
"Radiation Control for Health and Safety Act of 1968"
Amends the Public Health Service Act to provide for the protection of the public health from radiation emissions from electronic products. Available from Center for Devices and Radiological Health, HF2-44, Food and Drug Administration, 5600 Fishers Lane, Rockville, Md. 20857
10. Public Law 94-295
May 28, 1976
"Medical Device Amendments of 1976"
Amends the Federal Food, Drug, and Cosmetic Act to provide for the safety and effectiveness of medical devices intended for human use, and for other purposes. Available from Center for Devices and Radiological Health, HF2-44, Food and Drug Administration, 5600 Fishers Lane, Rockville, Md. 20857.
11. Title 21 Code of Federal Regulations
Parts 800 to 1299, Revised as of April 1, 1984
Food and Drugs

Comment: Contains the codification of rules published in the Federal Register. Title 21 is composed of nine volumes related to Food and Drugs. Parts 800 to 1299 include Subchapter H — Medical Devices, and Subchapter J — Radiological Health. Includes the Laser Product Performance Standard (21 CFR Part 1040).

Published by The Office of the Federal Register, National Archives and Records Service, General Services Administration. For sale by Superintendent of Documents, U.S. Government Printing Office, Washington, D.C. 20402.

12. Federal Register, Friday, September 14, 1984, Part VII
DHH, Food and Drug Administration
"Medical Device Reporting; final rule"
Comment: Final rule to require manufacturers and importers of medical devices to report to the FDA whenever the manufacturer or importer has information that reasonably suggests, or a person alleges and the manufacturer or importer is aware of the allegation, that a device has caused or contributed to a death or serious injury or that a device has malfunctioned, if a recurrence of the malfunction is likely to cause or contribute to a death or serious injury. Available from Center for Devices and Radiological Health, HFZ-44, Food and Drug Administration, 5600 Fishers Lane, Rockville, Md. 20857.

13. Federal Register, Friday, August 25, 1978, Part II
DHEW, Food and Drug Administration
Proposed Safety, Effective and Labeling Conditions
"Sunscreen Drug Products for Over-The-Counter Human Use"
Comment: A proposed rule based on recommendations of the Panel of Review of Topical Analgesics to establish conditions for the safety, effectiveness, and labeling of over-the-counter sunscreen drug products. Available from Center for Devices and Radiological Health, HFZ-44, Food and Drug Administration, 5600 Fishers Lane, Rockville, Md. 20857.

14. Suggested State Regulations for Control of Radiation
Volume II, Nonionizing Radiation, "LASERS", 1982
HHS Publication FDA 83-8220
Comment: These Suggested State Regulations for Lasers follow the laser classification scheme in the Federal Laser Products Performance Standard (21CFR Part 1040). The format involves three parts: Part A, general provisions; Part B, registration provisions; and Part C, the laser radiation protection provisions, such as maximum permissible exposure limits, specific safety regulations on safety eyewear, signs, labels, and posting. Special attention is given to medical surveillance. Available from Center for Devices and Radiological Health, Food and Drug Administration, 5600 Fishers Lane, Rockville, Md. 20857.

Appendix 2

POTENTIALLY PHOTOSENSITIZING AGENTS

Numerous products contain a drug or ingredient that is alleged, in the literature, to cause or result in a photosensitive skin reaction. Photosensitivity is a chemically induced change in the skin that results in the development of a rash or exaggerated sunburn on exposure to the sun or artificial radiation. The reaction is confined to uncovered areas of skin, providing a clue to the nature of its cause.

Photosensitivity reactions encompass both allergic (photoallergy) and nonallergic (phototoxicity) optical radiation-related skin responses. Photoallergy is the optical radiation-induced counterpart of contact allergy, whereas phototoxicity is the optical radiation-induced counterpart of irritation (primary irritation). The phototoxic skin response is likened to an exaggerated sunburn and is characterized by its nonimmunologic nature; the causative photochemical is usually activated by wavelengths longer than 320 nm; the absorption spectrum and the action spectrum of the photoactive chemical are in the same general range. The photoactive chemical may reach the skin by external application or indirectly via the bloodstream after ingestion or parenteral administration.

A listing of alleged photosensitizing drugs and agents has been made from an extensive body of literature (see Appendix 2 References). This list is not to be considered the total compendium of alleged photosensitizers, but rather an initial data base on the subject. Certain caveats must be considered prior to using the list, namely:

1. Not all persons using or taking these drugs or agents will experience a photosensitive reaction all the time or at any time.
2. Not all the drugs or agents will cause the same degree of skin reaction on all persons and, therefore, no attempt has been made to evaluate the drugs or agents as to reported incidence, i.e., incidence more frequent, incidence less frequent, or incidence rare.
3. Brand names of products are given only as examples of the generic drug and are not be considered as all inclusive of the availability of the product(s).
4. It is noteworthy that certain drugs may cause the skin to remain sensitive to optical radiation for several months after discontinuing use of the product. These are so designated in the listing.

As new information becomes available, this listing will require updating.

LIST OF ALLEGED PHOTOSENSITIZING DRUGS

Generic name	Example(s) of brand name(s)	Family/classification
Acetohexamide	Dymelor	Antidiabetic (oral) Sufonurea
Acetophenazine[a]	Tindal	Tranquilizer Antiemetic (phenothiazine)
Amitriptyline	Amitid Amitril Elavil Endep Etrafon[b]	Antidepressant (tricyclic)

	SK-Amitriptyline	
	Triavil[b]	
Amobarbital	Amytal	Sedative
	Dexamyl[b]	Hypnotic (barbiturate)
	Isobec	
	Tuinal[b]	
Azatadine	Optimine	Antihistamine
Bendroflumethiazide	Naturetin	Antihypertensive
		Diuretic (thiazide)
Benzthiazide	Aquastat	Antihypetensive
	Aquatag	Diuretic (thiazide)
	Exna	
	Hydrex	
Bromodiphenhydramine	Ambodryl	Antihistamine
Brompheniramine	TNTL[a]	Antihistamine
Butabarbital	Lotusate	Sedative
	Fiorinal[b]	Hypnotic (barbiturate)
Butaperazine	Repoise	Tranquilizer
		Antiemetic (phenothiazine)
Captopril	Capoten	Antihypertensive
Carbamazepine	Tegretol	Analgesic
		Anticonvulsant
Carbinoxamine	Clistin	Antihistamine
Carphenazine[a]	Proketazine	Tranquilizer
		Antiemetic (phenothiazine)
Chlordiazepoxide	TNTL[c]	Tranquilizer (benzodiazepine)
Chloroquine	Aralen	Antiprotozoal
		Antirheumatic
Chlorothiazide	Aldoclor[b]	Antihypertensive
	Diupres[b]	Diuretic (thiazide)
	Diuril	
	SK-Chlorothiazide	
Chlorotrianisene	Tace	Female sex hormone (estrogen)
Chloropheniramine	TNTL[c]	Antihistamine
Chlorpromazine[a]	Chloramead	Tranquilizer
	Klorazine	Antiemetic (phenothiazine)
	Promachlor	
	Promapar	
	Sonazine	
	Thorazine	
Chlorpropamide	Diabinese	Antidiabetic (oral)
		Sulfonurea
Chlorprothixene	Taractan	Tranquilizer (thioanthine)
		Antiemetic
Chlortetracycline	Aureomycin (ophthalmic only)	Antibiotic
Chlorthalidone	Hygroton	Antihypertensive
	Thalitone	Diuretic (thiazide)
Clemastine	Tavist	Antihistamine
Codeine	TNTL[c]	Analgesic
		Narcotic

Conjugated estrogens	Premarin	Female sex hormone (estrogen)
Contraceptives (oral)	TNTL[c]	Female sex hormone Contraceptive
Cyclobenzaprine	Flexgril	Muscle relaxant (skeletal)
Cyclothiazide	Anhydron Fluidil	Antihypertensive Diuretic (thiazide)
Cyproheptadine	Periactin	Antihistamine
Demeclocycline	Declomycin	Antibiotic
Desipramine	Norpramin Pertofrane	Antidepressant (tricyclic)
Dexchlorpheniramine	Polaramine	Antihistamine
Diethystilbestrol	DES Stilphostrol	Female sex hormone (estrogen)
Diltiazem	Carizem	Calcium-channel blocker Antiarrhythmic Antianginal
Dimenhydrinate	Dimate Dimentabs Dramaban Dramamine Dramilin Dramocen Dramoject Dymenate Eldoram Hydrate Marmine Motion-aid Reidamine Trav-arex Vertiban Wehamine	Antihistamine
Dimethindene	Forhistal Triten	Antihistamine
Diphenhydramine	Benadryl Bendylate Eldadryl Fenylhist Nordryl Nytol SK-Diphenhydramine Sominex Valdrene	Antihistamine
Diphenylpyraline	Diafen Hispril	Antihistamine
Doxepin	Adapin Sinequin	Antidepressant (tricyclic)
Doxycycline	Doxychel Doxy-lemmon Doxy-tabs	Antibiotic

	Vibramycin	
	Vibra-tabs	
Doxylamine	Decapryn	Antihistamine
	Unisom nighttime sleep aid	
Esterified estrogens	Amnestrogen	Female sex hormone (estrogen)
	Evex	
Estradiol	Delestrogen	Female sex hormone (estrogen)
	Estrace	
Estrogen	TNTL[c]	Female sex hormone (estrogen)
Estrone	Besterone	Female sex hormone (estrogen)
	Kestrin	
	Theelin	
	Theogen	
Estropipate	Ogen	Female sex hormone
	Piperazine estrone sulfate	
Ethinyl estradiol	TNTL[c]	Female sex hormone
Ethotoin	Peganone	Anticonvulsant (hydantoin)
Fluphenazine[a]	Permitil	Tranquilizer
	Prolixin	Antiemetic (phenothiazine)
Furosemide	Lasix	Diuretic
	SK-Furosemide	Antihypertensive
Gold salts (compounds)	Myochrysine	Antirheumatic
	Solganal	
Griseofulvin	Fulvicin P/G	Antibiotic (antifungal)
	Fulvicin-U/F	
	Grifulvin V	
	Gris-peg	
	Grisactin	
Haloperidol	Haldol	Tranquilizer (antipsychotic)
Hexobarbital	Sombulex	Sedative
		Hypnotic (barbiturate)
Hydrochlorothiazide	Esidrix	Antihypertensive
	Hydro-aquil	Diuretic (thiazide)
	Hydrodiuril	
	Oretic	
Hydrocodone	Dicodid	Analgesic (narcotic)
Hydroflumethiazide	Diucardin	Antihypertensive
	Saluron	Diuretic (thiazide)
	Salutensin[b]	
Hydroxychloroquine	Plaquenil	Antiprotozoal
		Antirheumatic
Imipramine	Imavate	Antidepressant (tricyclic)
	Janimine	
	Presamine	
	SK-Pramine	
	Tofranil	
Isocarboxazid	Marplan	MAO inhibitor
		Antidepressant
Mephenytoin	Mesantoin	Anticonvulsant (hydantoin)

Mephobarbital	Mebaral	Sedative Hypnotic (barbiturate)
Mesoridazine[a]	Serentil	Tranquilizer Antiemetic (phenothiazine)
Methacycline	Rondomycin	Antibiotic
Metharbital	Gemonil	Sedative Hypnotic (barbiturate)
Methotrexate	Mexate	Antineoplastic Antimetabolite Antipsoriatic
Methyclothiazide	Aqutensin Enduron	Antihypertensive Diuretic (thiazide)
Metolazone	Diulo Zaroxolyn	Antihypertensive Diuretic (thiazide)
Minocycline	Minocin	Antibiotic
Nalidixic acid	NegGram	Antimicrobial
Nortriptyline	Aventyl Pamelor	Antidepressant (tricyclic)
Oxytetracycline	Oxy-Kesso-Tetra Terramycin Tetramine	Antibiotic
Paramethadione	Paradione	Anticonvulsant
Pargyline	Eutonyl	MAO inhibitor Antidepressant
Pentobarbital	Nembutal	Sedative Hypnotic (barbiturate)
Perphenazine[a]	Trilafon	Tranquilizer Antiemetic (phenothiazine)
Phenelzine	Nardil	MAO inhibitor Antidepressant
Pheniramine	Combinations only (Triaminic, Panadyl, Triphenyl, Allerstat) Antihistamine	
Phenobarbital	TNTL[c]	Sedative Hypnotic (barbiturate) Anticonvulsant
Phenyltoloxamine	Combinations only (Naldecon, Sinubid, Sinutab, Percogesic, Tussionex)	Antihistamine
Phenytoin	Dilantin Diphenylan Diphenylhydantoin	Anticonvulsant (hydantoin)
Piperacetazine[a]	Quide	Tranquilizer Antiemetic (phenothiazine)
Piroxicam	Feldene	Anti-inflammatory Analgesic Antipyretic
Polythiazide	Renese	Antihypertensive Diuretic (thiazide)
Prochlorperazine[a]	Compazine	Tranquilizer Antiemetic (phenothiazine)

		Antiemetic (phenothiazine)
Promazine[a]	Sparine	Tranquilizer
		Antiemetic (phenothiazine)
Promethazine[a]	Historest	Antihistamine
	Phenergan	Tranquilizer (phenothiazine)
	Remsed	
	Numerous combinations	
Protriptyline	Vivactil	Antidepressant (tricyclic)
Pyrilamine	TNTL[c]	Antihistamine
	Numerous combinations	
Pyrvinium	Povan	Antihelminthic (antiworm medication)
Quinestrol	Estrovis	Female sex hormone (estrogen)
Quinethazone	Hydromox	Antihypertensive
		Diuretic (thiazide)
Secobarbital	Seconal	Sedative
	Tuinal[b]	Hypnotic (barbiturate)
Sulfacytine	Renoquid	Sulfa (sulfonamide)
Sulfaguanidine	Azulfidine	Sulfa (sulfonamide)
Sulfamethazine	S.A.S.-500	
Sulfamethoxazole	Gantrisin	
Sulfanilamide	Azo Gantanol[b]	
Sulfapyridine	Bactrim[b]	
Sulfasalazine	Cetamide	
Sulfathiazole	Septra[b]	
Sulfisoxazole	SK-Soxazole	
	Cotrim	
	Gantanol	
	Lipo Gantrisin	
	Methoxanol	
	Sosol	
	Sulfizin	
	Numerous combinations	
Talbutal (butalbital)	Fiorinal[b]	Sedative
	Lotusate	Hypnotic (barbiturate)
	Plexonal[b]	
	Sandoptal	
Tetracycline	TNTL[c]	Antibiotic
	Common examples:	
	Achromycin	
	Bristacycline	
	Cyclopar	
	Panmycin	
	Retet	
	Robitet	
	Sumycin	
	Tetracyn	
Thioridazine[a]	Mellaril	Tranquilizer
		Antiemetic (phenothiazine)
Thiothixene	Navane	Tranquilizer (thioxanthiene)
		Antiemetic

Tolazimide	Tolinase	Antidiabetic (oral) Sulfonurea
Tolbutamide	Orinase SK-Tolbutamide	Antidiabetic (oral) Sulfonurea
Tranylcypromine	Parnate	MAO inhibitor Antidepressant
Tretinoin	Retin-A	Antiacne (topical)
Triacetyldiphenolisatin	Phenisatin Trisatin Unilax	Laxative (cathartic)
Triamterene	Dyazide[b] Dyrenium	Antihypertensive Diuretic
Trifluoperazine[a]	Stelazine	Tranquilizer Antiemetic (phenothiazine)
Triflupromazine[a]	Vesprin	Tranquilizer Antiemetic (phenothiazine)
Trimeprazine[a]	Temaril	Antipruritic (sedative-hypnotic) Antihistamine
Trimethadione	Tridione	Anticonvulsant
Trimethobenzamide	Tigan	Antiemetic
Trimethoprim	Bactrim[b] Proloprim Septra Trimpex	Antimicrobial
Trimipramine	Surmontil	Antidepressant (tricyclic)
Tripelennamine	PBZ Ro-Hist	Antihistamine
Triprolidine	Actidil Actifed[b]	Antihistamine

[a] Skin may remain sensitive for 3 months after discontinuing taking of the drug.
[b] The brand is a mixture of two or more generic drugs.
[c] TNTL = too numerous to list.

OTHER PHOTOSENSITIZING AGENTS

Agent classification or use	Agent
Deodorant, bacteriostatic agents in soaps	Halogenated salicylanilides Halogenated carbanilides Halogenated phenols
Antimicrobial, antiseptic	Hexachlorophene Bithionol
Antifungal	Fentichlor Multifungin Jadit
Coal tar and coal tar derivatives for psoriasis and chronic eczema and in hair shampoos	Anthracene Phenanthrene Naphthalene Thiophene Many phenolic agents Pitch

Cosmetics and dyes	Fluorescein
	Rose bengal
	Eosin
	Erythrocine
	Trypaflavin
	Orange red
	Paraphenylenediamine
	Methylene blue
	Toluidine blue
	Trypan blue
	Acridine
	Methyl violet
In tattoos	Cadmium sulfide
Perfumes and toilet articles (essential oils)	Ethereal oils
	Oil of bergamot
	Oil of cedar
	Oil of citron
	Oil of lavender
	Oil of lemon
	Oil of lime
	Oil of rosemary
	Oil of sandalwood
Melanogenics (furocoumarins)	Psoralen
	Methoxypsoralens
	Petroleum products
Artificial sweeteners	Cyclamates
	Calcium cyclamate
	Sodium cyclohexylsulfamate
Fluorescent brightening agent for cellulose, nylon, or wool fibers	Blankophor
Sunscreen	*p*-Aminobenzoic acid (PABA)
	Esters of PABA
Perfumes or flavoring or as spices	Umbelliferae (plant)
	Rutaceae (plant)

It is important that physicians, as well as patients, be made aware that certain drugs have been reported in the literature to cause sensitivity of the eyes to optical radiation (photophobia).

Patients should be encouraged to report to their physicians and/or pharmacist any eye discomfort or visual changes that occur. This will allow for an appropriate evaluation of the situation, as well as any necessary corrective actions.

The following list is provided as examples of those drugs that are alleged to cause photophobia.

Anti-diabetic drugs	Mephenytoin
Atropine-like drugs	Methsuximide
Bromides	Mono-amine oxidase inhibitor drugs
Chloroquine	Nalidixic acid
Clomiphene	Oral contraceptives
Digitoxin	Paramethadione
Doxepin	Phenothiazines
Ethambutol	Quinidine

Ethionamide
Ethosuximide
Hydroxychloroquine
Quinine
Tetracyclines
Trimethadione

APPENDIX 2 REFERENCES

1. *Drug Information for the Health Care Provider,* Vol. 1, United States Pharmacopeial Convention, Rockville, Md., 1984.
2. Epstein, J. H., Chemical phototoxicity in humans, *J. Natl. Cancer Inst.,* 69, 265, 1982.
3. Fitzpatrick, T. B., Pathak, M. A., Harber, L. C., Seiji, M., and Kukita, A., An introduction to the problem of normal and abnormal responses of man's skin to solar radiation, in *Sunlight and Man, Normal and Abnormal Photobiological Responses,* Fitzpatrick, T. B., Pathak, M. A., Harber, L. C., Seiji, M., and Kukita, A., Eds., University of Tokyo Press, Tokyo, 1974, 3.
4. Fitzpatrick, T. B., Pathak, M. A., Magnus, I. A., and Curwen, W. L., Abnormal reactions to light, *Annu. Rev. Med.,* 14, 195, 1963.
5. Griffith, H. W., *Complete Guide to Prescription and Non-Prescription Drugs,* HP Books, Tucson, Ariz., 1983.
6. Harber, L. C., Armstrong, R. B., and Ichikawa, H., Current status of predictive animal models for drug photoallergy and their correlation with drug photoallergy in humans, *J. Natl. Cancer Inst.,* 69, 237, 1983.
7. Harber, L. C., Baer, R. L., and Bickers, D. R., Techniques of evaluation of phototoxicity and photoallergy in biological systems, including man, with particular emphasis on immunologic aspects, in *Sunlight and Man, Normal and Abnormal Photobiological Responses,* Fitzpatrick, T. B., Pathak, M. A., Harber, L. C., Seiji, M., and Kukita, A., Eds., University of Tokyo Press, Tokyo, 1974, 515.
8. Harber, L. C., Shalila, A. R., and Armstrong, R. B., Immunologically mediated contact photosensitivity in guinea pigs, in *Dermatotoxicology,* 2nd ed., Marzulli, F. N. and Maibach, H. I., Eds., Hemisphere Publishing, New York, 1983, 357.
9. Kligman, A. M. and Kaidbey, K. H., Human models for identification of photosensitizing chemicals, *J. Natl. Cancer Inst.,* 69, 269, 1982.
10. Long, J. W., *The Essential Guide to Prescription Drugs,* 3rd ed., Harper & Row, New York, 1982.
11. Maibach, H. I. and Marzulli, F. N., Phototoxicity (photoirritation) of topical and systemic agents, in *Dermatotoxicology,* 2nd ed., Marzulli, F. N. and Maibach, H. I., Eds., Hemisphere Publishing, New York, 1983, 375.
12. Pathak, M. A., Molecular aspects of drug photosensitivity with special emphasis on psoralen photosensitization reaction, *J. Natl. Cancer Inst.,* 69, 163, 1982.
13. *Photosensitivity from Drugs, Perfumes and Cosmetics,* Medical Letter, New Rochelle, N.Y., 1980, 64.
14. *Physician's Desk Reference,* 37th ed., Medical Economics, Oradell, N. J., 1983.
15. Schildkraut, M. L., Drugs that don't mix with summer sun and cosmetics, *Good Housekeeping,* 273, 1983.

REFERENCES

1. Sliney, D. H. and Wolbarsht, M. L., *Safety with Lasers and Other Optical Sources,* Plenum Press, New York, 1980.
2. Setlow, R. B. and Pollard, E. C., *Molecular Biophysics,* Addison-Wesley, Reading, Mass., 1962, chap. 7.
3. Barnes, F. S., Biological damage resulting from thermal pulses, in *Laser Applications in Medicine and Biology,* Vol. 2, Wolbarsht, M. L., Ed., Plenum Press, New York, 1974, chap. 6.
4. Wolbarsht, M. L., The function of intraocular color filters, *Fed. Proc.,* 35, 44, 1976.
5. Stellar, S., Polanyi, T. G., and Bredemeier, H. C., Lasers in surgery, in *Laser Applications in Medicine and Biology,* Vol. 12, Wolbarsht, M. L., Ed., Plenum Press, New York, 1974, chap. 8.
6. Buckley, W. R. and Grum, F., Reflection spectrophotometry. III. Absorption characteristics and color of human skin, *Arch. Dermatol.,* 89, 170, 1964.
7. McGinness, J. and Proctor, P., The importance of the fact that melanin is black, *J. Theor. Biol.,* 39, 677, 1973.
8. Wan, S., Anderson, R. R., and Parrish, J. A., Analytical modeling for the optical properties of the skin with in vitro and in vivo applications, *Photochem. Photobiol.,* 34, 493, 1981.
9. Parrish, J. A., Fitzpatrick, T. B., Tannenbaum, L., and Pathak, M. A., Photochemotherapy of psoriasis with oral methoxsalen and longwave ultraviolet light, *N. Engl. J. Med.,* 291, 1207, 1974.
10. Dougherty, T. J., Boyle, D. G., Weishaupt, K. R., Henderson, B. A., Potter, W. R., Bellnier, D. A., and Wityk, K. E., Photoradiation therapy — clinical and drug advances, in *Porphyrin Photosensitization, Advances in Experimental Medicine and Biology,* Vol. 60, Kessel, D. and Dougherty, T. J., Eds., Plenum Press, New York, 1983, 3.
11. Dougherty, T. J., Photoradiation therapy (PRT) of malignant tumors, *CRC Crit. Rev. Oncol. Hematol.,* in press.
12. Wan, S., Parrish, J. A., Anderson, R. R., and Madden, M., Transmittance of nonionizing radiation in human tissues, *Photochem. Photobiol.,* 34, 679, 1981.
13. King, G. W., *Spectroscopy and Molecular Structure,* Holt, Reinhart & Winston, New York, 1964, chap. 4.
14. Lamola, A. A., Fundamental aspects of the spectroscopy and photochemistry of organic compounds; electronic energy transfer in biologic systems; and photosensitization, in *Sunlight and Man,* Fitzpatrick, T. B., Pathak, M. A., Harber, L. C., Seiji, M., and Kukita, A., Eds., University of Tokyo Press, Tokyo, 1974, chap. 2.
15. Anderson, R. R. and Parrish, J. A., Selective photothermolysis: precise microsurgery by selective absorption of pulsed radiation, *Science,* 220, 524, 1983.
16. Tomberg, V. T., Non-thermal biological effects of laser beams, *Nature (London),* 204, 868, 1964.
17. Stern, R. H., Dentistry and the laser, in *Laser Applications in Medicine and Biology,* Vol. 1, Wolbarsht, M. L., Ed., Plenum Press, New York, 1971, chap. 5.
18. Steinert, R. F., Puliafito, C. A., and Trokel, S., Plasma formation and shielding by three ophthalmic neodymium-YAG lasers, *Am. J. Opthalmol.,* 96, 427, 1983.
19. Duke-Elder, S., Ed., *System of Ophthalmology,* Vol. 7 and 10, C.V. Mosby, St. Louis, 1967.
20. Frisen, L., Photography of the retinal nerve fibre layer: an optimized procedure, *Br. J. Ophthalmol.,* 64, 641, 1980.
21. Bone, R. A. and Landrum, J. T., Macular pigment in Henle fiber membranes; a model for Haidinger's brushes, *Vision Res.,* 24, 103, 1984.
22. Gallagher, P. N., MacRoberts, V. E., Edgerton, H. E., Wells, R. E., and Rees, S. B., Measurement of capillary blood flow using an electronic double flash light source, *Rev. Sci. Instrum.,* 36, 1760, 1965.
23. Wells, R. E., Schildkraut, E. R., and Edgerton, H. E., Blood flow in the microvasculature of the conjunctiva of man, *Science,* 151, 995, 1966.
24. Archer, D. B., Fluorescein angiography, in *Hereditary Retinal and Choroidal Diseases,* Vol. 1, Krill, A. E., Ed., Harper & Row, Hagerstown, Md., 1972, 73.
25. Cunha-Vaz, J., de Abreu, J. R. F., Dampos, A. J., and Figo, G. M., Early breakdown of the blood-retinal barrier in diabetes, *Br. J. Ophthalmol.,* 59, 649, 1975.
26. Wolbarsht, M. L. and Landers, M. B., Lasers in ophthalmology: the path from theory to applications, *Appl. Opt.,* 18, 1518, 1979.
27. American Conference of Governmental Industrial Hygienists, Ultraviolet radiation, in *Documentation of the Threshold Limit Values,* 4th ed., American Conference of Governmental Industrial Hygienists, Cincinnati, 1980, 471.
28. American Conference of Governmental Industrial Hygienists, Ultraviolet weighting factors, S, in Supplemental Documentation 1983, 472.1(83), American Conference of Governmental Industrial Hygienists, Cincinnati, 1983.

29. TLB Airborne Contaminants Committee, Light and near-infrared radiation, in *Threshold Limit Values for Chemical Substances and Physical Agents in the Workroom Environment with Intended Changes for 1985,* American Conference of Governmental Industrial Hygienists, Cincinnati, 1984, 89.
30. Sliney, D. H. and Wolbarsht, M. L., III, Radiation measurement and the development of standards, safety standards and measurement techniques for high intensity light sources, *Vision Res.,* 20, 1133, 1980.
31. Pitts, D. G., Cameron, L. L., Jose, J., Lerman, S., Moss, E., Varma, S. D., Zigler, S., Zigman, S., and Zuclich, J., Optical radiation and cataracts, in *Visual Health and Optical Radiation,* Waxler, M. and Hitchins, V. M., Eds., CRC Press, Boca Raton, Fla., in press.
32. Ederer, F., Long-term visual health and optical radiation, epidemiologic considerations, in *Visual Health and Optical Radition,* Waxler, M. and Hitchins, V. M., Eds., CRC Press, Boca Raton, Fla., in press.
33. Waxler, M., Long-term visual health problems: optical radiation risks, in *Visual Health and Optical Radiation,* Waxler, M. and Hitchins, V. M., Eds., CRC Press, Boca Raton, Fla., in press.
34. Fulton, A., Abramov, I., Allen, J., Gwiazda, J., Hainline, L., O'Neill, J., Raymond, P., and Varner, D., Optical radiation effects on visual development, in *Visual Health and Optical Radiation,* Waxler, M. and Hitchins, V. M., Eds., CRC Press, Boca Raton, Fla., in press.
35. Owsley, C., Kline, D. W., Werner, J. S., Greenstein, V., and Marshall, J., Optical radiation effects on aging and visual perception, in *Visual Health and Optical Radiation,* Waxler, M. and Hitchins, V. M., Eds., CRC Press, Boca Raton, Fla., in press.
36. Marshall, J., Greenstein, V., Kline, D. W., Owsley, C., and Werner, J. S., Optical radiation and the aged eye, in *Visual Health and Optical Radiation,* Waxler, M. and Hitchins, V. M., Eds., CRC Press, Boca Raton, Fla., in press.
37. Newsome, D. A., Berson, E., Bonner, R., Hyman, L., LaVail, M., Lawwill, T., Mainster, M. A., Sperling, H. G., and Tso, M. O. M., Possible role of optical radiation in retinal degenerations, in *Visual Health and Optical Radiation,* Waxler, M. and Hitchins, V. M., Eds., CRC Press, Boca Raton, Fla., in press.
38. Massof, R. W., Sykes, S. M., Rapp, L. M., Robison, W. G., Jr., Zwick, H., and Hochheimer, B., Optical radiation damage to ocular photoreceptors, in *Visual Health and Optical Radiation,* Waxler, M. and Hitchins, V. M., Eds., CRC Press, Boca Raton, Fla., in press.
39. Dayhaw-Barker, P., Forbes, D., Fox, D., Lerman, S., Megaw, J., McGinnis, J., Waxler, M., and Felten, R., Drug phototoxicity and visual health, in *Visual Health and Optical Radiation,* Waxler, M. and Hitchins, V. M., Eds., CRC Press, Boca Raton, Fla., in press.
40. Pitts, D. G., The ocular effects of ultraviolet radiation, *Am. J. Optom. Physiol. Opt.,* 55, 19, 1978.
41. Hull, D. S., Csukas, S., and Green, K., Trifluoperazine: corneal endothelial phototoxicity, *Photochem. Photobiol.,* 38, 425, 1983.
42. Lerman, S. and Borkman, R. F., A molecular model of lens aging, nuclear and cortical cataract formation, *Metab. Pediatr. Ophthalmol.,* 3, 27, 1978.
43. Kurzel, R. B., Wolbarsht, M. L., and Yamanashi, B. S., Ultraviolet radiation effects on the human eye, in *Photochemical and Photobiological Reviews,* Vol. 2, Plenum Press, New York, 1977.
44. Sykes, S. M., Robison, W. G., Jr., Waxler, M., and Kuwabara, T., Damage to the monkey retina by broad-spectrum fluorescent light, *Invest. Ophthalmol. Vis. Sci.,* 20, 425, 1981.
45. Ham, W. T., Jr., Ruffolo, J. J., Jr., Mueller, H. A., Clarke, A. M., and Moon, M. E., Histologic analysis of photochemical lesions produced in rhesus retina by short-wavelength light, *Invest. Ophthalmol. Vis. Sci.,* 17, 1029, 1978.
46. Waxler, M. and Hitchins, V. M., Eds., *Visual Health and Optical Radiation,* CRC Press, Boca Raton, Fla., in press.
47. Mainster, M. A., Ham, W. T., and Delori, F. C., Potential retinal hazards. Instrument and environmental light sources, *Am. Acad. Ophthalmol.,* 90, 927, 1983.
48. Rapp, L. M., Jose, J. G., Pitts, D. G., and Anderson, R. E., Evidence for DNA repair in the rat retina following exposure to UV light, *Invest. Ophthalmol. Vis. Sci.,* Suppl. 24, 180, 1983.
49. Henry, M. M., Henry, L. M., and Henry, L. M., A possible cause of a chronic cystic maculopathy, *Ann. Ophthalmol.,* 9, 455, 1977.
50. Bonner, R. F., Meyers, S. M., and Gaasterland, D. E., Threshold for retinal damage associated with the use of high-power neodymium-Yag lasers in the vitreous, *Am. J. Ophthalmol.,* 96, 153, 1983.
51. Wolbarsht, M. L., Editorial, *Invest. Ophthalmol. Vis. Sci.,* 19, 1123, 1980.
52. Pomerantzeff, O., Govignon, J., and Schepens, C. L., Indirect ophthalmoscopy: is the illumination level dangerous?, *Trans. Am. Acad. Ophthalmol. Otolaryngol.,* 73, 246, 1969.
53. Young, R. S. L., Goldberg, M. F., and Fishman, G. A., The minimum irradiance required for viewing human fundi in indirect ophthalmoscopy, *Invest. Ophthalmol. Vis. Sci.,* 20, 701, 1981.
54. Marshall, J., Light damage and the practice of ophthalmology, in *Intraocular Lens Implantation,* Rosen, E. S., Haining, W. M., and Arnott, E. J., Eds., C.V. Mosby, St. Louis, 1983, chap. 12.

55. Lerman, S., Jocoy, M., and Borkman, R. F., Photosensitization of the lens by 8-methoxypsoralen, *Invest. Ophthalmol. Vis. Sci.*, 16, 1065, 1977.
56. Hausser, K. W. and Vahle, W., Die Abhangigkeit des Lichterythems und der Pigmentbildung von der Schwingungszahl (Wellenlange) der erregenden Strahlung, *Strahlentherapie Sonderb.*, 13, 41, 1922.
57. Everett, M. A., Olsen, R. L., and Sayer, R. M., Ultraviolet erythema, *Arch. Dermatol.*, 92, 713, 1965.
58. Freeman, R. G., Owens, D. W., Knox, J. M., and Hudson, H. T., Relative energy requirements for an erythemal response of skin to monochromatic wavelengths of ultraviolet present in the solar spectrum, *J. Invest. Dermatol.*, 47, 586, 1966.
59. Parrish, J. A., Jaenicke, K. F., and Anderson, R. R., Erythema and melanogenesis action spectra of normal human skin, *Photochem. Photobiol.*, 36, 187, 1982.
60. Kaidbey, K. H. and Kligman, A. M., Cumulative effects from repeated exposures to ultraviolet radiation, *J. Invest. Dermatol.*, 76, 352, 1981.
61. Parrish, J. A., Zaynoun, S., and Anderson, R. R., Cumulative effects of repeated subthreshold doses of ultraviolet radition, *J. Invest. Dermatol.*, 76, 356, 1981.
62. Kahn, G., Photosensitivity from occlusion, *Arch. Dermatol.*, 103, 340, 1971.
63. Owens, D. W., Knox, J. M., Hudson, H. T., and Troll, D., Influence of humidity on ultraviolet injury, *J. Invest. Dermatol.*, 64, 250, 1975.
64. Boer, J., Schothorst, A. A., Boom, B., Hermans, J., and Suurmond, D., Influence of water and salt solutions on UVB irradiation of normal skin and psoriasis, *Arch. Dermatol. Res.*, 273, 247, 1982.
65. Kligman, A. M., Early destructive effect of sunlight on human skin, *JAMA*, 210, 2377, 1969.
66. Kligman, L. H. and Kligman, A. M., Reflections on heat, *Br. J. Dermatol.*, 110, 369, 1984.
67. Blum, H. F., *Carcinogenesis by Ultraviolet Light*, Princeton University Press, Princeton, N.J., 1959.
68. Forbes, P. D., Davies, R. E., and Cole, C. A., Photocarcinogenesis by UVA alone, *Am. Soc. Photobiol.*, 10(Abstr.), 66, 1982.
69. Van Weelden, H., de Gruijl, F. R., and van der Leun, J. C., Tumors induced by UV-A in mice, *Photochem. Photobiol.*, 37S, 79, 1983.
70. National Academy of Sciences, *Report of the Committee on Impacts of Stratospheric Change: Protection Against Depletion of Stratospheric Ozone by Chlorofluorocarbons*, National Academy of Sciences, Washington, D.C., 1979, Appendix F.
71. Michaelson, S. M., Standards for protection of personnel against nonionizing radiation, *Am. Ind. Hyg. Assoc. J.*, 35, 766, 1974.
72. Anderson, T. F., Waldinger, T. P., and Voorhees, J. J., UV-B phototherapy, *Arch. Dermatol.*, 120, 1502, 1984.
73. Rawls, R. L., Laser heart surgery teams chemistry and medicine, *Chem. Eng. News*, 62, 14, 1984.
74. News in medicine, *Lasers Appl.*, 3, 46, 1984.
75. Kessel, D. and Dougherty, T. J., Eds., *Porphyrin Photosensitization*, Vol. 60, Plenum Press, New York, 1983.
76. Pfister, J. and Kneedler, J. A., in *A Guide to Lasers in the O.R.*, Patterson, P., Ed., Education Design/Editorial Consultants, Aurora, Colo., 1983.
77. *Laser Safety Training Manual*, 6th ed., Rockwell Associates, Inc., Cincinnati, 1984.
78. Survey of otolaryngologists reveals high rate of laser complications, *Clin. Laser Mon.*, 2, 45, 1984.
79. Fried, M. P., A survey of complications of laser laryngoscopy, *Arch. Otolaryngol.*, 110, 31, 1984.
80. Ossoff, R. H. and Karlan, M. S., The CO_2 laser in otolaryngology, head and neck surgery: advantages, precautions and complications, Head and Neck Surgery, Inc., Training Course No. 4409, American Academy of Otolaryngology, Washington, D.C., 1982, 62E.
81. Morison, W. L., Parrish, J. A., and Epstein, J., Photoimmunology, *Arch. Dermatol.*, 115, 351, 1979.
82. Fox, I. J., Perry, L. L., Sy, M. S., Benacerraf, B., and Greene, M. I., The influence of ultraviolet light irradiation on the immune system, *Clin. Immunol. Immunopathol.*, 17, 141, 1980.
83. Elmets, C. A. and Bergstresser, P. R., Ultraviolet radiation effects on immune processes, *Photochem. Photobiol.*, 36, 715, 1982.
84. Morison, W. L., Photoimmunology, *J. Invest. Dermatol.*, 77, 71, 1982.
85. Morison, W. L. and Parrish, J. A., Photoimmunology, in *The Science of Photomedicine*, Reagan, J. D. and Parrish, J. A., Eds., Plenum Press, New York, 1982, 293.
86. Parrish, J. A., Morison, W. L., and Kripke, M., *Photoimmunology*, Plenum Press, New York, 1982.
87. *Experimental and Clinical Photoimmunology*, Vol. 1, Daynes, R. A. and Spikes, J. D., Eds., Vol. 2, Daynes, R. A. and Krueger, G., Eds., CRC Press, Boca Raton, Fla., 1983.
88. Parrish, J. A., Ultraviolet radiation affects the immune system, *Pediatrics*, 71, 129, 1983.
89. Hellman, K. B., Schuller, G. B., and Levis, W. R., Immunological defects in human radiation sensitive disorders, in *Topics in Photomedicine*, Smith, K. C., Ed., Plenum Press, New York, 1984, 143.

90. Kripke, M. L., Antigenicity of murine skin tumors induced by ultraviolet light, *J. Natl. Cancer Inst.*, 53, 1333, 1974.
91. Daynes, R. A., Spellman, C. W., Woodward, J. G., and Stewart, D. A., Studies into the transplantation biology of ultraviolet light-induced tumors, *Transplantation*, 23, 343, 1977.
92. Kripke, M. L., Latency, histology, and antigenicity of tumors induced by ultraviolet light in three inbred mouse strains, *Cancer Res.*, 37, 1395, 1977.
93. Daynes, R. A. and Spellman, C. W., Evidence for the generation of suppressor cells by ultraviolet radiation, *Cell Immunol.*, 31, 182, 1977.
94. Fisher, M. S. and Kripke, M. K., Suppressor T lymphocytes control the development of primary skin cancers in ultraviolet-irradiated mice, *Science*, 216, 1133, 1982.
95. Spellman, C. W. and Daynes, R. A., Modification of immunological potential by ultraviolet radiation. II. Generation of suppressor cells in short-term UV-irradiated mice, *Transplantation*, 24, 120, 1977.
96. Kripke, M. L. and Fisher, M. L., Immunologic parameters of ultraviolet carcinogenesis, *J. Natl. Cancer Inst.*, 57, 211, 1976.
97. Kripke, M. L., Lofgreen, J. S., Beard, J., Jessup, J. M., and Fisher, M. S., In vivo immune responses of mice during carcinogenesis by ultraviolet irradiation, *J. Natl. Cancer Inst.*, 59, 1227, 1977.
98. DeFabo, E. C. and Noonan, F. P., Mechanism of immune suppression by ultraviolet irradiation *in vivo*. I. Evidence for the existence of a unique photoreceptor in skin and its role in photoimmunology, *J. Exp. Med.*, 158, 84, 1983.
99. Toews, G. B., Bergstresser, P. R., and Streilen, J. W., Epidermal Langerhans cell density determines whether contact hypersensitivity or unresponsiveness follows skin painting with DNFB, *J. Immunol.*, 124, 445, 1980.
100. Germain, R. N., Accessory cell stimulation of T Cell proliferation requires active antigen processing, Ia-restricted antigen presentation and a separate nonspecific 2nd signal, *J. Immunol.*, 127, 1964, 1981.
101. Ahmann, G. B. and Ballas, Z. K., Whole body ultraviolet (UV) irradiation decreases natural killer (NK) activity, *Fed. Proc.*, 41, 601, 1982.
102. Hill, B. H. R., Immunosuppressive drug therapy as a potentiator of skin tumors in five patients with lymphoma, *Aust. J. Dermatol.*, 17, 16, 1976.
103. Green, M. H., Young, T. I., and Clark, W. H., Jr., Malignant melanoma in renal-transplant recipients, *Lancet*, 1, 1196, 1981.
104. Hoxtell, E. O., Mandel, J. S., Murray, S. S., Schuman, L. M., and Goltz, R. W., Incidence of skin carcinoma after renal transplantation, *Arch. Dermatol.*, 113, 436, 1977.
105. Poleksic, S. and Young, K. Y., Rapid development of keratocanthoma and accelerated transformation into squamous cell carcinoma of the skin, *Cancer*, 41, 12, 1978.
106. Nairn, R. C., Nind, A. P. P., Guli, E. P. G., Muller, H. K., Rolland, J. M., and Minty, C. C. J., Specific immune response in human skin carcinoma, *Br. Med. J.*, 71, 701, 1971.
107. Weimar, V. M., Ceilley, R. I., and Goeken, J. A., Cell-mediated immunity in patients with basal and squamous cell skin cancer, *J. Am. Acad. Dermatol.*, 2, 143, 1980.
108. Griffin, A. C., Methoxsalen in ultraviolet carcinogenesis in the mouse, *J. Invest. Dermatol.*, 32, 367, 1959.
109. Urbach, F. J., Modification of ultraviolet carcinogenesis by photoactive agents: Preliminary report, *J. Invest. Dermatol.*, 32, 378, 1959.
110. Stern, R. S., Thibodeau, L. A., Kleinerman, R. A., Parrish, J. A., and Fitzpatrick, T. B., Risk of cutaneous carcinoma in patients treated with oral methoxsalen photochemotherapy for psoriasis, *N. Engl. J. Med.*, 300, 809, 1979.
111. Kripke, M. L., Morison, W. L., and Parrish, J. A., Induction and transplantation of murine skin cancers induced by methoxsalen plus ultraviolet (320—400 nm) radiation, *J. Natl. Cancer Inst.*, 68, 685, 1982.
112. Lynch, D. H., Gurish, M. F., and Daynes, R. A., Relationship between epidermal Langerhans cell density ATPase activity and the induction of contact hypersensitivity, *J. Immunol.*, 126, 1892, 1981.
113. Morison, W. L., Parrish, J. A., Woehler, M. E., and Block, K. J., The influence of ultraviolet radiation on allergic contact hypersensitivity in the guinea pig. I. UVB radiation, *Br. J. Dermatol.*, 104, 165, 1981.
114. Kripke, M. L., Morison, W. L., and Parrish, J. A., Differences in the immunologic reactivity of mice treated with UVB or methoxsalen plus UVA radiation, *J. Invest. Dermatol.*, 78, 354, 1981.
115. Maguire, H. C., Jr. and Kaidbey, K. H., Photoallergic contact dermatitis: induction of hypersensitivity and of immunological tolerance, *J. Invest. Dermatol.*, 78, 356, 1982.
116. Strauss, G. H., Bridges, B. A., Greaves, P., Hall-Smith, P., Price, M., and Vella-Briffa, D., Inhibition of delayed hypersensitivity reaction in skin (DNCB test) by 8-methoxypsoralen photochemotherapy. Possible basis for pseudo-promoting action in skin carcinogenesis?, *Lancet*, 2, 556, 1980.

117. Makinodan, T., Immunity and aging, in *Handbook of the Biology of Aging,* Finch, C. F. and Hayflick, L., Eds., Van Nostrand Reinhold, New York, 1977, 379.
118. Odell, B., Jessen, R. T., Becker, L. E., Jackson, R. T., and Smith, E. B., Diminished immune response in sun-damaged skin, *Arch. Dermatol.,* 116, 559, 1980.
119. Fitzpatrick, T. B., Eisen, A. Z., Wolff, K., Freedberg, I. M., and Austen, K. F., Eds., *Dermatology in General Medicine,* 2nd ed., McGraw-Hill, New York, 1979.
120. Harber, L. C. and Bickers, D. R., *Photosensitivity Diseases,* W.B. Saunders, Philadelphia, 1981.
121. Der Kaloustian, V. M. and Kurban, A. K., *Genetic Diseases of the Skin,* Springer-Verlag, New York, 1979.
122. Magnus, I. A., The cutaneous porphyrias, *Sem. Hematol.,* 5, 380, 1968.
123. Andersen, F. A., Bockstahler, L. E., Chissler, P. G., Hellman, K. B., Levine, J. I., Lytle, C. D., Mills, L. F., and Segal, P., A review of biological effects considered in the development of a performance standard for sunlamp products, FDA By-Lines, No. 2, Food and Drug Administration, U.S. Department of Health and Human Services, Rockville, Md., 1981.
124. Lerman, S., *Radiant Energy and the Eye,* Macmillan, New York, 1980, 170.
125. Epstein, J. H., Chemical phototoxicity in humans, *J. Natl. Cancer Inst.,* 69, 265, 1982.
126. Risks associated with use of UV-A irradiators being used in treating psoriasis and other conditions, *Photochem. Photobiol.,* 30, 199, 1979.
127. Schram, P. J., Caloggero, J. M., and Tedesco, J. A., Eds., Health care facilities, in *The National Electrical Code Handbook,* 3rd ed., National Fire Protection Association, Quincy, Mass., 1984, article 517.
128. Standard for Health Care Facilities, No. 99, National Fire Protection Association, Boston.
129. Steere, N. V., Ed., *CRC Handbook of Laboratory Safety,* CRC Press, Boca Raton, Fla., 1971.
130. Stanley, P. E., Ed., *CRC Handbook of Hospital Safety,* CRC Press, Boca Raton, Fla., 1981.
131. Kern, H. E., Seasonal affective disorder and light, *Photochem. Photobiol.,* 39S, 2, 1984.
132. Kraft, M. C., Personal communication, 1985.

INDEX

A

B

C

D

E

F

G

H

I

J

K

L

M

N

O

P

Q

R

S

T

U

V